14th International Forum on Strategic Technology (IFOST 2019)

IOP Conference Series: Materials Science and Engineering
Volume 1019

Tomsk, Russia
14-17 October 2019

Editors:

Roman Ostvald
Daria Gvozdeva
Anastasia Ryabikina

ISBN: 978-1-7138-4550-8
ISSN: 1757-8981

Printed from e-media with permission by:

Curran Associates, Inc.
57 Morehouse Lane
Red Hook, NY 12571

Institute of Physics
Dirac House, Temple Back
Bristol BS1 6BE UK

Phone: 44 1 17 929 7481
Fax: 44 1 17 920 0979

techtracking@iop.org

Additional copies of this publication are available from:

Curran Associates, Inc.
57 Morehouse Lane
Red Hook, NY 12571 USA
Phone: 845-758-0400
Fax: 845-758-2633
Email: curran@proceedings.com
Web: www.proceedings.com

TABLE OF CONTENTS

IOP Conf. Series: Materials Science and Engineering 1019 (2021) 011002 doi:10.1088/1757-899X/1019/1/011002

Peer review declaration

All papers published in this volume of *IOP Conference Series: Materials Science and Engineering* have been peer reviewed through processes administered by the Editors. Reviews were conducted by expert referees to the professional and scientific standards expected of a proceedings journal published by IOP Publishing.

- **Type of peer review: Single-blind**
- **Conference submission management system: EasyChair**
- **Number of submissions received: 297**
- **Number of submissions sent for review: 297**
- **Number of submissions accepted: 182**
- **Acceptance Rate (Number of Submissions Accepted / Number of Submissions Received X 100): 61%**
- **Average number of reviews per paper: 2**
- **Total number of reviewers involved: 40**
- **Any additional info on review process:**
- **Contact person for queries: orpa@tpu.ru**

Detailed description of the peer review process

Authors needed to upload their papers to EasyChair. Conference organizers assigned at least two reviewers for each paper. Reviewers could reject or accept the invitation to peer-review a paper. If the invitation was accepted, reviewers were obliged to provide peer-review comments within 10 days. Peer-reviewers had three options: reject the article; accept the article; send comments to authors to edit the paper. If an article was poor and did not satisfy the requirements in the call for papers, peer-reviewers rejected that. If the article was good enough, peer-reviewers could accept the article without changes. The most popular option chosen by peer-reviewers was the one where they send comments to authors. Authors were obliged to edit the paper according to the comments within 10 days. Once authors submitted the edited files, editors checked if all changes were addressed. If so, the article was sent to a reviewer to approve the changes. If reviewers agreed with the changes, the article was considered as accepted for IFOST2019. If editors did not see that the article was changed according to the reviewers' comments, they sent the paper back to authors with a requirement to edit the paper.

IOP Conf. Series: Materials Science and Engineering 1019 (2021) 011001 doi:10.1088/1757-899X/1019/1/011001

14th International Forum on Strategic Technology (IFOST), October 14-17, 2019, Tomsk, Russia, organized by National Research Tomsk Polytechnic University (Russia).

International Forum on Strategic Technology (IFOST) is a nonprofit organization composed of member institutes located in Asia: University of Ulsan (Korea), Novosibirsk State Technical University (Russia), Tomsk Polytechnic University (Russia), Harbin University of Science and Technology (China), Mongolian University of Science and Technology (Mongolia), Chemnitz University of Technology (TUC) Germany.

The main purpose of the forum is to engage researchers into the discussion of strategic technologies, broaden their experience of participation in international events and giving an opportunity to find new colleagues and co-authors. IFOST is a meeting point for researchers from multiple countries. Plenary speakers of the forum were the distinguished experts in the field from universities of Germany, Brazil, and France.

IFOST-2019 proceedings cover a wide range of topics:
1. New Materials and Nanotechnologies
2. Information and Communication Technologies
3. Mechanical Engineering, robotics, and automation
4. Power Engineering and Renewable Energy
5. Ecology, Environmental Engineering, and Civil Engineering
6. Applied Engineering and Management
7. High Energy Physics
8. Chemistry and Biotechnologies
9. Aerospace Engineering

Committee

General Chair: Prof. Dr. Petr S. Chubik, acting rector of Tomsk Polytechnic University, Tomsk, Russia

Deputy Chair: Dr. Roman Ostvald, Deputy Vice-Rector for Research and Innovations, Tomsk Polytechnic University, Russia

Technical Program Committee Chair: Prof. Oleg Stukach, HSE Tikhonov Moscow Institute of Electronics and Mathematics, IEEE Senior Member, Moscow, Russia

Technical Program Committee Co-chair: Prof. Jayakody Dushantha Nalin Kumara, Research Laboratory for Information and Telecommunication Systems, Tomsk Polytechnic University, Tomsk, IEEE Senior Member, Russia

Editors

Roman Ostvald, Tomsk polytechnic university

Daria Gvozdeva, Tomsk polytechnic university

Anastasia Ryabikina, Tomsk polytechnic university

IOP Conf. Series: Materials Science and Engineering 1019 (2021) 012001 doi:10.1088/1757-899X/1019/1/012001

Effect of iron nanopowder on flammability of epoxy composites

O B Nazarenko, D S Lipchansky and I N Smirnova
School of Non-Destructive Testing, Tomsk Polytechnic University, Tomsk, Russia
E-mail: obnaz@mail.ru

Abstract. Reducing the flammability of polymeric materials is a serious problem that needs to be solved. The paper presents the results of a study of the effect of iron nanopowders, used as filler, on the flammability of epoxy polymers. Epoxy composites filled with 5 wt. % of iron nanopowder and 10 wt. % of boric acid separately, as well as in combination were prepared. The flammability of the prepared samples was evaluated by determining the ignition temperature and time-to-ignition.

1. Introduction

Epoxy resins are used in various industries and everyday life for the manufacture and repair of structural parts, as components of adhesives, paints, coatings, as well as sealing and impregnating compounds, etc. [1, 2]. The combination of unique properties, such as ease of curing, low viscosity, low shrinkage, high adhesive ability, high mechanical and electrical insulating properties, chemical resistance, promotes the widespread use of these materials. One of the main drawbacks of epoxy resins, limiting their field of application, is high fire hazard [3]. Therefore, the problem of reducing the flammability of polymers is topical.

To evaluate the flammability, an indicator such as the ignition temperature is used, by which is meant the lowest temperature of a substance, at which a substance emits combustible vapors and gases at such a rate that ignition is observed when exposed to an ignition source.

Reducing the flammability of polymers can be achieved by using such a method as the introduction of fillers, which have flame retardant properties [4, 5]. Fillers of different natures lead to changes in the character of the process of polymer destruction at the heating or blocking of the combustion process by non-combustible or inhibiting substances. One of the inorganic compounds with good flame retardant properties is boric acid. At the heating, boric acid releases water in the endothermic process that reduces polymer temperature and degradation. In our previous works, we found that the introduction of boric acid into the epoxy matrix at a percentage 10 wt. % has a positive impact on the thermal and thermo-oxidative degradation of the epoxy composites and does not lead to deterioration of the functional properties of the composites [6, 7]. In addition, boric acid can provide a greater effect on the improvement of the thermal stability of the epoxy composites when it is used in combination with other materials.

The disadvantage of traditional flame retardants is that their effectiveness in reducing the flammability of materials is achieved, as a rule, at relatively high concentrations (50 wt. % and more). At the same time, the structural and functional properties of polymer composites can significantly deteriorate. Nanodispersed materials, in contrast to conventional coarse flame retardants, can help to solve the problem of reducing the flammability of polymers at relatively low concentrations

(< 5 wt. %) [8, 9]. It is assumed that the mechanism of action of nanodispersed fillers is the barrier effect.

The aim of this work was to evaluate the effect of iron nanopowder separately and in combination with fine boric acid powder used as fillers on the flammability characteristics of epoxy polymers.

2. Experimental

2.1. Materials and preparation

The epoxy-diane resin grade ED-20 was used as a polymer matrix; polyethylene polyamine (PEPA) was used as a hardener for the preparation of the epoxy composites.

Iron nanopowder (Fe NP) was produced by the method of the electrical explosion of wires (EEW) [10–13] in Tomsk Polytechnic University. EEW is a process of explosive destruction of a metal wire under the action of a great density current ($>10^{10}$ A/m^2). To produce Fe NP, the argon was used as a working ambient in the discharge chamber. Fe NP produced by EEW is pyrophoric material, has high reactivity and ignites on contact with air. Therefore, air-passivation was carried out by Ar+0.1 vol. % air directly after the production to protect iron nanoparticles against oxidation.

Boric acid (BA) was milled in a mortar before the sample preparation. The particles with a size of fewer than 64 μm were used for the preparation of the epoxy composites.

For the preparation of epoxy composites, the required amount of the fillers was added into epoxy resin. We have used for the preparation of the composites the following fillers percentage: 1) 5 wt. % of Fe NP, 2) 10 wt. % of boric acid and 3) combination of 5 wt. % of Fe NP and 10 wt. % of boric acid. The formulations of the studied epoxy composites are given in table 1.

Table 1. Formulations of epoxy composites.

Sample	ED-20 (wt. %)	Fe NP (wt. %)	BA (wt. %)
E0	100	0	0
E/Fe	100	5	0
E/BA	100	0	10
E/Fe/BA	100	5	10

The epoxy resin with the fillers was mixed by hand for 5 min at room temperature. After that PEPA was added into the mixture and mixed again for 3 min. The ratio of epoxy resin and hardener was 10:1 by mass. The obtained mixtures were cured in the silicone molds at room temperature for 24 h. The images of the prepared samples are shown in Figure 1.

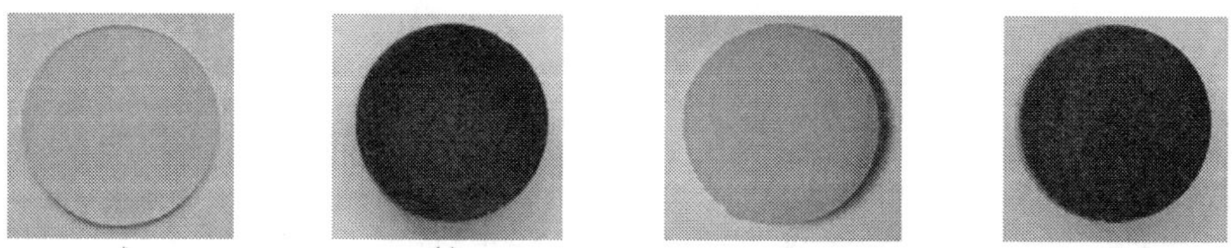

a) b) c) d)

Figure 1. Images of the prepared samples: a) E0; b) E/Fe; c) E/BA; d) E/Fe/BA.

2.2. Characterizations

The phase composition of the Fe NP was determined by X-ray diffraction (XRD) analysis using a diffractometer Shimadzu XRD-7000 with CuKα radiation. The size and morphology of the studied samples were investigated using a scanning electron microscope (SEM) TM-3000. The infrared spectra of the samples in the spectral range 4000–400 cm^{-1} were recorded using Nicolet 5700 Fourier transform infrared (FTIR) spectrometer.

The flammability of the prepared epoxy composites was evaluated by the standard method of experimental determination of the ignition temperature of solid substances and materials by GOST 12.1.044-89 "Occupational safety standards system. Fire and explosion hazard of substances and materials. Nomenclature of indices and methods of their determination". The ignition temperature of the epoxy composites was determined using the installation consisting of a vertical electric furnace with two coaxially located cylinders made of quartz glass, a container for placing the samples and a holder, an electric heater, a gas burner, and an air ejector. The principle of operation of this installation is based on setting the temperature mode in the reaction chamber and the effect of the burner flame and controlling the temperature parameters after adding the studied material to the reaction chamber. The epoxy samples in this test were cylindrical with a diameter of 45±1 mm. The weight of each sample was 3±0.1 g.

3. Results and discussion

3.1. Characterization of zeolite

Figure 2 shows the XRD patterns of the Fe NP nm. All the reflections correspond to metallic iron. The crystalline iron oxide phases as part of the protective layer formed during passivation were not detected in the XRD patterns of the Fe NP sample due to their amorphousness.

Figure 3 shows the SEM image of the Fe NP. The Fe NP consists of micron-sized and nanosized fractions. The average diameter of the micron-sized particles is about 1–3 μm and of the nanosized particles is about 80–150 nm. The particles of both powder fractions have the shape closed to spherical.

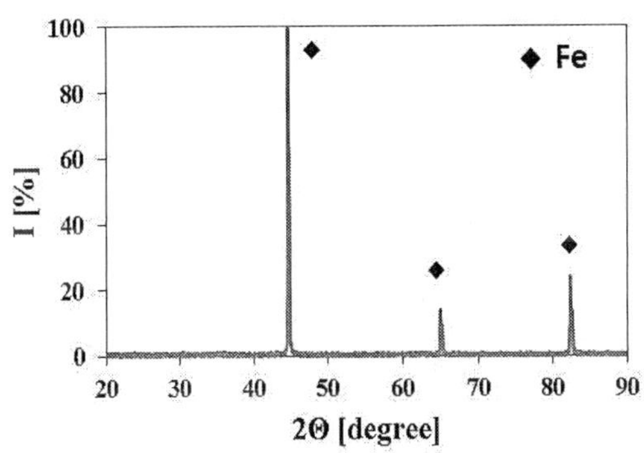

Figure 2. X-ray diffraction pattern of Fe NP.

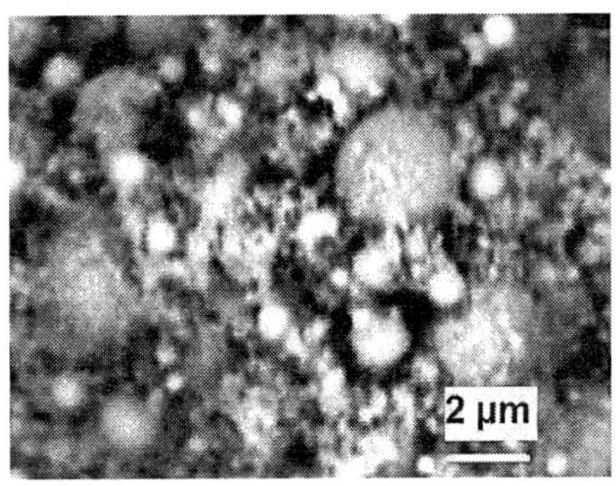

Figure 3. SEM image of Fe NP.

The chemical functional groups on the surface of the iron particles were identified using the FTIR spectroscopy. Figure 4 shows the FTIR spectrum of the Fe NP.

The several absorption bands in the range of 635–430 cm^{-1} corresponding to the Fe–O vibrations modes were observed in the FTIR spectrum of Fe NP. Thereby, it can be assumed that the surface layer of the Fe NP particles contains oxide phases γ-Fe_2O_3 and Fe_3O_4. The presence of two absorption bands at 1425 and 1483 cm^{-1} indicates that as a result of CO2 chemisorption, carbonate structures were formed on the Fe NP particles surface. The broad band in the range of 3300–3100 cm^{-1} and the band at 1564 cm^{-1} are due to the vibrations of hydrogen-bonded water molecules adsorbed on the Fe NP particles surface.

The results of the experimental determination of the ignition temperature and the time-to-ignition of the epoxy samples are shown in table 2. The time-to-ignition is defined as the minimum exposure time required for the sample to ignite and sustain flaming combustion.

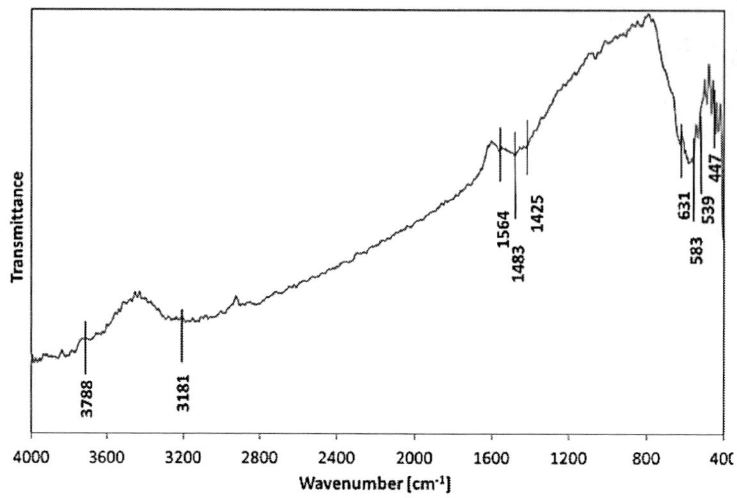

Figure 4. FTIR spectrum of Fe NP.

Table 2. Ignition temperature and time-to-ignition of epoxy composites.

Sample	Ignition temperature (°C)	Time-to-ignition (s)
E0	305	342
E/Fe	314	354
E/BA	317	343
E/Fe/BA	322	314

Figure 5 depicts the change of the ignition temperature of the studied epoxy composites. The incorporation of Fe NP and boric acid in epoxy resin individually, as well as of their combination leads to an increase in the ignition temperature of the samples compared to the control sample E0. Nanodispersed iron powder contributes to the formation during the burning of a strong and dense char layer on the surface of the epoxy sample that acts as a barrier to oxygen and combustible gases. In addition, formed iron oxide on the surface of the sample contributes to the protection of the epoxy sample from further decomposition as it was shown in [14]. The maximum ignition temperature was observed for the sample E/Fe/BA. The endothermic process of BA decomposition with the release of water leads to the cooling of the composite surface. Moreover, the boron oxide glassy layer on the surface of the epoxy matrix formed as a result of BA decomposition acts as a heat barrier [15] and increases the ignition temperature of the epoxy composites.

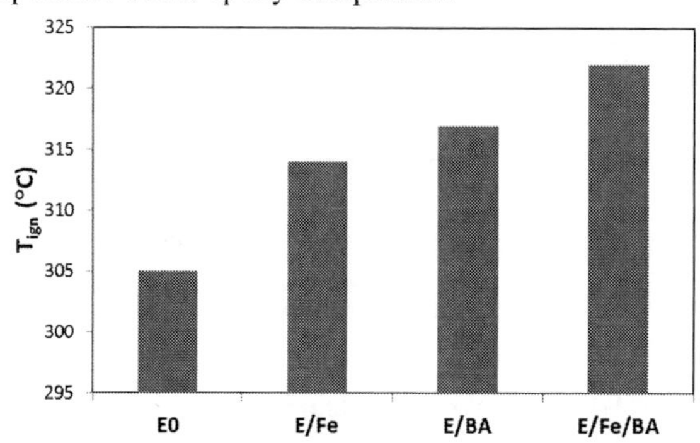

Figure 5. Ignition temperature of epoxy composites.

The sample E/Fe showed the maximum value of time-to-ignition. These results can be attributed to the active role of Fe NP in the formation of protective carbon chair residue.

4. Summary

In this work, the effect of iron nanopowder separately and in combination with fine boric acid powder used as fillers on the flammability characteristics of epoxy polymers was evaluated. The flammability characteristics of the produced epoxy composites were determined: the ignition temperature and time-to-ignition. It was found that in the case of the addition of 5 wt. % Fe NP to the epoxy matrix the ignition temperature increased by 2.9 % in comparison with that of the unfilled epoxy resin. In the case of a combination of 5 wt. % Fe NP and 10 wt. % boric acid the ignition temperature of the epoxy composite increased by 5.6 % compared to the unfilled epoxy resin.

Thus, the data obtained indicate the effectiveness of the use of nanodispersed iron powders separately, and especially in combination with boric acid as filler for epoxy resin in order to improve the fire hazard characteristics of the epoxy composites.

Acknowledgments

The research is carried out at Tomsk Polytechnic University within the framework of Tomsk Polytechnic University Competitiveness Enhancement Program grant.

References

[1] Lee H and Neville K 1967 *Handbook of Epoxy Resins* (New York: McGraw-Hill) p 960
[2] Jin F-L, Li X, Park S-J 2015 *J. Ind. Eng. Chem.* **29** 1–11
[3] Irvine D J, McCluskey J A, Robinson I M 2000 *Polym. Degrad. Stab.* **67** 383–396
[4] Wilkie C A and Morgan A B 2010 *Fire Retardancy of Polymeric Materials* (Boca Raton: CRC Press) p 856
[5] Dasari A, Yu Z Z, Cai G P, Mai Y W 2013 *Prog. Polym. Sci.* **38** 1357–1387
[6] Visakh P M, Nazarenko O B, Amelkovich Yu A, Melnikova T V 2015 *IOP Conf. Ser.: Mater. Sc. Eng.* **81** 012095
[7] Visakh P M, Nazarenko O B, Amelkovich Y A, Melnikova T V 2016 *Polym. Adv. Technol.* **27** 1098–1101
[8] Laoutid F, Bonnaud L, Alexandre M, Lopez-Cuesta J M, Dubois P 2009 *Mater. Sci. Eng. R.* **63** 100–125
[9] Arao Y. 2015 Flame retardancy of polymer nanocomposite *Flame Retardants: Polymer Blends, Composites and Nanocomposites* eds P M Visakh and Y Arao (Cham: Springer International Publishing) pp 15–44
[10] Kwon Y S, Ilyin A P, Tikhonov D V, Nazarenko O B, Yablunovskii G V 2003 *Proceedings of the 7th Korea-Russia International Symposium on Science and Technology KORUS'2003* vol 1 (Ulsan: University of Ulsan) pp 175–178
[11] Ilyin A P, Nazarenko O B, Tikhonov D V, Ushakov V Ya, Yablunovskii G V 2002 *Russian Phys. Journal* 45(12) pp 1176–1180
[12] Kwon Y S, Kim J S, Kim H T, Lee H K, Nazarenko O B, Ilyin A P 2004 *J. Ind. Eng. Chem.* **10**(6) pp 949–953
[13] Nazarenko O B 2005 *Glass Ceram.* **62**(11–12) 364–367
[14] Chen X L, Jiao C M, Wang Y 2009 *Express Polym. Lett.* **3**(6) 359–365
[15] Balci S, Sezgi N A, Eren E 2012 *Indian Eng. Chem. Res.* **51** 11091–11096

IOP Conf. Series: Materials Science and Engineering 1019 (2021) 012002 doi:10.1088/1757-899X/1019/1/012002

Application of software testing methodology based on quality criteria and expert assessments to mobile applications

E Y Galimova
Saint-Petersburg State University of Industrial Technologies and Design, Saint-Petersburg, Russian Federation
E-mail: galim81@mail.ru

Abstract. In this paper, we propose a technique for choosing between a manual, automated and semi- automated approaches to software testing. The great desire of each person to achieve maximum comfort in each of the spheres of life has also affected the international web. The user, wanting to always stay online, uses the phone as a communicator. This led to the emergence of mobile Internet. When staying away from home or during travel and business trips, you can connect using a tablet or mobile phone instead of a laptop. The efficiency and functionality of "mini" computers would not be brought to such a high level without specialized applications. Properly chosen testing methodology improves the quality of mobile applications.

1. Introduction

In the modern world, mobile devices are equipped with almost the same set of functions as stationary computers [1]. A distinctive feature of mobile devices is a more complex interaction with users. This fact makes their testing more resource-intensive [2], and organization should efficiently plan the testing process.

Quality assurance issues have been discussed in a number of research papers [3, 4]. The quality of data in software systems must comply with the characteristics described in the standards of the ISO 8000 group. Software testing is part of the quality assurance process for software systems. The quality management system for software systems should include all stages of the life cycle [5, 6]. This article proposes a formalization of the overall assessment of a software system for testing purposes. This is done by answering a list of questions that are close to unambiguousness to solve the issues of organizing testing of a given software system [7]. Experts are invited to fill out a questionnaire. The questions are based on the IEEE 830 standard, which forms the requirements for the software system. Examples of questions from the questionnaire are presented below.

2. Automated testing criteria

Do you plan to use cloud services for testing? Cloud-based testing platforms can be scaled up to many operation systems and devices. Mobile applications should work correctly on various combinations of platforms [8]. Cloud test automation tools are designed for automated test data generation, test case generation, test execution and test evaluation [9]. Consider the testing algorithm using a cloud farm with DevOps [10]. The last branch from the repository is loaded into the build system and the application is compiled. During the compilation process, you can run unit tests. Next the assembly is uploaded to the cloud device farm, where automated user interface testing (User Interface, UI) will be performed.

Is random testing important? Testing done without any plan; tests are run in random order. For example, clicking in a chaotic order on the first available buttons in the application. There are many applications for automation. For Android - UI / Application Exerciser Monkey, a command line tool that can be run on any emulator instance or device. It sends a pseudo-random stream of custom events to the system. The application has a number of settings: basic configuration parameters, operational limits, event types and their frequency, debugger settings.

Is penetration testing required? The purpose of this type of testing is to try to obtain confidential information; to use confidential information to commit malicious acts; to become a "superuser", get bonuses and privileges. Software protections must comply with the requirements of the Mobile Application Security Verification Standard (MASVS). There is a research open source community dealing with issues of application vulnerabilities (iMAS). Created automated tools for penetration testing of mobile environments (MobiSec, OWASP Zed Attack Proxy Project, IBM Application Security on Cloud) [11].

Is smoke testing planned? Post-build software testing to ensure critical functionality is working properly. Performed "before" any detailed functional or regression tests [12].

Will data quality testing be performed? The testing process verifies that incorrect data is processed correctly.

Do you plan to search for challenges of Transport Layer Security (TLS)? Starting with the Android PieE version, all incoming and outgoing connections will be made via the TLS protocol. It prevents unauthorized access and listening to packets when exchanging data in client-server applications. At the same time TLS does not allow lowering the version of the security protocol and weakening the encryption algorithm. Open source tool Nogotofail designed for testing TLS and works for Android, Windows, Linux, iOS, OSX, Chrome OS. The UniTEST automated testing technology can be used to verify TLS security protocol verification [13].

3. Manual testing criteria

Do you plan to use alpha testing? This is manual testing, one of the most common approaches to testing software systems. Used in the development of software systems, towards the end of the development stage. Consists of several iterations. Performed by the developers themselves, internal testing specialists. The advantage of alpha testing is that the found defects can be quickly eliminated by the developers. The testing process is carried out by performing tasks that a potential user can perform. Alpha testing can act as internal acceptance testing for the finished software system prior to beta testing. It is carried out using the white, gray or black box method.

Do you plan to use beta testing? This manual testing is done to get feedback from real users. Usually done by the black box method. Beta testing does not require a dedicated lab or test environment to be deployed. Testing is carried out in a real time environment. There are closed and open beta versions. The closed one is issued for a limited number of users, the open one is available to everyone interested.

Is installation testing important? Check that all previous data is saved after updates. Updates should be installed via GSM-connection, Bluetooth connection, via Wi-Fi. Possible installation options for mobile application shown in Figure 1.

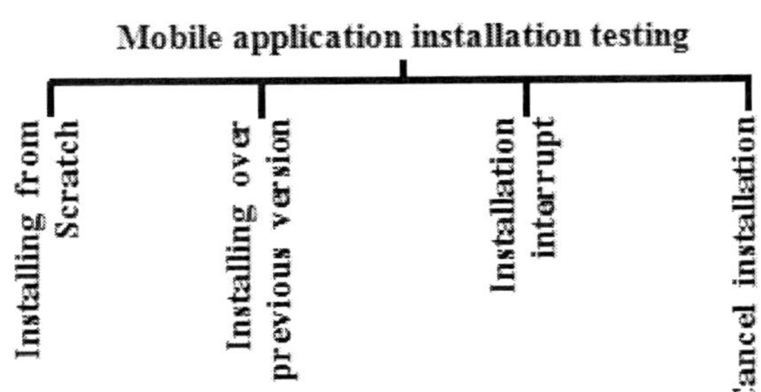

Figure 1. Types of mobile application installation testing.

Will the touch interface be tested? You need to check that all areas of the screen are equally sensitive to pressure. Check, how much effort you need to make to control the screen and whether light touches are registered.

Is accessibility testing important? The application should work correctly with at least some of the following functions that increase accessibility: High Contrast Display Modes, Voice Recognition, Screen Keyboards, Peripheral Keyboards, Blinking cursor, Peripheral Mouse, Visual interpretation of sound alerts.

Is interrupt testing planned? Different types of mobile interruptions shown in Figure 2.

Mobile interrupt testing

- Incoming/outgoing calls
- Incoming/outgoing SMS
- Incoming/outgoing MMS
- Battery full/medium/low strength
- Battery removal
- Network connection loss/restoration
- Disconnect/connect SD-card
- Battery charging

Figure 2. Basic test cases for mobile interrupt testing.

It is important to distinguish between third-party application and system notifications, since they are processed differently. Third-party applications allow you to quickly emulate the notification; system applications can notify about insufficient charge or availability of updates, and their notification is a bit more difficult to emulate. Tester must to check the transition to the standby mode after the user intentionally locks the screen. Like all checks, this should be done while working with the main functionality of the application, since locking can significantly affect the the processing of processes.

Is economic gain important in the short term? The cost of manual testing depends only on the salary of the tester. The company does not need to pay for automation tools. Manual testing is suitable primarily for small projects. A number of approaches have been developed for calculating the efficiency of automation applications [14].

4. Semi-automated testing criteria

Will we test the accelerometer? With the development of technology, almost all mobile devices are equipped with an accelerometer. The main purpose of this sensor is to provide information about the current acceleration of the device, more precisely, the difference between the acceleration of the device and the acceleration of gravity. At rest, the sensor readings coincide with the gravitational acceleration vector. The motion recognition algorithm using an accelerometer is described in [15].

Owing to the accelerometer, the mobile device responds to changes in orientation in space (turning the display, shaking, hitting), which will allow the screen to be rotated at the right time, as it would be convenient for the user. The accelerometer is used when operating the device as a pedometer. Calibration of the accelerometer can be done manually using the GPS Status & Toolbox for Android. You can record and play back the sequence of events from the accelerometer using the Android Debug Bridge (ADB) using the GetEvent / SendEvent commands.

At the device level, accelerometer testing is usually done using a microcontroller unit (MCU). Accelerometer and MCU are soldered to a printed circuit board. To verify the accuracy of the acceleration data received from the accelerometer, a special test program is developed for the MCU. Accelerations are read from each of the three axes. The obtained values are transferred to a computer for analysis.

Is regression testing planned? During testing, the existing functional and non-functional areas are checked after making changes to a part of the software or after adding new functions.

Is formal testing important? Standardized verification of software products, performed in accordance with the test plan and other formal documentation. During the testing process, final tests are performed for all the functionality of the software system. Formal mathematical methods are applied.

Is testing of responsive design planned? Adaptive design provides good perception when viewing pages. The site should adapt to the user mobile device. Adaptive design provides good perception when viewing pages. The site should adapt to the user gadget. Manual checks are carried out on existing mobile devices. Next, we use tools that mimic the screens of various mobile devices. For example, Am I Responsive allows you to drag and drop devices anywhere on the screen, you can embed your test site in a link and share it. Only four types of screens are available: desktop monitor, laptop, tablet, mobile phone. In another application, there are more device options and screen resolutions. There is a function to edit the background header. You can use the mobile emulator from Google DevTools. Without the need to install adb-drivers or something to configure, desktop Chrome detects all running instances of Chrome and WebView in Android and displays them on the screen, providing the ability to manage them both from the phone itself and from the desktop browser.

Is the test object a hardware and software complex? Usually, the main components are tested separately, then manually analyze the results obtained. Next is a semi-automated comprehensive testing of the entire system. Consider the software and hardware complex on the example of digital technological radio systems. Single special standard is used for rail transport. It is called the global mobile communications system for rail transport (GSM-R). There is also a backup radio standard Digital Mobile Radio (DMR) [16]. Scheme of digital technology radio system for GSM-R shown in Figure 3.

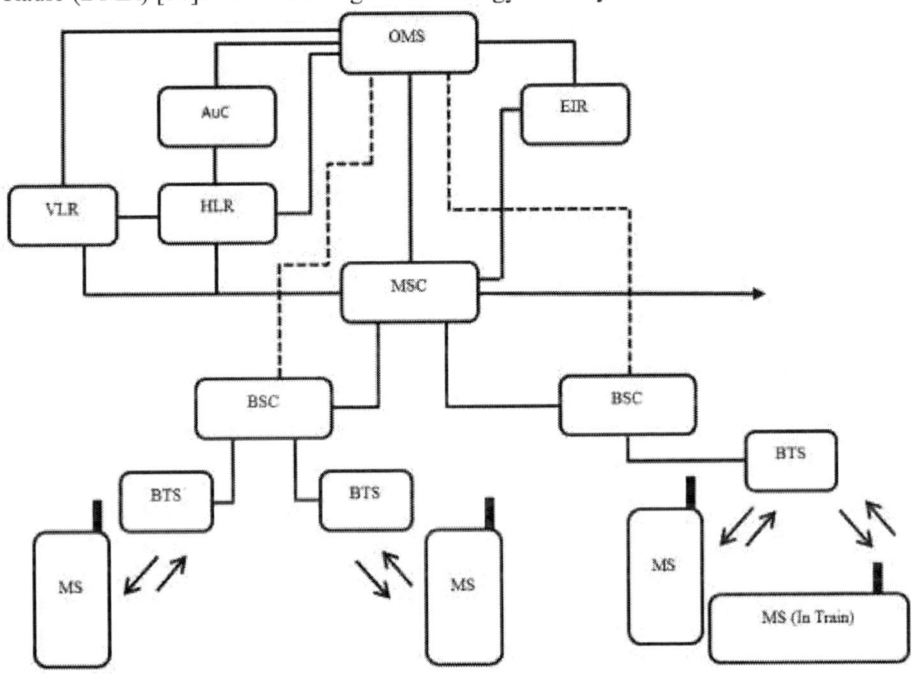

Figure 3. Digital technology radio system.

Mobile station (MS) is a mobile phone or other subscriber equipment that receives GSM communication services. MS consists of a mobile terminal and a SIM card. The mobile terminal has a unique identifier International mobile equipment identity (IMEI). The SIM card has a unique identifier International mobile subscriber identity (IMSI).

Base Transceiver Station (BTS) provides radio communication for a specific area. BTS installation options: on special masts, on the roofs of buildings, on boiler tubes. BTS functions: provision of interface with Base Station Controller (BSC); radio link management between the Mobile Station (MS) and the Base Station System (BSS). BTS is a distributed system. For testing of distributed systems we apply unit testing, system testing, integration testing. An additional approach is fault injection. In the process of the system using special programs, failures are added: network failures, failures inside components, disk failures. BTS must be resistant to these faults. Another complementary approach is formal verification. The description of internal algorithms and protocols is checked for correctness.

The main element of the base station system is the Base Station Controller (BSC). Functions: transport channel control with BTS; transport channel control with Transcoder (TC) or with Transcoding Rate and Adaptation Unit (TRAU); transport channel control between TC and Mobile Switching Center (MSC). We can use the BSC simulator for testing.

Handover - the procedure for transferring the active connection between cells. It allows subscribers to move within the network of the operator without breaking the connection. Many disconnections in GSM are due to Handovers, so very high requirements are placed on radio coverage in GSM-R. We can apply automated testing using handover selector role and handover client role. Client can open, close, reopen a connection with the server correctly.

TC (TRAU) performs the function of conversion of the speech signal from the main digital channel to the stream with a lower speed. For the choice of speed is responsible BSC. We can execute performance testing.

Home Location Register (HLR) stores the part of the location information of a mobile station that allows the central switch to deliver the call [17]. The HLR and MSC provide call routing and location changes (roaming) to the mobile station and contain all the administrative information of each subscriber registered in the corresponding GSM-R network, along with the current location of the mobile stations. We can apply automated load testing.

The Visitor Location Register (VLR) is the second main device that controls the movement of mobile stations from zone to zone. With its help, the operation of mobile stations outside the zone controlled by the register is achieved. When the mobile station moves from the zone of one BSC controller to the range of another, in the process of moving it is registered, and new information is entered into VLR. For network testing we can use Visitor Location Register Emulator.

Equipment Identity Register (EIR) is a database with information about subscriber equipment, indicating whether this equipment can be registered on the network or not. EIR stores three lists (white, gray and black) with IMEI. The presence of IMEI in the white list allows access to the network unconditionally. Equipment from the gray list will be allowed into the network, but will be continuously monitored while it is online. The black list is intended to store IMEI devices that are denied access to the network. In India, e.g., the cellular operators with the help of EIR can search out all mobile phones latched to their network [18]. Standard tools for testing databases are usually supplied by manufacturers. There are open source tools: Database Benchmark, Database Rider, DbFit, DbUnit, and more. If we are testing the white box, we need to test the database schema, including checking each object in the scheme; trigger testing. We can conduct stress testing. When testing the black box, we perform testing through the interface.

The Authentication Center (AUC) consists of several blocks and generates keys and authentication algorithms. With its help, the subscriber's authority is checked and his access to the network is provided. AUC makes decisions about the parameters of the authentication process and determines the encryption keys based on the database in the Equipment Identity Register (EIR). We can execute security testing.

Operations and Maintenance Center (OMC) provides management of network elements and the quality of its work. From OMS we execute control testing of the entire system as a whole. Triotrace Test Monitor can be used to monitor data and evaluate quality of service information for GSM-R. The Triotrace Test Monitor has Forcing functions for voice codec, channel codec, band codec. You can apply manual stress testing for failure of feeders (auxiliary lines between transformers and certain switches). The first cable, e.g., is considered the main one. We check that when the power is turned off, the magnetic starter commutes the electrical circuit to the backup feeder. In case of failure of two feeders at the same time, a mobile diesel generator should be connected.

5. Selection the software testing method using the multi-criteria task

The problem of choosing a testing method (automated, manual, mixed) is posed and solved as a formal multicriteria problem. In its formulation, this problem is nonclassical; it determines not the best element on a given set, but the best solution (alternative) for a given software system. At the same time, ideology and mathematical methods are used, similar to the solution of multicriteria problems.

To rank the questions in order of importance for making a decision on test automation, experts are invited to fill out tables where the degree of pairwise dominance of questions is assessed according to the scale developed by Thomas Saaty, which includes estimates from 1 to 9 with a single step [19, 20].

Using pairwise comparison of questions, it is possible to compose the most complete reflection of the subjective preferences of experts. With this approach, fewer restrictions are imposed on the choice compared to a number of other types of expert assessment. At each step, the expert makes a choice of only two alternatives. This task has an uncertainty level of 1 bit. The described approach facilitates the work of experts. The number of compared pairs for each questionnaire is $n \times (n-1) / 2$, where n is the number of questions. Questionnaires with tables of pairwise comparisons of questions from each expert are individually processed according to a special algorithm described in the work of V. V. Davnis [21]. Each questionnaire is a square n-by-n matrix A. The calculation is performed according to the iterative algorithm.

When selecting experts, it is not always possible to create a homogeneous group, therefore it is proposed to calculate the weighting coefficients of the competence of each expert. Let us apply an iterative algorithm that calculates in parallel the group preferences and competence coefficients of experts, considered in the work.

Let us construct a rectangular matrix P of size n by m, where n is the number of questions, m is the number of experts. Let's introduce additional designations: p_{i*} is the i-th row of the matrix P; p_{*j} is the j-th column of the matrix P; $p\ (p_1, p_2, ..., p_n)$ is the group evaluation vector; $v^0 = (v_1^0, v_2^0, ..., v_m^0)$ is a vector of initial competency weights.

In the first step of the algorithm:

$$v^0 = (1/m,\ 1/m, ...,\ 1/m).$$

Next, we calculate:

$$p^1 = v_1 p_{*1} + v_2 p_{*2} + ... + v_m p_{*m} = Pv^0.$$

We specify the weighting factors of experts' competence:

$$v^1 = p_{1*}^1 p_{1*} + p_{2*}^1 p_{2*} + ... + p_n^1 p_{n*}.$$

Next, we carry out the normalization and repeat the calculations using the iterative algorithm of parallel computations to the declared accuracy.

At the next step of the algorithm, a rectangular matrix Q of size n by m is constructed, where n is the number of questions, m is the number of experts, into which we write the experts' answers to 70 questions. The answer "yes" is denoted as 1, the answer "no" as 0. Next, calculate q_i, where i varies in the range from 1 to n:

$$q_i = q_{1i} * v_1 + q_{2i} * v_2 + ... + q_{mi} * v_m.$$

The We get the resulting vector $q\ (q_1, q_2, ..., q_n)$ – the weights of answers to questions, taking into account the coefficients of the experts' competence. Next, the convolution of the vector q and the vector of group assessment of questions $p\ (p_1, p_2, ..., p_n)$ is calculated. Depending on the result obtained, a recommendation is made to choose an automated, manual or mixed type of testing [22].

6. Conclusion

The development and testing of mobile applications have become one of the most popular tasks in the field of information technology due to the fact that the availability of mobile devices has made them an indispensable tool for solving a large number of tasks: data exchange, quick access to relevant information, mobile payment services, monitoring the flow of business processes, analytical reports and more. The method described in the article allows the testing engineer to make a decision on choosing an approach to testing a software product. In this paper, we approve the application of the selection method of testing type to mobile software product. Testing mobile applications is a separate direction in ensuring the quality of a software product. An engineer involved in a mobile product testing project should be aware of the approaches to the process, as well as special tools and specific checks. From the foregoing, the universality of the method proposed in the article follows, since it applies to mobile, desktop, and web applications. It also has a software implementation.

References

[1] Allen S, Graupera V and Lundrigan L 2010 *Pro Smartphone Cross-Platform Development* 1–14

[2] Choi W 2013 *Proc. of the 2013 ACM Workshop on Mobile Development Lifecycle ser MobileDeLi* 27–28
[3] Crosby P B 1978 *Quality is free* (US: McGraw-Hill Inc) 288 p
[4] Kshirasagar N and Priyadarshi T 2008 *Software testing and quality assurance: theory and practice* (US: A John Wiley & Sons) 616 p
[5] Kitchenham B and Pfleeger S L 1996 *IEEE Software* **January** 12–21
[6] Lewis W E 2000 *Software testing and continuous quality improvement* (US: Auerbach Publishers) 620 p
[7] Galimova E Y and Kovalenko A N 2016 *Vestnik of Don State Technical University* **4** 134–140
[8] Ridene Y and Barbier F 2011 *Proc. of the 5th European Conference on Software Architecture: Companion Volume* 9:1–9:7
[9] Incki K, Arı I and Sozer H 2012 *Proc. of IEEE Sixth International Conference on Software Security and Reliability Companion* 18–22
[10] Chernikov V *DevOps in the service of man: We automate the assembly, distribution and testing of new versions of the application.* Available at: https://xakep.ru/2017/02/17/devops-authomatization/
[11] Kumar V 2016 *Mobile Application Penetration Testing* (Birmingham: Packt Publishing) p 672
[12] Kumar V 2014 Smoke testing *International Journal of Scientific and Research Publications* **4** (2) 57–62
[13] Nikeshin A V, Pakulin N V and Shnitman V Z 2012 *Trudy ISP RAN* **23** 387–404
[14] Grebenyuk V M 2013 *Internet-zhurnal «NAUKOVEDENIE»* 13
[15] Jiayang Liu, Lin Zhong, Jehan Wickramasuriya and Venu Vasudevan 2009 *Pervasive and Mobile Computing* **5 (6)** 657–675
[16] Kazansky N, Lysuk P and Nemcov Y 2018 *Proc. of XV St. Petersburg International Conference Regional informatics "RI-2018"* 314–316
[17] Skachinsky E 2013 *Wireless Technologies* **3** 32–340
[18] Jog Y, Thenuan P, Khanna D and Chavan A 2016 *Indian Journal of Scientific Research* **5 (4)** 188–193
[19] Saaty T L 1985 *Analytical Planning: The Organization of Systems* (New York: Pergamon Press) 208 p
[20] Saaty T L 1987 *Decision Science* **18(2)** 157–177
[21] Davnis V V 2005 *Predictive models of expert preferences: monograph* (Voronezh: Voronezh State University Publishing House) 248 p
[22] Galimova E Y 2020 *E3S Web of Conferences* **177** 02008

Effect of local power systems integration on the distribution network functions

F L Byk, L S Myshkina and Ya A Frolova
Department of Automated Electrical Power Systems, Novosibirsk State Technical Universality, Novosibirsk, Russia
E-mail: Lsmyshkina@gmail.com

Abstract. The Russian energy industry today needs an "energy transition". The key link is the Regional Electric Network Companies as the basis of regional power supply system. There are significant changes in the industry, such as changes in legislation, the emergence of new requirements, changes in the regional fuel landscape. The mass appearance of distributed generation as the basis of local power systems has a great influence on the changes. This has led to new functions for Regional Electric Network Companies. The paper studies the effect of local power systems integration on the expansion of the functions performed by distribution network. Particular attention is paid to demand management for electricity; mutual reservation, multi-agent voltage regulation, functions of local energy systems integration with distribution networks.

1. Introduction

Raise of safety, reliability, quality, availability, environmental friendliness of power supply will ensure high competitiveness of domestic power engineering and its efficiency consequently. The new power engineering transition and digital transformation are aimed at this. The large Regional Electric Network Companies play an important role there as the basis of regional power supply.

Regional Electric Network Company (RENC) is a commercial organization that provides power distribution services using electrical network facilities related to united national (all-Russian) electric network. I.e., RENC provides the transmission and distribution of electricity in the regional power supply system. Wherein, [1] reflects that campaigns have the missions to providing services at an affordable price by carrying out the final consumer's power supplying functions:

- transmission and distribution of electricity;
- maintaining of structural and functional reservation;
- technological connection of new consumers.

These functions are provided by RENC since the creation of network companies. Nowadays the relevance and necessity of marked functions hasn't changed in the Russian power engineering. But new RENC functions have appeared due to the industry development, changes in law, emergence of new requirements, change of fuel-energy resources landscape and other factors. Implementation of them is accompanied by the extraction of new effects, both for network companies and for other subject of power engineering industry.

As result have new way – setting up local power systems based on distributed generation and using non-renewable (traditional) and renewable energy resources (alternative), integration which the greatest influence on the expansion of the functionality of region's network companies.

Thereby the actual tasks are:

- identifying and structuring new functions and effects related to the integration of the local and regional power systems;
- models development for assessing the technical and economic effects of integration.

The objective of the work is to identify new functions for regional electric network companies and their impact on efficiency of power supply.

2. Factors affecting new functions

2.1. Local power system

The efficiency of Russian centralized power supply system is reducing by its existing contradictions. That's why local power systems are emerging. The main factors causing this process are:

- technical and economic inexpediency of centralized power supply in isolated power systems while the availability of effective power plants usage of local or imported fuel;
- high growth rate of the electricity cost and the impossibility of its forecasting in region's systems power supply. In this case is not possible long-term planning of the business subjects` activities. On the other hand, there are modern technologies for efficient energy production on competitive price;
- appearance of new types of fuel, recycling of which is necessary by environmental and other requirements (associated gas, solid household and industrial waste, etc.) and fact the implementation of the total gasification strategy of the country are getting open up new ability to small business subjects. Important factor is availability technologies for using renewable energy sources;
- changes in the composition and structure of consumers while combined with the lack attract investments in the development of regional electrical networks, has led to a decrease in the reliability of electricity supply.

The above reasons have forced consumers to look for ways of reducing dependence of regional power supply systems. At the present time, about 10% of the installed capacity of power plants in Russia accounts for distributed generation of low power with an installed power source capacity up to 25 MW and a generator voltage of 10 kV.

The trend of appearing of local power supply systems is enhanced by the state policy in the energy conservation and energy efficiency. According to [2], the reconstruction of existing medium and large boiler houses should be carried out with the compulsory installation of cogeneration equipment. Consequently, the boiler house is turning into a mini-TPP (Thermal Power Plant) and in addition to the electrical supply of its zone; it will provide electricity to consumers, ensuring an increase in the energy efficiency of the heat and electricity production in the region.

The existing centralized power supply system has alternative options in modern conditions. It is caused by the possibility of making a choice for the consumer: to connect to a regional power supply system, to use its own power supply systems or to connect to local power systems.

In most cases, local power systems operate under conditions of self-balance and have excess (backup) capacities, which follow from the reasons of their appearance [3]. According to the genesis of the Russian power system development, which began in 1920 with known as the GOELRO plan, it may be concluded that the current development of the power industry in Russia is at a new stage of the development spiral (Figure 1).

The obvious next step will be the integration of local and regional power supply systems. "Pilot" integration projects are already being observed in Russia in a several regions. For example, the integration of local power system, providing electricity and heat supply for the "Berezovoe" residential community at Novosibirsk.

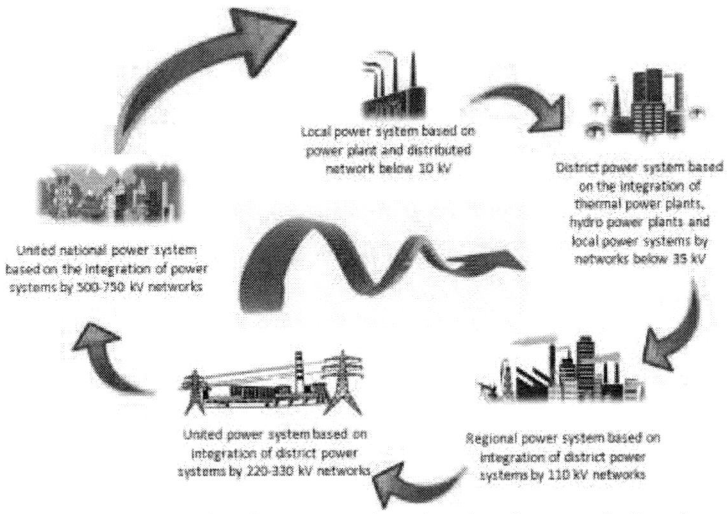

Figure 1. Genesis of power engineering development in Russia.

In general, power systems based on distributed generation are the solution of many problems for the consumer: starting from the need for reliable power supply to ensure the safety of technological processes during production, to fulfilling environmental requirements and solving the problem of reducing the expenses of electricity costs in the cost of their products. The last is most acute today for small and medium-sized businesses, where the cost of power supply makes their products noncompetitive, due to their unbearable cross-subsidizing in the power industry [4].

2.2. Demand Side Response Aggregator (DSR Aggregator)

In 2019 in Russia a new subject of electric power industry appeared due to the insufficiency of the controlled resource and the need to electricity demand management – Demand side response aggregator in the UPS [5]. DSR Aggregators are organizations that identify and coordinate the ability of a group of end users to manage their workload and convert it into goods and services in the electricity, power and system services markets. Essentially they are actively involved in maintaining the balance of production and consumption of electricity. The main objectives of managing the demand for electricity are to reduce the peak load in the power system, what should be accompanied by a decrease in prices on the electricity market and the prevention of excessive capital-intensive construction of power plants and electrical networks.

Conducted field experiments on the coordinated management of reduced electricity consumption have demonstrated their effectiveness, carried out in three regions of the Russian Federation with the participation of nine retail market consumers [6].

This subject is an analogue of operating abroad Demand response [7], where demand management has become a full-fledged tool to ensure the balance of supply and demand in power systems over the past decade. Nowadays, about 20 GW of controlled demand are involved in Europe, while the European Commission estimates the current potential at 160 GW in 2030. The usage of this resource in the US energy system allows avoiding investment of $ 270 billion (in prices of 2016) in the development of energy infrastructure [7].

In Russia, according to preliminary estimates, the potential of demand management can be 6–10 GW for the first price zone and 2–3 GW for the second price zone, which totals up to 13 GW.

The aggregator contracts the services providing with the subjects of the retail market for changing the load of their equipment by a specified amount a certain number of times according to a notification given in advance. Aggregator receives payment in the wholesale electricity and power market for reducing electricity consumption. In turn, the subject provides the aggregator with services for changing the load and receives payment from it.

The project of DSR Aggregators fully complies with the global trend of digitalization of the industry and the transition to a more sophisticated intelligent power engineering industry.

Any electric power industry entities or electricity consumers that have entered into an appropriate agreement will be able to become one of the aggregators. The important point is that as an adjusting resource, the aggregator can use the power of distributed generation and electrical energy storage systems. However, local power systems should be legislatively consolidated as a subject of the power industry, by amending the relevant legal acts. This status actualizes the begun processes of integration of local power systems with regional ones.

3. Functions acquired by the network company

Changes acquired by the emergence of local power systems and their integration allow the RENC to significantly expand the functions performed, including the functions of the energy supply company.

The subject of energy that receives the greatest number of functions, including the "hidden" effects from performing the function of a DSR aggregator, is precisely network companies. Performing the functions of an aggregator opens up the possibility for network companies to form a generalized adjusting resource and ultimately to obtain a significant effect due to participation in the retail and wholesale electricity market. And the most effective resource for managing the demand of electricity is generating capacity of local power systems connected to the networks of the network companies.

RENC will be able to determine in which places of the network it is advisable to emerge and connect local power systems from the standpoint of their interests, primarily to level the network load and control reliability and quality. Reducing the peak loads at the feeding centers make it possible to expect an increase in the durability of network equipment by increasing the irregularity factors and filling (density) of the load curve attached to the feeding center.

Creating an aggregator based on the network organization will allow solving a number of tasks specified in the concept of digital transformation of PJSC "ROSSETI" [8] related to improving the reliability characteristics of the power supply to consumers; increasing the efficiency of the company; increasing the availability of network infrastructure; diversification of the company's business through additional services.

In order to achieve maximum efficiency in the integration of local power systems, the RENC may receive an additional function - the integration of local power systems aimed at improving the efficiency and extracting new system effects that ensure the development of regional power engineering industry. Essentially it says about the emergence of regional power systems.

It is known that the connection of the local power system on the basis of distributed generation is accompanied by an increase of quality of power supply through the implementation of the principles of multi-agent voltage regulation [9]. Maintaining the voltage level in the nodes increases the stability of the motor load, both in normal conditions and in case of remote short circuit faults at high voltage classes.

Due to the adjusting capabilities of generators' automatic excitation control, it is possible to optimize the loss of active power in the networks to which the local power system is connected and to extract energy saving effects.

The integration of local power systems is generally associated with an increase of the reliability of power supply for both consumers of the local power system and consumers connected to the networks of the RENC. The main effect is the appearance of the function of mutual reservation. In [10] it's shown that with the integration of a self-balancing or redundant local power system, the power supply of consumers actually becomes two-way and the frequency and time of consumers' power-off are reduced.

4. Features of integration of local power systems

In accordance with certain functions, it is possible to highlight 3 main sections (cut sets) of networks of integrating the local power system (Figure 2):

1. used to separate the energy source according to the technological conditions and provide power supply to the consumers of the local energy system from the external network and control the energy produced by the energy source;
2. used to isolate the local power system for work in the island mode and control the energy supplied to the network;
3. used to regulate the requirements of the load aggregator in the mode of excess power of the local power system.

The technical conditions for connecting the local power system are different from the conditions for connecting distributed generation, in accordance with which a power plants' system of power output is being developed. That's why the connection of local power systems assumes the presence of appropriate automation, performing a number of functions, including:

- voltage regulation in parallel operation;
- prevention and elimination of interruptions of regime restrictions on the loading of network equipment at the feeding center in case of transformer failures at 110/10 kV substation and supplying it 110 kV grids.

An automatic control system for the parallel operation of low-power power plants with their direct connection to the networks of powerful power systems is developed by Novosibirsk State Technical University. Its production is carried out by two Novosibirsk companies: LLC "Modular Systems Tornado" and JSC "IAES".

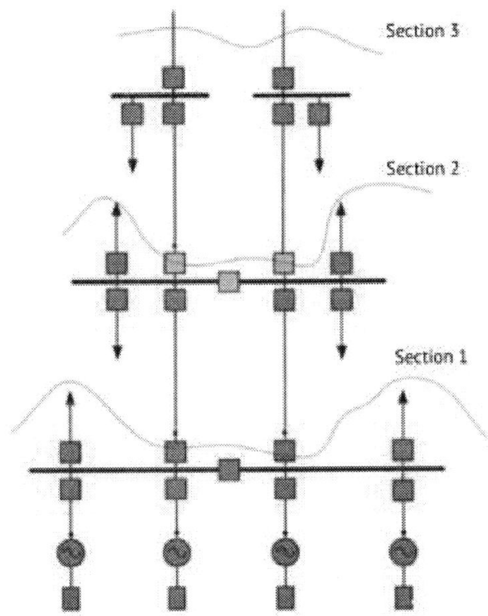

Figure 2. Controlled sections (cut sets) of network during integration.

The following idea and method are proposed for integration: advanced balanced division of the system according to a priori fixed cut sets of the network in interruption of the normal mode with the transition to the island mode with the subsequent automatic restoration of synchronism and normal mode with the required equipment load [9–11].

The proposed method of parallel operation with advanced balanced division of the system is aimed at limiting short-circuit currents, ground-fault currents. It helps to prevent interruptions of the stability of parallel operation with the emergence of asynchronized modes, eliminating shock moments on the shafts of synchronized machines and the need to coordinate external network protection with the protection and automation of the network being connected with plant [11,12].

The usage of this kind of automation for the integration of local and regional power systems will allow to fully implement the control and management of all concerned sections to perform new functions by the RENC.

5. Effects of energy saving and mutual reservation

A distribution network scheme (Figure 3), consisting of two energy districts, was selected to analyze the effectiveness of the implementation of new functions by the network organization. The right part of the scheme is the local power system connected to the external network of the RENC. There is mainly domestic consumption (70%) with an insignificant amount of non-production enterprises related to the service sphere (30%) in the structure of load of the local power system. The summary load is 3 MW. The generating capacities are represented by three 2 MW generators.

Section 1 allows to determine the output in the local power system. Cable communication lines between the local power system and external network form a section 2. Section 3 is formed by connections between the 110/10 kV power supply feeding bus and the external network power system region.

Proposals to assess the effect of mutual reservation of connecting local power systems were proposed by the authors at [10]. The values of structural and functional reliability of network are proposed to be determined on the basis of the failure frequency and the recovery time of the main network equipment, its structure and topology, incorporated in the engineering of reservation capabilities for determining the indicators according to the current order of incentives to improve reliability [13].

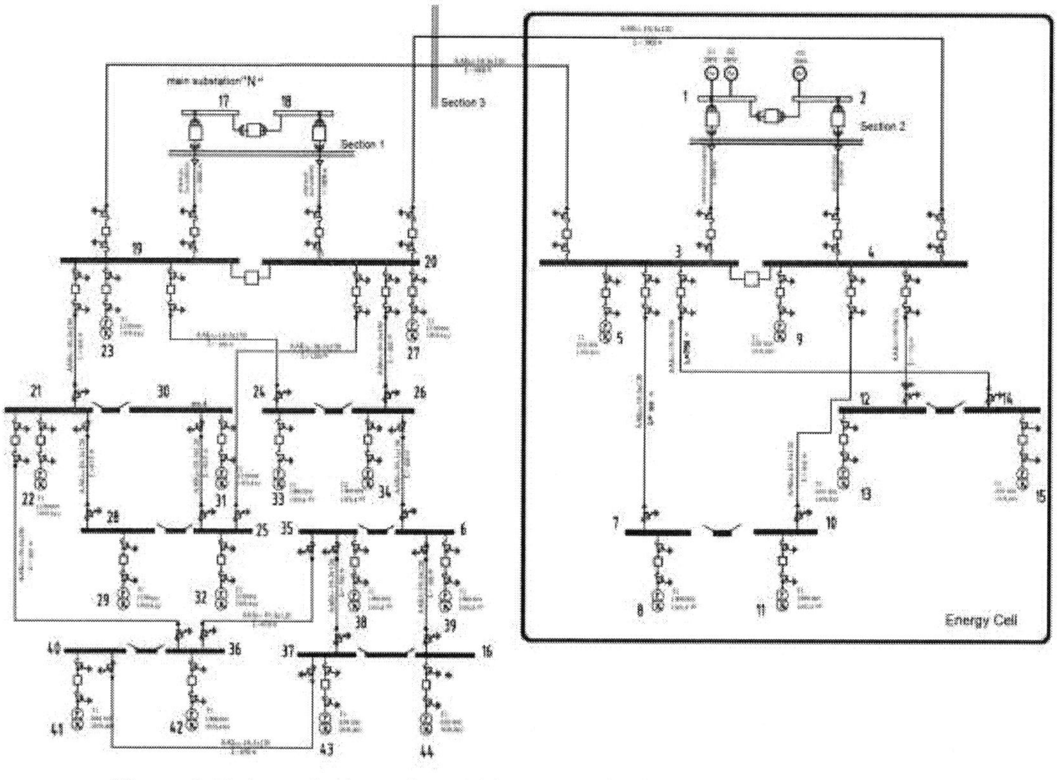

Figure 3. Estimated scheme for a 10 kV network of two power system regions.

Based on the values of reliability indicators and the restoration of network equipment, the structure and composition of the load, the values of indicative indicators of reliability of power supply of the areas before integration and after were determined (table 1). A multiple improvement in the reliability of power supply consumers is obvious.

Table 1. Values of indicative reliability indicators.

Region	Indicator	Before integration	After integration*
LPS	SAIDI	1.43E-03	5.01E-05
	SAIFI	1.78E-04	1.25E-06
DNet	SAIDI	4.42E-04	2.05E-06
	SAIFI	5.53E-05	5.12E-06

** - under the assumption of full mutual reservation*

Improving the reliability of power supply is an effect for RENC whose monetization mechanism is known and used in Russia. According to Resolution of the Government of the Russian Federation of 31.12.2009 N 1220, reaching certain values of reliability and quality indicators can lead to increase the tariff for electric power transmission by 2%.

The same monetization mechanism allows extracting the economic effect of improving the quality of power supply in the implementation of the principles of multi-agent voltage regulation. Calculations of electric modes show a decrease in the standard deviation of the voltage at the nodes of the 10 kV network as the result of multi-agent voltage regulation by attracting the aggregates' automatic excitation control in performing both the functions of mutual reservation and the DSR Aggregator, table 2.

Table 2. Mode indicators at peak load.

Indicator	*Before integration*		**Performing functions** *Mutual reservation*		*DSR Aggregator*	
	DNet	*LPS*	*DNet*	*LPS*	*DNet*	*LPS*
Standart deviation of the voltage at the nodes, %	1.418	4.014	1.441	0.611	1.214	1.002
Power losses in network, MW	0.142	0.054	0.139	0.062	0.122	0.116

It is modeled that the power reserve available for isolated operation in the local power system is used as a control resource when the network company is performing functions of DSR Aggregator in UPS. It's accepted that the removal of peaks is carried out for 4 hours per day (Figure 4).

Non-used power of distributed generation at the local power system is provided to the power district of the external network in cut set 2 to achieve energy efficiency and energy saving. It allows to reduce power losses in the network by 1.71 MW × h per day. The division of generating capacity of the local power system by purpose is shown in Figure 5.

The mode of peak load of generation equipment of distributed generation of the local power system is giving the highest efficiency in terms of improving energy efficiency, energy saving, quality and reliability of power supply. That's why this mode is recommended to consider as the main one in justifying technical solutions for integration with the regional power supply system [14].

It is obvious that the maximization of installed capacity ratios will accompany various economic effects. These effects are significant for owners of distributed generation and don't reduce the level of reliability of power supply to consumers in the local power network relative to available during isolated work.

IOP Conf. Series: Materials Science and Engineering 1019 (2021) 012003 doi:10.1088/1757-899X/1019/1/012003

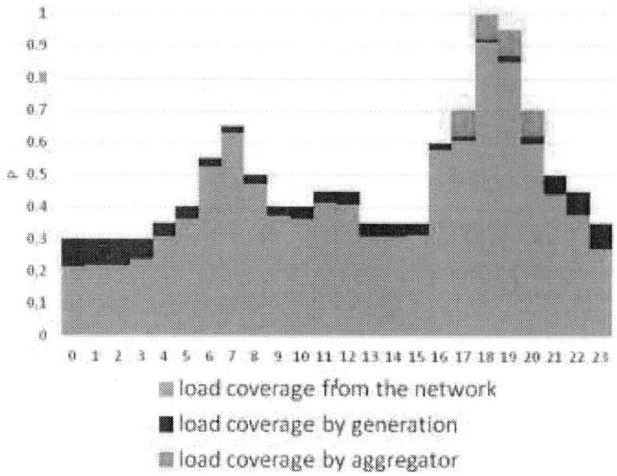

Figure 4. Load coverage curve of the external network power system.

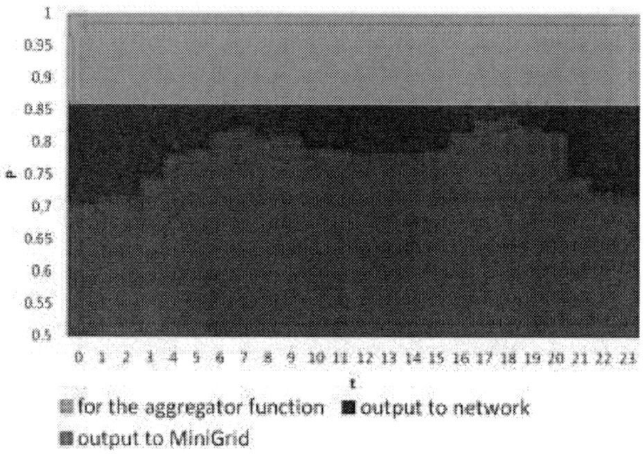

Figure 5. Objectives of loading the generating capacity at the local power system.

For the RENC it's accompanied by the redistribution of power flows in the 35–110 kV network, removal of restrictions from the "closed" feeding centers and new technological connection of consumers.

An important aspect of the integration process of local and regional power systems is the coincidence of interests of all subjects - owners of distributed generation and consumers of both regional and local power systems, and territorial grid organization.

6. Conclusion

Integration of local power systems with regional power systems is the pattern of development of the power industry in Russia, which coincides with global trends

RENCs are an integral element of the regional power supply system, traditionally performing the functions of transmission and distribution of electricity and power to end consumers. The development of the industry provides opportunities to expand the functions of network companies. New functions are emerging with the advent and connection of local power systems to the networks. These include: demand management for electricity; mutual reservation; multi-agent voltage regulation. It increases energy safety, energy independence, energy efficiency and energy saving, because the integration of local power systems leads to the emergence of a competitive regional power system based on distributed generation.

However, it is important to note that in order to implement most of the functions, local power systems should receive a fixed status of a power engineering subject and a retail market subject.

The most important function for territorial grid organizations, by existing commitments to implement digital network transformation, is the demand management function implemented by the DSR Aggregator in the UPS of Russia. Network organizations will be able to streamline the process of integrating existing local power systems and the emergence of new ones, determining the necessary place in the network, which will be reflected in the development of "Schemes and programs of the perspective development of the power industry of the regions"

Obviously, the integration will be accompanied by significant systemic technical and economic effects while reducing the expenses into region's networks development of higher and medium voltage, including in terms of improving the reliability and quality of power supply, reducing losses in networks, improving energy efficiency and energy saving. The number and size of system effects will be largely determined by the circuit-mode conditions in which new functions and their combinations are implemented.

Acknowledgment
The work was financially supported by Novosibirsk State Technical University (project C20-16).

References

[1] Federal Grid Company of Unified Energy System (PJSC FGC UES). Available at: https://www.fsk-ees.ru
[2] The federal law of 27.07.2010 *On heating* N 190-FL
[3] BateninaV, Bushuev V, Voropai N 2017 *Innovative electric power industry – 21* (Moscow: Energy)
[4] Zolotova I, Dolmatov I, Minkova V 2017 *Power engineering policy* **2** 51–59
[5] Byk F, Myshkina L, 2019 *E3S Web of Conferences* **139** 5 doi: 10.1051/e3sconf/201913901013
[6] Natural experiment of aggregator. Available at: http://www.so-cdu.ru
[7] Global Demand Response Capacity is Expected to Grow to 144 GW in 2025. Available at: https://www.navigantresearch.com/news-and-views/global-demand-response-capacity-is-expected-to-grow-to-144-gw-in-2025
[8] Conception "Digital Transformation 2030" PJSC "Rosseti". Available at: http://www.rosseti.ru/investment/Kontseptsiya_Tsifrovaya_transformatsiya_2030.pdf
[9] Kardzhaubaev N, Fishov A 2018 *Gazette of Irkutst State Technical University* **6** 183–195 doi:10.21285/1814-3520-2018-6-183-195
[10] Byk F, Myshkina L, Khokhlova K 2017 *Advances in Engineering Research* **133** 525–530 doi:10.2991/aime-17.2017.85.
[11] Fishov A, Marchenko A, Murashkina I *et al* 2018 *Procceding of 14th International Scientific-Technical Conference on Actual Problems of Electronic Instrument Engineering (APEIE 2018)* 108–114
[12] Fishov A, Marchenko A, Mukatov B 2018 *Russia Federation patent No. 2662728*
[13] Mitra J, Vallem V, Singh C 2016 *IEEE Trans. Ind. Appl.* **52** (3) 1989–1997 doi: 10.1109/TIA.2016.2517067
[14] Byk F, Frolova Ya, Myshkina L, 2019 *E3S Web of Conferences* **114** 6 doi: 10.1051/e3sconf/201911405007

IOP Conf. Series: Materials Science and Engineering 1019 (2021) 012004 doi:10.1088/1757-899X/1019/1/012004

Powerflow calculations in powersystems considering traction load

V V Shestakova, I M Kats and A V Darkhanova
School of Energy & Power Engineering, Tomsk Polytechnic University, Tomsk, Russia
E-mail: shestakova@tpu.ru

Abstract. Nowadays, the negative impact of traction load on power system operation poses a serious problem that may lead to false tripping or even failures of relay protection devices and reduce the quality of electrical energy. This situation is typical for the Zabaikal power system, in which in some areas the share of electricity consumption by railway reaches 70% of total consumption. A computing tool, that would allow simultaneous analysis of the modes, both in the utility's grid and in the traction network, do not exist. The inability to carry out such analysis often leads to inconsistency in the actions of control centers and railway transport authorities, especially when maintenance planning. Mathematic modeling in such software systems as "Mustang", "RastrWin3", "MathCAD", "Matlab/Simulink", "PSCAD", "Kortes". The developed model represents Zabaikal power system and contain detailed railway between substation Razmahnino – Shilka, Chita – Mogocha. Negative-sequence voltage unbalance factors were calculated for the case of train movement between substation Razmahnino and substation Shilka. Also, the necessity of back-up relays tripping values correction is stated. It was shown that for the powerflow calculations taking into account the traction load, it is rational to use a complex mathematical model, which uses compatible software systems with the ability of quick and easy data exchange.

1. Introduction

Nowadays, a negative impact of traction load on utility grid operation is widely observed in numerous publications [1–6]. This effect is strongly pronounced when railway consume more than 50% of total demand in the powersystem.

This is typical for Zabaikal powersystem (ZPS), where separate railway sections consume up to 70% from total demand. Traffic flow is taking off from year to year and this fact deteriorates the current situation. According to [3] it is planned to increase freight traffic through Transib railway, hence, through Zabaikal railway too until 2030.

Badly coordinated actions both of the control centers and railway authorities may often lead to mistakes when network calculations accounting the railway influence. As a consequence, equipment failures, supply interruption both of traction and non-traction consumers.

For instance, in ZPS trains were de-energized because of 220 kV overhead transmission line maintenance works. Besides, false tripping of back-up relays is also possible.

2. Problem formulation

The challenge is there are no any means that would facilitate utility system mode analysis considering traction systems impact. As it was mentioned before, this fact causes inappropriate responses of control centers and railway authorities, especially while maintenance planning. The aim of this paper is to develop a mathematic model, which is based on various software systems, to estimate mutual impact of supply and traction systems.

3. Traction systems power supply modeling principles

The main research method was mathematical modeling in such software systems as Mustang, RastrWin3, MathCAD, Matlab/Simulink, PSCAD, Kortes. Each one was used in accordance with theirs application area.

RastrWin3 has wide opportunities in powerflow calculations and steady-state modes optimization. Currently, this software is a major tool in all branches of System Operator of the United Power System. The RastrWin 3 database of the Zabaikal control area was a basis for our researchings. Should be noted that Zabakal power system is a "deadlock" as it operates apart from the Far-East power system.

Mustang was also main software for steady-state calculations and modeling electromechanial transients until recently. Now, it has been replaced by Eurostag, a more pioneering software package. Mustang has instruments to model automation components, for instance, automatic voltage regulation (AVR), automatic generation control (AGC) and so forth. Mustang as well as RastrWin3 (their databases are compatible) uses positive sequence schemes and this fact doesn't allow to determine electrical quantities in case of non-symmetric faults, besides, non-symmetric loads modelling is also impossible.

In Tomsk Polytechnic University Shemwizard, an add-on for Mustang, was developed. It automatically forms negative (NS) and zero sequence (ZS) schemes provided with positive sequence (PS) scheme. Shemwizard gave an opportunity to create the equilvalent sequence network, which is based on the RastrWin 3 PS scheme. This is achieved by varying the impedance of the branch connecting PS and NS schemes in accordance with the Fortescue method. The total number of nodes and branches in the PS scheme is 238 and 277 respectively. The model contains generators with AVR and AGC, transformers, loads and 110–220 kV transmission lines, the Gusinoozerskaya's power plant buses are taken as a swing node. A part of the 220 kV traction transit is showed on the Figure 1.

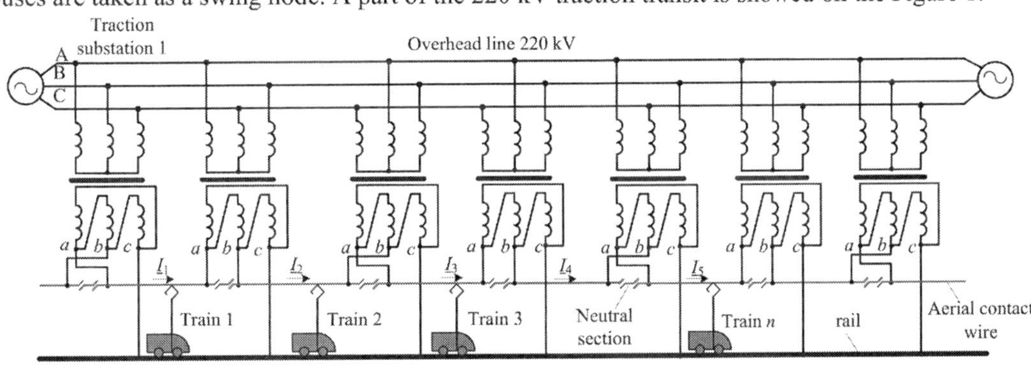

Figure 1. Connection scheme of traction substations to the longitudinal power supply line.

Mustang represents results as envelopes of complex PS and NS currents and voltages, which are time-varying. Such representation is not handy for evaluation, so by means of MathCAD an algorithm that returns instantaneous currents and voltage values for each phase was developed.

For more accurate modeling of 220 kV supply system between substation Darasun and Kholbon authors used Matlab/Simulink and PSCAD. This segment includes 4 traction substations. Initial voltage levels and generating power were based on steady-state mode parameters calculated by Mustang.

By means of "MatLab/Simulink" power circuits of a locomotive, including reversible conveter and traction motors were created. Modeling were performed for three phases.

When performing powerflow calculations in MatLab/Simulink, it is possible to take into the account the type of traction motor. Mostly on Zabaikal railway trains travel that driven by pulse control DC motors. Recently, locomotives with asynchronous traction motors have appeared on tracks of Zabaikal railway.

Supply system of the Trans-Siberian railway uses the 1×25 kV scheme implemented by three-winding transformers. Interfacing is conventional - two phase wires to the catenary, third phase is grounded. To reduce the non-symmetric effects in traction systems phasing combinations of traction transformers along the trajectory of a railway system was provided.

In author's opinion, Matlab/Simulink and PSCAD are not suitable for modeling ZPS, which consists of several hundred nodes, as these software packages implement advanced mathematical apparatus, hence this fact would slow down large schemes calculation. In addition, it may be problematic to tune AVR and AGC systems.

Kortes, a specialized software system for railway research centres was used to determine the power consumed by non-symmetric traction load. Kortes supposed to estimate modes only for railway systems (27.5 kV), however, it allows to take into consideration plane and profile conditions, train working diagram and mass of trains. It's admitted that the mass of trains is between 4000 and 8000 tons. Modern locomotives travelling along Zabaikal railway consume from 2 to 10 MW depending on factors mentioned above.

Should be stated that primary power supply systems are represented in Kortes as infinite buses, where constant voltage level is sustained.

To sum up, created model allows to estimate currents and voltages in each branch and node in cases of both unbalanced short-circuit and non-symmetric loading conditions considering protective relaying and remedial action schemes operation.

4. Findings

In lines of the research, situations of simultaneous train moving were modelled. In such cases a substantial voltage decreasing takes place in utility grid. When trains work accordingly to the schedule and supply system operates normally, trains don't stop due to power outages. If several sections 220 kV supply system voltage are being repairing, it becomes below critical level (18 kV) and trains stop. On the other hand, voltage rises up to 110–130% of nominal voltage if there are no locomotives on the track. This also affects on equipment condition.

The next research area is verification of back-up relays set-point values. Calculations were performed considering electric values changing during electromechanical transient processes, still initial conditions for repeated switchings were taken into account. Experiments showed that it is necessary to calibrate thresholds of back-up transmission lines protection. In some emergency and restorative modes it is needed to adjust setting values in back-up relays of Kharanorskaya power plant main assets.

Below simulation results when connecting a train to the buses of traction substations are given. The train consumes about 3 MW. 35 kV buses of substation Karymskaya and 11 kV buses of substation Urulga were chosen for voltage monitoring. Asymmetry of currents were controlled on lines Darasun-Karymskaya and Darasun-Urulga. In the time ranges 0.04 – 0.12 s and 0.2 – 0.28 s the train receives power from substation Karymskaya and substation Razmahnino respectively, which causes an increase in currents of PS and the occurring of the currents of NS.

In the Figure 2*a*, positive and negative sequence currents I1, I2 flowing through the Darasun-Karymskaya and theirs phase angles $\delta 1$, $\delta 2$ calculated in "Mustang" are displayed. The phase currents oscillogram was obtained using RMS values and is presented on the Figure 2*b*. These calculation results coincide with the experimental data obtained in [4].

The authors tend to believe that using a probabilistic approach in modeling trains movement is preferable for powerflow calculations considering railway demand. The key difference between the calculations of traction networks from the calculations of networks with static loads is continuous change in train`s location and consumed power (due to plane and profile conditions). For experiments it was considered that average power of train is 3 MW and traffic interval is 25 minutes. Also, the number of trains located near the substation zone at the same time may vary. Hence, this fact causes fluctuations of voltage and current asymmetry levels.

The results are given in Figure 3. Calculations showed that increasing of the unit capacity and reducing of the traffic interval leads to decreasing of voltage levels and has slight influence on negative sequence voltage unbalance (NSVU) factor. According to the simulation results change in the train unit power by 5 times causes factor deviations by less than 1 %. Monte-Carlo method was used. It was shown to achieve confidence coefficient of 0.95 and accuracy of 0.1 it's needed to perform about 600 iterations.

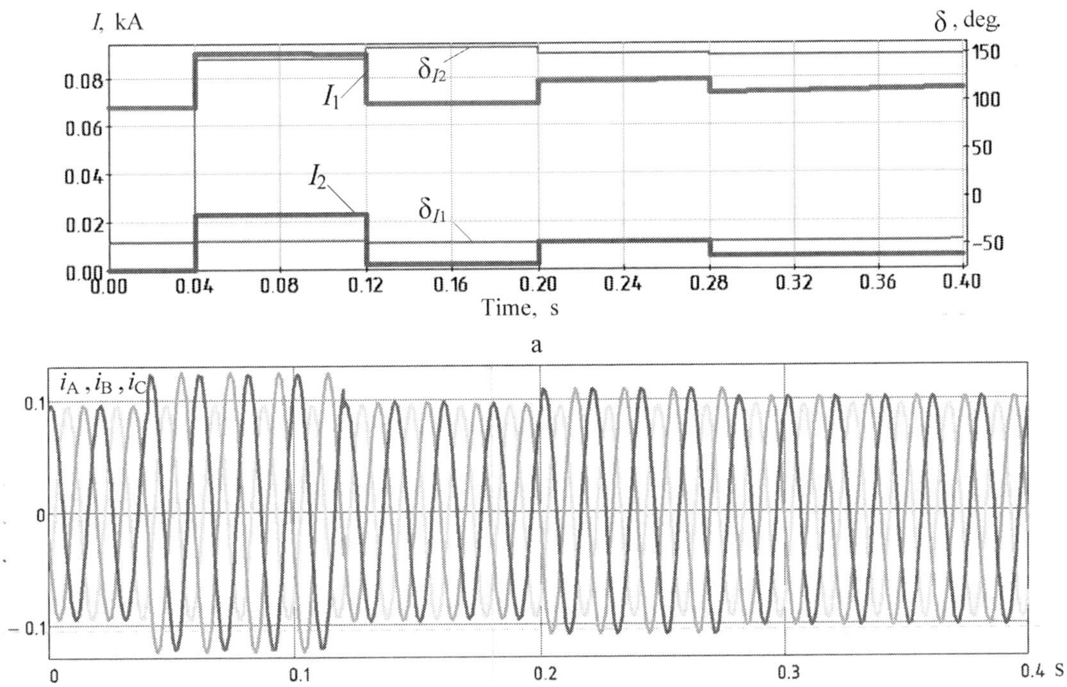

Figure 2. Connection scheme of traction substations to the longitudinal power supply line.

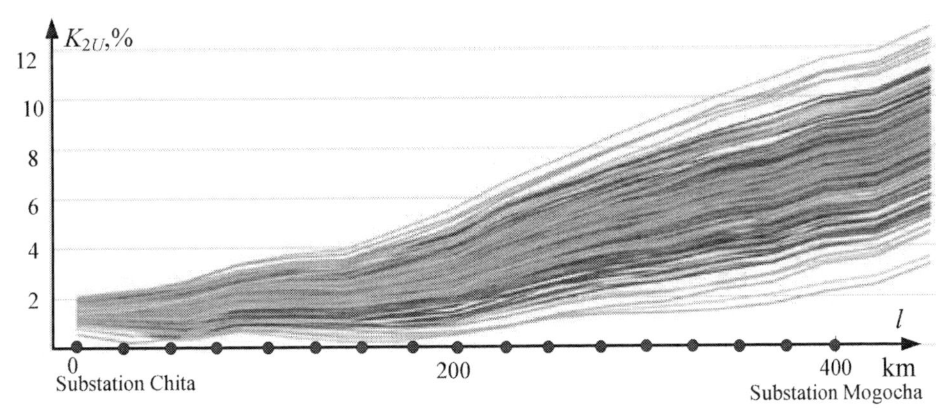

Figure 3. Change of negative voltage unbalance ratio (600 experiments).

As you can see the negative voltage unbalance ratio increases to the transit end. The value of the coefficient significantly exceeds the permissible level.

At present time, we are offered a range of balancing devices needed for current and voltage asymmetry compensation such as shunt compensation devices, series compensation devices, static var compensators (SVC), static synchronous compensator (STATCOM) and booster transformers [1–3].

According to [6] shunt compensation devices must be operated reliably and have rapid response of control systems. Series compensation devices possess positive regulation effect that obviously is a merit of such devices. However, series compensation condensers are vulnerable against ultraharmonics, which always take place in AC traction systems. SVC and STATCOM are devices of excellent perfomance due to relatively fast control systems, however tuning them properly may be a challenging process.

Installing booster transformer would be simpler and more reliable solution. It represents a three-phase transformer, which has its terminals connected in delta scheme to the catenary. Connection scheme of the booster transformer is shown on Figure 4.

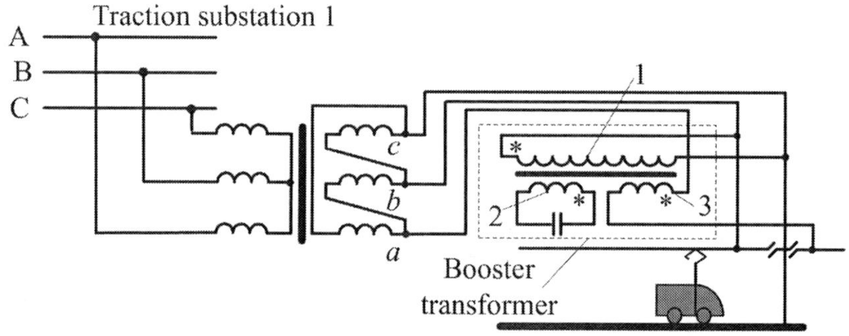

Figure 4. Connection scheme of the booster transformer: 1 – supply winding, 2 – compensating winding, 3 – auxiliary winding.

5. Conclusion

Complex mathematical model based on several software systems, that enable fast data exchange, is more advisable for powerflow calculations considering railway systems impact. It also obtains development dynamics of the processes in the integrated three phase - single phase electric scheme. This model may be applied for calculating parameters of booster transformer, relays set-point values correction and for maintenance planning.

References

[1] Milesevik B, Uglesic I and Filipovic-Grcic B 2016 *Electric Power Systems Research* **138** 172–179

[2] Zakaryukin V P, Kryukov A V and Cherepanov A V 2016 *Proceedings of Irkutsk State Technical University* **20** (11) 133–142

[3] Zakaryukin V P, Kryukov A V and Avdienko I M 2016 *Izvestiya Transsiba Journal of Transsib Railway Studies* **2**(26) 70–78

[4] Romanova V V and Hromov S V 2015 *Russian energy in the 21st century* **2**(3) 45–52

[5] Zupan A, Tomasovic Teklic A, Filipovic-Grcic B 2013 *Proceedings of IEEE EuroCon* (Zagreb) 844 – 849 doi: 10.1109/EUROCON.2013.6625081

[6] Albadi M, Al-Hinai A, Al-Badi A H 2014 *Proceedings of Int. Conference on Renewable Energies and Power Quality* 155–159 doi: 10/24084/repgj12.218

IOP Conf. Series: Materials Science and Engineering 1019 (2021) 012005 doi:10.1088/1757-899X/1019/1/012005

Transverse profile forming for high energy electron beams using 3D-printed plastic samples

A A Bulavskaya[1], Yu M Cherepennikov[2], A A Grigorieva[1], B M Gavrikov[3], A A Loginova[4], I A Miloichikova[2,5], N E Toropkov[6,7] and S G Stuchebrov[1]
[1] Research School of High-Energy Physics, Tomsk Polytechnic University, Tomsk, Russia
[2] School of Nuclear Science & Engineering, Tomsk Polytechnic University, Tomsk, Russia
[3] Moscow City Oncology Hospital №62, Moscow, Russia
[4] Radiotherapy Department Dmitry Rogachev National Research Center of Pediatric Hematology, Oncology and Immunology, Moscow, Russia
[5] Cancer Research Institute of Tomsk NRMC RAS, Tomsk, Russia
[6] Institute of Strength Physics and Materials Science of Siberian Branch of Russian Academy of Sciences, Tomsk, Russia
[7] School of Advanced Manufacturing Technologies, Tomsk Polytechnic University, Tomsk, Russia
E-mail: bulavskaya@tpu.ru

Abstract. The work presents application of previously proposed approach for transverse profile forming of high energy electron beams with plastic samples made using 3D printing technology. The collimator made of PLA (polylactide) plastic is designed and produced in a frame of this work. The results of transverse profile forming for 6 MeV electron beam of clinical linear accelerator Elekta Synergy using designed collimator is presented. Obtained results comparison with data from XiO dosimetry planning system shows good agreement of calculated and experimentally formed by plastic collimator electron beam dose fields.

1. Introduction

The wide range of applications where electron beams are used causes the presence of many different requirements for the form of their profiles [1, 2], that provides necessity to develop new approaches to their modification [3]. One of these approaches is to use new materials for the forming samples creation. In the framework of this study, it is proposed to form a transverse profile of high-energy electron beams with plastic samples made using 3D printing technology.

The use of rapid prototyping methods in medical physics is gaining more and more attention among researchers, since their advantages allow increasing the efficiency of patient treatment [4]. Additive technologies have found various applications in medicine. For example, 3D printing is used for surgical planning, where medical imaging data is used to make patient-specific phantoms [5]. Preliminary training with these phantoms lets surgeons to minimize risk in real procedure [4, 5]. Besides, samples made using additive methods is used for tutorial purposes [5], for structural implants production [6–8], and for tissue scaffolds [9]. Recent studies with clinical phantoms is dedicated to 3D printing technology application for boluses production, which provides percentage depth dose shifting in external beam radiation therapy sessions with photon and electron beam and in contact X-ray brachytherapy with gamma rays [10, 11].

Currently, metal is the most widely used material for manufacturing of samples for electron beam transverse profile forming [12–14], Difficult process of metal collimator manufacturing limits using of such devices since for working with metal specially trained and high qualified staff is necessary as well as specially equipped room, that is hard to provide in medical center [14]. An alternative approach is to use polymer samples produced with 3D printing technology [15,16]. This approach provides shorter manufacturing time of forming samples that is important because several oncology diseases are characterized by very high progression dynamics and high response to therapy. Besides, this approach has relatively low cost of equipment and consumables. The latter causes commercial prospects of the developed method [4].

To investigated prospects of the proposed approach to transverse profile forming for high energy electron beams using 3D printed plastic samples we produce plastic collimator allowing forming electron beam of linear accelerator [14].

2. Materials and methods

2.1. Radiation source.

As a source of radiation we use electron linear accelerator Elekta Synergy (Elekta Instrument AB, Stockholm, Sweden) [17]. This accelerator is designed for radiation therapy sessions and provides necessary parameters of the radiation field, particularly the uniformity of dose distribution. The test of proposed method to transverse profile forming for high energy electron beams is carried out with 6 MeV beam energy.

2.2. Production of plastic collimator.

The collimator made of PLA plastic by 3D printing was specially designed for this method [18, 15]. PLA plastic produced by Bestfilament (Bestfilament, Tomsk, Russian Federation) is chosen due to high availability and high strength of printed samples [18]. The shape of the collimation hole corresponds the data calculated with dose planning system XiO (Elekta Instrument AB, Stockholm, Sweden) [19]. The thickness of the plastic sample is chosen accordingly to the previous investigations [20] and equals 8 cm. This thickness is enough for total absorption of electron beam with energies up to 12 MeV.

To print plastic collimator with hole corresponding to coordinates calculated in XiO planning system, it's 3D model is designed using Autodesk Inventor software (Autodesk Inc., San Rafael, California, United States) in STL format (Figure 1), which can be used for sample printing by rapid prototyping systems [21, 15].

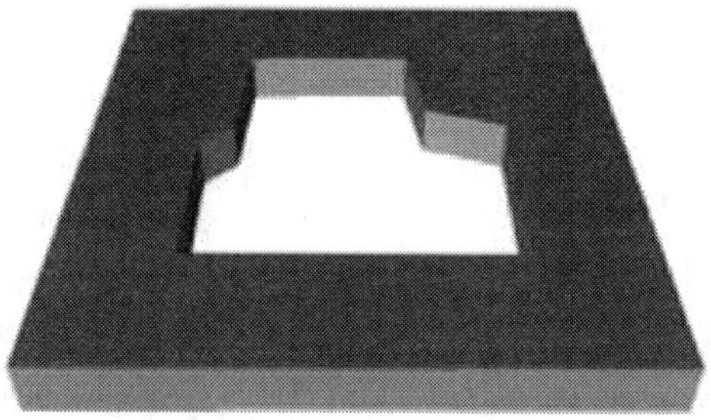

Figure 1. Three-dimensional model of plastic collimator.

The forming sample is produced with printer Prusa Mk2 (Prusa Research S.R.O., Prvniho Pluku, Praha) with modified printhead using fused deposition modeling approach [22]. The collimator was printed with 100% filling of sample volume by the material. Figure 2 shows process of plastic collimator production.

Figure 2. Production of plastic collimator using fused deposition method.

2.3. Experimental scheme.

In the experiment, conventional rectangular applicator of electron linear accelerator Elekta Synergy (10×10 cm^2 sized radiation field) is supplemented with a previously described custom plastic collimator [17]. Figure 3 demonstrates experimental scheme.

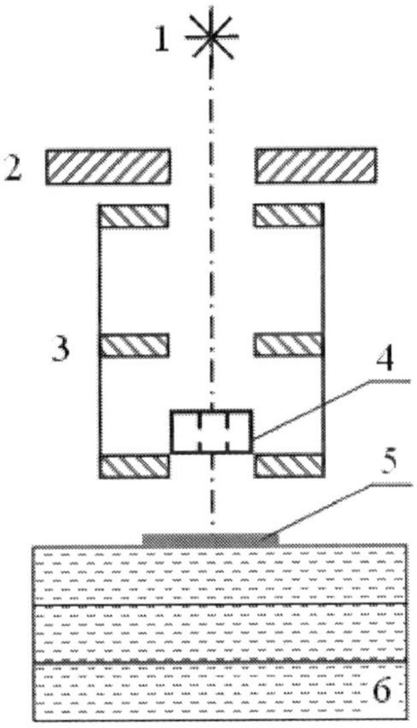

Figure 3. Experimental scheme: 1 – electron source; 2 – primary collimator; 3 – metal applicator; 4 – collimator; 5 – dosimetry film; 6 – tissue equivalent phantom.

To estimate dose field of formed electron beam we use pre-calibrated dosimetry film Gafchromic EBT3 (Ashland Inc., Covington, Kentucky, United States) [23], fixed to the surface of the solid tissue equivalent phantom SP34 (IBA, Louvain-La-Neuve, Belgium) [24]. The source-to-surface distance equals 100 cm, collimator front edge to the surface distance – 6.3 cm. Dosimetry film is fixed perpendicularly to the beam propagation.

3. Results and discussions

Experimental investigation of the applicability 3D printed plastic collimator produced by fused deposition method is performed in according with the experimental scheme shown in Figure 3.

The plan of irradiation is specially calculated in a frame of this work using dose planning system XiO. Electron beam with complex shape with 6 MeV energy falls on solid tissue equivalent phantom SP34 perpendicularly. Irradiation is normalized on the dose at a 1.1 cm depth of phantom. Results of calculation are 3D dose distribution in a phantom volume. Dose distribution on the phantom surface is a two-dimensional matrix contained dose value with resolution corresponding to the size of the voxel chosen for calculation. In the work this size equals 2 mm^3.

Experimental dose distributions of the high energy electron beam is observed by the dosimetry film Gafchromic EBT3 (Figure 4). Color flatbed scanner Epson Perfection V750 Pro (Seiko Epson Corporation, Suwa, Nagano, Japan) [25] is used to digitalize dosimetry film measurements.

Figure 4. Radiation field of the high-energy electron beam formed by 3D printed collimator.

Digitalized data observed with dosimetry films is processed with ad-hoc program code in MATLAB package (MathWorks Inc., Natick, Massachusetts, United States) [26]. The dose uncertainty is about 5% [27–29].

To compare calculated and experimental data the dose distributions from dose planning system XiO is overplayed with results observed with dosimetry film. Figure 5 shows comparison of the calculated in planning system and experimentally observed electron beam radiation field.

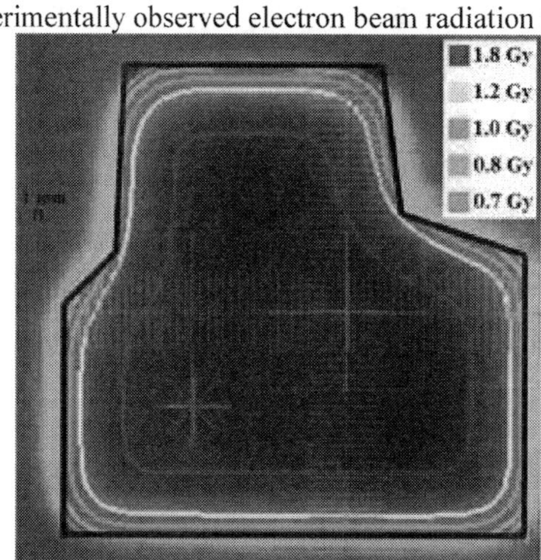

Figure 5. Comparison results of calculated in dose planning system XiO and formed by the 3D printed PLA plastic collimator electron beam radiation fields.

Figure 5 Demonstrates both quantitative matching of the calculated and experimental electron beam radiation field dose distribution, and quality matching of field shapes and collimation hole.

As it seen, the dose beyond the collimator hole is not higher than 0.1 Gy that proves effective absorption of the electron beam in the material of collimator.

The latter proves that samples made of PLA plastic using fused deposition modeling allows collimation of the high energy electron beam.

4. Conclusion

The work demonstrates, that samples made of PLA plastic using fused deposition modeling maintain both mechanical and absorption properties, after absorbing of radiation dose up to 1.5 kGy, that proves prospects of this material using for production of samples forming depth dose distribution of electron beams.

Acknowledgment

This work is supported by the Russian Science Foundation, project No. 18-79-10052.

References

[1] Hossain K, Maruthi Y A, Das N L, Rawat K P, Sarma K S 2018 *Appl. Water Sci.* **8** (1) 1–11
[2] Chaudhary N, Singh A, Debnath A K, Acharya D K 2015 *Solid State Phenom.* **239** 72–97
[3] Miloichikova I A 2016 *Phys. of Part. and Nucl. Lett.* **13**(7) 890–892
[4] Gibson I 2006 *Rapid Prototyping J.* **12**(1) 53–58
[5] McGurk M, Amis A A, Potamianos N M 1997 *Ann. of the R. College of Surgeons of England* **79**(3) 169–174
[6] Imanishi J, and Choong P F M 2015 *Int. J of Surgery Case Rep.* **10** 83–87
[7] Gopakumar S 2004 *Rapid Prototyping J.* **10**(3) 207–211
[8] Swann S 1996 *Rapid Prototyping J.* **2**(4) 41–46
[9] Melchels F et al. 2011 *Biofabrication.* **3**(3) 034114
[10] Su S, Moran K, and Robar J L 2014 J. *of Appl. Clin. Med. Phys.* **15**(4) 194–211
[11] Zhao Y et al. 2017 *Med. Dosim.*. **42**(2) 150–155
[12] Björk P, and Knöös P T 2002 *Phys. in Med. & Biol.* **47**(22) 4019–4041
[13] Delaperrière M 2003 *PRIMUS Basic Functional Description* (Erlangen: Siemens ag medical solutions CS TSC2*)* p 92
[14] Keyvanloo A 2016 *Phys. in Med. & Biol.* **61**(9) 3527–3539
[15] France A K 2013 *Make: 3D Printing: The Essential Guide to 3D Printers* (Sebastopol: Maker Media, Inc.) p 230
[16] I Miloichikova I A, Krasnykh A A, Stuchebrov S G 2016 *AIP Conf. Proc.* **1772** 060018
[17] Elekta Synergy Digital accelerator for advanced IGRT. Available at: https://www.elekta.com/radiotherapy/treatment-delivery-systems/elekta-synergy/
[18] Bestfilament: PLA plastic. Available at: https://bestfilament.ru/pla-1-1.75-black/
[19] XiO Comprehensive RTP system. Available at: https://www.elekta.com/dam/jcr:12b6794c-a8e8-4672-805a-94b964f4ac7a/XiO%C2%AE%20Brochure.pdf
[20] Miloichikova I, Cherepennikov Y, Gavrikov B, Krasnykh A, and Stuchebrov S 2019 *Radiother. and Oncol.* **133** S1008
[21] Autodesk Inventor 2018: Inventor API User's Manual. Available at: http://help.autodesk.com/view/INVNTOR/2018/ENU/?guid=GUID-5901102A-F148-4CD4-AF50-26E2AFDEE6A7
[22] 3D printing handbook: User Manual for 3D printers. Available at: https://prusa3d.com/downloads/manual/prusa3d_manual_mk2_en.pdf?5#_ga=2.105856451.171426022.1561708352-2049997063.1561708352
[23] GAFCHROMIC DOSIMETRY MEDIA, TYPE EBT-3. Available at: http://www.gafchromic.com/documents/EBT3_Specifications.pdf

[24] Phantoms for Absolute Dosimetry. Available at: https://www.iba-dosimetry.com/fileadmin/user_upload/rt-br-e-phantoms-for-ad_rev2_0813.pdf
[25] EPSON Perfection V700 Photo and V750 Pro: User's Guide. Available at: https://files.support.epson.com/pdf/prv7ph/prv7phug.pdf
[26] MATLAB. Available at: https://uk.mathworks.com/products/matlab.html
[27] Sipilä P, Ojala J, Kaijaluoto S, Jokelainen I and Kosunen A 2016 *J. of Appl. Clin. Med. Phys.* **17**(1) 360–373
[28] Devic S *et al* 2004 *Med. Phys.* **31**(9) 2392–2401
[29] Aldelaijan S, and Devic S 2018 *Phys. Med.* **49** 112–118

IOP Conf. Series: Materials Science and Engineering 1019 (2021) 012006 doi:10.1088/1757-899X/1019/1/012006

The development of post-processing algorithm for the ultrasonic evaluation by the application of automated robotic testing systems

D Dolmatov, V Zhvyrblya, D Sednev
School of Non-Destructive Testing, National Research Tomsk Polytechnic University, Tomsk, Russia
E-mail: dolmatovdo@tpu.ru

Abstract. The implementation of automated testing systems based on six degrees of freedom (DOF) robotic manipulators is the actual trend in ultrasonic testing equipment development. Such systems are able to provide fast ultrasonic evaluation with the respect of the surface of the testing object. In this work, the post-processing algorithm based on Synthetic Aperture Focusing Technique (SAFT) is suggested. Such algorithm allows presenting the results in the form of high-resolution imagery of the internal structure of testing objects. The suggested algorithm is applicable in the case of the utilization of automated testing systems based on six DOF robotic manipulators and takes into account all the features conditioned by such equipment application. Performance of the suggested algorithm was tested experimentally.

1. Introduction

Automation is one of the actual trends in ultrasonic nondestructive testing. In this regard, inspection systems based on six DOF robotic manipulators is of the big interest. It is conditioned by the ability of such equipment to perform fast automated ultrasonic scanning of a testing object with the respect of its surface.

Nowadays, the development of automated ultrasonic testing systems based on six DOF robotic manipulators is predominantly aimed at the composite materials inspection. For this purpose pulse-echo ultrasonic testing with focused transducers as well as the through-transmission evaluation was proposed [1–3]. Pulse-echo ultrasonic testing with focused transducers is effective in the case of thin objects evaluation. On the other hand, through-transmission testing can be performed in the case of access to the sample from two sides. Such conditions are not applicable to all areas of ultrasonic testing utilization. One of the examples where alternative approaches should be developed is the ultrasonic evaluation of metal castings. Such testing objects have several features which should be taken into account during the testing:

- generally complex shape of metal castings;
- complex form of the flaws in such objects which cause the strong scattering of acoustic waves;
- usually the coarse grain structure of the metal castings which cause the presence of high level of grain noise on the results.

In this regard, in order to increase the reliability of testing results the application of post-processing algorithms based on Synthetic Aperture Focusing Technique (SAFT) was proposed. Such an approach is able to provide a high resolution throughout the testing specimen and reduce the level of grain noise [4, 5]. The implementation of SAFT strongly depends on the conditions of ultrasonic testing which include:

- features of applied testing equipment;
- type of utilized acoustic contact;
- shape of the controlled object.

The features of SAFT implementation in the case of application of automated ultrasonic testing systems based on six DOF robotic manipulators are the following:

- the imaging should take place with the respect of the surface of the testing specimen;
- due to the fact that in automated systems immersion acoustic contact is used commonly, the presence of the areas with various acoustical properties should be taken into account;
- translation and rotational movement of the ultrasonic probe during the scanning should be considered during the post-processing.

Nowadays, ultrasonic imaging based on SAFT has found limited application in the framework of the utilization of automated ultrasonic testing systems based on six DOF robotic manipulators. In general, such an approach was found the application only for the surface reconstruction for the medical needs. Kerr et al. studied the application of ultrasonic system based on 6 DOF robotic manipulator and SAFT imaging for the determination of the 3D model of knee joint [6, 7]. Precise and accurate results demonstrate the applicability of this approach to the needs of knee arthroplasty. However, in surface reconstruction, it is not necessary to take into account the ultrasonic waves refraction phenomena. As a result, imaging for the ultrasonic testing by the application of 6 DOF robotic manipulators requires the development of special approaches.

2. Theory

The general idea of post-processing based on time-domain SAFT is the splitting of the volume of interest into the voxels and calculation of the time required for ultrasound propagation from the probe to the voxel and back. In the framework of 6 DOF robotic manipulators the following data is available and should be processed:

- the set of A-scans;
- the set of target points of the robotic manipulator in each point of scanning trajectory (x_T, y_T, z_T);
- the set of rotation angles of ultrasonic probe around the Z-axis (angle α), around the Y-axis (angle β) and around the X-axis (angle γ) in each point of scanning trajectory.

The target point is a point on testing specimen surface to which ultrasonic transducer is directed normally in the current point of scanning trajectory. In Figure 1 the imaging geometry in a single scanning position is presented.

In Figure 1 the imaging task is considered for one voxel of the internal structure of controlled object defined as a point C with coordinates (x, y, z). The ultrasonic transducer center is situated in point A. Point B is the refraction point on the interface between immersion liquid and testing specimen which provide the shortest propagation path between points A and C (according to Fermat principle).

In this regard the two-way travel time between points A and C can be calculated in the following way:

$$\tau_{xyz} = 2\left(\frac{|AB|}{c_1} + \frac{|BC|}{c_2}\right) \tag{1}$$

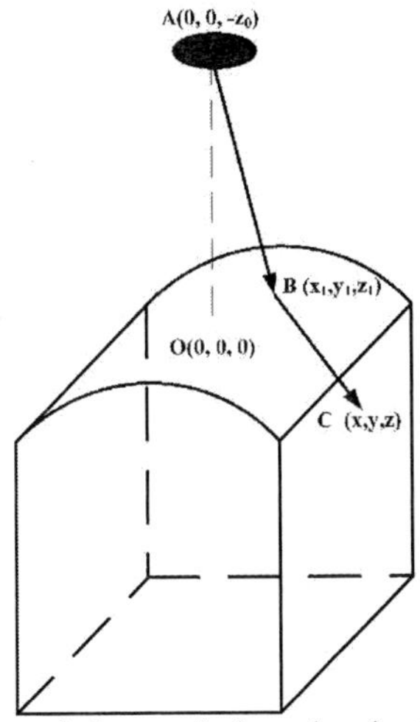

Figure 1. The imaging geometry of ultrasonic imaging in the single position of the probe.

In equation 1 c_1 and c_2 are the velocity of the ultrasonic waves in the immersion liquid and in the testing specimen respectively.

Therefore, the imaging task comes to the determination of point B. For this purpose, the following function can be introduced:

$$\delta(x', y', z') = \left|\hat{v}_r - \hat{v}'\right|, \tag{2}$$

where $\hat{v}_r$ is the normalized vector $\overline{v}_r$ which describes the propagation of the ray refracted on the point arbitrary point $B'(x', y', z')$ on the surface of the specimen whereas $\hat{v}'$ is the normalized vector $\overline{v}'$ which connects points $B'(x', y', z')$ and $C(x, y, z)$ (Figure 2).

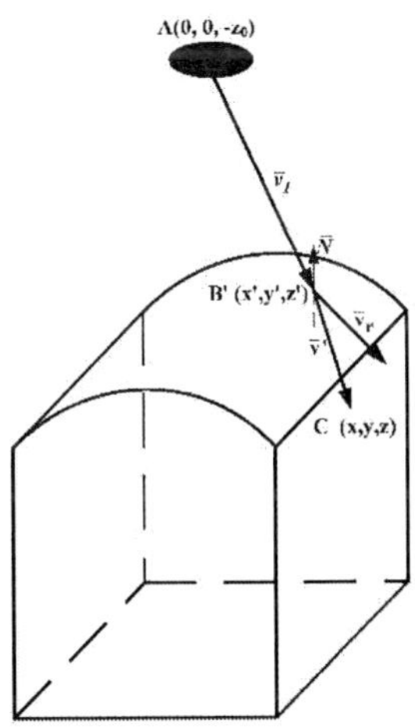

Figure 2. Vectors $\overline{v}_r$ and $\overline{v}'$.

The normalized vector $\hat{v}_r$ can be found via the Snell's law in vector form:

$$\hat{v}_r = \frac{c_2}{c_1}[\overline{N} \times (-\overline{N} \times \hat{v}_I)] - \overline{N}\sqrt{1 - \left(\frac{c_2}{c_1}\right)^2 (\overline{N} \times \hat{v}_I) \cdot (\overline{N} \times \hat{v}_I)}, \tag{3}$$

where $\hat{v}_I$ is the normalized vector $\overline{v}_I$ describing the propagation of the incident ray, $\overline{N}$ is the unitary normal vector to the surface in point B'.

In this regard, the point B can be found via the solving of minimization task of the function (e.g. gradient descent method):

$$(x_1, y_1, z_1) = \underset{x',y',z' \in \mathbb{R}}{\arg\min}(\delta(x', y', z')) \tag{4}$$

In the single position of the ultrasonic transducer, the propagation time should be determined for all voxels in the volume of interest. As a result, partial imagery in a single position of the probe is obtained:

$$I(x, y, z) = A(\tau_{xyz}) \tag{5}$$

where A is the A–scan sampled in current position.

Imagery obtained in different positions of scanning trajectory are defined by local coordinates relatively to the position of the probe and its orientation in current scanning point. In order to obtain total imagery the local coordination transform is necessary:

$$\begin{pmatrix} x_b \\ y_b \\ z_b \end{pmatrix} = R_{xyz} \begin{pmatrix} x \\ y \\ z \end{pmatrix} + \begin{pmatrix} x_T \\ y_T \\ z_T \end{pmatrix} \tag{6}$$

where R_{xyz} is the rotation matrix. It can be obtained via multiplication of the rotation matrixes along each of the axis:

$$R_{xyz} = R_z R_y R_x = \begin{pmatrix} \cos\alpha & -\sin\alpha & 0 \\ \sin\alpha & \cos\alpha & 0 \\ 0 & 0 & 1 \end{pmatrix} \cdot \begin{pmatrix} \cos\beta & 0 & \sin\beta \\ 0 & 1 & 0 \\ -\sin\beta & 0 & \cos\beta \end{pmatrix} \cdot \begin{pmatrix} 1 & 0 & 0 \\ 0 & \cos\gamma & -\sin\gamma \\ 0 & \sin\gamma & \cos\gamma \end{pmatrix} \tag{7}$$

Determination of partial imagery in the base coordinates of the robot for all positions of the ultrasonic transducer allows obtaining the total imagery of the testing object:

$$I_{total} = \sum_{i=1}^{N} I_i(x_b, y_b, z_b) \tag{8}$$

3. Experimental part

The suggested algorithm was implemented in Matlab R2016b. The effectiveness of the algorithm was examined through the in situ experiment. The experimental facility (Figure 3) consists of three parts:

- robotic manipulator;
- the immersion bath;
- the control cabinet.

The steel casting with plane surface was used as a testing specimen. The block contains seven flat bottom holes. Positions of the flaws are presented in Figure 4.

The transducer with the central frequency of 2.25 MHz and 6 mm in diameter was used for data sampling in the experiment. For the correct positioning of the transducer in the points of the testing specimen, the calibration procedure is necessary. Calibration procedure implies the determination of the coordinates of specific points of the testing object. Afterwards obtained parameters are juxtaposed with the STL model and appropriate scanning trajectory is built by the application of Cimstation software.

The sampled set of A-scans is used as the initial data for the developed algorithm. The output of the algorithm is the 3D imagery of the object. In order to perform the post-processing, the restoration of partial imagery occurred in each position of the ultrasonic probe. The dimensions of such imagery were 10x10x30 mm with a resolution of 0.5 mm. Therefore, the total number of voxels in imagery was 24000. Results obtained in each position of the ultrasound probe were coherently summed with the respect of probe orientation in order to get the total imagery. Imaging was performed on the PC with four-core and eight-threaded Intel Core i7-4790K processor and 16 GB of RAM.

Figure 3. The current experimental facility: 1 – robotic manipulator, 2 – immersion bath, 3 – control cabinet.

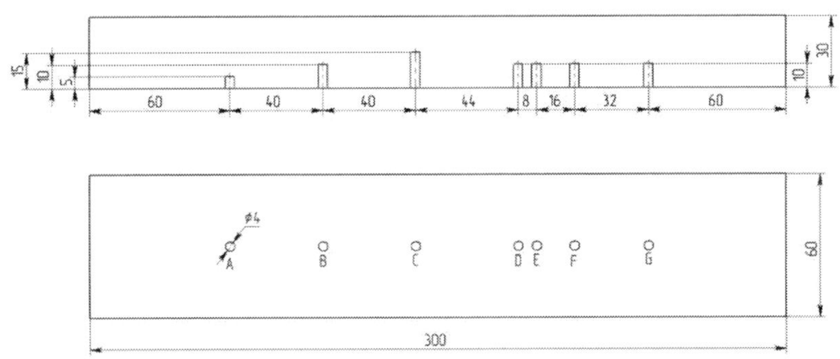

Figure 4. Position of the flaws in testing object.

On the Figure 5 the obtained volumetric result is presented whereas in Figure 6 the C-scan of obtained volumetric result is demonstrated.

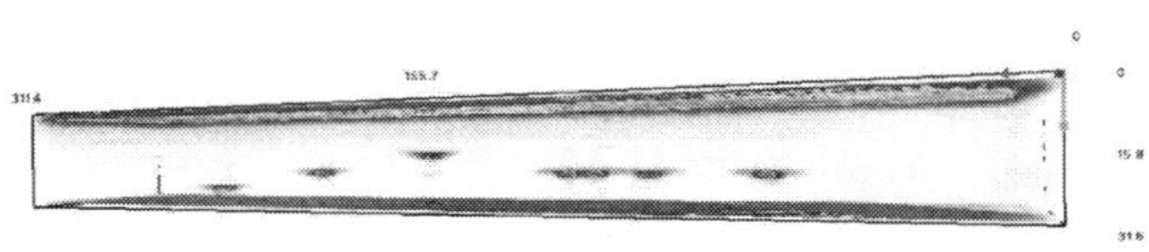

Figure 5. Obtained volumetric result.

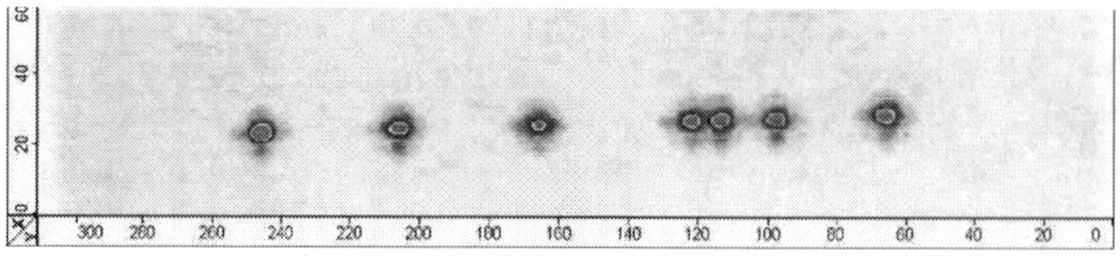

Figure 6. The C-scan of obtained volumetric result.

Results obtained by the application of the suggested algorithm could be evaluated via their comparison with the unprocessed ultrasonic data. It is possible to compose the volumetric imagery by the combining of A-scans with the respect of coordinates they were sampled. In Figure 7 the C-scan of volumetric imagery with unprocessed ultrasonic data is presented whereas Figure 8 contains the lateral profiles of imagery which contain raw and processed data.

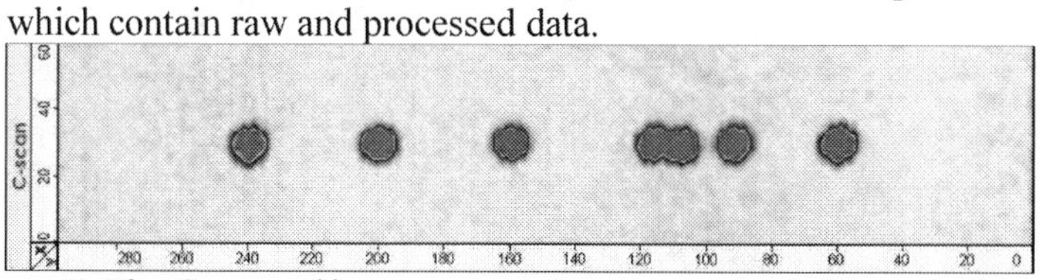

Figure 7. The C-scan of imagery contains unprocessed ultrasonic data.

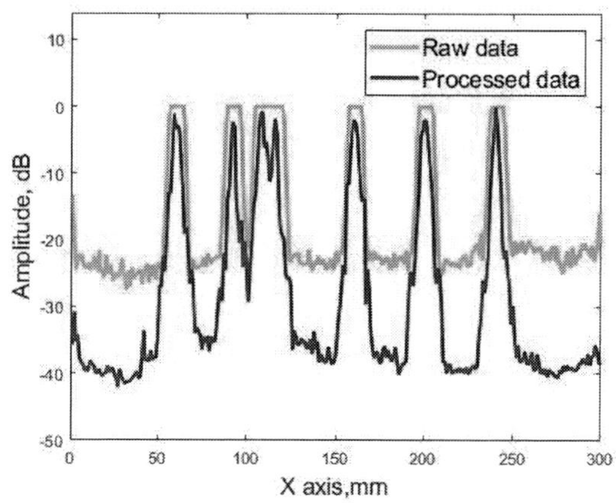

Figure 8. The lateral profiles of two imagery.

According to obtained experimental results and their comparison with unprocessed ultrasonic data several conclusions can be made. Firstly, the resolution of result increased in the comparison with the case when the post-processing is not applied. So, in the case of unprocessed data the flaws D and E were not resolved, whereas the application of the suggested algorithm allowed to resolve them. Secondly, the application of the suggested algorithm causes an increase in SNR. All of this demonstrates the imaging efficiency of the suggested algorithm.

However, industrial requirements for ultrasonic testing systems include the necessity to perform the ultrasonic imaging in real-time mode. For this reason in the framework of automated testing systems application, the post-processing time should be less than the time which is necessary for the object scanning. It was assumed that existed robotic testing system is able to provide quality ultrasonic scanning with an average speed of 5 cm/s. In this regard the time spent on scanning the testing object was 360 seconds. The main parameter which effects on the imaging speed is the number of voxels in imagery reconstructed in one position of the ultrasonic probe. In Figure 9 the dependence between the imaging time and the number of voxels is presented.

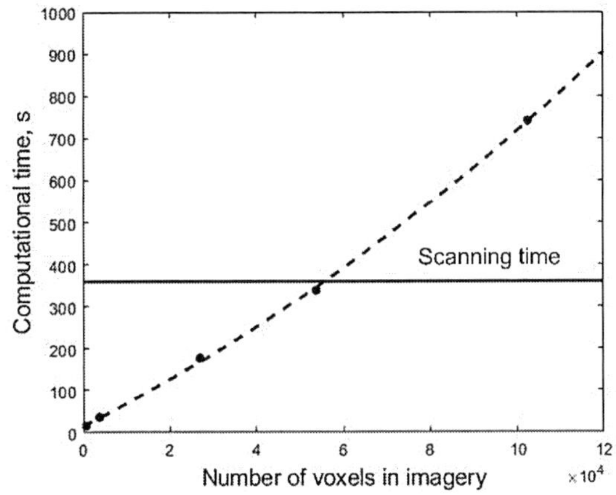

Figure 9. Dependence between the imaging time and the number of voxels.

According to obtained results, it is possible to conclude that imaging of testing object with the applied parameters can be done in real-time mode. The time which is necessary for the results obtaining (176.2 seconds) is less than the time which is necessary for the scanning of the specimen. However, increase the number of voxels (caused by the increase of resolution or volume of interest) in imagery restored in a single position of ultrasonic probe cause the increase of computational time. The critical number of voxels which make the real-time imagery was evaluated as 56641 on the given PC. However, it is expected that the suggested algorithm transfer on parallel computing units will increase the computational speed as well as a critical number of voxels.

4. Conclusion

In this paper, the ultrasonic post-processing algorithm in the case of testing systems based on six DOF robotic manipulators application was proposed. The effectiveness of the suggested algorithm was examined via the in situ experiment. Accurate and precise imagery, as well as high computational speed, demonstrates the effectiveness of the suggested algorithm. However, there is room for further work. Firstly, the effectiveness of the algorithm should be verified in the case of testing specimens with different curvatures. Secondly, the algorithm acceleration via its transferring on parallel computing units should be considered (CUDA, GPU). Thirdly, the possibility of algorithm adaptation in the case of phased array application should be studied.

Acknowledgment

The research is funded by the Governmental program "Science", research projects No. № FSWW-2020-0014. The research is carried out within the framework of Tomsk Polytechnic University Competitiveness Enhancement Program grant, the Project number is VIU-MNOL NK 187/2020

References

[1] Cuevas E, Hernandes S, Cabellos E 2016 *Proceedings of 19th World Conference on Non-Destructive Testing 1*-9

[2] Mineo C, MacLeod C, Morozov M 2016 *Proceedings of IEEE International Ultrasonics Symposium* 1-4

[3] Dolmatov D, Zhvyrblya V, Filippov G 2016 *IOP Conference Series: Materials Science and Engineering* **135**(1) 012010 doi:10.1088/1757-899X/135/1/012010

[4] Stepinski T 2007 *IEEE Transactions on Ultrasonics Ferroelectrics and Frequency Control* **54** (7) 1399–1408 doi: 10.1109/TUFFC.2007.400

[5] Dolmatov D, Abramets V 2016 *MATEC Web of Conferences* **48** 03004

doi: 10.1051/confmatec/20164803004
[6] Kerr W, Pierce S, Rowe P 2016 *Ultrasonics* **72** 165–176 doi: 10.1016/j.ultras.2016.08.007
[7] Kerr W, Pierce S, Rowe P 2017 *Computerized Medical Imaging and Graphics* **58** 23-32 doi: 10.1016/j.compmedimag.2017.03.002

The use of computer visualization in the analysis of breathing curves

O Berestneva[1], I Lyzin[1], A Yumasheva[1], O Marukhina[1] and A Trufanov[2]
[1]School of Computer Science & Robotics, National Research Tomsk Polytechnic University, Tomsk, Russia
[2]School of High Technologies, Irkutsk National Research Technical University, Irkutsk, Russia
E-mail: ogb6@yandex.ru

Abstract. Researchers in diverse domains use advanced computer techniques to describe complex entities and processes and impressively visualize those which are often unavailable for direct observation. However, one should emphasize that computer visualization conception implies more than just a convenient, impressive, or high rate information transfer. It also considers a problem of perception, processing, and further cultivation of such important personal qualities as intuition, professional "talent", and figurative thinking which are of value for experts in any domain. Computer visualization comprises traditionally such sophisticated techniques as computer graphics, animation and virtual reality. Traditionally, computer graphics has delivered strong instruments for creating, processing, and interacting with data representations. Interactive paradigm has led to emergence of a new scope within the problem of artificial intelligence, which is called cognitive computer graphics. The use of cognitive graphics allows physicians, analyzing modest volume of information, to draw significant conclusions. In whole, cognitive graphics forms a separate subfield in medical research. Visualization provides experts with data in current state of patients just to monitor their conditions continuously. In this article, we show the use of computer visualization to study characteristics (including network imprints) of such a common disease as bronchial asthma. The patients were grouped according to influence degree of psychological factors to the occurrence, progression and course stages of the disease. The study focuses on comparison and analysis of the patient' spirograms and demonstrate presence of physiological and psycho-physiological features among patients with diagnoses of bronchial asthma. In this respect computer visualization provides a solid platform for thorough research and deep analysis in spirometry.

1. Introduction

Bronchial asthma (from Greek. asthma – heavy breathing, smothering) – chronic lung disease affecting people of all age groups. It can occur in the form of single, episodic attacks or has a severe course with asthmatic status and fatal outcome. According to medical statistics, in recent years the incidence of bronchial asthma in most countries has increased significantly. The increase of the disease among young people indicates a continuing trend of increasing development of this disease. The pity fact is that, in spite of scientific advances in the sphere of etiology and the availability of new medicine, the incidence and mortality from bronchial asthma is constantly increasing. This is typical for most countries of the world.

The role of psychosocial, emotional factors in the development of bronchial asthma is estimated by various experts contradictory and the mechanisms are unclear. This is probably due to the fact that all patients with BA are regarded by them as a homogeneous population of people in terms of somatic status, but with different psychological states. In addition, clinicians (pulmonologists, therapists) do not always payattention to the fact that different emotional states and mental disorders bring different physiological reactions in healthy and sick asthma. So the variety of psychological effects causes a variety of psychological and somatic changes in different groups of patients with BA. Because of this it is necessary to study psychological (mental) and social factors in close relationship with clinical ones. On this basis, E. V. Nemerov classified bronchial asthma, taking into account psychological and social factors. The following classification was proposed [1]: BANP - Bronchial asthma is not psychogenic; BASP – Bronchial asthma somatopsychogenic; BAPI – Bronchial asthma psychogenically induced.

Employees of the Siberian State Medical University and Tomsk Polytechnic University during several years were making researches to detect significant differences between groups of patients (physiological and psychological indicators) with bronchial asthma, divided by the degree of influence of psychosocial factors on the occurrence, development and course of the disease [1–6]. To identify hidden patterns in the experimental facts, methods of structural data analysis were used, including methods of scientific visualization [5, 7–11].

The subject of this study is the patients' process of breathing with various forms of bronchial asthma. It is known, bronchial asthma is, first of all, a disease of the respiratory tract, because of it, a study of the patient's reayhing rhythm, the shape of the breathing curve, the presence of apnea, the duration of the inhalation-exhalation cycle, etc. It has great significance [3, 12–15].

Accounting for psychological and social factors, it has some importance [16–21].

2. Materials and methods of research

The experimental data consisted of patients' breathing curves with different types of bronchial asthma and in a group of conditionally healthy people registered with the "MONITOR" instrument (Figure 1) [1].

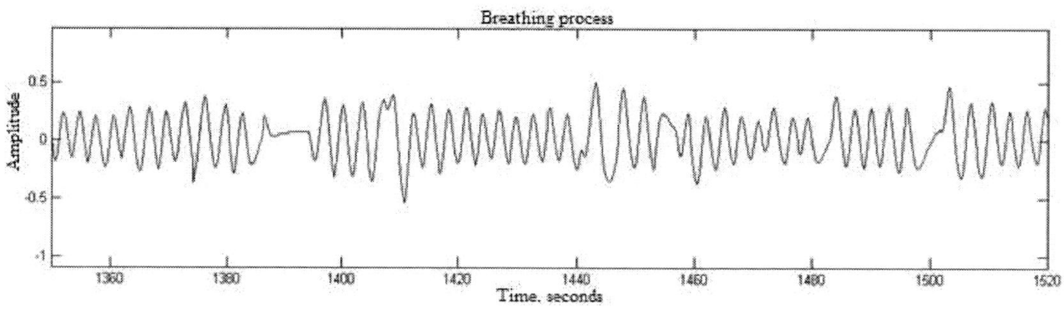

Figure 1. Breathing curve example.

Additionally, breathing curves were converted into networks for further analysis with powerful network instruments. First, the curves were put under digitizing by Engauge Digitizer 4.1 (Figure 2) [22].

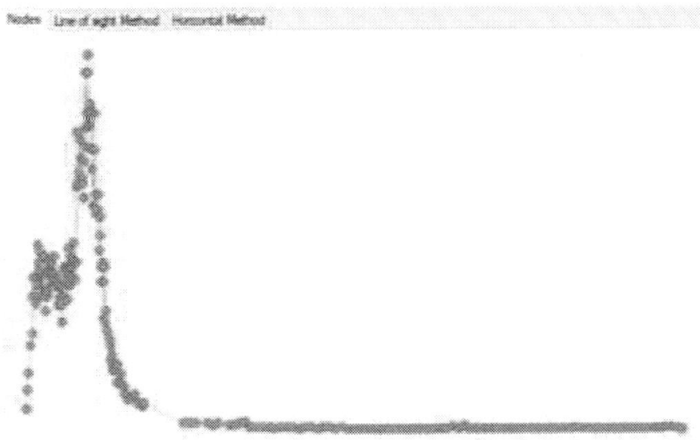

Figure 2. Example of digitized volume-time curve.

Second, a conversion code (C#) was developed and applied to realize natural visibility algorithm (NVG, [23]):

```
{
  class LineOfSightMethod
  {
    private List<Point> points;
    private int[,] adjencencyMatrix;
    public LineOfSightMethod(List<Point> points)
    {
      this.points = points;
    }
    public int[,] getAdjecencyMatrix()
    {
      return adjencencyMatrix;
    }
    public void buildAdjecencyMatrix()
    {
      adjencencyMatrix = new int[points.Count, points.Count];
            foreach (Point point in points)
        {
          testPoint(point);
        }
    }
    private void testPoint(Point point)
    {
        adjencencyMatrix[point.getId(), point.getId()] = 0;
        if (point.getId() < points.Count - 1)
        {
          adjencencyMatrix[point.getId(), point.getId() + 1] = 1;
          adjencencyMatrix[point.getId() + 1, point.getId()] = 1;
        }
        if (point.getId() < points.Count - 2)
        {
                  for (int i = point.getId() + 2; i < points.Count; i++)
        {
```

```
                if (testTwoPoints(point, points[i]))
            {
                adjencencyMatrix[point.getId(), i] = 1;
                adjencencyMatrix[i, point.getId()] = 1;
            }
            else
            {
                adjencencyMatrix[point.getId(), i] = 0;
                adjencencyMatrix[i, point.getId()] = 0;
            }
          }
        }
      }
      private bool testTwoPoints(Point point1, Point point2)
      {
        for (int i = point1.getId() + 1; i < point2.getId(); i++)
        {
          Point tempPoint = points[i];
                if (tempPoint.getY() >= (point1.getY() + ((point2.getY() - point1.getY()) *
(tempPoint.getX() - point1.getX()) / (point2.getX() - point1.getX()))))
            {
                return false;
            }
          }
          return true;
        }
      }
    }
```

Third, we used an effective free tool Gephi [24] to visualize and analyze network imprints. The tool delivers structure plots and diverese metrics for each node and link, and for an analyzing graph in whole : The following metrics were taken into account : size (number of vertexes N and edges L in a graph G(V,E), where V is set of vertexes, so that N=|V| and E is set of edges); diameter (the length max ij d(vi,vj) of the "longest shortest path" any two vertices vi,vj, where d(vi,vj) is a shortest path length, we examine the graph diameter D of the largest connected cluster); Clustering coefficient distribution and Average clustering coefficient (Ci for a node i, is a measure of the degree to which nodes in a graph tend to cluster together: $C_i=2e_i/\{k_i(k_i-1)\}$, where ki is the number of neighbours of the i'th node, and ei is the number of connections between these neighbours. Average clustering coefficient is noted as $\langle C\rangle = \Sigma C_i/N$). Average path length is $\lambda = \langle d(v_i,v_j)\rangle$.

3. Findings

The frequency spectrum of the breathing curve of each patient was evaluate (Figure 3).

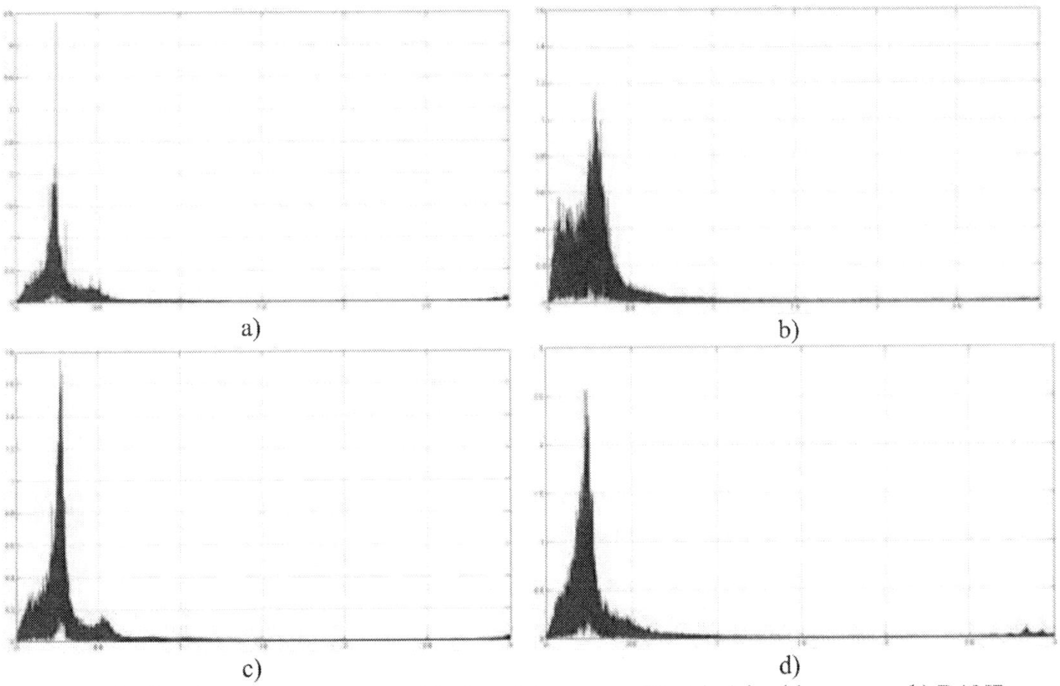

Figure 3. The frequency spectrum of the breathing curve (Hertz) a) healthy person; b) BANP; c) BAPI; d) BASP

The obtained results confirmed by the analysis of spectrograms (Figure 4).

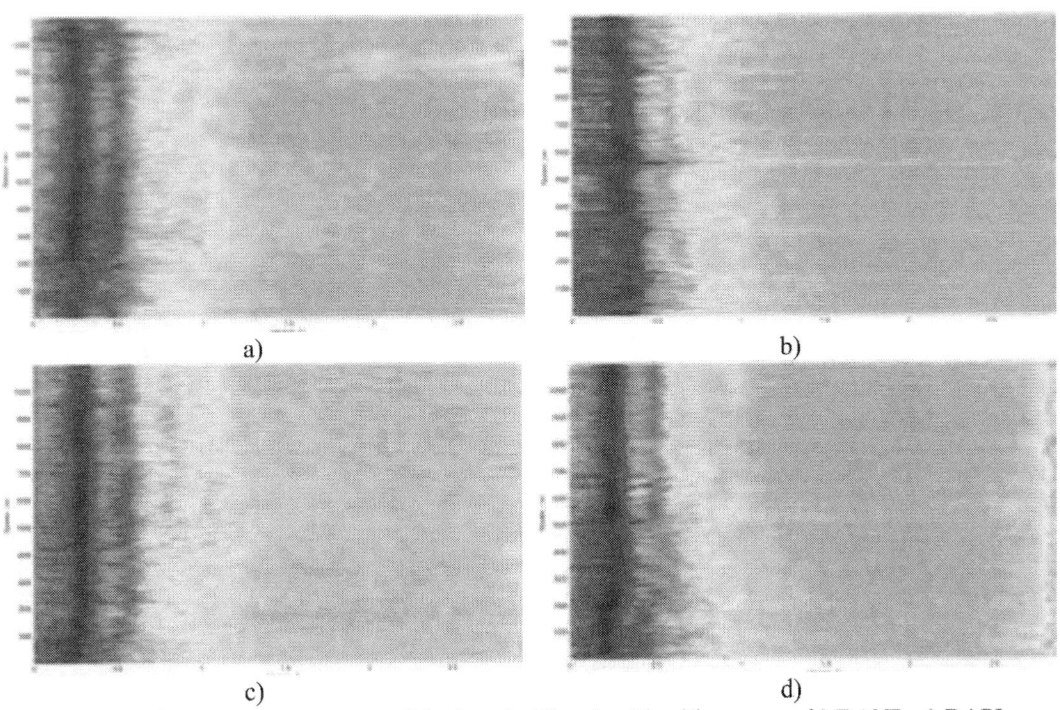

Figure 4. The spectrogram curve of the breath (Hertz): a) healthy person; b) BANP; c) BAPI; d) BASP

4. Discussion

After a careful study of the different groups patients' spectral images of the breathing curves, we managed to isolate several characteristic "single" graphic images (corresponding to one respiratory cycle) for patients with various forms of bronchial asthma and conditionally healthy people. The results are presented in table 1.

Network analysis conveys and enforces spectral image study (table 2).

Table 1."Single" graphic images for different patinen' group.

Diagnosis	Characteristical graphic for one breathing cycle
Almost healthy	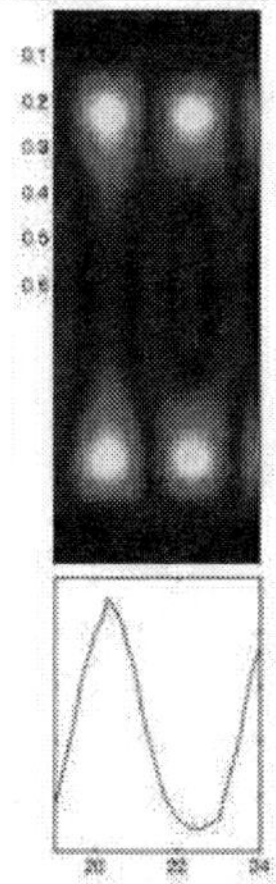
Non-psychogenic bronchial asthma (BANP)	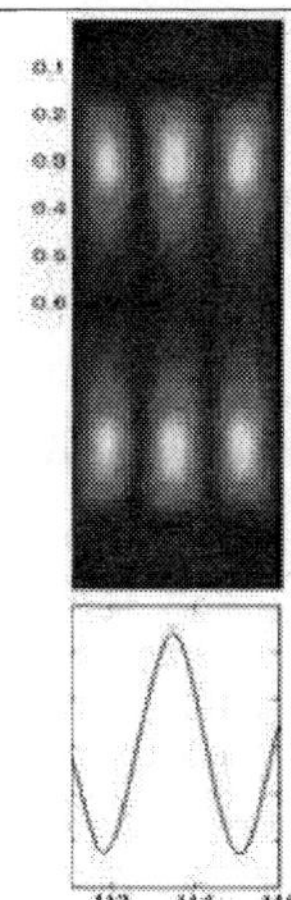

Psychogenic bronchial asthma (BASP)	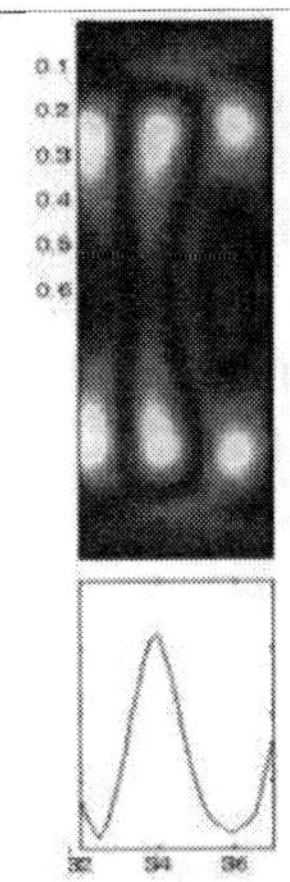
Inducted bronchial asthma (BAPI)	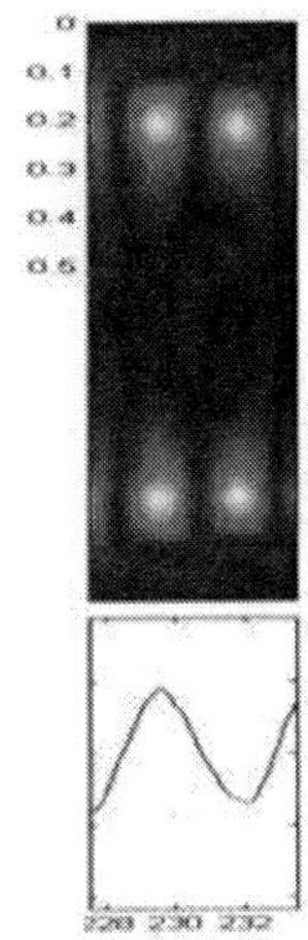

Table 2. Network for differen patient' groups.

NVG algorithm	Almost healthy	BANP	BAPI	BASP
	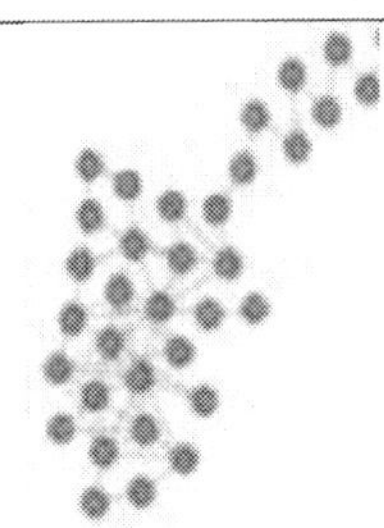	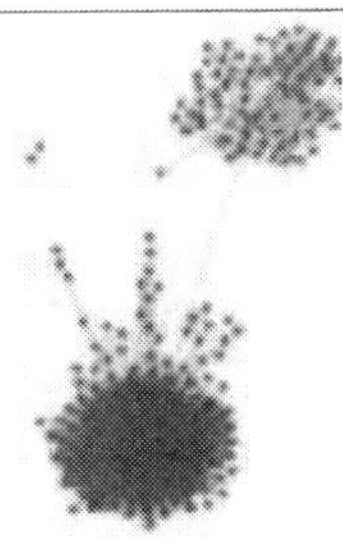		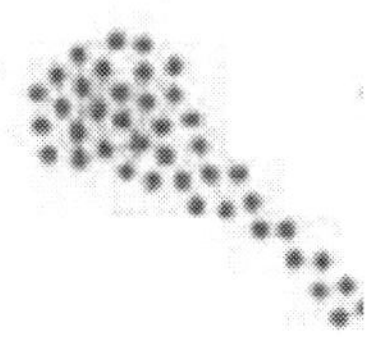
D	20	11	34	44
<C>	0.134	0.56	0.35	0.22
λ	7.1	3.7	12.	17.

5. Conclusion

The spectral-time analysis makes it possible to obtain characteristic "single" graphic images of patient' spirogtams with various forms of bronchial asthma. The table 1 portrays that a healthy person' spectral image of the volume time curve is noticeably different from almost all images that are typical for patients with various forms of asthma disease. However, patients from the BASP spirogram group are most similar to the spectral image characteristic for the group of "relatively healthy"; there is some similarity with the patient' spectral images diagnosed with BAPI. The patient' characteristical spectral image of the BANP breathing curve differs markedly from the others.

Strikingly the network imprints are in balance with the fact that is inferred above (see the table 2).

The results confirm the earlier conclusions [7, 3, 10, 11, 6, 14] about presence of characteristical physiological and psycho-physiological features among patients with diagnoses of BASP, BAPI and BANP.

In this respect computer visualization provides a solid platform for thorough research and deep analysis in spirometry.

Acknowledgment

The study was making with partial financial support from the Russian Foundation for Basic Research during the framework of research №18-07-00543.

References

[1] Nemerov E V, Yazykov K G, Berestneva O G and Zharkova O S 2015 *Proceeding of Information and mathematical technologies in science and management: works of the XX Baikal All-Russian Conference and School - Seminar of Scientific Youth, Irkutsk* **3** 241 – 246

[2] Berestneva O G, Zharkova O S, Nemerov E V, Yazykov K G, Fokin V A 2014 *Proceeding of Works of the Baikal All-Russian Conference Information and Mathematical Technologies in Science and Management* 139–147

[3] Karpenko P V, Berestneva O G and Stepanov D Y 2014 *Proceeding of Information technologies in science, management, socialsphere and medicine: Collection of scientific papers of the International Conference* 211–220

[4] Berestneva O G, Sharopin K A, Abdulkina N G, Stepanenko N P, Yumasheva A L and Zharkova O S 2015 *Mathematical methods and information technologies in problems of assessing the state of biosystems* (Tomsk: TPU publishing house)

[5] Osadchaya I A *et al* 2014 *Bulletin of Siberian medicine* **13** 89 –93

[6] Berestneva O G, Osadchaya I A, Nemerov E V 2012 *Bulletin of Siberian Science: electronic scientific journal* 333–338

[7] Osadchaya I A, Berestneva O G, Volovodenko V A and Marukhina O V 2014 *Communications in Computer and Information Science* **466** 568–576

[8] Osadchaya I A and Prokopev R O 2014 *Proceedings of Engineering messenger of Don* 2

[9] Berestneva O G, Osadchaya N A and Burtseva A L 2014 *Methods of analysis and visualization of multidimensional data* (Tomsk: Publishing House of Tomsk Polytechnic University)

[10] Volovodenko V A, Berestneva O G, Nemerov E V and Osadchaya I A 2012 *Bulletin of Tomsk Polytechnic University* **320** 125–130

[11] Kinash N A, Trufanov A I, Tikhomirov A A, Berestneva O G and Fisochenko O N 2016 *Proceedings of International Siberian Conference on Control and Communications (SIBCON)* 7491785

[12] Devjtykh D, Gerget O, and Berestneva O G 2014 *Communications in Computer and Information Science* **466** 556–567

[13] Vilunas Y 2013 *Crying breath cures bronchial asthma and other respiratory diseases*

[14] Karpenko P V and Nemerov EV 2019 *Proceedings of the VI International Student Electronic Scientific Conference "Student Scientific Forum".* Available at: http://www.scienceforum.ru/2014/482/2949

[15] Zulkarev R X 2007 *Cardiorespiratory variability in diseases of the respiratory organs. Thesis for the degree of Doctor of Medical Sciences*
[16] Shelekhov I L 2017 *Pedagogical Review* **2** 9–20
[17] Gadelshina T G and Shelekhov I L 2017 *Methodology and methods of scientific research: Tutorial* (Tomsk: Tomsk State Pedagogical University Publishing House)
[18] Shelekhov I L and Belozerova G V 2017 *Pedagogical Review* **3** 117–126
[19] Shelekhov I L, Belozerova G V and Berestneva O G 2018 *Proceedings of Information technologies in science, management, social sphere and medicine: Collection of scientific papers of the V International Scientific Conference* 160–167
[20] Shelekhov I L, Belozerova G V and Berestneva O G 2018 *Proceedings of Information technologies in science, management, social sphere and medicine: Collection of scientific papers of the V International Scientific Conference* 376–390
[21] Shelekhov I L and Belozerova G V 2016 *Personal aspects of adaptation in the study of images, symbols, plots of dreams* (Tomsk: Tomsk State Pedagogical University Publishing House)
[22] Engauge Digitizer 4.1 Available at: https://markummitchell.github.io/engauge-digitizer/
[23] Lacasa L, Luque B, Ballesteros F, Luque J, and Nuño J C, 2008 *From time series to complex networks: the visibility graph, Proc.Natl.Acad.Sci.U.S.A.* **105** 4972–4975. Available at: https://arxiv.org/pdf/0810.0920.pdf (accessed August,5, 2019).
[24] The Open Graph Viz Platform, Release 0.9.2/ 26 September 2017). Available at: https://github.com/gephi/gephi (accessed August,5, 2019)

IOP Conf. Series: Materials Science and Engineering 1019 (2021) 012008 doi:10.1088/1757-899X/1019/1/012008

Investigation of edge detection methods using the example of wave field pattern image

A A Yakimenko[1,2] **and A I Makfuzova** [1]
[1] Department of Computer Engineering, Novosibirsk State Technical University, Novosibirsk, Russia
[2] Laboratory of Geophysical Informatics, Institute of Computational Mathematics and Mathematical Geophysics SB RAS, Novosibirsk, Russia
E-mail: yakimenko@corp.nstu.ru

Abstract. This article explores the methods of detecting boundaries in images of wave field patterns. As a result, an application for image processing was developed using Sobel, Prewitt, Roberts, Gabor and Canny filters and experiments were carried out to detect the edge on images of wave field patterns. In conclusion, the quantitative and qualitative characteristics of the filters were presented, and a conclusion was made on the selection of the boundary detection filter. The results of this study will be used in the study of the neural network neocognitron for the recognition of geological and physical models of environments (GPME). This work was supported by the RFBR grant No. 19-07-00170.

1. Introduction

The search for a solution to the inverse problem of geophysics is currently the current direction. The inverse problem is to find the characteristics of the medium according to the physical parameters of the field, that is, the determination of the parameters of the medium from the existing pattern of propagation of the wave field [1]. Existing approaches to solving the inverse problem require a large amount of time for modeling proportional to the accuracy of the model.

The restore the GPME from the wave propagation pattern, it is proposed to use neural networks [2], which have already proven themselves in the field of image recognition. One example of a neural network for image recognition is the neocognitron network developed by K. Fukushima [3]. One of the features of the network is its insensitivity to shifts, resizing of the model or other distortions. The structure of the neocognitron contains a layer responsible for the selection of contrast, which will be modified to apply to this task.

This article examines the filters Sobel, Prewitt. Roberts, Gabor and Kanni's method for detecting the boundaries of images of wave field patterns by qualitative (visual) and quantitative (confusion matrix) characteristics.

2. Problem definition

It is necessary to develop an algorithm for determining the shape and location of the boundaries of the density drop in the geological-physical model of the environment. Consider and analyze filters for the edge detection method based on the difference in brightness. To develop software for the implementation of an algorithm for determining the shape and location of the boundaries of the density drop in the geological-physical model of the environment. Conduct a series of experiments with synthetic data.

3. Theory

Border detection filters are well designed for halftone images. A halftone image of a wave field pattern will be considered as a function of two variables (x, y).

3.1. *Filter Sobel*

The Sobel filter is used in image processing, especially in edge detection tasks.

Filter Sobel calculates the direction of the greatest increase in the brightness of the image at each point [4]. The Sobel filter is built on the calculation of the convolutions of the original image with the GX and GY cores, which provide the calculation of the first derivatives in the directions:

$$G_X = \begin{bmatrix} -1 & 0 & 1 \\ -2 & 0 & 2 \\ -1 & 0 & 1 \end{bmatrix} \times A \qquad (1)$$

$$G_Y = \begin{bmatrix} -1 & -2 & -1 \\ 0 & 0 & 0 \\ 1 & 2 & 1 \end{bmatrix} \times A \qquad (2)$$

where is $\times$ – dimensional convolution operation;

A – the origin image.

The use of the operator G_X allows to determine the approximate value of the first partial derivative of the intensity change in the horizontal direction, G_Y - in the vertical direction.

Two-dimensional operation of convolution is carried out according to the equation (3).

$$B(x, y) = \sum_i \sum_y F(i, j) \times A(x+i, y+i), \qquad (3)$$

where is $F(i, j)$ – core filter Sobel.

The magnitude of the gradient in each pixel of the image is determined by the equation:

$$G = \sqrt{G_X^2 + G_Y^2} \qquad (4)$$

Therefore, the Sobel filter is a discrete differential operator that calculates an approximate value of the brightness gradient at each point in the image.

3.2. *Filter Prewitt*

The Prewitt Filter is similar in how it works with the Sobel filter. It is an appropriate method for estimating the size and orientation of the boundary [5]. The Prewitt filter gives direction straight from the kernel with the maximum result. The method of detecting segments with its help along the boundary of the object localization:

$$G_X = \begin{bmatrix} 1 & 0 & -1 \\ 1 & 0 & -1 \\ 1 & 0 & -1 \end{bmatrix} \times A, G_Y = \begin{bmatrix} -1 & -1 & -1 \\ 0 & 0 & 0 \\ 1 & 1 & 1 \end{bmatrix} \times A \qquad (5)$$

The result of the Prewitt filter is:

$$\max\{P, Q\} \qquad (6)$$

where is P and Q – kernel responses G_X and G_Y .

3.3. *Filter Roberts*

The Roberts Filter is a non-linear contrasting method using two-dimensional discrete differentiation operations [6]. The method of detecting segments with its help examines two different convolution kernels:

$$G_X = \begin{bmatrix} -1 & 0 \\ 0 & 1 \end{bmatrix} \times A, G_Y = \begin{bmatrix} 0 & -1 \\ 1 & 0 \end{bmatrix} \times A \tag{7}$$

In the Roberts filter, a total vector of two diagonal differential vectors is used, which shows the largest value of the differential between the four covered points.

The transformation of each pixel by the Roberts cross filter can show the derivative of the image along a non-zero diagonal, and the combination of these transformed images can also be viewed as a gradient from the two upper pixels to the lower two. The response of this filter is determined by the equation:

$$S = \sqrt{P^2 + Q^2} \tag{8}$$

3.4. *Method Canny*

The method Canny consists of five steps [7]:

- Smoothing. It is used to suppress noise using a Gaussian image blur:

$$G(r) = \frac{1}{(2\pi\sigma^2)^{N/2}} e^{-r^2/(2\sigma^2)} \tag{9}$$

where is r – blur radius; N – number of measurements; σ – standard deviation of the Gaussian distribution.

- Search for gradients. At this stage, the Sobel filter is used, which is described in A.
- Suppression of non-maxima: pixels are declared as edges pixels in which a local gradient maximum is reached in the direction of the gradient vector.
- Dual threshold filtering. If the value of the gradient in some place on the fragment being viewed exceeds the upper threshold, then this element also remains the acceptable limit in those places where the value of the gradient considers two different convolution kernels;
- Trace ambiguity. Total boundaries are determined by suppressing all edges that are not bound to strong boundaries.

3.5. *Filter Gabor*

The core of the Gabor filter is the product of a Gaussian and harmonic function, and filters the signal based on the preferred spatial frequency parameter λ^{-1} [8]:

$$g(k,l) = \exp(-\frac{x'^2 + \gamma^2 y'^2}{2\sigma^2}) \cos(\frac{2\pi x'}{\lambda} + \psi), \tag{10}$$

where is x', y' – transition to polar coordinates;

k, l – position of the light pulse in the receptive area;

λ – the number of parallel excitatory and inhibitory zones in the receptive field;

σ – size of receptive field;

ψ – symmetry of excitatory and inhibitory zones;

γ – degree of ellipticity of receptive field.

Based on [9], the optimal value of the preferred spatial frequency can be calculated from the relation:

$$\frac{\sigma}{\lambda} = 0,56 \tag{11}$$

Gaussian serves to determine the receptive inhibitory and excitatory zones, the shape and dimensions of which determine the parameters σ, γ. The harmonic function generates a signal, when coinciding with a grayscale signal, the signal is amplified, or its complete attenuation otherwise.

4. Experiments

For experiments, an application was developed in C # to detect the boundaries of the image of a wave field pattern. The appearance of the application is shown in Figure 1.

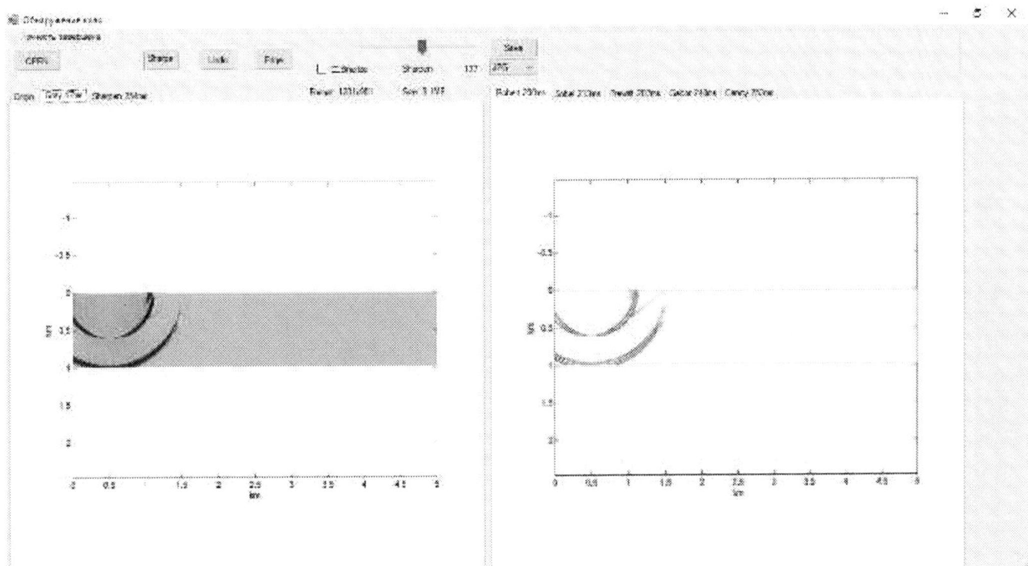

Figure 1. Edge Detection Application.

This application visually consists of two zones: processing the original image (Origin tab) into a halftone image (Gray tab), sharpening the image (Sharpen tab), and processing the original image with Roberts, Sobel, Prewitt, Gabor filters and Kanni to choose from.

It also provides for the binarization of the original image to improve the quality of border detection, and adjust the image sharpness (slider "Sharpen").

The image of the wave field pattern is dynamic, namely, the original images are made at equal intervals of time (Figure 2). Thus, experiments should be carried out not with one image, but with several.

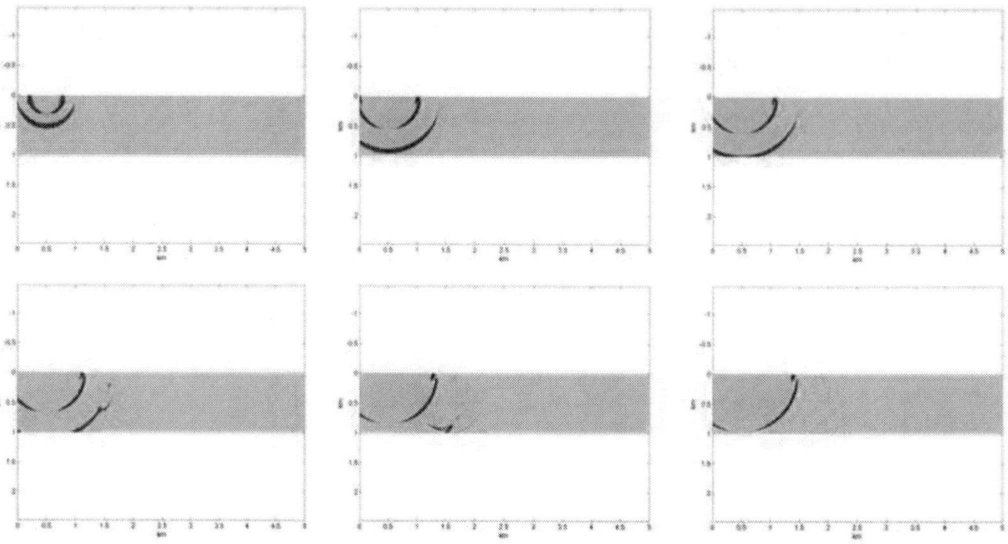

Figure 2. Original images of the wave field pattern.

In Figure 2, the initial data are synthetic, and they were obtained from the already existing GFME. For each model, 16 images of wave field patterns are presented. A total of 100 GFME with different positions in space and geometric characteristics.

4.1. Edge detection using a Roberts filter

The results of edge detection using the Roberts filter for various values of sharpness are presented in table 1.

Table 1. Edge detection with filter roberts

Sharpness	Original image	The resulting image
30	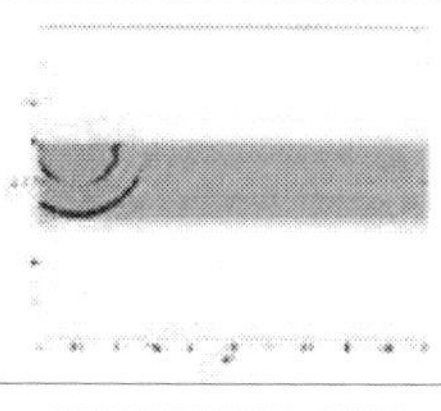	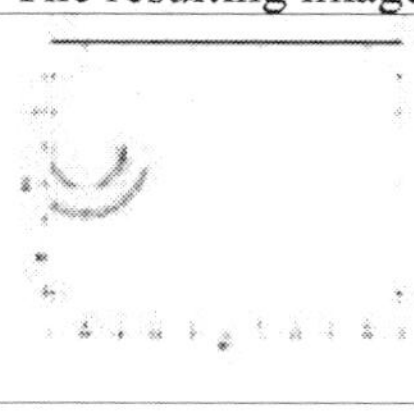
100	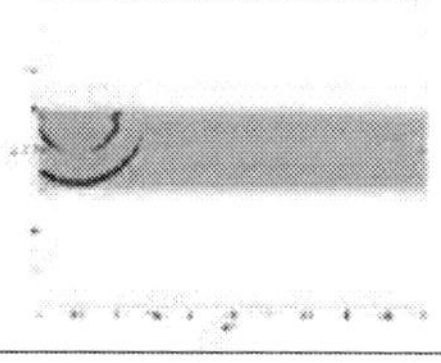	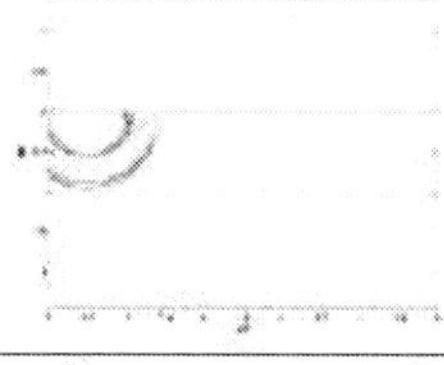
180	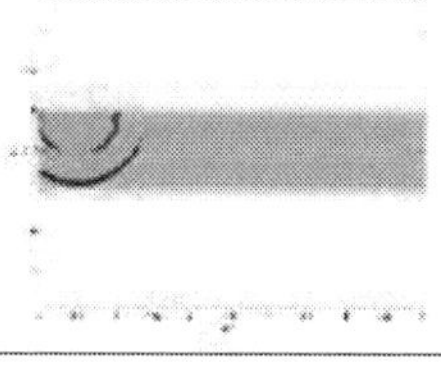	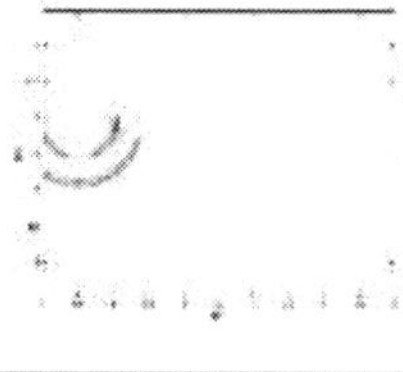

Based on the results, the recommended sharpness value for the Roberts filter is 45.

4.2. Edge detection using a Sobel filter

The results of edge detection using the Sobel filter for various values of sharpness are presented in table 2.

Table 2. Edge detection with filter sobel

Sharpness	Original image	The resulting image
30	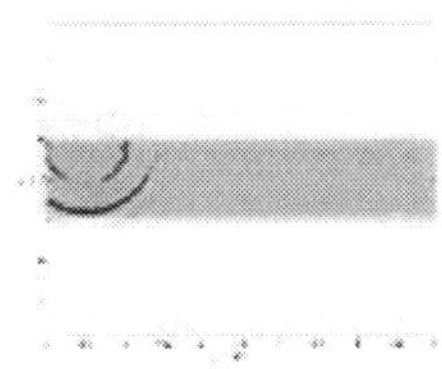	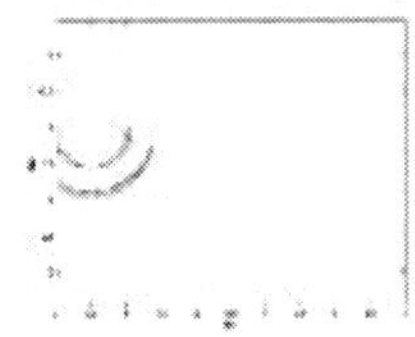
100	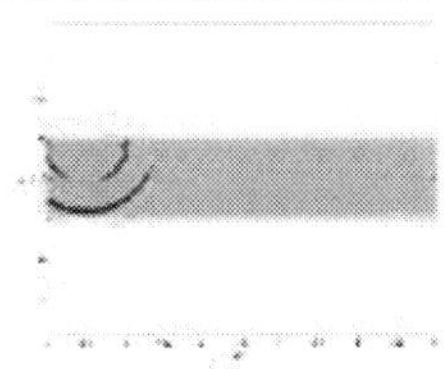	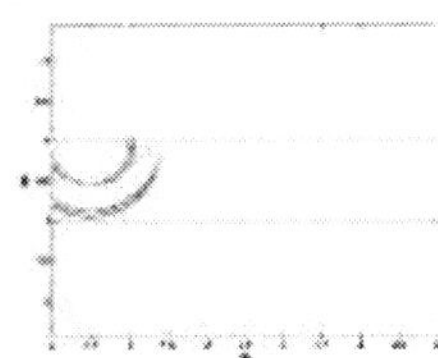
180	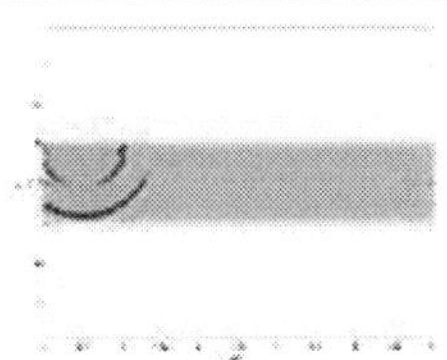	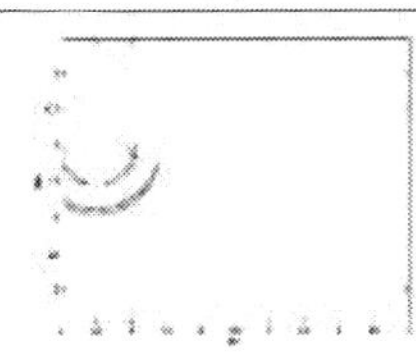

Based on the results, the recommended sharpness value for the Sobel filter is 55.

4.3. Edge detection using a Prewitt filter

The results of edge detection using the Sobel filter for various values of sharpness are presented in table 3.

Table 3. Edge detection with filter prewitt

Sharpness	Original image	The resulting image
30	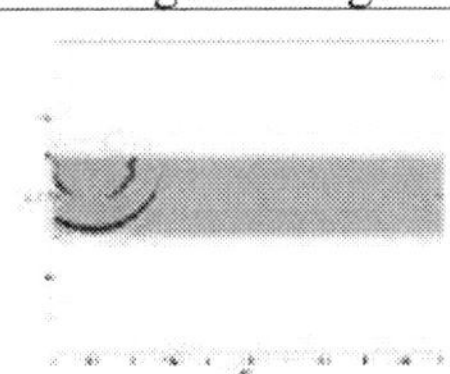	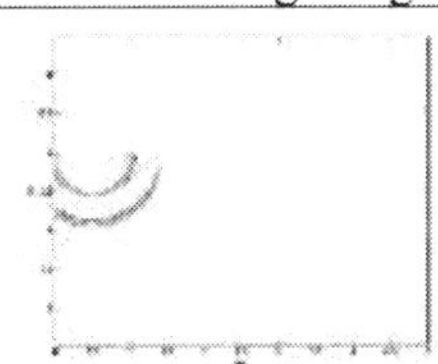
100	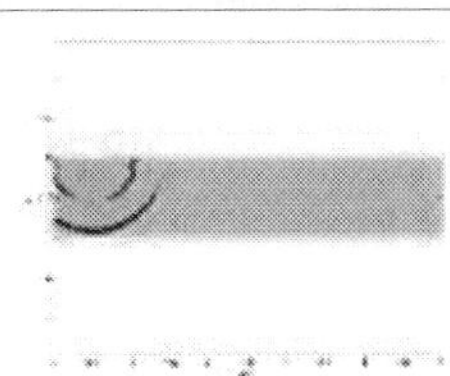	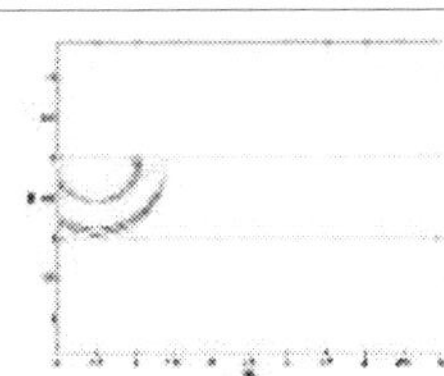

180	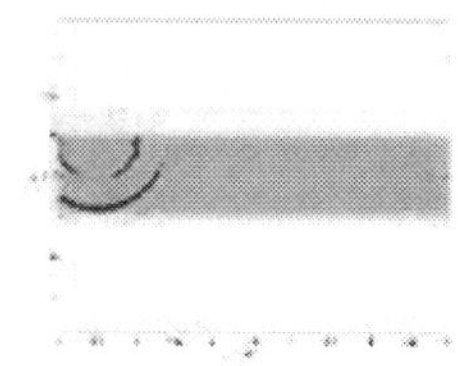	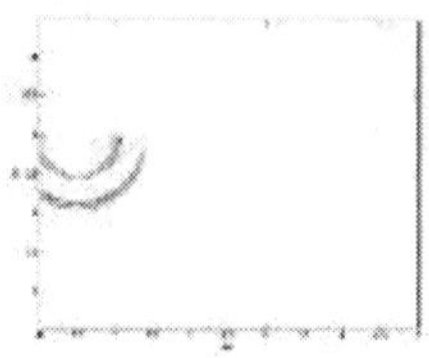

Based on the results, the recommended sharpness value for the Prewitt filter is 87.

4.4. Edge detection using a method Canny

The results of edge detection using the Sobel filter for various values of sharpness are presented in table 4.

Table 4. Edge detection with method canny

Sharpness	Original image	The resulting image
30	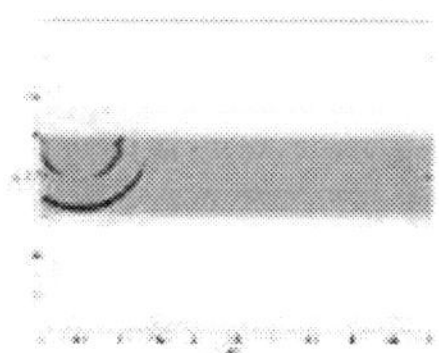	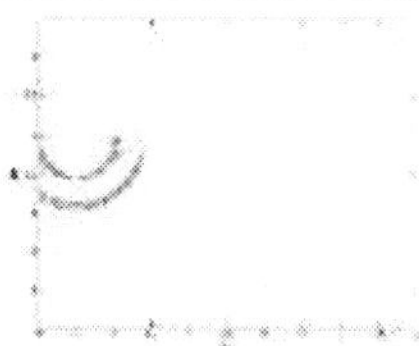
100	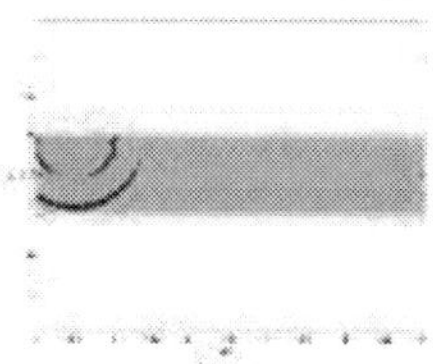	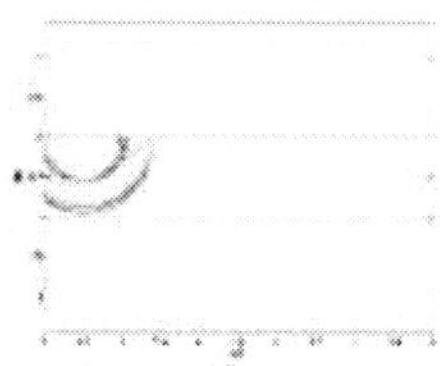
180	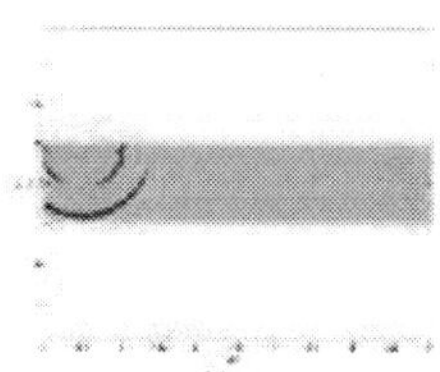	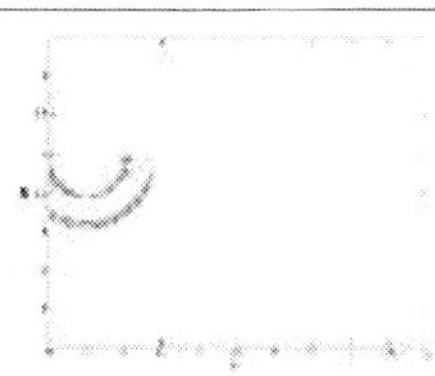

Based on the results, the recommended sharpness value for the method Canny is 32.

4.5. Edge detection using a Gabor filter

The results of edge detection using the Sobel filter for various values of sharpness are presented in table 5.

Table 5. Edge detection with filter gabor

Sharpness	Original image	The resulting image
30	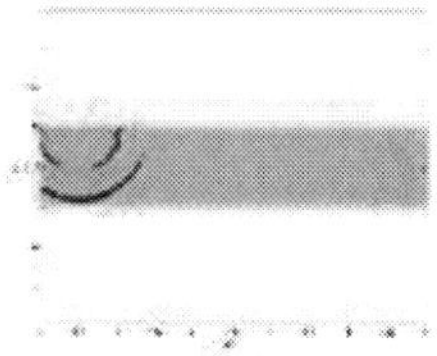	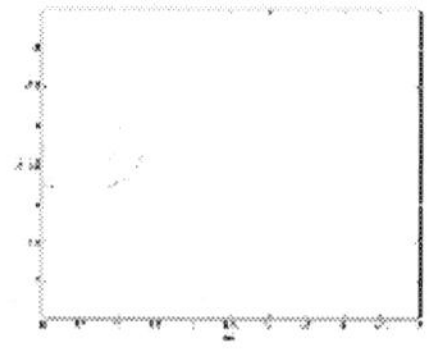

100	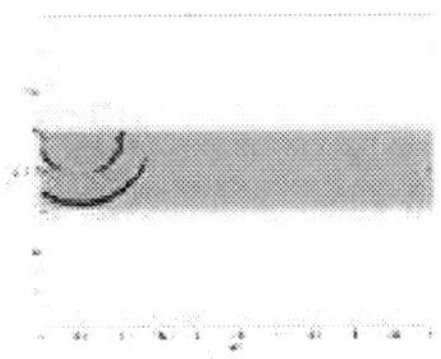	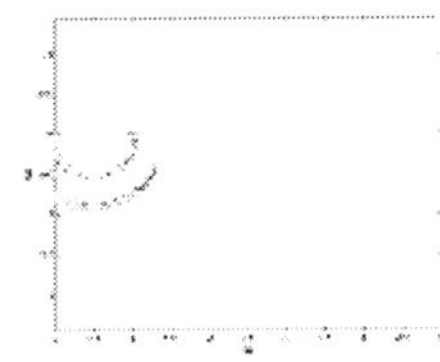
180	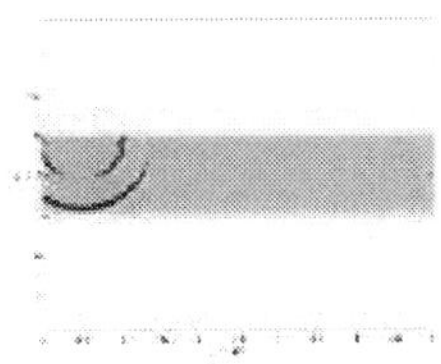	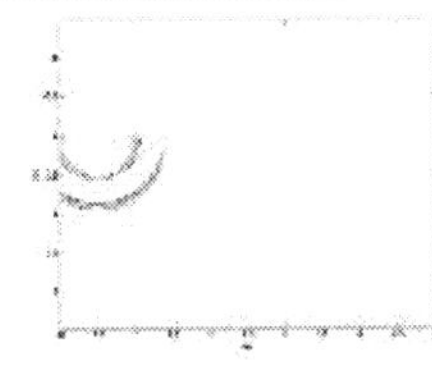

Based on the results, the recommended sharpness value for the Gabor filter is 100.

5. Choice of filter

5.1. Qualitative characteristics

Based on Chapter IV, it can be concluded that the detection of boundaries is affected by: the depth of field of the image, the filter parameters, and the binarization of the image. In consequence of this, for each filter, the parameters are selected separately, based on theoretical aspects. Thus, in experiments, the parameters of the filters did not change, but the amount of sharpness was varied. For the Roberts filter, the amount of sharpness is 45 based on qualitative analysis, 55 for the Sobel filter, 87 for the Prewitt filter, 32 for the Canny method, 100 for the Gabor filter.

5.2. Quantitative characteristics

To assess the quality and comparison of the above filters we will use metrics. But in order to go to metrics, it is necessary to introduce a concept to describe these metrics in terms of classification errors - confusion matrix [10].

Suppose that there are 2 classes and an algorithm that predicts that each object belongs to one of the classes, then the error matrix will look like this (table 6).

Table 6. Confusion matrix

	y=1	y=0
$y^*=1$	True Positive (TP)	False Positive (FP)
$y^*=0$	False Negative (FN)	True Negative (TN)

where is y* the response of the algorithm at the facility; y is the true class label on this object. Thus, classification errors are of two types: FN and FP.

5.2.1. Metric accuracy. A basic metric that measures the number of correctly classified objects relative to the total number of all objects.

$$accuracy = \frac{TP + TN}{TP + TN + FP + FN} \qquad (12)$$

5.2.2. Metric precision. The accuracy metric determines how many of all objects that are classified as positive are indeed positive, relative to the total number of positive labels received from the model.

$$precision = \frac{TP}{TP + FP} \qquad (13)$$

5.2.3. Metric recall. The completeness metric determines how many objects our model was able to correctly classify with a positive label from among a multitude of positive ones.

$$recall = \frac{TP}{TP + FN} \tag{14}$$

To select a filter for detecting borders on images of a wave field pattern, we use the accuracy metric, since the number of correctly detected borders on the image relative to the total number is important for us.

The calculation of the selected metric was carried out on the basis of the TensorFlow framework using the Keras library, where this metric is already included.

The results of the calculation are presented in table 7.

Table 7. Quantitative characteristics

	Sobel	Prewitt	Roberts	Gabor	Canny
accuracy, %	72	68	40	89	76

Thus, the Gabor filter showed the best result for the selected metric, and therefore, it is advisable to use it to detect boundaries in images of wave field patterns.

6. Conclusion

Experiments have shown that the Gabor filter is recommended to use as a detector for detecting boundaries in images of wave field patterns. This was confirmed by calculated quantitative and qualitative characteristics. During the study, a program was created to detect borders on the image, in which the parameters of the image can vary. This program can be implemented not only to solve the problem, but also for other experiments.

References

[1] Mikhailenko B G 1892 A Semi-Analytical Method to Calculate Nonstationary Wave Fields for Layered-Inhomogeneous Media Models *Mathematical Methods of Solving Direct and Inverse Problems of Geophysics* (Novosibirsk: Computing Center, Sib. Branch, USSR Acad. Sci.) pp. 92–104.

[2] A. A. Yakimenko, A. E. Morozov, D. A. Karavaev 2018 *J. Phys.: Conf. Series* **1015** 032148 doi: 10.1088/1742-6596/1015/3/032148

[3] Fukushima K 2003 *Neurocomputing* **51** 161–180

[4] Sobel I, Feldman G 1968 *A Talk at the Stanford Artificial Project* 271-272.

[5] Gonzalez R and Woods R 2006 *Digital image processing (3rd Edition)* (USA: Prentice-Hall)

[6] Guo Z, Xu W and Chai Z 2010 *Processin of Ninth International Symposium on Distributed Computing and Applications to Business, Engineering and Science* 169–171

[7] Canny J A 1986 *Processin of IEEE Transactions on Pattern Analysis and Machine Intelligence* **6** 679–698 doi: 10.1109/TPAMI.1986.4767851

[8] Huang L *et al* 2004 *Processin of Sixth IEEE International Conference on Automatic Face and Gesture Recognition* 397–402

[9] DeValois R L, Albrecht D G and Thorell L G 1982 *Vis. Res* **22** 545–559

[10] Gordienko N, Gang P, Gordienko Yu, ZengW, Alienin O, Rokovyi O, and Stirenko S 2018 *Processin of 25th International Conference on Neural Information* 14–16

Functioning of back-to-back high voltage direct current link model in Tomsk electric power system

A P Maltsev, V E Rudnik, Y Yu Malkova and R A Ufa
School of Energy & Power Engineering, Tomsk Polytechnic University, Tomsk, Russia
E-mail: maltsevaleksey@mail.ru

Abstract. In this article the research results of steady-state and dynamic performance of asynchronous back-to-back HVDC link in Tomsk electric power system are presented. It is proposed to use the back-to-back HVDC link to provide an asynchronous link for parallel operation of the North and South parts of Tomsk electric power system. The steady-state stability limit and maximum allowed power flow were defined. Based on this the more effective location and power of back-to-back HVDC link were determined. Therefore, it is shown the application of back-to-back HVDC link decreases the level of short-circuit current, power fluctuations and increases the stability of load operation, especially of motor load at substation.

1. Introduction

High-voltage direct current back-to-back link (back-to-back HVDC) based on static voltage converters with high-speed power fully-controlled semiconductor switches make it possible to effectively solve a number of urgent tasks in the electric power industry [1, 2]:

- integration of non-synchronous operating electric power systems (EPS) and their parts;
- DC power transmission;
- limiting short-circuit currents in the combined EPS and their parts;
- damping of low-frequency oscillations in an EPS and the solution of the related problem of oscillatory stability of the EPS operation.

According to the CIGRE report (Conseil International des Grands Réseaux Electriques – International Council for Large High-Voltage Electrical Systems), 15 back-to-back HVDC link are designed and already in operation, including two projects planned for implementation in Russian EPS [3].

The features and specifics of the functioning of these means significantly change and complicate the processes in the equipment and the dynamic properties of the EPS, respectively, the operating conditions of power equipment, relay protection and emergency automation system. So, the complete and reliable information about these processes is required to provide safety and effective operation of such modern EPS. Due to the inadmissibility of field experiments and the impossibility, due to the complexity of modern EPS, of their adequate physical modeling, the only way to obtain this information is mainly mathematical modeling.

2. Experimental studies of asynchronous back-to-back HVDC link

These studies were conducted on the example of Tomsk EPS (Figure 1). There are two asynchronous operated part of Tomsk EPS – North and South parts. The dividing point is located along the 220 kV transit: Substation (SS) Tomskaya (UES of Siberia) - SS Volodino - PS Parabel - PS Sovetsko-Sosninskaya - Nizhnevartovskaya power plant (UES of the Urals) about 800 km long [3].

In the northern part there are one of the largest load nodes - oil producing companies, in the southern part - one of the main sources of generation. If two chats of the Tomsk region are merged, it will increase the level of reliability in the electric power system, as well as increase the level of energy efficiency [3].

Operational switchings always leads to temporary shutdown of some consumers, and since the share of synchronous motors is large, a non-synchronous activation of the load node is impossible [2, 4].

The process of choosing the power and effective location of the back-to-back HVDC link is required to determine the following parameters:

• the value of the maximum possible emergency unbalance formed by the shutdown of large power units at the station or power lines;

• the value of the maximum permissible flow (MPF) in a controlled section.

The disconnection of a large unit in Tomsk EES is laid in the MPF Krasnoyarsk-Kuzbass-Tomsk and will not affect the power flow for the considered transit [3].

Accordingly, the installed power of the back-to-back HVDC limk will depend on the magnitude of the power flow from the UPS of the Urals to the UPS of Siberia, and vice versa, taking into account the communication MPF and the power of the loads of the transit substations [3].

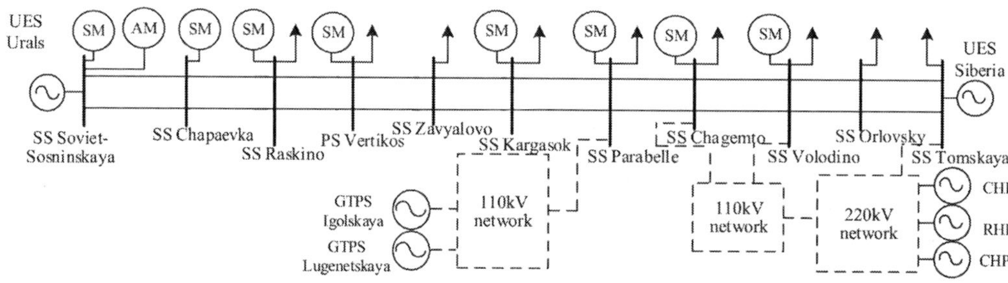

Figure 1. Test scheme of the power system of the Tomsk region: CHPP – Central Heating and Power Plant, RHPP – regional hydroelectric power plant, AM – asynchronous motor, SM – synchronous motor, GTPS –gas turbine power station.

1) Calculate the power of back-to-back HVDC link on Ss Parabel.

Define the MPF transit head sections:

• for Power lines (PL) 220 kV Nizhnevartovsk regional hydroelectric power plant - Ss Sovetsko-Sosninskaya for a normal scheme Pp1 = 247 MW (determined by ECL criterion (emergency current load) PL 220 kV Nizhnevartovsk regional hydroelectric power plant - Ss Sovetsko-Sosninskaya I or II PFM circuit (post-emergency mode) PL 220 kV Nizhnevartovsk regional hydroelectric power plant - Soviet-Sosninskaya II or I chain);

• for Power lines (PL) 220 kV Ss Tomskaya - Ss Volodino for a normal scheme Pp2 = 204 MW (determined by the criterion ECL PL 220 kV Tomskaya - Volodino with a tap on the Orlovka Ss I (II) circuit PEM PL 220 kV Tomskaya - Volodino with tap on Ss Orlovka II (I) chain);

Determination of load power:

• Consumption of the electric network from Ss Sovetsko-Sosninskaya to Ss Kargasok is Pn1 = 200 MW;

• Consumption of the electric network from Ss Tomsk to SS Parabel is Pn2 = 170 MW. At the same time, the total generation of gas turbine power plants averages 50 MW;

The required power of the back-to-back HVDC can be:

• to ensure the flow of active power from Ss 220 Kargasok towards Ss Parabel (from the Ural UES to the UES of Siberia): Pback-to-back HVDC = Pp1 - Pn1 50 = 247-200 = 47 MW

• to ensure the flow of active power from Ss Parabel towards Ss Sovetsko-Sosninskaya (from the UES of Siberia to the UES of the Urals): Pback-to-back HVDC = Pp2 - Pn2+50 = 204-170+50 = 84 MW

Accordingly, the recommended installed back-to-back HVDC link capacity is around 100 MW.

2) Calculate the power of back-to-back HVDC link on Ss Sovetsko-Sosninskaya.

Define the MPF transit head sections:

• for Power lines (PL) 220 kV Nizhnevartovsk regional hydroelectric power plant - Ss Sovetsko-Sosninskaya for a normal circuit Pp1 = 247 MW (determined by ECL criterion (emergency current load) PL 220 kV Nizhnevartovsk regional hydroelectric power plant - Ss Sovetsko-Sosninskaya I or II PFM circuit (post-emergency mode) PL 220 kV Nizhnevartovsk regional hydroelectric power plant - Soviet-Sosninskaya II or I chain);

• for Power lines (PL) 220 kV Ss Tomskaya - Ss Volodino for a normal scheme Pp2 = 204 MW (determined by the criterion ECL PL 220 kV Tomskaya - Volodino with a tap on the Orlovka Ss I (II) circuit PEM PL 220 kV Tomskaya - Volodino with tap on Ss Orlovka II (I) chain);

Determination of load power:

• the Ss 220 kV Sovetsko-Sosninskaya load is Pn1 = 125 MW;

• Consumption of a section of the electric network from Ss Tomsk to Ss Chapayevka is Pn2 = 250 MW

The required Pback-to-back HVDC may be:

• to ensure the flow of active power from SS 220 Sovetsko-Sosninskaya towards Ss Chapaevka (from the Ural Energy System of the Urals to the Siberia UES): Pp1 - Pn1 = 247 - 125 = 122 MW

• to ensure the flow of active power from Ss Chapaevka towards Ss Sovetsko-Sosninskaya (from the UES of Siberia to the UES of the Urals): Pp2 - Pn2 = 204 – 250+50= 4 MW. Accordingly, the front in this direction is not valid. Accordingly, the recommended installed capacity of the back-to-back HVDC link is ariund 125 MW.

Thus, after conducting a comparative analysis of the obtained calculations, it can be noted that to ensure the reversibility of power flow through the back-to-back HVDC link, as well as from the point of view of the minimum required power of the back-to-back HVDC link the most expedient is to install back-to-back HVDC link on Ss Parabel.

As a result of tests on a real power system, it was revealed that the scale of measuring the angle between the stress vectors (δ) in the northern and southern parts of the Tomsk EPS is -40^0 to 50^0. Here, the negative value of the angle corresponds to the advance of the voltage of the northern part of the Tomsk EPS, the positive - to the southern part [5].

Thus, the «safe» inclusion on Ss Parabel is possible with $\delta < 45^0$, otherwise dynamic stability is disturbed in nearby load nodes, In particular, the engines on Ss Parabel, Ss Chazhemto and Ss Kargasok (Figure 2).

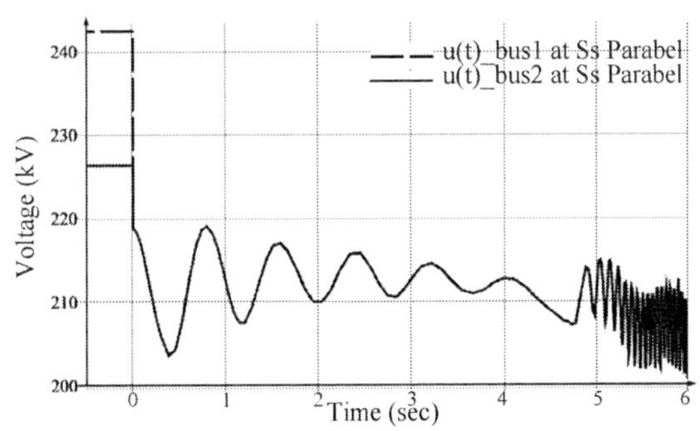

(a)

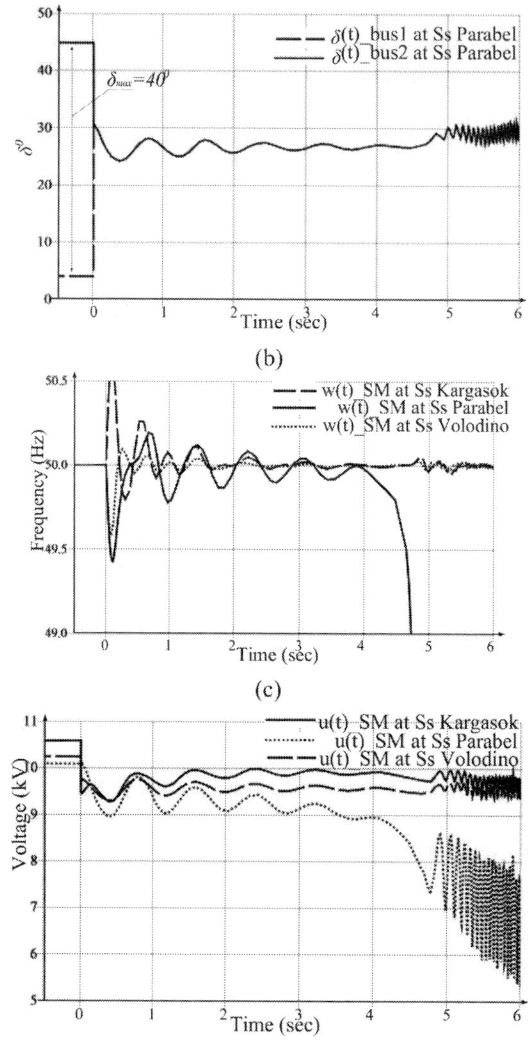

(b)

(c)

(d)

Figure 2. The oscillograms of voltage (a) at $\delta \approx 40^0$, voltage angle (b), rotational speed (c) and voltage of electrical machines (d).

Thanks to these conditions, it is possible to combine the parts of the Tomsk EPS by means of automatic reclosing with a synchrocheck function. However, in field tests it was found that such a combination for a long time cannot be performed without a specialized device [5].

According to the presented analysis of the properties and capabilities of the designated back-to-back HVDC link, it is possible to combine parts of the Tomsk EPS using this device, test fragments are presented below.

In Figure 3 the results of the simulation of the inclusion of the parallel operation through the installation of back-to-back HVDC link on Ss Parabel are presented. It is clearly demonstrated the effectiveness of the use of back-to-back HVDC link: even in the worst mode - $\delta \approx 180^0$ the union does not disrupt normal operation EPS.

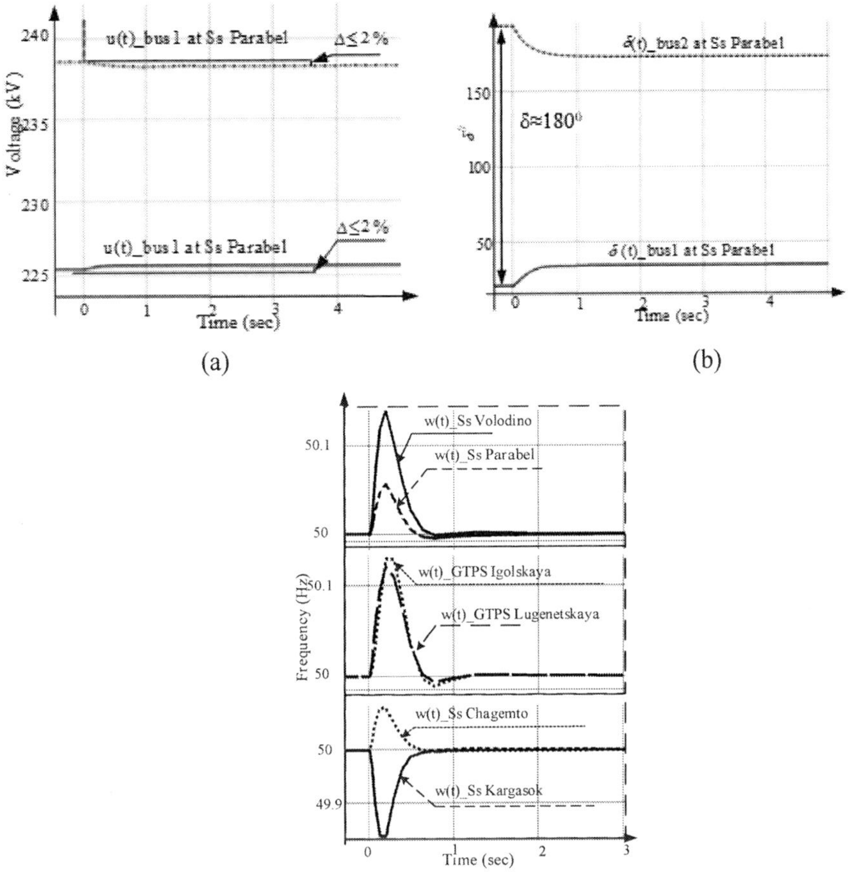

(c)

Figure 3. The oscillograms of voltage (a) and voltage angle (b), rotational speed (c).

Also, with the help of back-to-back HVDC link, step-by-step regulation and reversal of active power were carried out, the results of modeling the control processes are reflected in Figures 4: at 1 sec there was a step change in load (with 6 steps); at 6 sec reverse of the transmitted power occurred.

This study shows that back-to-back HVDC link can regulate and change power flow during few milliseconds [4, 6] in case of an accident at a power plant and the subsequent disconnection of the generator, thereby satisfying consumer demands for power without disturbing the parameters of EPS.

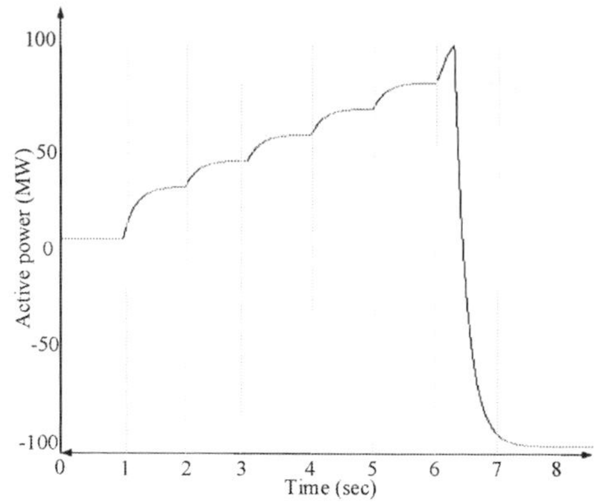

Figure 4. Active power curve, while controlling the transmitted power of the back-to-back HVDC mode.

The results of modeling the processes during short-circuit on the transit of 220 kV show the positive effect of back-to-back HVDC on emergency processes in the EPS (Figure 5). The voltage at the back-to-back HVDC connection point changes slightly, the frequency of oscillations in the system decreases.

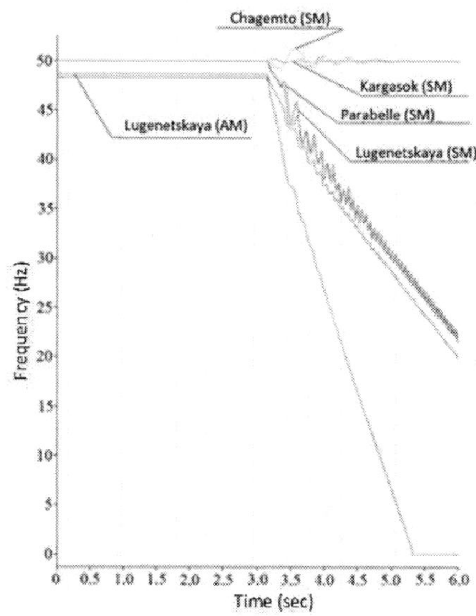

Figure 5. Oscillograms of the speed of rotation of the motors at a short circuit on the transit of 220 kV.

3. Conclusion

The considered option of combining the southern and northern parts of the Tomsk energy system using the back-to-back HVDC link is considered successful. The capabilities of this device allow you to combine at absolutely different angles of stress vectors. The presented results of modeling the processes at short-circuit on the 220 kV transit show the positive effect of the two-position direct current line.

The appropriate power of back-to-back HVDC link was determined in case for instalation at Ss Parabel (aroud 100 MW) and Ss Sovetsko-Sosninsk (aroud 125 MW). Howerver, after conducting a comparative analysis of the obtained calculations, it can be noted that to ensure the reversibility of power flow through the back-to-back HVDC link, as well as from the point of view of the minimum required power of the back-to-back HVDC link, the most effective installation is at Ss Parabel.

Acknowledgment
The work was supported by Ministry of Science and Higher Education of Russian Federation under the governmental grant "Science" № FSWW-2020-0017.

References

[1] ORDER No. 551-ra. «On the scheme and program for the development of the electric power industry of the Tomsk Region for the period 2014 – 2018». Available at: http://docs.cntd.ru/document/467907549
[2] Sellick R L and Åkerberg M 2012 *Proceeding of 10th IET International Conference on AC and DC Power Transmission* 1–6
[3] Ufa R A *et al* 2019 *Elektrichestvo* 33–41
[4] Deuse J, Stubbe M, Meyer B, Panciatici P 1995 *Proceeding of CIGRE Symposium on Power Electronics in Electric Power Systems* 1–6
[5] Markman G Z 2008 *Energoeffektivnost' preobrazovaniya i transportirovki elektricheskoj energii: Uchebnoe posobie* (Tomsk: Tomskij politekhnicheskij universitet)
[6] Wang H, Redfern M A 2010 *Proceeding of 45th International Universities Power Engineering Conference* 1–5

Assessment of wind power generation impact on the distance protection operation

N Yu Ruban, A B Askarov, A V Kievets and V E Rudnik
School of Energy and Power Engineering, Tomsk Polytechnic University, Tomsk Russia
E-mail: rubanny@tpu.ru

Abstract. In the article an experimental power system consisting of Russian equipment, as well as 220 kV power line distance protection, are simulated in the RTDS Simulator. The settings were calculated and the polygonal characteristics of distance protection were constructed. A sequential installation of wind turbines into various nodes of the power system was carried out and the correct operation of the protection was verified. As a result of the study, the impact of wind turbines on the distance protection when they are connected between the protection installation site and the point of the short circuit is revealed. The analysis of this impact is made. Possible ways to solve the identified problems are proposed, such as the use of adaptive distance protection, as well as numerous studies of the behaviour of wind turbines in various operating modes for further analysis and possible adjustment of guidelines for relay protection using new algorithms for calculating and configuring of distance protection. The results of this work can be used to improve and clarify the recommendations for tuning, produced by manufacturers of relay protection devices, which currently do not take into account the impact of wind power generation.

1. Introduction

Nowadays, alternative energy is gaining popularity. In modern power systems introduce new generating sources based on renewable energy. One of these are wind turbines. Their structure and behavior in different modes of operation of the power system differs significantly from the traditional synchronous generators existing in the power system. Therefore, it is necessary to investigate their behavior both in normal operating conditions of the power system and in emergency mode, that is, in case of various kinds of damages. It should be understood that wind turbines affect both the state of the power system as a whole and its individual parts. An integral part and the main type of electrical automation, without which normal and reliable operation of modern power systems is impossible, is relay protection. According to [1], the main protection of lines 110–220 kV is distance protection (mainly three-stage), based on monitoring changes in the calculated line resistance.

Thus, the purpose of this paper is to study the impact of wind power generation on the operation of distance protection of a 220 kV power line.

In this paper, it is proposed to simulate a wind turbine equipped with a squirrel-cage induction generator (SCIG) [2]. This generator is selected due to the prevalence in Russia and European countries. Although this model is a representative of the older generation, there are still a lot of such wind turbines, they are fully exploited and used everywhere.

In paper [3], as a review, different factors affecting distance protection in a power system with integrated renewable energy sources were considered, such as: (a) intermittent nature of a renewable source, (b) types of wind generators, (c) fault characteristics and nature of faults. After that, a generalized conclusion is made that the highly variability and intermittent nature of renewable energy sources in different network topologies causes changing in the levels of short-circuit currents and losing of coordinated operation of relay protection. It is also stated that renewable energy integration at the transmission level affect the reach of the distance relays and lead to their maloperation. The paper [4] summarizes many studies in which it was stated that: (a) fault characteristics of wind turbines have different behavior compared with synchronous based generators, (b) wind turbines do not provide sustained short-circuit currents during fault, (c) short-circuit current is pertinent to wind speed, penetration levels and number and type of wind turbine generator (d) relay selectivity is difficult to obtain with source impedance variation and bidirectional fault currents, (e) excessive fault-clearing times jeopardize distributed generator stability, and (f) traditional protection schemes limit distributed generation connection capacity. The paper [4] carried out studies using the RTDS Simulator. It can be concluded from the presented results that increasing wind penetration in distribution systems would have a detrimental effect on the distance relay operational characteristics. As a result, the operational characteristic of a distance relay expands and shrinks as a function of the wind penetration levels and wind speeds. In paper [5], the behavior of SCIG in case of short circuit on power lines is considered. Attention is drawn to the fact that the SCIG short-circuit current can exceed the rated current by more than six times. It is concluded that this type of wind generator introduces the greatest amount of short-circuit current, compared with other types of wind generators.

Based on the identified problems in papers [3–5], this paper proposes to simulate an experimental power system using Russian equipment and conduct appropriate studies to assess the effect of wind generators on the operation of distance protection that meets Russian standards.

2. Simulation in the RTDS simulator

2.1. Test scheme simulation

An experimental scheme of the power system has been developed and modeled in the RTDS Simulator. The structural view of the scheme is presented in Figure 1.

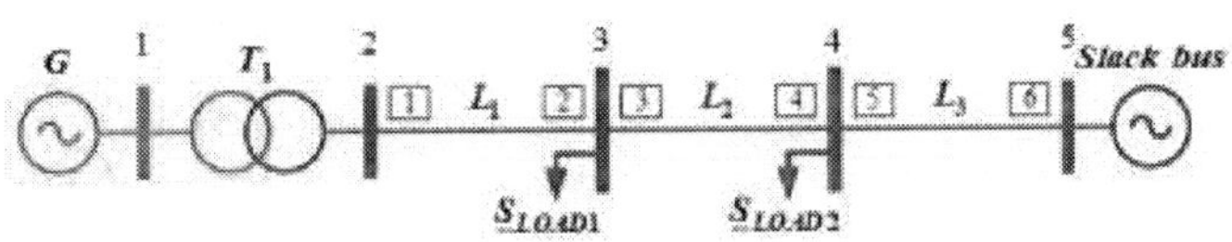

Figure 1. Experimental scheme.

On each power line in the scheme (Figure 1) two sets of protection are installed – at the beginning and at the end of the line. In this paper, a complete calculation was made only for the set number 1, located at the beginning of the power line L_1. Protection of the remaining lines is necessary for the coordination of the protection stages of the line L_1.

2.2. Simulation of distance protection

The simulation of distance protection in the RTDS Simulator is carried out using a special block, for which the input values are the measured values of the phase voltages on the buses at the beginning of each power line and currents in the corresponding lines and the output values are resistance and reactance.

2.3. Calculation of distance protection settings

The calculation results are summarized in table 1.

Table 1. Protection stages settings.

Stage Number	Settings
I	$\underline{Z}_{set}^{\mathrm{I}} = 37.535 \cdot e^{j78^{\circ}}$ Ohm
II	$\underline{Z}_{set}^{\mathrm{II}} = 72.56 \cdot e^{j78^{\circ}}$ Ohm
III	$\underline{Z}_{set}^{\mathrm{III}} = 116.365 \cdot e^{j77^{\circ}}$ Ohm

Thus, for set 1 of the distance protection of the power line L_1, the settings of three stages were calculated.

2.4. Plotting of Distance Protection Polygonal Characteristics

According to [6], when setting the distance protection polygonal characteristics, it is necessary to calculate the settings for the resistance, as well as one of the angles for the third stage characteristic, which takes into account the detuning from the load mode. When calculating the setting by resistance, the arc resistance at short-circuit at the end of the protected area is estimated. The electric field strength of the arc is assumed to be 2500 V/m, the arc length is taken to be equal to the distance between the phase conductors, which is 4.5 m for a selected section of a single-circuit power line of 220 kV [7].

Figure 2 shows the polygonal characteristic of all stages of the distance protection of the power line L1 on a scale of 1:1. It is based on the calculated settings and geometrical considerations in the Microsoft Visio software package. In Figure 2 there are following designations: Z_{set} is a setting of the protection stage; φ_1 is an angle of maximum sensitivity, assumed to be equal to the angle of setting; φ_2 is an angle of inclination of the face 0-1, assumed to be −15° for the first and second stages, for the third stage is calculated separately according to [6]; φ_2 is an angle of inclination of the face 0-4, is assumed to be 115° according to [6]; R_{set} is a setting for the resistance. The face 2–3 passes through the end of the straight line Z_{set} and is located parallel to the axis R. The characteristic of the first stage is shown in blue, the second – green, the third – lilac.

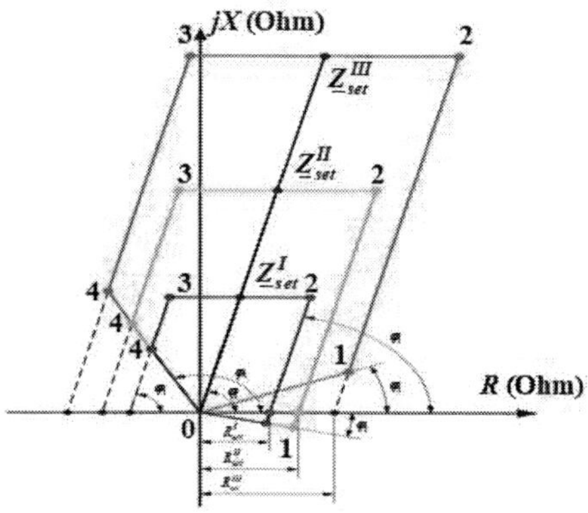

Figure 2. Distance protection polygonal characteristics.

Further characteristics were built in the RTDS Simulator.

3. Experiments in the RTDS Simulator

3.1. Verification of distance protection operation

When simulating three-phase-to-ground (3LG) faults and line-to-line (LL) faults in various nodes of the circuit, it was found that the impedance measured by distance protection falls into protection reach of the first, second and third stages. From which it can be concluded that the protection is correct and responds to all types of phase-to-phase faults.

3.2. *Wind Turbine Model*

The model of the first type of generator SCIG was chosen as the implemented wind turbine. This model consists of a windmill, a generator, a control system and its own transformers for connection to a 220 kV network. In the windmill model, it is possible to set the wind speed (input value) and remove the torque (output value) from the wind turbine shaft, which is transmitted to the rotor shaft of the asynchronous generator through the gearbox. It is known that the power generated by wind turbines depends on wind speed. Thus, in the presented model it is possible to change the values of the input wind velocity and obtain different values of the power generated. During the experiments, it was found that at a wind speed of 54 km/h = 15 m/s, the wind generator produces a maximum power of 6.36 MW.

3.3. *Installing wind turbines at the beginning of the protected line*

In previous experiments, one source was connected to the buses of the beginning of the protected power line L_1 – generator ***G*** (Figure 1) with a rated capacity of 100 MW. In this experiment, two sources were selected so that the total generation in the buses of the beginning of the power line L_1 remained approximately equal to 100 MW. One part of the power is generated by the newly installed generator ***G*** (Figure 3), and the other by several wind turbines. Thus, in addition to the generator ***G***, five wind turbines with a maximum capacity of 6.36 MW are connected to the buses of the beginning of the line L_1 (Figure 3), then the total capacity of the wind power plant (WPP) is 31.8 MW.

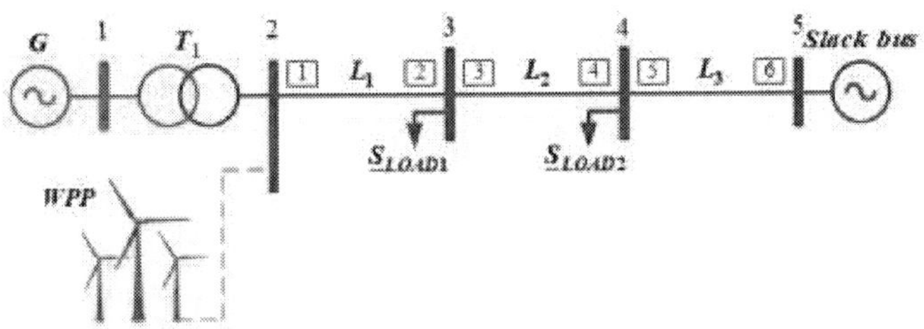

Figure 3. Experimental scheme with the installation of wind turbines at the beginning of the protected power line L_1.

After that, experiments with three-phase-to-ground (3LG) faults and line-to-line (LL) faults in various nodes of the circuit were conducted and the operation of distance protection was checked. The change in the level of generated power behind the protection, caused by the installation of a wind turbine at the beginning of the protected line L_1, does not affect the impedance measured by the distance protection during phase-to-phase faults.

3.4. *Installing wind turbines at the end of the protected line*

In this experiment, the power sources remained the same, but the wind farm, consisting of five wind turbines, is now connected to the buses at the end of the protected power line L_1 (Figure 4).

With this configuration of the circuit, recalculation of the distance protection settings is necessary, since a new power source has appeared in the node without generation. The calculation results are summarized in table 2.

In addition, when setting the distance protection polygonal characteristics, the settings for the resistance were recalculated. After that, experiments with 3LG faults and LL faults in various nodes of the circuit were carried out and the operation of distance protection with new characteristics was verified.

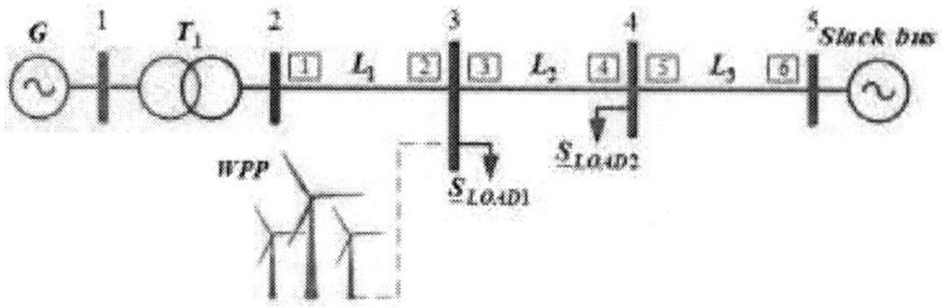

Figure 4. Experimental scheme with the installation of wind turbines at the end of the protected power line L_1.

Table 2. Protection stages settings.

Stage Number	Settings
I	$\underline{Z}_{set}^{\mathrm{I}} = 37.535 \cdot e^{j78°}$ Ohm
II	$\underline{Z}_{set}^{\mathrm{II}} = 75.088 \cdot e^{j78°}$ Ohm
III	$\underline{Z}_{set}^{\mathrm{III}} = 106.956 \cdot e^{j79.7°}$ Ohm

It is established that with phase-to-phase faults at the end of the protected power line L_1, distance protection measures the impedance, which is actually equal to the impedance of the line L_1, and connecting wind turbines to short-circuit buses does not affect the measured impedance.

The results of the simulation of phase-to-phase faults at the end of the adjacent line L_2 at various wind speeds are shown in Figures 5–7. As can be seen, with increasing wind speed and, consequently, power output by the wind turbine to the network, the hodograph of the measured impedance moves in the direction of decreasing the resistance and increasing the reactance. When installing the wind turbine at the end of the protected power line L_1 and simulating a LL fault on the buses at the end of the adjacent line L_2, the impedance hodograph does not fall into any of the distance protection reaches.

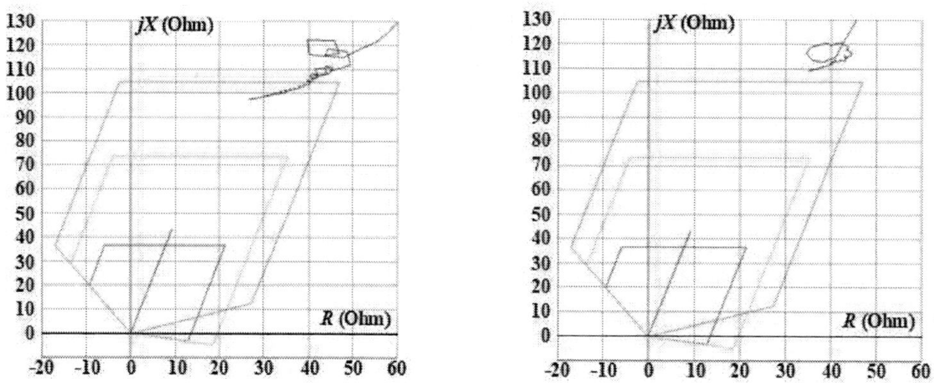

a b

Figure 5. Hodograph of (a) the phase A impedance in case of a 3LG fault and (b) the impedance between phases A and B in case of a LL fault at the end of the adjacent line L_2 at wind speed of *18 km/h = 5 m/s.*

IOP Conf. Series: Materials Science and Engineering 1019 (2021) 012010 doi:10.1088/1757-899X/1019/1/012010

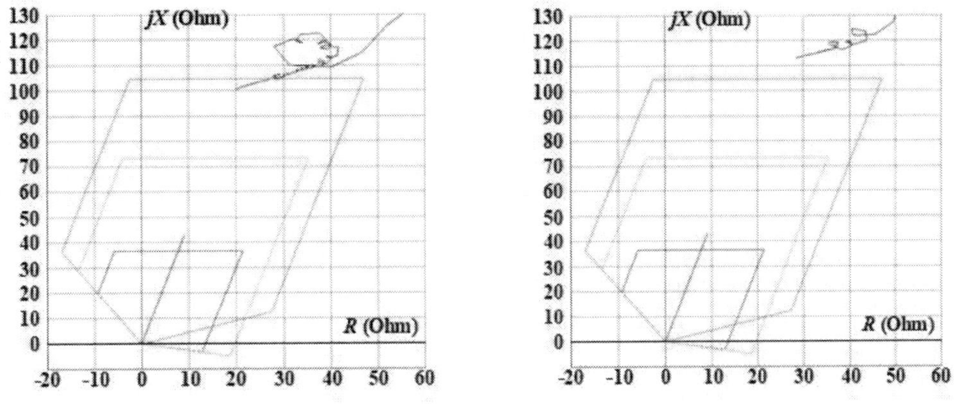

a b

Figure 6. Hodograph of (a) the phase A impedance in case of a 3LG fault and (b) the impedance between phases A and B in case of a LL fault at the end of the adjacent line L_2 at wind speed of *36 km/h = 10 m/s.*

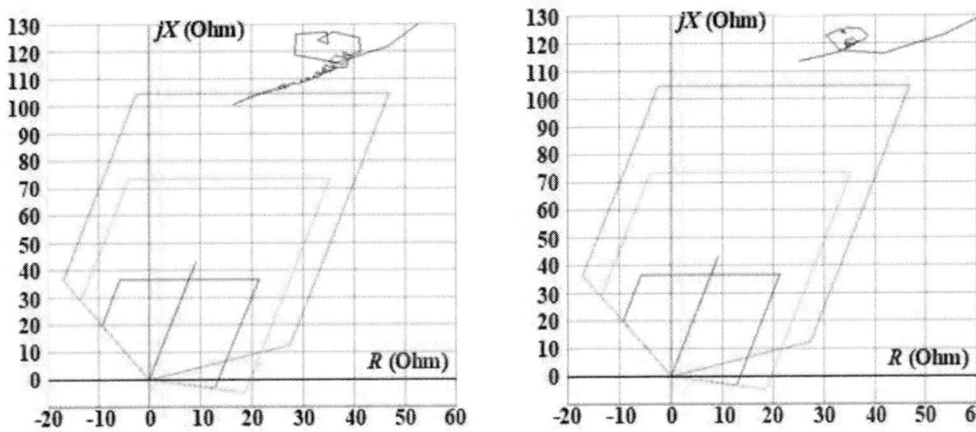

a b

Figure 7. Hodograph of (a) the phase A impedance in case of a 3LG fault and (b) the impedance between phases A and B in case of a LL fault at the end of the adjacent line L_2 at wind speed of *50 km/h ≈ 14 m/s.*

The connection of wind turbines to the buses in the beginning of the protected line and the change in the level of the power generated by them does not affect the impedance measured by the distance protection during phase-to-phase faults at all subsequent points on the protected and adjacent lines. This is due to the absence of additional generation between the protection installation site and the short-circuit point.

After connecting the wind turbine to the end of the protected line, distance protection settings were recalculated, as a result of which an increase of the protection reach of the second stage and decrease of the protection reach of the third stage were detected. When simulating phase-to-phase faults at the end of the protected line, it was established that the presence of wind turbines in the same node also does not affect the impedance measured by the distance protection, since the protection measures the impedance from its installation to the short circuit point. When modelling phase-to-phase faults at the end of an adjacent line (Figures 5–7), it was found that with increasing wind speed and, consequently, power output by wind turbines to the network, the hodograph of the measured impedance moves in the direction of decreasing resistance and increasing reactance. It is also noted that with 3LG faults, when wind turbines operate in the generation mode close to the maximum power, the point of the measured impedance is practically at the boundary of the protection reach of the third stage of distance protection. And with LL faults, the hodograph of the measured impedance does not fall into any of the protection reaches, which in this case indicates the failure of the remote backup function.

4. Introduction

According to the research, the effect of wind generators on the distance protection of a 220 kV power line has been revealed. The reasons for such deviations are: (a) a different behaviour of SCIG generators during a short circuit compared to traditional synchronous generators, which seems to be an important factor for determining the characteristics of distance protection, (b) the variation in wind speed and character significantly affects the reach of the distance relays set. This leads to changes in the apparent impedance seen by the protective relay and also causes changes in the reach setting of the relay.

The use of adaptive distance protection, as well as studies of the behaviour of wind turbines in various operating modes for further analysis and possible adjustment of guidelines for relay protection using new algorithms for calculating and configuring distance protection can be solutions to the identified problems.

Acknowledgments

The work was supported by Ministry of Science and Higher Education of Russian Federation, according to the research project № MK-1675.2019.8.

References

[1] 2005 *Electrical installation rules* (Novosibirsk: Sib. univ Publishing house)
[2] Ackermann T 2005 *Wind Power in Power Systems* (Hoboken: John Wiley & Sons, Ltd)
[3] Telukunta V, Pradhan J, Agrawal A, Singh M and Srivani S G 2017 *CSEE J. Power Energy Syst.* **3**(4) 365–379
[4] El-Arroudi K and Joós G 2018 *IEEE Trans. Power Deliv.* **33(2)** 620–629
[5] Muljadi E, Samaan N, Gevorgian V, Li J and Pasupulati S 2010 *Proc. Of IEEE Ind. Appl. Soc. Annu. Meet.* **1** 1–9
[6] 2015 STO 56947007-29.120.70.200-2015 *Guidelines for the calculation and selection of settings of microprocessor-based devices for relay protection and automation* (Moscow)
[7] Hondurov S A, Ilyukhin E V, Pirogov M G and Soloviev A L 2013 *Step distance protection of 35-220 kV transmission lines. The method of calculation* (St. Petersburg: PEIPK)

The influence of load distribution along a line on power and place of connection of a photovoltaic system

A V Bogdan[1], V A Bogdan[2], K A Garkavyi[3] and D S Netrebko[1]
[1]Department of electrical technology, Heat technology and renewable energy sources, Kuban State Agrarian University, Krasnodar, Russia
[2]Electrical regimes service, "Kubanenergo" PJSC, Krasnodar, Russia
[3]Department of physics and electrical engineering, Krasnodar Air Force Institute for Pilots, Krasnodar, Russia
konstantingarkavyi@gmail.com

Abstract. Distributed generation in the electrical network causes a redistribution of power and leads to a reduction in technological losses and a decrease in voltage deviations for consumers. The installation of additional generators of electric energy based on the use of wind generators and photovoltaic systems has become widespread in the world energy industry, and has now received development in the Russian Federation. It is shown that the connection of an additional source based on photovoltaic systems can reduce losses in a supply line. According to results of the analysis of a possible location of the connection of an additional power supply on a line with a uniformly decreasing or increasing load, it is proved that the greatest reduction in the level of losses in a line is achieved when additional power supply is connected to it in the case of increase in the load from the beginning to the end. Analytical expressions for determination of economically expedient capacities and a place of connection of an additional source for linearly decreasing or increasing specific load of a supply line are presented. There was determined the optimal power of an additional power source in the form of a photovoltaic system and the optimal point of its connection on a line to reduce power losses during power transmission.

1. Introduction

Power losses in power supply and distribution lines are one of the important criteria used to assess the efficiency of a network. Traditionally, to reduce energy losses, attention must be paid to the rational construction of a network circuit, increasing the cross section of conductors, reactive power compensation, etc. [1, 2]. Nevertheless, modern conditions dictate new tasks for design and operation of power supply networks, for which the government stimulates the development of distributed generation, placing ever higher demands on environmental and energy security [3].

Distributed generation, using a variety of energy sources, including photovoltaic systems, is one of the ways of energy development [4–6].

When connecting photovoltaic systems to the electrical network, it is necessary to solve problems with the parameters of the lines, such as determining a rational point of connection, voltage fluctuations, and power and energy losses. The photovoltaic system is usually connected to the low voltage side of the supply substation or to the consumer's supply input. In this case, there may be difficulties with the placement of a photovoltaic system or when it is split into a larger number of individual installations.

The source of stabilized supply is frequently used on the basis of the inverter with electric current output for connection to power distribution networks of a photoelectric system and [7]. It generates energy into the network, the parameters of the photovoltaic system are coordinated with the distribution network [8, 9]. The option of connecting a photovoltaic generating unit to power the load is shown in Figure 1.

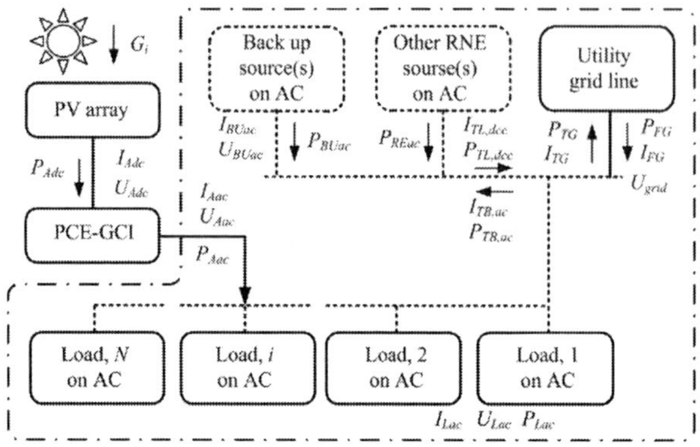

Figure 1. Block diagram for connecting a photovoltaic system to power the load.

In the diagram proposed in Figure 1, an additional source (photovoltaic system) is connected to one of the loads (Load, *i*), and not to the beginning of the power supply line for the loads Load, 1, Load, 2, ... Load, *N*. Such connection provides advantages: reduction of losses in the supply line and equalization of voltages at consumers connected to the line.

To determine the optimal power and the connection point of the additional source with the uniform distribution of the synchronously changing load along the line, a technique was proposed [3].

But the distribution of loads along the line can be uneven. In this paper we analyze the effect of load distribution, linearly decreasing or increasing, on power and place of connection of the photovoltaic system P_{Aac}, as shown in Figure 2.

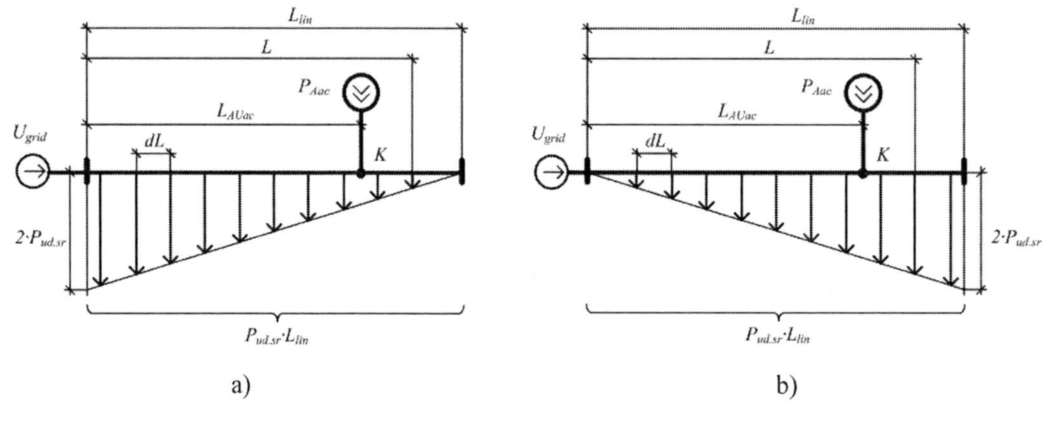

Figure 2. Rated scheme of a feeder with linear distribution of active power of consumers: a – decreasing load, b – increasing load.

We believe that the photovoltaic system is an additional source of energy of the active power P_{Aac} replaced by a current source and it needs to be installed at a distance of L_{Aac} from the beginning of the line. Let's determine the optimal value of the additional source of active power P_{Aac} and distance L_{Aac} if the efficiency of the installation of an additional source to estimate the amount of power loss reduction in the line.

IOP Conf. Series: Materials Science and Engineering 1019 (2021) 012011 doi:10.1088/1757-899X/1019/1/012011

2. Method

We have (Figure 2) for distribution of active loads on the line with connected additional P_{Aac} at a distance of L_{Aac} from the beginning of the line. We believe that the total load of the line is P_{lin}. Then the average specific load of the line is $P_{ud.sr}=P_{lin}/L_{lin}$. When distributing the specific load as shown in Figure 2, the law of its change along the line will be defined as:

- For decreasing load $P_{ud1} = 2P_{ud.sr}\left(1-\frac{L}{L_{lin}}\right)$,
- For increasing load $P_{ud,2} = 2P_{ud,sr}\frac{L}{L_{lin}}$.

And laws of change of active current along a line from its beginning to the point K at connection of an additional source

$$I_{lin.L_{Aac}1} = \frac{P_{ud,sr}}{U_{grid}}\left(\frac{L_{lin}-L}{2}\right), I_{lin.L_{Aac}2} = \frac{P_{ud,sr}}{U_{grid}}\left(\frac{L}{2}\right)$$

By integrating the entire length of the line for such changes in the value of current along the line we can obtain expressions for losses in the entire line with distributed load according to laws shown in Figure 2, in the absence of additional power supply

$$\Delta P_{0,1} = \frac{R_{den}}{U_{grid}^2}\int_0^{L_{lin}}\left[I_{lin,bL,1}\right]^2 dL = \frac{P_{ud,sr}^2 R_{den} L_{lin}^3}{5U_{grid}^2}$$

$$\Delta P_{0,2} = \frac{R_{den}}{U_{grid}^2}\int_0^{L_{lin}}\left[I_{lin,bL,2}\right]^2 dL = \frac{8P_{ud,sr}^2 R_{den} L_{lin}^3}{15U_{grid}^2}$$

where R_{den}– specific active resistance to the line.

If there is an additional source of active power P_{Aac} attached in the point K, the laws of current change along the line from the point to its beginning will have the form

$$I_{lin,0K,1} = \frac{P_{ud,sr}}{U_{grid}}\left(\frac{L_{lin}-L}{2}\right) - \frac{P_{Aac}}{U_{grid}},$$

$$I_{lin,0K,2} = \frac{P_{ud,sr}}{U_{grid}}\frac{L}{2} - \frac{P_{Aac}}{U_{grid}}$$

Losses in the line from active current generated by the load and additional source of active power are determined by the integration of its two sections with different current distribution along the line

$$I_{lin,0K,1} = \frac{P_{ud,sr}}{U_{grid}}\left(\frac{L_{lin}-L}{2}\right) - \frac{P_{Aac}}{U_{grid}},$$

$$I_{lin,0K,2} = \frac{P_{ud,sr}}{U_{grid}}\frac{L}{2} - \frac{P_{Aac}}{U_{grid}}$$

$$\Delta P = \frac{R_{den}}{U_{grid}^2}\left(\int_0^{L_{Aac}}\left[I_{lin,0K}\right]^2 dL + \int_{L_{Aac}}^{L_{lin}}\left[I_{lin,L_{Aac}}\right]^2 dL\right) \quad (1)$$

Take the keys for transition to relative units

$P_{lin} = P_{den}\cdot L_{lin}$ – general active load of the line;

$P_{pu} = P_{Aac}/P_{lin}$ – relative power of the additional source;

$L_{pu} = L_{Aac}/L_{lin}$ – relative length of a line to the place of connection of the additional source;

Relative power of losses

$$\Delta P_{pu} = \frac{\Delta P}{\Delta P_0}$$

where $\Delta P_0 = \frac{P_{ud,sr}^2 R_{den} L_{lin}^3}{U_{grid}^2}$– rated losses in the line from total active power of load connected to the end of the line at absence of the additional source of active power.

Then, in relative units for designation of losses in the line at different distributions of loads we obtain

$$\Delta P_{pu0,1} = \frac{1}{5}, \Delta P_{pu0,2} = \frac{8}{15}.$$

The calculation of the integrals in (1) allows one to obtain expressions for the losses in the line with the distribution of its loads according to Figure 2

$$\Delta P_{pu1} = \left[P_{pu}L_{pu}\left(2L_{pu} - \frac{2}{3}L_{pu}^2 + P_{pu} - 2\right) + \frac{1}{5}\right], \quad (2)$$

$$\Delta P_{pu2} = \left[P_{pu}L_{pu}\left(\frac{2}{3}L_{pu}^2 + P_{pu} - 2\right) + \frac{8}{15}\right]. \quad (3)$$

We define the extremum of functions of losses from equality to a zero of their derivate on L_{pu}. Initially we consider the type of the decreasing load

$$\frac{\partial \Delta P_{pu1}}{\partial L_{pu}} = P_{pu}\left(4L_{pu} - 2L_{pu}^2 + P_{pu} - 2\right) = 0 \quad (4)$$

From (4) we obtain the best value of L_{pu} by increasing of losses – relative remote from the place of the additional source from the beginning of the line for present load distribution. Name it economically reasonable point L_{puE1}.

$$L_{puE1} = 1 - \sqrt{\frac{P_{pu}}{2}} \quad (5)$$

When an additional source is connected to a point that is efficient for reducing losses, the dependence of the relative power losses in the line is obtained from (4) and (2)

$$\Delta P_{puPE1} = P_{pu}^2 - \frac{2P_{pu}}{3} - \frac{\sqrt{2P_{pu}^5}}{3} + \frac{1}{5}. \quad (6)$$

We can define the economically rational value P_{puE1} plugging in anywhere in the L_{pu} line, calculating the derivative on P_{pu} equality (2) equaled to zero.

$$\frac{\partial \Delta P_1}{\partial P_{pu}} = \frac{2}{3}L_{pu}\left(3L_{pu} - L_{pu}^2 + 3P_{pu} - 3\right) = 0. \quad (7)$$

We obtain the expression that shows the dependence of the economically feasible power of an additional source P_{puE1} in its place of connection

$$P_{puE,1} = \frac{L_{pu}^2}{3} - L_{pu} + 1, \quad (8)$$

If we put (8) in (2), we will find the dependence of power losses in the line at switching of the additional source in the economically cost-effective point of the line

$$\Delta P_{puLE,1} = \frac{L_{pu}^3}{3}\left(2L_{pu} - \frac{L_{pu}^2}{3} - 5\right) + 2L_{pu}^2 - L_{pu} + \frac{1}{5}. \quad (9)$$

3. Results

If (2) the objective function of the value of losses in the line, then to determine the minimum (2) we find its se-cond partial derivatives with respect to P_{pu} and L_{pu}

$$\frac{\partial^2 \Delta P_{pu1}}{\partial P_{pu}^2} = 2L_{pu}$$

$$\frac{\partial^2 \Delta P_{pu1}}{\partial L_{pu}^2} = 4P_{pu}\left(1 - L_{pu}\right)$$

and compound derivatives

$$\frac{\partial^2 \Delta P_{pu1}}{\partial L_{pu}\partial P_{pu}} = \frac{\partial^2 \Delta P_{pu1}}{\partial P_{pu}\partial L_{pu}} = 2\left(P_{pu} + 2L_{pu} - L_{pu}^2 - 1\right)$$

Taking together (4) and (7) we obtain two vectors of real roots of this system of equations

$$\begin{vmatrix} P_{pu} \\ L_{pu} \end{vmatrix} = \begin{vmatrix} \frac{\sqrt{21}+11}{25} & \frac{11-\sqrt{21}}{25} & 0 & 0 \\ \frac{9-\sqrt{21}}{10} & \frac{\sqrt{21}+9}{10} & 2 & 0 \end{vmatrix}$$

T By the sign of the determinant of the Hessian matrix for the system from equations (4) and (7), we find the required roots

$$\begin{Vmatrix} \frac{\partial^2 \Delta P_{pu1}}{\partial P_{pu}^2} & \frac{\partial^2 \Delta P_{pu1}}{\partial P_{pu}\partial L_{pu}} \\ \frac{\partial^2 \Delta P_{pu1}}{\partial L_{pu}\partial P_{pu}} & \frac{\partial^2 \Delta P_{pu1}}{\partial L_{pu}^2} \end{Vmatrix} = \frac{26\sqrt{21}-14}{125} > 0 \text{ only for the first pair of roots.}$$

Thus, it has been proved that the global minimum of function (2) exists. Point coordinates

$$\mathrm{P_{pu1}} = \frac{\sqrt{21}+11}{25} \approx 0.623,\ \mathrm{L_{pu1}} = \frac{9-\sqrt{21}}{10} \approx 0.442. \quad (10)$$

Substitution of the optimal coordinates (10) into (2) gives the value of the minimum power losses in the line with an additional power source and load distributed according to Figure 2*a*.

$$\Delta P_{pu.min1} = \frac{217-28\sqrt{21}}{3125} \approx 0.028, \quad (11)$$

$$\text{or } \frac{\Delta P_{pu1}}{\Delta P_{pu.min1}} = \frac{\frac{1}{5}}{\frac{217-28\sqrt{21}}{3125}} = 7.047.$$

Thus, connecting an additional source of optimal power to the optimal point reduces power losses in the line by 7 times with the considered law of distribution of loads along the line. The distribution of the value $\Delta P_{pu1}/\ \Delta P_{pu.min1}$ with changes in the values of L_{pu} and P_{pu} is shown in Figure 3.

Let's consider the option of increasing load (Figure 2*b*). The derivative of the loss expression (3) for L_{pu} in relative units with the base ΔP_0

$$\frac{\partial \Delta P_{pu2}}{\partial L_{pu}} = P_{pu}\left(2L_{pu}^2 + P_{pu} - 2\right) = 0. \quad (12)$$

From (12) we obtain the economically feasible value of the relative removal of the connection point of the additional source from the beginning of the line

$$L_{puE2} = \sqrt{\frac{2-P_{pu}}{2}}. \quad (13)$$

When an additional source is connected to a point that is expedient for reducing losses, the dependence of the relative power losses in the line is obtained from (13) and (3)

$$\Delta P_{puPE2} = P_{pu}\sqrt{\frac{2-P_{pu}}{2}}\left(P_{pu} + \frac{2-P_{pu}}{3} - 2\right) + \frac{8}{15}. \quad (14)$$

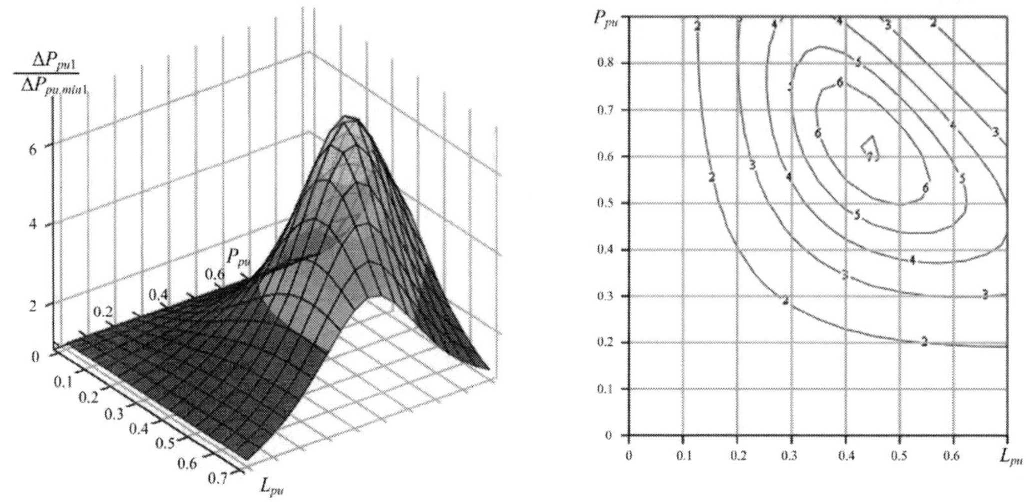

Figure 3. Dependence profiles $\Delta P_{pu1}/ \Delta P_{pu.min1}$ of relative power values of the additional source and the place of installation.

Economically rational value P_{puE2} , plug-in anywhere in the L_{pu} of the line,, determine from the equality to zero of the derivative of the expression (3) on P_{pu}.

$$\frac{\partial \Delta P_{pu2}}{\partial P_{pu}} = 2L_{pu}\left(\frac{L_{pu}^2}{3} + P_{pu} + 1\right) = 0. \tag{15}$$

From (15) we obtain the expression

$$P_{puE2} = 1 - \frac{L_{pu}^2}{3}, \tag{16}$$

to determine the power, an additional source, which is advisable to reduce power losses, from the point of its connection.

If put (16) in (3), we will obtain the dependence of power losses in the line at connection of the additional source in economically rational place of the line

$$\Delta P_{puLE2} = \frac{L_{pu}^3}{3}\left(2 - \frac{L_{pu}^2}{3}\right) - L_{pu} + \frac{8}{15}. \tag{17}$$

Expressions (14) and (17) for a given variant of load distribution along the line can be found economically feasible power or economically feasible place of connection if the power and place of connection of the additional source are known.

If (3) the objective function of the value of losses in the line when distributing the load according to Figure 2, b, then to determine the minimum (3), we will find its second partial derivatives with respect to P_{pu} and L_{pu}

$$\frac{\partial^2 \Delta P_{pu2}}{\partial P_{pu}^2} = 2L_{pu}, \frac{\partial^2 \Delta P_{pu2}}{\partial L_{pu}^2} = 4P_{pu}L_{pu}$$

and compound derivatives

$$\frac{\partial^2 \Delta P_{pu2}}{\partial L_{pu} \partial P_{pu}} = \frac{\partial^2 \Delta P_{pu2}}{\partial P_{pu} \partial L_{pu}} = 2\left(L_{pu}^2 + P_{pu} - 1\right).$$

Solving together (12) and (15) we will obtain two vectors of real roots of this system of equations

IOP Conf. Series: Materials Science and Engineering 1019 (2021) 012011 doi:10.1088/1757-899X/1019/1/012011

$$\begin{vmatrix}P_{pu}\\ L_{pu}\end{vmatrix} = \begin{vmatrix}\sqrt{3} & -\sqrt{3} & \sqrt{\frac{3}{5}} & -\sqrt{\frac{3}{5}} & 0 & 0\\ 0 & 0 & \frac{4}{5} & \frac{4}{5} & 0 & 2\end{vmatrix}.$$

The Hessian matrix determinant for the system of equations (12) and (15)

$$\begin{Vmatrix}\frac{\partial^2 \Delta P_{pu2}}{\partial P_{pu}^2} & \frac{\partial^2 \Delta P_{pu2}}{\partial P_{pu}\partial L_{pu}}\\ \frac{\partial^2 \Delta P_{pu2}}{\partial L_{pu}\partial P_{pu}} & \frac{\partial^2 \Delta P_{pu2}}{\partial L_{pu}^2}\end{Vmatrix} = \frac{16}{5} > 0 \text{ only for the third pair of roots.}$$

Therefore, the target function (3) has a global minimum at the point with coordinates

$$P_{pu2} = \sqrt{\frac{3}{5}} \approx 0{,}775,\ L_{pu2} = \frac{4}{5} = 0.8. \quad (18)$$

Substitution of the optimal coordinates (18) into (3) gives the value of the minimum power losses in the line with an additional power source and load distributed according to Figure 2*b*.

$$\Delta P_{pu.min2} = \left(\frac{8}{15} - \frac{16\sqrt{15}}{125}\right) \approx 0.038, \quad (19)$$

$$\text{or} \frac{\Delta P_{pu2}}{\Delta P_{pu.min2}} = \frac{\frac{8}{15}}{\frac{8}{15} - \frac{16\sqrt{15}}{125}} \approx 14.2.$$

From where, it can be concluded that the connection of an additional source to the line can reduce the loss of transmission of electrical energy by 14 times under the law of load distribution along the line (Figure 4). The distribution of the value $\Delta P_{pu2}/\ \Delta P_{pu.min2}$ for the range L_{pu} and P_{pu} is shown in Figure 4.

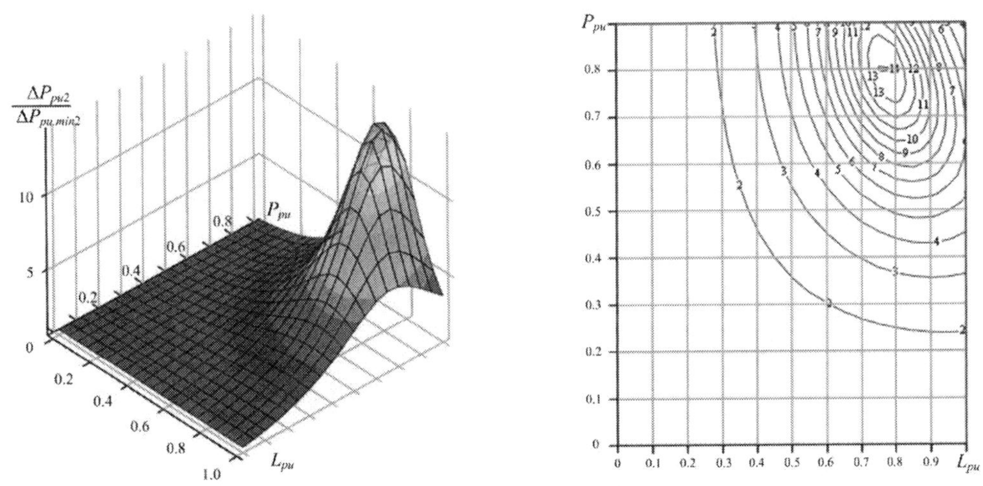

Figure 4. Profile of dependences $\Delta P_{pu2}/\ \Delta P_{pu.min2}$ of relative values of an additional source power the place of connection.

4. Discussion of results

As a result of the mathematical analysis of the possible location of an additional power supply on the line with unevenly distributed load (decreasing or increasing), it is found that the highest level of losses in the line will be with increasing load from the beginning to the end. Taking into account the results obtained by the authors earlier for the unevenly distributed load [10], the relative losses in the supply line before and after installation of an optimal additional power supply can be characterized as follows (table 1).

Table 1. Values of relative losses in the line at different load distributions

Type of load	Uniform distribution	Decreasing distribution (Figure 2*a*)	Increasing distribution (Figure 2*b*)
additional source absent	$\frac{1}{3}$	$\frac{1}{5}$	$\frac{8}{15}$
optimal additional source	$\frac{1}{27}$	$\approx \frac{1}{35}$	$\approx \frac{8}{212}$

The optimal point of connection of an additional source for different load distributions is in the range of 0.44 – 0.8 of the line length, and the optimal power of the additional source ranges from 0.66 – 0.77 of the total active power of consumers connected to the line.

5. Conclusions

There were obtained the analytical expressions for calculation of economically rational powers and points of connection of the additional source for linearly decreasing or increasing specific load of the supply line.

There were found the optimal points of connection of additional sources providing a minimum of losses in the transmission of energy along the supply line with irregular distributed load.

It is shown that in a line with the considered distribution of loads, connecting an additional source of optimal power to the optimal point of the line reduces losses by 7–14 times.

References

[1] Willis H L 2004 *Power distribution planning reference book. 2nd ed.* (New York: MarcelDekker)

[2] Kersting W H 2006 *Distribution system modeling and analysis 2nd ed.* (Boca Raton: CRC press)

[3] Bogdan A V, Bogdan V A and Garkavyi K A 2018 *Proceeding of International Ural Conference on Green Energy (UralCon)* 57–62

[4] Wang C and Nehrir M H 2004 *IEEE Trans. Power Syst* **4** 2068–2076

[5] Gozel T and Hocaoglu M H 2009 *Elect. Power Syst. Res.* **6**(79) 912–918

[6] Singh A K and Parida S K 2015 *Sustainable Energy Technologies and Assessments* **10** 1—11

[7] Eltawil M A and Zhao Z 2010 *Renewable and sustainable energy reviews* **1** 112–129

[8] Lazarte J C *et al* 2018 *Proceeding of North American Power Symposium (NAPS)* 1–6

[9] Bayrak G and Yılmaz A 2019 *Sakarya University Journal of Science* **23** 35–42

[10] Bogdan A. V., Bogdan V. A. and Garkavyi K. A. 2020 *International Ural Conference on Electrical Power Engineering (UralCon)* 441–448

Heating index for combined-cycle

P A Shchinnikov, O V Borush, A S Voronkova and P V Belyavskaya
Federal State Budget Educational Institution of Higher Education, Novosibirsk State Technical University, Novosibirsk, Russia
E-mail: voronkova.nastya@mail.ru

Abstract. The technique of the analysis of combined-cycle power plants with binarity coefficient which is less than one is offered. The technique enters a heating indicator. The index characterizes warmth anneafling by combined-cycle power plant. Plant is equipped with the steam turbine with heating selection of steam. Plants with afterburning and dual-fuel plants are considered. Interference of again entered index, degrees of binarity, various efficiencies is at each other shown in this article. The heat balance diagrams and the power streams's integrated charts of combined-cycle plants on the basis of the gas-turbine unit V64.3 are submitted. The calculated quantitative values of an efficiency on power generation on thermal consumption, coefficient of binarity, a heating index, the relative annealing of warmth and other characteristics for some of the actual or being in a stage of the project combined-cycle plants are given.

1. Introduction

Increasing the efficiency of fuel using in power plants is a relevant problem. One of the ways to increase the energy generation efficiency is using of combined-cycle power plants (CCPP) with binarity coefficient which is less than one. Through this, using of solid fuel in combined-cycle is possible [1– 4]. Binarity coefficient is determined by the following equation:

$$\gamma = \frac{Q^{\mathrm{GTU}}}{Q^{\mathrm{GTU}}+Q^{\mathrm{STU}}} = \frac{Q^{\mathrm{GTU}}}{Q^{\mathrm{CCPP}}}, \quad (1)$$

where Q^{GTU}; Q^{STU}, MW – energy supplied to the combined-cycle plant through the combustion chamber of the gas turbine and the steam boiler, respectively.

This index characterizes the amount of energy supplied to the installation through the gas turbine relative to the total energy supplied

2. Combined-cycle power plant with a coal afterburning

In contrast to the gas turbine combustion chamber where gas is burned, for generating steam of high parameters in the steam boiler can be used any type of fuel, including coal. In the combined cycle technology coal is involved with the efficiency of using unattainable in other technologies of energy generation where coal is used.

The thermodynamic feature of a combined-cycle with binarity less than one is described below. Energy is supplied in two places (Figure 1*a*). The first one – through the combustion chamber of the gas turbine $Q^{\mathrm{GTU}} = Q_{\mathrm{gas}}$) to ensure work of the gas turbine cycle (1-2-3-4-1, Figure 1*a*). The second one – through the steam power boiler ($Q^{\mathrm{CCPP}} = Q_{\mathrm{coal}}$ to ensure high parameters of the generated steam in the steam turbine cycle. The heat flow worked in the gas turbine ($Q^{\mathrm{GTU}}_{\mathrm{out}}$) is also transferred to the water-steam circuit through the waste-heat boiler. Useful product in the form of electricity are obtained both in the gas turbine and in the steam turbine units of the combined cycle. In addition, the steam turbine unit has a heat selection that provides heat supply to the consumer Q_h (Figure 1*a*).

The technological scheme describing thermodynamic approach is presented in Figure 1*b*. The air enters the compressor of a gas turbine unit (GTU), is compressed and enters with high pressure into the combustion chamber, where its temperature is increased due to the combustion of gas. After this the hot air and the products of combustion of the fuel enter the gas turbine, where they provide the electricity output (N_{GTU}). After gas turbine the exhaust gases is sent to a waste-heat boiler (WHB), where it is used to generate steam of average parameters for a steam turbine unit (STU) with power generation (N_{STU}). At the same time, high parameters of a sharp steam ensure the combustion of fuel (in particular, coal) in a steam boiler (SB), and an additional stream combined with reheating steam (process R-RR, Fig. 1a) and sent to the middle part of the steam turbine unit is generated in the waste heat boiler. Thus, the steam turbine has a wheel space where with steam expansion a flow of steam increases, even if there are bleeds (not shown in Figure 1*b*). The exhausted steam after the steam turbine enters the condenser (C), where it weeps (process K-K ', Figure 1*a*) and then is transferred to a waste-heat boiler and a steam boiler by the condensate pump (CP) and the feed-water pump (FWP). The regeneration system on the cycle arrangement is represented by a deaerator. In addition, the steam turbine has heat extraction, steam from which is sent to the heating system (HS) to provide heat supply Q_h to the consumer. The presence of the heat recovery boiler and steam boiler in the thermal scheme, which provide parallel steam generation of high and average parameters, determine the name of the installation, as a parallel-type CCPP [5].

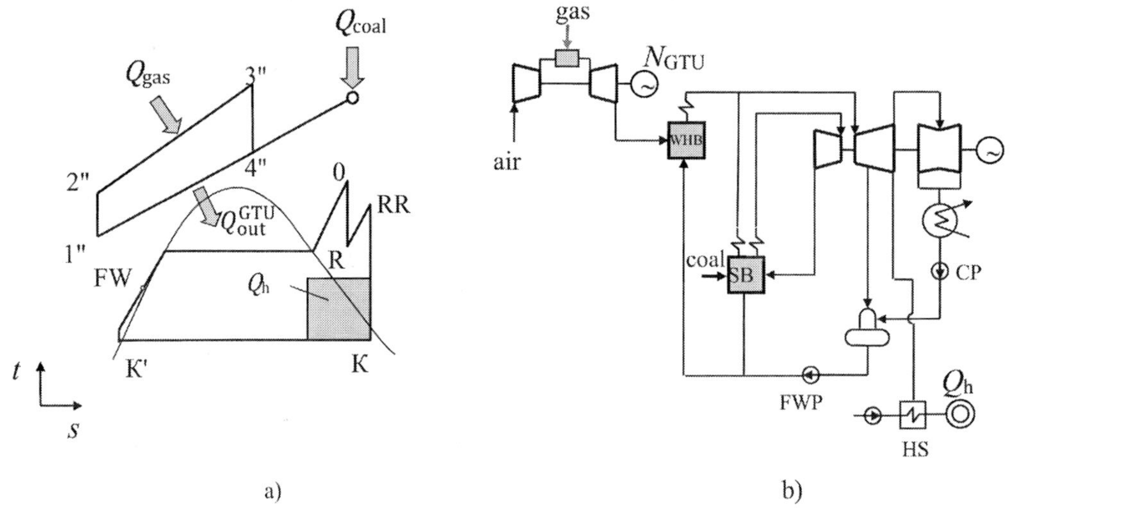

Figure 1. Thermodynamic diagram (a) and thermal scheme (b) of a combined-cycle power plant with a coal afterburning and a binary coefficient less than one.

Heat (energy) supplied to the combined-cycle plant (Q^{GTU}) is provided by the combustion of fuel and is determined by the following equations:

$$\begin{cases} Q^{GTU} = B_{GTU} \cdot Q^r_{i(GTU)}; \\ Q^{STU} = B_{STU} \cdot Q^r_{i(STU)}; \\ Q^{CCPP} = Q^{GTU} + Q^{STU}. \end{cases} \quad (2)$$

where B is the fuel consumer (kg/s; m^3 s) and Q^r_i is the lower calorific value (MJ/kg; MJ/m^3) of the fuel introduced into the GTU and the STU, respectively. In the case when the gas for a GTU and a steam boiler is a gas with one calorific value ($Q_{i(\mathrm{GTU})}^{r} = Q_{i(\mathrm{STU})}^{r}$), the equation for determining the binary coefficient:

$$\gamma = \frac{B_{\mathrm{GTU}}}{B_{\mathrm{GTU}}+B_{\mathrm{STU}}} . \quad (3)$$

Obviously, the binarity coefficient is less than one (γ <1) provided by the combustion of additional fuel in a steam boiler relative to the combustion chamber of a gas turbine. Such units are called combined-cycle power plants with afterburning of fuel. And units where gas is used as fuel in the combustion chamber of a gas turbine and coal in a steam boiler are called dual-fuel units.

The efficiency of the gas turbine unit of the CCPP can be determined by the following equation:

$$\eta_{\mathrm{GTU}} = \frac{N_{\mathrm{GTU}}}{Q^{\mathrm{GTU}}}. \quad (4)$$

Taking into account the energy conservation law and using the approaches [6], the amount of heat, which is lost in GTU, will be:

$$Q_{\mathrm{out}}^{\mathrm{GTU}} = Q^{\mathrm{GTU}} - N_{\mathrm{GTU}}. \quad (5)$$

After substitution:

$$Q_{\mathrm{out}}^{\mathrm{GTU}} = Q^{\mathrm{GTU}} - \eta_{\mathrm{GTU}} \cdot Q^{\mathrm{GTU}} = (1-\eta_{\mathrm{GTU}}) \cdot Q^{\mathrm{GTU}}. \quad (6)$$

This heat flow is directed to the steam turbine unit of the plant and it will provide the heating temperature of the steam in the waste-heat boiler equal to the exhausted gases temperature from the GTU (in the case there are no losses). In fact, the GTU output heat ($Q_{\mathrm{out}}^{\mathrm{GTU}}$) becomes part of the input heat flow for the Renkin steam-water cycle, which is realized in the steam turbine unit. The other part of the input heat flow for the Rankine cycle is heat (Q^{STU}) introduced with the fuel into the steam boiler. The efficiency of the steam turbine unit without of heat extraction ($Q_{\mathrm{h}} = 0$) is determined by the equation:

$$\eta_{\mathrm{STU}} = \frac{N_{\mathrm{STU}}}{Q_{\mathrm{out}}^{\mathrm{GTU}}+Q^{\mathrm{STU}}}. \quad (7)$$

Therefore

$$N_{\mathrm{STU}} = \eta_{\mathrm{STU}} \cdot \left(Q_{\mathrm{out}}^{\mathrm{GTU}} + Q^{\mathrm{STU}}\right) = \eta_{\mathrm{STU}} \cdot Q_{\mathrm{out}}^{\mathrm{GTU}} + \eta_{\mathrm{STU}} \cdot Q^{\mathrm{STU}} =$$
$$= \eta_{\mathrm{STU}} \cdot (1-\eta_{\mathrm{GTU}}) \cdot Q^{\mathrm{GTU}} + \eta_{\mathrm{STU}} \cdot Q^{\mathrm{STU}}. \quad (8)$$

Useful energy obtained plant:

$$N_{\mathrm{CCPP}} = N_{\mathrm{GTU}} + N_{\mathrm{STU}} =$$
$$= \eta_{\mathrm{GTU}}Q^{\mathrm{GTU}} + \eta_{\mathrm{STU}}Q^{\mathrm{STU}} - \eta_{\mathrm{GTU}}\eta_{\mathrm{STU}}Q^{\mathrm{GTU}} + +\eta_{\mathrm{STU}}Q^{\mathrm{STU}} = \eta_{\mathrm{GTU}}Q^{\mathrm{GTU}} + \eta_{\mathrm{STU}}(Q^{\mathrm{GTU}} +$$
$$Q^{\mathrm{STU}}) - -\eta_{\mathrm{STU}}\eta_{\mathrm{GTU}}Q^{\mathrm{GTU}} = \eta_{\mathrm{GTU}}(1-\eta_{\mathrm{STU}})Q^{\mathrm{GTU}} + \eta_{\mathrm{STU}}Q^{\mathrm{CCPP}} \quad (9)$$

The efficiency of CCPP can be determined by the equation:

$$\eta_{\mathrm{CCPP}} = \frac{N_{\mathrm{CCPP}}}{Q^{\mathrm{CCPP}}}, \quad (10)$$

$$\eta_{\mathrm{CCPP}} = \eta_{\mathrm{GTU}}(1-\eta_{\mathrm{STU}})\gamma + \eta_{\mathrm{STU}}. \quad (11)$$

At the same time, the combined-cycle power plants can work not only by electrical but also by thermal load curve ($Q_{\mathrm{h}} > 0$). In this case the efficiency of the steam turbine unit is:

$$\eta_{\mathrm{STU}} = \frac{N_{\mathrm{STU}}}{Q_{\mathrm{out}}^{\mathrm{GTU}}+Q^{\mathrm{STU}}-Q_{\mathrm{h}}} . \quad (12)$$

Therefore

$$N_{\mathrm{STU}} = \eta_{\mathrm{STU}}\left(Q_{\mathrm{out}}^{\mathrm{GTU}} + Q^{\mathrm{STU}} - Q_{\mathrm{h}}\right), \quad (13)$$

$$N_{\mathrm{STU}} = \eta_{\mathrm{STU}}(1-\eta_{\mathrm{GTU}})Q^{\mathrm{GTU}} + \eta_{\mathrm{STU}}\left(Q^{\mathrm{STU}} - Q_{\mathrm{h}}\right). \quad (14)$$

As a result, useful energy obtained plant:

$$N_{\mathrm{CCPP}} = N_{\mathrm{GTU}} + N_{\mathrm{STU}} = \eta_{\mathrm{GTU}}Q^{\mathrm{GTU}} + \eta_{\mathrm{STU}}Q^{\mathrm{GTU}} - \eta_{\mathrm{STU}}\eta_{\mathrm{GTU}}Q^{\mathrm{GTU}} + \eta_{\mathrm{STU}}Q^{\mathrm{STU}} - \eta_{\mathrm{STU}}Q_{\mathrm{h}} =$$
$$= \eta_{\mathrm{GTU}}Q^{\mathrm{GTU}} + +\eta_{\mathrm{STU}}(Q^{\mathrm{GTU}} + Q^{\mathrm{STU}}) - \eta_{\mathrm{STU}}\eta_{\mathrm{GTU}}Q^{\mathrm{GTU}} + \eta_{\mathrm{STU}}Q^{\mathrm{STU}} - \eta_{\mathrm{STU}}Q_{\mathrm{h}} =$$
$$= \eta_{\mathrm{GTU}}(1-\eta_{\mathrm{STU}})Q^{\mathrm{GTU}} + \eta_{\mathrm{STU}}\left(Q^{\mathrm{CCPP}} - Q_{\mathrm{h}}\right). \quad (15)$$

Determining the theoretical efficiency of the CCPP as:

$$\eta_{\mathrm{CCPP}} = \frac{N_{\mathrm{CCPP}}}{Q^{\mathrm{CCPP}}-Q_{\mathrm{h}}}, \quad (16)$$

will get

$$\eta_{CCPP} = \frac{\eta_{GTU}(1-\eta_{STU})Q^{GTU}}{Q^{CCPP}-Q_h} + \frac{\eta_{STU}(Q^{CCPP}-Q_h)}{Q^{CCPP}-Q_h}. \quad (17)$$

After substitutions and generations:

$$\eta_{CCPP} = \eta_{GTU}(1-\eta_{STU})\beta + \eta_{STU}. \quad (18)$$

In this equation the heating degree of CCPP:

$$\beta = \frac{Q^{GTU}}{Q^{STU}-Q_h}. \quad (19)$$

Heating degree β characterizes the heat supply by the combined-cycle power plant:

$$\beta = \frac{\gamma Q^{CCPP}}{Q^{CCPP}-Q_h}. \quad (20)$$

Where from after generations

$$\frac{Q_h}{Q^{CCPP}} = \frac{\beta-\gamma}{\beta}. \quad (21)$$

Considering that theoretically, in the studied installation $0 < \frac{Q_h}{Q^{CCPP}} \le 1$, the Heating degree β can take the values $\beta < 0$ (Figure 2).

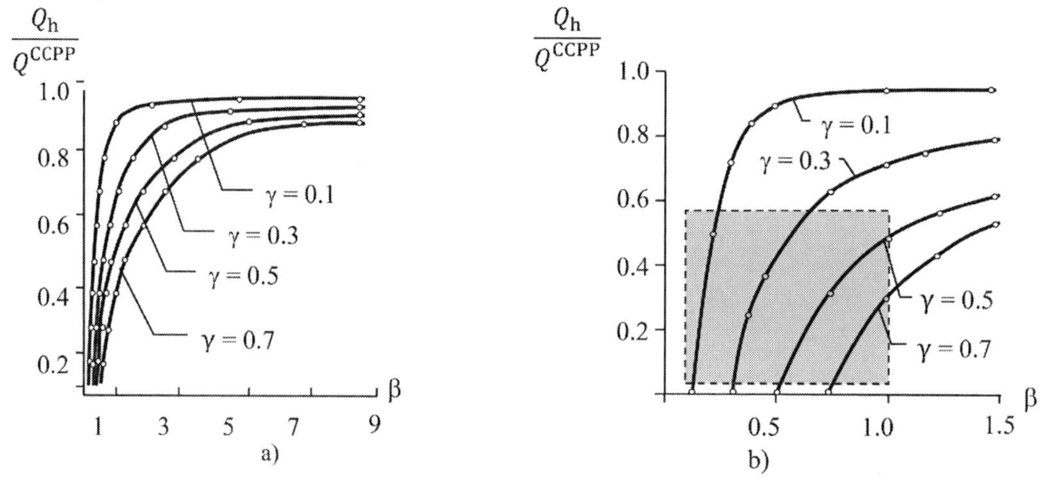

Figure 2. Relative heat supply by the combined-cycle power plant with afterburning of fuel, depending on the heating degree β (a) and the binary coefficient γ (b).

For technically implemented combined-cycle power plants the heating degree β depends on the binary coefficient γ and for plants close to the binary type can take values $\beta > 1$. Due to heat flows can be obtained for the needs of heating only in this case (Figure 2*a*).

For plants with a low binary coefficient, in particular for plants with predominant combustion of coal with $\gamma \le 0.5$, β may take values (Figure 2*b*, shown by hatching)

$$0 < \beta \le 1. \quad (22)$$

The relationship between the heating degree (β) and the binary coefficient (γ) of the CCPP is linear and depends on the relative heat supply of the CCPP (the value of the heat extraction). In general, β grows with γ (Figure 3).

Heating CCPP is characterized by the operation of a steam turbine unit in a heat-extraction mode. It is apparent that with the growth of combined electricity generation, characterized by the efficiency of a steam turbine unit increasing (η_{STU}), the heating degree (β) of the CCPP (Figure 4) also grows. At the same time, to ensure high efficiency of the CCPP, should not tend to high β values.

The effect of the gas turbine unit efficiency on the CCPP efficiency is less noticeable than the effect of the heating degree β. The greater the quantity of heat extraction (β grows), the higher the

CCPP efficiency (Figure 5). In general, we can say that for the high thermodynamic efficiency of the dual-fuel CCPP, there is should be a powerful gas turbine unit with a relatively small steam turbine, but at the maximum heat extraction. Questions of economic efficiency are not considered in this paper.

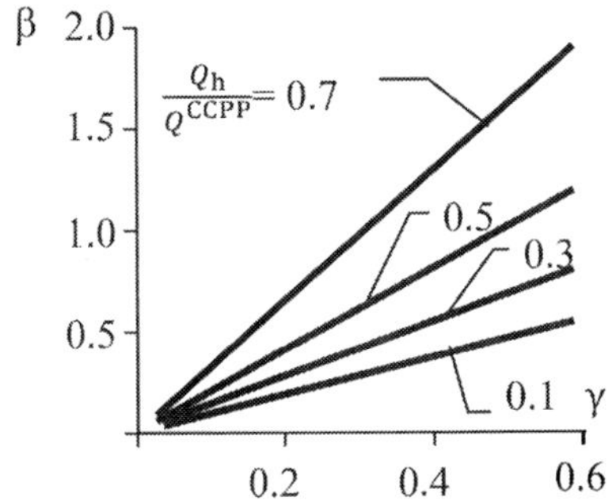

Figure 3. Dependence of the heating degree (β) on the binary coefficient (γ) of the CCPP for different relative output of heat extractions.

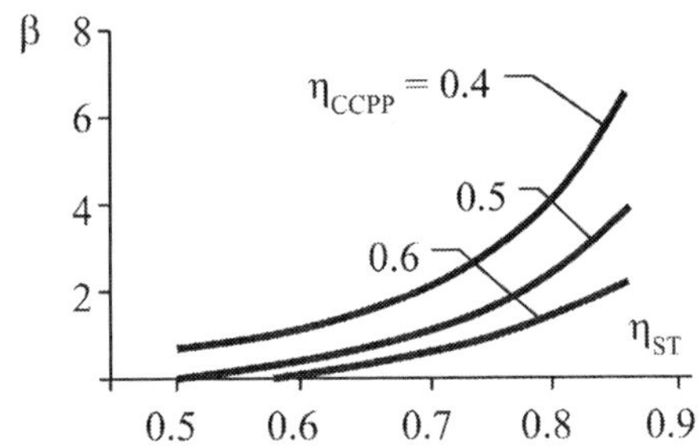

Figure 4. Influence of the heating degree (β) and efficiency of a steam-turbine unit operating according to the thermal load curve (η_{STU}) on the efficiency of the CCPP (η_{CCPP}).

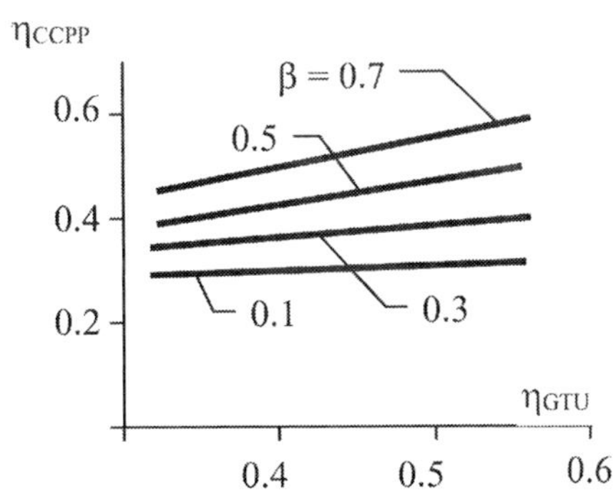

Figure 5. Influence of the gas-turbine unit efficiency (η_{GTU}) on the CCPP efficiency (η_{CCPP}) at different values of the heating degree (β).

3. Calculation results

The following are the results of some CCPP schemes calculations with heat extractions (Figures 6–9).

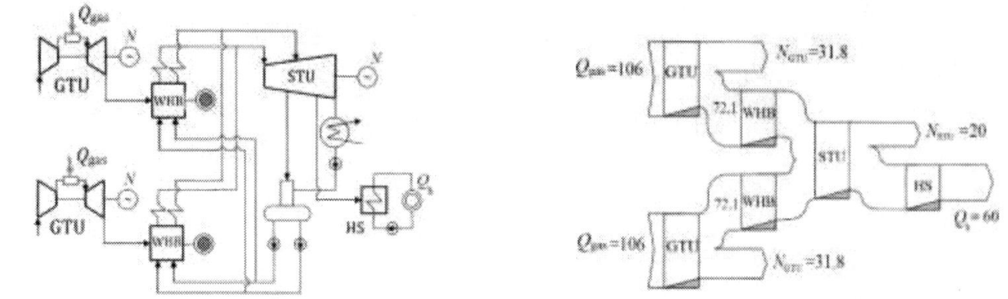

a) b)

Figure 6. The principle thermal scheme (a) and the enlarged diagram (b) of energy flows of a binary type combined-cycle power plant based on GTU LM2500 + G4 and STU T-14/23.

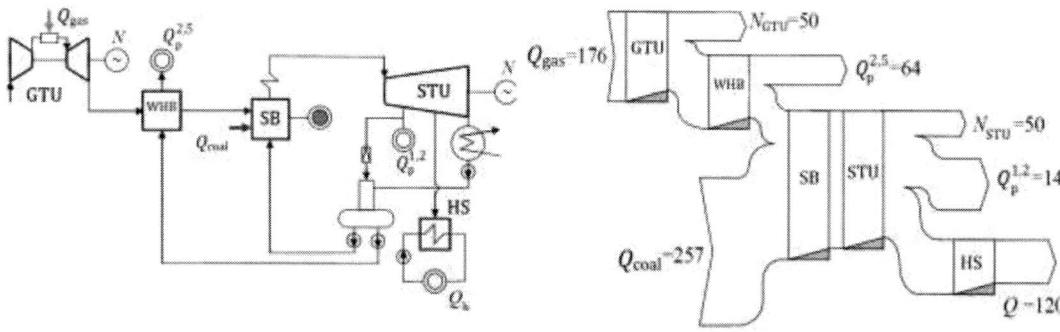

a) b)

Figure 7. The principle thermal scheme (a) and the enlarged diagram (b) of energy flows of the combined-cycle power plant based on GTU V64.3 Siemens and STU T-65 / 75-130 on the development of JSC "VTI" with afterburning.

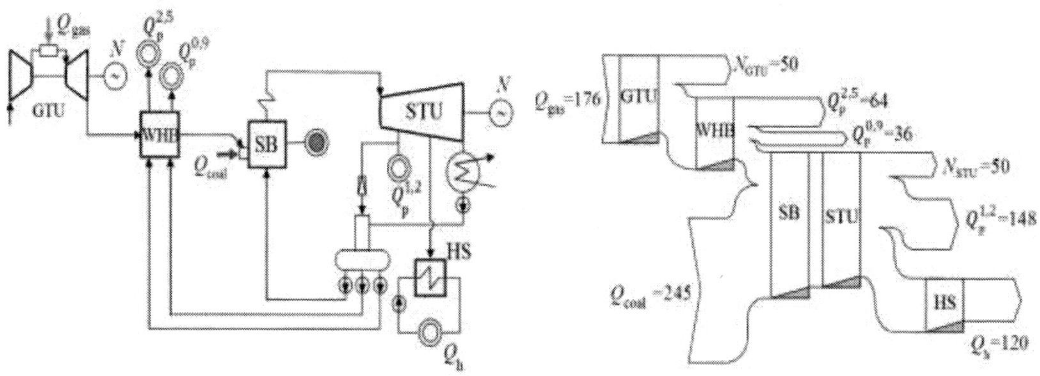

a) b)

Figure 8. The principle thermal scheme (a) and the enlarged diagram (b) of energy flows of the combined-cycle power plant based on GTU V64.3 Siemens and STU T-65 / 75-130 for the development of JSC "ENKO-center" with afterburning.

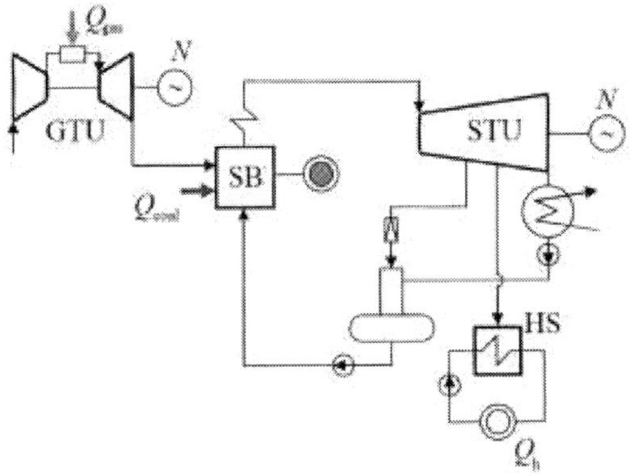

a)

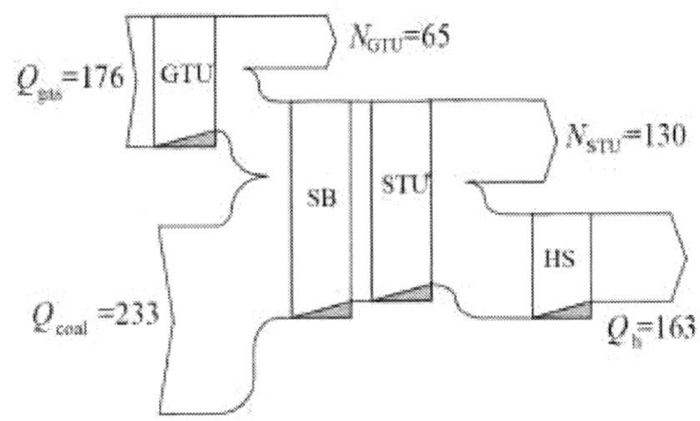

b)

Figure 9. The principal thermal scheme and the enlarged diagram of energy flows of the combined-cycle power plant based on GTU V64.3 Siemens and STU T-130 / 160-12.8 with afterburning.

During calculations according to the JSC "VTI" and JSC "ENKO-center" schemes it was taken into account that heat is supplied to the consumer after the waste-heat boiler does not enter the steam turbine unit (table 1).

In general, examples of calculations show that the heating degree can be used for the analysis of CCPP with the heat supply to the consumer.

Table 1. Characteristics of CCPP (using data [3, 5, 7, 8])

	Indicator name	CCPP-90 Omsk	CCPP-230 project JSC «KOTES»	CCPP-145 project JSC «VTI»	CCPP-145 project JSC «ENKO-center»
1	Installed capacity, MW	90	230	145	145
2	GTU type	LM2500	V64.3[a]	V64.3[a]	V64.3[a]
3	GTU capacity, MW[b]	32 (31.8)	70 (65)	70 (65)	70 (65)
4	STU type	T-14/20	T-130/ 160-12.8	ПТ-65/ 75-130	ПТ-65/ 75-130
5	STU capacity, MW[b]	20 (20)	160 (130)	75 (50)	75 (50)
6	Heat extraction, MW	60	163	120	120
7	Fuel, GTU/STU	gas	gas/gas	gas/gas	gas/gas
8	Industrial steam	no	no	yes	yes
9	Industrial extraction pressure, MPa	–	–	1.2	1.2
10	Industrial steam through WHB[c]/extraction, MW	–	–	64/148	100/148
11	GTU efficiency, relative unit	0.3	0.36	0.36	0.36
12	STU efficiency of electricity generation with heat extraction, relative unit	0.23	0.7	0.46	0.49
13	CCPP efficiency of electricity generation with heat extraction, relative unit	0.55	0.79	0.67	0.88
14	Binary coefficient, γ	1	0.44	0.4	0.41
15	Heating degree, β	1.39	0.72	0.67	0.49
16	Ratio Q_h/Q_{CCPP}	0.28	0.38	0.61	0.7

[a] manufacture Siemens;
[b] in parentheses are the values of the design mode;
[c] for the project JSC «VTI» through WHB is ensured industrial steam with P = 0.9 MPa and t = 350 °C; for the project JSC «ENKO-center» – P_1 = 2.5 MPa and t_1 = 350 °C, P_2 = 0.9 MPa and t_2 = 250 °C.

4. Calculation results

1. The energy research method of combined-cycle power plants with a binary coefficient less than one was developed and presented, the heating degree β was introduced.
2. It is shown that, in the general case, the heating degree
β > 0. For dual-fuel plants, in particular with predominant use of coal, 0 < β ≤ 1. For the binary type plants β > 1.
3. It is shown that the effect of the heating degree β on the combined-cycle power plant efficiency is generally more noticeable than the effect of the GTU efficiency.

References

[1] Peternya Y K *The history of the combined cycle cycle in Russia.* Available at: http://www.combienergy.ru/stat/900-Istoriya-parogazovogo-cikla-v-Rossii-Perspektivy-razvitiya
[2] Shchinnikov P A, Nozdrenko G V, Borodihin I V *et al* 2007 *Energy issues* 13–22
[3] Shchinnikov P A, Nozdrenko G V, Lovcov A A 2002 *Thermophysics and aerodynamics* **9** 445–449
[4] Burov V D, Konakotkin B V, Canev S V 1998 *Energy Saving and Water Treatment* 37– 43
[5] Canev S V, Burov V D, Remezov A N 2002 *Gas turbine and combined cycle plants of thermal power plants* (Moscow: MEI) p 584
[6] Kirsanova N I 2016 Thermodynamics of combined cycles *Energy and heat engineering: collection of scientific papers* (Novosibisrk: NGTU) pp 183–188.
[7] Shcheglyaev A V 1993 *Steam turbines. Theory of the thermal process* (Moscow: Energoatomizdat) **1** 384 p
[8] Grigor'eva V A, Zorina V M 1989 *Thermal and nuclear power plants. Directory* (Moscow: Energoatomizdat) 608 p

IOP Conf. Series: Materials Science and Engineering 1019 (2021) 012013 doi:10.1088/1757-899X/1019/1/012013

The analysis of power system steady-state aperiodic stability with electronic generation

A Fishov, I Murashkina, A Marchenko, E Enkhsaikhan and E Ivkin
Department of Automated Electric Power Systems, Faculty of Power Engineering, Novosibirsk, Russian Federation
E-mail: murashkinainna@yandex.ru

Abstract. One of the main directions of electric power industry development all over the world is the use of distributed small generation, both based on carbon fuel resources with synchronous communication between sources when they are connected to electric networks, and renewable energy sources working in the electric network through frequency-converting devices (electronic generation). The latter leads to the inevitable mass use of inverters in existing AC electric networks. The objectives of this investigation are to study the impact of electronic generation on the modes and stability of existing electric networks and power systems, the formation of requirements for the characteristics of electronic generation control, which minimize the coordination of relay protection and automation with electronic generation in the electric network, as well as increase the reliability of the general electrical mode. The article presents the results study of the steady-state aperiodic stability of the power system with electronic generation, the requirements for their steady-state characteristics and control in the power system.

1. Introduction

Currently Smart Grid based technologies are being developed differently than in Europe, USA and other countries. Small scale distribution generation abroad is based mainly on renewable energy sources (RES) operating in electric grid alone or through converters. Up to date Russia considering its climate and geographic features is mainly developing for cogeneration plants of small scale fuel distribution generation [1, 2]. Power-generating units synchronized with electric grid of power system, based on smart automated control system allows to fully use system effects from large scale concentrated and small scale distribution generation. However operation in an AC network is associated with a using of inverters in the form of current or frequency converters for most generators using RES [3].

A new concept of "Electronic Generation" has been introduced for such power sources.

Electronic generation (EG) – electrical energy sources supplying the load based on alternate or direct current via converters. One kind of electronic generation is a virtual synchronous machine.

The positive environmental effect of the using of RES is accompanied by many negative technical consequences due to the fundamental features of inverters when converting DC to AC and the lack of requirements for the characteristics of EG from electrical networks [4].

2. Inverters in the composition of electronic generation

Inverters in composition of EG perform the following main functions:

- Conversion of direct current to alternating current.
- Synchronization in frequency, voltage and angle with the power system.
- Stabilization of the output voltage.
- Limitation of current overloads in case of short circuit [5].

The Figure 1 shows typical connection schemes of inverter transformer substation (TS) to main switchgear. Radial connection schemes of inverter transformer substation to main switchgear are applied for low capacity power plants which used RES (2-3 MW) (Figure 1*a*). For power plants with a capacity of 4–8 MW, a reliable and inexpensive scheme is a circuit, where each feeder contains one or two pass-through inverter transformer substations and one dead-end (Figure 1*b*). For power plants with a capacity above 9 MW, the most suitable scheme is in the form of open ring sections (one ring for every 8–12 MW), where all inverter transformer substations are through-feed (Figure 1*c*).

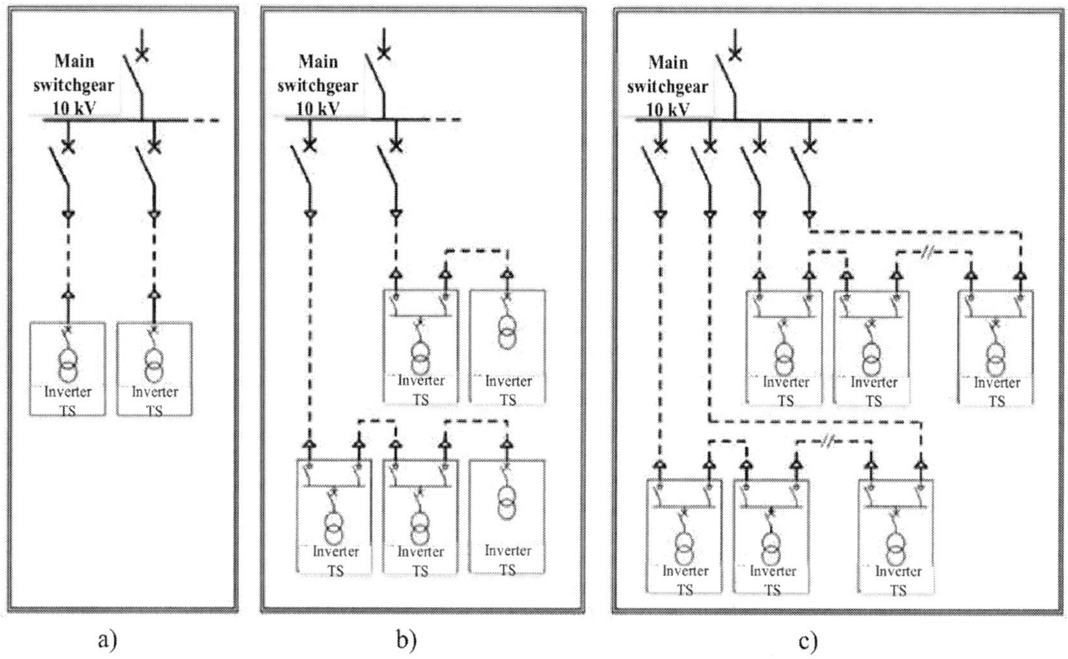

Figure 1. Typical connection schemes of inverter transformer substation to main switchgear.

Inverters operate either in Slave mode or Master mode.

The frequency and voltage are regulated depending on the power in Master mode (voltage source). In this mode it is possible to maintain a given voltage and adjust the voltage and frequency with statism (Figure 2*a*).

The output current is regulated depending on the voltage at the point of connection of the inverter to the network in Slave mode (current source). In this mode it is possible to stabilize the output current and adjust the output power (Figure 2*b*) [6].

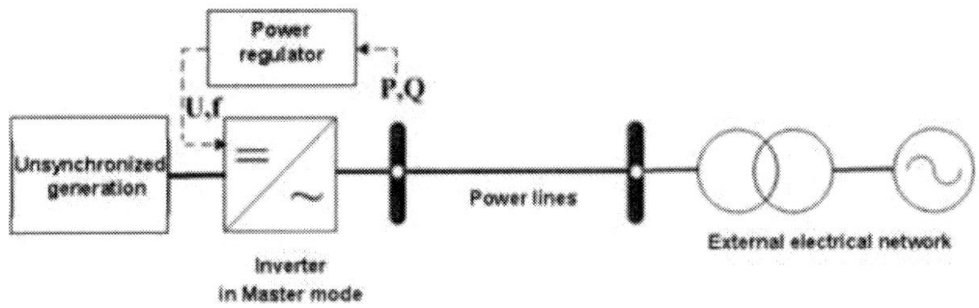

a)

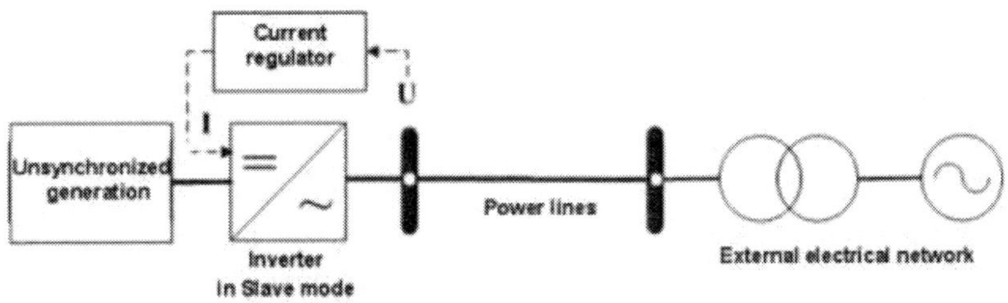

b)

Figure 2. Connection scheme of electronic generation in the mode: a) Master (voltage source); b) Slave (current source).

Modern converters allow creating a variety of electrical network architectures. The MicroGrid architecture, shown in Figure 3, makes it possible to connect consumers into a unified system, powered by different types of currents from power plants with different natural energy sources (sun, wind, fuel).

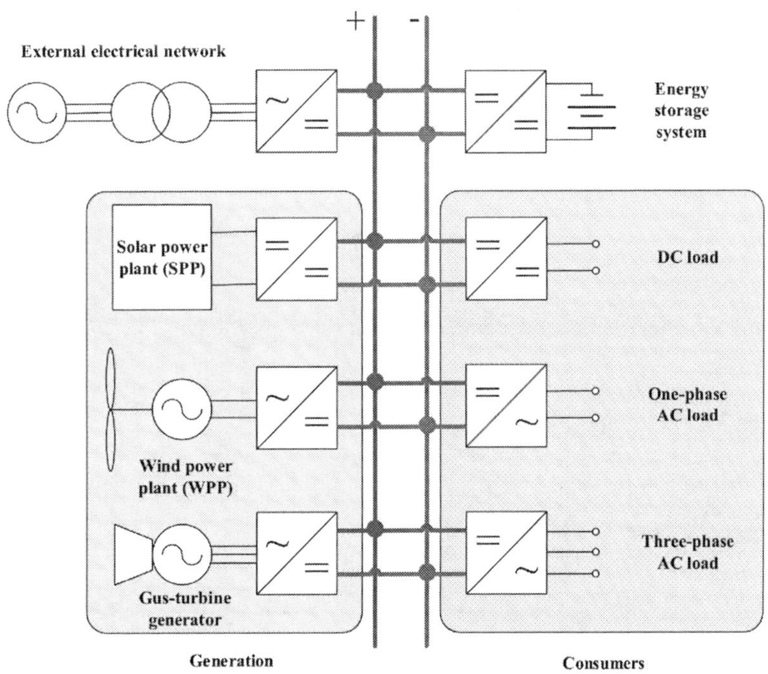

Figure 3. MicroGrid network architecture with different types of generation and consumers.

For systems with a high proportion of energy cells (Mini- and MicroGrid) and distributed RES, which have unguaranteed generation of power in their composition, it is advisable to introduce an additional concept of stability of the power system mode:

The systemically stable mode of the power system, which includes energy cells and sources of unguaranteed generation of power, is the mode of the power system in which small or large disturbances, as well as disturbances of the overall balance, the stability of the mode in the system is maintained as a whole. But stability disturbances can occur in cells division from the general network in which disturbances arose, without their propagation to other parts of the power system.

3. The modes and characteristics of EG influencing the steady-state aperiodic stability of power system

The limits of the steady-state aperiodic stability of power system are significantly dependent on the steady-state characteristics of the loads and generation, as well as electronic generation.

It is considering the basic modes and primary characteristics of EG inverters.

3.1. The mode A1

The inverter of EG operates in the Slave mode with the EG in the Master mode with synchronous generation or with synchronous electric network. The control system maintains a constant output current.

Steady-state characteristics for mode A1 are defined:

$$\begin{cases} I = const, \text{ if } |I| \leq I_{permissible} \text{ and } |U| < U_{max} \\ I = 0, \text{ if } |U| \geq U_{max} \end{cases}$$

Figures 4 and 5 show the dependence of currents on voltage at the connection point of EG and dependence of power on voltage, where $U_{lose\ synch}$ – voltage below which the inverter can lose synchronization.

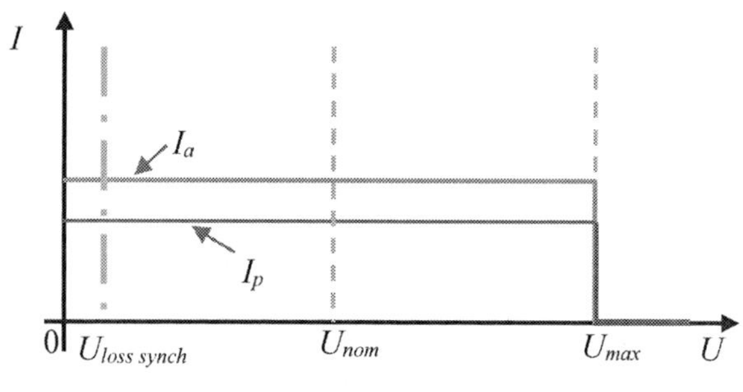

Figure 4. Characteristics of dependence of currents on voltage, $I = const$, the mode A1.

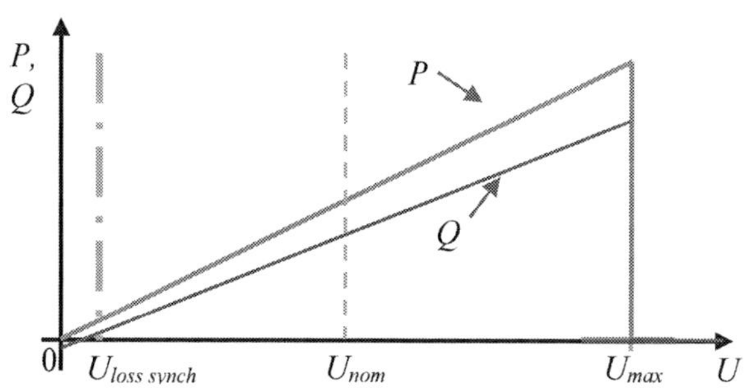

Figure 5. Characteristics of dependence of active and reactive powers on voltage, $I = const$, the mode A1.

3.2. The Mode A2

The inverter of EG operates with synchronous generation or with synchronous electric network similar to mode A1. But the control system maintains the constancy of the issued active and reactive powers in this mode.

Steady-state characteristics in the mode A2 are defined:

$$\begin{cases} |S| = const, \text{ if } |I| \leq I_{permissible} \text{ and } |U| < U_{max} \\ S = I_{permissible} \cdot U, \text{ if } |I| \geq I_{permissible} \text{ and } |U| < U_{max} \\ S = 0, \text{ if } |U| \geq U_{max} \end{cases}$$

Figure 6 shows the dependences of the active and reactive power on the voltage with changes in the behaviour of the inverter at critical points ($U_{min} = S/I_{permissible}$, $|U| = U_{max}$).

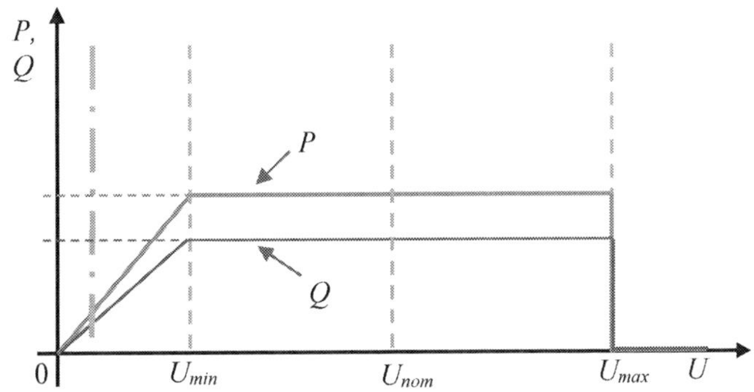

Figure 6. Characteristics of dependence of active and reactive powers on voltage, $|S| = const$, the mode A2.

3.3. The Mode B

In the mode B, the inverter of EG operates in the Master mode. This inverter is possible to work in parallel with other electronic generation(s), but provided that other electronic generation(s) operates in A1 or A2 mode (Slave mode). The control system automatically maintains constant frequency, phase and module of the output voltage.

Steady-state characteristics in mode B are defined:

$$\begin{cases} |U| = const, (U, \delta, f) = const, \text{ if } U < U_{max} \\ U = 0, \text{ if } U \geq U_{max} \\ U = 0, \text{ if } |I| \geq I_{permissible} \end{cases}$$

Figure 7 shows the dependence of voltage on inverter current.

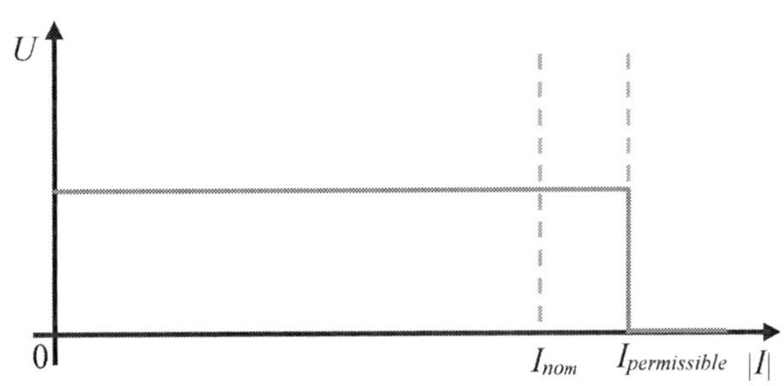

Figure 7. Characteristics of dependence of voltage on inverter current in mode B.

3.4. Steady-state characteristics of the inverter as a virtual synchronous machine (generator)

A virtual synchronous machine (generator) means electric power source connected to an alternating current electrical network through an inverter simulating dynamic characteristics, voltage and frequency statism of a traditional electric generator [6, 7]. The inverter as a virtual generator operates in mode A2. Moreover the active power set point changes depending on the frequency and reactive power set point changes depending on the voltage. The steady-state characteristic for active power is defined as $P_{steady} = P0–k \cdot \Delta f$ with the following limitations:

$$\begin{cases} P_{steady} = P_{max}, \text{ if } P_{steady} > P_{max} \\ P_{steady} = P_{min}, \text{ if } P_{steady} < P_{min} \\ P = 0, \text{ if } f > f_{max} \text{ or } f < f_{max} \text{ (simulation of disconnecting the generator from the network)} \end{cases}$$

Figure 8 shows the dependence of active power on the frequency of the current.

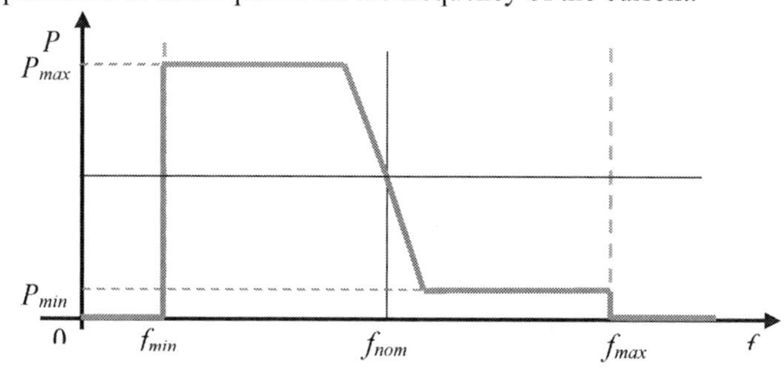

Figure 8. Characteristics of dependence of active power on the frequency of the current.

The steady-state characteristic for reactive power is defined as $Q_{steady} = Q_0–k \cdot \Delta f$ with the following limitations:

$$\begin{cases} Q_{steady} = Q_{max}, \text{ if } Q_{steady} > Q_{max} \\ Q_{steady} = Q_{min}, \text{ if } Q_{steady} < Q_{min} \\ Q = 0, \text{ if } U > U_{max} \\ Q = I_{permissible} \cdot U, \text{ if } |I| \geq I_{permissible} \end{cases}$$

Figure 9 shows the dependence of reactive power on the voltage.

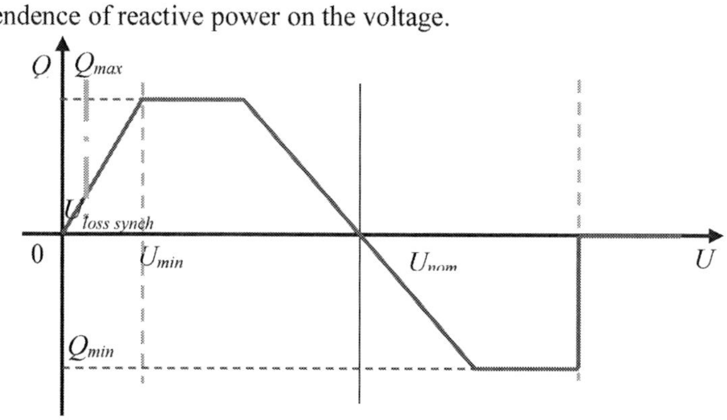

Figure 9. Characteristic of reactive power on the voltage (Virtual generator mode).

The disadvantage of the presented steady-state characteristics is piecewise differentiability. Because the derivatives of the quantities at critical points are equal to infinity, which makes it difficult to calculate the electric power mode and stability control.

IOP Conf. Series: Materials Science and Engineering 1019 (2021) 012013 doi:10.1088/1757-899X/1019/1/012013

4. Effect of EG on steady-state stability in the electric transmission scheme with its connection

For assess the effect of the EG on the limits of the steady-state aperiodic stability of the power system we turn to the simplest 35 kV power transmission scheme shown in Figure 10.

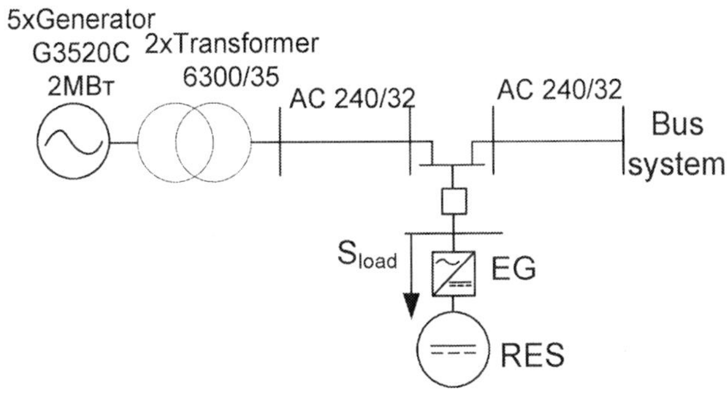

Figure 10. Power transmission scheme with electronic generation.

We define the angular characteristics of synchronous generation in the scheme with the on/off switch in the load node (load and EG of the same power). The load is represented by a typical steady-state characteristic and the EG will operate either in the mode of maintaining a constant current or in the mode of maintaining a constant generated power.

Typical steady-state load characteristics are given in table 1.

Table 1. Load characteristics.

№	Type of characteristics	a_0	a_1	a_2
1	Typical load characteristics of active power	0.83	-0.3	0.47
	-	b_0	b_1	b_2
2	Typical load characteristics of active power on the side of 6-10 kV	4.9	-10.1	6.2
3	Typical load characteristics of active power on the side of 110-220 kV	3.7	-7	4.3

To determine the angular characteristics and the limit on the steady-state stability of the synchronous generator, the mode was loaded by increasing the load of the synchronous generator and by increasing the load of the all scheme in the Mustang Program. Characteristic dependencies are shown in Figures 11–13.

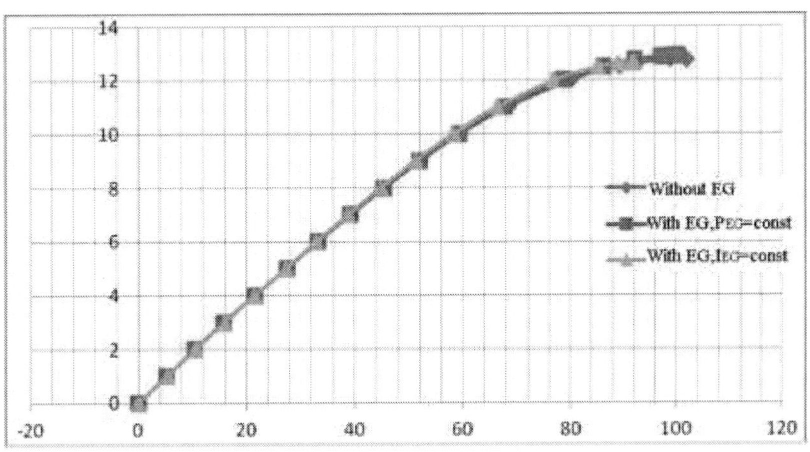

Figure 11. Angular characteristics of the synchronous generator with and without electronic generation (EG) power of 10 MW.

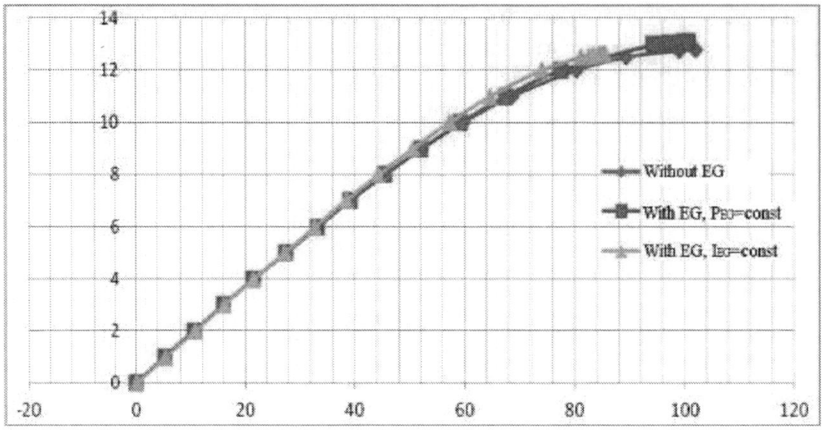

Figure 12. Angular characteristics of the synchronous generator with and without electronic generation (EG) power of 20 MW.

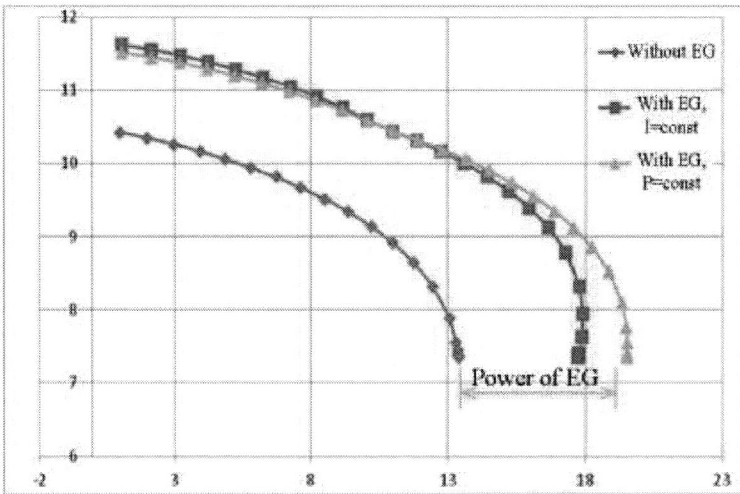

Figure 13. The dependence of the voltage at the point of load connection on the load value.

As follows from the presented graphs, the electronic generation has a positive effect on the steady-state stability of power system. At the same time for increase the parallel operation stability of generators in power system, EG modes are preferable with maintaining a constant output power and to prevent critical voltage drops - maintaining a constant output current.

5. The effect of electronic generation on the stability of Macrogrid

The study of the effect of EG on steady-state aperiodic stability was conducted on the example scheme of the electric power system of Mongolia. To calculate the limit value of steady-state aperiodic flow stability in Macrogrid design section it is required to load the mode (increased flow) along the lowest power limit. Both the steady-state characteristics and the places of EG connection in the power system were different.

As a calculation scheme, to determine the effect of EG on the limits of transmitted power over the main sections of the electric network, the Mongolian power system scheme was used (Figure 14).

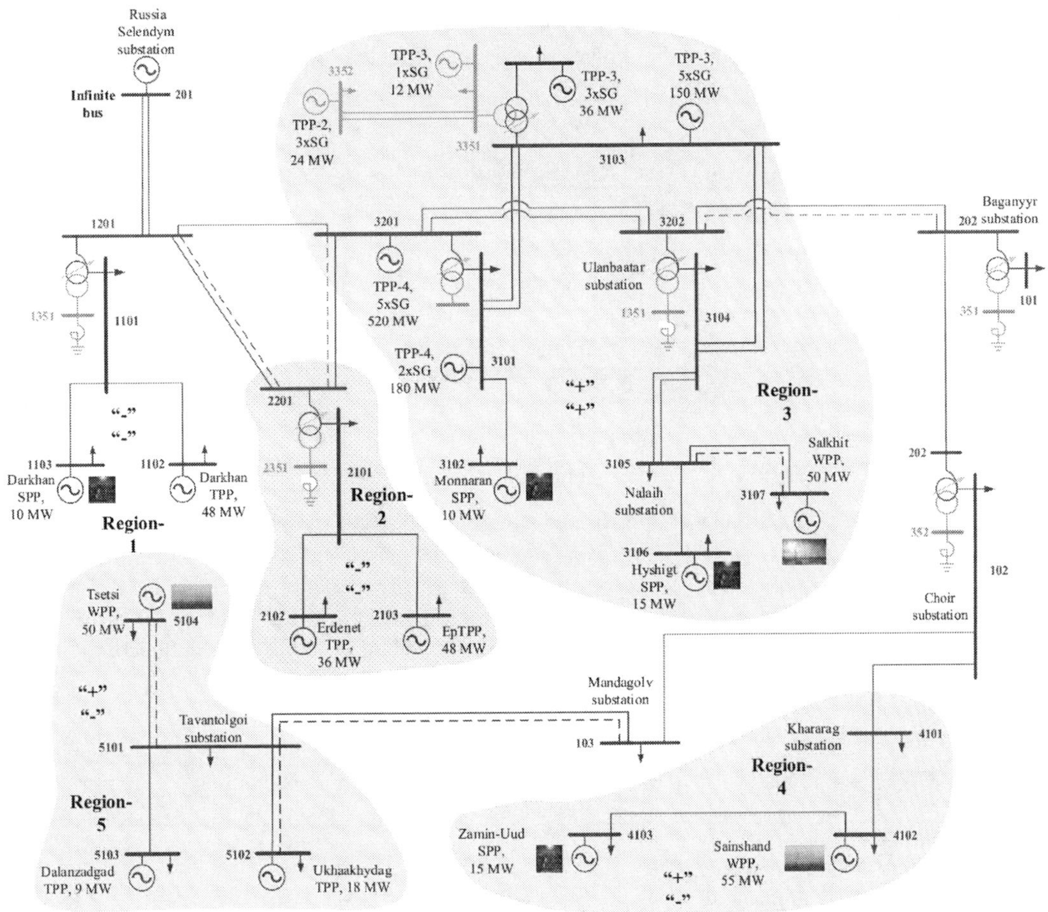

Figure 14. Calculated equivalent scheme of Mongolia power system.

The design scheme of the Mongolian power system contains the main network voltage (35 – 220) kV, distribution network 35 kV. Red color indicates a nominal voltage of 220 kV, blue – 110 kV, green – 35 kV. Electric loads are determined on the basis of the power and electricity energy system balances of Mongolia. The load of power plants is load of maximum mode.

Below are results for two design sections of the network.

Design section № 1 is the main connection of power system between Mongolia and Russia (Figure 15). The feature of this section is maintaining a power balance of the Mongolia electric power through the exchange of power with the Russian Federation. The average power transfer is 110 MW.

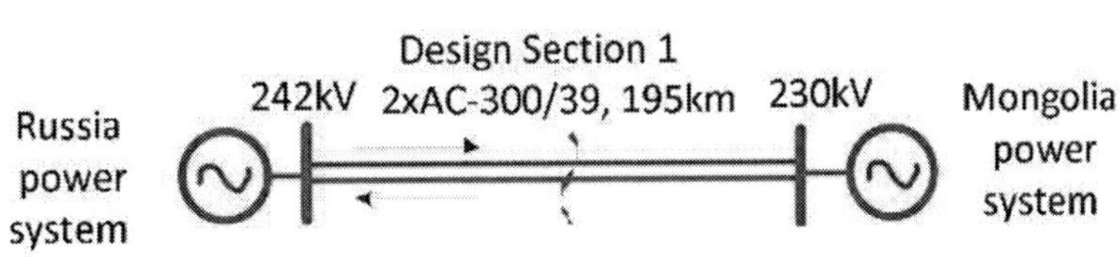

Figure 15. Design section of Russian Federation connection with Mongolia.

Design section 2 (Figure 16) refers to the critical part of the power system that is deficient in power and electricity (Region 2). Thus the power flows are directed towards the Erdenet substation (Mongolia) in this section.

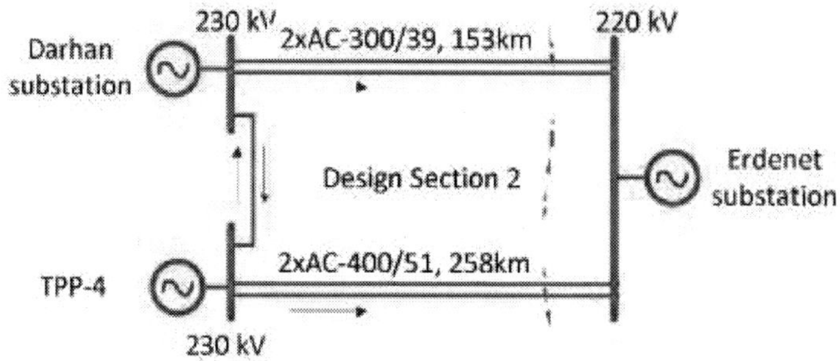

Figure 16. Design section of Mongolia with Region 2.

Flow distribution calculations were performed using the ANARES program.

5.1. Calculation of flow limits in steady-state aperiodic stability for design section № 1

Loading was carried out by increasing the load in region 3. The initial load in the load node is 185+j82MVA. The flow limits as a result of loading for section 1 with two connection lines are presented in Table 2.

Table 2. Results of limit modes for design section 1.

№	Variants	Increasing of general load in a load area, MWA	Section loading in initial normal mode, MWA	Limit of transmitted power, MW
1	Without EG	259+j116	190 – j24	527
2	With existing EG distribution	376+j169	55-j12	514
3	The concentrated placement of EG	380+j171	63-j13	540

5.2. Calculation of flow limits in steady-state aperiodic stability for design section № 2

Loading was carried out by increasing the load in region 3. The initial load in the load node is 165+j55MVA. The flow limits as a result of loading for section 2 are presented in Table 3.

Table 3. Results of limit modes for design section 2.

№	Variants	Increasing of general load in a load area, MWA	Section loading in initial normal mode, MWA	Limit of transmitted power, MW
1	Without EG	282+j127	146 – j38	459
2	With existing EG distribution	282+j127	146 – j42	460
3	The concentrated placement of EG	443+j199	22-j41	498

The limit of the transmitted power has not changed significantly when calculating the limiting modes and RES simulation that distributed throughout the energy system of Mongolia. The maximum power flow increased in the first calculated design section by 2.5%, and in the second by 8% when placing concentrated renewable energy sources in the loaded area.

It follows that the placement of EG in the power system has a positive effect on network limitations, and the stability limits depend not only on steady-state characteristics, but also on the location of the connected EG.

IOP Conf. Series: Materials Science and Engineering 1019 (2021) 012013 doi:10.1088/1757-899X/1019/1/012013

6. Conclusion

The article considers the effect of the location, modes and characteristics of EG on the steady-state aperiodic stability of power systems using the example of simple and complex power systems. Currently the steady-state characteristics of electronic generation are set by equipment developers. This steady-state characteristic has not taken into account the conditions of electronic generation operation in the power system including the effect on the stability of the mode. It is shown that the modes of EG with maintaining a given power and a required statism, have a positive effect on the parallel operation stability of power system synchronous parts.

The modes of EG with maintaining a constant current are useful in areas where are dangerous due to the voltage drop that is unacceptable under the condition of stability of the power system mode. The concentrated placement of electronic generation in critical areas of the power system increases the stability of modes and the transfer capacity of the electric network. At the same time balance stability of the power system mode is reduced if this generation is stochastic behaviour.

References

[1] Marchenko A and Fishov A 2015 *"Applied Mechanics and Materials* **792** 248–254 doi: 10.4028/www.scientific.net/AMM.792.248
[2] Mukatov B and Fishov A 2017 *E3S Web of Conf.* **25** 03009 doi: 10.1051/e3sconf/20172503009
[3] Fishov A, Marchenko A, Murashkina I, Erdenebat E and. Ivkin Y 2018 *Proc. of the 14th Int. Scientific and Technical Conf.* **1**, part 5 pp. 108–114 doi: 10.1109/APEIE.2018.8545916
[4] Marchenko A, Denisov V and Murashkina I 2019 *Science bulletin of the Novosibirsk state technical university* 77–90 doi: 10.17212/1814-1196-2019-1-77-90
[5] Ismailov F, SHarifov B, Gayzinz B, Teregubov T and Babkina N 2016 *Bulletin of UGATU* **20** 71–79 [in Russian]
[6] D'Arco S, Suul J and Fossob O 2015 *Electric Power Systems Research* **122** 180–197 doi:10.1016/j.epsr.2015.01.001
[7] Khokhlov A, Melnikov Yu, Veselov F, Kholkin D and Datsko K 2018 *Distribution energy in Russia: development potential* (Moscow: Skolkovo Energy Center Publ.). Available at: https://energy.skolkovo.ru/downloads/documents/SEneC/Research/SKOLKOVO_EneC_DER_2018.10.09_Eng.pdf

Nonequidistant antenna arrays synthesis based on genetic algorithm

V I Oznobikhin[1], V B Romodin[2] and L V Shebalkova[2]
[1]JSC «Scientific Research Institute on Electronic Devices», Novosibirsk, Russia
[2]Autonomous Information and Control Systems dept., Novosibirsk State Technical University, Novosibirsk, Russia
E-mail: shebalkova@corp.nstu.ru

Abstract. A new approach to the synthesis of nonequidistant antenna arrays with uniform distribution based on the Genetic Algorithm of the global extrema search is introduced and analyzed. The technique is demonstrated for various uniform antenna array structure. The radiation pattern of the nonequidistant antenna array is formed using the Genetic Algorithm to reduce a side lobe level. A mathematical formula is presented to highlight the operation of objective function minimization.

1. Introduction

In radar applications, the antenna has to be designed and manufactured to fulfil all requirements of size, weight, manufacturability, and of signal-processing performance. The low sidelobe level (SSL) of antenna array radiation pattern is one of most important requirements.

As is known [1], equidistant spaced uniformly exited arrays have a side lobe level of about -13 dB. In order to reduce the side lobe levels, the array must be designed to radiate more power towards the center, and much less at the edges. This can be achieved through tapering (windowing) the current distribution over the face of the array. There are many possible tapering sequences that can be used for this purpose. However, as known from spectral analysis, windowing reduces side lobe levels at the expense of widening the main beam. Thus, for a given radar application, the choice of the tapering sequence must be based on the trade-off between side lobe reduction and main beam widening. For example, Dolph-Tschebyscheff tapering leads to an antenna aperture efficiency decreasing, respectively, a directivity decreases, and the efficiency of the antenna decreases at the same aperture dimensions [1].

Many analytical techniques exist for synthesizing low-sidelobe amplitude tapers for arrays. Among the optimization methods, the Genetic Algorithm (GA) is one of the most popular optimization techniques used for SSL reduction. A binary GA is appropriate to use when the amplitude weights are quantized. The GA method is a global and random search algorithm that simulates natural selection and evolution. It searches through the total solution space and can find the optimal solution globally over a domain. However, the global nature of the algorithms and the lack of derivative information cause them to converge very slowly compared to other nonglobal methods [2].

The analysis of non-uniform linear arrays based on GA optimization techniques of elements locations and the SSL suppression is presented in [3–5]. The synthesis of linear and planar arrays was considered in [6]. The results show the flexibility of GAs to solve complex problems of classical optimization techniques.

GA optimization of large aperture nonuniform array antenna element's positions was proposed in [7], there are three parameters that can be changed in the development process, one is the spacing element arrangement, the other is the excitation weight of the array elements, and the third is the phase. Note, that SSL reduction doesn't cause the mainlobe width broadening.

Planar sparse array antennas with square shapes and variable number of elements are analyzed and synthesized using GA [8]. The radiation pattern of non-uniform elements distances and non-uniform element currents distribution optimization process is discussed. The GA technique was applied for distance and current optimization, the results show the advantage of this approach such as doubled SSL seduction and lower number of elements for similar performance.

This article discusses the possibility to reduce the SLL of linear array with uniform current distribution by interelement spacing GA optimization.

2. Mathematical formulation

A linear array far field and a normalized far field can be written as follows:

$$U(\theta) = F(\theta)\cdot\sum_{n=1}^{N} I_n e^{ikd_n(n-1)(\cos\theta-\cos\theta_{0t})}$$
$$U(\theta) = 20\lg\frac{|U(\theta)|}{|U_{\max}(\theta_{0p})|} \quad (1)$$

Where $F(\theta)$ is the far field of an element, i is the imaginary unit, k *is* the wave number ($k = 2\pi/\lambda$), N *is* the element's number, I_n is nth element excitation current, and $I_1 = I_2 \ldots = I_n = 1$, d_n is the distance to the nth element from the array origin or interelement spacing.

It can be seen from Eq. (1) that the array far field is related to the position of the array element, so the optimization of the array elements location can improve the antenna radiation pattern and reduce the SLL. So the objective function that minimized the maximum SLL can be determined from array factor.

The main goal is to find d_n to minimized a SLL. Herewith, the first and last radiation elements remain in their own place. That is, the 'N' element's array length is invariable. To solve the task, the functional is minimized:

$$f = c_1\cdot|U_{1t}-U_{1p}|/U_{1t} + c_2\cdot|U_{2t}-U_{2p}|/U_{2t} \quad (2)$$

Where $U_{1t,}U_{1p}$ are assigned and calculated SLL in the front radiation zone, U_{2t},U_{2p} are assigned and calculated SLL in the backward radiation zone, c_1, c_2 are cost coefficients (Lagrange coefficients).

In the cost function given in Eq. (2) interelement spacings d_n are unknown, and GA global extremum search is applied for their minimization [2].

The linear antenna schematic radiation zone arrangement is showed on Figure 1. The front radiation zone lies in angle range of 0-90 degrees, and the backward radiation zone lies in angle range of 90-180 degrees.

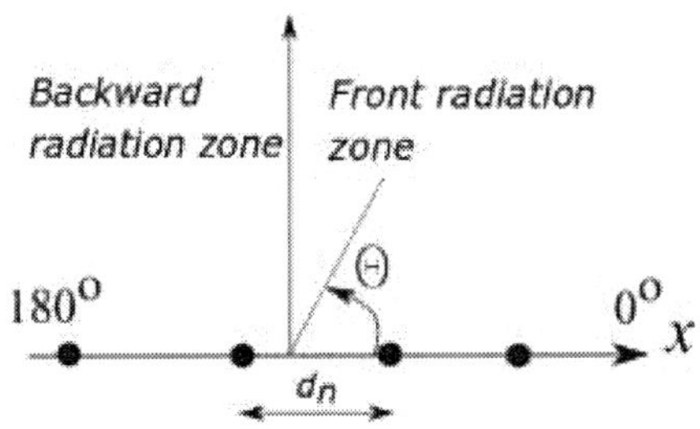

Figure 1. Radiation zones.

If the required SLL in the front and back radiation zones is the same as in cophasal array, the sum in Eq. (2) can be replaced by one term with the maximum calculated SLL.

Linear arrays directivity can be reduced and written as follows

$$D = \frac{2}{\int_0^{\pi} |U(\theta)|^2 \sin(\theta) d\theta} \quad (3)$$

3. The proposed algorithm

The main goal of this study is to design a low sidelobe radiation pattern for nonequidistant uniform linear antenna arrays with a fixed halfpower beamwidth. A considerable amount of discussion in the literature has been devoted to the problem of using GAs to optimize array element spacing. This is primarily because this class of array synthesis problems has been extremely challenging for traditional optimization methods due to the large number of variables that are typically involved. Moreover, conventional aperiodic array synthesis techniques based on statistical methods fall far short of optimal configurations.

Assume the array has an even number of elements. With this assumptions, the objective function is written as a function of the amplitude weights [9].

The GA is one of the evolutionary algorithms used to find the global extrema of the function of many variables. The principle of GA operation is based on modeling of some mechanisms of population genetics: manipulation of chromosomal set in the formation of a new biological individual by inheritance of sites of chromosomal sets of parents, accidental change of genotype (mutation). Another important mechanism is the natural selection procedure aimed to improve the fitness of the population members from generation to generation by making them more adaptable to the "survival" of individuals with certain characteristics. The function update population assigns ranks to individuals in the population generated by the union of parents and offspring. This is in order to hold the best individuals in each generation.

A GA population-based stochastic procedure was applied. The procedure for the used GA technique is described as follows.

The implementation of the basic GA can be represented as an iterative process that includes several stages:

1. The generation of the initial population.
2. The reproduction of offspring:
(a) choice of parent,
(b) selection and implementation of one of the crossover operators,
(c) mutation.
3. The establishment of reproduction of the group.
4. The procedure for the selection and formation on its basis of a new generation.
5. If the stop condition is not met, the transition to paragraph 2 is made.

According to GA terminology, a specific antenna array with a given amplitude current distribution over the radiation elements is called an individual, and the set of the latter is represented as a finite population of individuals. Each individual in the population has a measure of adaptation to the environment, that is, has a certain value of the objective function. The process of finding the optimal solution is described by the process of simulated "evolution", the purpose of which is to find an individual (or a set of individuals) with the maximum fitness, that is, an individual corresponding to the optimal value of the controlled parameters (d_n).

For a more convenient GA application, we present these parameters as

$$d_n = d + d_m g\mathbf{X}_n$$

Where d is the equidistant array spacing, d_m is the maximum element spacing, $\mathbf{X}_n$ is the parametric vector whose components must be found in the optimization process. These components lie within the range [-1;1]. In fact, the objective function depends on the parametric vector, and the main goal is to find its optimal value.

At the first stage, 'm' antenna arrays with definite d_n are formed randomly by the Monte Carlo method. Then, also by accident, 'm1' of antenna pairs is selected from 'm' antenna arrays. In the GA language, at this stage the initial population is formed, and then parent pairs are formed too.
At the next stage, the numerical values of d_n are converted from the decimal system to binary (transition to the genotype) and then GA algorithms are applied, that is, using the crossover operator (recombination of the bits of the corresponding pairs) and the mutation operator (replacement with some probability of each bit by its opposite value), six 'm1' new sets of currents are obtained. Then these currents are transferred back from the binary system to the decimal (the transition from the genotype to the phenotype), for each variant the objective function is calculated and a rank objective vector is formed, the components of which are arranged by the numerical value ('m' best options). If none of the selected options does not give the required value of the objective function ($f \geq 0.99$), then the cycle repeats. 'm' best variants of the previous cycle are used only as an initial population. If the condition ($f \geq 0.99$) is satisfied for at least one set of d_n, the selection of the optimal variant is completed.

4. Optimization results

The results of the objective function optimization given in Eq. (2) for three cases (N = 4, 8, 10) are shown in table 1. Interelement spacings d_n are given in fractions of wavelength. Initial spacing is equal 0.5, and arrays are cophased. Consider that the linear array has N antenna elements distributed on it randomly. The element was set at each end of the array to guarantee the fixed array aperture; the remaining 'N-2' elements were randomly assigned to the other locations.

For N=4 SLL is reduced to -23 dB, and for N=8, 10 SLL is reduced to -21 dB only. Since the length of the antennas remains invariant, the halfpower beamwidth does not change during optimization. The Figures 2–4 depict radiation patterns of optimized arrays compared with equidistant arrays one. The pattern peaks are normalized to 0 dB for easy comparison of SLL. There is no cross-polarized radiation in the principal planes. GA patterns generally exhibit lower levels of sidelobe radiation.

The table 1 shows that some radiation elements are shifted to the right (from the array origin), and some of them are shifted to the left.

Table 1. The interelement spacing for cophasal arrays.

n	Interelement spacing of nth element, d_n		
	$N = 4$	$N = 8$	$N = 10$
1	0	0	0
2	0.62683	0.78099	0.73037
3	0.90835	1.2648	1.2472
4	1.5	1.7132	1.7689
5	–	2.0005	2.2166
6	–	2.5267	2.5294
7	–	2.8179	3.0332
8	–	3.5	3.2317
9	–	–	3.8991
10	–	–	4.5

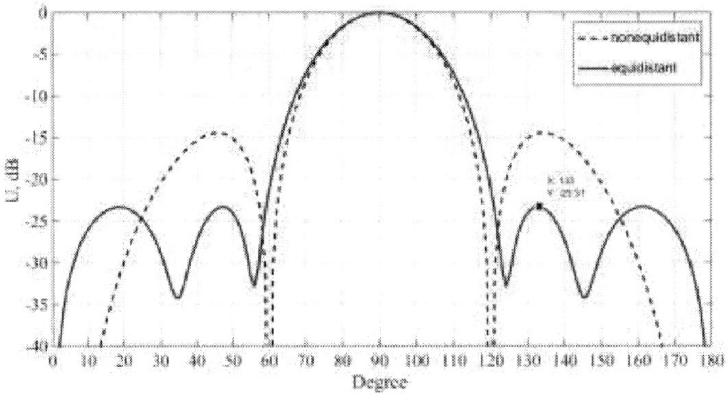

Figure 2. Radiation pattern of 4-elements arrays.

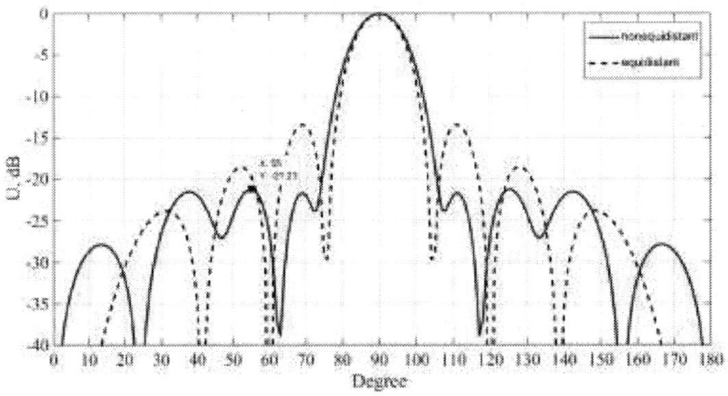

Figure 3. Radiation pattern of 8-elements arrays.

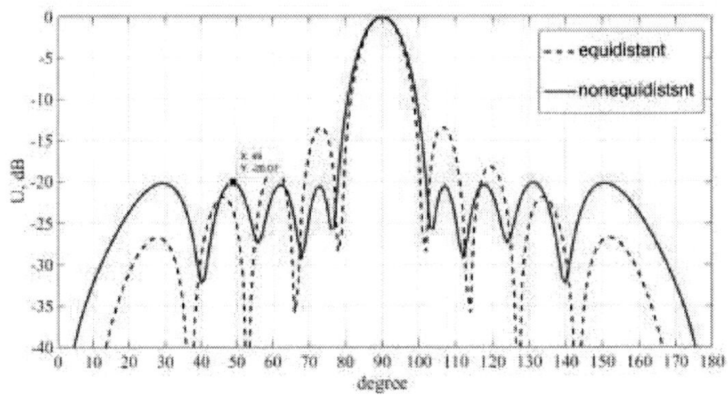

Figure 4. Radiation pattern of 10-elements arrays.

For linear phase arrays, SLL can also be reduced by varying the spacing of the elements. The element spacing and radiation pattern of the array that has a mainlobe position of 60 degrees are shown in table 2 and Figure 5 correspondently. The GA radiation pattern is compared with the equidistant array one. It can be seen that the mainlobe doesn't change its width.

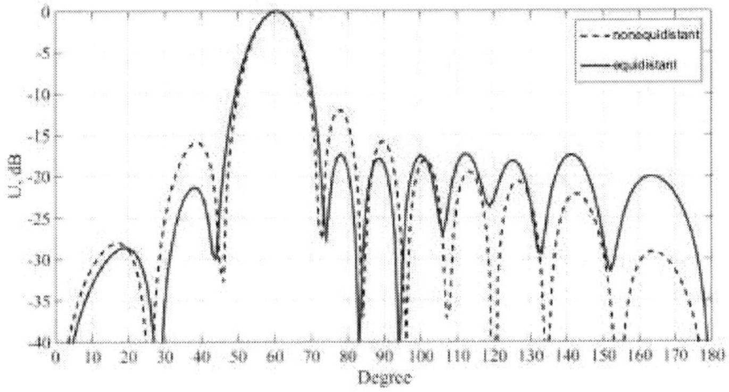

Figure 5. Radiation pattern of 10-elements linear phase arrays.

Table 2. The interelement spacing for linear phase array.

n	Interelement spacing of nth element, d_n
1	0
2	0.68298
3	1.2444
4	1.7444
5	2.1199
6	2.5645
7	2.9866
8	3.3853
9	3.9326
10	4.5

Note that the last radiation element will not be fixed, and the SLL can be further lowered. But the length of the array is reduced, respectively, directivity and antenna aperture efficiency are reduced too. For example, table 3 shows the interelement spacing d_n for 3-element arrays that has -25 dB and -30 dB of SLL, and halfpower beamwidth of 42 and 44 degrees accordingly. The Figure 6 depict their radiation patterns. For d_n=0.5, the halfpower beamwidth is equal 34 degrees (a dotted line).

Tables should have only horizontal rules and no vertical ones. Generally, only three rules should be used: one at the top of the table, one at the bottom, and one to separate the entries from the column headings. Table rules should be 0.5 points wide.

Table 3. The interelement spacing for 3-element array

n	Interelement spacing of nth element, d_n	
	SLL=-25 dB	*SLL=-30 dB*
1	0	0
2	0.41009	0.39848
3	0.81714	0.75697

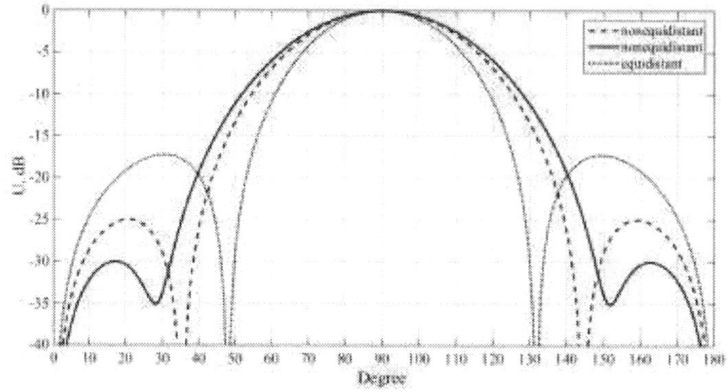

Figure 6. Radiation pattern of 3-elements arrays.

The directivity of nonequidistant and equidistant antenna arrays with same radiator number can be compared by Eq. (3) MATLAB evaluation. The results are showed in table 4. The directivities difference is negligible.

Table 4. The array's directivity.

N	Directivity	
	Nonequidistant array	*Equidistant array*
4	4.2961	4.2326
8	8.2907	7.8966
10	10.2896	9.7724

5.Conclusion

In this paper, we showed that a low SLL nonequidistant arrays can be well evaluated in MATLAB using the *GA*. It is possible to significantly reduce the SLL by nonequidistance introduced in the uniform array's element positions without compromising the energy efficiency of the antenna, and optimal interelement intervals were obtained using GA. It should be noted that the optimization time for small arrays (N < 15) is not large enough. For arrays with a large number of elements, this time increases significantly, which makes the method less suitable.

References

[1] Balanis C A 2005 *Antenna Theory Analysis and Design* (New York: John Wiley & Sons)
[2] Goldberg D E 1989 *Genetic Algorithms in Search, Optimization and Machine Learning* (Reading, MA, USA: Addison-Wesley)
[3] Li Y, Hong W, Li M, Lu G 2018 *International Workshop on Antenna Technology (iWAT)* 1–4
[4] Ridwan M, Abdo M, Jorswieck E 2011 *Proceeding of 13th International Conference on Advanced Communication Technology (ICACT2011)* 422–427
[5] Mandal D, Ghoshal S K, Das S, Bhattacharjee A K 2010 *Proceeding of International Conference on Recent Trends in Information Telecommunication and Computing* pp 126–129
[6] Marcano D, Duran F 2000 *Proceeding of IEEE Antennas and Propagation Magazine* **42** 12–20
[7] Tuan S C 2018 *Proceeding of 12th International Symposium on Antennas Propagation and EM Theory (ISAPE)* pp 1–3
[8] Enache F, Popescu F, Depărăteanu D, Enache A, Oroian T 2016 *Proceeding of 11th International Conference on Communications* pp 121–124
[9] Haupt R L, Werner D H 2007 *Genetic algorithms in electromagnetics* (Hoboken, NJ, USA: John Wiley & Sons)

IOP Conf. Series: Materials Science and Engineering 1019 (2021) 012015 doi:10.1088/1757-899X/1019/1/012015

Preparation and properties of composite material based on hafnium diboride and aluminum

S A Belyakovich[1], A V Mostovshchikov[2, 3], A P Ilyin[4] and F A Gubarev[5]
[1]School of Advanced Manufacturing Technologies, National Research Tomsk Polytechnic University, Tomsk, Russia
[2]School of Nuclear Science & Engineering, National Research Tomsk Polytechnic University, Tomsk, Russia
[3]Department of Physical Electronics, Tomsk State University of Control Systems and Radioelectronics, Tomsk, Russia
[4]School of Core Engineering Education, National Research Tomsk Polytechnic University, Tomsk, Russia
[5]Research School of Chemistry & Applied Biomedical Sciences, National Research Tomsk Polytechnic University, Tomsk, Russia
E-mail: sab41@tpu.ru

Abstract. Hafnium diboride (HfB_2) in aluminum matrix (Al) material was prepared by diffusion sintering of a mixture of hafnium diboride and aluminum micropowders. The effect of the hafnium diboride content (HfB_2) in the aluminum matrix (Al) on sintering characteristics and properties of the samples is presented. The effect of microwave activation of the aluminum powders has been established. The diffusion sintering behavior was concluded. The material properties and structure were defined.

1. Introduction

Due to the nuclear industry development the new materials design process is a crucial task to protect against ionizing radiation and neutrons [1]. New materials that can absorb or scatter ionizing radiation, at the same time having a low density and mass-dimensional characteristics of the final products are needed to be design. Moreover, using of lead screens and their subsequent disposal are a complex environmental problem; therefore, replacing such screens with new materials makes it possible to solve not only technical problems, but also to improve the environmental situation [2].

Hafnium diboride (HfB_2) should be noticed as a perspective material because of its nucleus high neutron capture cross section and scattering properties of boron nucleus. Aluminum has a relatively low density and low induced radioactivity [3]. Hafnium diboride physicochemical and radiation properties are well–studied, it is also known that it has high strength covalent bonding and low self-diffusion coefficient which indicates the need for high temperatures and pressure to create a strong non-porous structure. HfB_2 is a ceramic material, which suggests its fragility [4, 5]. Taking into consideration the known features of sintering ceramic materials with aluminum, leading to low-strength crack-sensitive systems (different thermal expansion coefficients of metals and ceramics, thermal stresses at the metal-ceramic interface) it was decided to create a system based on microwave radiation activated aluminum micropowder. Microwave effect on metal powders is known to an increase in the oxidation heat affect and a decrease in the oxidation onset temperature [6]. Microwave effect also leads to an increase of the aluminum nanopowder reserve energy, which is the result of the double electric layer formation and stabilization. Thereby microwave emission leads to a change in the substructural characteristics of metal-containing high-energy materials and results in an increase in the micro-stress in the metal component of aluminum particles [7].

Solid phase sintering was chosen because, in addition to the required lower energy costs, this solid phase interaction process is a diffusion process. A metal layer deposited on ceramics partially oxidized from the side facing the ceramics. The metal crystal lattice in the surface layer varies, at the same time, the chemically reorganized cells are not separated from the internal cells with the original crystal structure [8].

The aim of the work was to carry out exploratory researches of hafnium diboride diffusion (solid phase) sintering using microwave activated aluminum micropowder.

2. Experimental techniques and powder characteristics

To prepare material by diffusion sintering, mixtures of HfB_2 and Al micropowders were prepared by heating. Information about the initial micropowders are given in table 1.

Table 1. Initial powders.

HfB_2	TU No 6-09-03-418-76; particle shape is irregular; particle size – 40 μm
Al	ASD – 6M produced by OOO «SUAL-PM» (Shelekhov): the average surface diameter of the powder particles is 2.3 μm, the shape of the particles is close to spherical. The powder was produced by spraying molten aluminum in a chamber with a special atmosphere [9, 10]. The aluminum content in the powder after storage in conditionally sealed containers is 86 wt. %

The powder was irradiated according to the conditions and methods described in [6]. Figure 1 shows the thermograms of the initial (a) and irradiated aluminum powder (b).

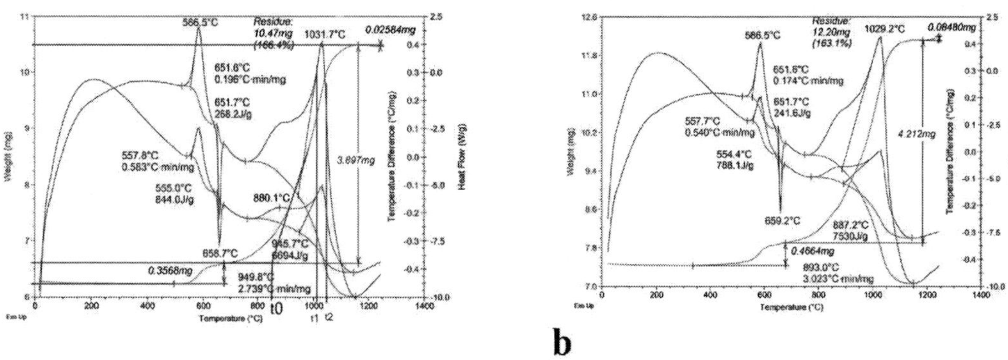

Figure 1. Thermograms of aluminum powder ASD-6M: a – thermogram of the initial powder; b – powder thermogram after microwave irradiation.

To prepare material samples HfB_2: Al mixtures were used, the composition and mass of which are presented in table 2. Similar mixtures were obtained on the basis of microwave irradiated aluminum [11, 12]. Tungsten nanopowder (W) was added to sample 4 in order to increase the sample adsorbing ability since W possesses high shielding properties due to high x-ray density. In addition, W is characterized by extremely high melting point, good machinability and high corrosive resistance [13, 14].

The initial and the microwave irradiated micropowders were checked for pyrophorosity by determining four activity parameters [6]. After the powder irradiating, its oxidation onset temperature decreased by almost 100 °C, remaining above 400 °C (Figure 1), i.e. microwave radiation does not give a pyrophorous property to the previously passivated powder [15, 16].

Table 2. Composition of mixtures of powders HfB2 and Al

Item No	HfB_2 content, wt. %	Al content, wt. %	The additive nanopowder W content, wt. %	The mixture mass, g.
1	90	10	0	5.5
2	50	50	0	10.0
3	10	90	0	9.5
4	50	20	30	8.0

The powders sample weights were mixed by grinding them on the tracing paper with a cork wrapped in tracing paper until a uniform color was obtained. The resulting mixtures were pressed using a hydraulic press. Compacting pressure – 1.5 MPa. Compressed tablets sintering was carried out in a muffle furnace at 550 °C in an anoxic environment for 4 hours.

X-ray phase analysis (XRD) of sintered samples was performed using a Diffrey-401 diffractometer (X-ray tube radiation F_{K_α}, $\lambda = 1{,}93$ Å).

3. Experimental results

Figure 2 shows photographs of the obtained samples. Initial aluminum based samples have splits and cracks. Microwave-irradiated aluminum based samples are characterized by higher mechanical characteristics. According to the obtained results, the sintered samples phase composition is consistent (table 1) with the data of tables 2 and 3: the composition of the sintered samples corresponds to the additivity concept. The results of the XRD are given in tables 3 and 4.

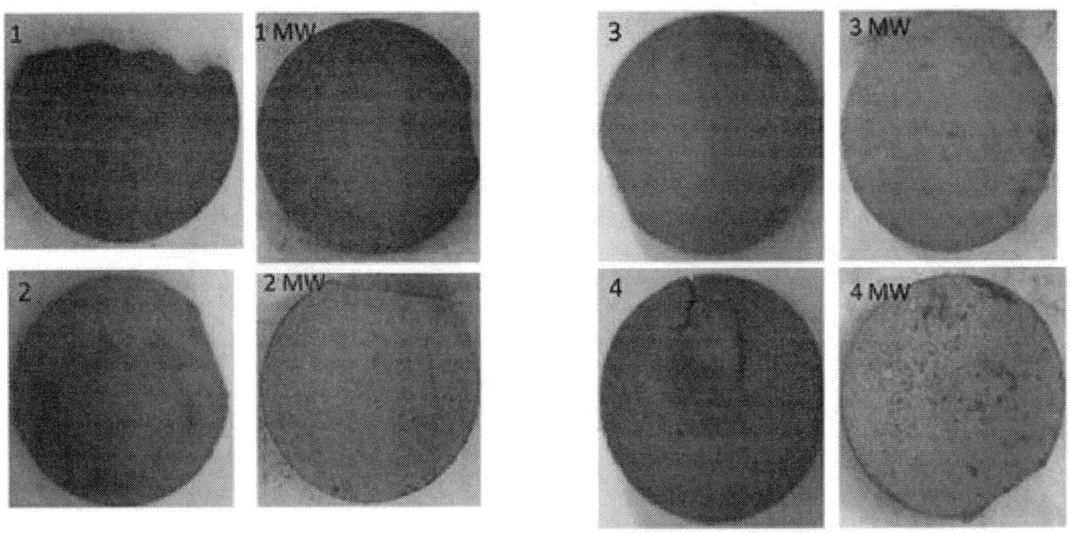

Figure 2. Photos of the samples: 1,2,3,4 – samples of the corresponding composition based on the initial aluminum powder and 1 MW, 2 MW, 3 MW, 4 MW – samples of the corresponding compositions based on microwave-activated aluminium.

Table 3. Sintered samples phase composition of material based on HfB_2 and initial aluminium.

Item No	HfB_2 content, wt. %	Al content, wt. %	The additive nanopowder W content, wt. %	The mixture mass, g.
1	90	10	0	5.5
2	50	50	0	10.0
3	10	90	0	9.5
4	50	20	30	8.0

Table 4. Sintered samples phase composition of material based on HfB2 and microwave irradiated aluminium.

Item No	HfB_2 content, rel. %	Al content, rel. %	Sintering products contents, rel. %
1	97	3	0
2	75	25	0
3	48	52	0
4	68	13	18 (B_2O_3)

Boron oxidized to B_2O_3 was found on the sample surface. Al_2O_3 on the surface was not detected. According to the XRD results, there are significant differences in the composition of samples 3 and 4 (tables 2 and 3). For sample 3 the differences can be explained by the structure homogeneity imperfection, when most of the aluminum is in the sample surface layer. For sample 4 the difference in the content of sintering products is related to the oxidation possibility of the additive (W) in the presence of sample defects noted for samples based on initial aluminum (there is a cleavage and crack in the sample volume).

The apparent density of samples without taking into account closed and open pores was measured in this work [17, 18]. The apparent density of initial Al based samples are given in table 5, microwave-activated Al based samples in table 6.

Table 5. The apparent density of HfB_2 based samples and initial Al powder

Characteristics	Item No (table 2)			
	1	*2*	*3*	*4*
Density ρ, g/cm3	1.97	2.05	2.36	2.79
Shrinkage,%	-0.41	-0.23	-0.50	-3.95

Table 6. The apparent density of HfB_2 based samples and microwave-activated Al powder

Characteristics	Item No (table 2)			
	1	*2*	*3*	*4*
Density ρ, g/cm3	3.98	1.94	1.97	3.14
Shrinkage,%	-0.31	-0.35	-0.35	-0.31

There are a high porosity and negative shrinkage for all samples, which indicate oxidation processes. At the same time, comparing the obtained data, it can be said that the microwave-activated Al based samples structure is more stable.

Figure 3 shows photographs obtained using scanning electron microscopy (SEM).

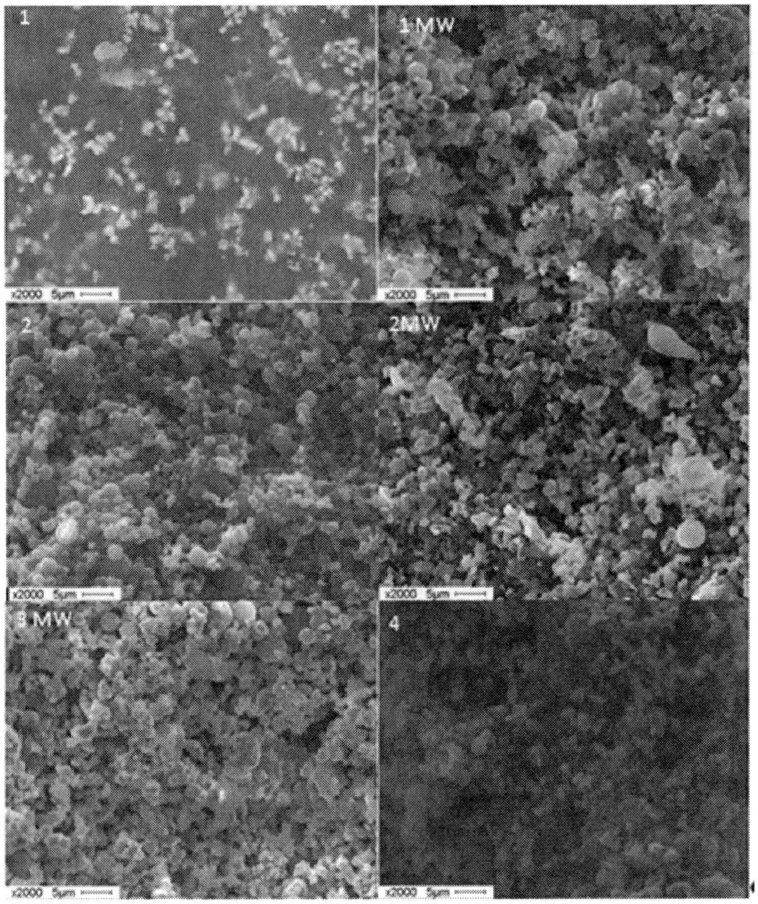

Figure 3. Samples photos obtained by diffusion sintering: 1,2,4 are photographs of the initial aluminum powder based samples respectively; 1 MW, 2 MW, 3 MW – photographs of the microwave-activated aluminum powder based samples respectively.

In photo 1 (Figure 3) spherical particles with an average size of 2.1 μm are observed, while in photo 1 MW a specific geometric shape is not observed and particle size is about 2.5 μm. The high light areas are hafnium diboride particles, characterized by a higher density and occupy about 11% of the volume in photograph 1 and about 28% in photograph 1 MW. Fragments are combined into associations, which indicates partial sintering. According to microanalysis, there are aluminum atoms in sample 1 detected by the electrons emission from the Kα orbitals at their mass ratio of 74.01%, and hafnium atoms in a mass ratio of 25.99% by electrons emission from the L orbitals. There are aluminum atoms in sample 1 MW detected by the electrons emission from the Kα orbitals at their mass ratio of 79.47 %, and hafnium atoms in a mass ratio of 20.53 % by electrons emission from the L orbitals. Boron atoms were not detected, since its content in the sample is below the microscope detection limit. For samples 2 and 2 MW, the average grain size is 2 and 2.6 μm, respectively. The mass ratio of aluminum and hafnium in the sample is 2 61.92 % and 38.02%, respectively. For sample 2 MW is 67.05 % and 32.95%. Boron atoms were also not detected. In sample 3 MW, the average grain size is 2.5 μm, the mass fraction of aluminum is 28.4%, the mass fraction of hafnium is 71.6%., The tungsten content is defined as 26.1 wt. % in sample 4, the hafnium content is 40.84 wt. %. The images determine structure of the samples as porous.

According to additionally carried out experiments using a spectrophotometer [19], the aluminum nanopowders burning samples emission spectra were recorded (Figure 4).

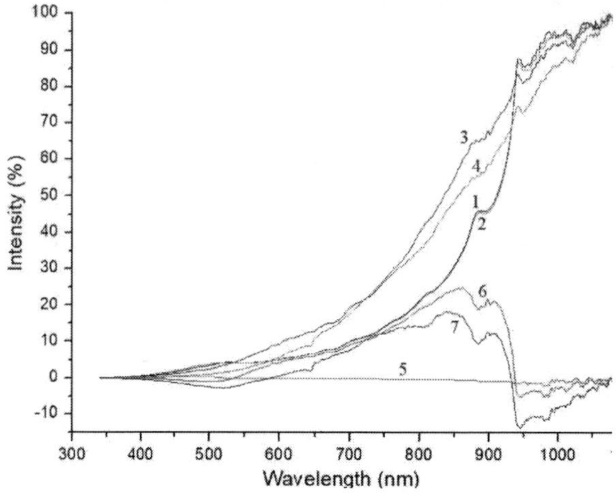

Figure 4. Aluminum powders emission spectra: 1 – Initial aluminum powder; 2 – Microwave-activated aluminum powder; 3 – Irradiated by 4 MeV electron beam (dose 31 x-ray); 4 – Irradiated by 7 MeV electron beam (dose 100 x-ray); difference spectra: 5 – Microwave-activated aluminum nanopowder and initial aluminum nanopowder; 6 – Irradiated by electron beam (dose 31 X-ray) aluminum nanopowder and initial aluminum nanopowder; 7 – Irradiated by electron beam (dose 100 X-ray) aluminum nanopowder and initial aluminum nanopowder.

Microwave-activated aluminum nanopowder during combustion emits a difference spectrum (spectrum 5 in Figure 4), as well as an internal nanopowder. At the same time, the irradiated by electron beam aluminum nanopowder has an additional emission band in the range of 600–950 nm (difference spectra 6 and 7 in Figure 4), which significantly improves the combustion characteristics of such aluminum in comparison with unirradiated powders [20, 21].

4. Conclusions

1. The method of diffusion sintering of a ceramic material in an aluminum matrix was tested in this paper. It was concluded that the obtained samples properties are not correspond to the required strength properties of the material. Nevertheless, it is noted that microwave-activated aluminum based samples have better sintering properties and a complete structure in general.
2. Due to the insufficient hardness of the obtained compact samples, this sintering method can be recommended for the material used to fill hollow protective partitions manufacturing.

Acknowledgments

The reported study was funded by RFBR, Project No. 19-03-00160.

References

[1] Kozlov V F 1991 *Handbook of radiation safety* (Moscow: Energoatomizdat) p 180
[2] Baratova A N and Korolchenko Y A 1990 *Fire and explosion hazard of substances and materials and their means of extinguishing* (Moscow: Chemistry) p 325
[3] Kircher J and Richard E 1964 *Effects of radiation on materials and components* (New York: Reinhold Publishing Corporation)
[4] Sheka I A and Karlysheva K F 1973 *Chemistry hafnium* (Kiev: Naukova Dumka) p 459
[5] Innocent J, Portehault D, Gouget G *et al* 2017 *Materials for Renewable and Sustainable Energy* **6** 6

[6] Mostovshchikov A, Il'in A, Chumerin P *et al* 2016 *Letters to the J. of Tech. Phys.* **42** 17–22
[7] Ilyin A, Root L, Mostovshchikov A 2012 *J. of Tech. Phys.* **82** 140–142
[8] Gordon W *et al* 2018 *The Eur. Phys. J.* **72** 140–142
[9] Zmanovsky S, Arkhipov V, Zhukov A *et al* 2014 Patent 2559080 Russian Federation IPC B22 F9 / 08. The method of obtaining metal powders by spraying melts / No. 2014109340/02; declare 03.11.2014; publ. 08/10/2015 Bull. No. 22
[10] Dubkova Y, Mostovchikov A, Kuznetsov 2018 *MATEC Web of Conferences* **194** 01014
[11] Ya-Wei Huan, Wen-Jun Liu, Xiao-Bing Tang *et al* 2019 *Nanoscale Research Letters* **14** 181
[12] Obstarczyk A, Kaczmarek D, Mazur M *et al* 2019 *J. Materials Science: Materials in Electronics* **30** 6358–6369
[13] Xiao-Ying Zhang *et al* 2019 *Nanoscale Research Letters* **14** 83
[14] Demirskyi D, Suzukib T, Grassoc S *et al* 2019 *J. Eur. Ceram. Soc.* **39** 898–906
[15] Demirskyi D, Solodkyi I, Nishimura T *et al* 2017 *J. Am. Ceram. Soc.* **100** 5295–5305
[16] Monteverde F, Melandri C, Failla S *et al* 2018 *J. Eur. Ceram. Soc.* **38** 2961–2970
[17] Yu M, Grasso S, Mckinnon R *et al* 2017 *Adv. Synthesis Catal.* **116** 24–60
[18] Bulanov Vand *et al* 1983 *Diagnostics of metal powders* (Moscow: Science) p 278
[19] Dilip Srinivas Sundaram *et al.* 2016 *Combust. Flame* **169** 94–109
[20] Li L, Mostovshchikov A, Ilyin A *et al* 2020 *IEEE Transactions on Instrumentation and Measurement* **69** 457–468
[21] Anand Bellam Balajia *et al* 2017 *Rad. Phys. and Chemistry* **141** 179–189

IOP Conf. Series: Materials Science and Engineering 1019 (2021) 012016 doi:10.1088/1757-899X/1019/1/012016

The Location estimation of seismic wave propagation layer velocities

E V Rabinovich, N Y Filipenko and G S Shefel
Faculty of Automation and Computer Engineering, Novosibirsk state technical university, Novosibirsk, Russia
E-mail: erabinovich1952@gmail.com

Abstract. On the basis of the original location-based technology developed by the authors for constructing seismic images of the deep structure of the earth's interiora method is proposed for estimating the seismic wave propagation layer velocities. The technology ensures the unambiguity and stability of seismic imaging by successively solving direct kinematic problems for all reflecting boundaries. It consists in locating the reflection pulse's sources from the boundaries of the deep section's layers and restoring the kinematic parameters of the layers through them. To assess the next layer velocity, information is used on the parameters of the preceding boundaries and the direct kinematic problem's solution with the additional equality's condition of local velocities at the reflection's points from the underlying boundary.

1. Introduction

The actual seismic exploration's task is the construction of seismic images, such as seismic maps of vertical and horizontal seismic sections, time sections, fracture zones arising from hydraulic fracturing, data cubes, etc.

The method of reflected waves plays a major role in seismic prospecting, and the methods of seismic imaging using this method possess high accuracy and reliability of the information obtained.

Currently, most seismic imaging technologies rely on the solution of the inverse kinematic seismic exploration problem (IKP) on reflected waves. The solution of the direct kinematic seismic exploration problem (DKP) is much less frequently used.

Solutions of these problems are sought within the geometric seismic framework of longitudinal reflected waves. It uses the temporal field's kinematic parameters of seismic waves (the shape of the rays and the reflection isochron for all seismic traces) propagating in the geological medium from oscillation's point source, to estimate the velocity and geometric parameters of this medium

This approach is a limiting case when in the wave equation the wavelength tends to zero, which schematizes the process of wave propagation, highlighting, above all, the energy's movement of the seismic pulse as a whole.

IKP consists in finding the spatial characteristics of the deep layer's boundaries, on the assumption that the propagation trajectory's type of the fronts, the velocity and arrival times of seismic waves on the seismic receivers, and the curvature's characteristics and roughness of these boundaries are specified.

The IKP solution is possible only within the framework of some a priori assumptions regarding the structure and the studied medium's properties. The system of such assumptions, which includes empirical data and known physical laws that control the deposit's formation and the formation of their properties' formation, is used to create a seismic-geological the environment model.

Modern methods of solving IKP use a layer model of the environment. It is believed that the structure model of the Earth is a layers' set of rocks with different elastic properties, the velocity of which, as well as the boundaries' properties separating them, are functions depending on parameters.

Solving the IKP means choosing a medium model and estimating its parameters so that the result of the DKP solution coincides with the received arrival reflection pulses' times from the deep layers' boundaries of the medium. In fact, the initial arrival times of seismic pulses are known not precisely, but with some errors. Therefore, as a criterion for the solution of the IKP, it is used not the best, but acceptable coincidence with the observed times [1].

Inverse problems are incorrect, they violate at least one of the conditions for the solution to be correct - these are the existence conditions, unambiguityand stability of the solution. Therefore, even an approximate solution of IKP is possible only within the framework of some a priori assumptions regarding the structure and properties of the studied medium.

There are many algorithms for approximate IKP solution. By approximate solutions, parameters of only integral temporal model's characteristics, such as average, effective velocity, average velocity gradient over depth and / or distance, can be estimated.

However, the experience of using these algorithms shows that with the same observations, the solution's results may differ significantly from each other, which are a lack's consequence of unambiguityand / or structural instability of the IKP solution [2].

To determine the kinematic parameters of the studied geological environment, various the hodographs' approximations of the target seismic waves for the selected reflecting boundaries' model are used. In order to restore individual seismic boundaries, compensation is required for the errors of these approximations, which is implemented by introducing kinematic corrections into the seismograms.

These procedures are uncertain and their implementation is associated with subjective decisions.

Modern studies of the deep environment structure are aimed at the continuation of the wave field obtained on the observation's surface, in the direction opposite to the waves' propagation - the migration of seismic wave fields.

A variety of complex computational procedures for seismic migration are performed in the temporal or deep region. But all of them are based mainly on the simplest horizontally layered model of the medium and the averaged parameters of its characteristics.

With more complex models, the results of constructing the migrated depth-velocity environment model are also subjective, especially when the migrated data are trying to be linked with the well data to better fit.

The presence of the proposed model parameters leads to the incorrect usual solution problem - there are many models that, satisfying the registered data, can lead to the same result.

For these reasons, IKP solutions can lead to poor quality estimates of the depth-velocity model's parameters of the environment.

DKP is to find the pulses' arrival times to seismic receivers, if the medium's depth model, the propagation path's type of the wave front, the receivers' position and the source are specified, the moment of seismic oscillations is known. In most cases, the direct problem is quite simple and reduces to an explicit system's solution of parametric optimization equations.

Solutions of direct problems are unambiguous and, within the limits of permissible errors, are stable. When solving the DKP, the propagation velocity of seismic waves from an oscillations' point source is not a presumed parameter of the medium model, but becomes a secondary calculated parameter.

2. Problem definition

The technology's basis of seismic imaging is information collected on the observation's surface about the temporal field of seismic waves reflected or refracted from the deep layers' boundaries. This field is recorded by seismic receivers at a constant position of the excitations' source (explosion or vibrator).

Reflected from the deep boundary of the medium seismic waves are recorded at the points of the profile or observation's area. They are presented in the form of seismic traces (time series) that define the waveform in the form of seismic pulses. Each seismic trace is the result of the waves' interference arriving at the seismic receiver.

In [3], an original location-based seismic imaging technology, free from the impropriety arising from the IKP, is proposed. It uses unambiguous and structurally stable DKP solutions based on the coordinates' location of the seismic pulses' sources reflected from the deep layers' boundaries of the medium under study.

Compared to the main seismic survey method - the common depth point method, location technology requires a significantly smaller number of vibration excitation points. It does not need to calculate the reflection time-distance curves, select the kinematic corrections and conduct the migration procedure.

The location is based on the superresolution detection's idea of reflected pulses used in radio and sonar [4, 5]. The location is made by the arrival times of seismic pulses found on the seismic traces. It consists in detecting these impulses' sources lying in the nodes of a given location grid covering the reflecting boundaries of the studied geological environment's layers.

The grid locations' coordinates, in which sources of seismic pulses were found, are used to build reflective boundaries of the layers.

Thus, the joint use of the reflected waves' time field obtained from the observed data and solving the problem's result of detecting the reflected pulses ensures the correct and seismic images' stable construction.

Estimate of the propagation velocity of elastic waves in seismic exploration is given much attention.

Using the well-known form of the propagation trajectory of the wave front in the medium, one can calculate the velocity with which the seismic pulse moved from its source to the seismic receiver, penetrating through all the studied medium's overlying strata.

Without knowledge of the seismic waves' propagation velocities, neither a geometric nor a dynamic interpretation of seismic data, as well as their geological interpretation, is possible. Data on velocities in real environments can be used in solving many geological and geophysical problems, since the dependencies of velocity on geological factors and the velocity's relationship with other physical properties of the studied medium are known [6].

The purpose of this work is to unambiguously and robustly find the propagation rates' estimates of seismic waves in each layer of the medium deep model under study.

3. Estimation of layer velocities

The paper uses an example of model data for the simplest layered 2D seismic survey model consisting of four reflecting homogeneous isotropic layers' boundaries (Figure 1).

In this model, monotypic reflected plane waves propagate from a point source of oscillations along rays in the form of straight lines, which corresponds to a constant seismic pulse velocity. The parameters associated with the dynamics of the excitation and propagation of waves, as well as changes in the shape and seismic pulses' interference, are considered ideal.

According to field data, subjected to pre-treatment in the form of cleaning from noise and interference, gain control, identifying the type of seismic pulse and much more, a seismogram (set of seismic trace) is formed that characterizes the time field of reflected waves created by a point source of vibration excitation. It is believed that these procedures are performed and the arrival times of seismic pulses are already known with a one millisecond's accuracy.

Locating sources of seismic pulses is carried out by placing excitation points (EP, several units) and reception points (RP, several dozen) - seismic receivers located on the earth's surface in a straight line at equal distances from each other. This set is used as a seismic antenna array (SAA). To improve the range resolution, the total number of SAA seismic receivers is divided into groups.

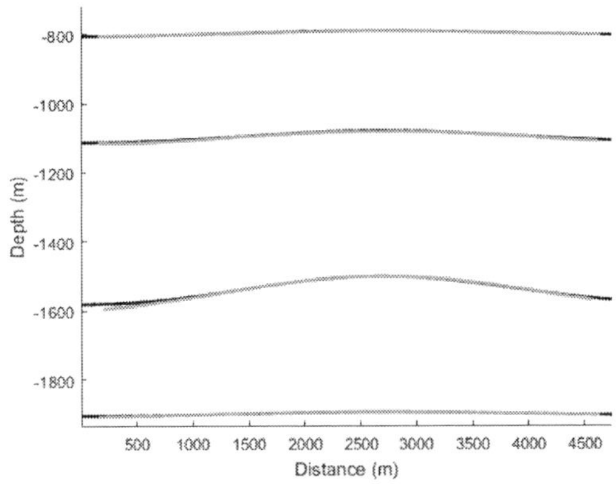

Figure 1. Depth model and its approximation (green lines).

The basis of the location technology of seismic imaging [3] is based on two algorithms. The first, designated as A1, is a parametric optimization algorithm that solves the DKP in order to determine the possible sources' coordinates of reflected seismic pulses on a given fragment of the location grid. It calculates the rays and travel times along them from the excitation source to the SAA seismic receivers, in turn for each layer. This calculation requires estimating the propagation velocity of seismic waves in each layer. The velocity in the first layer V_1 is considered known.

The location grid covers the reflecting boundaries of the medium and, in the simplest case, has the "beams" form of straight lines with multiple angles of inclination, whose centers are located on given horizontals and verticals of the deep environment. These straight lines serve to simulate a segment of the desired reflective border. In Figure2 given a simplified (not shown a lot of straight lines degrading the visual representation, reduced to 8 the number of seismic receivers in the SAA group) grid fragment's scheme location of the third border. It contains three "beams" located on three horizontals and one vertical (2350 m), and an image of the true boundary, represented by a black curved line. For all seismic receivers in the SAA group, 8 circles are plotted on each straight line depicting possible sources of seismic pulses.

The second A2 algorithm is the detection algorithm itself, which selects the coordinates of the best source using the idea of Capon's angular superresolution [7]. The idea is that the source is not detected by focusing the entire SAA or its group on a specific node of the location grid, but uses focusing the SAA group simultaneously on all the elements of the node vector to select the best element.

Algorithm A1 gives the coordinates of the elements of the node vector, and the A2 algorithm calculates the values of the decision function in them. The graph of this function is called direction-finding relief. In Figure 3 shows an example of three direction-finding reliefs belonging to three horizontals (different colors of lines) and straight lines with different angles of inclination. The blue color represents the best relief, and the best straight line has a slope of +4°.

The vector's element of nodes with the maximum relief value determines the inclination angle of the straight line, which is considered to be selected. The procedure for selecting the best node is performed for each seismic receiver in the SAA group. For the selected nodes, indicated by green stars, an approximation of a rectilinear boundary's segment is constructed. It is represented by the magenta color; squares are drawn on it, the coordinates of which correspond to the DKP solution for each seismic receiver of the current SAA group. Later, from the set of such approximations, a polynomial approximation of the entire boundary is constructed – the green line on Figure 4. A fragment of the true border is depicted in black.

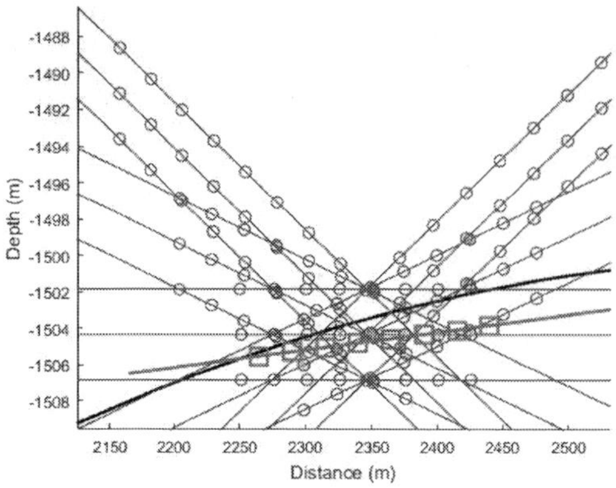

Figure 2. Mesh fragment of location.

The accuracy of the approximation of the obtained boundaries was estimated by theequation

$$SD = \sqrt{\frac{\sum_{i=1}^{n}(y_i - f(x_i))^2}{n-1}},$$

where $y_i$is the value of the i-th sample of the true boundary, $f(x_i)$ is the value of the i-th sample of the boundary's polynomial approximation, $x_i$is the coordinate of the i-th samples, n is the number of samples. The values are as follows:SD_1 = 0.88 m, SD_2 = 1.86 m, SD_3 = 10.84 m, SD_4 = 0.01 m.

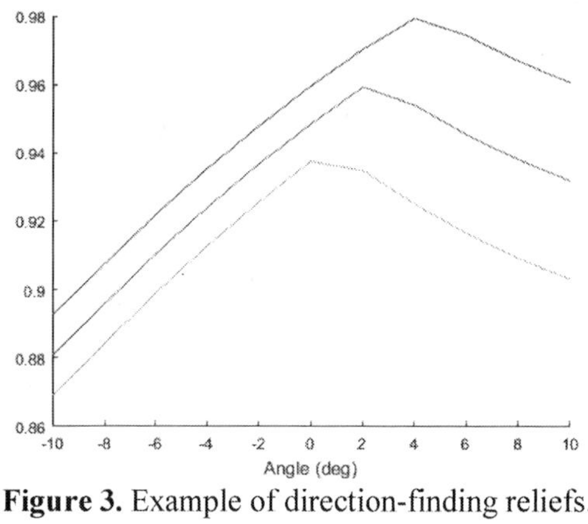

Figure 3. Example of direction-finding reliefs.

In this way, it is possible to detect the sources' coordinates of seismic pulses and the shape's approximation of the reflecting boundaries of all the layers that precede the one in which the layer velocity is estimated.

To construct the next layer's reflecting boundary of the medium's depth model under study, it is necessary to estimate the propagation velocity of seismic waves in it. For this purpose, several deployments of EP and RP are used. In Figure5, in order to facilitate the visual presentation, only two EPs and four RPs are shown, used to locate the underlying boundary of the fourth layer. In the Figure, green stars have the same meaning as in Figure4.

Polynomial approximations of all overlying boundaries and estimates of the layer wave propagation velocities have already been calculated. To estimate the fourth rate, the modified A1 algorithm is applied to all layouts. The modification is that the optimization is performed with the additional condition that the local velocities are equal at each selected node in the location grid for the current placement. The final estimate of the layer velocity is calculated as the median velocity obtained at different arrangements of EP and RP.

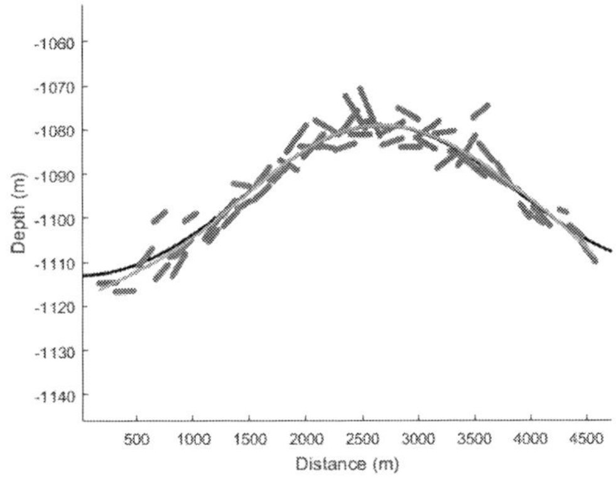

Figure4. Approximation of the second boundary.

Estimates of all layer velocities, produced in this example for two arrangements, and their absolute errors are as follows: V_1 = 2000.85 m/sec, ΔV_1 = 0.86 m/sec; V_2 = 2208.95 m/sec, ΔV_2 = 8.95 m/sec; V_3 = 2546.34 m/sec, ΔV_3 = 46.34 m/sec; V_4 = 2686.96 m/sec, ΔV_4 = -13.04 m/sec.

With a larger number of arrangements, the error in estimating stratum velocities can be reduced by calculating their average value for different arrangements.

4.Summary

The article describes a method for consistent assessment of propagation's layer velocities of seismic waves propagation reflected from several model's boundaries of the geological environment under study. The method is based on the technology of seismic imaging, which uses DKP solutions for all reflecting boundaries. The technology locates the reflection pulses' sources from the boundaries of the deep section layers and restores the kinematic parameters of the layers through them.

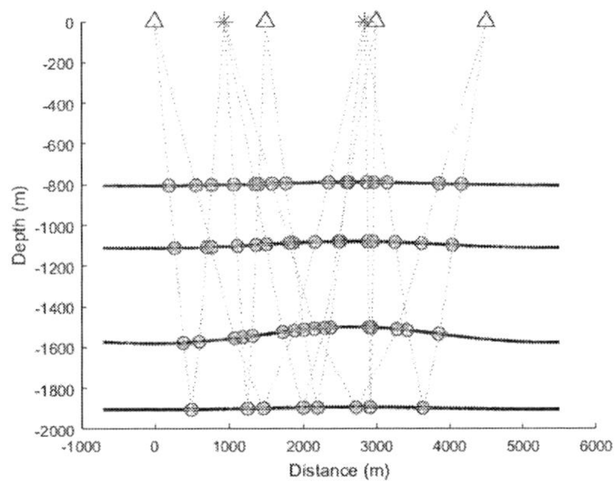

Figure5. Ray pattern for determining the fourth layer velocity.

The obtained errors of layer velocities estimate for the given model's example of the geological environment under study do not exceed 2%.

References

[1] Glogovskiy V, Langman S 2009 *Technologies of seismic prospecting* **1** 10–17

[2] Glogovskiy V*etal*1982 *Secondscientific seminar of the CMEA member countries on oil geophysics* **1** *Seismic prospecting* 327–331
[3] Rabinovich E, Shefel G, Jukov A 2018 *Proceedings of 14th International Scientific-Technical Conference on Actual Problems ofElectronic Instrument Engineering Proceedings APEIE-2018* 8545830
[4] Ratynskiy M V 2003 *Adaptation and superresolution in antenna arrays* (Moscow: Radio and communication) p 200
[5] Malyshkin G S *et al* 2014 *Acoustical Physics* **60** 570–587
[6] Urupov A K 1966 *Seismic velocity studies* (Moscow: Nedra) p 225
[7] Capon J 1969 *Proc. IEEE* **57** 1408–1418

Principles of manufacturing of DC on-board harness for a spacecraft based on 3D technology

S Efremov, N Karpov and A Leonov
Division for Power and Electrical Engineering, Tomsk Polytechnic University, Tomsk, Russia
E-mail: efremovv2005@yandex.ru

Abstract. This article describes in detail principles of manufacturing of DC on-board harness for spacecraft based on a new 3D design technology. This technology puts together all activities of on-board harness production, starting from electrical design and up to the final product readiness. Based on the analysis performed we proposed: necessary software, methods of wires and connector contacts connection, applicable wires and connectors available in Russian and European markets, shielding and insulating materials, test methods. The article discusses benefits of lightweight supply buses technology implementation (as compared to bundle of wires), which are: mass reduction, electrical performances stability improvement, lesser capacity between supply lines, better interference immunity, better thermal performances.

1. Introduction

Spacecraft (SC) – is a complex technological device comprising many components and systems. SC is composed of two major modules: payload module and service module.

Service module ensures nominal operation of all service equipment of a SC in a space orbit. Service module comprises:

- on-board computer;
- power supply system;
- thermal control system;
- propulsion subsystem;
- attitude determination and control subsystem.

Payload module ensures nominal operation of all SC mission equipment as per its function and application field (broadcasting, internet, navigation). This module includes only mission equipment.

The task of ensuring correct operation of a satellite in orbit is accomplished by connecting all units into an integrated electro technical system, and this requires a transmission medium between payload module equipment and service module. Such data transmission medium is on-board DC harness which provides electrical interfaces for on-board equipment. On-board DC harness connects not only SC modules, but also interlinks equipment within separate modules, and provides almost all electrical connections of a SC. That is why design and development of on-board harness is one of priority tasks of a SC design. As of today mass of on-board harness is 5-10 % of total SC mass. This is much lighter as compared to previously designed satellites, but reduction of on-board DC harness remains one of top-priority challenges of rocket-space industry [1, 2].

IOP Conf. Series: Materials Science and Engineering 1019 (2021) 012017 doi:10.1088/1757-899X/1019/1/012017

2. Improvements of on-board harness design and manufacturing

Today a SC on-board DC harness is designed, manufactured and tested by a new end-to-end technology, which puts together all technological activities of an on-board DC harness creation, starting from electrical design and up to final product readiness. This technology was introduced with a purpose to improve reliability, reduce mass and production lead-time of on-board harness. Prior to 2008 SC on-board harness was designed as a set of cables in the amount needed, and only connectors with solder contacts were used. While the new approach is to design on-board harness as an integrated three dimension network of electrical interfaces, featuring a complex configuration of branched and precisely located in space cables and having up to 12000 electrical connections with crimped removable contacts (Figure 1).

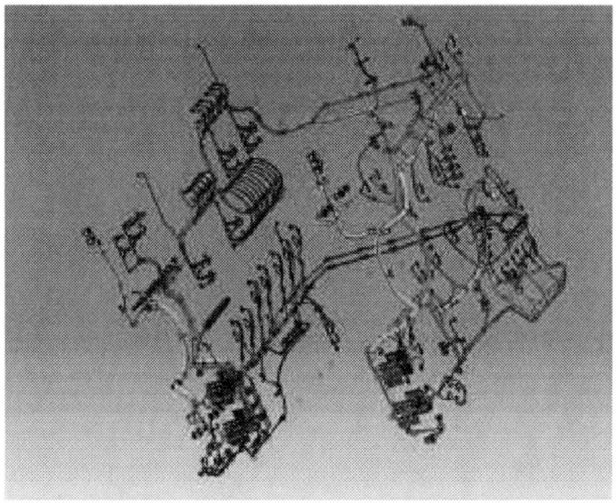

Figure 1. DC harness, 3D-model.

This technology allows reducing manpower input and shortening lead-time (from design and to product delivery) from 12 months down to 4–6 months. To put that in context: older technology allowed producing on-board harness for one, maximum two, SC per year.

Process of SC on-board DC harness creation consists of the following stages: design, manufacturing and test.

The following software can be used for design activities based on 3D-model of the satellite:

- Automated design system Catia PLM Express – this product is used to create harness topology (cables routing along the SC 3D-model).
- ECAD – electronic computer-aided design systems, such as E3.Series.
- PDM SmarTeam – electronic document management system.
- Program complex ALCAB – a set of programs and data bases, whose main function is to create packages of design documentation.

Apart from this the complex features other functions, such as:

- interaction with Catia system by means of interface files;
- storage of design solutions in data base;
- verification of description tables against specified data format;
- verification of final result against initial connections tables prior to design documentation release.

Manufacturing of on-board DC harness is done using two techniques of cables to contact connection: soldering and crimping.

Soldering:

- is used for Russian-made connectors (Figure 2);
- soldering technology is qualified, corresponds to Russian standards and is flight proven.

Figure 2. Soldering is used for Russian-made connectors.

Crimping:

- corresponds to the European Space Agency standard ECSS-Q-ST-70-26C and to the Russian regulatory document on rocket and space equipment standardization RD 134-0209-2013 (Figure 3);
- electrical connections are made with removable crimp contacts and crimp splices, which allows timely rework of cables when necessary. The process of cables crimped connection is more reliable and ecologically friendly as compared to traditional soldering.

Figure 3. Crimping is used for ESA-made connectors.

The following connectors are used for on-board DC harness:

- Russian-made solder connectors of such types as RS, MR1, 2RMDT, RP10, RRM46, RRM33, SNP339-T (Figure 4).

Figure 4. Russian-made connectors.

- European-made solder and crimp connectors as per ESCC 3401/001, ESCC 3401/002, ESCC 3401/008, ESCC 3401/029, ESCC 3401/052 specifications (Figure 5).

Figure 5. ESA-made connectors.

The following wires are used for on-board DC harness:

- Russian-made: radiation resistant wires with polyimide insulation type MS 16-15, MS 16-35, MS 26-15, MSE 26-15, MSEO 26-15 (specifications TU16.K76-160-2000, TU16.K76-011-88), wires with combined two layer insulation of fluoroplastic coating and polyimide varnish type MK26-15, MKE 26-15, MKEO 26-15, MK 26-35, MKE 26-35, MKEO 26-35 (specification PBMI.358200.002-2013TU) (Figure 6).

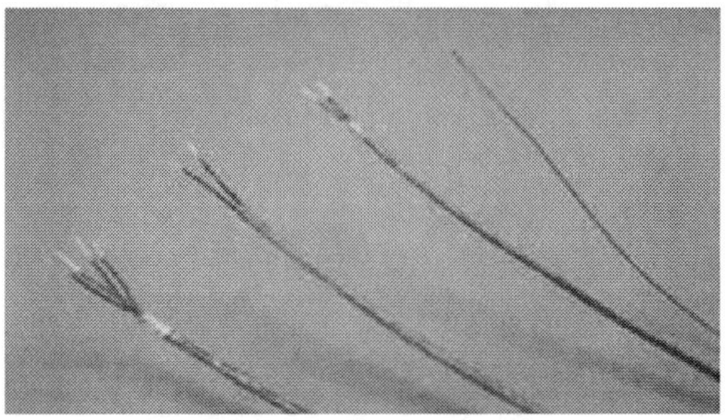

Figure 6. Russian-made wires.

- European-made: wires as per specification ESCC 3901 012 with radiation-modified fluoropolymer insulation (Figure 7).

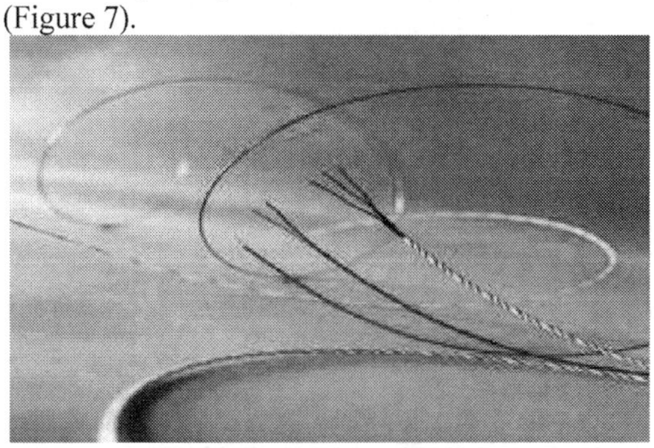

Figure 7. ESA-made wires.

The following cables are used for high-speed data transmission lines:

- cables Raychem specification 1200 with wave impedance 77 Ohm for data exchange channel via MIL-STD-1553B.
- cables ESCC 3902/002 with wave impedance 120 Ohm for data exchange channel via OBDH.
- cables ESCC 3902/003 and ESCC 3902/004 with wave impedance 100 Ohm for high-speed data exchange channel via SpaceWire.

Development of aluminum power-supply bus

One of the solutions to improve electrical performances and to reduce mass and size of a SC is to implement technology of lightweight supply bus into a SC power-supply on-board harness creation.

Lightweight supply bus is a rigid structure made of two aluminum buses (one with positive, one with negative polarity) minimal distance between which is fixed so that capacity and inductance are maintained at a certain level (Figure 8). Structure of supply bus features dedicated brackets which ensure minimal gap between them and SC structure when the bus is mounted [3, 4].

Figure 8. Aluminum bus.

Aluminum bus can be used instead of a copper wire sets with conductive cores. Such sets contain 240 wires with total cross-section of 120 mm^2 (maximal cross-section of a Russian-made space-application wire is 0.5 mm^2), each wire is individually insulated and, in order to meet electromagnetic compatibility requirement wires with positive and negative polarities are twisted in twos, thus the length of each wire increases by 5%. Position of wires in a cable bundle is not rigidly fixed, thus the distance between wires varies along the length of a cable, causing electrical parameters instability and capacity and inductance increase. Such cable is mounted onto the areas of the SC structure, that are not occupied by SC equipment, and the routing has many bends, which also impacts electrical parameters instability and apart from that creates possibility to damage wires during assembly activities. Lightweight supply bus uses connectors with allowable current load levels for connection with SC on-board equipment. Contacts inside connectors are soldered or welded to aluminum bus at a minimal possible distance using wires with silver plated copper cores.

Lightweight aluminum supply buses allow modernization of the following segments of standard cables in electrical power supply (EPS) system of a SC:

- segment between solar array (SA) and power conditioning unit (PCU);
- battery and PCU;
- PCU and power consuming devices.

The benefits of implementation of lightweight supply buses technology based on rigid structure of aluminum buses in production of SC supply on-board harness instead of standard cables made of wire sets with copper cores, are as follows [5, 6]:

- reduction of supply on-board harness by up to 30 %;
- improved stability of electrical parameters due to structural rigidity of the bus;
- lesser capacity between supply lines;
- improved interference immunity along supply lines;

- optimised allocation of SC mounting areas.
- increased heat removal of supply busses due to flat surface of the structure, which ensures heat removal with minimal gradient between conductive surfaces. This is particularly useful in space environment, where only conductive heat removal is possible.

The following materials are used for on-board DC harness:

- shielding materials:
- braid and synthetic tinsel tape wound in flattened metallic yarn;
- copper silver-plated braid;
- aluminum foil with conductive adhesive.
- electrical insulation materials:
- self-adhesive heat-resistant insulating tape;
- thermoplastic polyurethane cable sheath;
- polyvinylidene fluoride heat shrink tubing.

All mentioned materials are radiation resistant, qualified for usage in space environment for 15.5 years of a SC lifetime.

3. Conclusion and recommendations

In the frame of manufacturing phase, computer models are created and cable network are produces using with help of dummy representative of real structure. Such technology reduces usage of wires and materials, ensures bending radiuses, optimizes length of electrical lines and thus reduces cables mass.

On-board DC harness shall be manufactured in clean area of class 8 as per ISO 14644-1. Specified particles size: 0,5 μm (3520000 particles/m^3). Temperature of the environment shall be (22±3) °C. Relative humidity shall be (55±10) %.

Integrated three-dimensional on-board harness is manufactured by a paper-free technology, which requires working stations with large-sized monitors. During manufacturing process paper drawings are not used and workers consult screens directly, where they can project 3D configuration of separate cables or whole cabling system. Manufacturing process may further be facilitated by implementation of interactive electronic technical guides. Such guides make it easier for personnel to understand complex 3D structure of on-board harness without involvement of expensive computer-aided design systems.

On-board harness manufacturing process involves usage of certain tools for soldering and crimping of wires inside connectors, as well as certain equipment for crimping quality control: wire crimp pull test equipment, metallographic examination equipment, and voltage drop test equipment.

Test phase verifies quality of the manufactured harness by means of such tests as verification of conformance with design documentation, electrical tests, temperature cycling.

Insulation resistance between disengaged cable segments shall be less than 50 MOhm in normal environmental conditions;

Insulation between electrically disengaged cable segments shall be sufficient to prevent breakdowns.

Temperature range during tests (10 cycles) shall be from minus 50 to plus 60 °C.

Automated system of mounting verification in a matter of minutes checks all electrical lines and provides conclusion on the quality of manufactures on-board harness. Automated system of mounting verification called "GPI" is used for electrical checks of high-voltage circuits up to 1000 V (measurement limits of test voltage for alternating current is 5000 V and for direct current 6000 V).

At the final stage as soon as cable network manufacturing is completed, it shall be packed (without removing it from the dummy) and transported to the assembly area for integration onto SC. This approach helps to exclude any damage or deformations of the harness.

References

[1] Nawawi A, Tong C F, Yin S, Sakanova A, Liu Y 2017 *IEEE Transactions on Industry Applications* **53** 1168–1176
[2] Zhang H, Yang L, Wang S, and Puukko J 2017 *IEEE Transactions on Industrial Electronics* **64** 7563–7573

[3] Caponet M C, Profumo F, De Doncker R W and Tenconi A 2002 *IEEE Transactions on Power Electronics* **17** 225–231
[4] Bao H 2012 *IEEE Trans. Plasma Sci.* **40** 324–333
[5] Lawal Lasisi S, Chatwin Chris R 2009 *Proceedings of the 2012 Second International Conference on Electric Information and Control Engineering* **04** doi: 10.1109/ICEICE.2012.1164
[6] Jones M and Kelly A W 2000 *Proceeding of 15th Annual IEEE Applied Power Electronics Conf* (APEC-2000) **2** 839–845

Emission properties of YAG: Ce ceramics with barium fluoride flux

S A Stepanov, D T Valiev, V A Vaganov, A M Shrayber and V D Paygin
School of Advanced Manufacturing Technologies, National Research Tomsk Polytechnic University, Tomsk, Russia
E-mail: stepanovsa@tpu.ru

Abstract. In this work, we investigated the luminescent properties of YAG, Ce ceramics with an addition of barium fluoride flux. The ceramic samples were sintered from obtained YAG: Ce, BaF_2 phosphor powder. Morphological, luminescent and decay time characteristics of YAG: Ce ceramics were investigated. The luminescence decay kinetics analysis of the ceramics showed that the decay time in phosphors and ceramics is different in the visual spectral range.

1. Introduction

Ceramics are of great interest for present photonic devices. These optical materials occupy an intermediate position between crystal materials and glass, and combine the best properties of crystals such as a high mechanical and thermal durability and glassy materials (the possibility of pressing, molding, and forming elements of different geometry). One of the last directions in technologies of converting emission in construction of white LED are luminescent ceramics application [1, 2].

Using luminescent ceramics as an emission converter makes it possible to rank LEDs by color temperature with a high precision. This is achieved by more precise proportioning of phosphors. Another advantage of ceramic materials is to produce LEDs with high color homogeneity, a thermal durability, and an elevated temporal stable.

Yttrium-aluminum garnet doped with trivalent cerium ions (YAG: Ce^{3+}) is often used as an inorganic phosphor in the manufacture of lighting and photonic devices. The maximum of emission spectrum of YAG: Ce^{3+} phosphors are in the visible spectral region from 530 to 560 nm. This is due to radiation transitions between 5d and 4f energy levels of Ce^{3+} ions [3].

The solid-state reaction method (SSR) has gained great popularity for the industrial production of YAG phosphors [4–6]. At the same time, the SSR method requires a rather long treatment of phosphors at the high temperatures (> 1600°C) in an atmosphere of a gas mixture (N_2/H_2). A long-term high-temperature exposure leads to a partial sintering of the powder, and to the necessity for re-grinding the resulting solid-phase mixture.

Various sintering types of fluxes are used to improve the luminescent and morphological properties of phosphor powders obtained by SSR method [7, 8]. Fluoride materials (such as MgF_2, CaF_2, SrF_2, BaF_2, AlF_3) are often used as fluxes, which contribute well to the production of phosphor powders with a garnet structure [9, 10]. The authors of [9–13] are shown a positive effect of these fluxes on the ion diffusion, the crystallization process as well as on the formation of YAG host with a good crystallinity in the SSR method [9–13]. Obviously, under a radiation influence processes in ceramic materials differ from occurring processes in well-studied crystalline materials. In this regard, there is a growing of interest in the investigation of electron processes of excitation energy transfer (optical and radiation) to the active center in these materials. Therefore, understanding the mechanisms of the relaxation processes in the ceramic materials, directly interacting with the ionizing emission in the optoelectronic devices, is the key to the development and creation of new materials with given luminescent properties. The recombination processes will eventually determine the light output of the radiating medium, the degradation resistance of optical materials. This makes possible to control the properties of optical materials by creating efficient non-radiative channels or radiative energy transfer to the various luminescence centers at the stage of development of new radiation-resistant or radiation-sensitive materials, prognostication changes in their properties during extreme impacts.

There is a need to study the processes of energy transfer in materials between dopant ions of various types. Obviously, the size effects will make a significant contribution to the processes of energy exchange and the defect formation processes under radiation influence, which is especially important for the luminescent and optical materials.

The purpose of this work is to study the structural and luminescent characteristics of YAG: Ce, BaF_2 ceramics.

2. Materials and methods

The synthesis of phosphor powders more detail described in [14]. YAG phosphors powders with a cerium content of 0.06 wt. % were used for ceramics preparation. Barium fluoride was used as a flux. The concentration of barium fluoride in samples was 0; 0.05; 0.1; 0.15 wt. %.

The ceramic samples were sintered from obtained YAG: Ce, BaF_2 phosphor powder.

For pressing the powder we used a uniaxially automatic press IP-500 AVTO ZIPO. In the result the samples with a diameter of 9 mm and a thickness of 1 mm after sintering in the air were obtained. The photographs of ceramic samples are shown in Figure1.

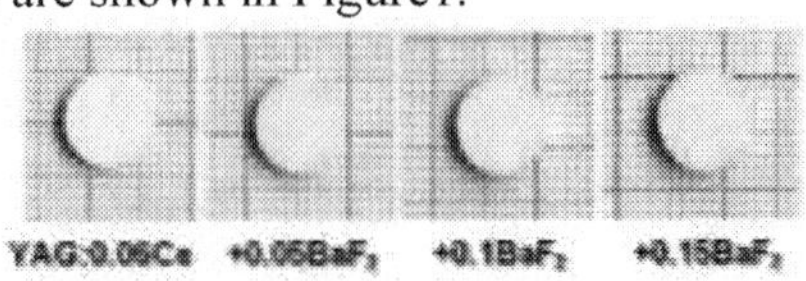

Figure 1. Sintered ceramic samples.

Integral spectral efficiency was measured by integrating sphere and spectrometer AvaSpec-3648 (integration time was 1 s, 200 – 1100 nm spectral range, 2.1 nm spectral resolution).At the first stage, the spectrum and emission flux of excitation source (LED) was measured. The next measurements were devoted to the emission spectrum of the investigated YAG:Ce^{3+} ceramics measurements.

The using of this method (Figure 2) made possible to subtract the measured spectrum of LED from the luminescence ceramic spectrum with a high precision. It allowed us to determine the reflected flux, the absorbed flux and the emission flux of the ceramic samples. The integral spectral efficiency determined as ratio of emission flux and absorption flux of sample.

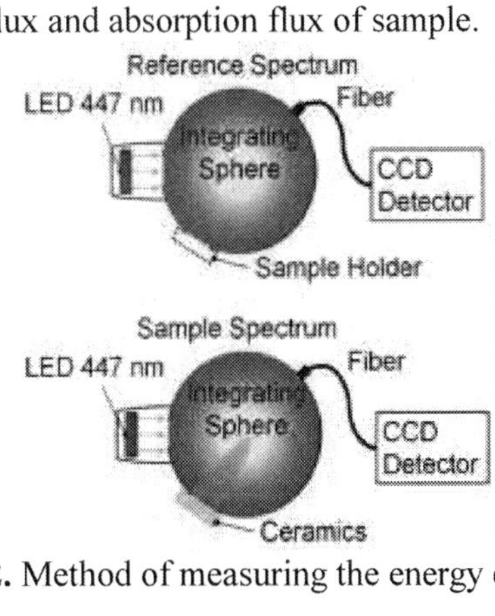

Figure 2. Method of measuring the energy efficiency.

IOP Conf. Series: Materials Science and Engineering 1019 (2021) 012018 doi:10.1088/1757-899X/1019/1/012018

For measuring pulse photoluminescence the LED (pulse duration $t_{1/2}$ ~15 ns, λ = 447 nm, FWHM-20 nm) as an excitation source was used. The luminescence decay kinetics were recorded with the monochromator MDR-204, and Hamamatsu H10720-20 photomultiplier tube and Tektronix DPO3034 (300 MHz) digital oscilloscope.

3. Results and discussion

3.1. SEM micro-images.

The SEM micro-images analysis of the investigated YAG:Ce^{3+} ceramics are shown in Figure3. The consolidation of particles with the formation of a high-density structure did not occur. Sintered ceramics are a porous monolith, which is confirmed by the results of scanning electron microscopy.

The sintering process is not complete until the end, consequently. The grain sizes remain at the level of the particle size of the initial powder, and the ceramics have a low density. YAG: 0.06Ce ceramics prepared with addition of 0.1 wt. % BaF_2 flux turned out with a smaller number of pores and more homogeneous than others one.

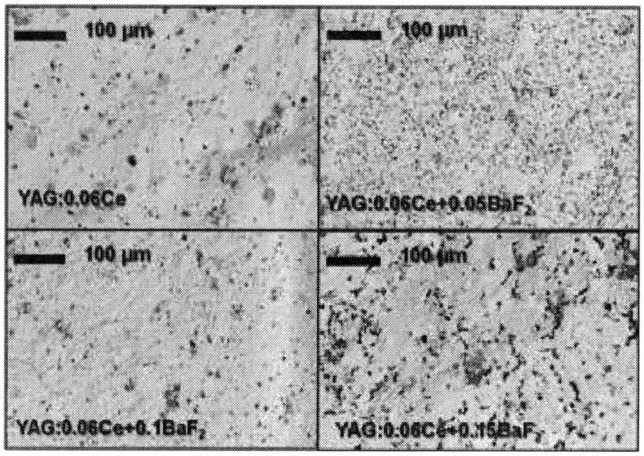

Figure 3.SEM micro-images.

3.2. Photoluminescence spectra and decay kinetics

There are two bands at 440 and 540 nm are recorded in the photoluminescence spectra excited by LED (λ_{ex} = 447 nm) ceramic samples. The results are represented in Figure 4. The photoluminescence spectra of synthesized YAG: Ce^{3+} ceramics with and without BaF_2 do not differ qualitatively.

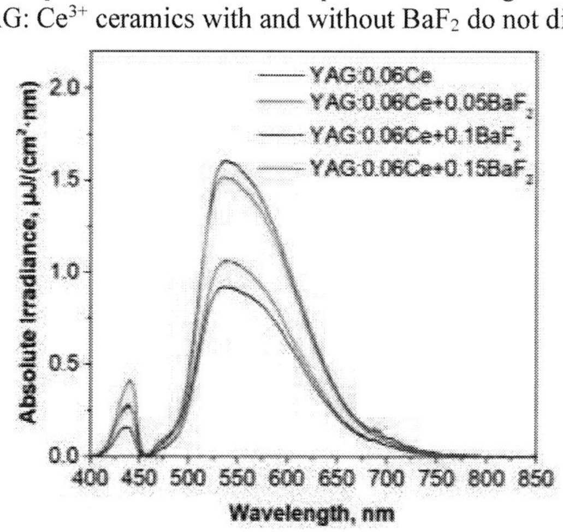

Figure 4. Photoluminescence spectra (λ_{ex}= 447 nm).

We suppose that the spin-orbit split ground state of Ce^{3+} ions determines the shape and the luminescence band position. The influence of YAG crystal field leads to the split of 5d energy level of Ce^{3+} ion into several sublevels. The emission occurs from the lower 5d sublevel into two splits states of the 4f level. This explains the luminescence double character of three valence cerium ion in YAG ceramics [15, 16].

Usually it is made experiments with the excitation by pulse laser radiation or by electron flux, but there are not results where as the excitation source used LED. According to [17], for YAG:Ce^{3+} nanocrystals short decay times (4 ± 3 ns and 17 ± 5 ns at 3.0 eV band) with excitation at 266 nm laser pulse is recorded. In our measurements we did not observe such short decay times (Figure 5). It can be explained by the pulse duration of excitation source. In our case the pulse duration was 30 ns and everything what was shorter than this time we could not observe.

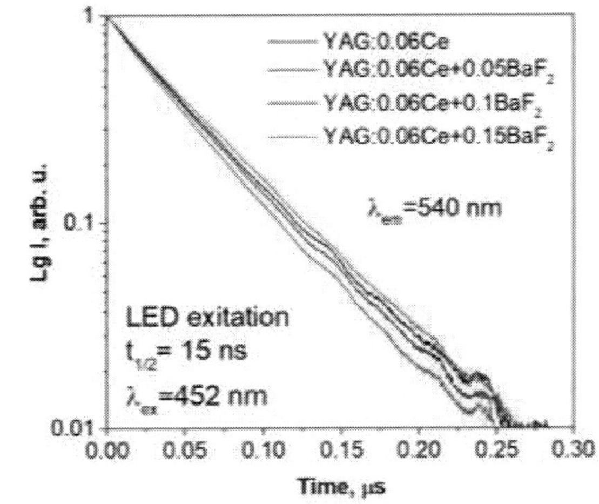

Figure 5. Luminescencedecaycurves.

Figure 5 shows the decay curves for 540 nm luminescence band. The character of the luminescence decay kinetics is explained by intra-center transitions. The radiative transition in each center occurs independently of the state of others and is determined only by its probability. The luminescence decay kinetics described by the single exponential function.

The results of measuring the decay kinetics showed that there is an effect of BaF_2 concentration on the luminescence decay time of YAG: Ce^{3+} ceramic samples (Figure 5).

Figure 6 shows the dependence of the decay times from barium fluoride flux. It can be seen that the lowest luminescence decay time corresponds to a concentration of 0.5 wt. % BaF_2. Also, it should be noted that the characteristic decay time for phosphors of similar composition, according to [14], averaged 60 ± 2 ns, while for ceramics 49 ± 2 ns.

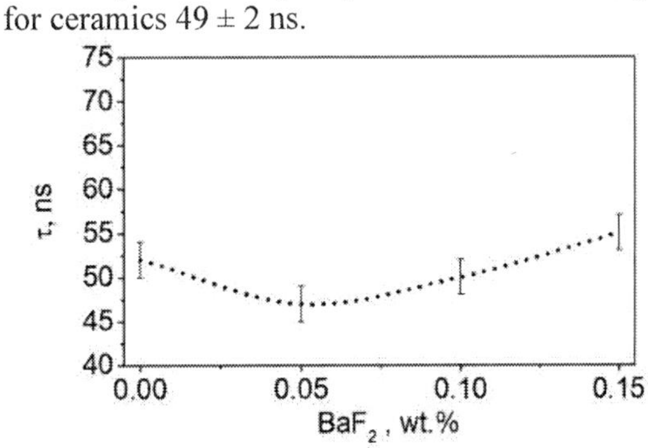

Figure 6. Dependence of decay times from barium fluoride flux.

3.3. Reflectance values and energy efficiency of YAG:Ce with barium fluoride flux

The results of a reflection measurement of the YAG:Ce^{3+} ceramic samples are shown in the Figure 7. The reflection values were measured for the band at 447 nm. It is shown the change of the reflection from 2 to 3% for the samples with different BaF_2 concentration. Often, the reflection of the materials is associated with the surface properties of ceramics. At the same time, we did not find a correlation between the dependence of the reflection on the concentration of barium fluoride with SEM images of the obtained ceramic samples.

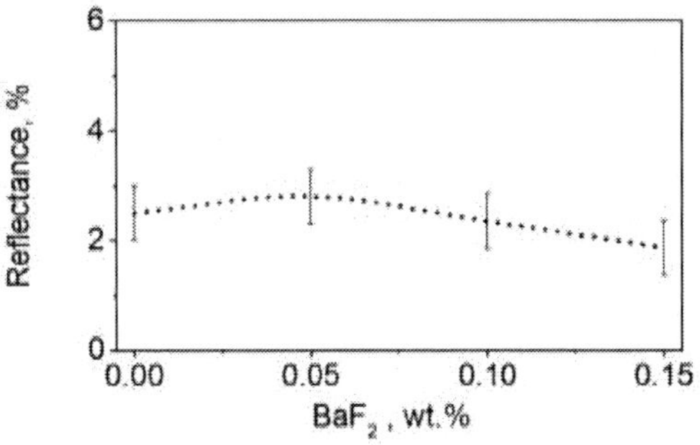

Figure 7. Dependence of reflectance values from barium fluoride flux.

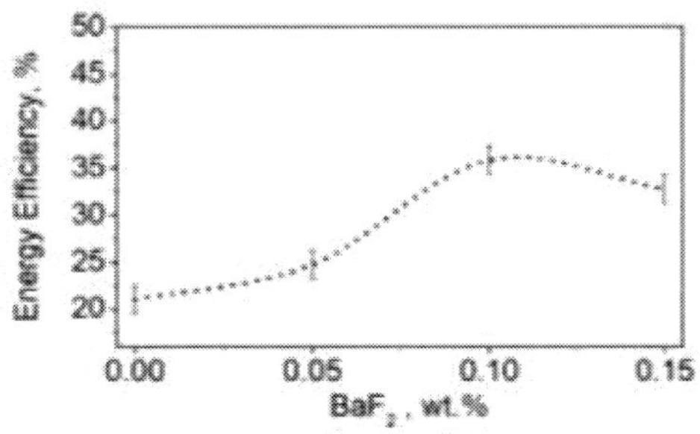

Figure 8. Dependence of energy efficiency from barium fluoride flux.

Therefore, the change in the reflection values with the increasing barium fluoride concentration in the main absorption band of the obtained ceramics is apparently associated with a change of properties grain boundaries and changes in the processes of sample recrystallization.

The results of energy efficiency measuring are shown in Figure 8. Energy efficiency increases from 22 to 36% with increase of BaF_2 concentration from 0 to 0.1%, respectively. Then, with the increasing fluoride flux concentration the decrease of efficiency is observed. It should be noted that the increase of energy efficiency have a fairly good correlation with a surface quality of ceramic samples (an existence of pores, a homogeneity) (Figure 3). The maximum efficiency is obtained for the sample, which matches to the best quality of the surface (a few of the residual pores).

4. Conclusion

Such properties of samples as luminescent, optical and morphological have been studied. Also the influence of barium fluoride flux on the luminescent and decay characteristics was investigated.

From the obtained results was enucleate that by synthesis ceramics on the basis of YAG: Ce, BaF_2 phosphors the spectrum of photoluminescence did not change. However, an increase in the luminescence intensity is observed. It was experimentally shown that the optimal concentration of BaF_2 is 0.1 wt.%. Upon this concentration value the maximum of luminescence energy efficiency is achieved. This is probably related to the small number of pores and the high homogeneity of the obtained ceramics.

As a result of analysis the luminescence decay kinetics it was shown that the luminescence decay time in ceramics is significantly lower than in the phosphor of a similar composition.

Acknowledgments

The reported research was funded by the Russian Foundation for Basic Research and the government of the Tomsk region of the Russian Federation, grant 18-43-703014.

References

[1] Chen Yi-C and Y-T Nien 2017 *Journal of the European Ceramic Society* **37** 223
[2] Bodišová K, Klement R, Galusek D, Pouchl′y V, Drdlík D and K Maca 2016 *Journal of the*

European Ceramic Society **36** 2975
[3] Xia Z and A Meijerink 2016 *Chemical Society Reviews* **46** 1
[4] TucureanuV, Matei A and A M Avram 2015 *Opto-electronics review* **23** 239
[5] Jang S, ChoiY H, Wu S, Lim T G and J S Yoo 2016 *Journal of Information Display* **17** 117
[6] Ohno K and T Abe 1994 *Journal of the Electrochemical Society* **141** 1252
[7] Shiqing X U , Liuzheng S, Ying Z, Haidong J, Shilong Z, Degang D, Huanping W and W Baoling 2009 *Journal of Rare Earths* **27** 327
[8] Lee S H, Jung D S, Han J M, Koo H Y and Y C Kang 2009 *Journal of Alloys and Compounds* **477** 776
[9] Chiang C-H, Liu T-H, Lin H-Y, Kuo H-Y and S-Y Chu 2013 *Journal of Applied Physics* **114** 243517
[10] Won C W, Nersisyan H H, et al 2011 *Journal of Alloys and Compounds* **509** 2621
[11] Shiqing X U , Liu S Z, Zhang Y, et al. 2009 *Journal of Rare Earths* **27** 327
[12] Kim H, Kim J, Han J S and K Park 2016 *Journal of of Ceramic Processing Research* **17** 405
[13] Song Z, Liao J, Ding X, Liu X and Q Liu 2013 *Journal of Crystal Growth* **365** 24
[14] Valiev D, Han T, StepanovS, Vaganov V and V Paygin 2018 *Materials Research Express* **5** 096201
[15] Wiatrowska A, Keur W and C Ronda 2017 *Journal of Luminescence* **189** 9
[16] Valiev D, Han T, Vaganov V and S Stepanov 2018 *Journal of Physics and Chemistry of Solids* **116** 1
[17] Pankratov V, Grigorjeva L, Millers D and T Chudoba 2007 *Radiation Measurements* **42** 679

IOP Conf. Series: Materials Science and Engineering 1019 (2021) 012019 doi:10.1088/1757-899X/1019/1/012019

Remote Reconfiguration for hardware-accelerated integrated Circuits

M Nagler[1], G G Kotre[2], W Hardt[1]
[1]Computer Engineering Department, Chemnitz University of Technology, Chemnitz, Germany
[2]Electrical Engineering Department, Chemnitz University of Technology, Chemnitz, Germany
E-mail: nagm@hrz.tu-chemnitz.de

Abstract. When it comes to hardware-accelerated solutions, Field Programmable Gate Arrays (FPGA) are often used to have a good trade-off between performance and the conservative cost restrictions the systems FPGAs are used in commonly have. With reconfiguration or partial reconfiguration the amount of needed resources can be reduced. In this work we show a solution to change the functionality of the system remotely. This means no physical connection is required. The different configurations can either be stored on the board in advance or can be sent to it just in time.

1. Introduction

Embedded systems in current world scenarios have to meet very high standards of performance and other requirements. The challenge is to get the needed computational resources while satisfying strict cost restrictions, which gets more complicated over time due to increasing consumer demands. This applies especially for automotive engineering, aircraft construction, and space industry. Systems in these domains are very complex, so they have to be separated into different subsystems [1–3].

In the field of autonomous driving or unmanned aerial vehicles, the cost requirements, especially regarding size, weight, energy consumption, and price are very conservative. On the other hand these systems need a high reliability as well as real-time capabilities. To be able to meet all these requirements Field Programmable Gate Arrays (FPGA) are used more often in the last years as current chips available on the market provide more functionalities, peripherals and essentially faster processing speeds as compared to their older counterparts. In contrast to software-based solutions they also provide a higher performance while consuming less energy. On the other side they have a higher flexibility than application specific integrated circuits.

For autonomous vehicles the situation the system is currently in, has a high impact on the needed functions [2, 4]. E.g. a car only needs the parking assistant while driving within a city whereas the lane assistant is only needed while driving on rural roads or highways. The same applies for aerial vehicles where the flight control has to be more precise during take-off and landing whereas the system for evaluating the sensor data needs more resources while flying over the target area (e.g. to find the important points in the target area). With this, the available resources can be shared during the runtime of the system.

The flexibility to share or re-allocate resources during runtime is also beneficial in the scenario described in [5]. Here a distributed system, which consists of several computational nodes, is to be developed. On one hand it tries to satisfy the conservative cost restrictions, that autonomous embedded systems nowadays have, and on the other hand it wants to provide as much reliability as possible. In contrast to regular solutions there are no redundant resources available that can be used in case of a failure. This full utilization of the system resources puts special requirements to the system when a computational node fails. In this case the *Quality of Service* of the application running on the system will be reduced. By e.g. reducing the resolution or the frame rate of an image processing application, resources can be freed up so the application will fit again into the reduced amount of available resources. This is achieved by having each module of the application available in different versions with a different *Quality of Service* which will be used for re-allocation. To be able share or to re-allocate resources, the system needs to change its functionality based on the current situation. In software-based solutions the functionality of a system can be changed by unloading not needed functions and replacing them with different ones. In FPGA-based systems the functionality of a system cannot be changed that easily. There is only the possibility to reprogram the complete chip with a different configuration or, if it is supported by the chip, reprogram a part of the chip while the remaining chip stays untouched. The latter one will not require to restart the chip. In general both cases need physical access to the chip to load the configuration via a programming interface. Nevertheless, there is also the possibility to write the configuration into a nonvolatile memory where it can be loaded from on system startup.

In this paper we show a solution to change the functionality of an FPGA remotely, without the need of physical access to the system. Different configurations will be stored inside the non-volatile memory and can be loaded dynamically during runtime. For this an internal configuration port is used to reboot the FPGA with one of the pre-stored configurations. In case a configuration is needed, that was not placed in the memory in advance, the configuration can be sent to the FPGA and a configuration slot in the memory is overwritten so that the new configuration can be loaded.

The structure of this paper is as follows. In section 2 solutions for reconfiguration in a software- and hardware-based systems are discussed. Section 3 describes the conceptual system setup as well as the different functionalities the proposed solution offers. Section 4 covers the technical realization of the system, its distribution into different modules and their division into subroutines, that are needed for the different functionalities. Section 5 evaluates the proposed solution. Next to the resource consumption of the different modules, measurements of the required time for the different operations are discussed. Section 6 concludes this paper with a short summary followed by the next steps which will be done to further improve the proposed solution.

2. State of the Art

System reconfiguration has to be done, based on the used components, either in software or hardware. This section provides a short overview about possible solutions for software and hardware-based systems.

2.1. Software-based System Reconfiguration

To change the functionality of a software-based system, not needed functions have to be unloaded to free up resources. These resources can be used to load new functions that suit to the current situation of the system. In general this is handled by the operating system. The challenge is to find an appropriate mechanism to select the necessary functions based on the inputs of the system.

Hanti proposes in his work [2] a solution that is based on a hierarchical scheduling algorithm for multi-core processors (HAMS). The HAMS scheduler has two abstraction levels. The first level assigns the functions to the different cores of the processor and the second level schedules the functions for every core independently. To find a suitable scheduling, all use cases for every situation are calculated apriori and stored in a so called *Knowledge Base*. With this, always the best scheduling can be selected, based on the current situation of the system.

Another approach are *Expert Systems*. One of such systems is discussed in [6]. It uses pre-defined rules to switch between different system states and draws decisions in real-time. The behavior is similar to a finite state machine. In contrast to [2] not all cases have to be specified in advance, so this rulebased decision making algorithm can react dynamically on the current situation.

2.2. Hardware-based System Reconfiguration

In hardware-based systems the change of their functionality has to be done by changing the configuration of the chip. Depending on the used device family, this can only be done completely (with rebooting the device) or it can also be done partially (without reboot). When targeting Xilinx FPGAs, not all devices support this so called partial reconfiguration. Looking at FPGAs from the 6th generation, there is only support for the *Virtex* family, not for FPGAs from the *Spartan* family. The same applies for FPGAs from other companies like Intel (formerly Altera).

Although officially not supported, there is research going on that provides partial reconfiguration for unsupported devices. References [7] and [8] propose a design flow that can generate partial bistreams for *Spartan 6* FPGAs from Xilinx. Here a new tool, called *GoAhead*, is introduced. It replaces the *PlanAhead* tool from the Xilinx toolchain. Based on userdefined areas, this tool removes everything from the complete bitstream, that should not be changed. With this a partial bitstream, that can be used to flash the FPGA, is generated.

To get access to the configuration memory, there are primitives available on the FPGAs. For Intel devices they are called *Partial Reconfiguration IP (PR-IP)* [9] and Xilinx has the so called *Internal Configuration Access Port (ICAP)* [9, 10]. Based on the used device, the functionality of these primitives is different. Devices that support partial reconfiguration can use it to update the configuration directly. If partial reconfiguration is not supported, it can only be used to reboot the FPGA with one of the configurations that is stored in a non-volatile memory. The selection of the according configuration is done by providing the start address of it.

The non-volatile memories can either be connected to a serial or a parallel port. A *Nexys 3* board from *Digilent* [11] with a *Spartan 6* FPGA on it, shall be shown as an example. This board contains three external memories, while two of them are non-volatile phase-change memories (PCM) [12]. Both have a size of 16 Mbyte. The serial memory is connected to a quadmode SPI bus whereas the parallel memory is connected to a 16-bit bus that only supports 16-bit operations. The latter is divided into 128 individually erasable 128 Kbyte blocks where one of these blocks is sub-divided into four individually erasable 32 Kbyte sub-blocks.

3. System Setup

The solution proposed in this work shall be realized on the previously mentioned *Nexys 3* board from *Digilent*. As the *Spartan 6* FPGA on it does not support partial reconfiguration, the method to reboot the board with a configuration that was stored in the non-volatile memory is used. This configuration can either be placed in the memory in advance or it can be sent to the board during runtime. The communication channel that is used to send a new configuration to the board or to trigger an reconfiguration shall be arbitrary, so an easy to use interface has to be designed, that can be adapted easily to different communication channels.

Furthermore, the proposed solution shall be as generic as possible, so there is only a low overhead to port it to a different FPGA with a different memory. To achieve this, all functionalities will be placed in separate modules: the *System Control* will handle the communication channel, extract the received data, and forward it to the according sub-components to call the different commands these components implement. The subcomponents will be the *Configuration Select* module, that can be used to reboot the chip with a different configuration, and the *Memory Controller* module, that can erase and write the non-volatile memory.

3.1. Memory Controller

This module can handle the data that is stored in the memory and will access the parallel memory. Based on the size of a configuration bitstream, the amount of possible configurations in the memory can be determined. A configuration for the *Spartan 6* FPGA used on the *Nexys 3* board has a size of

IOP Conf. Series: Materials Science and Engineering 1019 (2021) 012019 doi:10.1088/1757-899X/1019/1/012019

approximately 453 Kbyte. To be able to exchange a configuration inside the memory, the according part has to be erased first. As this is done blockwise, it has to be ensured that two different configurations are not stored in the same block. This requires that every configuration starts at a new block. We can calculate the amount of blocks per configuration by dividing the size of a configuration by the size of a block:

$$blocks = 453Kbyte/128Kbyte = 3.54$$

This means every configuration will need four blocks in the memory, which is equivalent to 512 Kbyte. By dividing the size of the memory with this value, we get the amount of configurations, that can be stored in the memory:

$$\begin{aligned} configs &= 16Mbyte/512Kbyte \\ &= 16384Kbyte/512Kbyte \\ &= 32 \end{aligned}$$

Based on this the memory can store up to 32 configurations. One of these configurations is reserved for the design that is developed in this work and must not be overwritten. This design is placed in *Slot 0* and will be called the *Golden Image*. It is the default configuration of the FPGA that is loaded on power-on. As Figure 1 shows, *Slot 0* covers *Block 0* to *Block 3*. After this, the user configurable slots follow. As the last slot would use *Block 127* which is sub-divided into four 32 Kbyte sub-blocks, it cannot be used here and leaves *Block 124* to *Block 127* unused. This means in total there are 31 slots in the memory for different configurations where 30 of them can be used for user designs.

The following commands will be implemented by this component:

- Write Data – *write one 16-bit word*
- Write Block – *write up to 32 16-bit words*
- Read Data – *read one 16-bit word*
- Read Status Register – *read status register of memory*
- Clear Status Register – *reset error flags in status reg.*
- Block Lock Status – *check if block is locked for writing*
- Unlock Block – *unlock block for writing*
- Erase Block – *erase data in block*

3.2. Configuration Select
This module can trigger the reboot of the chip with a different configuration. To do this, the *Internal Configuration Access Port (ICAP)* [10] primitive is used to issue a so called *IPROG* command. The instruction sequence is described in [10]. As input the module gets the number of the configuration that shall be loaded. Based on the size of a configuration the address in the memory, that has to be provided to the ICAP interface, is calculated.

The following command will be implemented by this component:

- Reboot – *load a different configuration*

3.3. System Control
This module is the main control unit in this work. It integrates the communication channel and handles the two sub-components *Memory Controller* and *Configuration Select*. The received operations will be converted into commands, the sub-components implement.

The complete system structure as well as the memory division is illustrated in Figure 1. The *Golden Image* is placed in *Slot 0*. The other slots can be used for configurations, that contain user logic. All configurations need to implement the *Configuration Select* module as well as a connection to a communication channel to be able to change the functionality of the system remotely. As the *Configuration Select* module is realized as independent component, it can be easily integrated into the user logic.

4. Implementation

The different components described in section 3 are realized as separate modules using the hardware description language VHDL. To show their functionality an example design with a serial interface as communication channel is implemented.

4.1. Memory Controller

This component is implemented as a finite state machine (FSM). The challenge with this FSM was the fact, that there is no clock signal, which can be used to synchronize the FSM with the memory. The memory latches the data on a rising edge of the according control signals. To implement this behavior, all states to write instructions or data were doubled so a rising edge could be modeled during the transition of these states. Nevertheless, the timing restrictions do not allow clock rates higher than 20 MHz.

The commands that were discussed in section 3 were implemented accordingly. Table 1 lists the input and output for every command. The start for every command will be triggered by the *System Control* module. After the completion of the command, a finish signal will be emitted.

Table 1. Memory Controller Commands.

Command	Input	Output
Write Data	Address, Data	–
Write Block	Address, Data(N)	–
Read Data	Address	Data
Read Status Register	–	Status Register
Clear Status Register	–	–
Block Lock Status	Block Base Address	Lock Status
Unlock Block	Block Base Address	–
Erase Block	Block Base Address	–

4.2. Configuration Select

This module was implemented as FSM as well by realizing the instruction sequence mentioned in [10]. It uses the same clock frequency of 20 MHz as the *Memory Controller*. The input for the reboot command is the slot number in the memory, that shall be loaded. Generics for the memory size, the size of a block, and the size of the configuration bitstream are used to calculate the start address of the slot. This makes the module reusable for different FPGAs.

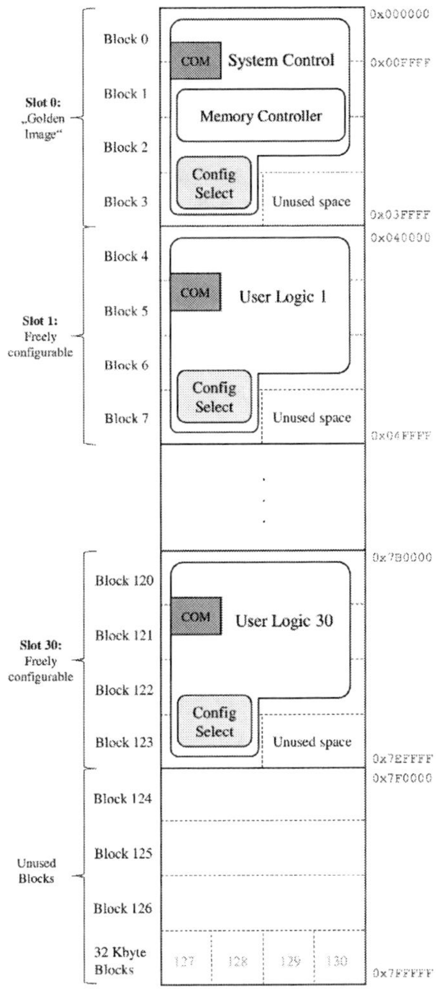

Figure 1. Structure of the non-volatile Memory.

4.3. System Control

This module triggers the different commands of its submodules. For this it also implements a FSM that starts a command and waits until its finish signal is emitted. Furthermore, it handles the data writing process. Internal count registers are used to keep track on how much data was already written or how much blocks were already erased. This module uses the same generic values as the *Configuration Select* module to calculate the addresses accordingly.

4.4. Example Design

An example design was used to show the functionality of all implemented modules. It integrates the *System Control* together with its sub-modules *Memory Controller* and *Configuration Control*. The communication channel was realized using a serial interface. Furthermore a script that converts a configuration bitstream into a binary file, which can be written to the non-volatile memory, was realized.

5. Results

The example design could show the proper functionality of the implemented modules. The generic address calculation of the *Configuration Select* and the *Memory Controller* module were able to select the different slots in the memory correctly. With this, any configuration could be changed by erasing it and writing a new bitstream to its memory location. After selecting a new configuration and rebooting the FPGA, it was loaded successfully. Additionally, the memory content written with the developed solution was compared to the memory content written with the *Digilent Adept* tool, that is provided by the manufacturer, to verify the correctness of the writing process.

Furthermore, the performance of the *Memory Controller* module was evaluated by measuring the time to complete the different operations. This was done by counting the clock cycles it took to perform an operation. The operation was applied to the complete slot. With the clock frequency of 20 MHz the elapsed time was calculated. The results show, that the time depends on the data that is stored in the memory or the data that is to be written. Two methods were tested for writing: *wordwise* and *blockwise* writing. The measurement was done with four different values:

- Zero – *writing/erasing the value 0x0000*
- One – *writing/erasing the value 0xFFFF*
- Count – *writing/erasing an up-counting value*
- Alternating – *writing/erasing a value where consecutive bits have a different value: 0x5555*

The results are shown in table 2. Every test was repeated 10 times. The table shows averaged values. As expected, the blockwise writing is much faster than wordwise writing. It outperforms wordwise writing by a factor of 11 to 19 (50 for 0xFFFF). Furthermore, it can be seen that writing and erasing 0xFFFF is much faster than the other operations. The reason here is the way the memory works. All entries have the value 0xFFFF after erasing. With this the controller of the memory chip does not have to change the value when writing or erasing, which results in a faster execution of the according operation.

Table 2. Performance of Controller Commands (Time to erase 4 Blocks or Time to write a Configuration of 453 Kbyte)

Value	Erase	Write Word	Write Block
0x0000	1634 ms	5025 ms	442 ms
0xFFFF	113 ms	2240 ms	44 ms
"Count"	1726 ms	5029 ms	313 ms
0x5555	1729 ms	5032 ms	268 ms

Additionally, the resource consumption of the design was evaluated. The *Golden Image* uses less than <5% of the FPGA resources. Considering that the *Spartan 6* on the *Nexys 3* board is a comparatively small chip from its device family, the resource consumption can be considered as very low. This makes it possible to integrate the complete design into the user configurations, not only the *Configuration Select*. If there are not enough resources available to integrate the complete design, at least the *Configuration Select* module has to be integrated to support the remote reconfiguration. Without *System Control* and *Memory Controller*, this module consumes approximately 1% of the FPGA resources and will fit in nearly every user design.

6. Conclusion and Future Work

In this work, a method to remotely reconfigure hardware-accelerated integrated circuits was shown. The design was realized for a *Spartan 6* FPGA mounted on a *Nexys 3* board from *Digilent*. With the independent modules, the implementation can be ported easily to different boards and memory chips. The low resource consumption makes it possible to integrate this solution into any design. The time to erase a configuration and write a new one into the memory is comparatively high with approximately 2000 ms. But with the low resource consumption of the proposed solution, the complete design can be integrated into the user designs, too. With this, it is possible to replace a configuration during runtime of a user design.

To further improve the proposed solution, the memory usage shall be optimized. Currently every slot uses with 3.54 out of 4 blocks only 88.5% of its size. Additionally, *Block 124* to *Block 127* at the end of the memory is unused, too. With this, approximately 2337 Kbyte (14%) of the complete memory is unused. By optimizing the memory usage up to 5 more configurations can fit into it.

Furthermore the here proposed solution shall be tested with the partial bitstreams generated by the *GoAhead* tool proposed in [7] and [8].

Also the time it takes to reboot into a different configuration has to be evaluated in the future. As this needs additional external logic, it was not considered in this work. Finally, the design proposed in this work will be integrated into the system discussed in [5] to be able to react on unforeseen situations as they can occur in autonomous distributed embedded systems.

References

[1] Watkins C B, Walter R 2007 *Proceefing of the 26th Digital Avionics Systems Conference* 2.A.1–1–2.A.1–10
[2] Hanti T, Frey A, Hardt W 2015 *Proceefing of the 34th Digital Avionics Systems Conference* (DASC) 7A3–1– 7A3–11
[3] Moens H, Famaey J, Latre S, Dhoedt B, De Turck F 2011 *Proceefing of the 12th IFIP/IEEE International Symposium on Integrated Network Management (IM 2011) and Workshops* 137–144
[4] Schneider S, Meisel A, Hardt W *Proceefing of the 2009 IEEE/IFIP International Symposium on Rapid System Prototyping* 61–67
[5] Nagler M, Blokzyl S, Hardt W 2017 *Proceefing of the IEEE Sensors Applications Symposium* 85–87
[6] Halbig J, Windisch A, Kingsbury P, Oswald N, Hardt W 2012 Autonomous Unmanned Systems 119–128 doi: 10.1007/978-3-642-32217-4_13
[7] Koch D, Beckhoff C, Tørrison J 2010 *Proceefing of the International Conference on Field-Programmable Technology* 361–364
[8] Beckhoff C, Koch D, Torresen J 2012 *Proceefing of the 20th International Symposium on FieldProgrammable Custom Computing Machines* 37–44
[9] Vipin K, Fahmy SA. *FPGA Dynamic and Partial Reconfiguration: A Survey of Architectures, Methods, and Applications.* doi:10.1145/3193827g.
[10] Spartan-6 FPGA Configuration User Guide. 2100 Logic Drive, San Jose, CA 95124; 2019.
[11] Nexys 3 FPGA Board Reference Manual. 1300 Henley Court Pullman, WA 99163; 2016.
[12] 128Mb: P8P Parallel Phase Change Memory (PCM); 2012

Parallel solution of 3D forward and inverse problems of airborne electromagnetic survey

M G Persova, Y G Soloveichik, D V Vagin, D S Kiselev, O S Trubacheva, N S Kondratyeva
Department of Applied Mathematics, Novosibirsk State Technical University, Novosibirsk, Russia
E-mail: mpersova@mail.ru

Abstract. The paper is devoted to the comparison of different approaches to parallel solution of 3D forward and inverse problems of airborne electromagnetic (EM) surveys when multidimensional processing of data obtained on large areas is needed. The mathematical models for the solution of 3D forward and inverse problems and different methods of their parallelization are considered. The results of numerical experiments for the airborne EM survey problem, which is typical of prospecting the kimberlite bodies in Eastern Siberia, are presented. It is shown that the combination of different techniques of parallelization in a distributed computational system, taking into account the specifics of the numerical schemes implementation for forward and inverse problems solution, makes it possible to achieve the speedup close to linear.

1. Introduction

Nowadays, the airborne EM survey methods are widely used for prospecting solid mineral deposits [1–3]. The investigated geological media, as a rule, are of a complex 3D structure, and in order to recover them, the 3D data inversion methods must be applied [4–9]. In this case, due to high density of observation systems, the number of 3D electromagnetic fields calculations can come to about several thousands even for the forward problem solution. Due to the fact that the modeling is performed in the time domain, the finite element system of the linear algebraic equations (SLAE) must be solved approximately 100–200 times for each airborne system position, depending on the time channels range used in the airborne survey. When the 3D inversion is carried out, this number is multiplied by the number of parameters, which, in its turn, can reach several hundreds or even thousands. Therefore, within one iteration of the non-linear 3D inversion, several hundreds of thousands or even millions of forward problems can be solved (as regards the number of finite element SLAEs to be solved, it can reach dozens or hundreds of millions).

In order to make the computational time acceptable for the operating cycle of airborne EM surveys, it is necessary to use parallelization for solving forward and inverse airborne survey problems. The effectiveness of parallelization essentially depends on the approach to its implementation, taking into account the numerical modules specifics.

In this paper, we consider several techniques of parallelization for the numerical solutions of airborne survey problems and perform their comparative analysis.

2. Methods

2.1. The Mathematical Models for Solving the Forward and Inverse Airborne EM Survey Problems

In order to solve the forward airborne EM survey problems, the vector finite element method on the non-conforming meshes [10,11] with hexahedral cells is used. This allows smooth approximation of the Earth surface relief and borders between geological layers. The electromagnetic field induced by the induction loop of an airborne system is described by the intensity $\vec{\mathbf{E}}^t$, which can be found as a sum of two summands $\vec{\mathbf{E}}^t = \vec{\mathbf{E}}^p + \vec{\mathbf{E}}^s$. Here, $\vec{\mathbf{E}}^p$ is the intensity of the primary field, which is defined by the change of the current in the transmitter, and $\vec{\mathbf{E}}^s$ is the intensity of the secondary field, which is defined by the 3D geological medium influence.

The field $\vec{\mathbf{E}}^s$ can be found as $\vec{\mathbf{E}}^s = -\partial\vec{\mathbf{A}}^s/\partial t$, and the vector potential $\vec{\mathbf{A}}^s$ is sought from the solution of the vector differential equation

$$(1/\mu_0)\nabla\times\left(\nabla\times\vec{\mathbf{A}}^s\right)+\sigma\partial\vec{\mathbf{A}}^s/\partial t=\sigma\vec{\mathbf{E}}^p \tag{1}$$

where μ_0 is the magnetic permeability of vacuum, σ is the conductivity of the 3D medium. On the remote boundaries $\partial\Omega$ of the 3D computational domain Ω, the zero tangential components of $\vec{\mathbf{A}}^s$ are set.

The magnetic field induction $\vec{\mathbf{B}}^s$ is can be calculated as $\vec{\mathbf{B}}^s = \nabla\times\vec{\mathbf{A}}^s$.

In order to speed up the forward problem calculation, the grouping [4,9,11] of airborne system positions and time steps is used. This enables us to use one and the same matrix of the finite element SLAE obtained as a result of the space and time approximation of (1) for large groups of airborne system positions. In its turn, this makes it possible to effectively apply the direct solvers of SLAE (for example, PARDISO [12]), because the matrix factorization, which is the most computational cost operation, is carried out only once for all problems of a group, and in order to obtain the solution of each problem, it is necessary to solve only the SLAE with the triangular matrices.

The inverse problem solution is based on the minimization of the functional of the residual between the observed and numerical data

$$\Phi(\mathbf{b})=\sum_{i=1}^{K}\left(\omega_i\delta\mathcal{E}_i(\mathbf{b})\right)^2+\sum_{m=1}^{M}\alpha_m\left(b_m-\bar{b}_m\right)^2\rightarrow\min_{b_m} \tag{2}$$

where K is the number of the observed data, $\delta\mathcal{E}_i$ is the difference between the observed and numerical values, M is the number of parameters, b_m are the components of the vector of parameters, $\bar{b}_m$ are the components of the vector of parameters obtained at the previous iteration, α_m are the regularization parameters.

The vector of the unknown parameters is constructed depending on the chosen geoelectrical model parameterization. We use the geometric inversion [7–9,13,14], where the vector of the unknown parameters contains both the electrophysical characteristics of geological medium objects and the coordinates of inhomogeneities boundaries. The total number of parameters used in processing the data, which are observed on the relatively small area of 5×5 km, can come to about several thousands.

The minimization of functional (2) is performed on the basis of the Gauss-Newton method. In this case, the functions $\delta\mathcal{E}_i(\mathbf{b})$ are linearized in the vicinity of parameters $\mathbf{b}$, and to calculate the values of $\partial(\delta\mathcal{E}_{il})/\partial b_m$, the equations similar to (1) are solved for each parameter b_m and each airborne system position.

Thus, as we noted above, the number of forward problems, which must be solved in the inverse problem, can reach several hundreds of thousands or even millions. Therefore, parallelization is important in implementing numerical methods for solving airborne EM survey problems.

2.2. *Distributed Computations*

Parallelization is performed in the computational system consisting of the multicore personal computers connected in a local network (further, we name them the computational nodes-servers).

Taking into account the specifics of the mathematical models used and the numerical schemes of their implementation, several techniques of parallelization can be applied.

- The first technique is the parallel calculation of one problem at one node: the parallelization of the SLAE solver, including the matrix factorization and the solution of systems with the triangular matrices for several right hand side vectors.
- The second technique is the parallel calculation of several problems on one computational node: for example, the simultaneous electromagnetic field calculation for several groups of the airborne system positions in the forward problem solution or simultaneous calculation of several fields of influence of the geoelectrical model structural parts (the derivatives with respect to parameters) in the inverse problem solution.
- The third technique is the parallel solution of problems on several computational nodes.

The first technique of parallelization is implemented using the OpenMP on one computational node. The second technique is performed by running several executable modules simultaneously on one computational node.

For the third technique implementation, the control program is required. It consists of the client and server applications. These applications are installed on the client-computer (where the forward or inverse problems are run by means of the graphic user interface) and on computational nodes-servers, respectively. The input data of the client application of the control program include the file "tasks.json", which contains the list of "tasks" in the form of the schedule of files, which should be sent to the corresponding computational node, and the information files, which are necessary for running electromagnetic field calculation for tasks (further, the calculation of electromagnetic fields for group of airborne system positions will be named task). The input data of the control program also include the file with the list of IP-addresses of the servers on which the calculations should be performed as well as the option defining the number of tasks, which are simultaneously sent to one computational node. Also, the input data include the number of repeated tries to access the server if it does not respond (before this server is assumed as inaccessible) as well as the pause between them.

The control program orders the tasks with respect to the number of airborne system positions in a group (at first the tasks corresponding to the groups with the most of the positions should be calculated), inquires the servers which are on the list of IP addresses, and sends the first part of the tasks. After receiving the task, the server application runs the corresponding software modulus on the computational node. During the calculation the client application of the control program controls the tasks completion, takes the data, and marks the corresponding task as completed or reruns the task. The task restarting is performed if there are no more computational resources on the node (due to any reasons), or if the connection with the computational node terminates for a long time. In the first case, the server is marked as inaccessible only for the task, which is completed with error, and in the second case, the server is marked as inaccessible for all tasks. In the latter case, if the connection with the computational node gets back up, then it is included in the list of accessible nodes again. If the results of the completed task are received by the client successfully, then the client application of the control program sends the following task to the free node. The process is completed after all tasks from the file "tasks.json" are solved successfully.

3. The analysis of the speedup factor by means of different parallel methods

We investigate the computational effectiveness of parallel techniques considered above and their combinations. For the calculations, we use the geoelectrical model shown in Figure 1, which is typical of airborne EM survey problems aimed to prospecting the kimberlite bodies in Eastern Siberia.

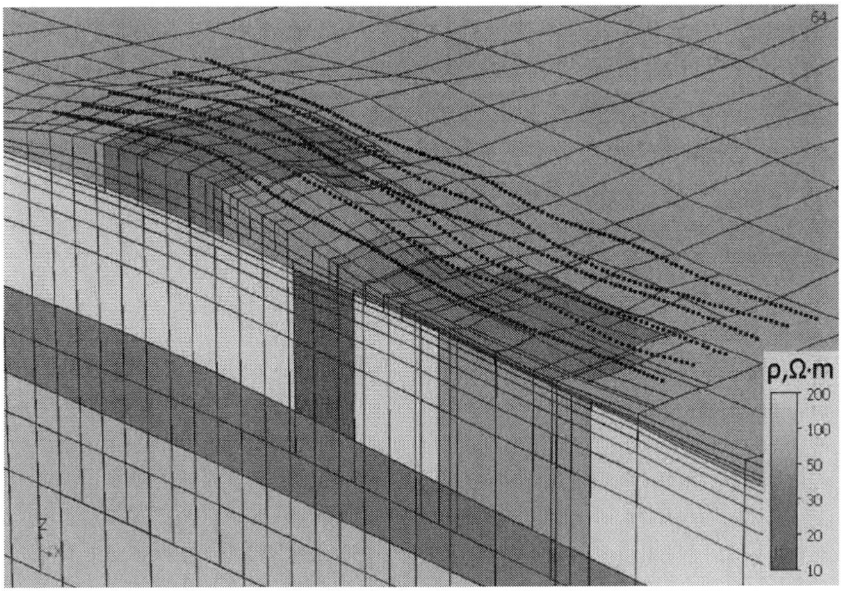

Figure 1. Geoelectrical model and finite element mesh.

In order to perform the numerical experiments, we use six-core computers with the processor Intel Core I7-8700 3.7 GHz and 32 GB of memory and four-core computers with the processor Intel Core I7-7700 3.6 GHz and 16 GB of memory. The calculations are carried out for the task in which the groups contain about 50 positions of the airborne system, and the finite element mesh contains about 25 thousand nodes. Below, the tables present the computational time for different techniques of parallelization when different numbers of threads are used. The computational time is given in seconds.

Tables 1–2 show the computational time required for computing one task (i.e. for calculating fifty 3D non-stationary fields for one group of airborne system positions) when different numbers of threads are used for the first techniques of parallelization on six- and four-core computers, respectively.

Table 1. The computational time when using different numbers of threads for the first technique of parallelization on a six-core computer.

Number of threads	Run Time, s		
	Matrix factoriz ation	Solution of SLAE with triangular matrices	Total time
1	41.4	76.6	282.3
2	26.2	40.8	185.2
4	15.7	19.1	137.6
5	17.6	16.2	145.5

Number of threads	Run Time, s		
	Matrix factorization	Solution of SLAE with triangular matrices	Total time
6	18.0	23.2	165.8
8	17.8	27.8	165.8
11	19.6	28.0	172.4

Table 2. The computational time when using different numbers of threads for the first technique of parallelization on a four-core computer.

Number of threads	Run Time, s		
	Matrix factorization	Solution of SLAE with triangular matrices	Total time
1	28.9	51.4	154.9
2	17.9	29.0	118.6
3	14.1	21.5	107.1
4	13.1	19.7	104.3
5	13.0	19.9	104.9
6	12.8	19.6	104.0
7	13.2	20.2	104.7

Note that except for the time required for the matrix factorization and solving the SLAE with the triangular matrices, the total time includes the time for generating meshes, global matrices and right hand side vectors, as well as the output of finite element solutions using special smoothing procedure.

As seen from the data given in tables 1–2, the use of four threads enables us to obtain the solution in minimal time. In this case, the speedup factor is only 1.5–2 and little different from the factor obtained when using two threads.

Let us consider the second technique of parallelization. To do this, we run different numbers of tasks (one task includes the fields calculations for fifty airborne system positions) on one computational node and analyze the total time required for their performance. The computational time for six- and four-core computers is given in tables 3–4, respectively.

Table 3. The computational time when using different numbers of threads for the second technique of parallelization on a six-core computer.

Number of tasks	Run Time, s					
	Matrix factorization		Solution of SLAE with triangular matrices		Total time	
	All tasks	Per task	All tasks	Per task	All tasks	Per task
1	41.4	41.4	76.6	76.6	282.3	282.3
2	40.3	20.2	82.2	41.1	255.3	127.7
4	31.1	7.8	58.3	14.6	200.6	50.2
5	32.8	6.6	57.3	11.5	204.1	40.8
6	35.9	6.0	65.5	10.9	218.4	36.4

Number of tasks	Run Time, s					
	Matrix factorization		Solution of SLAE with triangular matrices		Total time	
	All tasks	Per task	All tasks	Per task	All tasks	Per task
8	44.9	5.6	81.5	10.2	243.9	30.5
11	59.3	5.4	118.0	10.7	334.5	30.4
12	67.9	5.7	120.2	10.0	359.0	29.9
13	68.1	5.2	120.0	9.2	389.0	29.9

Table 4. The computational time when using different numbers of threads for the second technique of parallelization on a four-core computer.

Number of tasks	Run Time, s					
	Matrix factorization		Solution of SLAE with triangular matrices		Total time	
	All tasks	Per task	All tasks	Per task	All tasks	Per task
1	28.9	28.9	51.4	51.4	154.9	154.9
2	30.4	15.2	53.9	27.0	162.8	81.4
3	31.3	10.4	55.6	18.5	167.7	55.9
4	34.4	8.6	61.9	15.5	182.8	45.7
5	40.0	8.0	73.6	14.7	211.8	42.4
6	47.6	7.9	85.1	14.2	245.0	40.8
7	48.2	6.9	99.1	14.2	271.7	38.8
8	60.1	7.5	111.9	14.0	311.5	38.9
9	57.3	6.4	104.6	11.6	347.2	38.6

The results presented in these tables show that the minimal time it takes for the solution of one task is achieved when one runs in parallel from 8 to 13 tasks on a six-core computer (see table 3). If a four-core computer is used, the minimal time is achieved when the number of tasks computed in parallel is 7 to 9 (see table 4). Further increase in the simultaneously computing tasks on a computational node is impossible due to memory limitation.

Thus, the second technique of the parallelization allows speedup the solution of one task by 3–4 times in comparison with the first technique.

The following numerical experiments are performed when the first and second techniques of parallelization are used simultaneously. The calculations are carried out for the cases, when one task is computed using two and four threads. The results of these numerical experiments are given in tables 5–6.

Table 5. The computational time requred for simultaneous use of the first and second techniques of parallelization on a six-core computer when one task is computed with two threads.

Number of Tasks	Run Time, s					
	Matrix factorization		Solution of SLAE with triangular matrices		Total time	
	All tasks	Per task	All tasks	Per task	All tasks	Per task
1	26.2	26.2	40.8	40.8	185.2	185.2
2	20.2	10.1	30.8	15.4	151.1	75.6
4	24.3	6.1	42.9	10.7	168.2	42.1
5	28.8	5.8	51.5	10.3	181.0	36.2
6	33.6	5.6	57.6	9.6	202.3	33.7

Table 6. The computational time requred for simultaneous use of the first and second techniques of parallelization on a six-core computer when one task is computed with four threads.

Number of Tasks	Run Time, s					
	Matrix factorization		Solution of SLAE with triangular matrices		Total time	
	All tasks	Per task	All tasks	Per task	All tasks	Per task
1	26.2	26.2	40.8	40.8	185.2	185.2
2	20.2	10.1	30.8	15.4	151.1	75.6
4	24.3	6.1	42.9	10.7	168.2	42.1
5	28.8	5.8	51.5	10.3	181.0	36.2
6	33.6	5.6	57.6	9.6	202.3	33.7

The results given in the tables show that, if a large number of tasks is necessary to solve, the increase in the number of threads does not speed up the computations but slightly slows them. The simultaneous use of the first and second techniques of parallelization is appropriate only if the number of tasks sent to one computational node does not exceed six tasks.

Finally, the last experiment is connected with the analysis of the speedup factor using the second and third techniques of parallelization simultaneously. The calculations for 18 groups (the computation of 18 tasks) are performed on three computational nodes on each of which the second technique of the parallelization is used.

The computational time required for all 18 tasks on three computational nodes is 4 min 19 s, and on one computational node, this time is 12 min 8 s.

The result obtained shows that the simultaneous use of the second and third technique of parallelization requires the minimal cost for data transfer, and the speedup achieved is almost linear.

4. Conclusion

The study of the computational effectiveness of parallel solution of 3D airborne EM survey problems shows the following. The combination of techniques of parallelization in a distributed computational system, taking into account the specifics of the numerical schemes implementation for forward and inverse problems solution, makes it possible to achieve the speedup close to linear.

Acknoledgement

This work is supported by the Russian Science Foundation under grant 20-61-47072.

References

[1] McMillan M S, Schwarzbach C, Haber E and Oldenburg D W 2015 *Geophysics* **80** K25–36
[2] Kaminski V, Prikhodko A and Oldenburg D 2011 a *SEG Tech. Progr. Expand. Abstr.* **30** 1262–1266
[3] Fullagar P K, Pears G A, Reid J E and Schaa R 2015 *Explor. Geophys.* **46** 112–117
[4] Haber E and Schwarzbach C 2014 *Inverse Probl.* **30** 28
[5] Liu Y and Yin C 2015 *Geophysics* **81** E401–408
[6] Yang D, Oldenburg D W and Haber E 2014 *Geophys. J. Int.* **196** 1492–1507
[7] Persova M G, Soloveichik Y G, Koshkina Y I, Vagin D V and Trubacheva O S 2016 *Proceedings of First Conference on Geophysics for Mineral Exploration and Mining* 1–5
[8] Persova M G, Soloveichik Y G, Vagin D V, Domnikov P A, Kiselev D S, Koshkina Y I and Simon E I 2018 *Proceedings of Engineering and Mining Geophysics* 1–10
[9] Persova M G, Soloveichik Y G, Vagin D V., Koshkina Y I and Trubacheva O S 2018 *Proceedings of 14th International Scientific-Technical Conference on Actual Problems of Electronic Instrument Engineering* 226–231
[10] Persova M G, Soloveichik Y G, Trigubovich G M, Vagin D V. and Domnikov P A 2014 *Geophys. Prospect.* **62** 1193–1201
[11] Soloveichik Y G, Persova M G, Domnikov P A, Koshkina Y I and Vagin D V 2018 *Geophys. J. Int. Geophys. J. Int* **212** 2159–2193
[12] Schenk O and Ğartner K 2004 *Future Generation Computer Systems* **20** 475–487
[13] Persova M G, Soloveichik Y G, Trigubovich G M and Tokareva M G 2013 *Izv. Phys. Solid Earth* **49** 329–343
[14] Mogilatov V, Goldman M, Persova M, Soloveichik Y, Koshkina Y, Trubacheva O and Zlobinskiy A 2016 *J. Appl. Geophys.* **135** 17–31

IOP Conf. Series: Materials Science and Engineering 1019 (2021) 012021 doi:10.1088/1757-899X/1019/1/012021

Analysis of the opportunities for improving energy efficiency in public buildings

A Terziev[1], V Stoyak[2], V Ushakov[3], D Brazhanova[2], S Suleimanov[3]
[1]Department of Power Engineering and Power Machines, Technical University of Sofia, Sofia, Bulgaria
[2]Department of Scientific and innovation activities, Almaty University of Power Engineering and Telecommunication, Almaty, Kazakhstan
[3]School of Power Engineering, National Research Tomsk Polytechnic University, Tomsk, Russian Federation
E-mail: dana_b.k@bk.ru

Abstract. Reducing energy consumption and maintaining an appropriate indoor air quality in buildings is a global trend. This is achieved in several ways – improving the building envelope thermal characteristics; fuel base replacement; introduction of smart energy consumption tracking systems. The aim of this paper is an analysis of the impact of energy efficiency measures introduced in buildings that provide health services. Measures have been taken to improve the energy performance of the envelope, as well as to modernize the system for the production and distribution of thermal energy. One complete year after energy savings measures implemented the total energy savings of 37.9% have been achieved. Further economic assessment has been made regarding the profitability of the applied measures.

1. Introduction

Issues related to increasing energy efficiency in buildings have been particularly relevant over the past three decades. The need arises both from the rise in energy carriers' prices and the provision of an adequate indoor climate. Globally, around 30% of energy consumption in buildings is attributable to municipal and public buildings [1]. The paper points out that the high share of energy consumption is due to the poor thermal performance of the enclosures, the inefficient operation of HVAC systems (Heating, Ventilation, & Air Conditioning) and the lack of or poor management of energy consumption.

Some papers [2–5] examine the effect from some of the most commonly applied energy-efficient measures in municipal and public buildings. For this purpose, energy audits were carried out using different mathematical approaches to model energy consumption. In some of the cases near-infrared-reflective architectural coatings was suggested to reduce heat losses. In [6], the effect of energy saving from replacing the lighting system has been examined in compliance with the regulatory requirements.

The choice of energy-efficient measures is most often in line with the calculated energy savings, respectively the financial parameters of the measure. In some cases, when the cost of the energy used is low and the investment is significant, the measure is unprofitable [7]. In this case, different co-financing mechanisms (whether national or international) are available to promote energy efficiency improvement in buildings.

In a number of cases, the regulatory framework and a grant scheme are the driving force behind the energy targets set in a given country [8].

2. Problem identification

The subject of the present survey are four different buildings designed to provide health cares (public buildings), which are situated on a single site. Summary of the four buildings under analysis is presented below.

2.1. Fthysiatric ward (Building 1)

The first building is the Fthysiatric ward. The building is public state-owned and was built in 1889. The surveyed part of the hospital is a brick building with a monolithic structure. The facades are lime-cement plaster in a relatively good condition. The exterior walls are masonry of solid bricks with different thickness without thermal insulation. The building is equipped with contemporary aluminum frames with double glazed windows. The roofs are tiled without thermal insulation, with the same height throughout the attic spaces. The floors of the building are made of concrete slab without thermal insulation with laminate and terracotta coating. All parts of the building are heated. The working regime of the hospital is 24 hours and 7 days a week. Figure 1 shows the design and orientation of the building. Thermal characteristics of the building are presented in table 1.

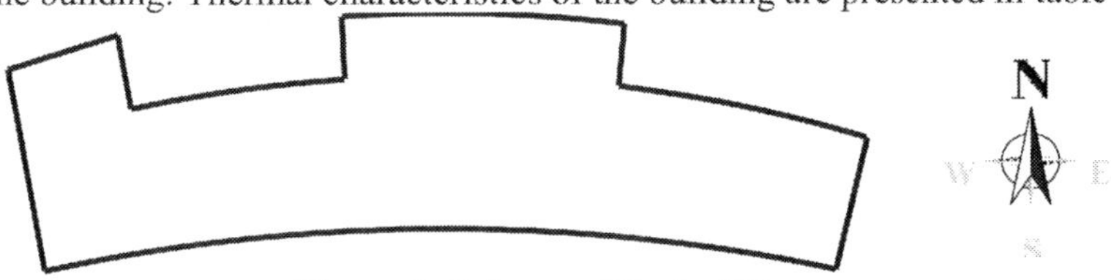

Figure 1. Scheme of the building.

Table 1. Thermal characteristics of the building envelop 1.

Type of envelope	Thermal resistance (R), m^2K/W	
External wall		
№ 1. Inner plaster $\delta_l = 0.03$ m; $\lambda_l = 0.70$ W/(mK)	**0.043**	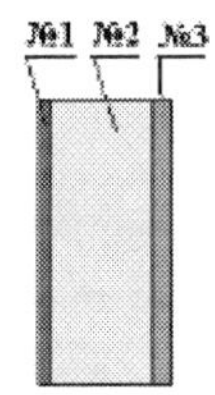
№ 2. Brick wall $\delta_l = 0.38$m; $\lambda_l = 0.79$ W/(mK)	**0.437**	
№ 3. External plaster $\delta_l = 0.04$ m; $\lambda_l = 0.87$ W/(mK)	**0.046**	
Overall heat transfer coefficient U=1.436 W/(m^2K)		
Floor		
№ 1. Terracotta $\delta_l = 0.015$ m; $\lambda_l = 1.28$ (W/mK)	**0.012**	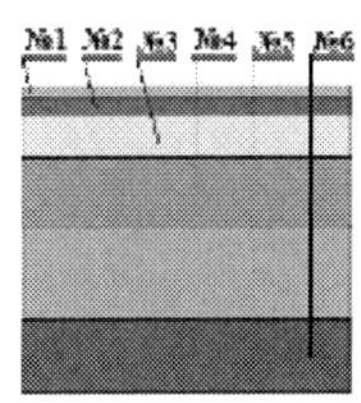
№ 2. Dressing cement plaster $\delta_l = 0.02$ m; $\lambda_l = 0.93$ W/(mK)	**0.021**	
№ 3. Reinforced concrete slab $\delta_l = 0.10$ m; $\lambda_l = 1.63$ (W/mK)	**0.061**	
№ 4. Timber structure $\delta_l = 0.20$ m; $\lambda_l = 0.61$ (W/mK)	**0.33**	
№ 5. Air cavity $\delta_l = 0.3$ m; $\lambda_l = 0.19$ (W/mK)	**1.58**	
№ 6. Pudding $\delta_l = 0.20$ m; $\lambda_l = 1.50$ (W/mK)	**1.58**	

Type of envelope	Thermal resistance (R), m^2K/W
External wall	
Overall heat transfer coefficient U=0.264 $W/(m^2K)$	
Roof	
№ 1. Roof tiles $\delta_1 = 0.03$ m; $\lambda_1 = 0.99$ (W/mK)	**0.03**
№ 2. Air cavity $\delta_1 = 1.55$ m; $\lambda_1 = 1.99$ (W/mK)	**0.778**
№ 3. Lacing	-
№ 4. Water-repellent plaster $\delta_1 = 0.04$ m; $\lambda_1 = 0.87$ (W/mK)	**0.046**
№ 5. Solid bricks $\delta_1 = 0.26$ m; $\lambda_1 = 0.79$ (W/mK)	**0.329**
№ 6. Timber structure $\delta_1 = 0.18$ m; $\lambda_1 = 0.61$ (W/mK)	**0.295**
№ 7. Inner plaster $\delta_1 = 0.03$ m; $\lambda_1 = 0.70$ (W/mK)	**0.043**
№ 8. Suspended ceilings $\delta_1 = 0.0125$ m; $\lambda_1 = 0.052$ (W/mK)	**0.240**
Overall heat transfer coefficient U=0.695 $W/(m^2K)$	

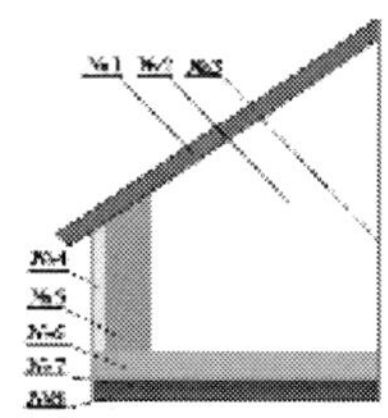

Heat supply is provided by local heating installations (boiler and substation) located in the underground part of the building. The boiler room consists of one boiler with power of 116 kW that is installed 2004 – 2005.

The heat supply of the building is not regular. In the winter, the building is heated up to the end of the heating season, with the boiler running in mode - switching on and off at the temperature of the heat carrier at the boiler inlet (return water). During holidays and bank holidays the building is also heated. The circulation of the heat supplier is forced through a single circulation pump. Collectors (collector and distributor) and pipelines in the boiler room and substation premises are insulated, their fittings are maintained and work well. The heating installation is two-pipe with forced circulation, with parameters of the heat carrier 90/70° C. The radiators are cast iron, installed in all heated rooms.

2.2. Administration of hospital (Building 2)

The building is public state-owned, built in 1889 and used as Administration. The surveyed part of the hospital is a brick building with a monolithic massive structure. The facades are lime-cement plaster, in a relatively good condition. The outer walls are masonry of solid bricks of different thickness without thermal insulation. The building is equipped with contemporary aluminium frames with double glazed windows. The roofs are tiled without thermal insulation, with the same height throughout the attic spaces. The floors of the building are made of concrete slab without thermal insulation with wooden flooring (laminate) and terracotta. All parts of the building are heated. An outline of the building is presented in Figure 2.

The thermal performance characteristics of the building are presented in table 2.

2.3. Pneumology 1 (Building 3)

The structure of the building is similar to that of Building 2. A general view is presented in Figure 3.

The thermal performance characteristics of the building are presented in table 3.

Figure 2. View of the hospital administration.

Figure 3. View of pneumology 1 hospital building.

Table 2. Thermal characteristics of the building envelop 2.

Type of envelope	Heat transfer coefficient (U), W/(m^{2}K)
External wall 1	**1.436**
External wall 2	**1.823**
Floor 1	**0.24**
Floor 2	**0.291**
Floor 3	**0.318**
Floor 4	**0.401**
Roof	**0.692**

Table 3. Thermal characteristics of the building envelop 3.

Type of envelope	Heat transfer coefficient (U), W/(m^{2}K)
External wall 1	**1.436**
Floor 1	**0.241**
Roof	**0.694**

2.4. Pneumology 2 (Building 4)

The structure of the building is similar to that of Building 2. A general view is presented in Figure 4.

The thermal characteristics of building 4 are presented in table. 4.

The heat supply of all buildings is the same - the supply of heat is supplied trough a convection system from a boiler located in the basement of the building.

Figure 4. Scheme of the building.

Table 4. Thermal characteristics of the building envelop 4.

Type of envelope	Heat transfer coefficient (U), W/(m²K)
External wall 1	**1.436**
Floor 1	**0.249**
Floor 2	**0.305**
Roof	**0.682**

3. Analysis of the current energy consumption of the buildings

During the survey period, two different energy carriers were used — electricity and diesel fuel. Electrical energy is used to ensure the operation of the lightning system in the hospital rooms, the operation of medical equipment, and the auxiliary equipment to the heating system - pumps and fans. Diesel fuel is used solely to maintain the microclimate in the buildings and to produce hot water for household purposes.

Figure 4 presents the energy consumption in the buildings for the period 2009–2011. The difference in consumption is no more than 15% and depends mainly on the use of the available equipment as well as the external weather conditions.

Table 5 and Figure 5 show the energy consumption by years for the period, as well as the number of heating day degrees. From Figure 5 it is evident that the share of electricity consumption during the surveyed period is between 31% and 35% and the consumption of the thermal energy – 65–69%.

The relative annual energy consumption, as shown in Figure 6, is calculated based on this data. It is clear that the relative annual energy costs do not significantly change over the last two years and are less than the costs in 2009. This is due to the fact that in 2009, there has been a replacement of the window frames of the building, as a result of which the specific energy consumption has decreased.

Table 5. Annual energy consumption of the buildings.

Year	Electricity, kWh	Thermal energy, kWh	Total energy, kWh	Day degrees (DD)
2009	246 270	543 210	789 480	2 714
2010	234 239	437 638	671 877	2 815
2011	222 256	477 538	699 794	2 989

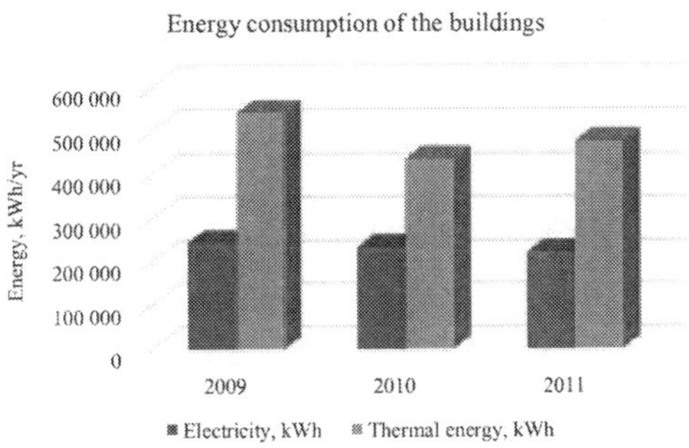

Figure 5. Yearly annual energy consumption of the buildings.

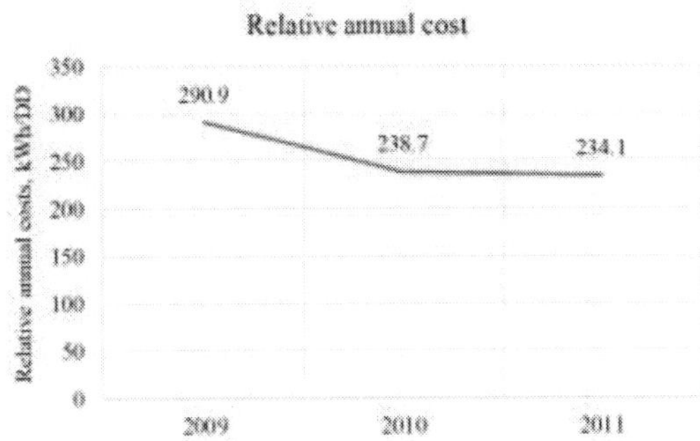

Figure 6. Relative annual costs, kWh/DD.

4. Modeling of energy consumption of the buildings

A survey of the energy consumption in the buildings was carried out. The results of the survey are presented in Figures 7–10. The normalization of the energy consumption was needed, in compliance with the normative requirements for maintaining the necessary temperature comfort. For all four buildings, the normalized energy consumption is increased, indicating that at the time of the survey the buildings were under-heated.

The survey shows that the energy consumption of the building is significantly higher than the standard for this type of buildings, therefore measures are recommended to increase the energy efficiency.

As a result of the audit, three energy saving measures have been introduced as reasonable for the buildings:

- Introduction of thermal insulation on the external walls of the buildings;
- Laying thermal insulation on roofs / attic spaces;
- Modernization of the heating system.

Given the architectural features of the building, it was found that laying thermal insulation is impossible on the external walls of the building, which is why its application is chosen to be on the inside of the external walls.

The modernization of the heating system consists of replacing the local boilers with natural gas. The suggested boilers are equipped with the corresponding control systems, characterized by a high efficiency.

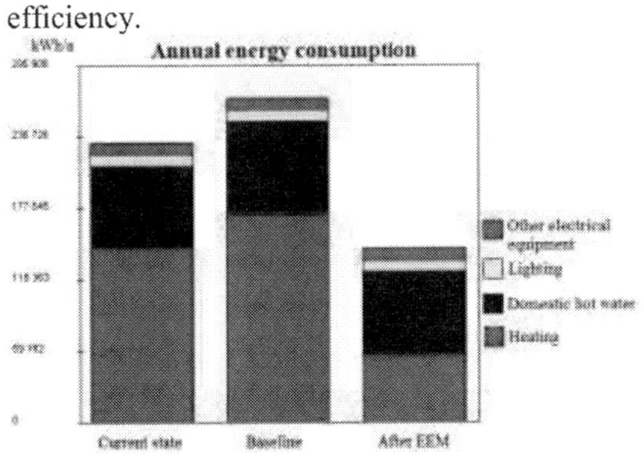

Figure 7. Outputs from the energy modeling of building 1.

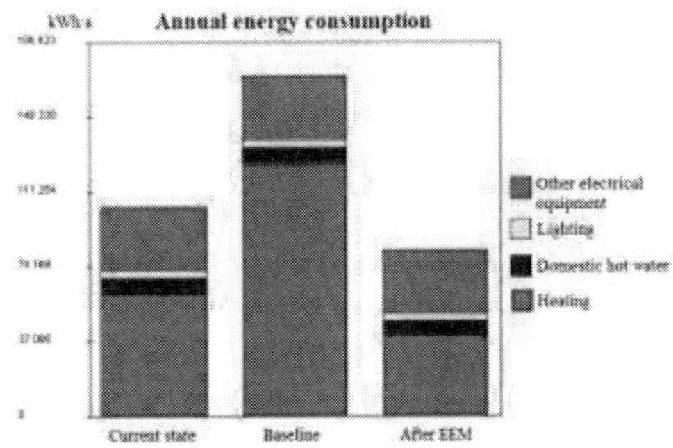

Figure 8. Outputs from the energy modeling of building 2.

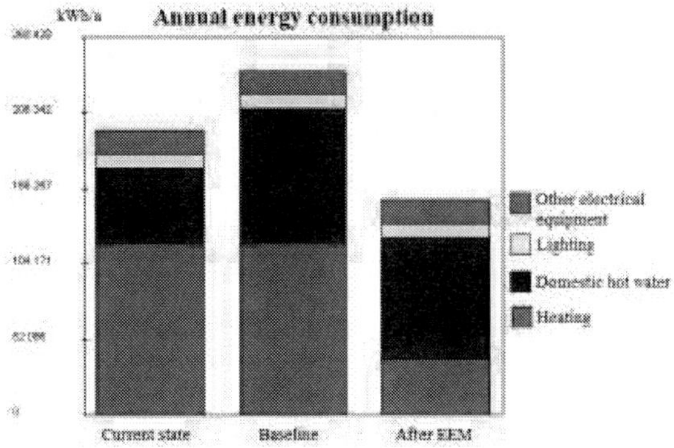

Figure 9. Outputs from the energy modeling of building 1.

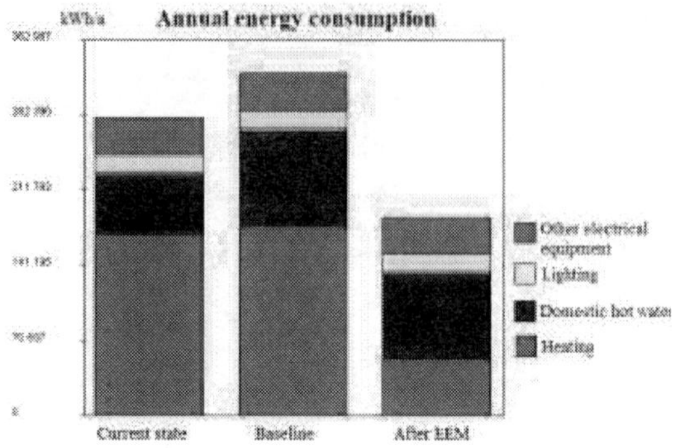

Figure 10. Outputs from the energy modeling of building 4.

5. Energy consumption after implementation of measures

The measures were put into operation in 2014, with three full years of since the implementation.

Energy savings from the implementation of the measures are calculated as the difference between the baseline energy consumption (or normalized) and consumption after the introduction of the measures. However, it should be noted that the baseline must be adjusted to the outdoor conditions with which it is compared. Thus the energy consumption of the buildings before the implementation of the measures in operation is according to table 6.

Table 6. Energy consumption of the buildings before energy efficiency measures implementation

Month	**External temperature**	**Heating days**	**Day degrees**	**Baseline, kWh**
January	2.4	31	546	87 162
February	2.1	28	501	80 069
March	5.4	31	453	72 305
April	16.6	16	54	8 691
October	13.8	0	0	0
November	5.3	30	441	70 452
December	1.5	31	574	91 619
Total		**167**	**2 568**	**410 297**

Table 7. Energy consumption of the buildings before energy efficiency measures implementation

Month	Baseline, kWh	Real consumption, kWh	Savings, kWh	CO2 emission reduction, t
January	87 162	52 434	34 728	12.7
February	80 069	46 404	33 665	12.0
March	72 305	47 902	24 403	9.6
April	8 691	7 453	1 237	0.8
October	0	0	0	0.0
November	70 452	42 747	27 705	10.2
December	91 619	57 961	33 658	12.8
Total	410 297	254 901	155 396	58.1

Table 7 presents information on energy consumption (natural gas) after the introduction of the measures, as well as the corresponding savings. It can be seen that the energy saving is 37.9% of the

IOP Conf. Series: Materials Science and Engineering 1019 (2021) 012021 doi:10.1088/1757-899X/1019/1/012021

normalized baseline, with savings of 155,396 kWh / year or cash equivalent of EUR 13 433. The carbon savings achieved amount to 58.06 t / yr. The investments costs for the implementation of the measures are EUR 114 137, which sets a payback period of 8.49 years.

6. Financial analysis

A financial analysis summary of the cost-effectiveness of the proposed energy-efficient measures is shown below. Table 8 provides information for the achieved savings, both in natural units and in currency. It can be seen that the savings from the introduction of measure 1 - about 43%, followed by the heating rehabilitation measure - 36% are the most significant ones. About 21% are the savings from increasing the roof's energy efficiency of the buildings.

Table 8. Net incomes from implementation of the measures.

Energy efficiency measure	Energy savings achieved, kWh	Total savings, EUR
Thermal insulation of walls	66 323	5 733
Thermal insulation of roof	33 814	2 923
Rehabilitation of heating system	55 259	4 777

Table 9 provides information on the financial parameters of the implemented measures: Internal rate of returns (IRR), Net present value (NPV), simple payback period (PB) are shown. It is clear that the highest value of IRR and NPV is achieved from the measure for thermal insulation of the external walls. The total simple payback period of the measures is 8.5 years and the IRR value is 11.3%. The real interest rate is 5%, which includes 0.5% inflation rate and 5% nominal interest rate.

Table 9. Main financial indicators

Energy efficiency measure	Investment, EUR	PB, yrs	IRR, %	NPV,EUR
Thermal insulation of walls	39 947	7.0	13	31 499
Thermal insulation of roof	28 534	9.8	8	7 893
Rehabilitation of heating system	45 654	9.6	6	3 930

7. Conclusion

The paper presents a feasibility study for the implementation of energy-efficient measures related to the improvement of energy efficiency in public buildings. The proposed measures are rehabilitation of the external walls, rehabilitation of roof spaces and modernization of the heating system. Energy modelling of the building was carried out, and based on the determined savings, the energy efficiency measures were prioritized. After one year of operation of the building, the actual energy consumption was also recorded. The comparison between baseline and actual consumption shows an energy saving of 37.9%. The internal rate of return for the selected measures is between 6-13%, with the net present value being a positive figure. This qualifies the measures as profitable.

References

[1] Jankovic A, Andjelkovic M and Minic S 2016 *Proceedings of 2016 4th International Symposium on Environmental Friendly Energies and Applications (EFEA)* 1–4

[2] Xu Y, Zhang L 2011 *Proceedings of International Conference on Multimedia Technology* 6058–6061

[3] Wu Y 2009 *Proceedings of International Conference on Information Management, Innovation*

Management and Industrial Engineering 619–622
[4] Yashuang L, Yisheng L 2015 *Proceedings of 2015 International Conference on Logistics, Informatics and Service Sciences (LISS)* 1–4
[5] Stimoniaris D *et al* 2014 *MedPower 2014* 1–6
[6] Coelho D, Valdez M and Oliveira C 2016 *Proceedings of 51st International Universities Power Engineering Conference (UPEC)* 1–6
[7] Kerimray A, Bektineyev R and Rojas-Solórzano L R 2016 *Proceedings of 4th IET Clean Energy and Technology Conference (CEAT 2016)* 1–6
[8] Talamon A, Almási I, Nagy A, Czinger E and Babina Z 2013 *Proceedings of 4th International Youth Conference on Energy (IYCE)* 1—5.

IOP Conf. Series: Materials Science and Engineering 1019 (2021) 012022 doi:10.1088/1757-899X/1019/1/012022

AREIOM: adaptive research multicopter platform

R Harradi, B Battseren, A Heller and W Hardt
Professorship of Computer Engineering, Faculty of Computer Science, Chemnitz University of Technology, Chemnitz, Germany
E-mail: reda.harradi@informatik.tu-chemnitz.de

Abstract. The applications of unmanned aerial systems have dramatically increased in the late days, in particular within dull, dirty and dangerous missions, where it has been clearly shown that unmanned systems are better suited than manned aircrafts. The reduction of human involvements leads to an increasing demand for autonomous and operationally adaptive airborne systems. In addition, unmanned aerial systems share similar limitations with most computer embedded systems, e.g., limited space and power resources, and increasing computation requirements and complexity of the applications running onboard. To fulfill these requirements and achieve the fundamental features of autonomy and adaptability, the AREIOM—Adaptive Research Multicopter Platform—along with its layered modular architecture and the specific details of its underlying hardware and software components are presented.

1. Introduction

Unmanned Aerial Systems (UAS) have been proving their utility over the few past years. Not only are they better suited for dull, dirty or dangerous (3D) missions, but also are considered as to be a suitable platform for various tasks such as disaster assistance, telecommunication relay, surface reconnaissance, agricultural surveying, power line monitoring or inspection tasks. A simple example would be the usefulness of the UAS systems during the earthquake and the following Tsunami that damaged the Fukushima Daiichi nuclear facility in Japan on March 11th, 2011, and which urged the local authorities to advise the citizens to avoid the area and to have recourse to UAS systems to inspect the states of the nuclear power plant [1]. In fact, reducing the involvement of humans leads to a rush of autonomously operating airborne systems. However, it is necessary to distinguish between an automatic and an autonomous system [2, 3]. An automatic system is a system that is pre-programmed to execute a specific task without the need of an intervention from the controlling operator but cannot promptly react to the changes of the surrounding environment from its own, whereas an autonomous system would imply that the system is adaptive and is able to derive and take decisions by itself when confronted with a variety of choices that are required for performing the specific task it has been developed for. A considerable effort has been made to define the automatic and the autonomous levels of a system [4], which are based on the six autonomous vehicle navigation levels stated by the National Highway Traffic Safety Administration of the U.S. Department of Transportation.

Due to the technological improvements such as the miniaturization of the necessary sensory devices, the interest of carrying additional computational devices for data processing and storage or using new wireless-communication standards; such as the 5G network standard, is increasing. This enhances the functionalities and opens new horizons for unmanned aerial systems, which therefore brings the need for improving the various aspects pertaining to adaptability, autonomy, efficiency, reliability and safety [5]. In fact, the Adaptive Research Multicopter Platform (AREIOM) which is

IOP Conf. Series: Materials Science and Engineering 1019 (2021) 012022 doi:10.1088/1757-899X/1019/1/012022

based on hardware and software components for unmanned aerial systems, in particular for multicopters, is presented to serve as a base for the research on adaptive copter missions for civilian visual inspections. As a first use case, the inspection of high voltage power lines and the detection and recognition of faulty insulators using electro-optical sensors is described. It is based on research results on autonomous and adaptive flying copter missions, where the flying control flow of the copter has been fully automated. This allows the copter to autonomously take off, detect and recognize the insulators of interest, as well as to take snapshots of the insulators, and to finally land back to its initial position from where it started, without any needed intervention from the pilot.

2. Requirements

Generally, unmanned aerial systems share the same limitations with most computer embedded systems: limited space, limited power resources, increasing computation requirements and complexity of the applications [6]. Thus, several requirements need to be fulfilled in order to achieve the fundamental features of autonomy or adaptability. The most essential part is the onboard real-time decision making that helps changing the flight plan according to the current situation. Therefore, the acquisition of data by multiple sensors, the onboard real-time data fusion and exploitation, the automatic detection of relevant features, the definition of significant facts and dependent rules to trigger the corresponding actions, and the closed loop system action notification as well as the notification of the system's status to the user figure among the aspects that need to be addressed.

In addition, it is necessary to consider the health monitoring of the system, in combination with its safety features, in order to ensure a maximized reliability and safety in critical situations. Several counter-actions could be undertaken depending on each different use-case scenario. For example, using distance measurement sensors, such as ultra-sonic sensors, lasers or radars, could ensure not to go closer than the minimal safety distance, in order to avoid the electro-magnetic interferences around the power lines and power poles. Another example could be the monitoring of the battery voltage and the health statuses or the trespassing of predefined no-fly zones, e.g., geo fences around the power poles, which could help ensuring a prompt reaction to critical situations by using appropriate counter-measures, such as an emergency landing or a "return-to-home" action. In this sense, the AREIOM platform introduces a layered modular architecture, as depicted in figure 1.

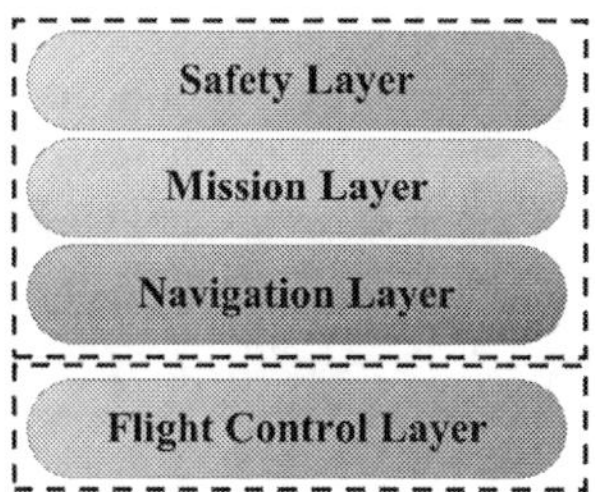

Figure 1. The abstraction levels of the AREIOM platform's layered architecture.

The Flight Control Layer represents the basic layer and consists of the flight controller that is based on an unaltered open source firmware. Several sensors like gyroscope, accelerometer, GPS, or barometer are providing various information about the surrounding environment to the flight controller for executing a predefined flight plan. Within the AREIOM platform, the flight controller is used as it comes out of the box, with no changes or modifications being conducted. Thus, any commercial off-the-shelf autopilot product that supports the Micro Air Vehicle Link (MAVLink) communication protocol is possible, as it is discussed in chapter 3. Nowadays, no commercial system that supports the actual mission exists on the market. Hence, a layer for navigation, mission and safety are specified within the AREIOM platform and are located on an additional computational device—the companion board. The communication between the flight controller and the companion board is ensured through a defined interface and a layered modular approach that enables the exchange of data and information between the different inter-process communication protocols. Moreover, the usage of such a companion board provides the possibility of using multicore and multithreading approaches, in order to further enhance the performance speeds.

Moreover, the Navigation Layer is in charge of handling and processing all the information parts that are needed to reach the required safety distance to the object of interest, and also to track the target of interest during the mission's flight. Furthermore, the Mission Layer includes all the components that are necessary for the on-board decision-making process according to the predefined mission's objectives. In fact, the application or the task that the UAS needs to perform is meant by the term "mission", e.g., the mission of inspecting the power pole insulators to detect defects like burn marks. Finally, the Safety Layer ensures that a set of predetermined behavioral actions that the UAS needs to undertake in critical and emergency situations. Not to mention that the safety of the human factor involved within the mission has to always be taken into account, where the actions undertaken by the UAS should be ensured never to put human lives in danger.

3. Architecture

The modular architecture of the AREIOM platform introduces an adaptive approach for designing a multi-component inspection platform that satisfies the needs and requirements of the different safety and criticality aspects, while offering a high performance throughput and fully autonomous flight and mission controls.

3.1. Hardware Architecture

The proposed architecture for the AREIOM platform is implemented based on six main hardware components: electro-optical sensors whose orientation is controlled by a gimbal, a companion board, a flight controller along with a remote controller, a ground control station and an access point, as it is designated in figure 2. All of the software components implemented within the AREIOM platform, going from the multiple sensors data acquisition to the onboard real-time data fusion and exploitation, as well as the automatic detection and extraction of relevant features can rely on this hardware architecture, while satisfying the aforementioned requirements for each architectural layer.

3.1.1. Navigation and Inspection Cameras. In order to satisfy the requirements of the system, two high definition cameras are installed on-board. Both are high resolution industrial cameras. While the first camera is meant to serve the navigation purpose, the second camera is mounted to help perform the inspection process. The navigation camera is a FLIR Blackfly S 3.2 Megapixel industrial camera, equipped with a 4-13 mm focal length wide angle lens. It provides a video stream with a resolution of 1024×768 pixels to the image processing algorithm which eventually processes the data and provides a real-time decision input for the rest of the processing chain. Moreover, the second; i.e., inspection camera, is required due to the need of sharp and highly detailed images for the inspection mission. In the case of the presented platform, a FLIR Blackfly S 12 Megapixel industrial camera with a 25 mm focal length prime lens provides high resolution sensor data for running detailed inspection on the object of interest. Both cameras use the USB 3.0 interface, and are able to achieve high throughput rates.

3.1.2. Gimbal. In order to acquire clear and sharp imagery data, the system's vibrations have to be accounted for and have to be canceled as much as possible at the level of the attached cameras. Moreover, the inspection object needs to be constantly tracked and focused at, during the complete inspection process of the flight. For these intentions, the HD Air Studio Infinity MR-S 3-axis gimbal was chosen, and delivers both camera stabilization and object tracking capabilities at the same time. The gimbal uses three brushless motors to level and adjust the position or viewing angle of the cameras. It is equipped with an AlexMos controller, that is controlled by the decision making algorithm for the tracking and the following of the inspection object, and an IMU sensor that is used to measure the instant alignment of the gimbal and provide the feedback loop input to a closed-loop control, in order to achieve the desired angular movements in all three axes. In fact, the gimbal enables the system to take stabilized imagery sets of the inspection objects and ensures the datasets are free of any unwanted blurriness or fuzziness artefacts.

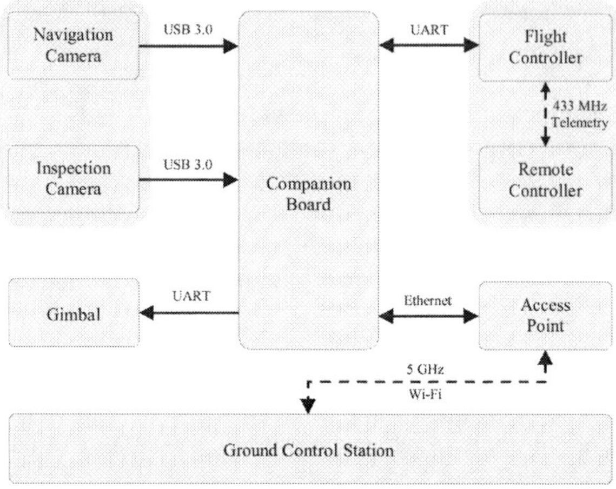

Figure 2. The hardware architecture of the AREIOM platform.

3.1.3. Flight Controller and Remote Controller. The flight controller is a dedicated embedded system that consists of a microprocessor and a set of sensors, and is designed for controlling the UAS' flight. In the proposed platform, the PixHawk 1 flight controller that is based on the second Flight Management Unit (FMUv2) was preferred, due to its seamless compatibility with ArduPilot's open source flight control stack. As mentioned previously, the flight control stack is used without any modification. All connections and configuration schemes are following the flight controller's own specifications. Besides that, the flight controller is connected to the companion board over an UART interface enabling the sending and receiving of MAVLink messages and commands to control the behavior and attitude of the multicopter.

Note here that the remote controller is not in any way involved in the autonomous mission's control chain. However, since no system on earth is perfect, it has been deliberately decided to keep the possibility to the pilot of taking over the system's control and securely taking any appropriate safety measures, when needed in particular unforeseen emergency cases.

3.1.4. Companion Board. The companion board is an on-board embedded computer, which runs the digital signal processing algorithms based on the sensors data and makes the real-time decisions to control and regulate the flight controller's actions, in order to run the missions in a safe and faultless manner. To execute this complex task, a number of software components have been implemented. Due to the considerable computational power needed by these software components and the narrow power, space and weight restrictions, the ODROID-XU4 from Hardkernel has been carefully selected as a companion board to the UAS. It is has a 32-bit octacore ARM processor powered with a Quad Cortex-A15 core clocked at 2 GHz and a Cortex-A7 core clocked at 1.3 GHz, and showcases a small (83×58×20 mm) and light weight (60 g) factor. The single board is equipped with both an USB 3.0 host and a Gigabit Ethernet interface, where the cameras are connected through the USB ports, the flight control and the gimbal controller through the serial ports, and the access point is connected over the Ethernet port.

3.1.5. Ground Control Station. The pilot or the team of operators working on the ground should have access to a real-time monitoring and controlling station. This station has two main functions—monitoring the flight behavior of the UAS, as well as the overall progress of the mission's process. For monitoring the status of the UAS while in operation, a telemetry hardware device, i.e., a 433 MHz radio telemetry (TLM), and the Mission Planner software (MSP) are needed. Moreover, the monitoring and supervision of the mission's operation can be achieved through the usage of the proposed 5 GHz wireless network in order to enable a bi-directional transmission of the mission's status and other sensor data along with the video stream between the UAS and the Ground Control Station (GCS).

3.1.6. Access Point. All the gathered data are saved and processed on-board during the flight's mission. Several Inter-Process Communication (IPC) techniques are used for sharing the video and bit stream data between the software components within the companion board. On the other hand, as noted before, the different raw and processed video streams, sensor data and processed information about the states of the UAS are transmitted to the GCS for the purpose of monitoring the mission. The current setup uses the Bullet BM5HP access point from the Ubiquity Networks company, which provides the necessary wireless communication between the UAS and the GCS. It is offering a maximum transmitted output power of 20 dBm, and is able to achieve a satisfying outdoor range performance of up to 50 km with a 135 Mbit/s point-to-point data rate at a 5 GHz frequency. While the companion board connects to the access point through its Ethernet port, the GCS is connected wirelessly.

3.2. Software Architecture

In order to execute the several mission tasks that the UAS has to fulfill, a specific software architecture has been developed, allowing not only an efficient communication between each component, but also ensuring a real-time execution feasibility of the required inspection and navigation tasks on the companion hardware. As it has previously been mentioned, the communication between the components has been achieved through the appropriate usage of multiple IPC mechanisms, such as Named Pipes, Shared Memory and Unix Domain Sockets.

Table 1 summarizes the categorization of the architecture's building blocks in accordance with the defined abstraction layers and defines the criticality assignment of each of the different architectural components.

Table 1. The layering of the AREIOM architecture's building blocks.

	CTH	EXS	FPD	MAL	CGC	IMR	FLC
Safety	X	X					
Mission	X	X	X		X	X	
Navigation	X	X		X			
Flight Control							X

The first decisive building block within the proposed architecture is the Feature Point Detector (FPD). It accesses the direct feed captured by the on-board mounted cameras, and separates internally the navigation from the inspection visual processing objectives. The component applies both low level image processing filters and feature extraction and description mechanisms, on the one hand in order to localize the power lines using edge filtering, line detection and line clustering techniques, and therefore determine the resulting intersection point and deduce the heading direction of the UAS, and on the other hand in order to detect and analyze the insulators and assess the defects and damages, such as burn marks or missing insulator disks on the insulators strings.

As it is depicted in figure 3, all the components within the architecture exhibit a connection form to the Control Handler (CTH). The tasks of the CTH range from handling the different communications between the software components, to the evaluation and analysis of the data packets received from the FPD, as well as the generation of the knowledge base necessary for the Expert System to fire the appropriate rule, based on each appropriate scenario.

IOP Conf. Series: Materials Science and Engineering 1019 (2021) 012022 doi:10.1088/1757-899X/1019/1/012022

It is good to note here, that some of the connections to the CTH are on the one hand unidirectional, when the CTH sends some data to or receives some data from a component, and on the other hand bidirectional, when the CTH and the component constantly exchange data. Once the flight parameters have been extracted from the Flight Controller (FLC), the CTH analyzes the received packets from the FPD and arranges the corresponding Working Memory Elements that would trigger the appropriate software component to take the required action, based on the defined rules in the Expert System. The triggered component can then adjust for instance the orientation of the camera to allow the following of the objects of interest, start the recording and the transmitting of the inspection mission's stream and mission and sensors data to the Ground Control Monitor (GCM) and the Mission Planner (MSP) in the case of the Inspection Mission Recorder (IMR), regulate the climbing rate to smoothly achieve a certain elevation, or control the lateral movements of the UAS to reach a desired destination.

The whole decision making mechanism that handles the situational behavior of the UAS relies on the Expert System (EXS) component. It is based on the C Language Integrated Production System (CLIPS) shell and implements the RETE forward-chaining pattern matching algorithm [7], which consists of a rule-based system that efficiently represents the production rule sets in a directed acyclic graph; called the RETE network, and asserts a given set of patterns with their appropriate facts, in order to determine the possible matches that define the rules to be fired in each match-resolve-act cycle. In fact, the Expert System is directly linked to the Control Handler, in such a way that the CTH handles all the messages from all the other different components, and regularly updates and prepares the list of facts that need to be added to the Working Memory. Once this is done, the expert system's inference engine is used to activate the matching production rules whose left-hand side constraints have been satisfied, and store them within the Conflict Set. Different conflict resolving strategies could then be used in order to pick the right rule to be fired, based on several factors such as the priority, the recency or the complexity of the activated rules [8]. In fact, the EXS notifies the CTH about which rules have been satisfied, and suggests then the appropriate commands and messages that need to be sent by the MAL to the FLC, in order to achieve the necessary UAS' maneuvers.

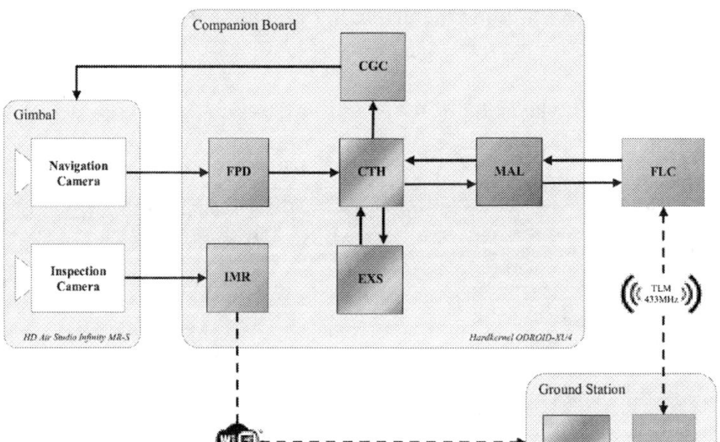

Figure 3. The mapping of the software architecture with the hardware architecture within the AREIOM platform.

Once the FPD has detected the insulators, the Camera Gimbal Controller (CGC) controls; as its name implies, the gimbal and orients the camera's field of view towards the detected insulators' string. The process is triggered by the CTH and is done repeatedly, in such a way that the insulators are always appearing within the center of the video frames captured by the inspection camera. Note here

IOP Conf. Series: Materials Science and Engineering 1019 (2021) 012022 doi:10.1088/1757-899X/1019/1/012022

that the insulators could be replaced by any other inspection object, depending on each use case application. In fact, the CGC receives the x- and y-coordinates of the object's relative origin within the frame, as well as its width and height. Based on these parameters, the CGC then sends signals to the gimbal controller to direct the gimbal in both its yaw and pitch axes and stabilize it, such as the middle of the object of interest is always centered and well-focused, while the UAS is pursuing its inspection mission.

Furthermore, the communication with the flight controller takes place within the MAVLink Abstraction Layer (MAL). The MAL implements the second version of the MAVLink protocol, and is compatible with most commercial off-the-shelf autopilots. It itself is based on a client-server architecture, where the clients get connected to the server-side and request a command to be executed. The server-side is then responsible of sending the commands to the FLC through the UART serial interface, and send a confirmation message back to the clients, when the commands have been acknowledged. In fact, the output rules that are activated by the EXS and sent to the CTH, trigger the appropriate MAL clients that request the execution of the actions needed for the fulfilment and pursuing of the inspection mission.

4. Use case

The initial use case of the AREIOM platform is the Automated Power Line Inspection (APOLI) project. The goal of this project is the automated vision-based inspection of High Voltage Transmission Lines (HVTL) and Power Poles. Due to the strong electro-magnetic radiation of the HVTL, conventional positioning and orientation sensors are not properly working. Therefore, the UAS maintains a safety distance and uses the vision-based approach for its navigation. As a matter of fact, the main focus of the inspection is the damage detection and visual assessment of the insulators, in particular the electrical short cuts visible by the so called burn marks. A combination and modification of state-of-the-art image processing approaches providing real-time and onboard data analysis for insulator detection and tracking to ensure the navigation of the UAS regarding the mission target is applied using the navigation camera [9]. A second high-resolution camera is used to take images of the insulators that would need to be detected and later assessed on the presence of burn marks through a post-processing step within a standalone program. However, hardware acceleration [10], has a sufficient potential to reliably detect the damages on-board and in real-time, and is considered as a future work. The main software component within the Mission Layer is the online decision making part that is based on the predefined expert system rules and facts. Furthermore, the communication between the companion board and the flight controller is based on the MAVLink protocol, where an abstraction layer has been developed for this purpose.

An UAS-based fully automated inspection system significantly improves the reliability and the quality of the power line inspection process, and decreases the risks which are caused by the human skill limitations and faults. As a result, no experienced pilots would be required anymore, and the electrical experts would be able to do the inspection tasks self-dependently and more conveniently by themselves, thanks to the reduced amount of workload and fatigue [11]. Beside this presented use case, the AREIOM platform architecture can be also applied to vision-based inspection of wind turbines or photovoltaic systems, aerial photogrammetry, e.g., 2D/3D buildings and objects reconstruction, areas documentation, exploration and surveillance of large-size areas or regions with hard accessibility, disaster and emergency relief, e.g., flooding, forest fire, and transportation of lightweight goods and payloads, e.g., small parcels, medicines.

5. Conclusion

The applications for unmanned aerial systems are actually increasing from day to day and range from inspection tasks to transportation of lightweight goods, and the exploration and surveillance of areas like construction sites. However, until today, no commercial system that supports the actual needs of the mission exists. Thus, the focus of the presented research was to develop a generalized platform for unmanned aerial systems, demonstrated with the example of a multicopter, that can be used to fulfil different missions. As a result, a layered modular architecture is achieved with specific software components. Moreover, the key features for the underlying hardware were defined. The mission relevant components are mainly located within the companion board that offers the additional needed computational power, while the flight controller stack remained unmodified. The developed platform was firstly applied to the use-case of vision-based inspection of insulators within HVTL. The proposed layered modular architecture with the relating software components and hardware requirements has been used within the development of the UAS-based insulator inspection.

References

[1] Ackerman E 2011 Japan earthquake: global hawk UAV may be able to peek inside damaged reactors
[2] Stenger A, Fernando B and Heni M 2012 *61st Deutscher Luft- und Raumfahrtkongress (Berlin)* (Bonn: DGLR) pp 1–10
[3] McCarthy N 2009 *Autonomous systems: social, legal and ethical issues* (London: The Royal Academy of Engineering) pp 2–7
[4] Kang D and Cha Y J 2018 *Comput-Aided Civ. Inf.* **33** 885–902
[5] Boubeta-Puig J, Moguel E, Sánchez-Figueroa F, Hernández J and Preciado J C 2018 *IEEE Internet Comput.* **22** 6–15
[6] Pastor E, Lopez J and Royo P 2006 *IEEE-AESM* **22** 3–8
[7] Forgy C L 1982 *Artif. Intell.* **19** 17–37
[8] Halbig J, Windisch A, Kingsbury P, Oswald N and Hardt W 2012 *Proc. 22nd AMS Conf. (Stuttgart)* ed P Levi *et al* (Berlin/Heidelberg: Springer) pp 119–28
[9] Tudevdagva U, Battseren B, Hardt W, Blokzyl S and Lippmann M 2017 *Proc. 12th IFOST Conf. (Ulsan)* (Ulsan: University of Ulsan)
[10] Blokzyl S, Vodel M and Hardt W 2014 *Proc. 9th SAS Symp. (Queenstown)* pp 59–64
[11] Tudevdagva U, Battseren B, Hardt W and Troshina V G 2018 *Proc. 14th APEIE Conf.* pp 579–83

IOP Conf. Series: Materials Science and Engineering 1019 (2021) 012023 doi:10.1088/1757-899X/1019/1/012023

Study of the laws of oxidation of biodiesel

S A Yudaev[1], E N Ivashkina[1], V N Sapunov[2] and M C Voronov[2]
[1]Chemical Engineering Department, Tomsk Polytechnic University, Tomsk, Russia
[2]Department of Technology of Basic Organic and Petrochemical Synthesis, Mendeleev University of Chemical Technology of Russia, Moscow, Russia
E-mail: udgin92@mail.ru

Abstract. The paper discusses in detail the first stage of the catalytic oxidation of fatty acid methyl esters with atmospheric oxygen. According to the data obtained as a result of processing the kinetic data, it was concluded that the reactivity of the components of biodiesel. It is established that polyunsaturated and monounsaturated compounds have different reactivity from each other. The kinetic pattern of consumption of unsaturated components of biodiesel was set up and kinetic constants were calculated, which are equal to: oxidation of C18/1 $k_1 = 0.106 \pm 0.008$ h^{-1}, and oxidation of C18/2 $k_2 = 0.269 \pm 0.005$ h^{-1}, respectively

1. Introduction

In the world there is a growing need for obtaining products from renewable raw materials. One of the main sources of raw materials are fats of plant origin [1–3]. From vegetable fats, a product is obtained that is in demand both on the energy market and on the market of raw materials for chemical technology of organic chemistry products (plasticizers, stabilizers, corrosion inhibitors, etc.). Such a product is methyl esters of fatty acids (FAME) or biodiesel (BD).

The main direction of use of FAME is the fuel direction. In the countries of the European Union and the United States, FAMEs are used as an environmentally friendly additive to diesel fuel. But there are a number of ways to obtain modified FAMEs [4–6]. Modification of FAME involves the introduction of functional groups in the fatty acid residue. Basically, the modification is carried out through several stages, the first of which is the formation of an epoxy cycle at the location of the double bond in the fatty acid residue. There is a method for producing epoxy compounds through radical aerobic oxidation. The radical mechanism of oxidation implies a static expenditure of components, but in our experiments it was established that the rate of consumption of components depends on the reactivity and stability of the radicals.

2. Experimental

To obtain FAME used commercial refined sunflower oil (GOST 1129-2013). The initial FAMEs were obtained by the standard method of alkaline transesterification of sunflower oil [7–10] in the laboratory of the Department of Chemical Technology of the Basic Organic and Petrochemical Synthesis of the Mendeleev University of Chemical Technology of Russia. The obtained FAME - crude was distilled at the molecular distillation unit KDL 1 BASIC VERSION 2007, manufactured by UIC GmbH at 170–175°C and pressure 0.2 bar. Analysis of distilled chromatograph chromatograms showed that they represent a mixture of esters of acids: palmitic (C16:0), stearic (C18:0), oleic (C18:1), linoleic (C18:2), linolenic (C18:3) and arachidic (C20:0), in the ratio 6.5: 6.0: 29.0: 53.2: 1.4:

1.9 mass%, respectively. The kinematic viscosity of the obtained product at 40°C was 4.6 mm^2 / s, determined in accordance with GOST 33-2000 (ISO 3104-94) "Oil products. Transparent and opaque liquids. Determination of kinematic viscosity and calculation of dynamic viscosity". Dry compressed air was used as an oxidizer.

A. Oxidation of methyl esters of fatty acids

The bubbling reactor was loaded with biodiesel of 50 ml, after which the supply of dry compressed air was switched on, its flow rate was 2÷5 ml/sec and the reaction system was heated. After reaching the reaction mass of the desired temperature (110°C), a catalyst was added to the solution (0.12 g MoO_2PrGl_2, the molybdenum concentration in the reaction mass is 0.12 mmol. Mo/l). The process was followed by the analysis of the samples taken from the reactor for the content of hydroperoxides, unreacted remaining and formed epoxides. The duration of the experiments was 15 hours. After the end of the experiment, a sample was taken for analysis – gas chromatography.

B. The method of preparation of catalyst (MoO_2PrGl_2)

In a three-neck flask with a volume of 50 ml, 0.1766 g of ammonium molybdenum acid was placed. Dissolved in 10 ml of distilled water at 50–70°C. After complete dissolution, the liquid was heated to 90°C. Then 3 g of purified propylene glycol was added dropwise to the reaction mass. Then the solution was heated to 105°C. With intensive stirring, within 10 minutes the solution acquired a blue-green color. After that, another 7 ml of propylene glycol was added. After 1.5 hours with intensive stirring and a temperature of 105°C, the mixture acquired a dark green color.

Then the water was driven away, 1.5 hours were intensively mixed at 120°C. the Reaction mixture acquired a green color. Under vacuum (6÷10 mm Hg art.) drove propylene glycol, after which the solution began to acquire a dark green color, and then – green-brown. The total mass of the catalyst was 8.06 g.

C. Methods of analysis of the reaction mixture

The analysis of raw materials and reaction products was carried out by various methods of instrumental analysis, firstly, on the gas-liquid chromatograph "Crystal 4000 Lux" with a flame ionization detector on a capillary column 1 m long, 3 mm in diameter, phase – SE-30 (5%)

Chromatographically conditions: flow rate of nitrogen is 25 ml/min, hydrogen flow of 50 ml/min, airflow of 450 ml/min, T detector – 250°C, T appetites – 275°C, T speakers – 170°C – 200°C, analysis time – 40 min the Volume of injected sample was 0.1 μl. The device and the method of analysis was controlled by the concentration of the components of raw materials and products in the course of the experiments.

3. Results and discussion

The feedstock in its composition has methyl esters of oleic and linoleic acids. These esters have an increased reactivity in oxidation reactions to hydroperoxides, since they have double bonds in their composition. Previously, the reactivity of these compounds in different ratios in the feedstock was not studied. In the presented systems, oxidation reactions are presented in Figure 1.

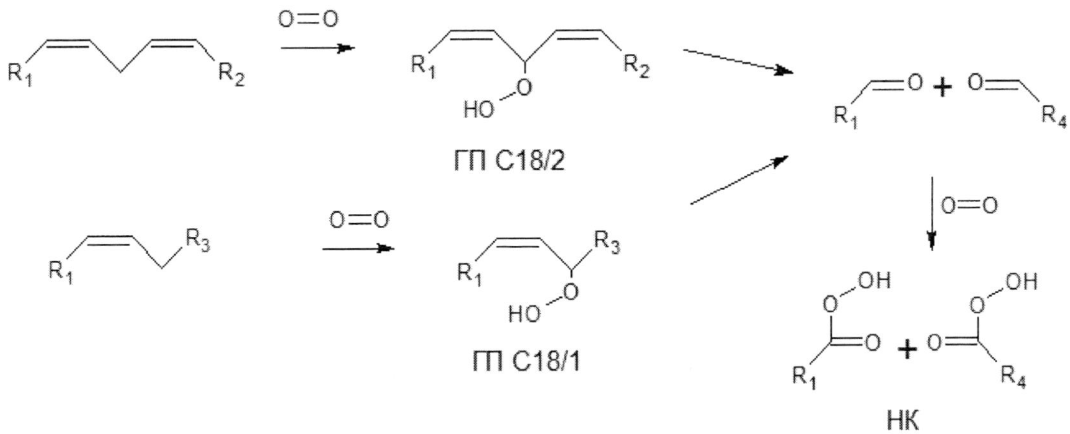

Figure 1. Oxidation scheme.

Aerobic oxidation of compounds, having in its composition of olefinic sites, such as methyl esters of oleic and linolevoy acids occurs by a radical chain mechanism. In the study of oxidation on different types of raw materials, it was found that the oxidation of methyl ester of linoleic acid (C18/2) occurs at a higher rate than the oxidation of methyl ester of oleic acid (C18/1) and does not depend on the initial ratio of acids in the raw material. To determine this pattern, a number of experiments were conducted on 6 different types of raw materials presented in Figure 2.

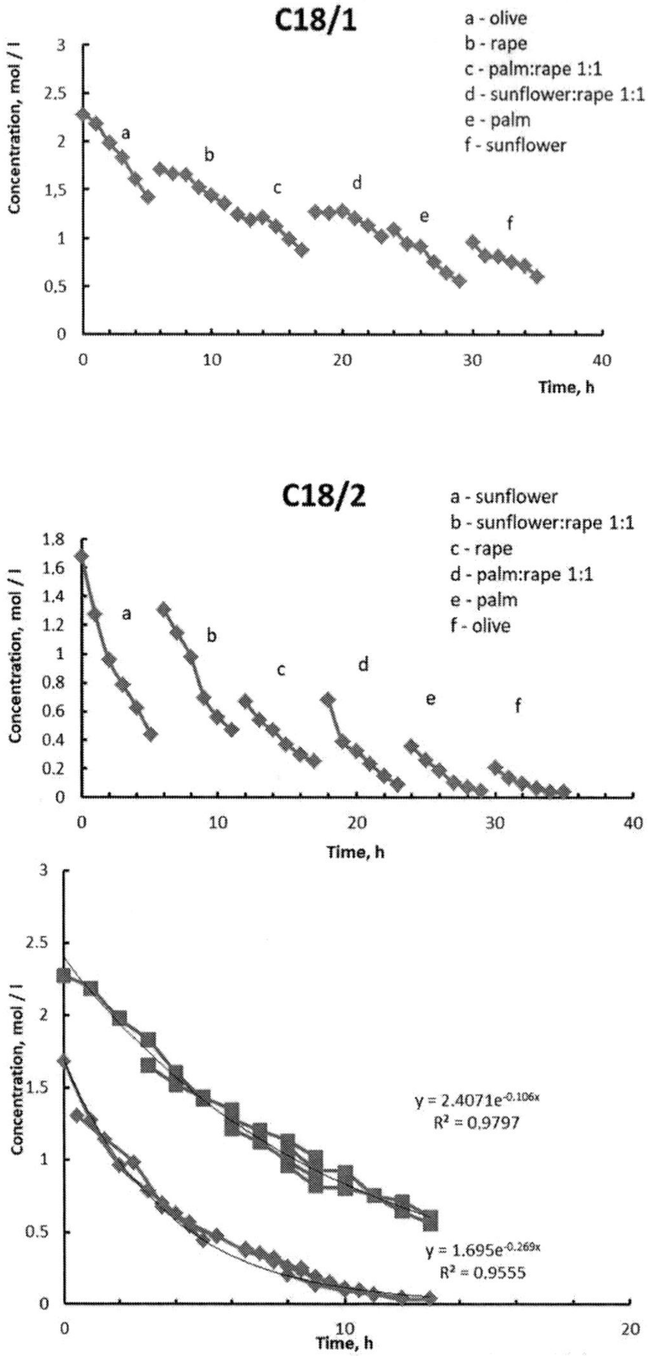

Figure 2. Kinetic curves of consumption of initial components of various raw materials

For processing the experimental data, a method of time kinetic curve shift was used. Found that the oxidation of the initial components of C18/1, C18/2 occurs according to independent kinetic laws. These laws are described by the first order kinetic reaction law. It was calculated that the oxidation rate constant C18/1 $k_1 = 0.106 \pm 0.008$ h^{-1}, and the oxidation rate constant

C18/2 $k_2 = 0.269 \pm 0.005\ h^{-1}$. the standard deviation of the calculated data from the experimental data is no more than 5%.

The increased consumption rate of C18/2 compared with C18/1 can be explained by the fact that the radical C18/2 formed when the hydrogen radical is detached has a more stable structure than the radical C18/1. The stability of the radical is determined by the number of carbon nuclei involved in stabilization, which is shown in Figure 3.

All the patterns shown demonstrate that the radical oxidation processes of various raw materials can be controlled. From a technological point of view, this makes the process controllable. Managed oxidation processes can form the basis of the industrial process for the oxidation of FAME.

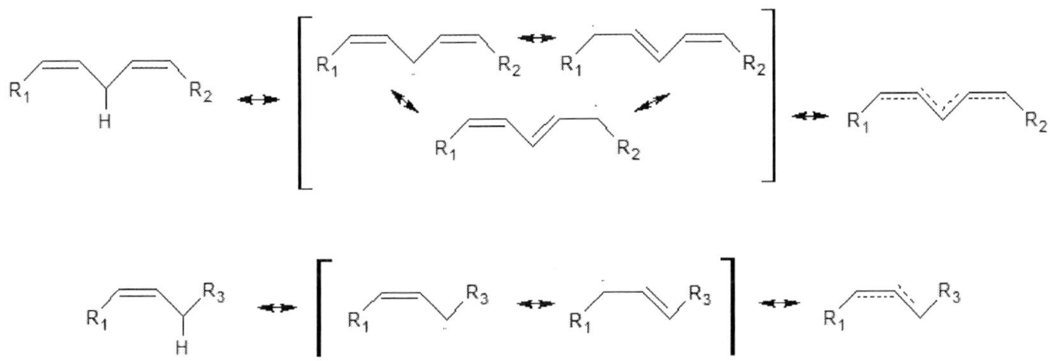

Figure 3. Stability of the radicals C18/2 and C18/1

4. Conclusions

As a result of the work done, it was established:

1. Component C18/1 has a less stable radical than component C18/2. In this connection, this radical has a higher reactivity, which is proved by the behavior of kinetic curves;
2. Oxidation rate constant C18/1 k_1=0.106±0.008 h^{-1}, and the oxidation rate constant C18/2 k_2=0.269±0.005 h^{-1}. The standard deviation of the calculated data from the experimental data is not more than 5%.

References

[1] Kennedy E *et al* 2012 *Fire Science Reviews* **1**(1) 1–36
[2] Sapunov V N *et al* 2017 *Russian Journal of Applied Chemistry* **90** 1598–1604
[3] Giua L *et al* 2013 *Food Chemistry* **140** 680–685
[4] Natarajan E 2012 *International Journal of Energy Science* **2** 152–155
[5] Morales A *et al* 2014 *Food Research International* **62** 1080–-1086
[6] Jakeria M *et al* 2014 *Renewable and Sustainable Energy Reviews* **30** 154–163
[7] Fattah I *et al* 2014 *Renewable and Sustainable Energy Reviews* **30** 356–370
[8] Jain S *et al* 2010 *Renewable and Sustainable Energy Reviews* **14** 667–678
[9] Tiozzo C *et al* 2013 *Eur. J. Lipid Sci. Technol.* **115** 86–93
[10] Jankovic M *et al* 2014 *Ind. Eng. Chem. Res.* **53** 9357–9364

IOP Conf. Series: Materials Science and Engineering 1019 (2021) 012024 doi:10.1088/1757-899X/1019/1/012024

Development of impulse method for transformer winding condition control technology

A Mytnikov, A Lavrinovich, V Strugov and M Saqib
Power Engineering School, National Research Tomsk Polytechnic University, Tomsk, Russia
E-mail:mytnikov66@mail.ru

Abstract. Mechanical shift of windings is serious problem of power transformers which belong to the very significant components of power engineering system. Mechanical displacements and following insulation failure due to short circuit and inrush currents can lead to an emergency situation. That's why the effective control technology of windings is an relevant task of a current power engineering. A new approach to winding state control technology is described in the paper. The described technology is based on nanosecond probing impulse duration with short (no more than 20 ns) rapidness of front impulse. The experimental results of sensitivity growth at decreasing front pulse duration are shown. The experimental equipment and research are described. It is established that duration of probing impulse has optimum value around 100ns. It is shown that it is possible to do a condition control under operating voltage. It is shown that both shorter probing impulse and front of probe pulse duration allows upgrading common effectiveness of the diagnostic technology.

1. Introduction

The Electrical strength of windings insulation is a necessary condition of a power transformer stable operation. It is impossible to realize without sensitive and reliable condition control technologies. The appropriate technology to control a mechanical state of transformer windings was proposed in 1966 [1]. The principle of the method consists in applying probing standard lighting impulse of 1.2/50 microsecond with the amplitude of 200 – 300 V to one of the windings. Other windings were short-circuited and the shunt which gave a response to a probing impulse was installed in them. The response represents the signal corresponding to transient, arising in windings, as a reaction to a probing impulse. At first, it is necessary to measure the normogram. This is a response from the winding of the working transformer. The process of sounding is repeated at the next test. Comparison of normograms and current sounding (defectograms) allows making a conclusion about winding state. The difference in the normogram and defectograms represents the problem in the winding. Further, this method was very popular in European, American countries [2] and was called the Low-Voltage Impulses method (LVI).

Furthermore, in Canada, the LVI method was modified and transformed into a method of measurement of amplitude-frequency characteristics. It was done to avoid at the measurement errors due to strict rules to measurement procedure. The principle of amplitude-frequency method consists in measuring the amplitude-frequency characteristics from one of the transformer windings when applying to the other winding a sinusoidal signal with the amplitude around 10 V of various frequencies. Then the amplitude-frequency characteristics are compared with the normograms, received on the serviceable transformer. This allows to avoid strong influence of circuit parameters to measurement results. Nowadays this method is called a Frequency-Response Analysis method or a

FRA Technology and it is widely used around the world [3–7]. The FRA interpretation that is based on the analysis of physical electrical parameters is not feasible since one of the most important parameters, the winding series capacitance, cannot be determined in transformer bulk [5]. Special generators for the diagnostic procedure by the FRA technology are being produced. Frequencies for diagnostics of high-voltage transformers are in range from 1 Hz to 2.5 MHz.

Both the LVI and the FRA methods not always provide exact diagnostics and sometimes have low sensitivity. We consider that it occurs due to rather narrow frequency range of probing signals. The standard storm impulse has the top frequency about 500 kHz, the FRA method – 2.5 MHz.

Our approach is to upgrade common efficiency of the LVI method by means of a probing pulse using the frequency range up to 50 MHz. It can be implemented by applying to one of the transformer windings a probe pulse with the amplitude of 300 V and a short front duration of and pulse duration of 20...500 ns. A short probe impulse is a way to increase sensitivity of the diagnostic procedure. But to determine optimal parameters of probing impulse the experiments both on model and real transformer should be done. Main goal of research is to determine probing impulse parameters which provide maximum efficiency of winding condition control.

2. Experimental research of condition control on transformer model

Experimental facility consists of a generator of nanosecond impulses, physical model of three-phase two-winding transformers and oscilloscope Tektronix TDS 2024. The model of three-phase transformer includes three low voltage coils inserted in high voltage ones. The turn number at the low voltage coil is 20, at the high voltage coil – 120. A nanosecond impulse of rectangular form with front duration 5-25 ns was supplied to enter of the high voltage coil. The response signal was measured on exit of the low voltage coil. Oscilloscope Tektronix TDS 2024 was used to control the parameters of a probe impulse as well as response signal. The electric circuit of the facility is shown in Figure 1.

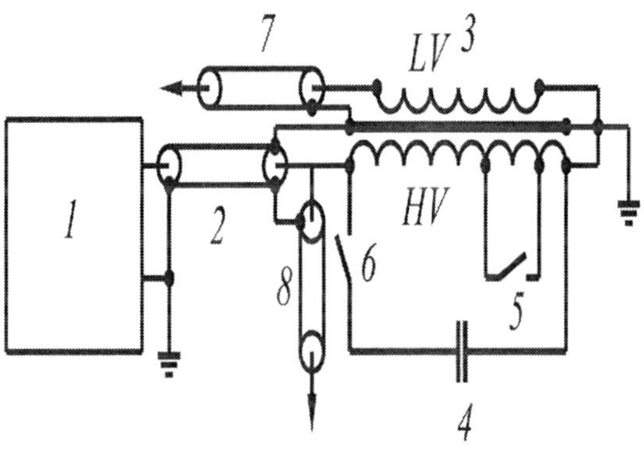

Figure 1. Electric scheme of the transformer model experiment: 1 – the generator of probing impulses, 2 – the connecting coaxial cable impedance of 75 Ohms, 3 – the diagnosed transformer (HV—high voltage, LV – low voltage), 4 – the condenser for change of the front of a probing impulse, 5 – short-circuit of turns (common turn number is 120), 6 – the switch for connection of the condenser 4; 7, 8 – probes for connection to an oscilloscope TDS 1024.

Some changes of windings geometry due to tilting, bending, axial or radial shifts, faults as a result of short-circuit currents, cause changes of capacities and inductances along winding. As result, a influence of probing pulse lead to changes in response essence as a winding reaction. Preliminary experiments proved the prospect and common opportunity and good efficiency of the nanosecond approach as a diagnostics method [8–10].

The impulse of rectangular form generated by the pulsed scheme is put to a high-voltage winding, the impulse-response is fixed on a low-voltage winding of the transformer. Windings had no loads (a case of completely switched-off transformer). Fail type turn to turn short circuit including three turns was simulated at the high voltage winding of phase A. The front of a probing impulse was changed and had meanings durations 25 and 400 ns respectively. Results of winding control for two cases with turn to turn fail and without one are described below. A comparison of an impulse of the normogram and an impulse of the response for both given cases is presented in Figure 2. It is established that duration of a probing impulse front of 25 nanoseconds the response at a fail winding differs strongly enough in comparison with the response at duration of a probing impulse of 400 nanoseconds.

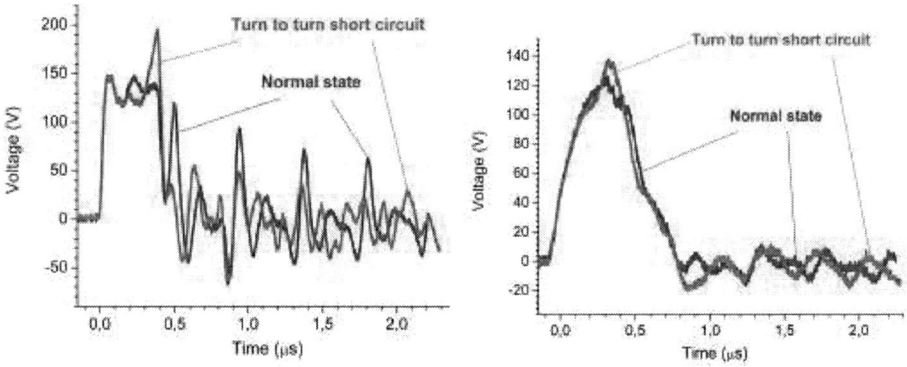

a) b)

Figure 2. Comparison of impulses of the normogram and responses for probing impulses with the different duration of the front and identical defects in a transformer winding: a) duration of a probing impulse front of 25 nanoseconds, b) duration of a probing impulse front of 400 nanoseconds.

3. Experimental research of condition control on real equipment

Next research run was carried out on real high voltage transformer. Main parameters of the transformer are as follows: total power is 160 000 VA, voltage of HV winding is 10 kV, voltage of LV winding is 400 V, three phases, winding connection type is star, insulation is transformer oil. The schemes of experiment of monitoring the condition of windings including the generator of probing impulses, oscilloscopes measuring parameters of probing impulse as well as response signals, are shown in Figure 3. Scheme of condition control is traditional for low voltage impulse method. Probing impulse is supplied to a high-voltage winding of the transformer A, the impulse-response is measured on a low-voltage winding of the transformer a. Oscilloscope 1 is for measure of probing impulse parameters. Oscilloscope 2 is for response signal registration. To research a dependence of winding condition control efficiency versus probing impulse duration, special pulsed generator was designed and engineered. Picture of this is shown on the Figure 4.

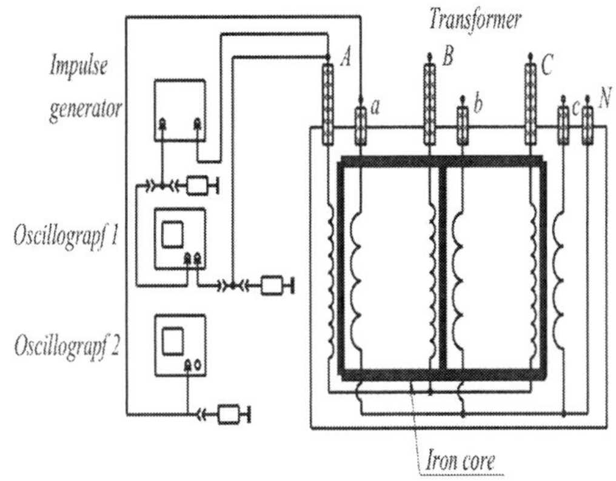

Figure 3. Scheme of experiment at the condition control by the nanosecond pulsed method.

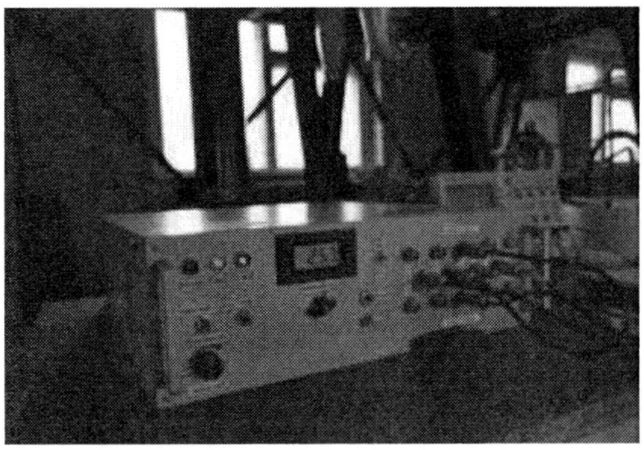

Figure 4. Common view of probing pulsed generator.

The generator produces probing impulses with different pulse duration, namely, 520, 260, 100, 60 and 20 ns. Two types of fail were simulated on the high voltage winding of real high voltage transformer. Defect number 1 is turn short circuit of three rounds. Defect number 2 is axial shift of turns at the bottom of winding (lodging conductors). Main goal of experiments on real high voltage transformer is research of influence of both pulse and pulse's front durations to effectiveness of winding condition control at two different fail types. Views of fails are shown on the Figures 5.

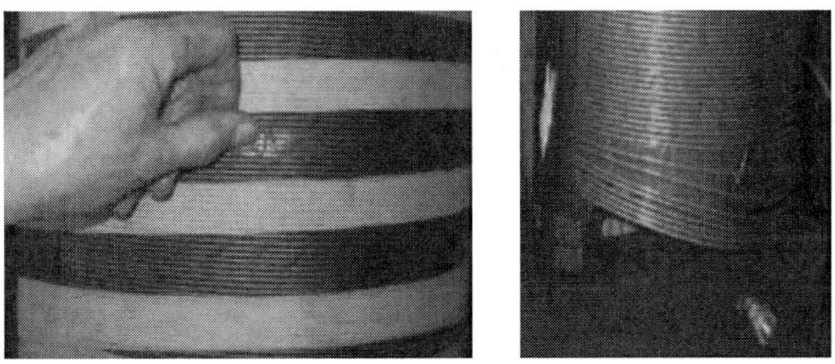

a) b)

Figure 5. View of faults: a) fail turn to turn short circuit, b) fail axial shift of turns type

To research a dependence of winding condition control efficiency versus probing impulse duration probing impulse with five different durations is applied to winding. Three cases were researched – normal winding condition, winding with defects number 1 and 2. To research a dependence of winding condition control efficiency versus probing impulse front rapidness, a capacitor is placed in parallel between probing pulse generator and HV winding of phase A. 4 different value of capacitance were used – 330, 2200, 4700 and 9100 pF. The more capacitance value the more front duration is. For probing pulse duration 520 ns, front has following values: 5 ns without capacitor, 66 ns for capacitance 330 pF, 512 ns for capacitance 9100 pF. In last case front duration is practically equal to pulse duration. The same measurements were carried out for all pulse durations – 520, 260, 100, 60 and 20 ns.

To determine an effectiveness of condition control process, special program for response comparison were design and used. The results of comparison of normograms (a blue curve) and response signals for both defects (a red curve) are given in Figures 6 and 7.

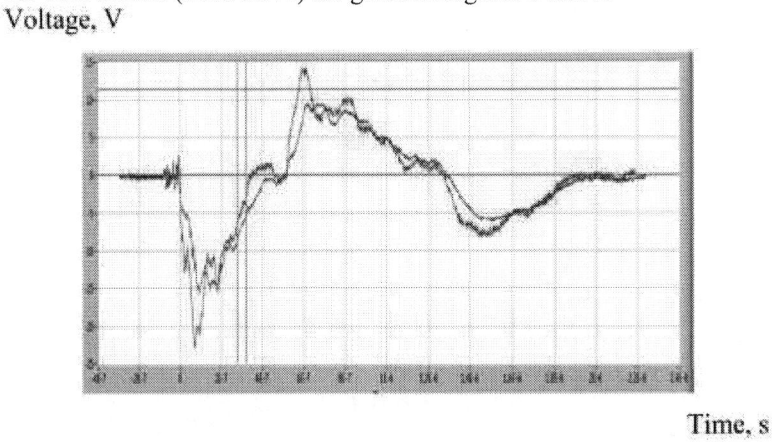

Figure 6. Result of two signal comparison. The blue curve is the normogram; the red curve corresponds to "turn to turn short circuit" type made in HV winding. The probing impulse is applied to HV winding of phase A, the response is recorded in LV winding of phase a.

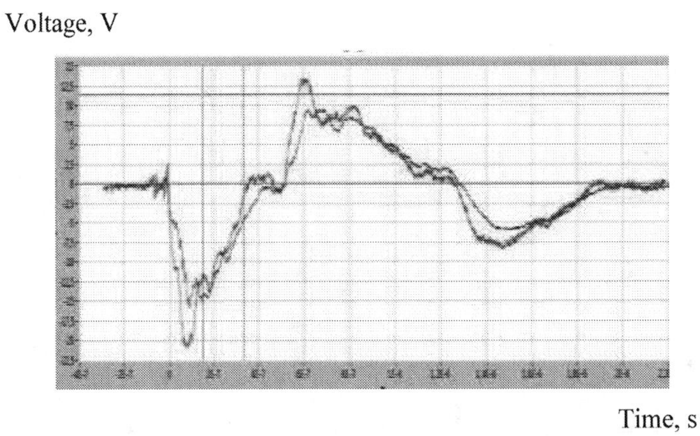

Figure 7. Result of two signal comparison. The blue curve is the normogram; the red curve corresponds to "axial shift of turns" type made in HV winding. The probing impulse is applied to HV winding of phase A, the response is recorded in LV winding of phase a.

The calculation by the comparison program for the case in Figure 7 shows the difference integral of 3.4×10^{6} Wb (in the range of $0–2{,}2\times10^{-6}$ s). The integral of the normogram (in the range of 0 to

2.2×10^{-6} s) equals 1.0114×10^{-5} Wb. The difference integral between the normogram and defectogram of response integral normogram is 34%. The calculation by the comparison program for the case in Figure 8 shows the difference integral of 9.3243×10^{-7} Wb (in the range of 0–2.2×10^{-6} s). The integral of the normogram (in the range of 0–2.2×10^{-6} s) equals $1{,}9740\times10^{-6}$ Wb. The difference integral between the normogram and defectogram percentage of response integral is 47%. So, efficiency of condition control is expressed in percentages.

Result of detailed research of influence of pulse and front of probing pulse duration to effectiveness of winding condition control is shown on Figures 8 and 9.

It is established that optimum pulse duration is in 100 – 110 ns. At this parameter condition control efficiency has maximum value for both types of different defects. The decrease in efficiency to the left side of the maxim is due to the fact that a short pulse does not penetrate into the winding. The decrease in efficiency to the right of the maximum, in our opinion, is explained by a decrease in the degree of excitation of the winding circuits as the duration of the probing pulse increases. According to above experiments it is established that the less front duration the more condition control efficiency. This statement is valid for defects of both types and for whole range of probing impulse duration.

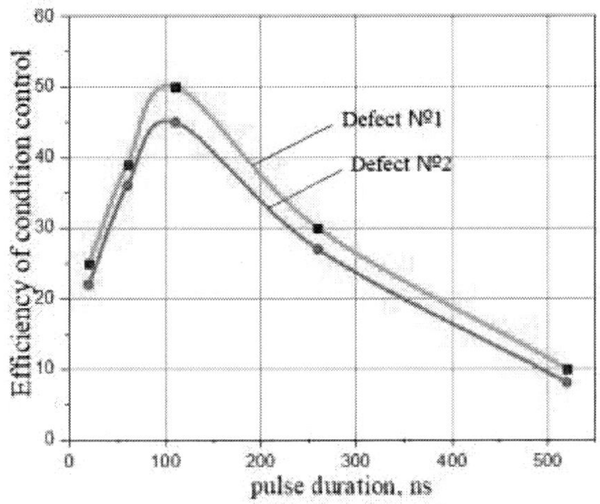

Figure 8. The dependence of the efficiency of monitoring the winding condition from probing impulse duration for two defects.

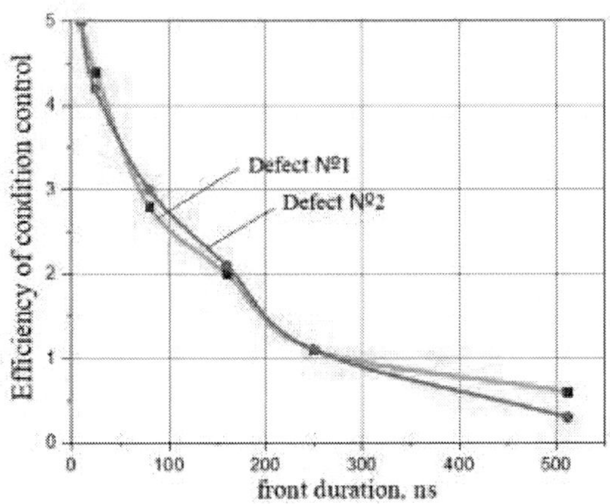

Figure 9. The dependence of the efficiency of monitoring the winding condition from front of probing impulse duration for two defects.

Optimum point for efficient diagnostics around 100 ns pulse duration could be explained as following. Impulses with duration more than 100 ns can't penetrate to winding without big losses. These losses influence sensitivity of diagnostics negatively. Impulses, with duration less than 100 ns, due to skin effect come to surface of investigated winding and stop to participate in condition control process. As for the front of probing impulse duration, it is established that high frequency part of common spectrum depends on it. The shorter front rapidness the more high frequency part comes to winding. Namely high frequency part is responsible for initial fail reveal. That fact makes short front more appropriate for sensitivity increasing.

4. Experimental research of winding condition control under operating voltage

In order to investigate the fundamental possibility of monitoring the state of the transformer windings at an operating voltage between the probe pulse generator and one of the phases of the transformer, an isolating coupling capacitor was installed. The experiments were carried out for model as well as real high voltage transformer. Separator condenser parameters: type K15-4; rated voltage 40 kV, capacity 2200 pF. In the measuring circuit used the same capacitor to obtain response signal under operating voltage. Thus, the probe pulse generator and measuring equipment were isolated from the transformer windings, which can be applied in practical work to monitor the state of the transformer windings in the operating mode. In Figures 10 and 11 are shown probing impulse and response for cases of without and under operating voltage.

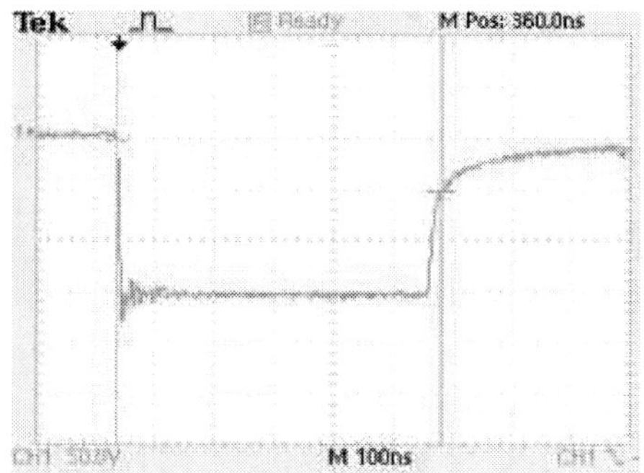

Figure 10. View of the probing impulse applied to the transformer winding

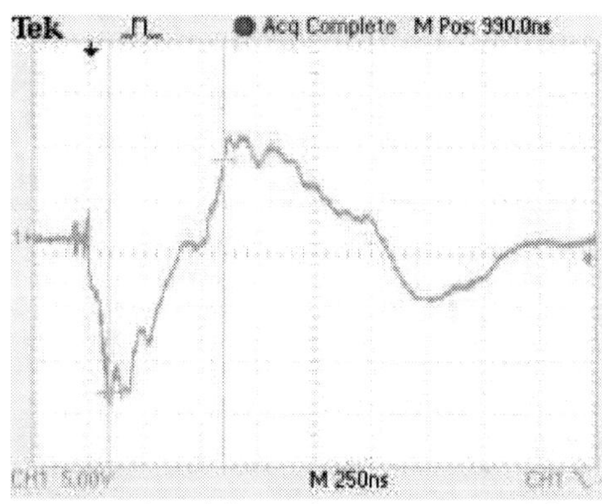

Figure 11. Oscillogram of the response signal from phase LW "*a*" of the transformer winding when the probing impulse is applied to HV "*A*" phase.

The signals have been taken in the same experimental conditions in the presence and absence of a coupling capacitor. Operating voltage was in 5–10 kV range. The signals in both cases are completely identical, which indicates the absence of extraneous signals and distortions in the measuring circuits.

The results of the experiments clearly demonstrate the fundamental possibility of implementing a diagnostic procedure under operating voltage.

The established fact that it is possible in principle to carry out a diagnostic procedure under operating voltage is important and has a significant prospect, as it allows monitoring the status of transformer equipment without shutting down and balancing, which is an undesirable (and often impossible) factor in the process of electric power systems.

5. Discussion

It is shown that optimum pulse duration is in 100–110 ns. At this parameter condition control efficiency has maximum value for kind of different defects. The decrease in efficiency to the left side of the maxim is due to the fact that a short pulse does not penetrate into the winding. The decrease in efficiency to the right of the maximum, in our opinion, is explained by a decrease in the degree of excitation of the winding circuits as the duration of the probing pulse increases.

The short probe pulse measurement also allows determining not only the initial stage of the defect, but also the place of the defect.

The shorter impulse, the wider its spectrum and the more relative contribution of the pulse high – frequency components and less noise level contribution in the signal response at the same time. Transition to impulse durations of 50…100 nanoseconds will allow marking out the defects in the most initial stage and thus avoiding the excessive fluctuations arising in contours with a smaller own frequency. It is possible since short impulses have a wider spectral structure which is in the range of frequencies which are required for excitation of contours arising from just initial defects (defects type like the most initial stage of turn to turn short circuit or axial shift).

6. Conclusion

Insulation destruction, deterioration of dielectric properties of the main and turn insulation of the transformer, electrical defects and mechanical deformations are influence of consequences of electro-dynamic forces of short circuit currents. To prevent all these consequences, the reliable methods of diagnostics should be used.

Despite many positive characteristics, in some cases sensitivity of most popular technology for these goals FRA makes a lot mistakes during the transformer diagnostics. One of the possible reasons for that is the fact that upper boundary of the frequency range is restricted by 2 MHz of frequency. Appropriate way to solve a problem is using a probing impulse with duration 100 ns with a rise time of 5 ns. Experiments confirm the effectiveness of this approach.

The established fact that it is possible in principle to carry out a winding condition control under operating voltage or ON-LINE regime is important and has a significant prospect, as it allows monitoring the status of transformer equipment without switching down. This fact opens new prospects before pulse detection method.

References

[1] Lekh W, Tyminskiy L 1966 *Electrichestvo* **1** 77–81
[2] Avetisov V, Levitsckaya E, Popov B 1978 *Electrotechnika* **1** 53–57
[3] Wang M, Vandermaar A, Srivastava K 2013 *IEEE Transactions on power delivery* **20** 1930–1938
[4] Bagheri M, Phung B, Blackburn T 2013 *IEEE Transactions on Dielectrics and Electrical Insulation* **20** 1962–1970
[5] Pham D, Pham T, Borsi H, Gockenbach E 2013 *IEEE Transactions on Dielectrics and Electrical Insulation* **20** 2026–2036
[6] Senobari R, Sadeh J, Borsi H 2018 *Electric Power Systems Research* **155** 172–183
[7] Cheng Q, Zhao Z, Tang C, Qian G, Islam S 2019 *Measurement* **140** 197–206
[8] Lavrinovich V, Lavrinovich A, Mytnikov A 2012 *International Journal on Technical and Physical Problems of Engineering* **13** 149–153

[9] Lavrinovich V, Mytnikov A 2015 *IEEE Transaction Dielectrics and Electrical Insulation* 22 2041-2045
[10] Lavrinovich V, Mytnikov A, Hongda Li 2016 *Resource-Efficient Technologies* 3 111–117

High-resolution phonocardiogram parameters

I Yu Zemlyakov[1,2], D S Zdanov[1,2], Y V Kosteley[1,3,2], E V Golobokova[1,2], E Yu Dikman[4] and A Sh Bureev[1,2]
[1] LLC Diagnostica, Tomsk, Russia
[2] Research Laboratory, Instrumentation National Research Tomsk State University Tomsk, Russia
[3] Department of Economic Mathematics, Informatics and Statistics, Tomsk State University of Control Systems and Radioelectronics, Tomsk, Russia
[4] Division for Electronic Engineering, National Research Tomsk Polytechnic University, Tomsk, Russia
E-mail: d_s_zhdanov@mail.ru

Abstract. The article describes the results of studying and analyzing phonocardiograms (PCGs) obtained during a physiological experiment with Blu-ray standard equipment. It provides the findings of a spectral and spectral-time analysis for signals with a sampling frequency of 10, 44.1 and 192 kHz. It shows that the differences in the PCG spectra of identical signals are unreliable. The article specifies the onset and disappearance moments of the harmonic components of heart sounds. It also provides recommendations on the sampling frequency and bit resolution of digitized PCG signals for telemetric systems.

1. Introduction

The World Health Organization reports that, in 2016, 17.9 million people died of cardiovascular diseases (CVDs) in the world. It is about 31% of all fatal cases. 85% of these fatal cases were caused by heart attacks and strokes. Many experts believe that up to 60% of the fatal cases caused by a premature stroke and infarction could be prevented if deceased patients from the risk groups were provided with professional medical care in due time. Apart from unhealthy living, one of the primary mortality causes is late diagnostics, including the lack of easy-to-use, effective and reliable home control devices, which is important to people with limited mobility and senior persons [1]. The development of equipment utilizing up-to-date cardiologic signal analysis approaches, including artificial intelligence elements, requires a test signal library. Such a library should include not only normal, physiological signals but also potential pathological changes in electrocardiograms (ECGs) and phonocardiograms (PCGs). The development of such a library itself is a serious challenge requiring the involvement of high-cost experts and a sufficient amount of records to choose the most illustrative examples. Such libraries are often used for educational purposes: for instance, PCG records meant for medical students are freely available (e.g., [2]). These records contain all known heart sounds and pathological side tones.

Unfortunately, the PCG records from such libraries are available in the widespread MP3 audio signal coding format, with a loss of quality and sampling frequency of 10–22 kHz. The authors believe that it is sufficient to represent natural sounding. However, the MP3 format uses the physiological features of our sense of hearing masking the intentional loss of the “fine signal structure” with loud sounds [3]. Unfortunately, the modern digital signal processing (DSP) approaches are sensitive to such signal structure "simplifications." That is why the developers of CV diagnostic equipment need libraries of high-quality cardiologic signals with no loss in quality or tools to generate reliable signals.

As it has already been mentioned, the development of such libraries requires too considerable resources. A simpler solution, the development of software-based signal generators, is based on corresponding mathematic models that are able to imitate high-quality signals. Normal and pathological ECG modeling is quite well-developed. However, the authors of this article have not found the description of realistic PCG models in available literature. Indubitably, the initial development of such a model requires a thorough study of the structure of high-quality PCGs registered in laboratory settings. This article describes the preliminary findings of several experiments of this kind.

2. Materials and Methods

2.1. Signal registration

PCGs were recorded at the laboratory of clinical diagnostics with the participation of patients aged 18–36 years old. During the recording process, the patients were in a lying position. Recording was started after a 5–7-minute adaptation period. In total, 300 records with a length of 19–56 minutes were obtained. PCGs were recorded at Botkin point via a contact piezoelectric sensor with an electric capacity of 27 nF. This sensor is able to register signals within a range of 10 Hz to 2.2 kHz. Signals were recorded by means of the ASUS X75A Intel CORE i3 6GB laptop in an on-battery mode to prevent electric shock accidents.

Signals were transmitted from the sensor to a charge-to-voltage converter based on the AD8610 low-noise operational amplifier [4] via a coaxial cable. Then signals were transmitted from the output of the converter through the buffer amplifier of the OP07 integrated circuit [5] to the input of the Creative SBX Prostudio sound card connected to the laptop via a USB channel. To reduce the level of noise, the converter was shielded and powered (±9 V) from two Duracell 6LR61 batteries. Signals were registered in the Windows WAVE 24 bit 192 kHz sound file format with no loss in quality (Blu-ray standard level). The experts, i.e. the leading clinical employees, claim that the records did not differ phonetically from reference samples from verified MP3 signal libraries. The sampling frequency of 192 kHz and the window width of 512 samples allowed obtaining the time resolution of spectrogram signals amounting to 2.67 ms. It would be impossible to ensure such a resolution with the "standard" sampling frequency of reference files amounting to 44.1 44.1 kHz and especially 10–12 kHz.

2.2. Digital signal processing (DSP)

The obtained PCG signals were processed and analyzed by means of a computer equipped with the Intel CORE i5 8600 processor, 16GB RAM and 2GB hard drive under Linux Debian 9 64 bit. All the used programs are free software available under the GNU license.

Preliminary signal processing was performed via the Ardour 5.1 and Audacity 2.3 audio editors. This processing implied the elimination of sounds under 16 Hz. The upper limit of the working frequency range was chosen individually in each case depending on the level of noise and interference and varied from 180 to 450 Hz. For further analysis, individual heart sound complexes were randomly singled out from the PCG records and saved as WAVE files with the same resolution.

The processed signals were analyzed via the Scilab 6 system by writing command scripts in a high-level internal language. The same system was used to create the charts and graphs introduced in this article. Standard Scilab function libraries were used to build spectrograms. The data were formed into blocks with 512 samples via the *window*() function with the *beta* parameter = 6, which made its properties close to those of the Kaiser Window according to the Scilab Help system.

3. Research results

The experimental stage of this work took long time, so some patients were examined several times. An expert carried out the analysis of experimental data by a detailed study of the results obtained after constructing spectrograms. The findings of these examinations were analyzed separately: first of all, the authors analyzed the stability and repeatability of spectral and time PCG properties. In total, four episodes of this kind were registered during the experiment: two patients underwent tests twice, and two patients underwent them thrice.

3.1. Influence of the signal sampling frequency on PCG characteristics

The primary question determining the need for further research endeavors was the influence of the signal sampling frequency on the reliability of their assessment. It was planned to carry out comparative tests of PCGs with the sampling frequencies of 192 kHz (Blu-ray standard), 44.1 kHz (music CD standard) and 10 kHz (the frequency used in specific electronic telemedicine stethoscopes, e.g., CMS-VESD, Contec Medical Systems). To do this, the authors created the spectrograms of one and the same randomly chosen PCG segment containing 10 heart sound cycles with the initial sampling frequency and spectrograms of this segment after the procedure of resampling to a necessary frequency via the Audacity audio editor. All the output signals were saved in the WAVE format with no loss in quality. The spectrograms of one and the same signal for the sampling frequencies of 192 kHz, 44.1 kHz and 10 kHz are shown in Figure 1.

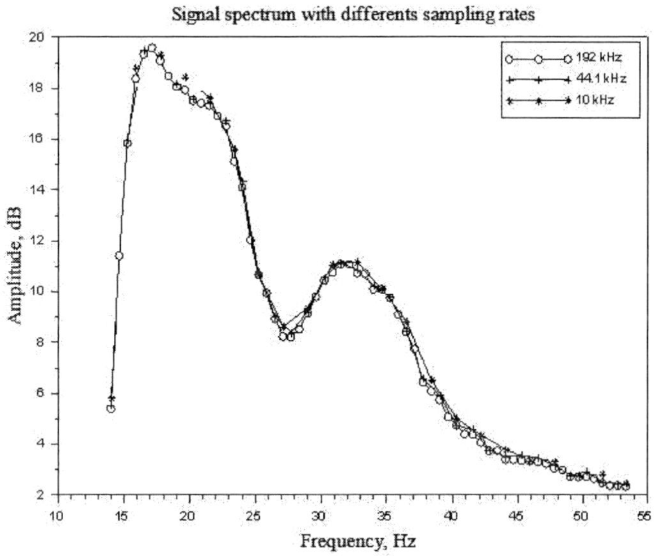

Figure 1. Spectrogram family for one and the same PCG signal with the sampling frequencies of 192 kHz (a), 44.1 kHz (b) and 10 kHz (c).

Figure 2 shows corresponding differential spectrograms for randomly chosen signals with the sampling frequencies of 192 kHz and 44.1 kHz (Figure 2*a*), and 192 kHz and 10 kHz (Figure 2*b*).

As one can see from these spectrograms, the differences in the signal energy are insignificant. The comparison of the findings of the tabular integration for the spectra of the same PCGs at different sampling frequencies shows that the difference in the signal energy was insignificant and amounted to 0.43% ±0.57% (P>0.1, N=18) for the sampling frequencies of 192 kHz and 44.1 kHz and 0.88% ±0.68 (P>0.1, N=18) for the sampling frequencies of 192 kHz and 10 kHz.

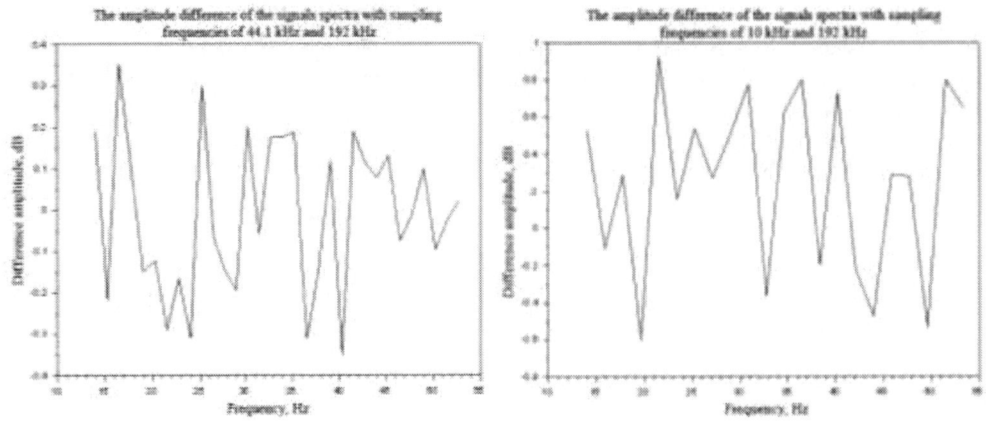

a) b)

Figure 2. Differential spectrograms of the signals with the sampling frequencies of 192 kHz and 44.1 kHz (a), and 192 kHz and 10 kHz (b).

3.2. Spectral-time PCG characteristics

The moments of emergence, duration and extinction of the harmonic components of heart sounds are essential to understand the fine PCG structure. Previously, the authors already reported on the preliminary research findings obtained during the analysis of signals with the sampling frequencies of 10 kHz and 11.6 kHz [6,7]. In their research study [7], the authors carried out a spectral-time analysis within a frequency band of 14…78 Hz at 2Hz intervals over a period of 350 ms with an accuracy of 10.93 ms, which appeared to be absolutely insufficient to obtain precise time parameters for the characteristic marks of heart sound harmonics. Increasing the signal sampling frequency up to 192 kHz allowed enhancing the time positioning accuracy of harmonics approximately by 3.5 times, up to 2.67 ms with the fast Fourier transform (FFT) data window width of 512 samples. As a result, the analysis of the PCG segments being 340 ms long allowed building a 128×128-point surface. A relevant area was a 32×128-point segment representing a signal spectrum within the range of 16–80 Hz over the entire duration of heart sound existence: the energy of the higher-frequency components was at the level of noise (Figure 3).

Table 1 shows the fine time profile of the harmonics of heart sounds based on the findings of 48 measurements. The table 1 demonstrates time marks for the harmonics that form 95% of the spectrum energy. The boundaries of harmonic existence were set as the time period when its amplitude was higher than 5% of the maximum value. Lower levels were taken as 0.

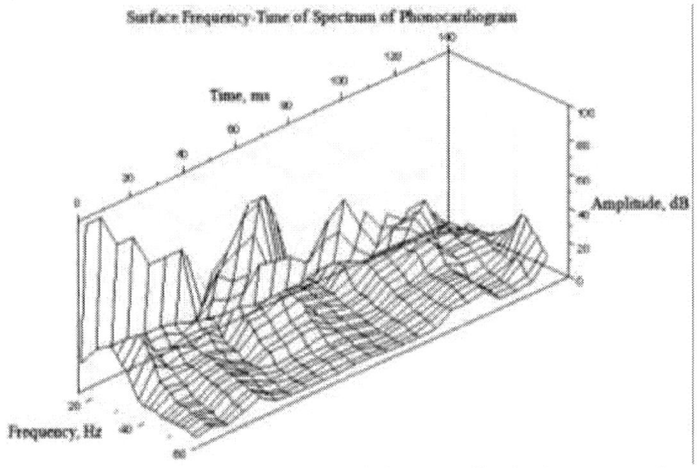

Figure 3. Typical high-resolution spectral-time profile for heart sounds I and II.

Table 1. Fine Time profile of the harmonics of heart sounds.

Frequency, Hz	**Time marks (N=48)**		
	Start, ms	Maximum, ms	End, ms
16	0	1.44±0.88	178.2±43.3
18	0	1.76±0.89	158.2±22.4
20	1.74±0.90	2.14±0.93	160.2±42.8
22	1.96±0.64	3.48±1.01	248.2±63.6
24	1.94±0.92	4.74±0.92	228.2±44.8
26	1.96±0.43	5.65±1.73	248.2±93.4
28	1.98±0.84	6.91±1.34	231.7±33.9
30	2.08±0.76	17.45±1.95	198.8±51.6
32	2.14±0.96	19.7±2.16	174.7±43.11
34	2.35±1.02	21.33±2.17	194.4±49.66
36	2.88±1.14	23.92±2.65	188.2±49.13
38	2.94±1.12	3.84±1.99	197.8±41.3
40	3.42±1.1	5.78±2.92	183.2±91.15
42	4.16±1.41	5.66±2.89	208.2±83.16
44	4.74±1.53	11.14±4.12	196.2±73.17
46	6.52±1.59	13.48±7.46	218.2±91.6
48	7.64±2.11	15.23±8.1	285.2±109.2
50	8.02±2.41	16.32±6.15	278.2±133.2
52	10.1±2.96	21.12±6.12	306.4±147.1
54	10.58±2.71	27.47±5.53	348.2±143.3
56	11.1±1.18	38.74±10.8	392.2±190.8
58	13.74±4.91	71.95±11.19	377.2±181.2
60	17.4±8.11	101.43±16.6	378.2±140.5

As compared with the data introduced in [7], it was found out that not all low-frequency harmonics emerged at the first moment: fluctuations within the range of 26…32 Hz emerged earlier than others, while fluctuations with other frequencies emerged only 8…10 ms after that. Overall, the high-resolution spectral-time profile matched with the profile obtained from the signals with a sampling frequency of 44.1 kHz with minor deviations.

4. Discussion

The reliable absence of differences in the spectra of identical signals with the sampling frequencies of 192 kHz, 44.1 kHz and 10 kHz confirms that the sampling frequency of 10 kHz is sufficient to digitize heart sounds with no loss in quality. If the goals of the experiment and data analysis were expanded, it would be possible to determine the minimum permissible frequency of PCG signal digitizing, when the level of spectrum energy would reduce vs. that of the reference 192 kHz signal by a substantial value, e.g., 2% or 5%. Probably, it will be the goal of one of the subsequent research studies.

Such issues arise during the development of equipment meant for telemedicine applications and autonomous operations: sampling frequency reduction definitely leads to a reduction in the volume of collected and transferred data. However, at the same time, it is necessary to ensure a balance between this system parameter and requirements to input signals on the part of a software signal detection complex. Excessive data transferred by wireless communication channels slow down system operations, and low-resolution signals can cause detection failures and unpredictable errors [8].

Another important issue connected with the selection of sampling frequencies is the elimination of noise and interference where the frequency is higher than the doubled sampling frequency. This is an additional parameter of ADC-based signal digitizing and a very efficient filter. For instance, digitizing signals having the frequency of 500 Hz suppresses all interferences with the frequencies of over 250 Hz [9].

Speaking about the time resolution of signal spectra, the following can be noted. Without doubt, there is a specific boundary PCG sampling frequency that makes a further increase in this parameter senseless. In the authors' opinion, the sampling frequency of 192 kHz, a standard option for the hi-fi Blu-ray audio reproduction, will also be optimal for detection systems. As for the Blu-ray format, music experts cannot differentiate between "live" music and reproduced audio fragments. Therefore a further increase in the time resolution of signals can hardly be justified [10].

The same approach can be used to determine the number of bits during signal digitizing. Being part of inexpensive mass market digital data collection systems, analog-to-digital converters with the number of bits of 10…12 theoretically cover the dynamic range of ECG signals. However, digitizing errors and limitations typical of this procedure make digital signals either artificial, nuance-free (coarsening) or having a high level of noise due to a high sampling interval. The authors believe that the best option for ECG signals is ADCs with a conversion capacity of no less than 14 bits or, optimally, 16 or 18 bits with a sampling frequency of 192 kHz.

5. Conclusion

In the authors' opinion, for the recognition of physiological PCG heart tones recorded signals must satisfy the conditions Nyquist theorem. For conducting a full and qualitative analysis of PCGs, the sampling frequency should be at least 8 kHz, the resolution of the ADC should be 16 bits.

At the same time, while placing high demands on systems for recording and collecting data parameters of the acoustic sensors should be taken into account as well. So, the best models of measuring electret microphones, for example, Panasonic WM-61A, have a dynamic range of 62 dB with a sensitivity of -35 dB [11]. Contact microphone CM-01B for medical purposes with a piezoelectric transducer has an equivalent dynamic range of 44 dB at a sensitivity of -32 dB [12]. Preference should be given to modern small-sized acoustic sensor MEMS type: they combine a wide frequency band and a large dynamic range with high sensitivity. For example, MEMS microphone ADMP441 [13] with a built-in 24-bit ADC has a high sensitivity -24 dB bandwidth of 60 Hz to 15 kHz and a dynamic range of 61 dB. Built-in microphone allows the driver to directly connect it to a digital input of the microcontroller via the I2C bus. Thus, all necessary conditions currently exist for the registration and analysis of high-resolution PCG using inexpensive and available technical means.

Acknowledgment
The research has been carried out with the financial support from the Russian Foundation for Basic Research (RFBR) under Scientific Project No.18-31-20012 mol_a_ved.

References
[1] 2017 *Fact sheets: Cardiovascular diseases (CVDs)* (Geneva: WHO) p 8
[2] Heart Sounds. Multimedia reference guides and courses. Available at: https://www.practicalclinicalskills.com/heart-sounds
[3] Rassol Raissi 2002 The theory behind MP3. Available at: https://www.semanticscholar.org/paper/The-Theory-Behind-Mp3-Raissi/ed14d9a452c4a63883df6496b8d2285201a1808b
[4] 2008 *AD8610 Precision Very Low Noise, Low Input Bias Current, Wide Bandwidth JFET Operational Amplifiers* (Norwood: Analog Devices, Inc) p 24
[5] 2011 *OP07 Ultralow Offset Voltage, Precision Operational Amplifier: Data Sheet* (Norwood: Analog Devices, Inc) p 16
[6] Zemlyakov I, Zhdanov D, Bureev A and Kostelei Y 2019 *Pren. Méd. Argent* **105** 644–651
[7] Zemlyakov I, Zhdanov D, Bureev A, Kostelei Y and Golobokova E 2019 *IJEAT* **8** 2168–2173
[8] Caldarola P, Gulizia M M 2017 *European Heart Journal Supplements* **19** 229–243 doi:10.1093/eurheartj/sux028
[9] Mneney S H 2008 *An Introduction to Digital Signal Processing: A Focus on Implementation*, (Aablog: River Publishers ApS) p 284
[10] Subair N 2008 *Blu-ray Technology: A Seminar Report* (KOCHI: COCHIN University) p 43
[11] 2008 *Omnidirectional Back Electret Condenser Microphone WM-61*: *Data Sheet* (Panasonic, Inc.) p 2
[12] 2011 *CM-01B Contact Microphone* (Hampton: Measurement Specialties, Inc.) p 3
[13] 2015 *Omnidirectional Microphone with Bottom Port and I2S Digital Output ADMP441: Data Sheet* (Norwood: Analog Devices, Inc) p 16

A system-inno-diversified approach to the development of technologies for the effective transformation of organizations of the agro-industrial complex into a digital economy

M K Chernyakov[1] and M M Chernyakova[2]
[1]Accounting and Finance Department, Novosibirsk State Technical University, Novosibirsk, Russia
[2]Siberian Institute of Management, Novosibirsk State Technical University, Novosibirsk, Russia
E-mail: mkacadem@mail.ru

Abstract. The paper examines the most important trends and concepts of the modern stage of digitalization and the end-to-end technologies offered by the "Digital Economy" program. The necessity of creating a technology of regulation by the agro-industrial complex as an important component of the modern digital economy is substantiated. The aim of the research is to develop technologies for the effective transformation of the organizations of the agro-industrial complex into a digital economy. According to the definition of the digital economy proposed in the work, it should be built as an in-diversified, self-developing system. The concept of technology for regulating the digital economy of the agro-industrial complex should be formed on the basis of a systematic approach and have a hierarchical 5-tier network structure. The possibility of implementing the structure using the mathematical apparatus of the theory of Petri nets is substantiated. It has been established that model design of technology is the most optimal, if not the only possible way to improve the structure of production systems and technological processes of organizations of the agro-industrial complex, as well as to manage its structures.

1. Introduction

The development of the state regulation practice of the agro-industrial complex occurs both under the influence of digitalization and as a result of the development of economic and mathematical theories and their influence on the national economy through the formulation of recommendations and proposals in the field of government regulation. Today, leading economists agree that the economic policy of the state was shaped by the development of economic and mathematical theories. However, there are disagreements about the scale, mechanisms and objectives of such an impact. Therefore, we need to consider the peculiarities of the formation of the main theoretical approaches to the problems of state regulation of the agro-industrial complex. Successful digital transformation requires organizations to consider a wide range of opportunities that will vary depending on the importance in the business context and the needs of the particular organization [1].

The concept of "digital economy" does not yet have an unambiguous, clear definition due to the large number of its diverse interpretations. Based on the definitions and concepts cited in various sources, we can formulate a generalized definition of the term "Digital economy". We suggested using the following definition: "Digital economy is an environment that includes a combination of digital infrastructure and information and communication technologies for doing business" [2]. Intensive digitalization and the Internet of things penetrating into the agro-industrial complex, however, are able to transform it, to a lesser extent than other areas subject to greater influence of IT, into a high-tech business through accelerated increase in productivity and reduction of unproductive costs.

The aim of the research is to develop technologies for the effective transformation of organizations of the agro-industrial complex into a digital economy. According to the proposed definition of the digital economy, it should be built on the basis of the innovation diversification model proposed by us [3] and should be a self-developing system. The term "innovation diversification" was introduced by us in 2016 [3] and describes the penetration of innovations in new areas.

2. Literature Review

In the work of A. Melnikov, V. Sidorenko, P. Mikhailushkin, O. Makarevich [4]. the role of the state in the sustainable development of the agro-industrial complex of Russia is stressed. The need to develop a concept of a more advanced system of state regulation and support of the agricultural sector of the country's economy, increasing the competitiveness of the industry and integration into the global agro-food system was noted. Studies have shown that the refusal of state planning, state order for produced agricultural products, unreasonable pricing policy in the industry, insufficient state support, unfounded privatization in some cases, especially enterprises processing and selling products, restrain the growth rate of the rural economy, led to a huge the debt of the Russian farmers. There is no doubt that the strengthening of state regulation today is the basis of the agrarian policy and the shortest way of reviving the village.

In the explanatory note of the direction "Digitization of agriculture" [5] the goal was set to provide the participants with the opportunity to use broadband, mobile, LPWAN communications, information technologies (small and big data, artificial intelligence, control platforms) of domestic instrument manufacture (tags, controllers, sensors, controls). Unfortunately, the program lacks a methodological substantiation and recommended platforms for achieving this goal. Therefore, to achieve this goal, it is necessary to provide a methodological substantiation for digitalization, in particular, state regulation of the agro-industrial complex, which we should begin with an analysis of the composition of the industry.

Many scientists, such as N. A. Volkova, O. A. Stolyarova, E. M. Kosterin [6], N. A. Popov [7] and a number of others propose to consider the agro-industrial complex on the basis of the systematic approach that unites the national economy sectors related to the production, procurement, processing and sale of food, as well as machines for agriculture. The agro-industrial complex is a 3-tier, interconnected and interdependent system consisting of three subsystems: production, agriculture and services. The leading place in this system belongs to agriculture, which is characterized by special conditions of influence on other subsystems and on the whole system as a whole. The spheres of production and services in the agro-industrial complex play, although an important, but still supporting role. G.M. Gritsenko [8] in addition to the 3 main subsystems proposes to include the 4th subsystem which represents scientific and educational institutions (SEI), information centers (IC) and monitoring services (MS). The authors agree that in the conditions of digitalization of the economy, the development of telecommunications means, it became necessary to increasingly introduce innovative ICTs in the production of the agro-industrial complex, to carry out personnel training of specialists who are able to work in the new conditions [5]. In our opinion, for the interaction and development of these subsystems and the system of the agro-industrial complex itself, it is necessary to include a regulatory subsystem (Figure 1).

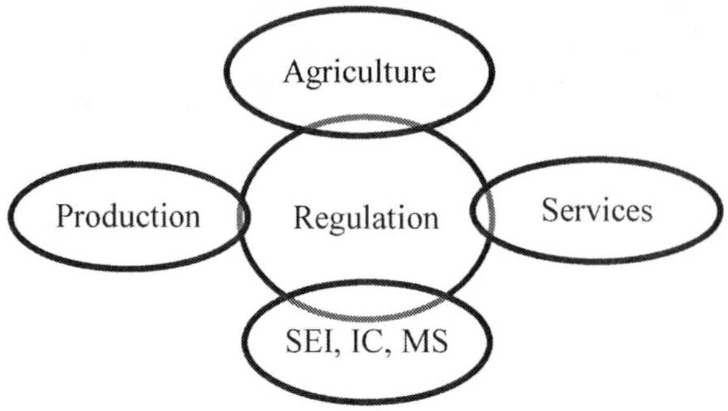

Figure 1. The structure of the agro-industrial complex, as a system.

3. Research Methodology

Figure 1 shows that the methodological substantiation of the digitization of state regulation of the agro-industrial complex requires the use of the systematic approach. "The systematic approach is, first of all, the integrated approach, involving a comprehensive account of the specific characteristics of the selected object, determining its essence, structure and, therefore, organization" [9]. Accordingly, the methodological basis of the research in this work is the systematic approach to the theory of systems and system analysis [10] to the problems of developing the state regulation system of the agro-industrial complex and analyzing the resource potential of agricultural organizations, taking into account the conditions of transformation into digital economy and forming its own point of view on the processes taking place while studying the topic.

The difficulties of analyzing and synthesizing such complex objects as the organizations of the agro-industrial complex led to the need to develop special scientific tools. Such areas as systems analysis, cybernetics, synergetics act as a methodological toolkit for the study of complex systems. Their commonality lies in the fact that they are a tool for researching systems. However, there are differences between them, and they are significant enough which reflects their specificity and uniqueness (Figure 2) [11].

The system, being a multidimensional object, is the subject of research from the perspective of multidimensional statistical analysis. Methodologies for multivariate statistical analysis are based both on the basis of heuristic methods [10] and on the basis of analytical methods [12]. The toolkit of multivariate statistical analysis is more adequate in the study of multidimensional objects, such as the regulation of organizations of the agro-industrial complex, with a combination of heuristic and analytical methods of system analysis.

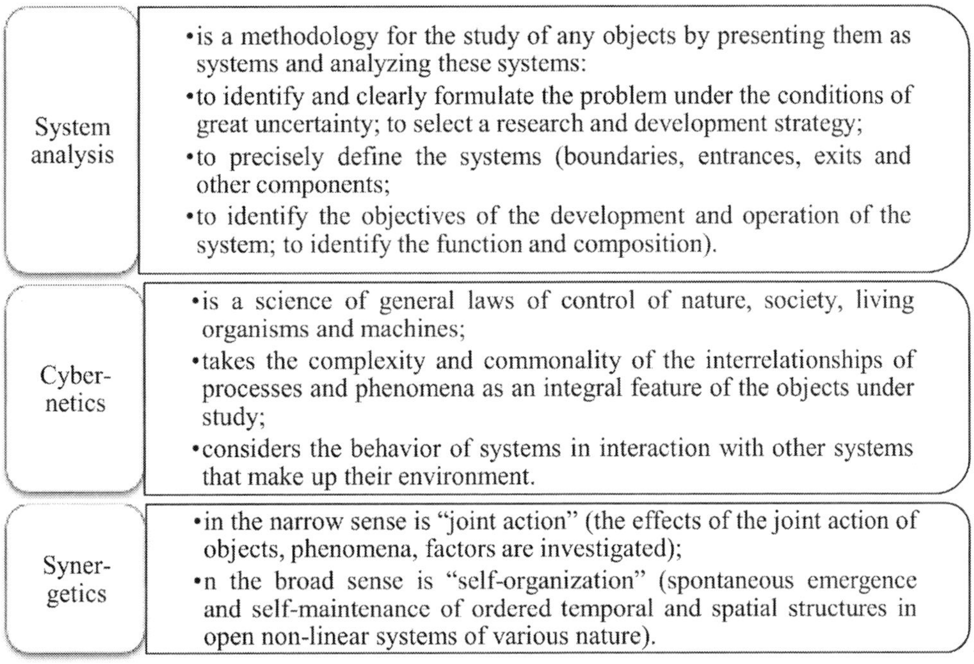

Figure 2. Methodological toolkit for the study of complex systems.

4. Results

The system-inno-diversified approach to the formation of a technology for efficiently transforming organizations of the agro-industrial complex into a digital economy should have a strict structure based on the theory of government regulation. The proposed version of such a structure is shown in Figure 3.

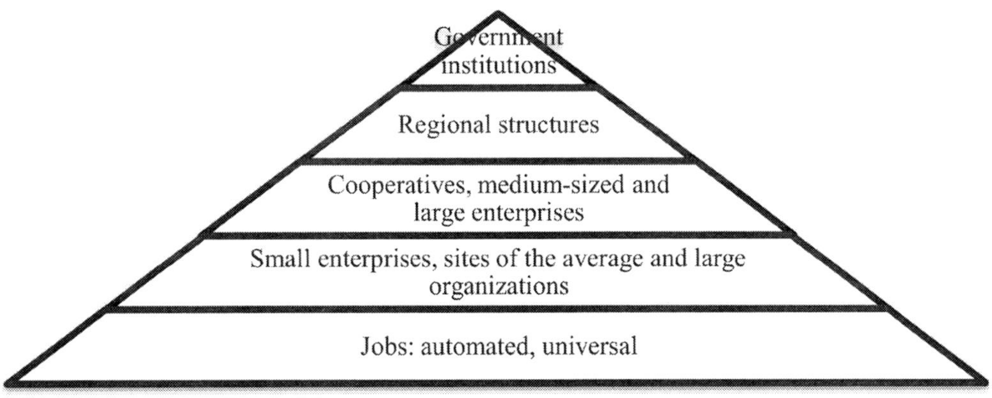

Figure 3. The structure of the platform for regulating the transformation of the digital economy of the agro-industrial complex.

The system-inno-diversified approach to the technological process of organizations of the agro-industrial complex is characterized by a complex dynamic system consisting of various units: equipment, control and management means, auxiliary and transportation equipment, processing tools or environments which are in continuous motion, interaction and change, production facilities, operators (people, robots, manipulators) carrying out processes and controlling them. Analysis of complex technological processes involves the decomposition of the production system into subsystems and sub-subsystems of various depth levels. As a result of the decomposition of the system into subsystems, a hierarchy of the structure of the production system can be built which will allow to consider it at various levels of its detail.

Due to the high complexity and labor intensity of the design processes of agricultural technological processes and systems, it is advisable to use digital technologies.

Predicting the results of designing technological processes and systems using ICTs is best done using simulation modeling methods.

The objective function of the concept of in-diversification simulation modeling of processes and systems of the agro-industrial complex involves the design calculations of the characteristics of productivity and efficiency taking into account the analysis of various options for the structure of organizations, risk levels of the digital economy [10], as well as the impact of disturbing factors of the surrounding and internal spheres. To achieve this goal, it is necessary to solve the following problems of model design:

- predicting the main characteristics of production systems and technological processes;
- obtaining statistical indicators and other characteristics of technical and economic efficiency;
- using the obtained results of model design to find the best option for production systems or technological processes;
- studying the optimal variants of the structure of operations of the technological process using the tools of the digital economy model.

With the help of the model design tools it is possible, for example, to describe the work of an agricultural producer's workplace, a farm, an agricultural holding or the industry as a whole. Model design of production systems or technological processes will allow reproducing parallel, sequential or parallel-sequential schemes of functioning, taking into account stochastic events and their influence on systems and processes [13]. Model design will allow a detailed analysis of the designed operating and route structures and the influence of various factors on productivity, load factor and other economic indicators necessary for making management decisions considering the unforeseen risks of the digital economy [10].

The algorithm for implementing the system-inno-diversified approach for regulating the systems of the agro-industrial complex is shown in Figure 4. In the algorithm, the system is understood as the structures of the platform for regulating the transformation of the digital economy of the agro-industrial complex shown in Figure 3.

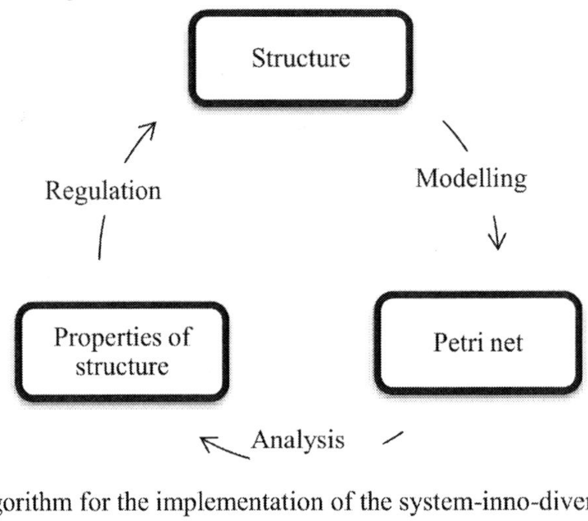

Figure 4. Algorithm for the implementation of the system-inno-diversified approach.

Structures of any level are modeled using the Petri nets. Petri nets in the algorithm are considered as an element of the analysis. The classification of Petri nets [14] and their characteristics are shown in Figure 5. Each type of Petri nets has a number of advantages and disadvantages. Combined networks allow you to reduce or eliminate the disadvantages and enhance the benefits. Therefore, for greater efficiency, it is necessary to use a combination of all known networks and refer to such a network as an inno-diversified network (Figure 5).

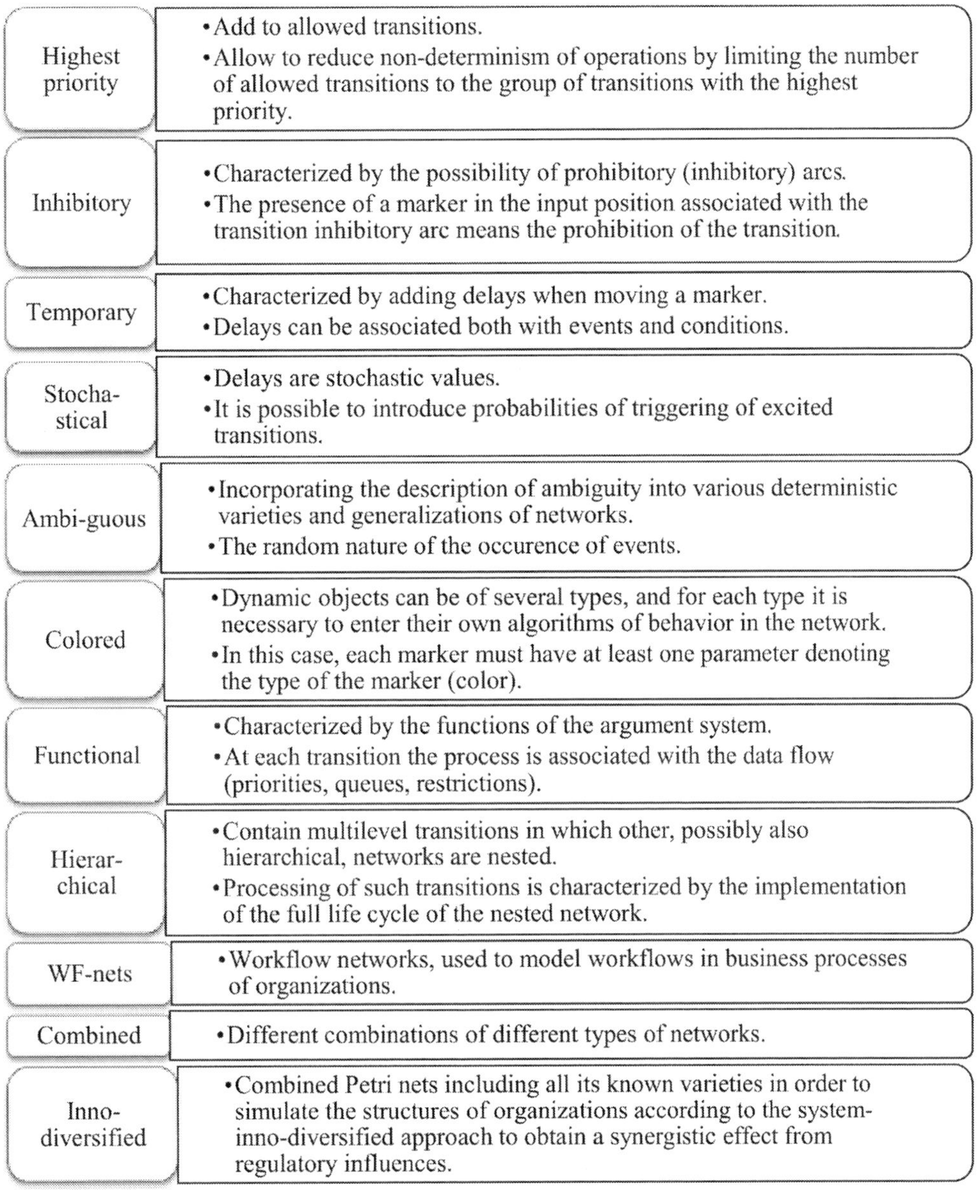

Figure 5. Characteristics of Petri nets.

5. Discussion of the results

Given that the methodology of the system-unique diversification approach is recommended to build the structures of organizations of the agro-industrial complex, it is better to use inno-diversified types of Petri nets for its implementation. At the first stage, the structure of the system or process is constructed as a Petri net (Figure 6), then the structure model is analyzed using it. The process of designing structures and determining its characteristics is carried out in terms of Petri nets (table 1). The challenge is to convert the Petri net vizualization into a real working structure.

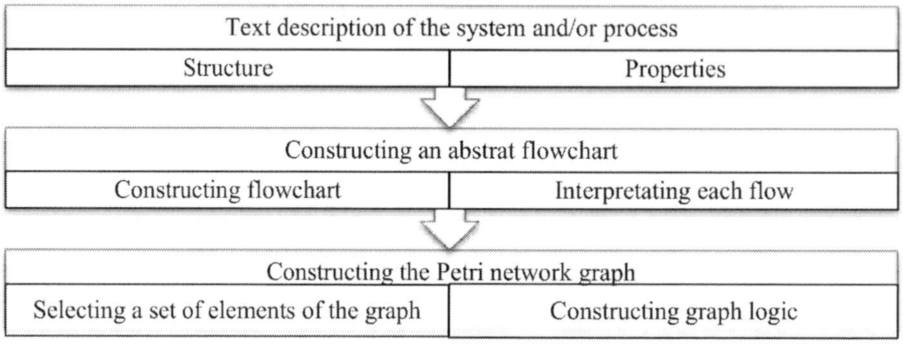

Figure 6. Algorithm for constructing a Petri net.

Table 1. Graphic Structure of Systems and Technologies in Modeling.

	Name	Notation	Presentation	Remarks
C	Petri net: C=(P,T,I,O)	$C_{i,j}$ (box)	Final set of networks: $C=(C_{1.1},C_{1.2}... Ci, j)$ where i is number of hierarchical levels, j - number of networks on i-volume level	Understand networks, I will stay the hierarchies of regulation which are located at various levels representing or hierarchical network
T	Event	▮	Finite set of transitions: T=(T1,T2,...Ti), where i is a number of transitions	Sowing, collection, processing, loading, unloading
P	Condition	○	Finite set of positions: P=(P1,P2,...Pj), where j is a number of positions	Units, operators, accumulators, equipment
A	Relations	⤴	Finite set of relations: A=(A1,A2,..Ak) where k is a number of relations	Sequence of implementing a manufacturing process
M	Tags (tokens)	•	Tag (token) in the condition means its execution	Marking of a set positions
I	Input function	P(A)→T	Finite set of inputs: I(j,k)	How many times the position P(j) is an input of the transition T(k)
O	Output function	T→P(A)	Finite set of outputs: O(j,k)	How many times the position P(j) is an output of the transition T(k)
Z	Actuation of transitions		T(k)→P(j)×M(i)→A	Transition is allowed when there are places for tags and if the number of tags is larger or equal to the number of relations

The textual description of the system and/or process implies:

1. A description of the structure of a process or a structure includes a list of events that should occur during its implementation. For example, in a dairy farm in the workplace in the technological

process of milk production, at least two events are necessary: to begin the process of milking cows (T1) and to complete it (T2).

2. In order for the events to occur, it is necessary to create conditions. In order to start the milking process, at least cows (P1), equipment for this operation realization (P2) and milking operators (P3) are required. These conditions contribute to the process of starting the milking (T1) and creating the conditions for the implementation of the milking process itself (P4).

3. In order to complete the milking process (T2), it is necessary at least to have a completed state of the milking operation (P4) and operator (milker). These conditions contribute to the implementation of the milking completion process (T2) and the creation of conditions for the production of finished products (placing milk in a store of a certain capacity) (P5). As a result, operators and inventory are freed from participation in the technological process.

As all operations do not occur instantaneously, a delay duration in the same units of measurement is needed to be assigned to them.

The number of cows, sets of equipment and operators can be any, depending on the specific workplace. You can simulate both manual and machine milking processes. In the case of manual milking, the operator is not released after loading but accompanies the milking process and is released only after unloading. Not only humans can be used as operators but also robots or manipulators.

Constructing an abstract flowchart to display the logical structure of the system consists of a graphic illustration of the stages of the process (Figure 7). Simple flowcharts are easy to create in Microsoft Word, Excel, PowerPoint and easy to read because of the simplicity and clarity of the figures. Automatically create a simple flowchart based on data that is more efficiently designed using data visualizer in Visio which is an addition to the standard Microsoft application package. The "Simple flowchart" template in Visio contains figures that can be used to visually represent a variety of processes, including processes in organizations in the agro-industrial complex. The display of simple processes, such as the process of milking cows is shown in Figure 7.

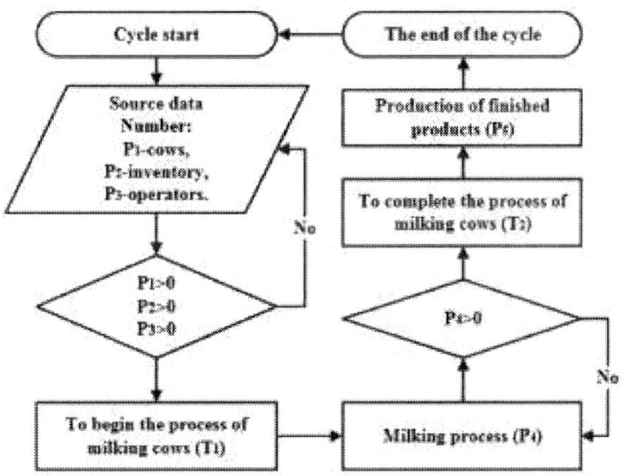

Figure 7. Flowcharts for displaying the logical structure of a cow milking process.

In Petri nets events and conditions are designated by abstract symbols from two non-intersecting alphabets named accordingly as the transition set and the position set. There are several formal representations of Petri nets: set-theoretic; graphic – bichromatic (bipartite) graph; graphical; matrix [15]. The algorithm for constructing the Petri network graph is shown in Figure 8.

The simulation modeling is based on the structure of operations which clearly defines the order of carrying out: in which sequence which transitions in which position are implemented (table 1).

The construction of a Petri network graph is carried out in three steps.

1
- Simulated processes (phenomena) occur in the system:
- described by the set of events (T) and conditions (P) that these events determine,
- cause-and-effect relations (A) established on the set of "events-conditions".

2
- Events-actions are defined, their sequence of occurrence is controlled by the state of the system.
- System states are set by a variety of conditions.
- Conditions are formulated as predicates.
- Quantitative conditions are characterized by capacity.
- The capacity of the conditions is expressed by the the natural numbers.

3
- Conditions (predicates) can be fulfilled or not fulfilled.
- Only the fulfillment of the conditions ensures the possibility of the occurrence of events (preconditions).

4
- After the occurrence of the event, other conditions that are in a cause-and-effect relationship with the preconditions are fulfilled (postconditions).
- After the event occured, postconditions are fulfilled which are the preconditions for the next event, etc.

Figure 8. Algorithm for constructing a Petri network graph.

At the first step of modeling design, a graphical model of modules (Figure 9), mathematical models in the form of a graph represented by a matrix of binary relations (table 2) and a set of two-component corteges are created:

$$\mathbf{M_1} = \begin{pmatrix} \langle P_1, T_1\rangle, \langle P_2, T_1\rangle, \langle P_3, T_1\rangle, \langle T_1, P_4\rangle, \\ \langle P_4, T_2\rangle, \langle T_2, P_2\rangle, \langle T_2, P_3\rangle, \langle T_2, P_5\rangle \end{pmatrix}.$$

At the second step, on the basis of the graphical model (Figure 9) a table for inputting the initial data is formed (table 2). Source data is processed using the "Simulator" application package [3].

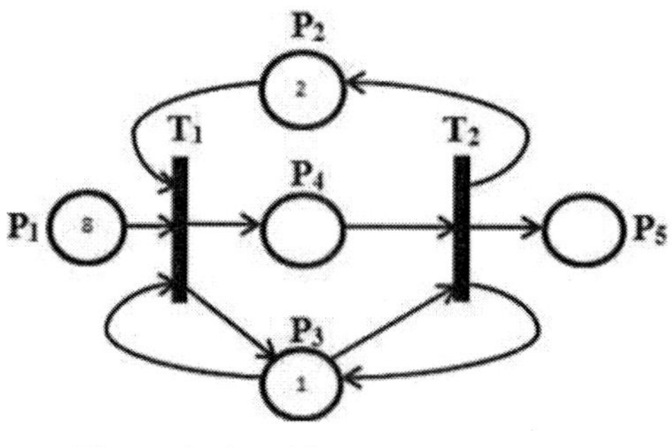

Figure 9. Graphic model of modules.

Table 2. Binary Relation Matrix.

		T1	T2	P1	P2	P3	P4	P5
	T1	0	0	0	1	0	1	0
	T2	0	0	0	0	1	1	1
	P1	1	0	0	0	0	0	0
G=	P2	0	1	0	0	0	0	0
	P3	1	0	0	0	0	0	0
	P4	1	1	0	0	0	0	0
	P5	0	0	0	0	0	0	0

Petri nets consist of three components: the set of positions (places) P, the set of transitions T and the connections (incidence relations) A (table 1). The state of the system is characterized by chips or marks (tokens), which are indicated by black dots and placed in the positions of P. If large quantity of them, they can be replaced by numerical symbols.

At the third step, it is required to perform a simulation of a simulation model at a minimum of five points for a qualitative analysis of modeling design. The obtained results of the calculations are presented in a summary table, on the basis of which a graph is constructed for clarity (Figure 10).

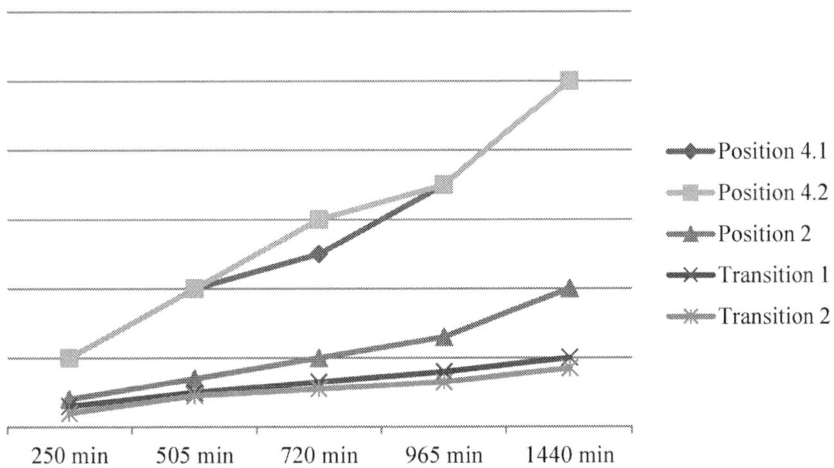

Figure 10. Efficiency of equipment and operations loading.

Analyzing the calculation results, based on the initial data, the module structure is optimized and management decisions are taken to modernize the production system or technological processes: reduce or increase equipment, operators, to change their technical and economic characteristics: increase or decrease performance indicators (duration of delays in work positions and in transitions), changes in storage capacity (delay arrows), etc.

6. Conclusion

In conclusion, it should be noted that as a result of the a priori research and analysis of various sources of information, the most rational, according to the authors of the theory, were identified: government regulation, risks, modeling, mathematical logic, sets, graphs, Petri nets and system-inno-diversified approaches; design-calculated, statistical-economic, economic-mathematical, sociological and other research methods can be involved as well. As a result, a new direction for the development of a system-inno-identified approach for simulation modeling of state regulation of the agro-industrial complex in a digital economy was formulated.

System-inno-diversified approach will allow to obtain the most accurate description of technological processes in organizations of the agro-industrial complex (field processing time or harvest, equipment utilization or loading coefficients, technical performance, operation delays, etc.). To effectively use the system-inno-diversified approach, preliminary development of algorithms for its implementation is necessary.

Given that system-inno-diversified approach is recommended to build the structures of organizations of the agro-industrial complex, it is more expedient to use a combined type of Petri nets for its implementation. The process of designing structures and determining its characteristics should be implemented in terms of Petri nets. The task is to transform the Petri nets visualization into a real working structure. Initially, the structure of the system or process is built in the form of a Petri net, and then, using the proposed algorithms, it is necessary to analyze the structure model for making decisions on regulatory influences.

Acknowledgments
This research is achieved with the financial support of the Novosibirsk State Technical University (project C19-12).

References

[1] Carcary M, Doherty E, Conway G 2016 *Proceeding of 10th European Conference on Information Systems Management: Univ Evora* 20–-28
[2] Chernyakov M K, Chernyakova M M 2018 *Journal of Corporate Finance* **12(4)** 99–109
[3] Chernyakov M K 2018 *Proceedings of the 2nd International Scientific conference on New Industrialization: Global, national, regional dimension (SICNI 2018)* **240** 262–267
[4] Melnikov A, Sidorenko V, Mihailushkin P, Makarevich O 2017 *Mezhdunarodnyy sel'skokhozyaystvennyy zhurnal* **4** 37–40.
[5] Chernyakov M K et al 2020 *E3S Web of Conferences* **175** 13024 doi: 10.1051/e3sconf/202017513024
[6] Volkova N A, Stolyarova O A, Kosterin E M 2005 *Ekonomika sel'skogo khozyaystva i pererabatyvayushchikh predpriyatiy [Economics of Agriculture and Processing Enterprises]. Ed. N.A. Volkova* (Moscow:Koloss)
[7] Popov N A 2008 *Consultation in the field of optimization of management of modern agricultural production* (Moscow: RUDN)
[8] Gritsenko G M 2016 *Modern trends in the development of science and technology* **11** 48–52
[9] Kostusenko I I, Kosyakova L N, Smekalov P V 2014 *System analysis of investment and innovation processes in the agro-industrial complex* (Saint Petersburg: Prospekt Nauki) p176
[10] Babeshko V N *et al* 2018 *Mezhdunarodnyy zhurnal perspektivnykh issledovaniy* **(1-2)** 15–20
[11] Shalanov N V 2012 *Mathematical methods for the study of systems* (Deutschland: Palmarium academic publishing) p 520
[12] Shalanov N V 2012 *Vestnik Sibirskogo universiteta potrebitel'skoy kooperatsii* **2(3)** 109–119
[13] Chernyakov M K *et al* 2018 *Advances in Engineering Research* **157** 124–128
[14] Simankov V S, Tolkachev D M 2012 *Vestnik Adygeyskogo gosudarstvennogo universiteta* **4** 202–209
[15] Kudj S A, Loginova A S 2015 *Vestnik MGTU MIREA* **1(6)** 10–22

Application of neural networks in problems of determining geometrical properties of objects placed in geophysical elastic media

A E Morozov[1], **A A Yakimenko**[1,2] **and D A Karavaev**[2]
[1] Department of Computer Engineering, Novosibirsk State Technical University, Novosibirsk, Russia
[2] Laboratory of Geophysical Informatics, Institute of Computational Mathematics and Mathematical Geophysics, Novosibirsk, Russia
E-mail: yakimenko@corp.nstu.ru

Abstract. The use of a neural network approach is associated with high requirements for software implementation. The main criteria is a high degree of accuracy in the recovery of the desired geological-physical model of the environment (GPME) and processing speed. The paper presents a description of the developed neural network architecture for solving the inverse problem of geophysics for determining the geometric properties of GPME objects. The performance of a single neural network and an ensemble of neural networks (NN) has been evaluated. The results are presented comparing the operating time of the NN when restoring models on various computing devices: CPU and GPU. The results of experiments on the restoration of various GPME using the developed NN based on the LSTM layer and the U-Net architecture are presented. This work was supported by the RFBR grant No. 19-07-00170.

1. Introduction

The aim of the work is the development and study of a methodology that uses basically neural networks. This approach is used to solve the problem of reconstructing geological and physical models of media from wave field patterns [1]. To this end, we developed software for the implementation of the NN architecture and tested the developed neural networks. The results showed a good match with the original pictures of the models of GPME. Thus, the paper shows the high practical significance and applicability of the neural network approach to solving the inverse problem of geophysics. The solution of such a problem is connected with the restoration of 2D GPME and the determination of the geometric characteristics of the object in the form of a cavity located in the medium. Under the cavity should be understood as an oval region with non-zero values of elastic parameters. The elastic medium model is represented by a rectangular area in a 2D coordinate system. The case of an isotropic inhomogeneous medium described by three parameters is considered: two velocities of elastic waves and density. To train a neural network, it took the results of solving a direct geophysical problem of modeling the propagation of elastic waves in a 2D environment. As a result of a series of numerical experiments, a set of initial GPME models and a set of 2D field images were obtained. The presented calculation results are used as test data for testing the performance of the NN. 2D modeling of the propagation of elastic waves was performed using the computational resources of the Siberian Supercomputer Center of the Siberian Branch of the Russian Academy of Sciences: cluster НКС-30Т and НКС-1П.

Analysis of the wave field images showed that for high-quality NN work it is necessary to store information about the sign of the wave amplitude. The use of a three-channel image at the input of the NN leads to a significant increase in the amount of memory occupied by the model in the learning process. Therefore, when working with a neural network, it is necessary to convert a three-channel snapshot of the wave field into a single-channel one, while preserving the information about the wave amplitude.

It should be noted that the simulation results presented in the images are interrelated. The observed pattern of wave propagation of the current image depends on the picture in the previous image (iteration step in the calculation time). So The neural network must store certain information about previous images to form a qualitative opinion on the structure of the GPME.

It can be concluded that the target model of the neural network is faced with the task of analyzing the sequence of images for working with the wave field, as well as the task of segmentation to restore the position and size of the cavity.

2. Neural network development

2.1. Neural network training

To obtain a trained neural network model, it is necessary to solve the problem of optimizing the function defined by this model. First of all, it is necessary to determine the loss function, depending on the task set for the NN. The loss function is a function whose arguments are the correct answer to the input data set and the answer received by the neural network. When teaching NN its parameters tend to lead to such a form that the loss function takes the smallest value. In other words, when training a neural network, it is required to choose its weights so that the loss function takes the minimum value [2].

In order to apply gradient descent for training an NN, a backpropagation method was developed. It consists in propagating error signals from the network outputs to its inputs, in the opposite direction. This is achieved at the expense of the chain rule - the rule of differentiation of a complex function, which allows us to calculate the derivative of the composition of several functions on the basis of individual derivatives.

Using the method of back propagation of error, it becomes possible to study deep NN. Such NN have 2 or more layers in their structure. At the expense of the rule of the chain, the error during training extends from the output of the NN to its initial layers.

When teaching neural networks, stochastic gradient descent is widely used, since it offers an advantage in speed compared to standard gradient descent. When used, the value of the gradient is approximated by a gradient of the cost function calculated on one element of training [3, 4].

2.2. Data preparation

The size of the GPME and the obtained wave field images for teaching the NN was originally 929×186 pixels. It should be noted that the process of training the NN model to restore images of this size requires a large amount of computational resources — RAM and processor time. Therefore, to perform the work it was necessary to change the size of the input data in the form of images. First of all, the wave field and GPME were trimmed to sizes 736×184. It turned out that such a pruning did not spoil and did not remove the basic information, which reflects the presence of a cavity. Then the size of the images of wave field images and GPME was reduced to 256×64. This is for ease of operation and the need to use MaxPooling's operation. This operation is used to quickly increase the end-neurons of the field of perception. During this operation, the size of the incoming image is divided into two. Thus, the most advantageous dimensions of the sides of the input image for a model are multiples of 2. If one of the sides of the image is not such, at a certain stage of the model's direct passage, information about a row or column of pixels may be lost, since the division will be integer and the remainder of the division will be discarded.

For high-quality training of any neural network that works with images, first of all, the requirement is to have a large amount of training data. During the study, there was a problem with a small number of GPME for training. In total, there were 107 models, each of which had 14 snapshots of the wave field propagation.

To increase the amount of training data, noise from input images with normally distributed noise was used. For each GPME several copies were created, where the wave field patterns were noisy randomly. The mathematical expectation of the distribution of such noise was zero, and the dispersion was as small as possible and ranged from 10^{-8} to 5×10^{-8}. Using this augmentation method, the number of images in the training sample was increased from several dozen to 4–5 thousand. On such a number of images, the target model of the neural network could already be trained and highlight any patterns in the data to solve the problem.

2.3. LSTM-based architecture

To implement the model based on the LSTM layer, it was first necessary to create a neural network unit. Such a block is necessary to encode information from each snapshot of the wave field into a numerical vector. Then the vector can be transferred to the input of the recurrent layer. Simultaneous learning of such a model takes too much memory. The computing devices available for the job do not have the necessary resources. Therefore, the learning process of such a model consists of two stages. The first is the training of a wave field image encoder. The second is the training of the GPME restorer from a sequence of numerical vectors obtained using an encoder.

In order to get a quality encoder, he was trained on the principle of auto-encoder. The autocoder is a neural network that has a narrowing in its architecture and is trained to play its own input. Autocoders are widely used in solving various problems. In particular, when solving the problem of generating semantic numerical vectors for words (word2vec).

Figure 1 shows the structure of the simplest auto-encoder. In the place of narrowing of such an autocoder during training, such a numerical vector is formed, from which the model can then easily restore the original data. In this case, the appearance of the most representative numerical vector for the input data is said. After learning such an autocoder, its first part, before narrowing, can be used to encode the input data into a vector of smaller dimension. The first part of the autocoder is called the encoder.

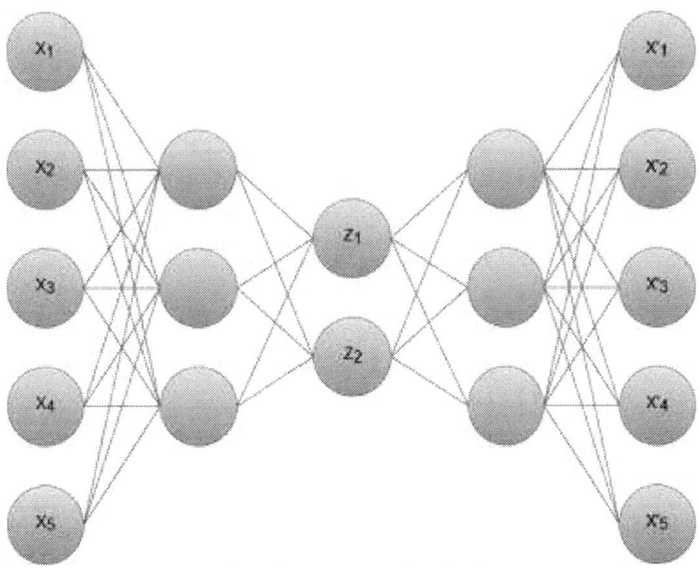

Figure 1. The simplest auto-encoder with a narrowing in the center on the example of NN from fully connected layers.

In the context of this task, the auto-encoder was assembled from convolution and pulling operations (encoder, first part) and sweep and anpuling (decoder, second part). The architecture of the auto-encoder was almost completely analogous to the architecture of the segmentation model of the neural network from [5], however, this technology was not applicable in this task due to the lack of data transmission over time within the model itself.

2.4. Reducer Architecture

The reducer of the GPME should have taken as input a sequence of numerical vectors created by the encoder and synthesize the GPME in which, according to his assumption, this is precisely the propagation of the wave field. Obvious was the need to use the operations of unpilling and scanning to realize the possibility of generating images based on numerical data coming from the LSTM layer [5]. During the direct passage of the model. Figure 2 shows the final architecture developed by the NN.

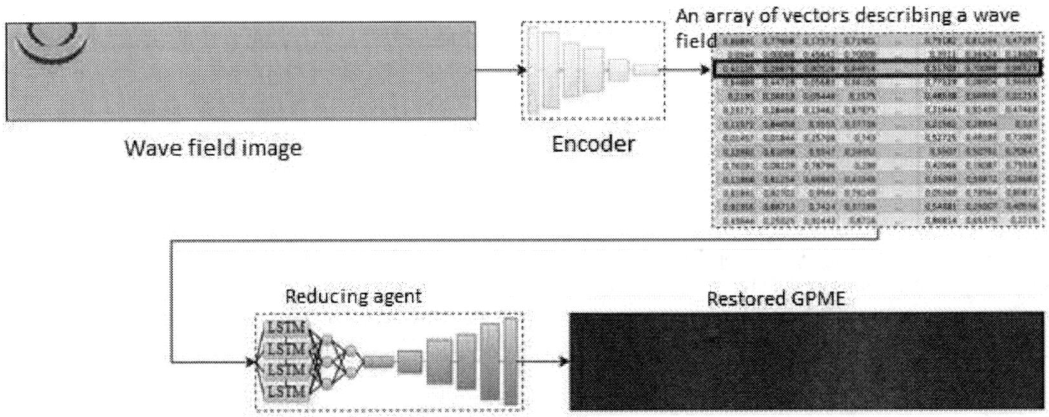

Figure 2. Developed NN architecture based on LSTM layer.

2.5. Restored GPME Architecture based on U-Net model

To build a neural network model based on the U-Net architecture [6], it was necessary to solve the problem of storing information in time from all incoming wave field images. To solve this problem, it was decided to add the output information from each level of the model after running each snapshot of the wave field. Each shot at the same time passed through the same weight. In other words, one encoder was used for all wavefield images, instead of separate encoders for each image. Subsequently, the sum of the encoded properties from each level was processed by subsequent layers of the model at the same level. Thus, from each snapshot that came to the input of the model, features of different levels were extracted, which were then used by the decoding part. Figure 3 shows the architecture of such a model.

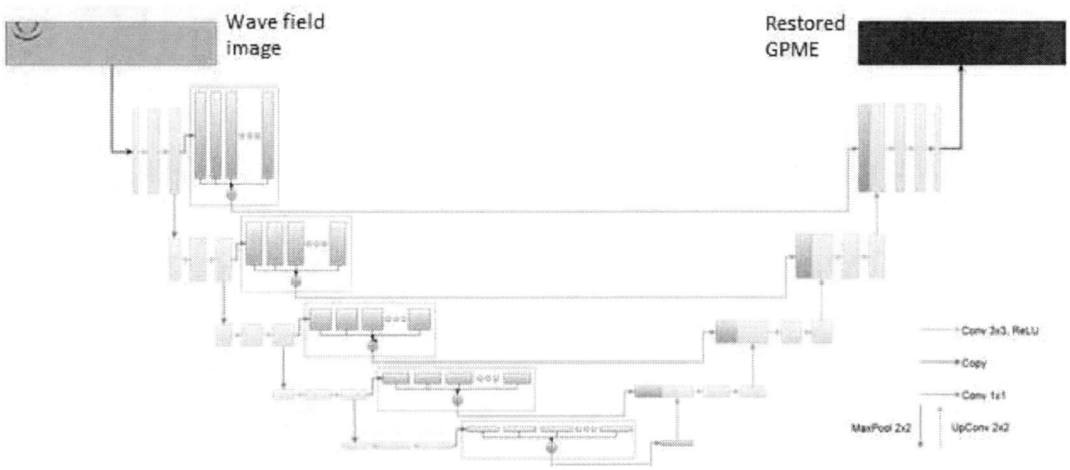

Figure 3. Designed by NS based on U-Net architecture.

3. Experimental results

3.1. Recovery examples

To confirm the quality of the recovery of the GPME with the developed models, experiments were carried out on the direct processing of wave field images developed by the NN, followed by visualization of the results.

The results of the restoration of the GPME on one wavefield pattern based on the LSTM layer (Figure 4) and the U-Net architecture (Figure 5) are presented. GPME models with different sizes and position of a cavity in an isotropic medium were considered.

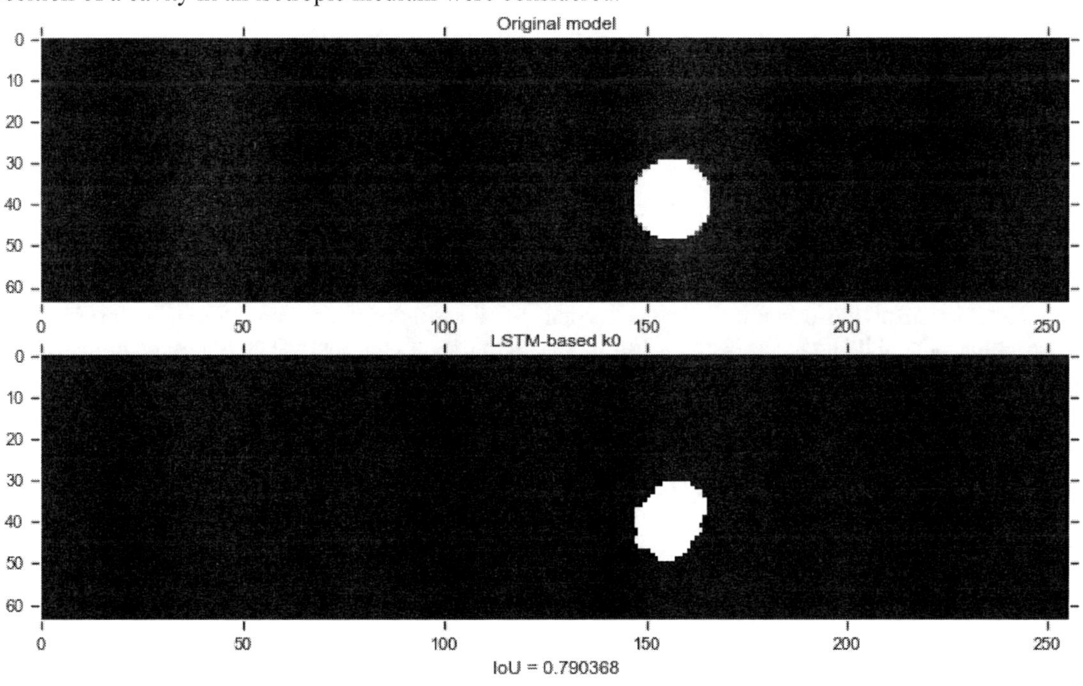

Figure 4. Result of restoration of the GPME based on the LSTM layer.

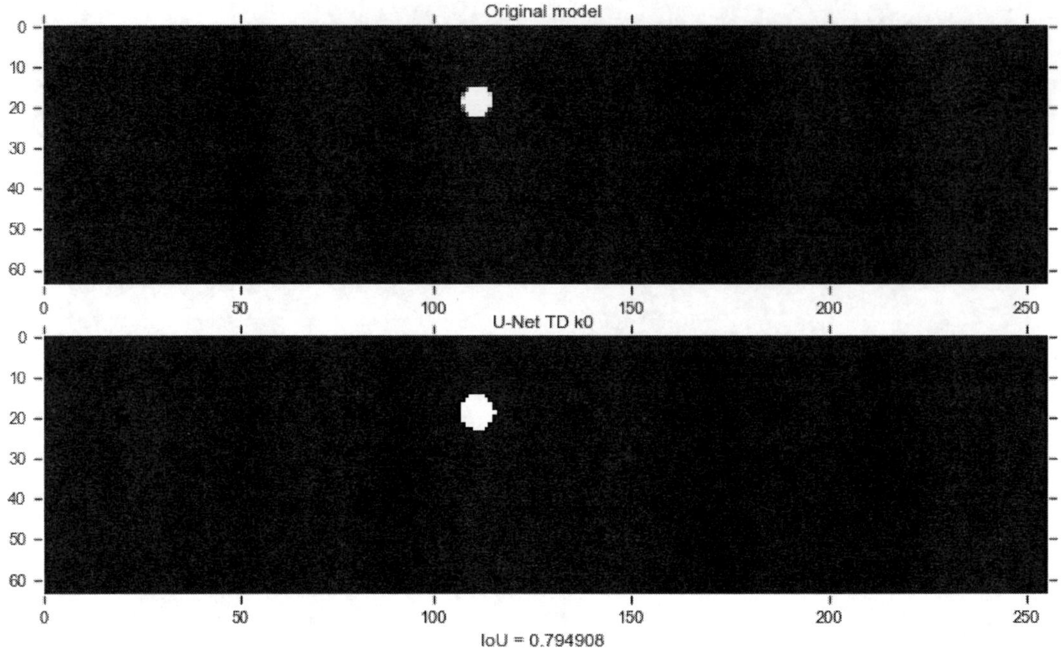

Figure 5. Result of restoration of the GPME based on the U-Net architecture.

Thus, when comparing the results presented in Figures 4 and 5, it can be seen that the model based on the U-Net architecture did much better than the model based on the LSTM layer. You can evaluate the quality of the restored model compared to the original. When using the U-Net, the oval area of the cavity is more fully restored.

3.2. Comparison of the working hours of the NN

The process of developing an NN is associated both with the quality of the results obtained and with speed. To compare the work of the presented NN architectures, experiments were carried out to restore models on different computing devices: a CPU and two GPUs. The main difference used by the GPU in the type and size of memory. The Nvidia GTX 850M is equipped with 4 GB DDR3 memory. Nvidia Titan X is equipped with 12 GB GDDR5 memory. The core frequency is also different. While the GTX 850M has a core frequency of 876 – 936 MHz, the Titan X has a core frequency of 1000 – 1089 MHz. All this gives Titan X an advantage in the processing speed of the NN model.

Evaluation of the execution time is carried out for single models and for their ensemble. An ensemble should be understood as a set of models that process the same data, whose readings are then averaged to improve the quality of their work.

In Figures 6, 7 shows the results of measuring the processing time of models based on the LSTM layer and the U-Net architecture on various devices.

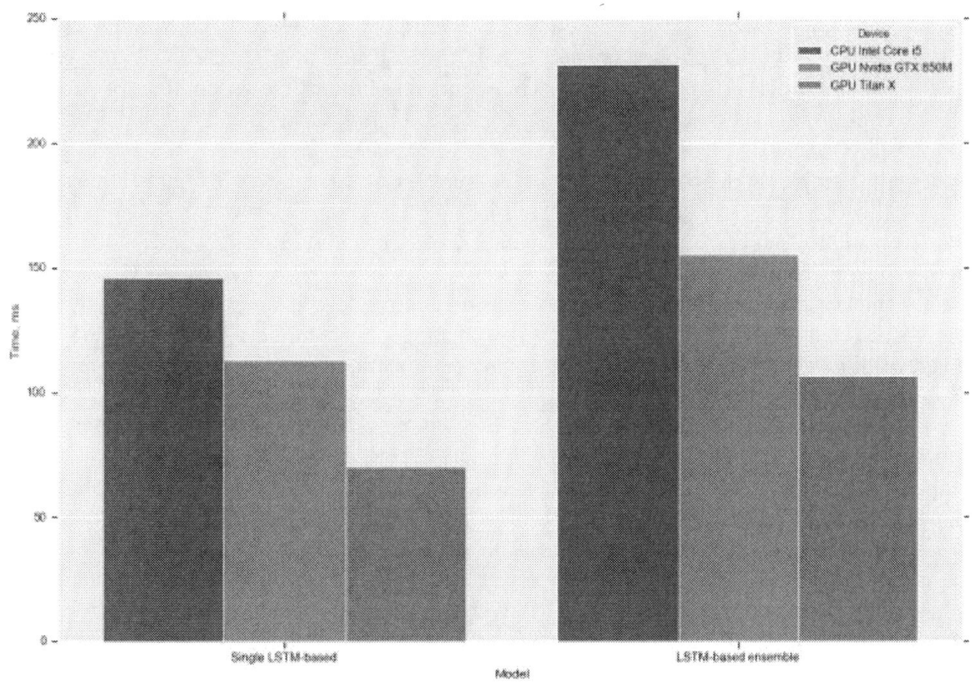

Figure 6. Measurement times for processing models based on the LSTM layer on various devices.

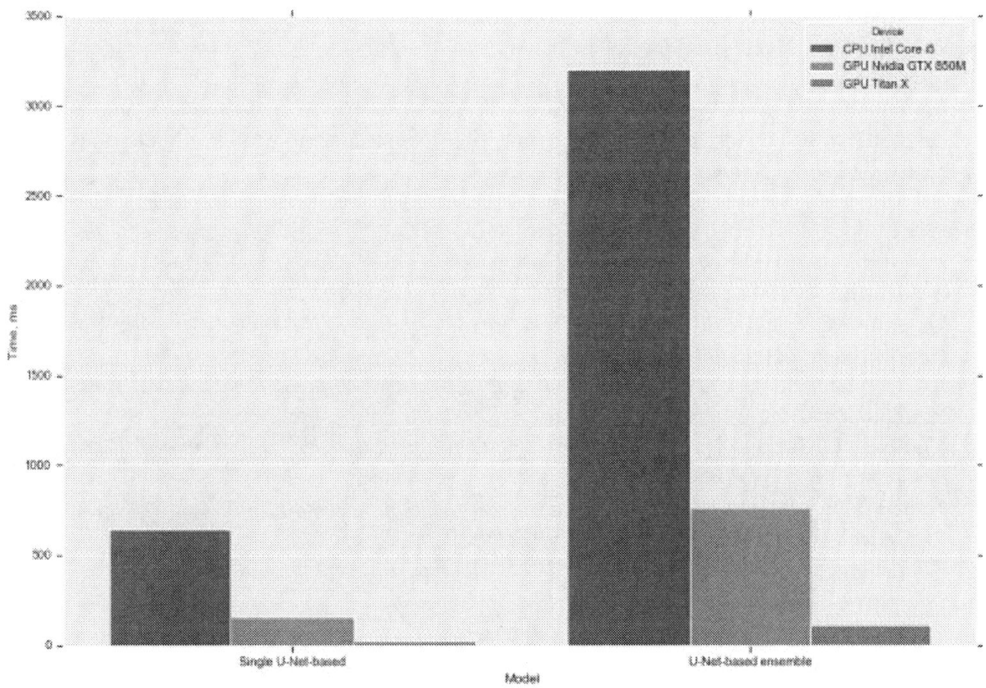

Figure 7. Measurement times for processing models based on the U-Net architecture on various. Devices.

Processing a single model based on the LSTM layer runs about twice as fast on the Titan X as on the Intel Core i5 CPU and about one and a half times faster than the GTX 850M GPU. When processing the ensemble, Titan X also wins. The processing time of the ensemble of models based on the LSTM layer on it is about 2 times less than on the CPU Intel Core i5 and one and a half times less than on the GTX 850M.

Processing a single model based on the U-Net architecture takes about 15 times longer on an Intel Core i5 CPU and 6 times longer on a GTX 850M than on a Titan X. Processing an ensemble of such models takes about 22 times more time on an Intel CPU Core i5 and 5 times more time on the GTX 850M than on the Titan X.

When comparing the results obtained between the two models, it was concluded that, on average, models based on U-Net architectures are processed longer than models based on the LSTM layer. However, the time difference is compensated by the demonstrated quality of the work of models based on the U-Net architecture.

4. Conclusion

The paper presents the architecture of the developed NN. The results of practical application of the developed NN architectures based on the LSTM layer and U-Net are shown. The features of the construction of such NN are described. The results of practical application of the developed software for the implementation of NN on test examples are presented. According to the obtained results, we can conclude that models based on the LSTM layer are processed faster than models based on the U-Net architecture. Also the ensemble of these models is processed much faster. This is achieved due to the fact that the coding part of this architecture only needs to process the wave field images that came to the model input once. After that, you can analyze the resulting sequence of vectors with different models of reducing agent to get the result of the ensemble. Models based on the U-Net architecture have no divisions with common parts, therefore, to analyze the wavefield pattern, complete processing is required for each U-Net-based model.

Despite the serious superiority of LSTM-based models in terms of processing time, both types of models are applicable to the on-site inspection task, since the processing time of a single model does not exceed 800 milliseconds, and the processing time of an ensemble of 5 models does not exceed 3300 milliseconds on a conventional dual-core CPU.

References

[1] Yakimenko A A, Morozov A E, and Karavaev D A 2018 *IOP Conf. Series: Journal of Physics: Conf. Series* **1015** 032148 doi: 10.1088/1742-6596/1015/3/032148

[2] Gavrilov A V and Maliavko A A 2017 *Proc. of the 2 Russian-Pacific conf. on computer technology and applications* 64–67

[3] LeCun Y and Bengio Y 1995 *Convolutional Networks for Images, Speech and Time-Series* (The Handbook of Brain Theory and Neural Networks MIT Press) pp 3361

[4] Yue-Hei Ng, Joe and Hausknecht 2015 *Proceedings of the IEEE conference on computer vision and pattern recognition* 4694–4702

[5] Noh H, Hong S, Han B 2015 *Proceedings of the IEEE International Conference on Computer Vision (ICCV)* 1520–1528 doi:10.1109/ICCV.2015.178

[6] Ronneberger O, Fischer P, Brox T 2015 *Medical Image Computing and Computer-Assisted Intervention* 234–241

Mathematical model of power supply system for remotely operated underwater vehicle with dc power transmission line and load voltage feedback

V Chekh[1], D Lyapunov[1,2], A Pravikova[1]
[1]Research Institute of Automation and Electromechanics of Tomsk State University of Control Systems and Radioelectronics
[2]National Research Tomsk Polytechnic University
E-mail: chekh1993@gmail.com

Abstract. The paper deals with a mathematical model of the closed loop power supply system for a remotely operated underwater vehicle with a DC transmission line via a rope-cable and the load voltage feedback. The state-space method is used to develop the mathematical model. Using differential equations in the form of Cauchy alleviates significantly the mathematical description. The results for both simulation and mathematical models of the closed loop system are compared, thus, confirming the adequacy of mathematical description and prospects for further application in the design of a versatile tool for calculating and adjusting the parameters of the control system for the power supply system under study.

1. Introduction

Currently, the use of remotely operated underwater vehicles (ROV) contributes significantly to the development of the oil and gas exploration and production industry in the shelf zones. Moreover, carrying out various underwater works in the seas, oceans and inland waters such as hydrographic and biological researches is an up-to-date demand. Likewise, providing rescue and search operations on sunken objects is a fruitful topic for underwater engineers and scientists. The ROV performance is largely determined by the power supply system (PSS).

According to the review of PSS versions for ROV of a research type [1], the transfer of DC energy over a cable-cable results in a reduction of the weight and size of the underwater part. That is due to the absence of a transformer and voltage converters within the subsea equipment.

The open-loop model of the PSS is considered in [2]. We will use this model to develop a closed-loop model of the PSS to ensure the stable load voltage within ±10% around its nominal value, and to provide appropriate dynamic performance of the system.

The purpose of the present paper is to develop a mathematical model of PSS for ROV with DC power transmission line over a rope-cable and a load voltage feedback.

2. A block diagram of the power supply system

The adopted block diagram of the PSS for ROV is portrayed in Figure 1.

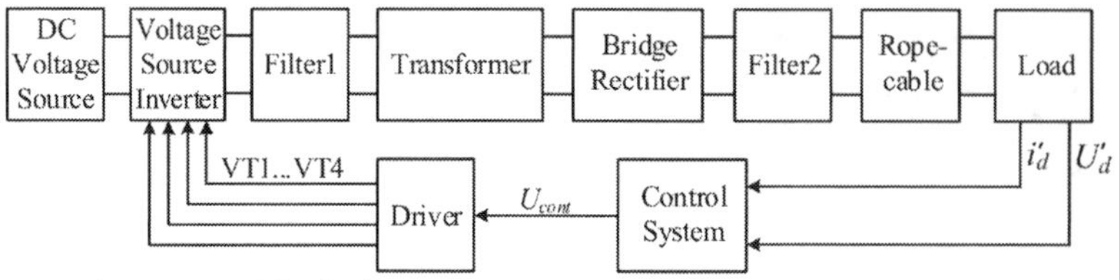

Figure 1. A block diagram of the closed-loop power supply system.

Accepted designations: i'_d and U'_d are the reduced to the primary winding of the transformer load current and voltage, accordingly; U_{cont} is the output signal of the voltage controller; VT1...VT4 are control signals to drive the gates of the transistors.

The diagram shows a DC voltage source connected to a voltage source inverter (VSI) with an output filter (Filter1). The output voltage of the filter is applied to the primary winding of the step-up transformer. The secondary winding of the transformer is linked to a single-phase bridge rectifier with its output filter (Filter2), providing the high voltage applied to the load via a rope-cable.

The basic principles of each PSS block operation are considered in detail in [3].

To enable the feedback we need to provide a set of equations for the power part of the PSS schematic depicted in Figure 2.

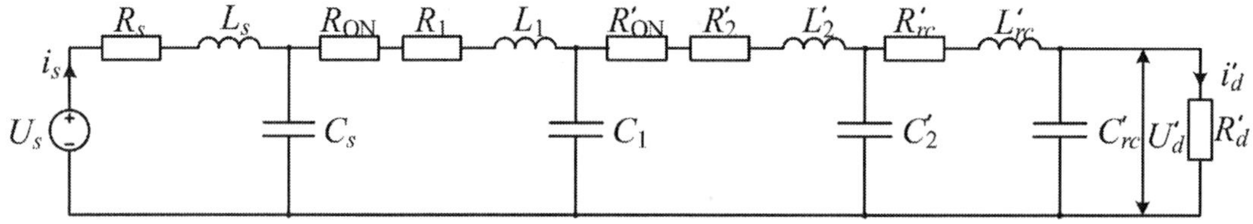

Figure 2. Equivalent electric circuit of the power supply system.

Accepted designations: U_s and is are DC source voltage and current, accordingly; R_s, L_s and C_s are the resistance, inductance and capacitance of the DC source circuit, accordingly; R_{ON} and R'_{ON} are the equivalent on-state resistances of the VSI and the bridge rectifier, accordingly; R_1, L_1 and C_1 are the resistance, inductance and capacitance of the VSI output filter, accordingly; R'_2, L'_2 and C'_2 are the resistance, inductance and capacitance of the rectifier output filter, accordingly; R'_{rc}, L'_{rc} and C'_{rc} are the resistance, inductance and capacitance of the rope-cable, accordingly; R'_d is the load resistance; the prime mark related to the parameters and values reduced to the primary winding of the transformer.

3. Mathematical description of the power supply

A set differential equation for the input circuit of the voltage source inverter (Figure 2) can be written in the following form:

$$\begin{cases} \dfrac{dU_{Cs}(t)}{dt} = \dfrac{1}{C_s}\cdot\left(i_s(t) - i_{\text{in}}(t)\right), \\ \dfrac{di_s(t)}{dt} = \dfrac{1}{L_s}\cdot\left(E - R_s\cdot i_s(t) - U_{Cs}(t)\right), \end{cases} \tag{1}$$

where $U_{Cs}(t)$ is the voltage on the input circuit capacitor C_s, $i_s(t)$ is the current of the voltage source. The VSI input current $i_{\text{in}}(t)$ is defined by the expression:

$$i_{in}(t) = K_m(\xi)\cdot i_{tr1}(t), \tag{2}$$

where $i_{tr1}(t)$ is the primary current of the transformer, $K_m(\xi)$ is the switching function for the VSI, the difference function ξ is defined by the difference between the reference signal $U_{ref}(t)$ and the sawtooth time-base voltage $U_{st}(t)$:

$$\xi = U_{ref}(t) - U_{st}(t). \tag{3}$$

The output VSI voltage represents the primary voltage of the transformer $U_{tr1}(t)$ and is described by the following relationship:

$$U_{tr1}(t) = K_m(\xi) \cdot U_{Cs}(t). \tag{4}$$

Within the framework of this paper we neglect the leakage resistances and inductances of the transformer and represent it as an ideal component with the transformation ratio K_{tr}.

The input current of the rectifier represents the secondary current of the transformer expressing by the equation:

$$i'_{tr2}(t) = (K_{1,3} - K_{2,4}) \cdot i'_2, \tag{5}$$

where i'_2 is the output current of the rectifier reduced to the primary winding, $K_{1,3}$ and $K_{2,4}$ are the switching functions of the rectifier.

All the issues concerning the control signal formation in the control system under study are considered in [4, 5]. Generally, the control signal is generated by comparing the reference signal to the periodical sawtooth one. The result is a set of the voltage pulses applied to the gates of the VSI transistors with a definite pulse width.

For the output filter of the rectifier the differential equation are as follows:

$$\begin{cases} \dfrac{dU'_{C2}(t)}{dt} = \dfrac{K_{tr}^2}{C'_2} \cdot (i'_2(t) - i'_{rc}(t)), \\ \dfrac{di'_2(t)}{dt} = \dfrac{1}{L'_2 \cdot K_{tr}^2} \cdot (U'_2(t) - U'_{C2}(t)), \end{cases} \tag{6}$$

Where $U'_2(t)$, $U'_{C2}(t)$, $i'_2(t)$ and $i'_{rc}(t)$ are reduced to the primary winding output voltages of the rectifier and its output filter, the output current of the rectifier and the current in the rope-cable, accordingly.

The electrical balance equations for the rope-cable acquire the following form:

$$\begin{cases} \dfrac{dU'_d(t)}{dt} = \dfrac{K_{tr}^2}{C_{rc}} \cdot (i'_{rc}(t) - i'_d(t)), \\ i'_d(t) = U'_d(t) / (R_d \cdot K_{tr}^2), \\ \dfrac{di'_{rc}(t)}{dt} = \dfrac{1}{L_{rc} \cdot K_{tr}^2} \cdot (U'_{C2}(t) - R_{rc} \cdot K_{tr}^2 \cdot i'_{rc}(t) - U'_d(t)), \end{cases} \tag{7}$$

where $U'_d(t)$ and $i'_d(t)$ are reduced to the primary windings load voltage and current, accordingly, C_{rc} and L_{rc} are the capacitance and inductance of the rope-cable, accordingly, R_d is the load resistance.

4. Simulation of the power supply

To simulate the model of the PSS with closed loop structure we used Matlab Simulink 2013b environment to create a simulation model using blocks. Using the Matlab Simulink software for developing the simulation model represents a versatile tool to create both the power circuit and control system in a single file [6]. Moreover, programming in C++ language directly [7] using Matlab C++Builder block gives the opportunity to create a program suitable for the PSS controller.

We demonstrate in Figure 3 the simulation model in Matlab Simulink obtained using S-Function block named `contr`.

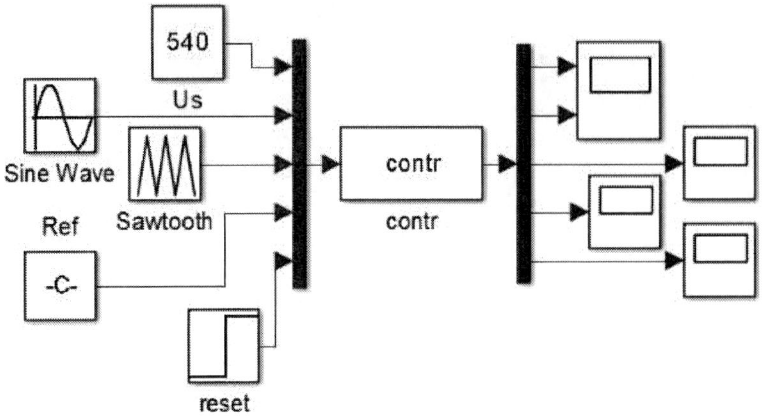

Figure 3. Simulation model of the power supply system in Matlab Simulink using S-Function.

Table 1 provides the realization of the block `contr` in C++, where the left column illustrates the control signals generation and the right one describes the electric circuit of the PSS. PI-controller is tuned optimally [8] to ensure 1000 V across the load within the error band of ±1%.

Table 1. Realization of the block `contr` representing the power supply system in C++

// Signals from Simulink	//Rectifier output voltage
E=*uPtrs[0]; si=*uPtrs[1];	if (Uab>0) U2=((Uab/Ktr));
pila=*uPtrs[2]; gamma=*uPtrs[3];	else if (Uab<0) U2=((-Uab/Ktr));
if (*uPtrs[4]==0) err=0,err_pre=0,e_reg=0,	else U2=0;
Ua=0,Ub=0,K_13=0,K_24=0,itr2=0,itr1=0,	//Secondary current of transformer
i2=0,irc=0,id=0,U2=0,UC2=0,Ud=0,	itr2=(K_13-K_24)*i2;
Km=0,Kma=0,Kmb=0;	//Primary current of transformer
//Controller	itr1=itr2/Ktr;
u_set=*uPtrs[3];	//VSI input current
k_fbv=0.00035507;	ivh=Km*itr1;
Kp=301.616;	Us=Us+dt*(1/Cs)*(is-ivh);
Ki=33478.406427854034148;	is=is+dt*(1/Ls)*(-Rs*is-Us+E);
err=u_set-k_fbv*UC2;	//Filter2 current
e_reg=e_reg+(Kp+dt*Ki)*err-Kp*err_pre;	i2=i2+dt*(1/L2)*(U2-UC2-i2*R2);
err_pre=err;	//Output filter2 voltage
if (e_reg>1) e_reg=1;	UC2=UC2+dt*(1/C2)*(i2-irc);
else if (e_reg<0) e_reg=0;	//Current in rope-cable
else e_reg=e_reg;	irc=irc+dt*(1/Lrc)*(UC2-(Rrc*irc)-Ud);
//VSI transistors	//Load voltage
Ua=e_reg*si;	Ud=Ud+dt*(1/Crc)*(irc-id);
Ub=-Ua;	//Load current
if (Ua>pila) Kma=0.5;	id=Ud/Rd;
else Kma=-0.5;	
if (Ub<pila) Kmb=0.5;	
else Kmb=-0.5;	
Km=Kma+Kmb;	
//VSI output voltage	
if (Km==1) Uab=(Us);	
else if (Km==-1) Uab=(-Us);	
else Uab=0;	
//Switching functions for diodes	
if (Uab>0) K_13=1, K_24=0;	
else if (Uab<0) K_13=0, K_24=1;	
else K_13=0, K_24=0;	

The software is written in C++ language and solves the set of equations (1)–(7). Here we note that the reference signal, sine wave signal, sawtooth signal, and DC source voltage are obtained directly from Simulink blocks (Figure 3). This could also be provided within the C++ file *contr.c*. To provide the appropriate model convergence we used Euler method for solving the differential equations with the step size dt=0.000001.

Thus, we obtained the transient response plot of the output voltage of the PSS applying a step reference signal at the input $U_{ref} = 0.9$ (Figure 4).

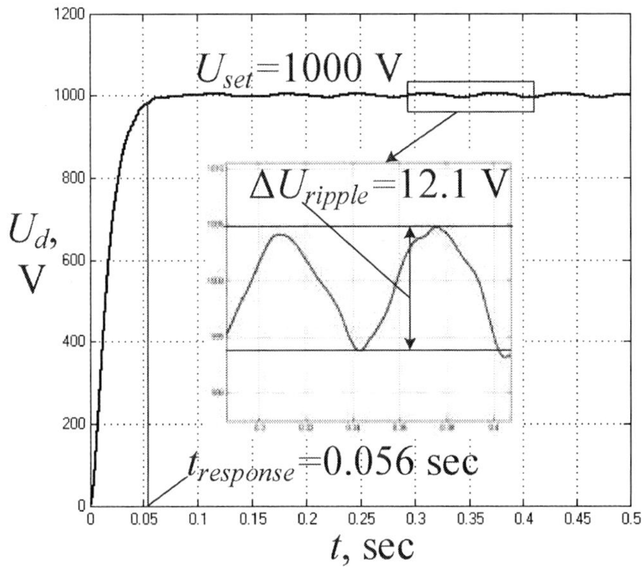

Figure 4. Transient response for the closed-loop power supply system

5. Conclusion

The mathematical model of PSS for ROV allows investigating both static and dynamic performance of the system under development. Using load voltage feedback along with the controller ensures the output voltage of 1000 V within the prescribed error band ±1% as well as the optimal response time. The developed model will presumably be taken as a basis for the development of an adaptive system that would take into account possible parameter changes during the PSS operation. This will serve as a versatile tool for calculating and adjusting the parameters of the control system Moreover, the model created will allow investigating all the operating modes of the PSS as well as the effect of the load on the power supply.

References

[1] Christ R, Wernly R 2007 *The ROV Manual: A User Guide for Observation-Class Remotely Operated Vehicles* (Waltham: Elsevier Ltd) p 679

[2] Rulevskiy V, Chekh V, Shurygin V and Pravikova A 2018 *IOP Conference Series: Materials Science and Engineering* **327** (2) 022018

[3] Rashid M 2018 *Power Electronics Handbook* (Cambridge: Elsevier) p 1496

[4] Kobzev A Mikhalchenko G Andriyanov A and Mikhalchenko S 2007 *Non-linear Dynamics of Semiconductor Converters* (Tomsk: Tomsk State University of Control Systems and Radioelectronics Press) p 224

[5] Dorf R and Bishop R 2014 *Modern Control Systems* (Hoboken: Pearson Education Limited)

[6] German-Galkin S 2013 *Virtual Laboratory of Semiconductor Systems in Matlab Simulink* (Saint-Petersberg: Lan) p 448

[7] Shaffer R 2007 *Fundamentals of Power Electronics with Matlab* (Boston: Charles River Media) p 384

[8] Rulevskiy V, Chekh V, Bukreev V, Mescheryakov R 2018 *Bulletin of the South Federal University, Technical Sciences* **1** 155–167

The Joule–Thomson effect and the non-equilibrium thermodynamics of sliding nano-contact

V S Deeva[1] and S M Slobodyan[2]
[1]Department of Engineering Entrepreneurship, National Research Tomsk Polytechnic University, Tomsk, Russia
[2]Department technology metals & materials, Tver State Technical University, Tver, Russia
E-mail:veradee@mail.ru

Abstract. Dynamical systems, having the space-time-varying behavior, takes place in the range of processes. Especially interesting are those which affect the structure and properties complex materials during its formation because of thermal communication between the different parts of the system. One of the most important features of the study is the temperature behavior in the contact area under sliding interaction. This behavior results from the large contribution of the heat capacity to the entropy. We investigate the influence of the Joule-Thomson effect on the behavior and thermal characteristics of contact area formed by debris and lubricant between the sliding surfaces when they are brought into contact with one another in commutator-and-brush assembly as an example. The model reflecting the influence of the change in the third body temperature on the material properties, geometry of the debris and the surrounding environment is proposed. The result shows the strong correlations between Joule-Thomson effect and mechanical fluctuations of the wear particles that are responsible for the size of the contact area.

1. Introduction

Dynamic contact between two surfaces that are moving over each other is usually accompanied by many physical phenomena [1–6]. From a physical point of view, the dynamic contact is an impulse process having the space-time-varying behavior in contact area, so called spatio-temporal processes [4–13]. The probabilistic nature of the dynamics and the structure of material leads to the strong correlations between interaction and physical properties that are responsible for the dynamic equilibrium [2–5, 9–13]. This takes place in the range of processes, including sliding contact [14]. However, despite of the secular study of the sliding contact [1, 14–17], many research questions of the pure equilibrium of the dynamical interaction between moving bodies remain unexplored, in particular, the influence of surface properties on thermodynamics of the sliding contact layer. The structure and properties (physical, thermal, mechanical, etc.) of complex materials during its formation process have strong correlation with the surrounding medium [2–5, 14–18]. Each of the material particles interacts with surroundings randomly, independently of each other, resulting from a probabilistic nature. This fact leads to the rich material properties with small changes in parameters [18–20].

An electrical contact pair is of interest [4,5,14,17] from the point of view of its structure due to material multivariance (carbon, metal, nanostructure, and others), which have a correlation [3–5] with environment formed by wear particles or lubricant under sliding interaction. These elements exemplify the strong correlations [7–11] between random mechanical fluctuations of the wear particles and temperature behavior of the contact area (air with wear particles or lubricant), in particular, the non-equilibrium thermodynamics [5,6,8–12,20–21]. The random mechanical fluctuations of the wear particles during interaction with the environment cause the Joule-Thomson effect [14,21–27] on the contact area. We could optimize the choice of material and evaluate the contacting pair durability taking into account the thermodynamics of interaction at the electrical contact.

The aim of this article is to investigate the measure of influence of the Joule-Thomson effect on the behavior and thermal characteristics of contact area between the sliding surfaces when they are brought into contact with one another. To achieve our research outcomes, we create the model reflecting the influence of the change in the contact area temperature on the material properties, geometry and surrounding environment.

2. Methods of research

We assume that the dynamic contact area of surface sliding interaction is constant. Thereafter, one surface slides on another surface under the action of the normal loading force with a constant linear speed, for example [17], when the electric brush slides on a slip ring (collector) of electrical machine (Figure 1). We ignore the destruction of the brush lateral surfaces inside the brush-holder because it has no real impact on the thermodynamics of sliding contact layer and could be negligible.

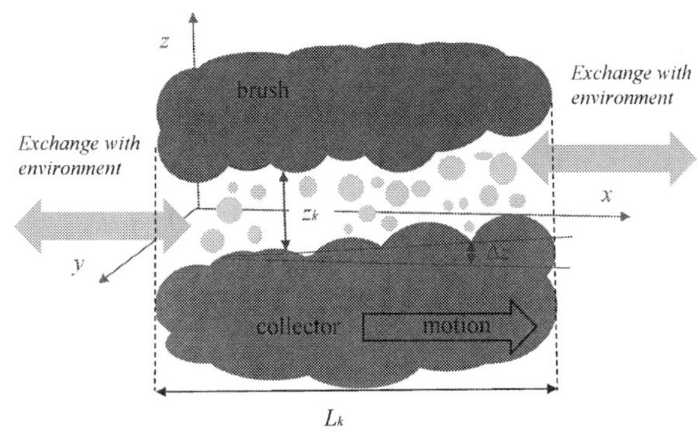

Figure 1. A schematic of sliding contact area (z_k is an height of the contact area Δz is the change of the inclination).

Any surface texture is rough; it is not ideally smooth on the nanoscale [5–8,18–20]. The surface roughness causes the wear damage processes, whilst the abrasive action is probably the most important factor affecting the surface wear. The geometry and size of the wear particles formed by the product of surface destruction (so called "the third body") are the random variables [14,17,28,29]. The random nature of the height of the layer between the surfaces is specified by the random geometric size and dimension of the wear particles and depends on theirs. Therefore, we assume that the average height of the contact area (z_k) is affected by the specific volume of the randomly packed particles between the surfaces: $V = s_{ca} \cdot v_{ca} \cdot t$, where s_{ca} is the cross-section through the contact area, v_{ca} is the average linear velocity of the particle stream along the surface (longitudinal) and t is the time period [14–17]. One thing that is evident from the existence of surrounding is that the wear particles dynamically interact with surrounding air or lubricating medium on the contact area. Meanwhile, the changing in height of the contact area leads to random mechanical fluctuations of the wear particles as a consequence of non-equilibrium state which can be altered by parameters such as pressure or temperature and correlate well with their values. Among the reasons for changes in the temperature are the mechanical fluctuations of the bodies during interaction which can result in the Joule-Thomson effect on the contact area [14,20–23]. This is exactly what stipulates the temperature-jump or the pressure-jump correlating with non-equilibrium state.

In the Joule-Thomson process a gas is forced through porous plug. In this paper, we consider the abundance of wear particles as the porous plug. To study this throttling process, we shall concern with the stream as a two-phase system, in particular, gas medium and wear particles in the sliding contact. This throttling process in the isenthalpic sliding contact layer can lead to either an increase (the Joule-Thomson coefficient is negative) or a decrease (the Joule-Thomson coefficient is positive) in temperature [14,20–26]. The Joule -Thomson coefficient (D_T) is defined as the ratio limit of the temperature change (ΔT) to the pressure change (ΔP): $D_T = lim\ [\Delta T / \Delta P] = D_T(T;\ P)$. It follows, that with choice of the surrounding gas, and suitable provided conditions, it should be possible to cool contact area for positive Joule-Thomson coefficient or to heat it for negative Joule-Thomson coefficient.

3. Equations

According the First law of thermodynamics the total amount of the heat is calculated through the formula $\Delta Q = \Delta Q_1 - \Delta Q_2 = dI - V\ dp$, where ΔQ_1 is the external interchange of heat; ΔQ_2 is the internal interchange of heat; dI is the sliding contact layer (third body) enthalpy; $V\ dp$ is the (third body) potential energy; p is the pressure in contact area; V is the contact area volume. This formula applies to the quasistatic process [14] in a homogeneous system. To calculate the external interchange of the heat ΔQ_1 we can use the formula

$$\Delta Q_1 = -k_{mo} L_K (t - t_0) dx, \tag{1}$$

where k_{mo} is the heat transfer coefficients between the contact area and an external medium; L_k is the size of contact area surface; $(t - t_0)$ is the temperature difference.
After due calculation, the equation (1) takes the form

$$k_{mo} L_K (t - t_0) dx = c_p (mdt - mD_T (\Delta p / l) dx) + mg (\Delta z / l) dx + md (v2 / 2) \tag{2}$$

where c_p is the specific heat capacity of the contact area; m is the mass flow rate; $\Delta p / l$ is the pressure variation per length l of the sliding surface; Δp is the distinction between the pressure at the beginning and at the end of the sliding contact area; $d(v^2/2)$ can be neglected due to the rate of negligible speed. Furthermore, $dz \equiv (\Delta z / l) dx$, where Δz is the change of the inclination between two ends of a length l of the contact area. The contact layer has a random orientation and, thuswise, the coordinates of the entry and outlet ends of a length l can be described by the arbitrary orientation coordinates in a three-dimensional space (Figure 1).
Integrating (2) [30–32], we receive

$$T = T_0 + (T_b - T_0) e^{-k_c x} - D_T (\Delta p / l)(1 - e^{-k_c x}) / k_c - g k_c c_p (\Delta z / l)(1 - e^{-k_c x}), \tag{3}$$

where T_0 is the background temperature; T_b is the temperature at the beginning of the sliding contact area; $l \geq x \geq 0$ is x coordinate of the contact layer cross-sectional; a component $k_c = k_{mo} \cdot L_k / m$ denotes the proportion of surrounding gas (medium) leaving the contact area/

4. Results of research and discussion

Let us denote the non-dimensional variation for a change of the relative inclination of the layer $a = \Delta z / l$. To calculate the temperature change ΔT for $k_c \cdot l < 1$ due to the Joule-Thomson effect, we need to estimate

$$\Delta T / g c c_p = k_c (\Delta z / l)(1 - e^{-k_c x}) = k_c \cdot a \left[k_c l + 0.5 (k_c l)^2 \right] = k_c^2 \cdot a l (1 + 0.5 k_c l), \tag{4}$$

The expression (4) shows that a measure of the strength of association between the Joule-Thomson effect and the surrounding medium follows a parabolic law, likewise under a temperature gradient, the temperature in contact area is not constant, and it varies linearly with the coordinate of the beginning of a length l of the contact area. Specifically, the contribution of the Joule-Thomson effect (normalized on the component of the specific heat capacity $g{\cdot}c_p$) to the changing in temperature depends on the coefficient k_c.

In addition, we offer to provide insight into the coefficient kc, which is used in (3) to determine the heat exchange with the environment, and material removal factors and it depends on the size of the contact area L_k. To analyze the role of the coefficient k_c, we compare the changes in temperature (normalized on $g{\cdot}c_p$) occurring in the contact area (4) considering two cases $k_c = 1$ and $k_c = 0.2$ for the small change of the relative inclination of contact area $a = 0.1$, and rather more $a = 0.2$. The coefficient kc is a function of the size of the contact area. In the Figure 2 one can see significant differences between the thermal response in the entry and outlet ends of the contact length l even though for the small change of the relative inclination $a = 0.1$, and more significantly for $a = 0.2$. The first is carried out on porous material composed of a slightly packing of the wear particles, the latter – at the material composed of close-packed particles [14,22–29]. Thus, we have a clear increasing trend in the influence of the Joule-Thomson effect in the both cases [21–27]. The comparison indicates a maximum difference of 30 percent of the thermal effect between the entry and outlet ends of the contact length l. This means that the strong correlations that are responsible for mechanical fluctuation leads to the increase in the dispersion of the temperature fluctuations in the outlet end of the contact area caused by the corresponding susceptibility to the surrounding medium [17,32–36]. It is also reasoned that the heat transfer to the environment and contact area extension can result in the gain in influence of the Joule-Thomson effect [21–27].

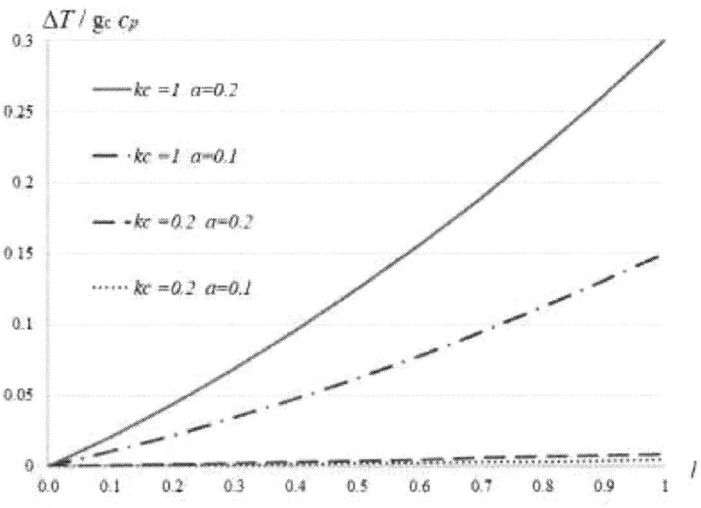

Figure 2. The temperature change in the contact area as a function of $k_c = f(L_k)$ for the change of the relative inclination $a = 0.2$ and $a = 0.1$

Due to the facts that k_c is a measure of heat output from the sliding contact, and degree of compaction is usually assessed through the density or specific volume reduction of the particles, the case of $k_c = 1$ corresponds to the maximum porosity for close-packed particles or, in other words, the lake of surroundings in the contact layer. In our opinion, in this case there is accumulation of heat due to the rolling and sliding friction under zero heat current [14,17].

As noted earlier, the changes in the contact area temperature under Joule-Thomson effect has resulted from mechanical fluctuations of wear particles which directly depend on the size of the wear particles, associated with the change of the relative inclination a of the sliding contact. The Figure 3 shows the correspondence between the change of the relative inclination of the contact layer and its thermodynamics, normalized on $g{\cdot}c_p$, for $k_c = 1$ and $k_c = 0.2$. Both lines depict reasonably linear correlations of these parameters, and moreover the temperature 8-fold increases in the average with increasing coefficient a. This fact means that the correlation between thermal effects in the sliding contact and wear damage caused by the different type of abrasion must be strong [3,37–39].

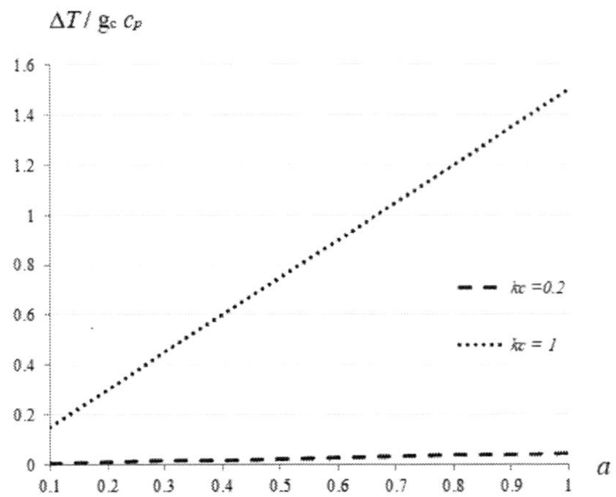

Figure 3. The influence of the change of the relative inclination of the contact layer on its thermodynamics for $k_c = 1$ and $k_c = 0.2$.

Figures 2 and 3 indicate that there is significant impact of the Joule-Thomson effect on the increase of the average temperature. It is interesting to note that larger wear particles have a greater temperature increase because of Joule-Thomson effect. It follows, that more porous material has a greater damage rate because of the large size of wear particles and possibly their mechanical fluctuations leads to the greater change in the temperature.

5. Conclusion

In this paper, we have attempted to explain the phenomenon underlying the occurrence of the Joule-Thomson effect in the consequence of differential change (decrease or increase) in the temperature of the contact area using the sliding contact in commutator-and-brush assembly as an example. The significance of the material properties dependency of temperature, in particular the Joule-Thomson effect was also evaluated. The coefficient which strong correlates with temperature changes in the contact area could have dramatic consequences on material properties due to the interaction or lack of interaction with a surrounding environment. Increasing the coefficient will decrease the influence of the Joule-Thomson effect. It can be expected that the more densely particles become closer packed, the less surrounding can penetrate into the contact area result in a reduction in the rate of heat exchange having as a consequence dramatic change in the temperature. The results presented in this paper show the strong correlations between Joule-Thomson effect and mechanical fluctuations that are responsible for the size of the contact area which leads to the conclusion that as the size of the contact area is significantly reduced, the Joule-Thomson effect could be negligible [14].

Acknowledgments

This work was supported by the Ministry of Education and Science of the Russian Federation in the framework of the implementation of the Program «Research and development on priority directions of scientific-technological complex of Russia for 2014–2020», and is funded from Tomsk Polytechnic University Competitiveness Enhancement Program grant. Assistance in the publication of the manuscript was provided by Tomsk Polytechnic University within the framework of TPU DRIaP-79/2018.

References

[1] Bhushan B 2013 *Introduction to Tribology* (New York: John Wiley & Sons)
[2] Lide D 2005 *Handbook of Chemistry and Physics* (Boca Raton: CRC Press LLC)
[3] Reinhart A 2018 *Statistical Science* **33** 299–318
[4] Bryant M and Khonsari M *et al* 2008 *Proc. R. Soc. A* **464** 2001–2014
[5] Deeva V and Slobodyan S 2018 *International Journal of Engineering & Technology* **7(2)** 167–170
[6] Li Q, Tullis T *et al.* 2011 *Nature* **480** 233–236
[7] Press W, Tolan M *et al* 1996 *Physica* B*: Condensed Matter* **221** 1–9
[8] Deeva V and Slobodyan S 2019 *Journal of Tribology* **141** 031602
[9] Brinkman H 1952 *J. Chemical Physics* **20** 571–581
[10] Kanafi M, Kuosmanen A *et al* 2015 *Int. J. Pavement Eng.* **16** 168–179
[11] Attia H, Gessesse Y *et al* 2007 *Wear* **263** 419–429
[12] Gattinoni C, Heyes D *et al* 2013 *Phys. Rev.* E **88** 10
[13] Pellegrini F, Franois P *et al* 2016 *Physical Review* E **94** 053001
[14] Deeva V and Slobodyan S 2017 *Tribol. Inter.* **105** 299–303
[15] Tzanakis I, Conte M *et al* 2013 *Wear* **303** 154–168
[16] Deng C and Zhang H *et al* 2017 *Materials Research Express* **4** 025602
[17] Deeva V, Slobodyan S 2017 *Proceedings of the 27th European Safety and Reliability Conference ESREL* 2577–2584
[18] Xie H, Mead J *et al* 2017 *Scientific Reports* **7** 44907
[19] Qianming G, Dan L *et al* 2013 *Tribology of polymeric nanocomposites. Friction and Wear of Bulk Materials and Coatings* (Berlin: Elsevier)
[20] Leff H 2012 *Part II Phys. Teach.* **50** 87–90
[21] Wright P 1966 *Nature* **209** 1125
[22] Chang H 2004 *Inventing temperature* (Oxford, Oxford University Press)
[23] Schmidutz T, Gotlibovych I *et al* 2014 *Phys. Rev. Lett.* **112** 040403
[24] Yang Y, Ding T *et al* 2016 *Heat Transfer Research* **47** 177–192
[25] Stauffer P, Lewis K *et al* 2014 *Transp. Porous Med.* **105** 471–485
[26] Pimentel L, Guerrieri Y *et al* 2016 *Ind. Eng. Chem. Res.* **55** 1117–1125
[27] Ng K, Xue H *et al* 2002 *Int. J. Heat & Mass Trans.* **45** 609–618
[28] Godet M 1984 *Wear* **100** 437–452
[29] Sanchez C, Chen Y *et al* 2017 *Tribol. Int.* **111** 66–72
[30] White R 2010 *Asymptotic Analysis of Differential Equations* (Singapore:World Scientific)
[31] Incropera F, DeWitt D *et al* 2011 *Fundamentals of heat and mass transfer* (New York: John Wiley & Sons)
[32] Morin D 2008 *Introduction to Classical Mechanics with Problems and Solution* (Cambridge: Cambridge University Press)
[33] Schlichting H 1979 *Boundary-Layer Theory* ed H Schlichting (New York: McGraw-Hill)
[34] Ewen J, Restrepo S *et al* 2017 *Tribol. Int.* **107** 264–273
[35] Xie H, Mead J *et al* 2017 *Scientific Reports* **7** 44907
[36] Zhang Y, Yu W *et al* 2017 *Scientific Reports* **7** 40507
[37] Zheng B, Huang Z *et al* 2016 *Tribol. Lett.* **62** 1–11
[38] Olomolehin Y, Kapadia R *et al* 2010 *Tribol. Lett.* **37** 49–58
[39] Hu C, Bai M *et al* 2014 *Appl. Surf. Sci.* **321** 302–309

IOP Conf. Series: Materials Science and Engineering 1019 (2021) 012030 doi:10.1088/1757-899X/1019/1/012030

Influence of modified natural zeolite on the thermal stability of epoxy based composites

Y S Murashkina[1], D V Martemyanov[2] and O B Nazarenko[1]
[1]School of Non-Destructive Testing, Tomsk Polytechnic University, Tomsk, Russia
[2]R&D School of Advanced Manufacturing Technologies, Tomsk Polytechnic University, Tomsk, Russia
E-mail: obnaz@mail.ru

Abstract. The effect of modified natural zeolite of the Shivyrtuysky deposits on the thermal behavior of epoxy composites at the heating was studied. The modification of the zeolite particles with aluminum oxyhydroxide (AlOOH) nanofibers was made. AlOOH nanofibers were obtained at the heating and oxidation of aluminum powder by water. The parameters of thermal-oxidative and thermal destruction of the samples with a filler concentration of 1, 2, 5, and 10 wt. % were determined. The limiting oxygen index was determined by the calculation method.

1. Introduction

Epoxy polymers have a range of unique properties, such as ease of curing, low shrinkage during curing, high mechanical and electrical insulating properties, adhesive ability, chemical resistance [1, 2]. Therefore, epoxy resins are widely used in construction, mechanical engineering, electrical engineering, and other areas of production. However, there is a disadvantage that complicates the use of epoxy resins: their high flammability and increased fire hazard [3, 4].

The flammability of epoxy resins can be reduced by the introduction of fillers with flame retardant properties into the epoxy matrix. The use of traditional halogen-containing flame retardants based on bromine and chlorine leads to the formation of highly dangerous and persistent toxic substances, such as phosgene, cyanide compounds, dioxins. Therefore, modern research is aimed at finding and developing environmentally friendly fire retardants. Among inorganic fillers, researchers are attracted by fillers based on natural minerals that are not highly toxic, they are relatively cheap and widely distributed in nature materials [5–7].

Natural zeolites belong to the group of crystalline aluminosilicate minerals. The structure of zeolites has channels and cavities occupied by cations and water molecules that can be freely removed and absorbed by the structure, due to which ion exchange occurs [8]. The structural features of the crystal lattice, the mineral composition, the presence of adsorbed water in the structure of natural zeolites, allow the use of zeolites to reduce the flammability of polymers.

The use of natural zeolites as flame retardant materials for protecting polymeric building structures from the fire was shown by Beyer H.K. et al. [9]. Lee J.-Y. et al. studied the effect of natural zeolite on the mechanical characteristics of epoxy resin [10]. They demonstrated that the mechanical properties of the composites were improved with decreasing particle size of the zeolite, and Young's modulus increased with the zeolite content raise in the epoxy polymer [10]. Bourbigot S. et al. studied the intumescent system based on a combination of ammonium polyphosphate and zeolite. They showed that the presence of zeolite in the studied intumescent system under investigation led to an enhancement in the flame retardant properties of polymeric materials due to the formation of phosphorus-containing carbon compounds at high temperatures [11].

The aim of this work is to determine the effect of the modified zeolite of the Shivyrtuysky deposit on the thermal properties of epoxy polymers.

2. Experimental

2.1. Objects of investigation

To obtain epoxy composites, the epoxy resin of grade ED-20 was used. The curing of the epoxy resin was performed using an amine hardener of polyethylene polyamine (PEPA). Fine powder (<50 μm) of zeolite rock from the Shivyrtuysky deposit of Russia was taken as a filler.

The zeolite particles were modified by precipitation of AlOOH nanofibers on their surface, which are formed by heating and oxidizing of the submicron aluminum powder in the chemical reaction with water under certain conditions in accordance with the method of hydrothermal synthesis [12, 13]. The containing zeolite particles suspension with a fraction size less than 0.1 mm and aluminum powder in distilled water was heated to 80 °C. In order to accelerate the aluminum-water reaction rate, an alkaline catalyst was used: 10% aqueous solution of sodium hydroxide NaOH. The addition of NaOH in water contributes to removing the aluminum oxide layer, and water can contact with the aluminum metal. Thus, the AlOOH nanofibers are deposited on the zeolite particles. The stirring of the reaction mixture was carried out using a magnetic stirrer. Then, the modified zeolite was washed to neutral pH, filtered and dried at room temperature.

According to the chemical composition, the zeolitic rock of the Shivyrtuysky deposit belongs to high-silica and is characterized by a high Si/Al ratio (> 4) [14]. The total content of SiO_2 and Al_2O_3 in the Shivyrtuysky zeolitic rock is 79.1%. The molar ratio of SiO_2/Al_2O_3 is 9.25. The content of zeolite in the rock of the Shivyrtuysky deposit can reach 95%. We used the samples which contain up to 60% of clinoptilolite.

2.2. Preparation

The procedure for the manufacture of epoxy-based composites consisted of the following steps.

The zeolite powder surface was treated with PEPA. Then, the required quantity of the zeolite filler was introduced into epoxy resin, and this composition was mixed by hand for 5 min without additional heating. The zeolite content in the epoxy compositions ranged from 1 to 10 mass %. After that, the hardener was added to the mixture in a ratio of epoxy resin to the hardener 10:1 by mass. Then, the composition was mixed for another 3 min. The resulting mixtures were cured in the molds made of silicone at room temperature for 24 h. The composition of the specimens is presented in table 1.

Table 1. Formulations of epoxy composites.

Sample	ED-20 (wt. %)	PEPA (wt. %)	Zeolite (wt. %)
E0	100	10	0
S1	100	10	1
S2	100	10	2
S5	100	10	5
S10	100	10	10

2.3. Characterizations

The size and morphology of the studied zeolite samples were investigated using a scanning electron microscope (SEM) TM-3000 and a transmission electron microscope (TEM) JEOL JEM 100 CX II. The Fourier transform infrared spectrum (FTIR) for the studied zeolite rock was recorded using spectrometer Nicolet 5700 in the spectral range 4000–400 cm^{-1}. The values of specific pore volume and specific surface area of modified zeolite were determined by the BET method using a Sorbtometr M instrument. The thermogravimetric analysis (TG) was made using SDT Q600 thermal analyzer. TG-curves were measured from 25 °C to 850 °C at a rate of heating 10 °C/min under argon atmosphere and air.

3. Results and discussion

3.1. Characterization of zeolite

Figure 1 shows the SEM image of the modified Shivirtuysky zeolite. Figure 2 shows the TEM image of the zeolite particles modified with AlOOH nanofibers. On the surface of the zeolite particles, several AlOOH nanofibers are observed; they have a width from 2 to 5 nm and a length from 50 to 150 nm.

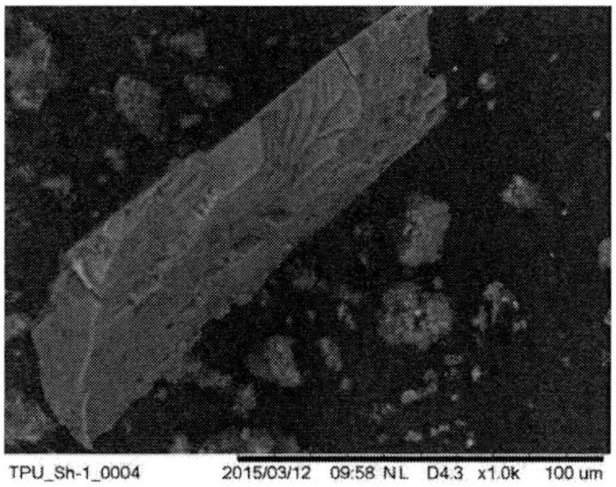

Figure 1. SEM image of zeolite.

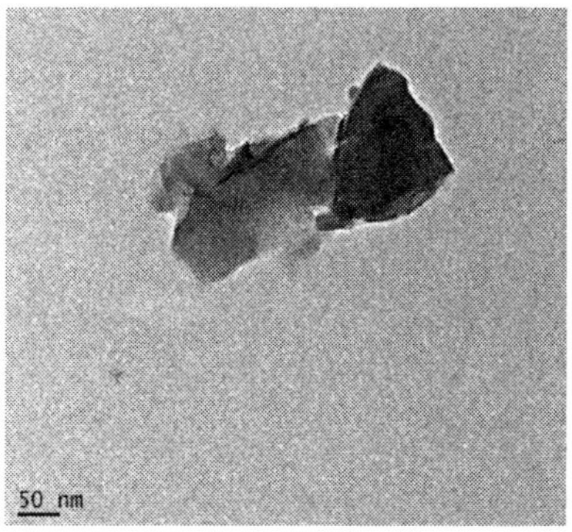

Figure 2. TEM image of zeolite.

The specific surface area of the modified Shivyrtuysky zeolite determined by the BET method is 69.6 m^2/g, and the specific pore volume is 0.030 cm^2/g. This corresponds to the AlOOH content of about 17 % in the zeolite.

The chemical functional groups on the zeolite particle surface were identified using the method of FTIR spectroscopy. The FTIR spectrum of the modified zeolite sample is given in Figure 3. The most intensive absorption band in the FTIR spectrum of the modified by AlOOH nanofibers zeolite is observed at 1058 cm^{-1}. This band corresponds to asymmetric stretching vibrations of Si-O-Al and Si-O-Si. The absorption band at 740 cm^{-1} corresponds to symmetric stretching vibrations of these bonds" [8]. The absorption bands at 442 and 482 cm^{-1} correspond to the bending vibration O-Si(Al)-O. The band at 1626 cm^{-1} corresponds to the bending vibration of zeolitic and adsorbed water H-O-H. In the range from 3000 to 3750 cm^{-1} there are strong absorption bands that are associated with the existence of zeolite water [15]. The peaks at approximately 3383, 3095, 1162, 1058, 740, and 482 cm^{-1} are typical absorption bands for AlOOH [15].

3.2. Characterization of epoxy composites

The results of TG analysis obtained at the heating of the zeolite filled epoxy composites in air atmosphere are given in Figure 4. From the TG results, the following data were obtained: T_5 – the value of temperature at which the mass loss was 5%; T_{50} – the value of temperature at which the mass loss was 50 %; T_{90} – the value of temperature at which the mass loss was 90 %; and residue at 600 °C. The results obtained are given in table 2.

According to the data obtained, the initial temperature of decomposition of the epoxy samples upon heating T_5 is depended on the concentration of filler. Thus, the temperature T_5 obtained for the unfilled epoxy resin E0 is 183 °C, and for other samples, T_5 is lower than for E0, except for the sample S1. The mid-point temperature at 50 % mass loss T_{50} increased for the samples S1, S2, S5, and S10 compared

IOP Conf. Series: Materials Science and Engineering 1019 (2021) 012030 doi:10.1088/1757-899X/1019/1/012030

to that of the unfilled epoxy resin E0. In addition to this, increasing the content of filler in epoxy composites leads to an increase in the residue at 600 °C.

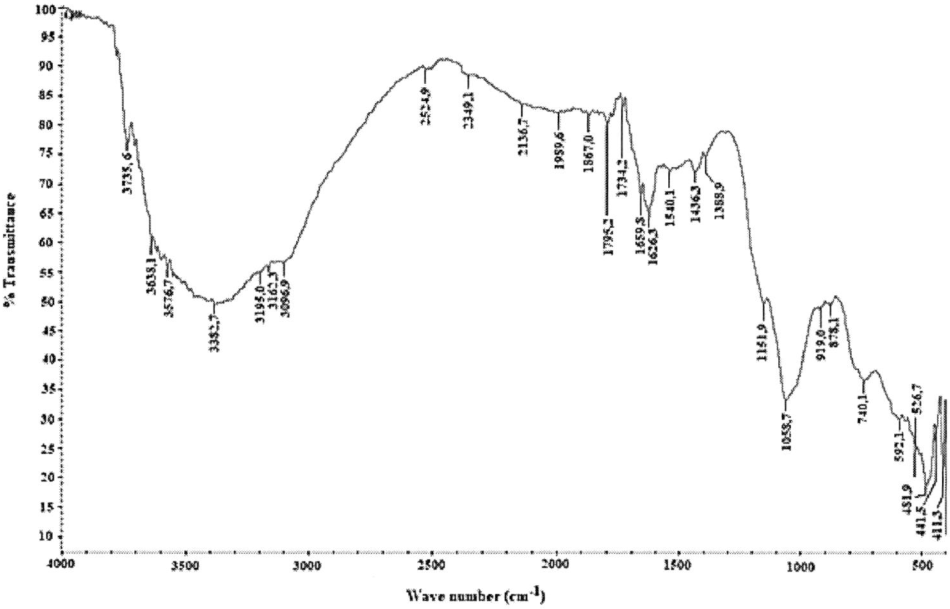

Figure 3. FTIR spectrum of zeolite.

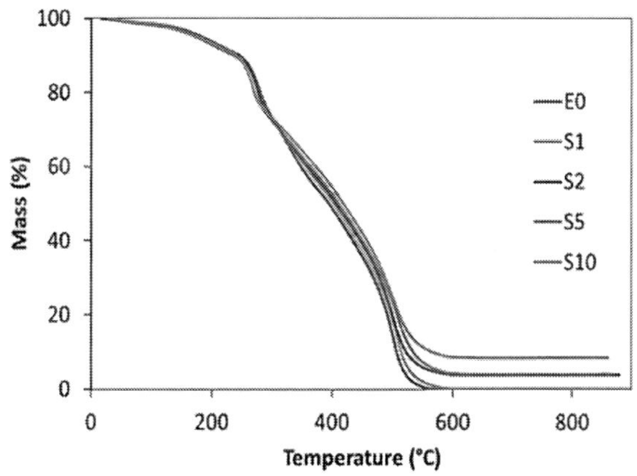

Figure 4. TG curves of epoxy composites filled with zeolite at heating under air.

Table 2. Thermal properties of zeolite based epoxy composites at heating under air.

Sample	T_5 (°C)	T_{50} (°C)	T_{90} (°C)	Residue at 600 °C (%)
E0	183	395	507	0
S1	186	402	514	0.1
S2	181	407	523	4.0
S5	174	412	535	4.3
S10	179	419	567	8.7

The results of thermal analysis of the epoxy composites received from TG data at the heating in argon are presented in table 3. Table 3 shows the values of temperature at which the mass loss of the epoxy composites was 5, 50, and 90% – T_5, T_{50}, T_{90}, respectively, and limiting oxygen index (LOI).

Table 3. Thermal properties of zeolite based epoxy composites at heating under argon.

Sample	T_5 (°C)	T_{50} (°C)	T_{90} (°C)	LOI
E0	192	376	554	20.8
S1	211	374	742	22.2
S2	186	373	588	20.9
S5	195	368	-	22.5
S10	184	368	-	23.3

The limiting oxygen index is the parameter that is used to characterize the flammability of polymeric materials. It can be determined as the minimum concentration of oxygen (in percent) in the mixture of oxygen and nitrogen that is needed to support the combustion of polymers. In our work, we calculated LOI from TG results made at the heating in argon according to the Van Krevelen and Hoftyzer equation [16]: LOI = 17.5 + 0.4Ch, where Ch is char yield at 850 °C.

According to the data obtained at the heating of studied samples in argon, the decomposition temperature T5 for the unfilled epoxy resin is 192 °C. The value T_5 is higher by 19 °C for the sample S1 and by 3 °C for the sample S5 in comparison with that of the sample E0. For the sample S2, the value T_5 is lower by 6 °C, and for the sample S10, is lower by 8 °C in comparison with that of the unfilled epoxy E0. The temperature corresponding to a 50% mass loss for all the studied samples is lower than that of the unfilled epoxy polymer. It can be seen from data presented in table 3 that the unfilled epoxy resin has LOI value of 20.8. It means that the sample E0 is referred to as flammable material according to classification [17] (the materials with LOI less than 20.95 %). The incorporation of the zeolite into the epoxy matrix resulted in an increase in LOI, except the sample S2. The zeolite based epoxy composites can be ascribed as slow-burning materials (the materials having LOI in the range of 20.95–28.0).

The results obtained can be explained by the influence of the release of water adsorbed by zeolite particles, as well as by the physical-mechanical approach. The particles of zeolite restrain the motion of macromolecular chains in the epoxy polymer and lead to the formation of the carbon residue upon heating. The carbonaceous residue acts as a heat barrier in the epoxy composites and prevents the heat flux into the polymer matrix and the destruction of the epoxy resin.

4. Summary

In our study, the zeolite based epoxy composites were fabricated with a content of the filler 1; 2; 5 and 10 wt. %. The zeolite particles were modified with nanofibers of aluminum oxyhydroxide. The parameters of thermal degradation and thermo-oxidative degradation of the produced epoxy samples were determined by the method of thermal analysis. The data obtained indicate the effectiveness of the use of zeolite particles as filler for the epoxy resin to enhance the thermal stability and reduce the flammability of the epoxy composites. It is shown in the work that the effect of improving the thermal stability depends on the content of the filler and can be explained.

Acknowledgments

The research is carried out at Tomsk Polytechnic University within the framework of Tomsk Polytechnic University Competitiveness Enhancement Program grant.

References

[1] Lee H and Neville K 1967 *Handbook of Epoxy Resins* (New York: McGraw-Hill) p 960

[2] Jin F-L, Li X, Park S-J 2015 *J. Ind. Eng. Chem.* **29** 1–11
[3] Irvine D J, McCluskey J A, Robinson I M 2000 *Polym. Degrad. Stab.* **67** 383–396
[4] Wilkie C A and Morgan A B 2010 *Fire Retardancy of Polymeric Materials* (Boca Raton: CRC Press) p 856
[5] Dasari A, Yu Z Z, Cai G P, Mai YW 2013 *Prog. Polym. Sci.* **38** 1357–1387
[6] Kiliaris P, Papaspyrides C D 2010 *Prog. Polym. Sci.* **35** 902–958
[7] Hewitt F, Hull T R 2017 *Fillers for Polymer Applications. Polymers and Polymeric Composites: A Reference Series* ed R Rothon (Cham: Springer) pp 329–354
[8] Breck D 1974 *Zeolite Molecular Sieves* (New York: Wiley) p 320
[9] Beyer H K, Borbely G, Miasnikov P, Rozsa P 1989 *Zeolites as Catalysts, Sorbents and Detergent Builders. Applications and Innovations* eds H G Karge and J Weitkamp (Amsterdam: Elsevier Science Publishers) pp 635–644
[10] Lee J Y, Shim M J, Kim S W 1999 *Polym. Eng. Sci.* **39** 1993–1997
[11] Bourbigot S, Bras M L, Delobei R, Tremillon J M 1996 *J. Chem. Soc., Faraday Trans.* **92** 3435–3444
[12] Zykova A, Livanova A, Kosova N, Godymchuk A, Mamontov G 2015 *IOP Conf. Ser. Materi. Sci. Eng.* **98** 012032
[13] Lerner M I, Svarovskaya N V, Psakhie S G, Bakina O V 2009 *Nanotechnologies in Russia* **4**(11–12) 741–757
[14] Melnikova T V, Nazarenko OB, Visakh P M 2017 *Proceedings of 11th International Forum on Strategic Technology (IFOST 2016)* Pt. 1 (Novosibirsk: NGTU) pp 36–38
[15] Salimi F, Abdollahifar M, Jafari P, Hidaryan M 2017 *J. Serb. Chem. Soc.* **82**(2) 203–213
[16] Van Krevelen D W, Hoftyzer P J 2009 *Properties of Polymers* (Amsterdam: Elsevier) p 1004
[17] Fenimore C P 1975 Candle-type test for flammability of polymers *Flame-Retardant Polymeric Materials* vol. 1 eds M Lewin, S M Atlas, E M Pearce (New York: Plenum) pp 371–397

IOP Conf. Series: Materials Science and Engineering 1019 (2021) 012031 doi:10.1088/1757-899X/1019/1/012031

Recognition of english and russian-language texts based on frequency characteristics

Y Kotov[1] **and O Sanina**[2]
[1]Department of Information Security, Novosibirsk State Technical University, Novosibirsk, Russia
[2]Department of Computer Engineering, Novosibirsk State Technical University, Novosibirsk, Russia
E-mail:kotov@corp.nstu.ru

Abstract. Distinction of texts in one language from texts in others is necessary to solve the problems of automated text analysis. The paper presents criteria and critical values for recognizing English-language and Russian-language texts. The obtained criteria are estimated by experiments. The paper describes the methods to estimate the size of character codes and to identify a space character in a text. The algorithm for recognizing texts in the English and Russian languages with arbitrary encoding is studied and its accuracy is estimated experimentally.

1. Introduction

Language recognition is required to solve the problems of multi-coding communication, steganography, cryptanalysis and other problems of formal text analysis [1–4]. Nowadays, to solve the problem of language recognition, neural networks and Naive Bayes Classifier [5–7] are used. However, if the text encoding is unknown, the mentioned methods cannot be applied.

The problem solution with an arbitrary encoding being used assumes comparing certain characteristics of the text under investigation with sample characteristics of texts in a known language. In general cases it means that an arbitrary character sequence must be studied to answer the question if the sequence is a text in a given language (languages) or otherwise *not a text* at all. Such definition of language recognition differs from the known [5–7] cases and requires forming a new approach to solve the problem considered in the paper.

2. Problem definition and solution steps

In the paper we assume that each letter is represented by k characters ($k \geq 1$) and consider texts in the Russian and English languages. In this regard, to recognize a text in an arbitrary character sequence it is necessary to tackle the following tasks:

1. To estimate the size of a character code k.
2. To make a primary distinction between English-language and Russian-language texts.
3. To substitute characters if $k > 1$.
4. To identify spaces in the texts.
5. To compare frequency characteristics of the texts with the language samples.
6. To identify letters in the texts.
7. To substitute characters of the texts for language letters.

3. The size of character codes and primary text distinction

If the size of character codes is known to be not greater than 2 ($k \leq 2$), the conjunction index can be used for estimating the size of character codes. The conjunction index is calculated based on the digram characteristics of a text [8]. If the value of the conjunction index tends to zero it can be concluded that combinations of two characters represent the letters of the alphabet; if it tends to one then one character stands for one letter. The critical value of the conjunction index for English-language texts is 0.816, and for Russian-language texts the index is equal to 0.821 [8]. If the current value is smaller than the critical, then two characters represent each letter in the text, otherwise one character represents one letter.

In the general case to estimate the size of a character code k, the alphabet cardinality of letters in a given language can be used as a criterion [9]. Measured for each point of the scale that represents the text size x, the minimal and maximal amount of the alphabet letters being used can be taken as critical values. Figure 1 shows the ranges of the letters' number being used in the English and Russian languages at each scale point.

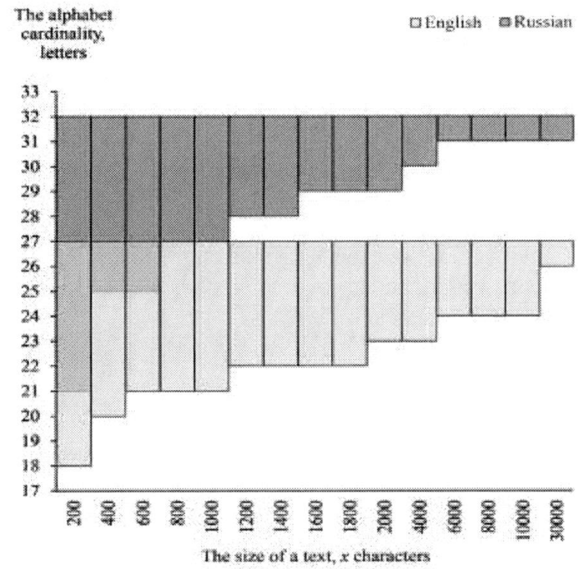

Figure 1. Ranges of the alphabet cardinality in Russian-language and English-language texts.

In this regard, the size of character codes can be estimated as follows:

1. To fragmentize a text – i.e. to represent a text as sequences of characters with the size k;
2. To compare the alphabet cardinality of the obtained code fragments with the sample number of the alphabet letters used in natural language texts (Figure 1).

If the alphabet cardinality of the obtained code fragments falls within the range presented in Figure 1, the code size is considered as accepted. There may be cases when several possible code sizes for one text fall within the range of the accepted number of letters. In particular, if the encoding sequences consist of the alphabet letters in the language under consideration or of single texts in a natural language, and the alphabet cardinality meets the criterion shown in Figure 1. Brute force can be employed to find all possible values for the size of the character code.

As a result, after all the possible values k from one up to $x/200$ being brute-forced, the vector with accepted values of character codes k is formed. In addition, if the cardinality of character codes fits the language, a flag L must be assigned to indicate this language. The flag L can have the following values:

- 0 – the cardinality of character codes does not fit any language (and the current value of k is rejected);

- 1 – the cardinality of character codes fits the range of the Russian language;
- 2 – the cardinality of character codes fits the range of the English language;
- 3 – the cardinality of character codes fits the range both of the Russian and English languages ($x < 1200$).

If $k > 1$, the text is reencoded so that one letter of the language is presented by one character, and the analysis moves to the next step.

Stating with the greatest, the accepted values of the obtained vector are checked. The first size of a character code for which a text is identified as a text in a natural language is taken as the size of character codes.

At this step, English-language texts can be distinguished from Russian-language texts based on the cardinality of used alphabet letters. The unambiguous distinction between Russian-language and English-language texts based on this criterion is possible for texts with size x more or equal to 1200 characters. The distinction of texts with the size up to 1200 characters is possible only if the number of letters being used in the text does not fall within the intersection area of the languages.

However, the sole knowledge of the alphabet cardinality is not enough to make a final decision if the text belongs to any language. For that, frequency characteristics of a text should be compared with sample frequency characteristics of the chosen language. Prior to obtaining the frequency characteristics, the presence or absence of a space character must be confirmed, and a space character should be identified.

4. Identification of space characters in a text

A variety of criteria can be employed to identify a space character in a text [10]. The most accurate is the criterion based on the deviation of the expected value and the variance of the Poisson distribution, since the distribution of words from dictionary over the length is close to it [10]. The critical values for Russian-language texts are determined as (1):

$$\delta(x)=\begin{cases}2.76\cdot 10^{-4}x+2.09, & 200\le x\le 6000\\ 3.75, & x>6000\end{cases} \quad (1)$$

For English-language texts, a constant value equal to 1.35 can be taken as a critical value for texts of any size (2):

$$\delta(x)=1.35 \quad (2)$$

If the index of words' length is used as a criterion, the critical values for Russian-language texts are considered as (3):

$$y(x)=\begin{cases}0.00087x+9.206, & 200<x<1200\\ 10.25, & x\ge 1200\end{cases} \quad (3)$$

For English-language texts, a constant value equal to 23 [10] can be assumed as a critical value for texts of any size (4):

$$y(x)=23 \quad (4)$$

5. Recognition of English-language and Russian-language texts

The following frequency characteristics should be used to recognize texts in a natural language: the coincidence index and the number of bigrams being used in a text [8].

The range of English-language text's size can be divided into 4 intervals, and in each interval, the following ranges for the critical values of the coincidence index can be accepted (with 95% of texts fall within the range) (5):

$$
\begin{array}{lll}
1) & 200 \le x < 600, & 0.0541 < I_c(x) < 0.0778; \\
2) & 600 \le x < 1400, & 0.0584 < I_c(x) < 0.0738; \\
3) & 1400 \le x < 30000, & 0.0592 < I_c(x) < 0.0729; \\
4) & 30000 \le x \le 350000, & 0.0613 < I_c(x) < 0.0700.
\end{array} \quad (5)
$$

The intervals for English-language texts (5) differ from the intervals (6) known for Russian-language texts:

$$
\begin{array}{lll}
1) & 200 \le x < 800, & 0.04950 < I_c(x) < 0.06830; \\
2) & 800 \le x < 8000, & 0.05187 < I_c(x) < 0.06587; \\
3) & 8000 \le x < 100000, & 0.05415 < I_c(x) < 0.06175; \\
4) & 100000 \le x \le 350000, & 0.05481 < I_c(x) < 0.06069.
\end{array} \quad (6)
$$

If the calculated coincidence index does not fall within the range of the critical values (5) or (6), the text is rejected and its analysis is over; otherwise the analysis continues in checking the number of bigrams being used in the text.

For Russian-language texts, six intervals and their corresponding critical values are determined [8] according to the relative number of used bigrams (7):

$$
y(x)=\begin{cases}
3.16\cdot 10^{-4}x+0.09, & 200\le x<800; \\
1.30\cdot 10^{-4}x+0.23, & 800\le x<2000; \\
5.70\cdot 10^{-5}x+0.38, & 2000\le x<4000; \\
2.19\cdot 10^{-5}x+0.52, & 4000\le x<10000; \\
4.86\cdot 10^{-6}x+0.70, & 10000\le x<30000; \\
3.01\cdot 10^{-7}x+0.83, & 30000\le x\le 350000.
\end{cases} \quad (7)
$$

The following intervals and their corresponding critical values are determined for the relative number of used bigrams in English-language texts (8):

$$
y(x)=\begin{cases}
3.39\cdot 10^{-4}x+0.12, & 200\le x<800; \\
1.25\cdot 10^{-4}x+0.29, & 800\le x<2000; \\
3.74\cdot 10^{-5}x+0.46 & 2000\le x<4000; \\
2.00\cdot 10^{-5}x+0.53, & 4000\le x<10000; \\
5.18\cdot 10^{-6}x+0.68, & 10000\le x<30000; \\
4.02\cdot 10^{-7}x+0.83, & 30000\le x\le 350000.
\end{cases} \quad (8)
$$

If the number of used bigrams normalized to 961 (for Russian-language texts) and 676 (for English-language texts) is smaller than the critical values, the text is accepted as Russian/English; otherwise, a text is considered as *not a text* and rejected.

The accuracy of the step for identification of English-language texts without considering the cardinality of the alphabet being used has been estimated by the experiment with the validation sample of the Russian-language and English-language texts. The results are shown in table 1, where N is the number of texts with the size x, m stands for the number of errors in texts with the size x, P_E is the error frequency, P_{AV} is the average error frequency in texts, α – type I errors (English-language texts are rejected), β – type II errors (Russian-language texts are identified as texts in English).

The average error is 7.9%. The minimal error is zero at several scale points. The maximal error is 38% at the scale point $x = 200$.

The amount of error reduces as the text size is increasing: at the point $x = 200$ the average error achieves 38%, at the point $x = 400$ – 25%, the error is 16.3% at the interval (400; 800], at the interval (800; 2000] – 11.3%, and at the interval (2000; 350000] – 1%.

The accuracy of the step for identification of Russian-language texts without considering the cardinality of the alphabet being used has been estimated by the experiment with the same validation samples. Results are provided in table 2, where α – type I errors (Russian-language texts are rejected), β – type II errors (English-language texts are identified as texts in Russian).

Table 1. The errors in identifying English-language texts.

x	α			β			P_{AV}	x	α			β			P_{AV}
	N	m	P_E	N	m	P_E			N	m	P_E	N	m	P_E	
Group 1								Total 2	500	2	0.004	500	38	0.076	0.04
200	100	4	0.04	100	72	0.72	0.38	Group 3							
400	100	9	0.09	100	41	0.41	0.25	10000	100	0	0	100	3	0.03	0.015
600	100	4	0.04	100	28	0.28	0.16	30000	100	0	0	100	0	0	0
800	100	2	0.02	100	31	0.31	0.165	50000	99	0	0	100	0	0	0
1000	100	1	0.01	100	24	0.24	0.125	70000	99	0	0	100	0	0	0
1200	100	3	0.03	100	23	0.23	0.13	90000	98	1	0.010	100	0	0	0.005
1400	100	1	0.01	100	18	0.18	0.095	110000	98	2	0.020	100	0	0	0.010
1600	100	0	0	100	21	0.21	0.105	Total 3	594	3	0.005	600	3	0.005	0.005
1800	100	4	0.04	100	17	0.17	0.105	Group 4							
2000	100	2	0.02	100	24	0.24	0.13	100000	30	0	0	100	0	0	0
Total 1	1000	30	0.03	1000	299	0.299	0.165	150000	30	1	0.033	50	0	0	0.017
Group 2								200000	30	1	0.033	50	0	0	0.017
2000	100	2	0.02	100	18	0.18	0.1	250000	30	0	0	50	0	0	0
4000	100	0	0	100	8	0.08	0.04	300000	29	0	0	79	0	0	0
6000	100	0	0	100	3	0.03	0.015	350000	27	0	0	21	0	0	0
8000	100	0	0	100	5	0.05	0.025	Total 4	176	2	0.011	300	0	0	0.006
10000	100	0	0	100	4	0.04	0.02	Total	2270	37	0.016	2400	340	0.142	0.079

Table 2. The errors in identifying Russian-language texts.

x	α			β			P_{AV}	x	α			β			P_{AV}
	N	m	P_E	N	m	P_E			N	m	P_E	N	m	P_E	
Group 1								Total 2	500	3	0.006	500	214	0.428	0.217
200	100	11	0.1	100	61	0.61	0.355	Group 3							
400	100	13	0.13	100	66	0.66	0.395	10000	100	0	0	100	0	0	0
600	100	8	0.08	100	78	0.78	0.43	30000	100	0	0	100	0	0	0
800	100	1	0.01	100	76	0.76	0.385	50000	100	0	0	99	0	0	0
1000	100	2	0.02	100	62	0.62	0.32	70000	100	0	0	99	0	0	0
1200	100	0	0	100	51	0.51	0.255	90000	100	0	0	98	0	0	0
1400	100	1	0.01	100	39	0.39	0.2	110000	100	0	0	98	0	0	0
1600	100	1	0.01	100	52	0.52	0.265	Total 3	600	0	0	594	0	0	0
1800	100	1	0.01	100	48	0.48	0.245	Group 4							
2000	100	1	0.01	100	47	0.47	0.24	100000	100	0	0	30	0	0	0
Total 1	1000	39	0.039	1000	580	0.58	0.310	150000	50	2	0.04	30	0	0	0.02
Group 2								200000	50	2	0.04	30	0	0	0.02
2000	100	1	0.01	100	52	0.52	0.265	250000	50	1	0.02	30	0	0	0.01
4000	100	0	0	100	57	0.57	0.285	300000	79	0	0	29	0	0	0
6000	100	0	0	100	52	0.52	0.26	350000	21	1	0.048	27	0	0	0.024
8000	100	0	0	100	51	0.51	0.255	Total 4	300	6	0.02	176	0	0	0.01
10000	100	2	0.02	100	2	0.02	0.02	Total	2400	48	0.02	2270	794	0.350	0.185

The average error is 18.5%. The minimal error is zero at several scale points. The maximal error is 43% at the scale point $x = 600$.

The amount of error reduces as the text size is increasing: at the interval [200; 1000] the error is 37.7%, at the interval (1000; 10000] – 22.9%, at the interval (10000; 350000] – 0.6%.

As the language of the text is identified, the text analysis can move to the next step, where the language letters are identified and substitution of character codes for letters of the language follows the identification. The techniques described, for instance, in [11] can be employed for letter identification.

6. Conclusion

The flowchart of the algorithm for recognizing English-language and Russian-language texts with unknown encoding is shown in Figure 2.

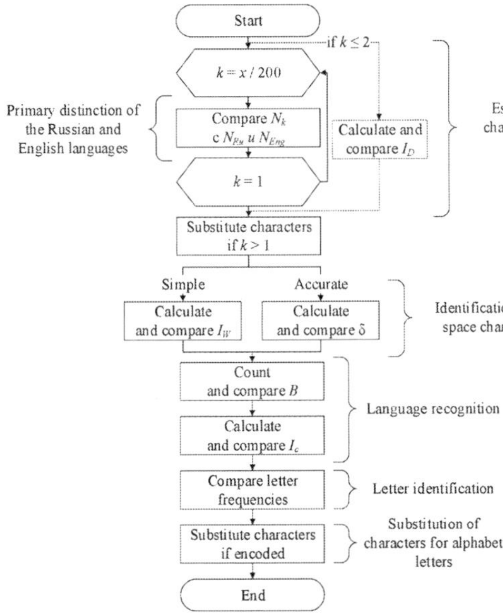

Figure 2. The flowchart of the algorithm for language recognition.

The accuracy of the suggested algorithm has been estimated experimentally with the validation samples of Russian-language and English-language texts. The results are provided in table 3, where N is the number of texts with the size x, P_E – the number of all errors (both type I and type II), P_{AV} is the average error frequency in texts.

According to the algorithm, the text is examined to check if it belongs to the English language and after that, if rejected, if it belongs to the Russian language. Examination of texts for Russian and then for English provides with worse results.

Table 3. The errors in language recognition.

x	*Russian texts*		*English texts*		P_{AV}	x	*Russian texts*		*English texts*		P_{AV}
	N	P_E	N	P_E			N	P_E	N	P_E	
Group 1						Total 2	500	0.008	500	0.008	0.008
200	100	0.45	100	0.04	0.245	Group 3					
400	100	0.13	100	0.09	0.11	10000	100	0.03	100	0	0.015
600	100	0.08	100	0.04	0.06	30000	100	0	100	0	0
800	100	0.01	100	0.02	0.015	50000	100	0	99	0	0
1000	100	0.02	100	0.01	0.015	70000	100	0	99	0	0
1200	100	0	100	0.03	0.015	90000	100	0	98	0.010	0.005
1400	100	0.01	100	0.01	0.01	110000	100	0.01	98	0.020	0.015
1600	100	0.01	100	0	0.005	Total 3	600	0.007	594	0.005	0.006
1800	100	0.01	100	0.04	0.025	Group 4					
2000	100	0.01	100	0.03	0.02	100000	100	0	30	0	0
Total 1	1000	0.073	1000	0.031	0.052	150000	50	0.04	30	0.033	0.037
Group 2						200000	50	0.04	30	0.033	0.037
2000	100	0.01	100	0.04	0.025	250000	50	0.02	30	0	0.01
4000	100	0.01	100	0	0.005	300000	79	0	29	0	0
6000	100	0	100	0	0	350000	21	0.048	27	0	0.024
8000	100	0	100	0	0	Total 4	300	0.02	176	0.011	0.016
10000	100	0.02	100	0	0.01	Total	2400	0.036	2270	0.018	0.027

The average error is 2.7%. The minimal error is zero at several scale points. The maximal average error is 24.5% at the point $x = 200$. Furthermore, the average error for Russian-language texts (3.6%) is found to be higher than for English-language texts (1.8%).

The greatest number of errors is found in texts with small sizes (up to 600 characters). The average error at this interval is equal to 13.8%. As the text size is increasing, the distribution of frequency characteristics is getting stable and the amount of errors in language recognition reduces.

References

[1] Rubinstein-Salzedo S 2018 *Cryptography* (Luxembourg: Springer) p 259
[2] Banik B G and Bandyopadhyay S K 2018 *IETE Journal of Research* **66:3** 384–395
[3] Bochkarev A I, Pepelyaeva M A and Skvortsova E B 2018 *Proc. XIV International Scientific-Technical Conference on Actual Problems of Electronics Instrument Engineering* 33–36
[4] Yang N and Mali A D 2016 *Proc. 28th International Conference on Tools with Artificial Intelligence* 165–168
[5] Zazo R, Lozano-Diez A, Gonzalez-Dominguez J, Toledano T D and Gonzalez-Rodriguez J 2016 *PLOS ONE* **11** 1–17
[6] Simonchik K, Novoselov S and Lavrentyeva G 2016 *Lecture Notes in Computer Science* **9811** 174–181
[7] Noraset T, Demeter D and Downey D 2018 *Proc. of the AAAI Conference on Artificial Intelligence* 5333–5341
[8] Kotov Yu and Sanina O 2018 *Proc. XIV International Scientific-Technical Conference on Actual Problems of Electronics Instrument Engineering* 175–179
[9] Kotov Yu and Sanina O 2019 *Proc. VIII International Scientific Conference of Graduate and Postgraduate Students "Progress through Innovations"* 69–71
[10] Kotov Yu and Sanina O 2020 *Vestnik SibGUTI* 60–72
[11] Alkazaz N R, Irvine S A and Teahan W J 2018 *Information Security Journal* **27** 57–75

IOP Conf. Series: Materials Science and Engineering 1019 (2021) 012032 doi:10.1088/1757-899X/1019/1/012032

Permanent magnet synchronous motor sensorless control based on rotor flux observer with low-pass filter adaptation

D V Bevz, V M Zavyalov
Department of Power Engineering, Tomsk Polytechnic University, Tomsk, Russia
E-mail: d.v.b@mail.ru

Abstract. This paper presents a strategy for a permanent magnet synchronous motor sensorless vector control based on rotor flux estimation considering a rotor saliency. The integration problem in the stator flux estimation was solved by replacing the integrator with a low-pass filter. The low-pass filter parametric adaptation algorithm with phase delay compensation, based on Bode diagrams was designed for the rotor angular position estimation accuracy improving. The proposed adaption algorithm allowed to expand a stable motor control with a high accuracy in the rotor position estimation down to 15 rpm. Also the developed rotor angular position estimation algorithm together with adaptive field weakening algorithm provides a sable motor operation with a high accuracy in the rotor position estimation at the high speed.

1. Introduction

The electric drives based on permanent magnet synchronous motor (PMSM) are in high demand in such industry areas as robotics, tracking and guidance systems, mining, servo systems, medical equipment, electric traction drives, etc. This is explained by that the PMSM has such advantages over another motors type as low values of the electromagnetic and electromechanic time constants, high torque overload capacity, small weight-size parameters compared with other motors type at the same power, wide speed control range, high efficiency (more than 90%). This allows to design high-dynamic electric drives and also makes the PMSM application attractive at the electric drives used in common industry. The PMSM acquired these advantages with new materials emergence for the permanent magnets manufacturing. These materials, as a rule, are Sm-Co, or Ne-Fe-B.

The PMSM control system is based on PMSM mathematical model in the d-q reference frame, fixed relative to the PMSM rotor. The information about rotor position is needed to control PMSM torque and speed. The rotor position is measured by the rotor angular position sensor. However, the rotor position can be calculated by the mathematical expressions describes the PMSM model. It allows to refuse the rotor position sensor using. Failure from the rotor position sensor using allows to reduce PMSM overall dimensions, to increase electric drive reliability, and also to reduce electric drive cost in general. Thus, sensorless control algorithms development is the topic problem to research in the PMSM control algorithms area.

Sensorless control algorithms use the rotor angular position observers based on different mathematical models and using various mathematical methods to estimate the rotor angular position.

PMSM rotor position observers can be divided into two categories: 1) observers using injection of high frequency test signal [1–3]; 2) observers using PMSM mathematical model. The first ones are based on high frequency or pulse signal generation by inverter and the rotor angular position searching by the reaction to this signal. The main disadvantages are additional losses and noise level increasing.

Observers based on PMSM mathematical model, which are called PMSM state observers directly use mathematical expressions of the model to calculate the rotor position [2, 4, 5]. The main disadvantage of these algorithms is the low accuracy of the angular position estimate. In order to improve the accuracy the following methods are used: sliding mode control [6]; Kalman filter [7]; neural networks [8]. The sliding mode control generates a high-frequency component in the estimated rotor angular position, that is required the output signal additional filtering. The algorithms based on Kalman filter or neural networks theory are accompanied by a large amounts of computation, which determines the high requirements for the microprocessor.

Another kind of state observers is observers with adaptive system. The most popular of the adaptive estimation methods is the model reference adaptive system (MRAS) method. This method use PMSM mathematical model equations in order to adapt it to real processes occurring in the motor and, as a result, reduce the orientation error to zero [9]. However, the are no methods for the regulating adapter parameters tuning.

In this paper, the PMSM rotor angular position estimation method that provides stable performances with a high accuracy of rotor position estimation in range from 2 Hz to 400 Hz is considered.

2. State observer for the rotor flux vector angle position estimation

The PMSM rotor angular position will be estimated with PMSM mathematical model in the stationary reference frame α, β.

The PMSM voltage equation in the stationary reference frame is:

$$\begin{bmatrix} v_\alpha \\ v_\beta \end{bmatrix} = \mathrm{R}\begin{bmatrix} i_\alpha \\ i_\beta \end{bmatrix} + \frac{\mathrm{d}}{\mathrm{d}t}\begin{bmatrix} \psi_{s\alpha} \\ \psi_{s\beta} \end{bmatrix}, \tag{1}$$

where:

v_α, v_β – stator voltage in the stationary α, β frame;

i_α, i_β – stator current in the stationary α, β frame;

$\psi_{s\alpha}$, $\psi_{s\beta}$ – stator flux linkage in the stationary α, β frame;

R – stator resistance.

The components of the stator flux linkage are determined by the following expressions:

$$\begin{bmatrix} \psi_{s\alpha} \\ \psi_{s\beta} \end{bmatrix} = \begin{bmatrix} \mathrm{L}_0 + \mathrm{L}_1\cos 2\theta & \mathrm{L}_1\sin 2\theta \\ \mathrm{L}_1\sin 2\theta & \mathrm{L}_0 - \mathrm{L}_1\cos 2\theta \end{bmatrix}\begin{bmatrix} i_\alpha \\ i_\beta \end{bmatrix} + \lambda_{\mathrm{PM}}\begin{bmatrix} \cos\theta \\ \sin\theta \end{bmatrix}, \tag{2}$$

where:

λ_{PM} – permanent magnet flux linkage;

θ – rotor position;

$\mathrm{L}_0 = \frac{\mathrm{L}_d + \mathrm{L}_q}{2}$, $\mathrm{L}_1 = \frac{\mathrm{L}_d - \mathrm{L}_q}{2}$,

L_d and L_q are the d- and q-axes inductances.

As the initial equations for estimating θ, we take the equations system (2). Express from them the components with angle θ and permanent magnet flux linkage λ_{PM}:

$$\lambda_{\mathrm{PM}}\begin{bmatrix} \cos\theta \\ \sin\theta \end{bmatrix} = \begin{bmatrix} \psi_{s\alpha} \\ \psi_{s\beta} \end{bmatrix} - \begin{bmatrix} \mathrm{L}_0 + \mathrm{L}_1\cos 2\theta & \mathrm{L}_1\sin 2\theta \\ \mathrm{L}_1\sin 2\theta & \mathrm{L}_0 - \mathrm{L}_1\cos 2\theta \end{bmatrix}\begin{bmatrix} i_\alpha \\ i_\beta \end{bmatrix} \tag{3}$$

Then:

$$\frac{\sin\theta}{\cos\theta} = \tan\theta = \frac{\psi_{s\beta} - (\mathrm{L}_0 - \mathrm{L}_1\cos 2\theta)i_\beta + \mathrm{L}_1\sin 2\theta i_\alpha}{\psi_{s\alpha} - (\mathrm{L}_0 + \mathrm{L}_1\cos 2\theta)i_\alpha + \mathrm{L}_1\sin 2\theta i_\beta} \tag{4}$$

Thus, the rotor angular position is:

$$\theta = \arctan\left[\frac{\psi_{s\beta} - \left(L_0 - L_1\cos 2\theta\right)i_\beta + L_1 \sin 2\theta i_\alpha}{\psi_{s\alpha} - \left(L_0 + L_1\cos 2\theta\right)i_\alpha + L_1 \sin 2\theta i_\beta}\right] \tag{5}$$

The values $\psi_{s\alpha}$ and $\psi_{s\beta}$ can be determined by integration of (1):

$$\begin{cases} \psi_{s\alpha} = \int_o^t \left(v_\alpha - Ri_\alpha\right)dt; \\ \psi_{s\beta} = \int_o^t \left(v_\beta - Ri_\beta\right)dt. \end{cases} \tag{6}$$

Integration in (6) accumulates error if there is constant component. In order to eliminate this drawback the integrators are replaced by 1st order low-pass filters with the following transfer function:

$$W(s) = \frac{T}{Ts+1}, \tag{7}$$

where T is low-pass filter cutoff frequency.

It is recommended to choose the cutoff frequency in the range 1 – 3 Hz [10]. Figure 1 shows Bode plots of the integrator and low-pass filter with cutoff frequency 3 Hz. Low-pass filter gain and phase shift will be identical to the integrator at the nominal motor speed ω_{nom}. The phase error will be increase up to 45° with a decrease in speed from ω_{nom} to ω_c. This will lead to stator flux linkage estimation error increasing.

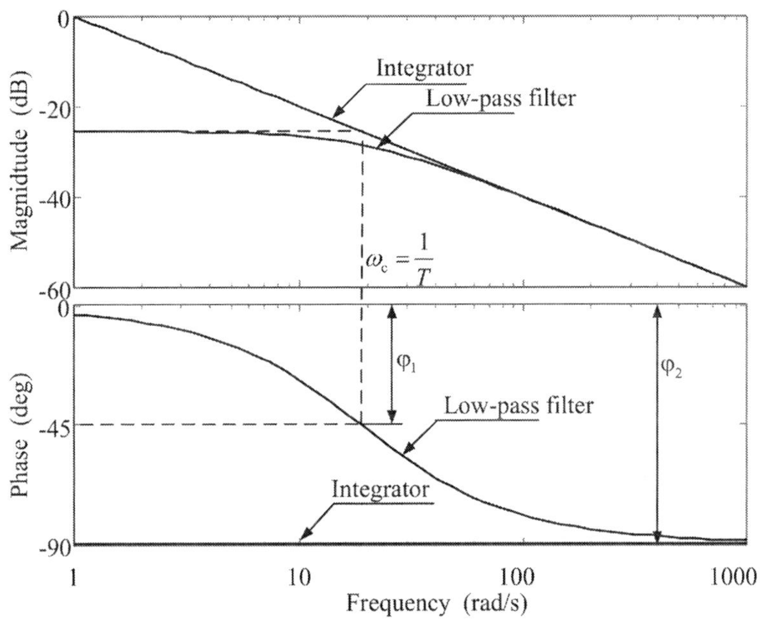

Figure 1. The low-pass filer and integrator Bode plots.

Figures 2 and 3 show that the phase error in ψ_s estimation have the greatest influence on the accuracy of the λ_{PM} angular position estimation, which determines the rotor angular position.

IOP Conf. Series: Materials Science and Engineering 1019 (2021) 012032 doi:10.1088/1757-899X/1019/1/012032

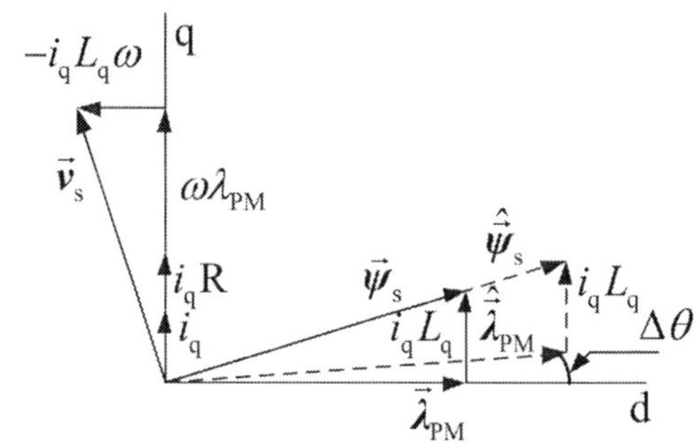

Figure 2. Error in stator flux linkage vector amplitude estimation on PMSM vector diagram.

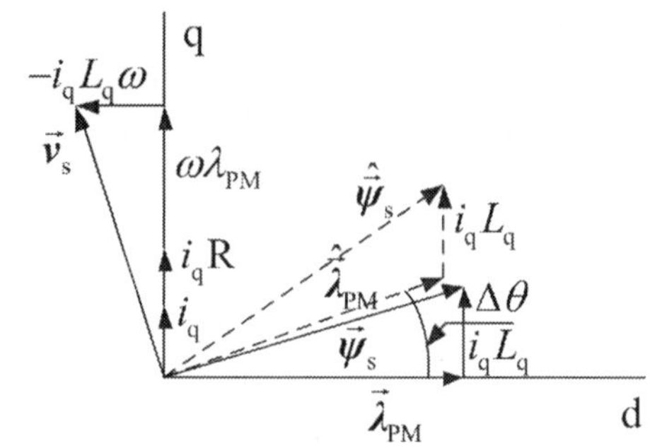

Figure 3. Error in a stator flux linkage vector angular position estimation on PMSM vector diagram.

3. Proposed low-pass filter adaptation algorithm developing

We will use low-pass filter time constant adaptive tuning depending on the motor shaft rotation frequency to ensure a constant phase shift of -45°. The adaptive tuning is to set the low-pass filter cutoff frequency equals to the current electrical rotation frequency. Thus, the time constant $T = f(\omega_{el})$. The low-pass filter Bode plots with adaptive tuning are shown in Figure 4.

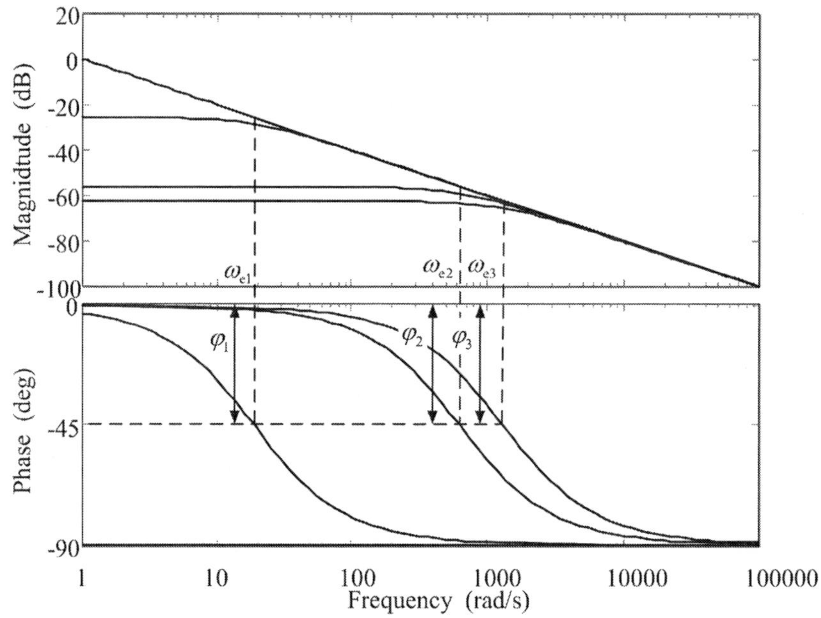

Figure 4. The low-pass filter Bode plots with proposed adaptation algorithm.

Then, to obtain a phase shift of -90° it is necessary to provide an additional phase shift of -45°at filter output. As a result, the low-pass filter using with angle correction to determine the stator flux linkage will provide a phase shift identical to the integrator, without error accumulation at the motor low speed operation. This will make it possible to obtain accurate rotor position estimation at the low speed.

4. Experimental results

The rotor angular position observer algorithm was developed for the fan electric drives based on PMSM sensorless control. Also, a frequency converter EC-F with a rated power of 3 kW and three-phase AC voltage of 380 V was designed to control a PMSM. PMSM parameters are presented in table 1.

Table 1. Parameters of the PMSM.

Number of poles pairs (p)	8
Rated power	2.2 kW
Rated speed	1500 rpm
Rated frequency	200 Hz
Rated torque	14 Nm
Rated phase to phase voltage	380 (rms)
Rated pahse current	4.2 A(rms)
Stator resistance per phase (R_s)	2.4 Ω
d-axis inductance (L_d)	101 mH
q-axis inductance (L_q)	133 mH
Rotor permanent – magnet (λ_{PM})	0.213 V s rad-1

The frequency converter EC-F is based on digital signal processor STM32F405G. The sampling frequency is set to 5 kHz and the PWM frequency is set to 20 kHz. The dead time of IGBT inverter is set to 1 μs. The experimental setup with a radial fan is shown in Figure 5.

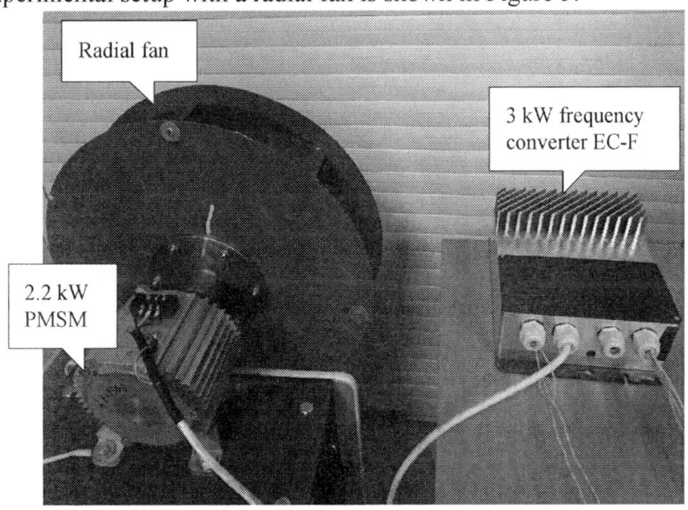

Figure 5. Experimental setup with a radial fan.

PMSM sensorless control is based on vector filed-oriented control, where the rotor angular position and the rotation speed are estimated by the observer. In parallel, the actual rotor angular position was measured by incremental encoder with 2500 pulses per revolution.

The electric drive start-up begins with initial rotor angular position identification. The identification is carried out when a stator current is formed along the d-axis and zero angle applied to Park transformations. During this this time the observer estimates the rotor angular position. Then the control structure switches to current control along the q-axis and current vector rotates on angle, which is estimated by observer.

The estimated rotor position $\hat{\theta}$ and the actual rotor position θ are shown in Figures 6–8, where 1 relative unit equals 360 electrical degrees.

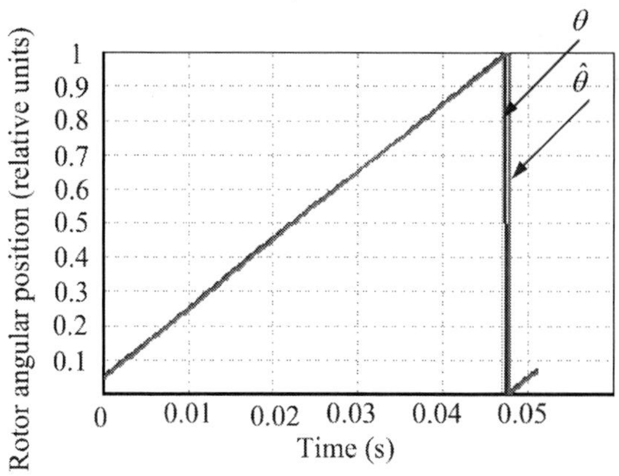

Figure 6. Rotor angular position at the rotation speed of 20 rpm.

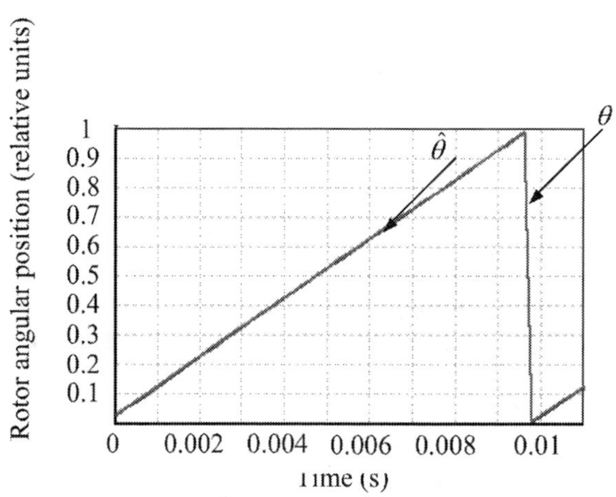

Figure 7. Rotor angular position at the rotation speed of 750 rpm.

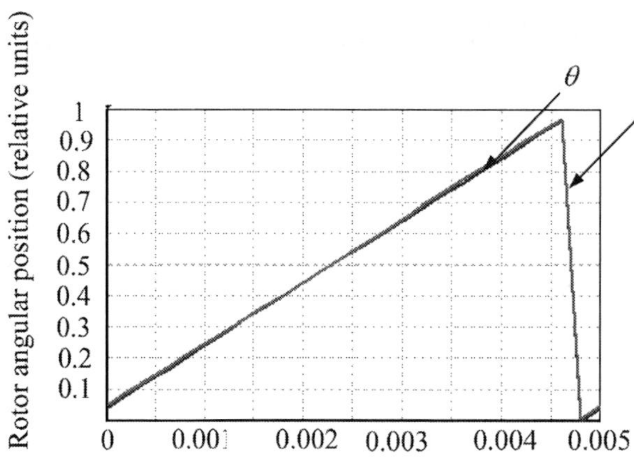

Figure 8. Rotor angular position at the rotation speed of 1500 rpm.

IOP Conf. Series: Materials Science and Engineering 1019 (2021) 012032 doi:10.1088/1757-899X/1019/1/012032

The information about error between estimated and actual angle rotor position is presented in table 2.

Table 2. The observer characteristic.

Speed reference ω, p.u.	The rotor position estimation error $\Delta\theta$,°	
	$f_c = 3$ Hz	$f_c = \frac{\omega_{el}}{2\pi}$
0.01	3.915	0.423
0.1	2.85	0.423
0.5	1.89	0.325
1.0	1.525	0.426
2.0	1.635	0.485

Transient performances at the electric drive accelerate are shown on Figures 9 and 10, where 1 relative unit for speed equals 1500 rpm and 1 relative unti for current equals 5 A.

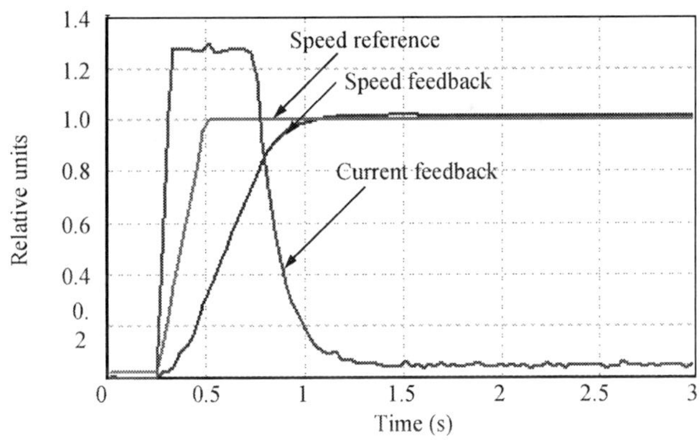

Figure 9. Electric drive speed and torque transient performances at speed reference of 1500 rpm.

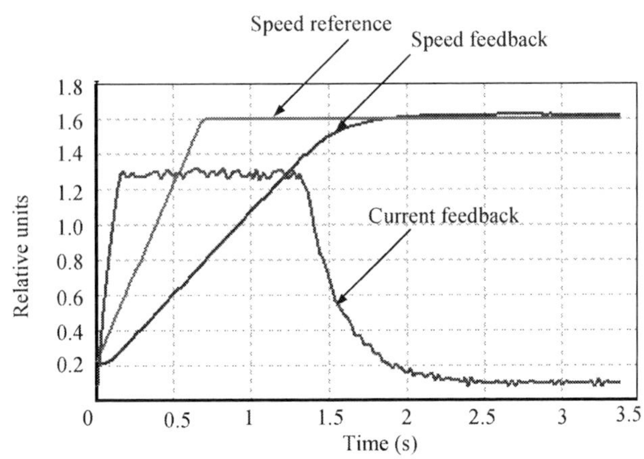

Figure 10. Electric drive speed and torque transient performances at speed reference 2400 rpm.

In order to expand speed control range, field weaking algorigm "Constant current – constant power" (CCCP) was applied.

Figure 11 shows experimental setup for proposed observer algoritm research with a different load on the PMSM shaft. The inducton motor of rated power 5.5 kW with control from Danfoss FC302 are used to provide a load.

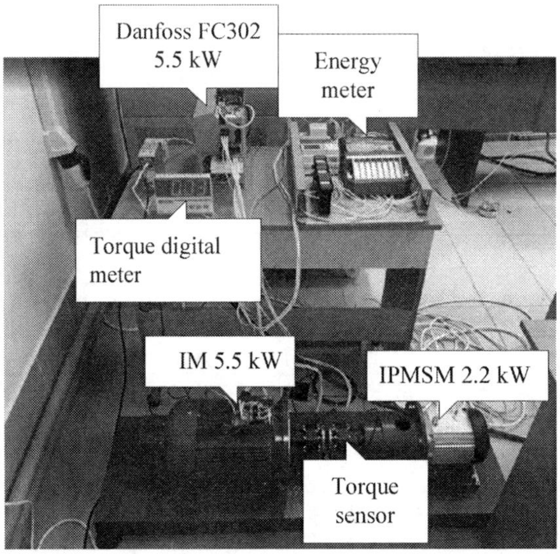

Figure 11. Experimental setup with a load machine.

The limiting electromechanical characteristic of PMSM when it is opertaed in sensorless mode is shown in Figure 12. Sensorless control provides a stable operation at the dobule electromechanical torque overload in speed range from 250 rpm to 1150 rpm.

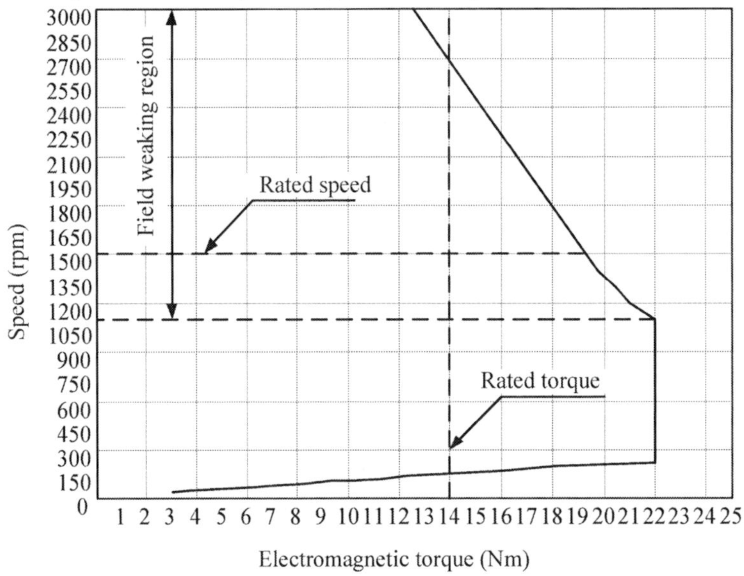

Figure 12. PMSM electromechanical characteristic with vector control based on rotor position observer.

5. Conclusion

The developed rotor angular position estimation method provides a stable motor control in closed speed loop at the speed range from 15 rpm to 3000 rpm. The error between actual rotor position and estimated rotor position no more 1 mechanical degree The low-pass filter time constant adaptive tuning is showed significant error decreasing in the rotor angular position estimation, especially at low speed operation. The observer algorithm provides a low computing load on the processor and can be integrated into any electric drive control system.

However, there is no ability to reverse. Also, the maximum electromagnetic torque is decreased with a speed decreasing form 150 rpm to 15 rpm. Analysis and improvement of the observer dynamic performances, expansion the load range at the low speed operation and the ability to revers are directions for further research.

References

[1] Briz F, Diez A and Degner M W 2000 *IEEE Transaction on Industry Applications* **36** pp 1360–68
[2] Sul S, Kwon Y and Lee Y 2017 *Ces Transactions on Electrical Machines and Systems* **1** pp 91–99
[3] El-Murr G, Giaouris D and Finch J W 2009 *Proceedings of the 2008 IEEE Reagion 10 and the Third international Conference on Industrial and Information Systems* pp 1–6
[4] Bedetti N, Calligaro S and Petrella R 2013 *IEEE Energy Conversion Congress and Exposition* pp 3496–3503
[5] Inoue T, Hamabe Y, Tsuji M and Hamasaki S 2018 *The 2018 Internattional Power Electronics Conference* pp 1276–81
[6] Lee H and Lee J 2013 *IEEE Transactions On Control Systems Technology* **21** pp 1394–99
[7] Termizi M S, Lazi J M, Ibrahim Z, Talib M H N, Aziz M J A and Ayob S M 2018 *IEEE Conference on Energy Conversion* pp 145–150
[8] Ren Y and Zhou L 2018 *Cyberworlds 2008 International Conference* pp 1823–29
[9] Kumar P, Kumar M and Dahiya S 2015 *Proceedings of the World Congress on Engineering* pp 198–204
[10] Holtz J 2002 *IEEE Electrical Machines and Drives Group* **90** pp 1359–94

IOP Conf. Series: Materials Science and Engineering 1019 (2021) 012033 doi:10.1088/1757-899X/1019/1/012033

Quantized lobe level theoretical estimates for phased array antennas applicability

A A Slobodyanenko, V G Edvabnik and L V Shebalkova
Autonomous Information and Control Systems dept., Novosibirsk State Technical University, Novosibirsk, Russia
E-mail:sepwood@gmail.com

Abstract. This paper raises the question of the applicability of some theoretical estimates of the quantization lobe level. Existing assessment methods and their features are considered. Based on numerical simulated method, statistical characteristics are constructed for the magnitude of the level of quantization lobes and, according to the proposed criterion, a conclusion is given on the consistency of theoretical estimates.

1. Introduction
To date, one of the areas attracting the attention of researchers and developers of airborne radar systems is the use of phased antenna arrays. Due to the ability of the antenna array to change the phase relationships between the currents in the emitters, you can control the direction of the beam in a given sector of angles, which contributes to solving the tasks set for the onboard radar system, namely: target detection, angular coordinates determining, velocities determining, moving objects tracking, and objects classifying.

The need to place such antennas on the aircraft board imposes a number of restrictions, one of which is the overall mass characteristics. The desire to minimize the size of the antenna array at the same time, while maintaining the ability to scan, has led to widespread use of discrete-switching phase shifters as phase control devices. These phase shifters are a set of n fixed delay lines with values $2\pi/2$, $2\pi/2^2$ ... $2\pi/2^n$, n is called the digitizer phase capacity. The ability to implement phases only multiples of the discrete $\Delta\varphi$ leads to errors in the phase distribution. This adversely affects the parameters of the pattern. There is a decrease in the directional coefficient, the increase in the level of side lobes, as well as the inaccuracy of the installation of the beam in space.

Studies on the effect of phase distribution errors on the characteristics of the antenna array are presented in [1–6]. One may notice that most of the previous works treated the effect of random errors on the average sidelobe level. In a modern radar system, the peak side lobe level (SLL), however, is more important. Operation requirements are usually specified in terms of peak sidelobe level, in particular for a low sidelobe system. The reason is that for jamming rejection, it is most important that a jammer noise comes from a spatial location at the peak of a sidelobe and this noise must not interfere with the radar detection. Average sidelobe level in this case does not help. An array may have a very low average sidelobe level, but at the same time, it may have a few high peak sidelobes located at the wrong spot, which renders the radar system vulnerable.

2. The theoretical estimates of quantization lode level
A linear array far field and a normalized far field can be written as follows:

$$\mathrm{AF}(\theta)=\sum_{n=1}^{N} A_n \mathrm{e}^{ikd(n-1)\cos(\theta)}\mathrm{e}^{\mathrm{i}\Delta\varphi_n} \tag{1}$$

where i is the imaginary unit, θ the angle measured from the direction of the array, k is the wave number ($k=2\pi/\lambda$), N is the elements number, A_n is n^{th} element excitation current, $\Delta\varphi_n$ is the phase error at the nth element d is the interelement spacing.

In order to steer the beam to a given direction, we must set up the phases computed according to the steering direction in the phase shifters. Although the array is required to produce a smooth phase taper, an N-bit phase shifters has phase states separated by the least significant bit: $\Delta\varphi=2\pi/2^N$. These can only approximate to the desired phase distribution across the array, and if a straightforward rounding off is used, a periodic phase error arises which causes a pointing deviation of the main beam and introduces a quantization lobe.

Figure 1 shows that this quantilization allows only a staircase approximation of the continuous progressive shift required for the array. The staircase phase front results in a periodic triangular phase error that produces the pattern with grating lobe- like sidelobes.

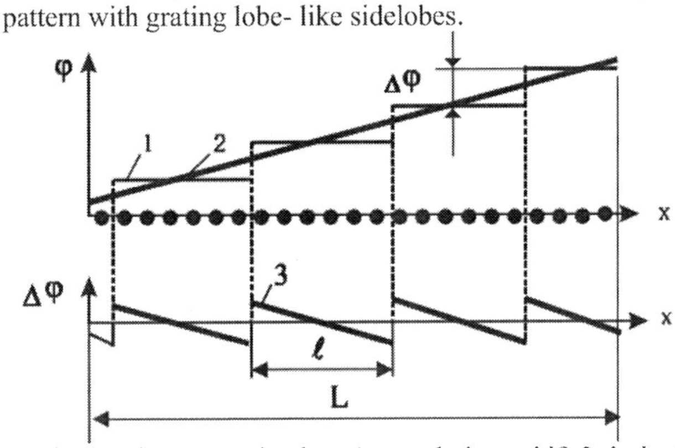

Figure 1. Phase error due to phase quantization: 1-actual phase shift 2-desired phase shift 3-phase error.

The phase errors are

$$\Delta\varphi(\mathrm{x})=\begin{cases}\Delta\varphi(1-\dfrac{(x-x_m)\mathrm{d}}{l}) \\ -\dfrac{\Delta\varphi(x-x_m)\mathrm{d}}{l}\end{cases} \text{ with probability } \begin{vmatrix} p(x) \\ q(x)=1-p(\mathrm{x})\end{vmatrix}$$

where $x_m=(m-1)l/d$ is the coordinate of the start point of the mth period; $l=Md$ is the length of a period; M is number of element in subarrays. As the variation of the current amplitude in a period is not large, we can approximately obtain

$$AF(\mathrm{u})=f_1(\mathrm{u})\cdot f_2(\mathrm{u})$$

where

$$u=kd(\cos\theta-\cos\theta_0)$$

$$f_1(u)=\frac{l}{d}\sum_{n=1}^{[L/l]} A_n \mathrm{e}^{\mathrm{i}\frac{(n-1)ul}{d}}$$

$$f_2(u)=\frac{d}{l}\int_0^{l/d}\left(p(x)\mathrm{e}^{\mathrm{i}\Delta\varphi(1-\frac{x\mathrm{d}}{l})}+q(\mathrm{x})\mathrm{e}^{-\mathrm{i}\frac{\Delta\varphi x\mathrm{d}}{l}}\right)\mathrm{e}^{\mathrm{i}ux}\mathrm{d}x$$

Using the above formulas to deduce the size of the quantization lobe for the rounding-off method, using the "probability" distribution:

$$p(\mathrm{t})=\begin{cases}1\,, t\geq 0 \\ 0\,, t<0\end{cases}$$

With this approximation, the first quantization lobe level [4] is given as

$$L_1 \cong 20\log\left(\frac{\delta}{\pi-\delta}\right) = 20\log\left(\frac{\pi}{2^n}\left(\pi-\frac{\pi}{2^n}\right)^{-1}\right) = -6n$$

To obtain an estimate of the average quantization lobes level due to a single quantization error is decomposed in equation (1) by the factor characterizing the phase error in the Maclaurin series, representing the sum in the form of the scalar product of two N-dimensional vectors.

As a result, the value of the level of switching lobes is written as

$$L_2 = \frac{1}{\sqrt{2}}\beta_{\text{mean}}$$

where β_{mean} is the RMS value of phase errors.

The mean square of the systematic phase error can be found by taking the integral of the square of the phase error magnitude over the entire length of the antenna. According to Figure 1 we have

$$\delta\varphi_{\text{stm}} = \frac{1}{l}\int_{-l/2}^{l/2}\left(\Delta\varphi\frac{x}{l}\right)^2 dx$$

Then

$$\left(\delta\varphi_{\text{stm}}\right)^2 = \frac{\pi^2}{3\cdot 2^{2n}}$$

Let the value of the random phase error introduced by the phase shifter be small compared with the value of the systematic phase error.

$$\beta_{\text{mean}} = \delta\varphi_{\text{stm}}$$

Then

$$L_2 = \left(\frac{\pi^2}{3\cdot 2^{(2n+1)}}\right)^{-\frac{1}{2}}$$

Miller [7] published analysis of the adverse effects of using phase shifters with discrete phase states for large phase arrays with uniformly illuminated contiguous subarrays.

The general expression for power at the peak of the quantization lobe [3] is written as:

$$L_3 \approx -20\log(M) + 9.94 - 6.02N$$

Theoretical estimates calculated for the antenna array with 3, 4, 5-bit phase shifters are summarized in table 1.

Table 1. The theoretical quantization lobe level estimates.

Phase shifter bitness	Values of theoretical estimates		
	L_1	L_2	L_3
3	-18	-15.9	-14.2
4	-24	-21.92	-20.1
5	-30	27.94	-26.2

3. Numerical simulated method and statistical approach for estimate quantilization lobelevel

Errors due to the discreteness of the phase shifter are systematic, and the level of quantization lobes for a particular spatial direction of radiation is determined (by the function of the initial phase distribution). We will consider the entire set of the level values of the switching lobe for all possible radiation directions of the main maximum. Then we can talk about the frequency of occurrence of the quantization lobe level in the scanning sector and thereby assigning the probability of its occurrence. Formulating the problem in this way, we use the apparatus of mathematical statistics. Based on the results obtained by the method of numerical simulations of a 32 element array antenna, the phasing of which is carried out by 3, 4, and 5-bit phase shifters, the corresponding statistical series of the realized quantization lobe levels are obtained. Some radiation patterns obtained by numerical simulation are presented on Figures 2–7. Phase shifts between radiating elements are set in the rounding-off method and are calculated by the following formula:

$$\Delta\varphi_n = \Delta\varphi \cdot \left[\frac{\varphi_{n,\theta_0}}{\Delta\varphi} + 0.5 \right]$$

where $\varphi_{n,\theta_0} = n \cdot \frac{2\pi d}{\lambda} \cdot \cos(\theta_0)$ is required phase shift on the nth element to form a beam in the direction of θ_0; $[a]$ the integer part of a.

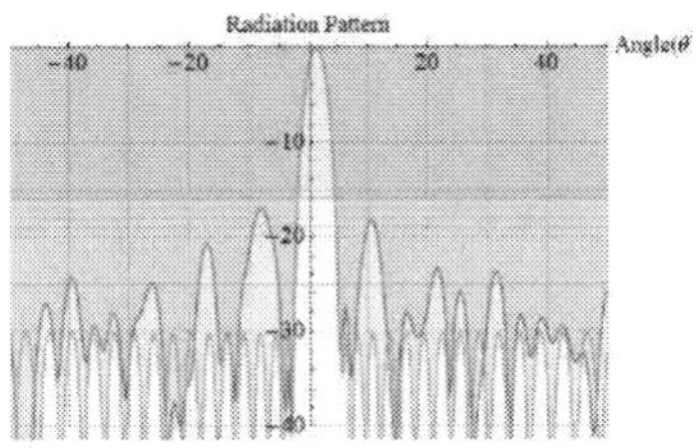

Figure 2. Radiation patterns for scan angles $\theta_0 = 1°$ (a) for a 3-bit phase shifter.

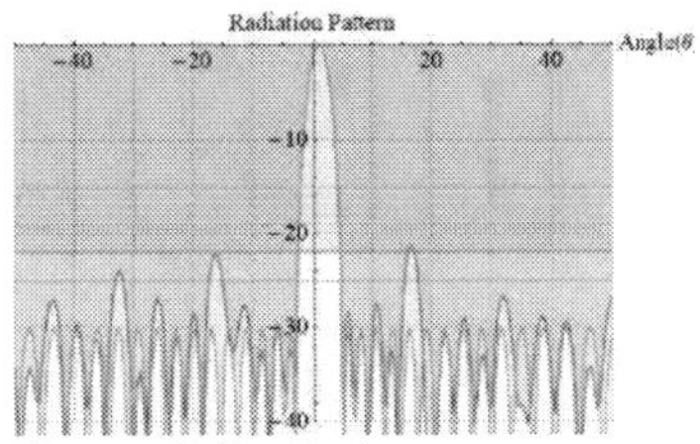

Figure 3. Radiation patterns for scan angles $\theta_0 = 1°$ for a 4-bit phase shifter.

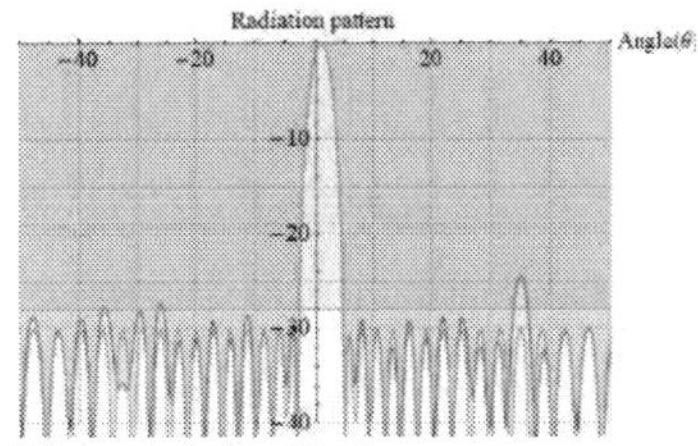

Figure 4. Radiation patterns for scan angles $\theta_0 = 1°$ for a 5-bit phase shifter.

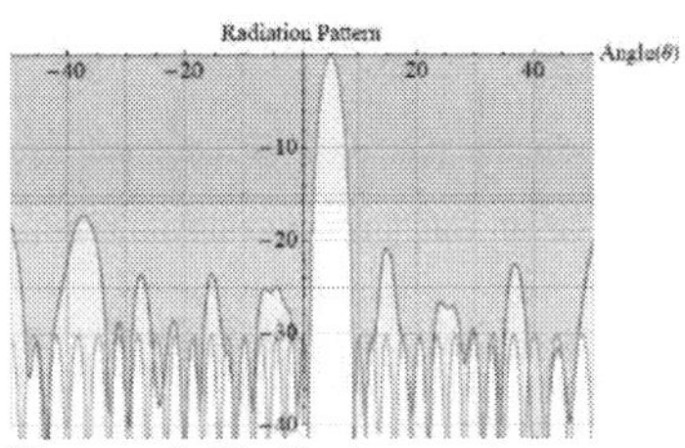

Figure 5. Radiation patterns for scan angles $\theta_0 = 5°$ for a 3-bit phase shifter.

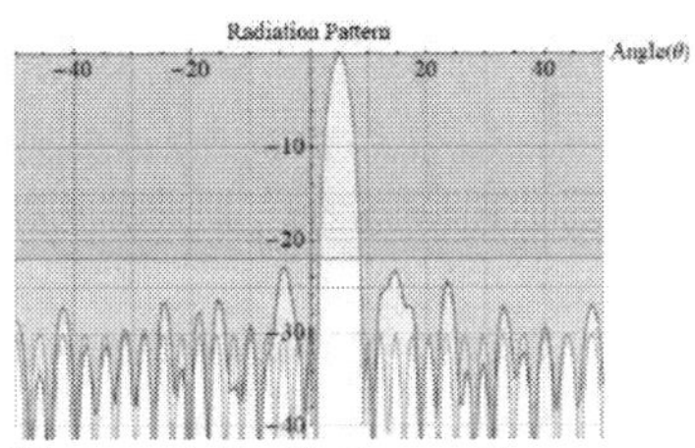

Figure 6. Radiation patterns for scan angles $\theta_0 = 5°$ for a 4-bit phase shifter.

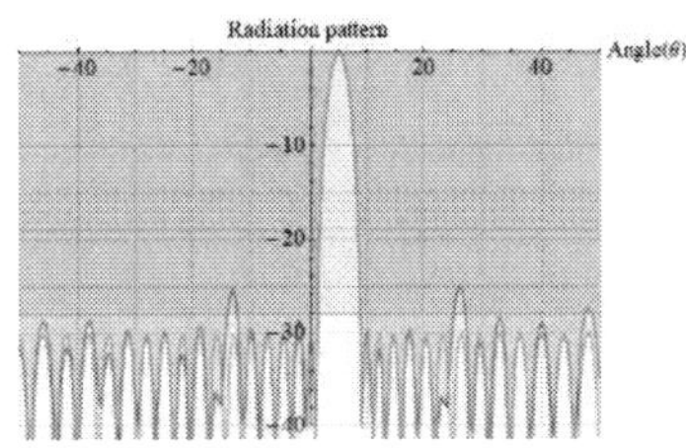

Figure 7. Radiation patterns for scan angles $\theta_0 = 5°$ for a 5-bit phase shifter.

Since the type of statistical distribution of the quantization lobe level is not known in advance, the Pearson curves, based on solving a differential equation expressing the general properties of the distribution function, was used to approximate the probability distribution density. To estimate the distribution parameters, the method of moments was used, the essence of which is to calculate the required number of distribution parameters from experimental data and equate them to theoretical values. As is known, the Pearson curve is uniquely determined by its first four moment givrn in table 2. Based on their values, the types of Pearson curves are determined that are suitable for specific distributions obtained by the method of numerical simulations and the approximated probability density functions of the distribution of Figures 8–10 are constructed. Probability distribution density satisfies the chi-square criterion with significance level $\alpha = 0.05$.

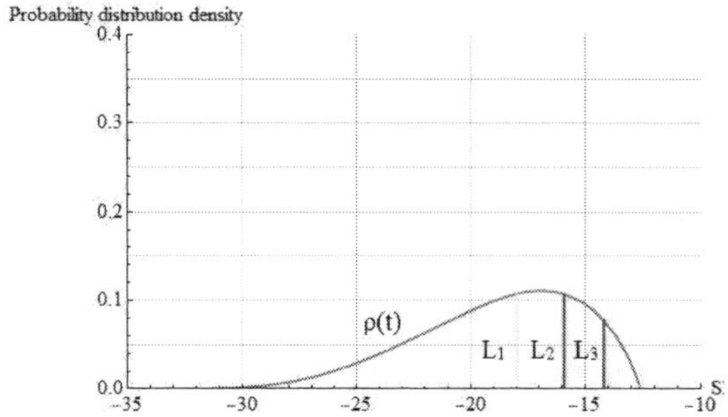

Figure 8. The normalized probability distributiondensity for a 3-bit phase shifter.

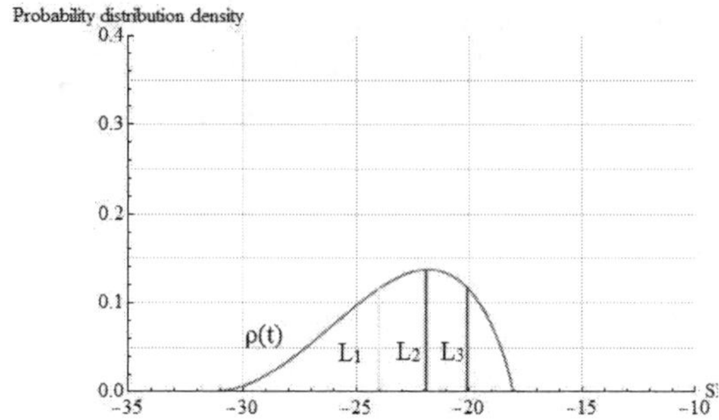

Figure 9. The normalized probability distribution density for a 4-bit phase shifter.

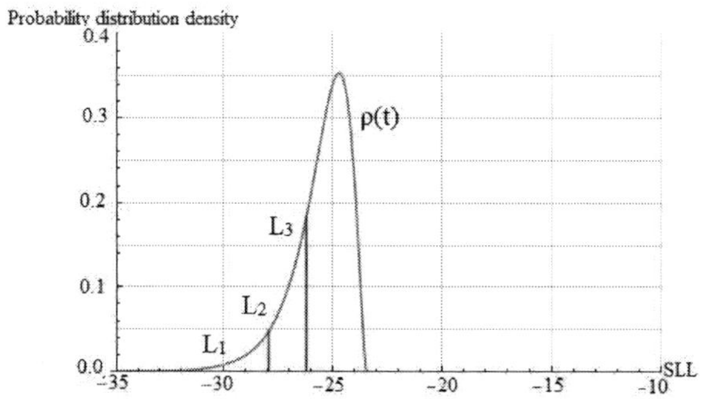

Figure 10. The normalized probability distribution density for 5-bit phase shifter.

Table 2. Values of statidtical moments.

Phase shifter bitness	Values of moments			
	m_1	μ_2	μ_3	μ_4
3	-16.83	4.47	-39.6	499.47
4	-21.82	3.56	-21.68	192.99
5	-25.59	1.96	-4.29	25.17

All probability distribution densities were centered relative to the expectation and were normalized according to the condition:

$$\mathrm{P}\left(-\infty < \mathrm{L} < +\infty\right) = \int_{-\infty}^{+\infty} \rho(\mathrm{t})\mathrm{d}t = 1$$

Since the method of numerical simulations allows to obtain the actual realizable values of the quantization lobe level, taking into account the accepted assumptions, it is possible to draw a conclusion on the reliability of the theoretical estimates from the obtained probability densities.
To do this, we introduce the validity criterion: we will consider the L_k estimate to be reliable if, with a probability of $\geq 90\%$, the quantization lobe level lies in the interval $(-\infty, \mathrm{L}_k)$. Using the obtained probability distribution densities, we find the probability that of the quantization lobe level does not exceed the theoretical estimate

$$F(\mathrm{L}_k) = \mathrm{P}\left(-\infty < \mathrm{L} < \mathrm{L}_k\right) = \int_{-\infty}^{\mathrm{L}_k} \rho(\mathrm{t})\mathrm{d}t \quad (2)$$

The probabilities calculated in equation (2) are given in the table 3.

Table 3. Probability quantization lobe level less then theoretical estimate.

Phase shifter bitness	Probability		
	$F(L_1)$	$F(L_2)$	$F(L_3)$
3	0.55	0.78	0.94
4	0.35	0.62	0.85
5	0.01	0.06	0.23

L_1, L_2 are not satisfactory for estimating the peak value of the quantization lobe. Evaluation of L_3 satisfies the criterion in the case of application in the antenna array of a 3-bit phase shifter, in the later cases, L_3 does not give a value close to the peak quantization lobe level. Since the estimate of L_3 takes into account a certain general symmetry of errors, its value is closest to the realizable peak value of the quantization lobes (for $n=3,4$). In the case $n=5$ where the influence of systematic errors increases, the consideration of the general symmetry of errors becomes insufficient.

Moreover, with an increase in the digitizer of the phase shifter, the probability tends to zero, which indicates a drop in the quality of the estimate and its insolvency. It is possible to draw conclusions about the insufficient consideration of only average phase errors over the lattice and about the predominant influence of their symmetries.

The proposed approach can be applied to assess the radiation patterns of the onboard radio frequency systems.

4. Conclusion

Various theoretical estimates of the level of quantization lobes are considered and analyzed. Based on statistical series, methods for calculating numerical symmetries are obtained using 3, 4, and 5-bit phase shifters. Based on the proposed reliability criterion, a conclusion was made on the applicability of theoretical estimates for antenna arrays as part of airborne radar systems.

References

[1] Mailloux R J 2005 *Phased Array Antenna Handbook* (Norwood, MA, USA:Artech House)
[2] Jacobs E 1980 *IEEE Tmns Antennas Propagat* **29** 243–246
[3] Mailloux R J 1984'*IEEE Trans. on Antennas and Propagation* **32** 1364–1368
[4] Smith M S and Guo Y C 1983 *IEEE Trans. on Antennas and Propagation* **31** 821–827
[5] Hansen R C 1986 *Linear Arrays Ch. 9 in The Handbook of Antenna Design* (London: Peter Peregrinus) **2** pp 104–106
[6] Keizer W P 2011 *IEEE Trans. on Antennas and Propagation* **59** 4520–4524
[7] Miller C J 1964 *Procceding of Symp. Electronically Scanned Phased Arrays and Applications* **1** 17–38

Application of standard ZET-algorithm for gaps filling in wide area measurement system data assets

N L Batseva and J A Foos
Power Engineering School, Tomsk Polytechnic University, Tomsk, Russia
E-mail: JuliaAlekseevna6797@gmail.com

Abstract. This article explores the application of Standard ZET-algorithm for gaps filling in Synchronized Phasor Measurement data assets received from Wide Area Measurement Systems. In spite of Synchronized Phasor Measurement data assets apparent advantages, they have gaps affecting the accuracy of solving power engineering tasks. Authors analysed the reasons of gap occurrence in data assets, provided experiments with voltage and current data assets and collected the key points about application of Standard ZET-algorithm.

1. Introduction

Synchronized Phasor Measurement (SPM) data assets received from Wide Area Measurement Systems (WAMS) are in widespread use for solving technological tasks in power systems [1]. These tasks cover the operational dispatch management and automatic emergency control:

a) improving accuracy of power system state estimation [2];
b) on-line identifying a point of power system emergency separation into parts and then their synchronization;
c) monitoring margin of power system stability;
d) choosing volume of action adjustments for emergency response.

Definitely, SPM data assets have unquestionable advantages:

a) obtained from one node, they allow calculating the voltage of neighbouring nodes by load current values of a line, increased power system observability;
b) they give an opportunity to control the phase imbalance, as the signal, when equipment tend to accident-caused failure [2];
c) high discreteness of regime parameters provides the possibility to analyse electromechanical transients for centralized and local automatics according to type of stability [3].

However, SPM data assets include noises and gaps inside, leading to errors in result data, especially, for technological tasks based on data analysis methods, and making a contribution to wrong conclusions and taking incorrect solutions [4, 5]. This fact underlines the relevance of the research.

2. The reasons of gap occurrence in SPM data assets

In order to understand the nature of gaps, we analyzed the collecting process of SPM data assets received from WAMS (Figure 1).

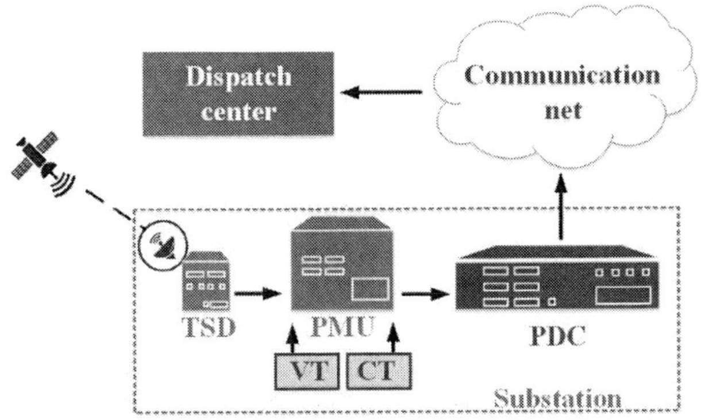

Figure 1. The collecting process of SPM data assets.

According to Figure 1, WAMS consists of three main devices: time synchronization device (TSD); phasor measurement units (PMU); and Phasor Data Concentrator (PDC).

PMU records SPM data assets received from current (CT) and voltage transformers (VT) set at a power station and/or substation. PMU assign the record time tag to every parameter in data assets. TSD generates time tags under synchronization of satellites [6]. After preprocessing SPM data assets in PDC, they are transmitted to the server of power system control center via communication net.

At the stage when data assets are recorded, gaps can be appeared due to PMU technical failures. It requires restarting PMU but it is not possible to restore PMU normal operation immediately since the additional adjustments are required. Moreover, the operating staff cannot fully complete these adjustments therefore required period of time for restoration is significantly increased.

Time synchronization malfunctions can also lead to gaps in data assets due to satellite communication channel failures. They cause by technical reasons or weather conditions in a location of power facilities. For example, low temperatures and high humidity provide icing transceiver facilities, consequently, the connection between a satellite and TSD temporary disappears. As a result the time tag shifts. Now this problem has short-term solution by integrating the special function to the program code of the system server. Unfortunately, this method of data synchronization cannot guarantee the absence of gaps in SPM data assets [4].

WAMS communication net needs to transmit SPM data assets from measurement and registration location to the server of power system control center. During transfer SPM data assets in on-line regime, they are often lost. These are because of connection loss between PDC and the server of power system control center; clients' reconnection; communication net reconfiguration; operation failures of net equipment – routers, hubs, modems, servers; aggregations of SPM data assets received from several PMU to one protocol packet [5].

Specialists of System Operators and Joint-stock companies identified the dependence of gaps amount from the time of day (Figure 2). They calculated the proportion of gaps in SPM data assets as percentage of total amount of received data in 10-minutes intervals per 24 hours period.

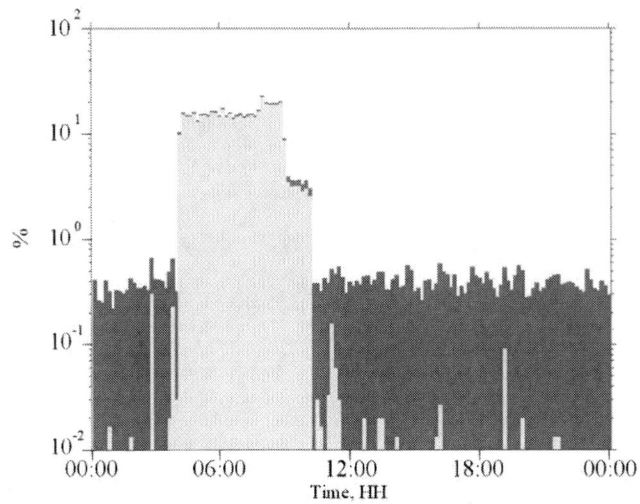

Figure 2. The proportion of gaps in SPM data assets.

Figure 2 shows total amount of gaps in assets: grouped (gaps come one after another) and single gaps (red color); only grouped gaps (yellow color).

The most part of gaps appear at the beginning of a working shift, when clients' reconnections and communication net reconfiguration take frequently place. At the peak load period about 20% of data are lost. Therefore, these data assets can be used for technological tasks only after their processing by special gaps filling algorithms [7].

3. The algorithm for gaps filling

Nowadays the algorithms for gaps filling are constantly developing [7, 8]. The simplest algorithms recommend deleting those rows and columns in data assets which have even one gap [9]. The complicated algorithms use methods of least squares and maximum likelihood [10], and EM-algorithm [11]. All of them are considered to be global ones. It means that the specified type of the dependence is relevant for the whole data assets.

Apart from global algorithms there are local algorithms which define the dependence type according to incomplete selection around a gap. These are standard ZET-algorithm and WANGA-algorithm [7, 8].

This article illustrates application of standard ZET-algorithm. The choice of ZET-algorithm is due to hypotheses which lay in the core of the algorithm and the possibility of application for value forecasting in "time-property"-type time series.

Standard ZET-algorithm consists of three main hypotheses:

a) the excessiveness: it visualizes that SPM data assets are redundant caused by conformity among rows and depending on each other columns;
b) local competence: it states that in order to fill the gap, the analysis of a whole data asset is not needed, but only "competent" part including rows similar to i-row and columns similar to j-column are taken into account;
c) linear dependency: a dependence among rows and columns must be linear.

Standard ZET-algorithm has the preliminary stage, and then three main stages are realized *l*-times.

At the preliminary stage, columns of SPM data asset are normalized according to dispersion for adjusting the properties to unified scale (1):

$$a'_{ij} = A_j^{-1} \cdot (a_{ij} - \bar{a}_j) \tag{1}$$

Where a'_{ij} is adjusted value; a_{ij} is initial value; $\overline{a}_j$ is value of mathematical expectation; A_j is mean square deviation.

Assumed that unknown gap has x and y coordinates (Figure 3).

At the first main stage, from the parent SPM data asset where columns have been normalized according to dispersion by equation (1), for an unknown gap (Saxon blue color), a sub multitude of "competent" rows (blue color) is selected. Then, for "competent" rows, the "competent" columns (blue color) are selected to produce a "competent" matrix. The matrix order is S_{ij}, when $i \in [1;p]$ and $j \in [1;q]$. We must select (p-1) "competent" rows for y-row with an unknown gap and (q-1) "competent" columns for x-column with an unknown gap.

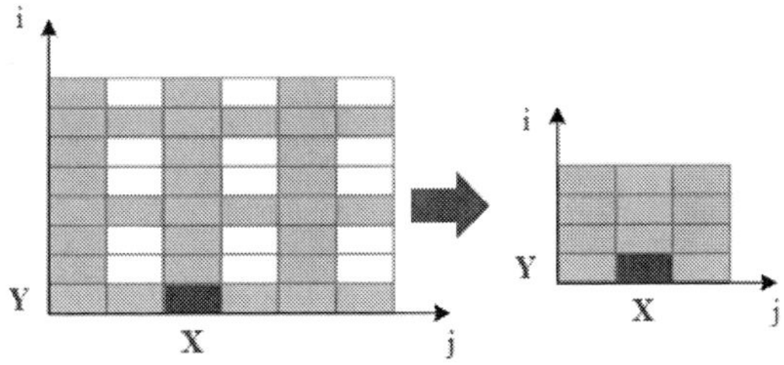

Figure 3. The first stage of ZET-algorithm: asset normalized according to dispersion and "competent" matrix.

The i-row "competence" L_{iy} in relation to the y-row is defined by (2):

$$L_{iy} = t_{iy} / r_{iy} \tag{2}$$

where i is "competent" row number; t_{iy} is completeness that is the number of values known for i- and y-rows; r_{iy} is Descartes' distance between rows.

The "competent" y-row must not contain the gap on the x-position.

The j-column "competence" L_{iy} in relation to the x-column is defined as (3):

$$L_{jy} = |k_{jx}| \cdot t_{jx} \tag{3}$$

where t_{jx} is completeness that is the number of values known for the columns j and x; k_{jx} is correlation coefficient for j- and x-columns. It should be emphasized that for k_{jx} calculation we must use those column values which belong to competent rows.

The "competent" column must not contain the gap on the y- position.

At the second stage, reasonable coefficients σ_p and σ_q are defined for the goal of row and column forecasting gap. This procedure is similar as for rows as for columns. Given coefficient a (a=σ_p for rows and a=σ_q for columns within stated limits and with stated step size, we minimize the function (4):

$$\sum |a_{ik} - b_{ik}| \to \min, \tag{4}$$

where a_{ik} is i-component real value of k-row (k-column) with gap; b_{ik} is i-component forecasting value calculated by (5):

$$b_{ik} = \sum_{j=1}^{p-1} bl_{jk} \cdot L_{ij}^{\sigma_p} \Big/ \sum_{j=1}^{q-1} L_{ij}^{\sigma_q}, \tag{5}$$

where bl_{jk} is k-row (k-column) forecasting value defined as simple regression by the least squares method.

At the last stage, we forecast the gap value: for "competent" rows by (6):

$$b_y = \sum_{i=1}^{p-1} bl_{iy} \cdot L_{iy}^{\sigma_p} \Big/ \sum_{i=1}^{p-1} L_{iy}^{\sigma_p}, \tag{6}$$

and for "competent" columns by (7):

$$b_x = \sum_{i=1}^{q-1} bl_{ix} \cdot L_{ix}^{\sigma_q} \Big/ \sum_{i=1}^{q-1} L_{ix}^{\sigma_q}. \tag{7}$$

The restored gap value is calculated by (8):

$$b_{xy} = 0.5 \cdot (b'_{xy} + b''_{xy}). \tag{8}$$

The error of the gap value is defined by (9):

$$e_{xy} = |B_{xy} - b_{xy}| / |b_{xy}|, \tag{8}$$

where B_{xy} is initially known value.

4. The application of standard ZET-algorithm for gap filling in SPM data assets

We researched the application of standard ZET-algorithm by SPM data assets of current and voltage modules and phases. It should be noted that the data of current and voltage phases play a key role when the task of power system state estimation is decided. Meanwhile, it is impossible to understand, whether the current and voltage phase values are trusted or not, since a phase values can dynamically change from $+\pi$ to $-\pi$ during the time (Figure 4).

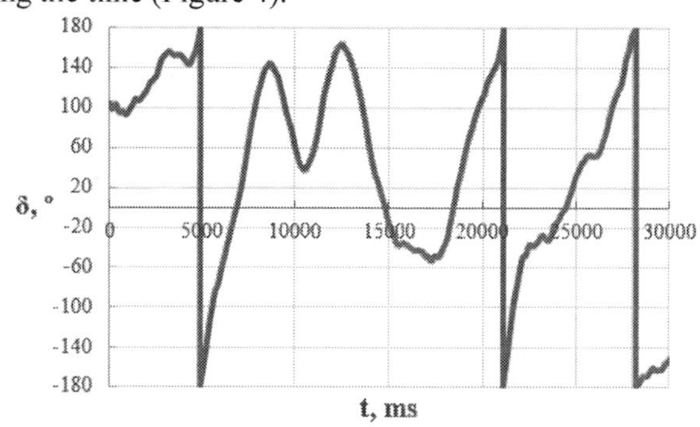

Figure 4. Dynamical changes of phase values during the time.

All researches are provided by SQL Server Management Studio.

SPM data assets are received from PMU installed at 500kV substation 1 in 500 kV power lines Substation 1 – Substation 2; Substation 1 – Substation 3, and in 220 kV power line Substation 1 – Substation 4 (Figure 5).

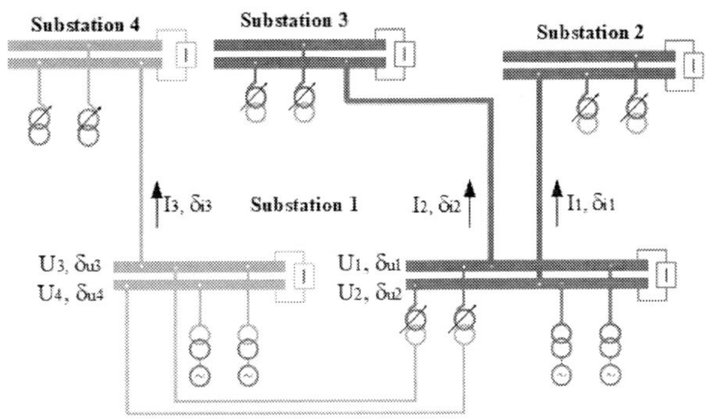

Figure 5. The scheme of power grid.

SPM data assets include 42 000 values of synchronized voltage and current vectors U_A, U_B, U_C, δ_{Ua}, δ_{Ub}, δ_{Uc}, I_A, I_B, I_C, δ_{Ia}, δ_{Ib}, δ_{Ic} per 10 minutes with sample spacing 5 values per a second.

Two test assets are created based on the real SPM data assets. The first test asset consists of 25% of single gaps and 5% of grouped gaps, and the second one includes 5% of single gaps and 25% of grouped gaps (in following tables see asset I and asset II respectively).

The values of the gaps are calculated by (1–8). The error for the gap value is defined by (9). Tables 1 and 2 demonstrate the results of calculations.

Table 1. Result of calculations for gap filling in assets of voltage and current modules.

Test asset	Type of gaps	Error, %
Asset I	single	0.078
	grouped	0.002
Asset II	single	0.033
	grouped	0.046

Table 2. Result of calculations for gap filling in assets of voltage and current phases.

Test asset	Type of gaps	Error, %
Asset I	single	0.335
	grouped	0.798
Asset II	single	0.280
	grouped	0.485

Tables 3 – 6 illustrate the real and calculated phase values for assets I and II with the lowest error of calculation for both single and grouped gaps.

Table 3. Result of voltage and current phase values calculations for asset I. Single gaps.

Real δ°	**ZET-algorithm δ°**	**Δ δ°**	**Error, %**
0.67252	0.57999	0.09253	15.953
-1.71186	-2.07783	0.36597	17.613
-1.54104	-2.01156	0.47052	23.39
2.04425	1.64436	0.39989	24.3188

Table 4. Result of voltage and current phase values calculations for asset I. Group gaps.

Real δ°	ZET-algorithm δ°	Δ δ°	Error, %
3.06494	1.92793	1.13701	58.975
3.22679	1.92827	1.29852	67.341
3.38853	1.93140	1.45713	75.444
-3.20492	-4.27932	1.0744	25.106
-3.03814	-4.24894	1.2108	28.496
-2.87093	-4.23185	1.36092	32.159
-35.77520	-36.79924	1.02404	2.782
-2.71039	-4.21641	1.50602	35.718

Table 5. Result of voltage and current phase values calculations for asset II. Single gaps.

Real δ°	ZET-algorithm δ°	Δ δ°	Error, %
-0.39773	-0.47312	0.07539	15.934
1.02544	0.92375	0.10169	11.008
1.34082	1.22206	0.11876	9.718
2.23318	2.11331	0.11987	5.672
-5.47433	-5.81228	0.33795	5.814

Table 6. Result of voltage and current phase values calculations for asset II. Group gaps.

Real δ°	ZET-algorithm δ°	Δ δ°	Error, %
9.16659	8.14266	1.02393	12.574
9.33588	8.15297	1.18291	14.509
9.51735	8.15548	1.36187	16.698
27.58634	26.22689	1.35945	5.183
27.71748	26.22881	1.48867	5.675
27.85212	26.22936	1.62276	6.186
9.16659	8.14266	1.02393	12.574
9.33588	8.15297	1.18291	14.509

Table 7 shows the results of gap filling for voltage and current phases if the phase values variate at the boundary of [-180°; +180 °] diapason.

Table 7. Results of phase values calculations. Phase values are changing at the boundary of [-180°; +180°] diapason.

Real δ°	ZET-algorithm δ°	Error, %
-179.77574	**144.92745**	**224.045**
-179.25549	-179.32759	0.0402
-179.15950	-179.27198	0.0627
-179.02856	-179.10581	0.0431
-179.70389	**89.97538**	**299.725**
-179.77574	**144.92745**	**224.045**
179.85089	179.77194	0.043
179.81882	179.66501	0.085
179.12233	179.03520	0.0486
179.41821	179.41525	0.001
179.31870	179.13311	0.103
179.37088	179.35614	0.0082
179.50395	179.43094	0.04

5. Conclusion

Proposed researches and results of experiments demonstrate that:

a) the excessiveness: it visualizes that SPM data assets are redundant caused by conformity among rows and depending on each other columns;
b) the main advantage of standard ZET-algorithm is the robustness against occurrence of grouped gaps in SPM data assets;
c) results of experiments from tables 1 and 2 illustrate that in assets when grouped gaps are dominant, they can be adequately filled by standard ZET-algorithm (Asset II);
d) the average error for gap filling in assets of voltage and current phases is higher than for assets of voltage and current modules;
e) results of experiments from tables 3 – 6 prove the high calculation accuracy for both single and grouped gaps as in asset of phases as in asset of modules. The difference between real and calculated phase values is less than 2°;
f) in case of phase value is changing near the boundary of [-180°; +180°] diapason, the gap value is calculated with low accuracy. Therefore, standard ZET-algorithm needs to be sophisticated by the algorithm for a "competent" matrix size correction.

References

[1] Phadke A G and Thorp J S 2008 *Synchronized phasor measurements and their applications* (New York: Springer) p 81
[2] Novosel D *et al* 2007 *Proceeding of IEEE 40th Annual Hawaii Intern. Conf. on System Sciences* 118.
[3] Phadke A and Kasztenny B 2009 *IEEE transactions on power delivery* **24** 89–95
[4] Crotti G *et al* 2019 *IEEE Transactions on Instrumentation and Measurement* **68** 1724–1731
[5] Chen Y *et al* 2016 *Dianli Zidonghua Shebei/Electric Power Automation Equipment* **36** 95–101
[6] Zhang Y *et al* 2010 *IEEE Transactions on smart grid* **1** 159–167
[7] Little R, Rubin D 1989 *Sociological Methods & Research* **18** 292–326
[8] Beale E and Little R 1975 *Journal of the Royal Statistical Society* **37** 129–145
[9] Gleason T, Staelin R 1975 *Psychometrica* ***40*** 229–252

[10] Dempster A, Laird N and Rubin D 1977 *Journal of the Royal Statistical Society* **39** 1–38
[11] Kaminskyi R, N. Kunanets N *et al* 2018 *Paper presented at the CEUR Workshop Proceedings* 108–118

Scientific-technical developments that ensure the growth of the resource potential of the regional oil and gas complex

A S Novikova[1], O V Eremenko[2] and S A Gancharov[3]
[1]Faculty of oil and gas field development, National University of Oil and Gas «Gubkin University», Moscow, Russia
[2]Orenburg branch, National University of Oil and Gas «Gubkin University», Orenburg, Russia
[3]LLC «Gazprom FLOT», PJSC «Gazprom», Murmansk, Russia
E-mail: annov106@mail.ru

Abstract. The Orenburg oil and gas complex is one of the key participants in the Russian fuel-energy sector of economy. However, the Orenburg fields have moved to the final stage of operation. In this regard, the authors in article attempt to formulate and justify scientific-technical developments that allow to stabilize gas production and increase the resource potential through: the introduction of innovative geological-technical measures; the use of «CO_2-technologies»; modernization of the gas collection and primary treatment system; changes in the technology of oil and condensate production; creating tools for the intellectualization of subsurface use, etc. The authors assessed the prospects of new technologies in terms of achieving sustainable gas production in the next 10 years in the amount of 10 billion m^3. For all the proposed innovations, a positive economic effect was obtained, however, some of them did not affect the increase in the coefficient of hydrocarbon recovery from the reservoir, ensuring only the maintenance of the volume of raw material production at the field. The total increase in production over the calculation period will amount to 33.4 billion m^3 of gas and over 0.9 million tons of stable condensate.

1. Introduction

Commercial operation of the Orenburg gas chemical complex began in 1974. At that time, it was the first regional complex in the USSR that included gas production and gas processing capacities. The development of the oil and gas condensate field was characterized by intensive drilling of the main Deposit with a grid of vertical wells and massive acid treatment of them, which made it possible to produce 45-48 billion m^3 of gas annually until 1985 [1]. The perfect nature and high degree of opening of the reservoir provided high working flow rates of wells. Thus, at the initial stage of field development, the gas flow rate for individual wells exceeded 1 million m^3/day.

In the first years of development, there was a progressive flooding of the field through highly permeable reservoirs. Therefore, the newly drilled wells were equipped with a production string with further perforation of the selected intervals. this, in turn, led to a decrease in well productivity. Thus, since 1986 [5], the Main Deposit has been operated under conditions of falling production with an annual decrease in gas extraction by 1.0–2.0 billion m^3.

To reduce the rate of decline in gas production, since 1990, the technology of drilling horizontal wells and side horizontal shafts has been introduced, allowing to return to operation wells that were stopped due to flooding; to increase the productivity of wells located in areas with degraded filtration-capacitance properties of reservoirs [2]. Geological-technical measures are also widely used, such as:

repair and insulation works with transfer to the upper horizons or attaching additional intervals; directed acid treatments; thermal acid and foam acid treatments; radial opening of the reservoir. The entire set of measures outlined is aimed at solving the main problem of the field, which is that half of the gas reserves (52.4%) do not participate in active development (Figure 1).

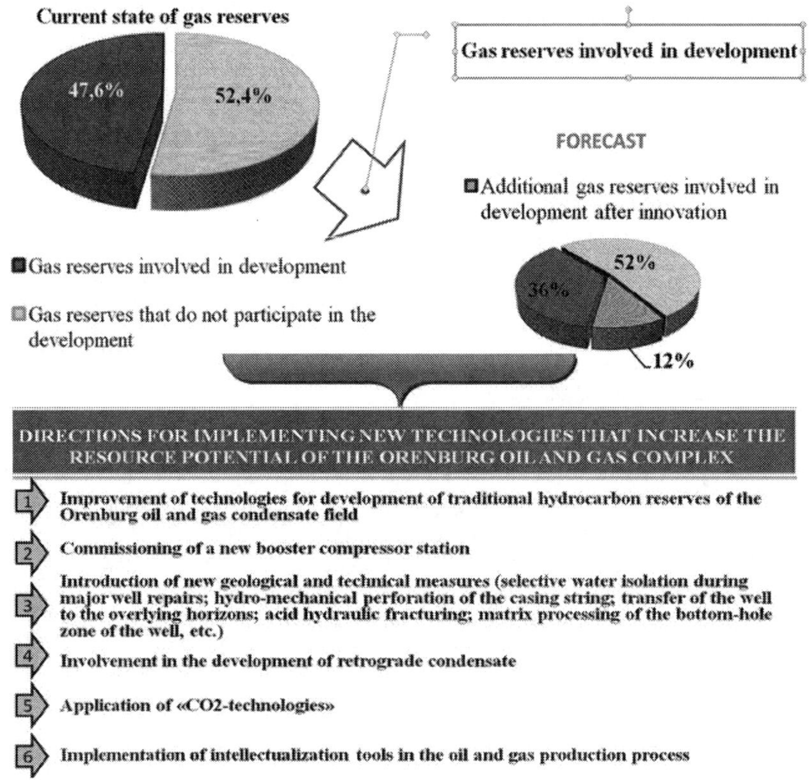

Figure 1. Current state of the field's reserves and directions for implementing innovations that ensure the growth of the resource potential.

Additional involvement in the development of gas reserves is achieved when implementing innovations in priority areas by full opening of the reservoir's productive thickness, using plunger-liftt, etc. Let's look at them in more detail.

2. Technologies and technical solutions offered for application at the Orenburg oil and gas condensate field

The technologies currently used at the Orenburg field can only partially compensate for significant annual losses in hydrocarbon production due to a sharp decrease in reservoir pressure and well stoppage (table 1). Irretrievable gas losses over the past seven years amounted to 1.4 billion m^3 of gas.

Table 1. Retirement of wells from the Orenburg field in 2013–2020.

	Total wells stopped, units	The reason for stopping the wells	The loss of gas, billion m^3
Wells stopped	82	Due to a drop in reservoir pressure	0.6
	44	As a result of wells flooding	0.8

Continuing this trend will increase losses and reduce the resource potential of the gas field by 8.5 billion m^3 of gas by 2030; will reduce the gas recovery rate by 0.6%, which is equivalent to 10.7 billion m^3. Mechanized production using submersible electric centrifugal pumps should have the greatest positive impact on reducing irretrievable losses of hydrocarbons. This will lead to a significant increase in the volume of reservoir water extracted from hydrocarbon raw materials (Figure 2). Testing of this method of production at well No.178 provided a 22-fold increase in the well flow rate (from 1–2 thousand m^3/day to 44 thousand m^3/day).

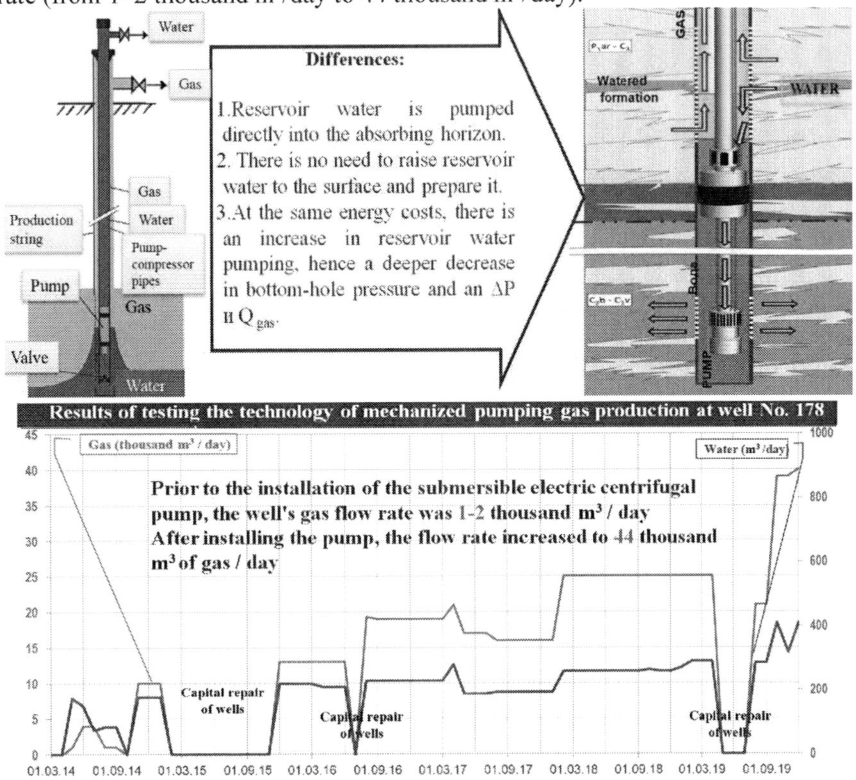

Figure 2. Introduction of mechanized gas production technology using submersible electric-centrifugal pumps.

To ensure the efficiency of pumping-production of gas in areas of mass flooding of wells, it is necessary to develop a technology with reverse injection of reservoir water into the absorbing horizon, so that production does not depend on the permissible volumes of reservoir water treatment in the field. At the same time, special attention should be paid to the justification of the permissible content of liquid hydrocarbons in the injected reservoir fluid. The value of the effect in waterlogged wells will be higher if reconstruction is carried out in neighboring wells transferred to higher horizons. This will allow creating areas with low reservoir pressure in areas of mass flooding, which will receive gas from areas with high pressure, which will ensure filtration of «pinched» gas reserves [3].

In addition, to increase the resource potential, it is necessary to improve the system for maintaining reservoir pressure in the oil and gas condensate deposits of the field. To do this, it is necessary to increase the capacity of booster compressor stations (which requires significant investment) and build injection wells in the areas of the expected location of oil rims. Today, the Orenburg field uses 3 booster compressor stations with a total capacity of 296.7 megawatt for pumping hydrogen sulfide-containing gas from complex gas treatment plants to a gas processing plant, but these capacities are insufficient [4].

The efficiency of the reservoir pressure maintenance system will increase more rapidly if carbon dioxide CO_2 is used as the working agent, provided that the displaced water is subsequently pumped into the reservoir. The use of «CO_2-technology» will ensure the replacement of reservoir gas with carbon dioxide with the displacement of hydrocarbon gas and liquid hydrocarbons (oil and condensate) to production wells. At the same time, the saturation of rocks with hydrocarbons will significantly decrease behind the displacement front.

Testing of this method at the field has shown that injection of CO_2 into the reservoir is accompanied by a significant increase in production of both oil and associated gas. Moreover, 83–87% of the injected carbon dioxide is extracted together with the produced products, and there is a rapid breakthrough of carbon dioxide to the producing wells. Already in the 4th year of CO_2 injection, its amount in the extracted products reached 92%. This result suggests that associated gas with a high CO_2 content can be reused as a working agent. These facts, as well as a preliminary assessment of the economic efficiency of work on the organization of the process of pumping carbon dioxide into a gas-oil Deposit, showed that the use of this technology is commercially feasible (table 2).

Table 2. Efficiency of various development options for the Assel Deposit.

	Option 1 (maintaining reservoir pressure by flooding)	Option 2 (CO_2 injection)	Change (+,-)
Accumulated oil production by the end estimated period, thousand tons	7431	8539	1108
Capital investment, mln. rubles	4710	8264	3554
Revenue, mln. rubles	64018	79015	14997
Operating costs, mln. rubles	50632	61944	11312
NPV, mln. rubles	1958	2271	313

As you can see, the introduction of «CO_2-technology» in the conditions of the Orenburg field looks quite attractive. There is an increase in NPV by 313 million rub. However, the issue of applying injection of carbon dioxide into the reservoir in order to increase the recovery coefficients of hydrocarbons requires additional research (theoretical, experimental and field-based) [5].

The undoubted advantages of this technology are: the possibility of increasing oil and gas production, production of gas products at the Orenburg gas processing plant; increasing the gas recovery coefficient by replacing reservoir gas with carbon dioxide (which will provide an increase in production by 100–150 billion m^3 of gas); increase in the oil recovery rate by 50%; creation of conditions for involving in the development of matrix oil in the amount of 2.6 billion tons, sorbed gas (360 billion m^3), retrograde condensate (60 million tons); slowing the rate of flooding wells by compensating for reservoir pressure during CO_2 injection; reduction in greenhouse gas emissions due to CO_2 utilization.

The main disadvantage of the technology is the high uncertainty of the degree of corrosion activity of the equipment used due to the lack of data.

No less significant in terms of increasing the resource potential are the tools for intellectualizing oil and gas production, which allow not only to increase the efficiency and accuracy of controlling the technological parameters of equipment, but also to save fuel and energy resources. The need for their implementation is also dictated by the complexity of energy technology systems for field development, which involve effective management using modern intelligent systems. They solve multi-factor optimization problems in real time, providing energy monitoring, calculation of equipment energy efficiency indicators, parametric diagnostics, storage, transmission of necessary information, etc. In this regard, the authors propose an information and analytical system for controlling the operation of a booster compressor station.

The main optimization tasks of using this system include: parametric diagnostics of the technical condition of the gas pumping equipment; selection of the optimal gas temperature behind the gas air cooling device in order to reduce hydraulic resistance during gas transportation to the gas processing plant; optimization of gas pressure at the station inlet; minimization of gas bypass from the injection at the input of the bypass system, and gas condensate from the injection at the input of the booster pump station; establishment of optimal parameters of network water temperature at the outlet of heat exchangers.

Among the information and management tasks solved by the system, it is worth noting: continuous monitoring of equipment energy efficiency indicators; determining the reasons for deviations from the optimal operating mode; formulation of recommendations for optimizing the technological mode based on the collection and processing of analytical information about the operation of the booster compressor station. The list of effects that occur when implementing this system is shown in Figure 3:

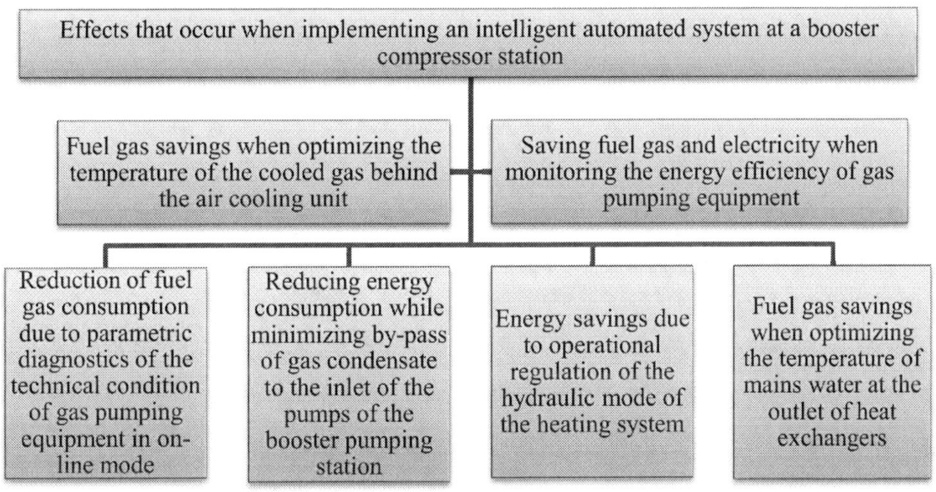

Figure 3. The effects of introducing a system of intellectualization of the management booster pump station.

The total annual energy savings with the introduction of an intelligent control system will amount to 3,497.4 thousand m3 of gas and 688.7 MWh of electricity. Taking into account the company's costs for the development and installation of the system at the level of 8191 thousand rubles, the return on investment will reach 11 months, which once again confirms the feasibility of this scientific and technical development.

3. Results

The priority of recommended technologies to ensure sustainable production of hydrocarbons until 2030 should be determined, first of all, based on the analysis of the development process and the current state of the main Deposit of the field, as the basis of the resource base of the Orenburg oil and gas complex. The mode of its development, despite the progressive flooding of wells, is characterized as a gas mode, which causes a natural decrease in reservoir pressure during development. A drop-in reservoir pressure in conditions where it is not allowed to reduce the wellhead pressure leads to a decrease in the depression on the reservoir, a decrease in the working flow rate of wells for gas, and, consequently, to a deterioration in the conditions for carrying liquid (reservoir water and liquid hydrocarbons) to the surface. As a result, during the operation of waterlogged wells, they are self-trapped by a column of liquid accumulating in the wellbore, which causes the wells to be liquidated or transferred to operation in higher-lying horizons with degraded filtration and reservoir properties. The need to solve these problems determines the significance of the technologies proposed by the authors, which will reduce the rate of decline in gas production and stabilize it at the level of 10 billion rubles. m3 over the next 5 years, as well as increase the resource potential of the field from 330 to 790 billion tons of conventional fuel (Figure 4).

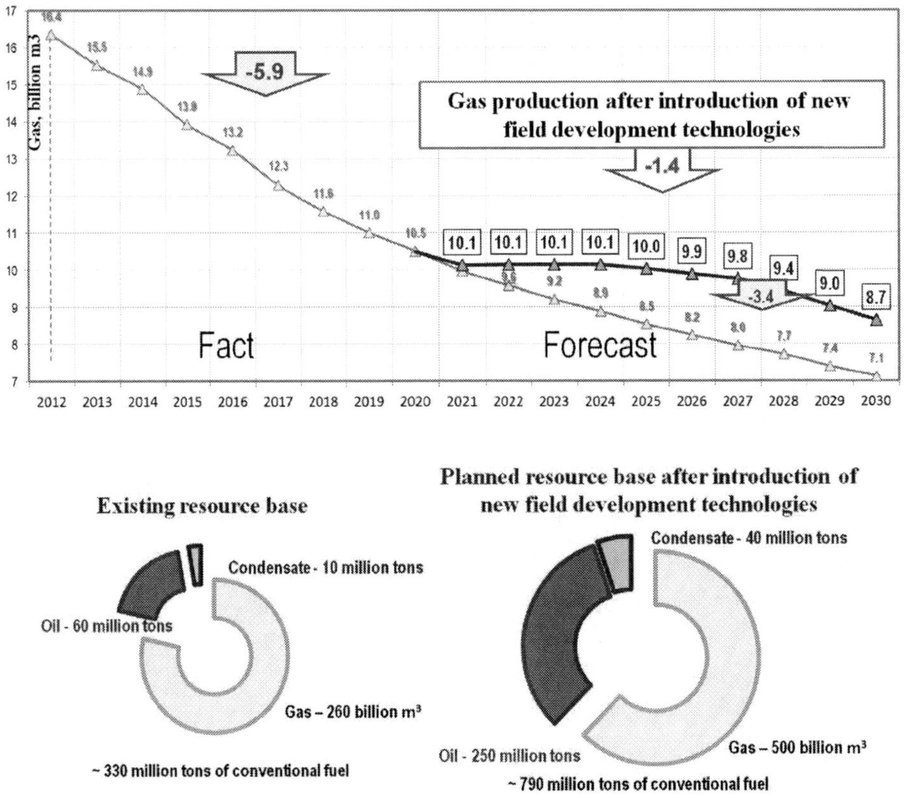

Figure 4. Forecast of changes in the resource potential of the Orenburg oil and gas condensate field.

4. Summary

Thus, when formulating innovative technologies and measures to increase the resource potential of the Orenburg oil and gas complex, the authors gave preference to scientific and technical developments that make it possible to ensure sustainable production of hydrocarbons and involve gas reserves that were previously out of development in active development. For this purpose, it is proposed to widely use mechanized and, above all, pumping operation of watered gas production wells, including reverse injection of reservoir water into the absorbing horizon. To return to active development of the lower part of the productive section of the Main Deposit, it is necessary to reconstruct low-yield wells, which will allow transferring wells to the operation of overlying horizons, as well as returning previously liquidated wells to the Current well stock. In the future, to extend the period of stable hydrocarbon production, technologies will be required that will reduce reservoir, bottom-hole and wellhead pressure in wells after reaching the minimum allowable pressure at the inlet of booster compressor stations (using mobile compressor units), as well as technologies aimed at maintaining reservoir pressure («CO_2-technology», geological-technical measures, increasing the volume of drilling horizontal wells, etc.).

References

[1] Golunov N N, Martinov V G 2020 *et al Neftegaz.RU 6 (102) pp 16-26*
[2] Dmitrievsky A N, Martinov V G 2016 *The fundamental basis of innovative technologies of prospecting, exploration and development of oil and gas and priority directions of development of resource base of fuel and energy complex of Russia pp 29-37*
[3] Eremenko O V, Novikova A S 2019 *Proceeding of 19th international multidisciplinary scentific geocoference SGEM2019* doi: 10.5593/sgem2019/5.3/S21.056
[4] Novikova A S, Eremenko O V 2019 *Proceeding of Tyumen 2019: 6th Conference* doi:10.3997/2214-4609.201900591
[5] Novikova A S, Eremenko O V 2019 *Proceeding of Tyumen 2019: 6th Conference* doi:10.3997/2214-4609.201900625

IOP Conf. Series: Materials Science and Engineering 1019 (2021) 012036 doi:10.1088/1757-899X/1019/1/012036

Road pavement crack detection using deep learning with synthetic data

I A Kanaeva and Ju A Ivanova
Division for Information Technology, Tomsk Polytechnic University, Tomsk, Russia
E-mail: iap15@tpu.ru

Abstract. The improvement of road system quality is a critical task. The mechanism to address such important issue is close monitoring of road pavement condition. Traditional approach requires manual identification of damages. Taking into account considerable length of road system it is essential to create an effective automatic pavement defects detection tool. This approach will extremely reduce time for monitoring of current road state. In this paper global experience in solution of detection issues of road pavement's distress is reviewed. The article includes information about the existing datasets of road defects, which are commonly used for detection and segmentation. The present work is based on deep learning approach with the use of synthetic generated training data for segmentation of cracks in driver-view image. The novelty of the approach lies in creating synthetic dataset for training state-of-the-art deep learning frameworks. The relevance of the research is emphasized by processing of wide-view images in which heterogeneous pixel intensity, complex crack topology, different illumination condition and complexity of background make the task challenging.

1. Introduction

Infrastructure in modern economic model is a sufficiently important key point. From the perspective of Russian Government statistical data analyses the 67.1% of all freights in 2018 were transported by road. Moreover in recent years the volume of carriage steadily increase as well as the amount of private vehicles. All of these factors considerably influence the road pavement condition. Deterioration of road system in turn lead to decrease in shipment transport speed and also increase the number of road accidents. This is precisely why the task of road system quality monitoring is practically essential.

Furthermore, every third car accident in Russian Federation occurs due to unsatisfactory maintenance and furnishing of road system. According to the data of World Health Organization (WHO) each year over 1.35 million people die on roads and more than 50 million get injures. Analogous research in European region shows that 120 thousand of road users die and 2.4 million get injures each year. In Russia 20 thousand traffic deaths and 221 thousand injuries were reported in 2016. The last data in 2019 shows that 17 thousand people died and 210 thousand were injured. Despite the steady decline of numbers of death accidents from 2012 till nowadays the death rate in Russian federation is still beyond than average in Europe. Recently, in the Russian Federation, the national project "Safe and high-quality roads" has been carried and now it is actively developing. The goal of this project is to improve the quality of significant regional roads and road network of urban agglomerations up to the standard state. The released program is directed not only to traffic improvement, but also to reduction of deaths numbers on roads in 3.5 times.

Considering that fact that Russian road system is 5th longest in the world, the road diagnostic system critically need to automate routine tasks such as defects detection. Due to the sufficient length of the road, some aspects of the diagnostic process require analysis of a considerable amount of video data. For this aim the computer vision may be effectively implemented.

Current progress in computer vision is based on deep learning and has been achieved mostly due to large image data sets with careful annotations. Such spheres as for example autonomous driving systems are actively developing due to accurate semantic segmentation datasets, such as Cityscapes [1], Wilddash [2] and KITTI [3]. Manually labeling process of such kind of data is expensive and labor-consuming.

Due to the need of computer, processing of high-quality video data of roads in the road industry there is requirement for developing automatic algorithm of road surface defects detection by images. The development of an effective algorithm of pavement defects identification is notably important task. On one hand, it is able to improve quality and speed of the monitoring process. On the other hand, repair roadworks carried in time not only allow to maintain the standard road condition, but also preclude significant amount of accidents, inclusively resulting in death.

2. Related works

Current methods of monitoring road pavement can be divided into three types: 3D scanning-based, vibration-based and 2D image vision-based. The first group of methods are based on different sensors such as inertial measurements, 3D scanners, and optical sensors. A high cost of LIDAR system and complexity of exploitation and data post-processing are limit their use. Accelerometer is a principal component of vibration-based methods. It measures vibrations caused in a vehicle due to hitting a pothole or other road damages.

Over the past decade, many different techniques of image processing have been proposed in the field of automatic detection, classification and segmentation of pavement distress. A. Mohan and S. Poobal in [4] reviewed 50 research papers in the area and provided the collective survey of the different image processing techniques used for the detection of the cracks in the engineering structures.

In a review [5] on automated pavement distress detection methods authors conducted that the highest need for further research is demanded from the surface defects: raveling, polished aggregate and bleeding. German researchers in paper [6] divided the algorithms developed for evaluation of the pavement surface into three major groups: crack threshold filtering, patch-based classification, and depth-based algorithms.

The first group of methods designed explicitly for crack detection, mostly by applying image processing methods that segment distress textures by threshold filtering. In order to reduce illumination artifacts image-preprocessing algorithms are preceded. Due to pixels of cracks have minimum intensity threshold filtering is applied after preprocessing. On last stage, the detection is refined by morphological image operations and by searching for connected components. Next papers present the aforementioned approach.

In [7] the research results were implemented as a CrackIT software tool for segmentation of cracks in an image taken directly above the road surface. The CrackTree [8] toolbox is based on the construction of probability map of pixels belonging to a crack on the image that previously was preprocessed by geodesic shadow-removal algorithm. Paper [9] proposes a new unsupervised multi-scale fusion crack detection algorithm that works on a series of images smoothed by different-scale Gauss filters and combines the resulting masks. Gabor filters in [10] are used for searching areas as cracks.

The algorithms of the second group apply various types of classifiers to patch of the image to determine distress areas or cracks. One part of the researchers initially selects a certain feature vector from the considered image region, and then use it as input of the classifier. The advantage of this approach is that the size of the region is not fixed, but initially you need to divide the image into a sufficient number of regions. In [11] regions are defined using the over-segmentation algorithm SLIC. Then support vector machines were used as a binary classifier. Despite the fact that the method does not have high accuracy, it allows to calculate the ratio of damaged and non-damaged pavement. In addition, this method can be expanded for different defects detection, such as pothole, manhole and road marking.

With the advent of public available datasets of road images with pavement distress, such as GAPs [5] and CRACK500 [12], many researchers have used deep learning approaches for the problem. For example, in [13] a truncated CNN VGG16 is used for extract the feature vector from the input image. Next, a neural network with one hidden layer of 256 neurons classifies the feature vector.

3. Existing datasets

Consider the most popular and public available datasets for road distress classification, segmentation and detection tasks:

- GAPs dataset [5]: includes total 1,969 gray valued images with resolution is 1,920×1,080 pixels. These images are divided into 64×64 patches and each patch is labeled as a crack or not. The pictured surface material contains pavement of three different German federal roads.
- CRACK500 dataset [14]: consists of 500 RGB images of pavement cracks of size around 2,000×1,500 pixels that were collected on main campus of Temple University using cell phones. Each crack image has a pixel-level annotated binary map.
- CrackTree200 dataset [8]: includes 206 pavement images of size 800×600 with various types of cracks that were annotated in pixel-wise level. The images have complex asphalt context with shadows, occlusions, low contrast and noise.
- CFD dataset [15]: have 118 images of annotated road crack of size 480×320 pixels. These images was captured on urban Beijing roads. The images contain a significant amount of noisy pixels like oil spots and water stains, and some of them are under the poor illumination condition.
- Road Damage dataset [16]: consist of 9,053 labelled road images of size 600×600, acquired from a smartphone camera installed on the dashboard of a car. The main aim of this is capturing general front view from driver's position, as opposed to capturing images above the road surface. It decreases difficulty of capturing image process and increases practical applicability of such images. The dataset has 15,435 bounding boxes of damages in total, annotated for the dataset.

The Road Damage dataset was collected in seven Japan municipalities and has eight classes of pavement distress: five types of cracks, two types with wear of road marking, one class that combines other damages like rutting, bump, pothole and separation. The dataset has a PASCAL VOC [17] format and was presented at the IEEE International Conference on Big Data Cup in 2018.

The dataset of Japanese scientists has revived interest in solving the challenge of automatic damage detection on wide-view road images using convolutional neural networks. This was facilitated by sufficiently large scale of the dataset and presence of different distress types, not only cracks. Disadvantages of the dataset is that bounding boxes may include a lot of redundant information due to various forms and sizes of damage on image, especially diagonal cracks on road. For maintains of road pavement condition pixel-level segmentation task is more appropriate and allows localize damage and estimate distress size.

Creating dataset for road distress segmentation with high quality pixel-level annotations is a laborious and time-consuming process due to manual labeling of object's pixels in the image. Another approach to deal with this problem is a synthetic data generation. In next section, we propose algorithm for generation instance-level synthetic dataset for crack segmentation based on well-known collections with a marked road, such as KITTI and Cityscapes dataset.

4. Related work

4.1. Synthetic dataset creation

At current stage of machine learning evolution, the formation of a set of training data have paramount importance for the successful solution of the tasks of detection and segmentation. However, the meticulous manual targeting of several thousands of images is an enormous and sufficiently labor-consuming process, so quite important task is to develop methods for obtaining a representative synthetic samples.

A synthetic dataset is a repository of data that is generated programmatically and cannot be collected by any real-life survey or experiment. For creating the synthetic road crack dataset, we decided to use three publically available sets: KITTI and Cityscapes dataset as images of the road scene, CFD as a source of cracks marked at the pixel level (Figure 1).

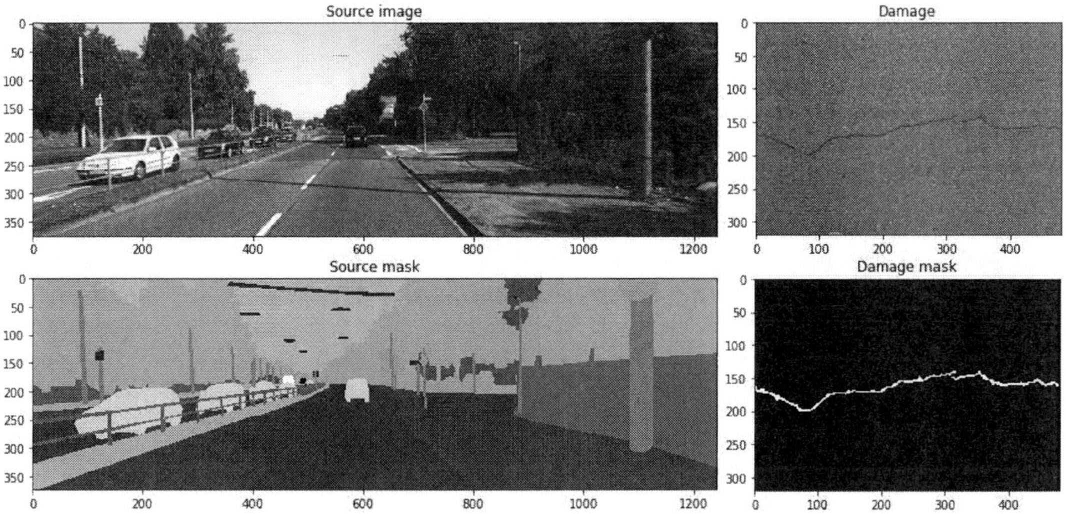

Figure 1. Road and crack images with masks.

To determine the part of the image corresponding to the roadway, all the pixels of the road mask are chosen. Then, the connected components algorithm with 8-connectivity is applied to the resulting binary mask for determine connected areas. As a result, the area with the maximum number of pixels is taken as the main road mask (Figure 2, main road mask is highlighted in gray).

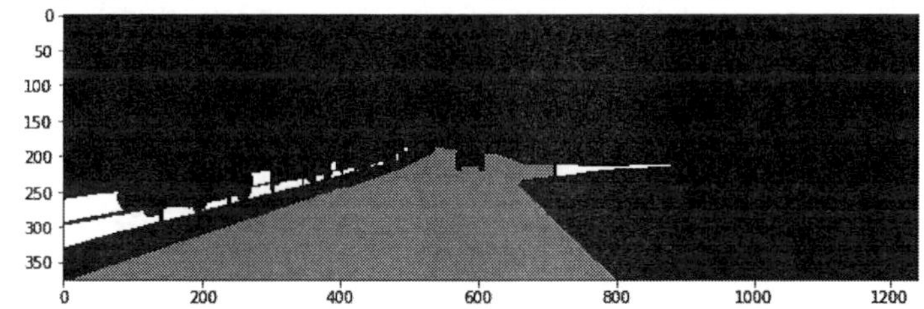

Figure 2. Maximal area of roadway.

Then the original image of the crack and its mask are cropped at the minimum bounding rectangular for the purpose of reducing following calculations. After that, the image of crack with the mask is scaled and rotated randomly. The result is an image of the crack D and its mask D^{mask} (Figure 3).

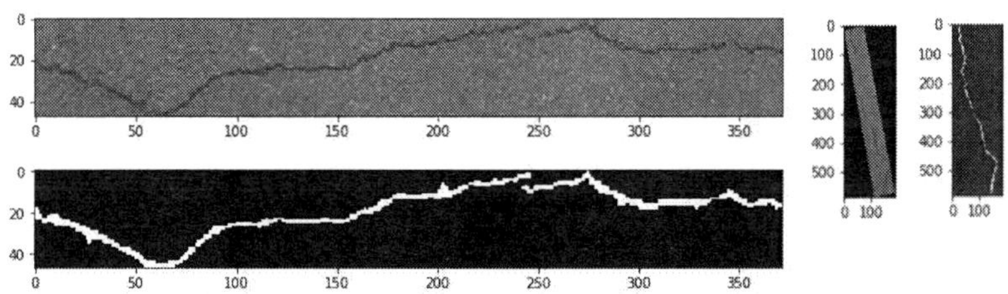

Figure 3. Cropping, rotating and scaling of crack.

In the next step a point, which is the center of the overlapping area, is selected inside the mask of the main area of the road, and the S^{mask} is cut out of the original image mask, equal in size to D^{mask}. To mix two images in mask area, the mean asphalt value $\overline{D_c}$ are calculated in each channel c for the image of the crack that does not lie under the mask:

$$\overline{D_c} = \frac{1}{k}\sum_p D_c(p)\cdot\left(1 - D^{mask}(p)\right), c \in \{R, G, B\}, \tag{1}$$

where $D_c(p)$ – pixel value p of crack image D in channel c, $D^{mask}(p)$ – pixel value p in binary crack mask, k – number of pixels for which $D^{mask}(p) = 0$. This calculation process is possible due to the homogeneous texture of the asphalt on the image of the defect. Then, when a crack is applied to an image of a road, only the values under the mask of the roadway and the cracks are considered:

$$M(p) = D^{mask}(p)\cdot S^{mask}(p) \tag{2}$$

The crack is added by changing the pixel values of the original image S. These pixels are located under the common mask M and their values are multiplied by the ratio of the calculated mean asphalt surface to the crack pixel values:

$$S_c(p) = S_c(p)\cdot M(p)\cdot D_c(p)/\overline{D_c} \tag{3}$$

The algorithm result is shown in Figure 4.

Figure 4. The result of synthetic crack generation.

For the purpose of increasing the informational capacity of the training image sample, from 1 to 5 cracks, which can intersect were generated on one image. As a result, the training set contains 1524 images, and the test dataset includes 505.

4.2. Object Segmentation System
To solve the crack detection problem in pixel level as segmentation task, we decided to use two modern convolution neural network systems: Mask R-CNN [18] and U-Net [19]. Consider its structures and main principles of operation.

4.2.1. Mask R-CNN is a state-of-the-art framework for detecting objects in an image while simultaneously generating a high-quality segmentation mask for each instance. Mask R-CNN extends object detection algorithm Faster R-CNN by adding a module for predicting an object mask. Simplify Mask R-CNN structure is illustrated in Figure 5. Mask R-CNN have a complex, flexible and powerful block architecture with two stage: generating object proposals and classifying proposals to generate bounding boxes and masks in parallel.

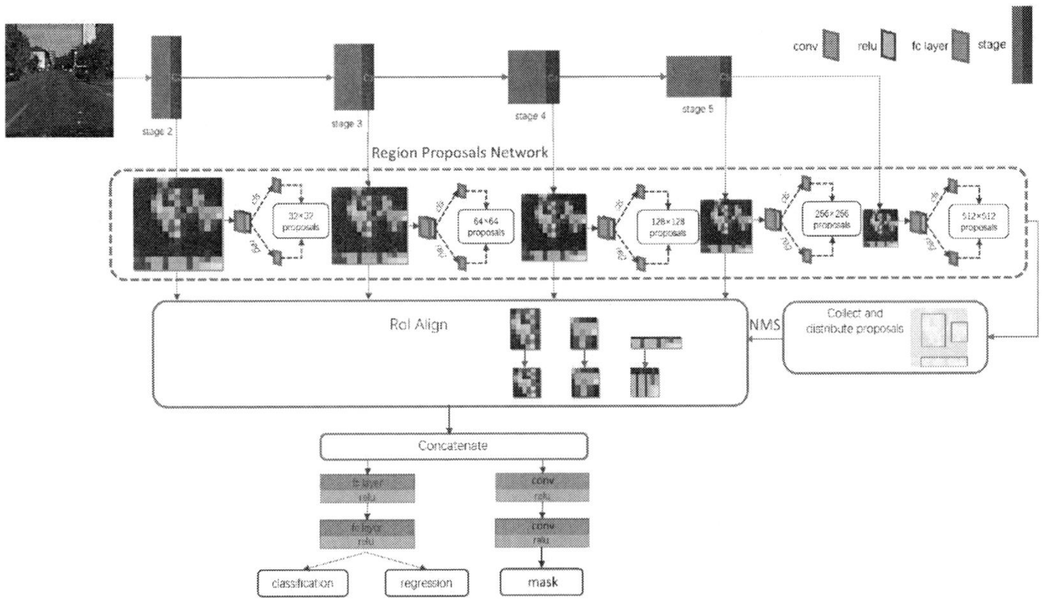

Figure 5. Mask R-CNN architecture.

Initially, the image is fed to the input of Mask R-CNN to produce a feature map, which often uses pre-trained VGG16 or ResNet50/101 with excluded layers responsible for classification and named backbone. One of the improvements in this framework is using Feature pyramid networks (FPN) for generation multi-scale feature maps. Sequential layers of FPN with decreasing dimension are considered as a hierarchical pyramid, in which the lower level maps have high resolution, and the upper level maps have high generalizing, semantic ability.

The resulting feature maps are processed in CNN Region Proposals Network (RPN), whose task is to create the regions of interests (RoIs) which may contain objects. For this purpose, each feature map is scanned by lightweight neural network with a 3×3-convolution layer. Output of RPN is based on k anchors - set of boxes with predefined locations and scales relative to images. For each anchor, RPN generates a probability of a proposal having the target object, and a refinement of the coordinates of the bounding box of the object. The purpose of this stage is to identify regions of interests that may contain objects. At the end, duplicate proposal regions are discarded due to non-maximum suppression operation.

Then the proposals are mapped from corresponding feature map levels, extracted from its and resized to the same size using the RoI Align operation. According to RoIs, the final operations of classification, refinement of the bounding box's coordinates and mask prediction are performed at the second stage. The output mask has a greatly reduced size, but contains real values, which allow to obtain sufficient accuracy by scaling the mask to the size of the selected object's bounding box.

4.2.2. U-Net model is a fully convolutional network that outputs a classification of each pixel in the image to generate a segmentation mask. U-Net architecture consists of a contracting path to capture context and of a symmetrically expanding path that enables precise localization. The contracting path follows the typical architecture of a convolutional network with alternating convolution and pooling operations and progressively down-samples feature maps, increasing the number of feature maps per layer at the same time. Every step in the expansive path consists of an up-sampling of the feature map followed by a convolution.

Typically, U-Net is trained from scratch starting with randomly initialized weights. In order to account that our training dataset is synthetically generated dataset, it only models plain cracks on road surface without perspective distortion and light aspects, we use transfer learning similar to TernausNet

[20]. We used U-Net type architecture improved by the using of the pre-trained VGG16 on ImageNet as the encoder. Initialized weights from the pre-trained network are frozen. This approach allows significantly reducing the number of trainable parameters and shows better performance than training from scratch. To construct an encoder, we kept only first four convolution blocks with last convolutional layer with 512 channels. This U-Net architecture is illustrated in Figure 6.

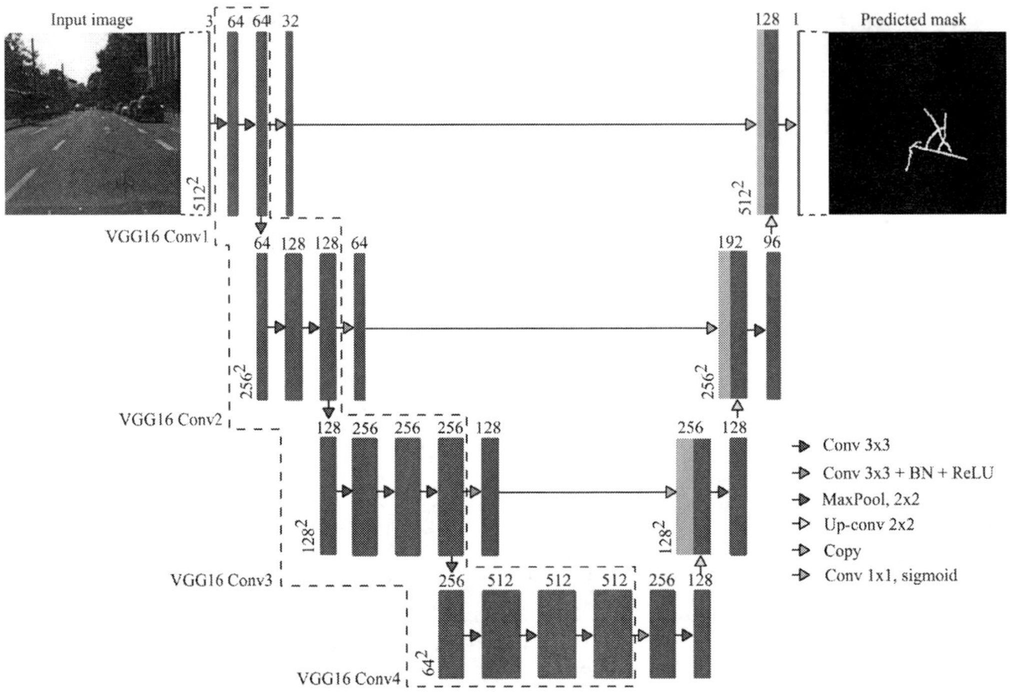

Figure 6. VGG16 + U-Net architecture.

5. Experiments

The obtained synthetic set of images with per-pixel labeled road cracks was used to train modern convolutional neural networks Mask R-CNN and U-Net. For our research we used Abdulla et al. [21] Tensorflow + Keras implementation of Mask R-CNN with modifications to perform our dataset. ResNet101 with FPN forms backbone for feature maps construction. Due to the fact that the synthetic dataset cannot contains all characteristics of real asphalt cracks, the transfer learning technology was used in combination with pre-trained model ResNet101 on the MS-COCO dataset. Transfer learning allows accelerating the training process and improving the performance of a model.

We use RGB images of size 1024×1024 as input of the Mask R-CNN. In addition, the following values are used as anchor scales: 0.33, 0.5, 1, 2, 3. The training took out 40 epochs of 400 iterations using mini-masks with size 56 × 56 pixels to optimize the used computer memory. We get the best result using learning rate annealing by starting with 0.001 and decreasing it by a factor of 10 every 10 epochs. Removing detection results of the same class happen if there is more than 0.7 value of IoU among bounding boxes.

VGG16 + U-Net realization in input uses RGB image with 512×512 resolution and output prediction mask of the same size. Pre-trained VGG16 on ImageNet is used as the encoder. Images from the synthetic dataset resized to 472×472 size and padded to the input size. This operation allows to avoid losing border pixel in the input due to convolution sequence. In these experiments, we use Adam optimizer with a batch size of 8, momentum of 0.9 and a learning rate of 0.001 with decay of 0.000001 and trained the models for 17 epochs.

All the experiments have been conducted on a workstation with a NVIDIA Tesla K80 GPU graphics card machine with 13 GB DDR5X memory on Google Colaboratory platform.

6. Results & evaluation

Instance segmentation using Mask R-CNN often is evaluated in PASCAL VOC style. The best Mask R-CNN's result estimated by the average precision (AP) metric with the value IoU = 0.5 on the validation synthetic subset is 78.1%.

Jaccard index (Intersection over Union) is chosen as evaluation metric for segmentation task. It reflects similarity measure between a finite numbers of sets. The measure between two sets A and B is defined as following:

$$IoU(A,B)=\frac{|A\cap B|}{|A\cup B|}=\frac{|A\cap B|}{|A|+|B|-|A\cap B|} \quad (4)$$

Pixel accuracy metric not used in the evaluation stage, because it can provide misleading results for the reason of small crack class representation within the image in contrast of negative case.

To evaluate the relevance of the training on our synthetic dataset we constructed a small dataset of 67 real images with cracks on carriageways that were manually labeled. These real images were obtained by digital camera mounted at a roof of a vehicle. The collected real-image dataset consists of consecutive frames of two street and has different cracks, potholes, shadows, road marking, manholes, road facilities and equipment. The presence of these elements on the images leads to the conclusion that the real-image dataset is representative.

Finally, the IoU score for segmentation results on synthetic and real image dataset was calculated and summarized in table 1.

Table 1. Intersection over Union evaluation

Model	Dataset		
	Synthetic		Real-image
	training	validation	
Mask R-CNN	0.79	0.59	0.46
U-Net	0.81	0.56	0.47

Figure 7 shows four examples from real-image dataset and corresponding results of manual human annotation, Mask R-CNN detection and VGG16+U-Net segmentation.

7. Conclusions

The automatization of road condition control is current trend for practical application of computer vision methods. The conducted analytical review of road distress detection problem has shown that crack detection in pixel level in the driver-view image is a priority and challenging task. We have proposed a novel deep learning approach based on synthetic training data generation for segmentation of cracks in the images with road pavement. The synthetic data generation algorithm allows to easily obtain crack training dataset of any size for instance segmentation task. It makes possible to use state-of-the-art Mask R-CNN-based and U-Net-based segmentation model. The models, trained on synthetic crack data, give acceptable outcomes over 47% of IoU metrics on real images with surfaces' cracks. It should be noted that the proposed approach is weak sensitive to shadows, road marking and light condition. Improving the detection results is possible with additional training of models with attention to road technical facilities, such as manholes and restoration patches.

a) b) c) d)

Figure 7. Segmentation results: (a) examples of pavement surface crack from real-image dataset, (b) their groundtruth labels, (c) instance-segmentation result by Mask R-CNN, (d) pixel-wise segmentation results by U-Net.

Acknowledgment
This research was supported by Tomsk Polytechnic University Competitiveness Enhancement Program and was funded by RFBR according to the research project № 18-08-00977 A.

References

[1] Cordts M, Omran M, Ramos S, Rehfeld T, Enzweiler M, Benenson R, Franke U, Roth S and Schiele B 2016 *IEEE Conf. on Computer Vision and Pattern Recognition (CVPR)* 3213–3223
[2] Zendel O, Honauer K, Murschitz M, Steininger D and Dom´ınguez G F 2018 *Proceeding of Computer Vision – ECCV 2018* 407–421
[3] Fritsch J, K¨uhnl T and Geiger A 2013 *Proceeding of 16th Int. IEEE Conf. on Intelligent Transportation Systems (ITSC 2013)* 1693–1700
[4] Mohan A and Poobal S 2018 *Alexandria Engineering J.* **57** 787–798
[5] Coenen T B and Golroo A 2017 *Cogent Engineering* **4** 1–23
[6] Eisenbach M, Stricker R, Seichter D, Amende K, Debes K, Sesselmann M, Ebersbach D, Stoeckert U and Gross H 2017 *Proceeding of Int. Joint Conf. on Neural Networks (IJCNN)* 2039–2047
[7] Oliveira H and Correia P L 2014 *Proceeding of IEEE Int. Conf. on Image Processing (ICIP)* 798–802
[8] Zhang L, Yang F, Zhang Y D and Zhu Y J 2016 *Proceeding of IEEE Int. Conf. on Image*

Processing (ICIP) 3708–3712

[9] Li H, Song D, Liu Y and Li B 2019 *IEEE Transactions on Intelligent Transportation Systems* **20** 2025–2036

[10] Salman M, Mathavan S, Kamal K and Rahman M 2013 *Proceeding of 16th Int. IEEE Conf. on Intelligent Transportation Systems (ITSC 2013)* 2039–2044

[11] Varadharajan S, Jose S, Sharma K, Wander L and Mertz C 2014 *IEEE Winter Conf. on Applications of Computer Vision* 115–22

[12] Liu Y, Yao J, Lu X, Xie R and Li L 2019 *Neurocomputing* **338** 139–153

[13] Gopalakrishnan K, Khaitan S K, Choudhary A and Agrawal A 2017 *Construction and Building Materials* **157** 322–330

[14] Yang F, Zhang L, Yu S, Prokhorov D, Mei X and Ling H 2020 *IEEE Transactions on Intelligent Transportation Systems* **21** 1525–1535

[15] Shi Y, Cui L, Qi Z, Meng F and Chen Z 2016 *IEEE Transactions on Intelligent Transportation Systems* **17** 3434–4345

[16] Maeda H, Sekimoto Y, Seto T, Kashiyama T and Omata H 2018 *Computer-Aided Civil and Infrastructure Engineering* **33** 1127–1241

[17] Everingham M, Eslami S M, Van Gool L 2015 *Int J Comput* **111** 98–136

[18] He K, Gkioxari G, Doll´ar P and Girshick R 2017 *Proceeding of IEEE International Conf. on Computer Vision (ICCV)* 2980–2988

[19] Ronneberger O, Fischer P and Brox T 2015 *LNCS* **935**1234–1241

[20] Iglovikov V and Shvets A 2018 TernausNet: U-Net with VGG11 Encoder Pre-Trained on ImageNet for Image Segmentation *Preprint* 1801.05746

[21] Abdulla W 2017 Mask r-cnn for object detection and instance segmentation on keras and tensorflow. Available at: https://github.com/matterport/MaskRCNN

IOP Conf. Series: Materials Science and Engineering 1019 (2021) 012037 doi:10.1088/1757-899X/1019/1/012037

Particularities of spatial kinetics of hybrid thorium reactor installation containing the long neutron source based on magnetic trap

I V Shamanin[1], S V Bedenko[1,3], I O Lutsik[1], S D Polozkov[1] and A V Arzhannikov[2]
[1]School of Nuclear Science and Engineering National Research Tomsk Polytechnic University, Tomsk, Russia
[2]Department Plasma Physics Budker Institute of Nuclear Physics SB RAS, Novosibirsk, Russia
E-mail: bedenko@tpu.ru

Abstract. In this work, we study the features of the spatial kinetics of installation as a hybrid thorium reactor with an elongated plasma neutron source based on a magnetic trap. The active zone of the installation under study consists of an assembly of hexagonal fuel blocks of a unified design and a long solenoid with a high-temperature plasma column passing through the axial region of the core. Combining engineering expertise in creating nuclear reactors with a physics-technical potential for obtaining high-temperature plasma in a long magnetic trap we ensure the solution of the multidisciplinary problem posed. These studies are of undoubted practical interest, since they are necessary to substantiate the safety of operation of such hybrid systems. The research results will allow optimizing the active zone of the hybrid system with leveling the resulting offset radial and axial energy release distributions. Results of our study will be the basis for the development of new and improvement of existing methods of criticality control in related systems such as "pulsed neutron source – subcritical fuel assembly"

1. Current state of the art

The study of physics of low-power high temperature gas-cooled thorium reactor unit (HGTRU) (Tomsk Polytechnic University (TPU), Tomsk) was started three years ago [1–3] and its authors have already chosen the most appropriate core configuration and a nuclear fuel material composition. The authors have stated [1] that the reactor being studied is capable of operating for not less than 3000 effective days at the power of 60 MW_{th}.

A special feature of the reactor studied in the work [1, 3] is that it is able to produce heat, electricity energy and hydrogen simultaneously, and its core can rather easily be modified for solving another task [4]. Besides, the reactor power can be adjusted according to the regional energy needs.

At present a series of experiments for the purpose of studying physical properties of the new generation nuclear fuel for HGTRU is being carried out at TPU. Dispersion fuel, which is being developed at TPU, also refers to the fuel of new generation. It is a fuel material with advanced mechanical, thermophysical and neutronic properties. As far as the reactor core with such fuel has not been studied earlier in neutronic experiments, on this reason it became necessary to create an installation to carry out required experiments. A special facility intended for studying neutronic properties of the dispersion $(Th,Pu)O_2$ fuel was suggested in the capacity of such installation by the employee group of Budker Institute of Nuclear Physics of SB RAS (Novosibirsk, Russia) [4–6]. This facility is an assembly of fuel blocks the axial part of which is substituted by a long magnetic trap [5,6] with high temperature plasma providing generation of thermonuclear neutrons.

In our paper, we present the results of computer simulation of neutronic processes in the core of the HGTRU for 30 different loading options. To provide stable reactor operation the fraction of the dispersion phase (ω_f) and the composition of the starting fissionable nuclide were chosen. The possibility to modify the axial part of the core was investigated in accordance with the concept suggested in the works [4].

From the perspective of advancement in the field of fundamental knowledge, the purpose of these researches is to expand and to deepen some understanding of the possibilities that become open due to the development of thermonuclear power energy technologies of the future. From the point of view of important applied tasks solution it can be said that the results will form the basis of organizing a stable operation of HGTRU in a long-term operating cycle (not less than 10 years) with a high burn-up degree of both thorium and plutonium components of the fuel.

2. Numerical investigation

2.1. Computational model of the core basic configuration

The core of the HTGRU [1, 3] consists of fuel blocks having the configuration of a graphite hexagon. The core is surrounded by two rows of graphite hexagons of the same configuration, but without holes for fuel. On the top and on the bottom the core is closed with graphite blocks too, but they are laid in one row there.

Schematic drawing of the fuel block and its cross section are presented in Figure 1*a, b*. One can see that there are the cylindrical channels along all length of the block. The fuel block contains 76 channels of small diameter for fuel pellets (red channels of small diameter) and seven channels of large diameter for helium flows (blue channels of large diameter).

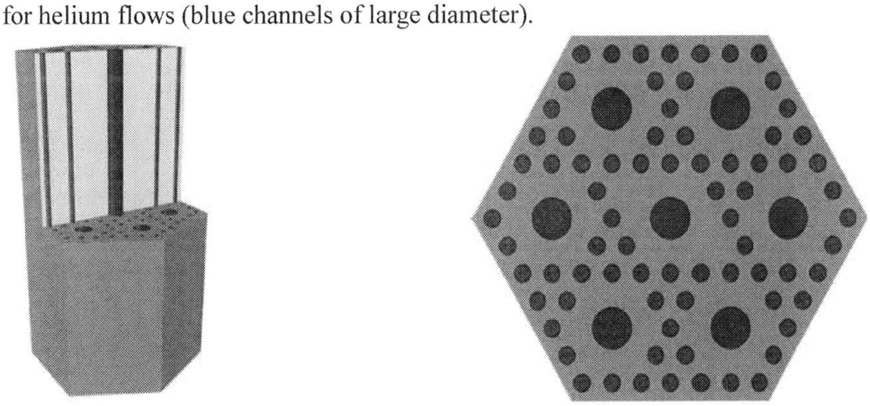

(a) (b)

Figure 1. The schematic of the core fuel block (a) and its cross section (b).

Microencapsulated fuel (microfuel) for fuel pellets of HGTRU [1–3] is a spherical fuel kernel from fissionable material $(Th,Pu)O_2$ with the diameter of 350 μm, covered successively by the layers of PyC and Ti_3SiC_2. These fuel kernels are dispersed into a graphite matrix of cylindrical fuel pellets, which are placed in the fuel block.

At the first stage of research presented in the works [1,3], three types of fuel pellets with reference designations 0817, 1017 and 1200 were used. The diameters of these fuel pellet types were different:

8.17×10^{-3}, 10.17×10^{-3} and 12.00×10^{-3} m, correspondingly. At the same time, the height of the fuel pellets (20×10^{-3} m) and thickness of the external SiC layer (0.3×10^{-3} m) were equal for all types.

In the work [1] it was shown that from the point of view of the campaign life the best reactor core loading can be achieved by using the fuel pellets of the type 1200. However, in the reactor with the fuel pellets of this type the increase of the disperse phase quantity ω_f ($\omega_f = V_{coated\ particle}\times N/V_{matrix}$) results in significant decrease of spent $((\eta(^{239}Pu) = (N(t_{start}) - N(t_{end}))/N(t_{start}))$ ^{239}Pu. Therefore, the fuel pellet of the type 1017 was chosen in the researches for better ^{239}Pu burnup.

2.2. Neutronic computations for different core loading configurations

At the second stage of the performed computer calculations the cylindrical reactor cell that is an equivalent system of Wigner-Seitz, is studied. The fuel part of the cell is homogenized; computations are performed in the program WIMS-D5B [7].

WIMS-D5B (Nuclear Energy Agency) is a universal program for computing cells of different reactor types. The WIMS-D code uses a 69-group system of constants on the basis of the evaluated nuclear data base ENDF/B-VII.0 (The basic library has been compiled with 14 fast groups, 13 resonance groups and 42 thermal groups), which allows computing reactors on fast and thermal neutrons. In WIMS-D, the accuracy of the computed functionals is set by the step equal to 10 days. To assess poisoning the reactor by xenon, the first 2 days are computed separately.

To determine effective K-factor (k_{eff}) an axial and radial geometrical parameter (B) is introduced. It is computed taking into account transition from the actual core size to the equivalent system of Wigner-Seitz. "White mirror" is used as boundary conditions on the side surface of the cell and "translation symmetry" is used on the cell faces. The computation results of 30 core-loading options with fuel pellets of the type 1017 are presented in Figure 2.

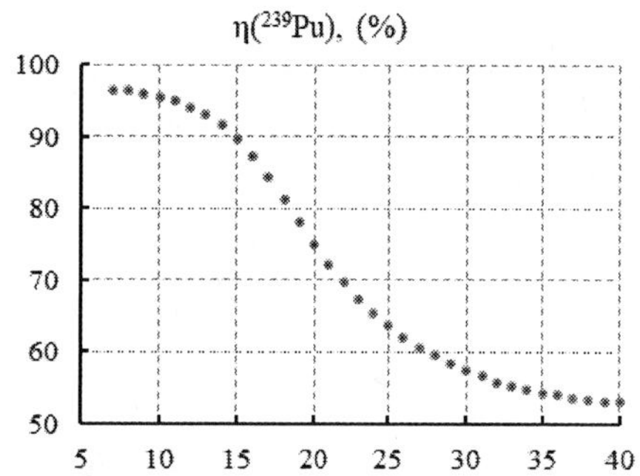

(a)

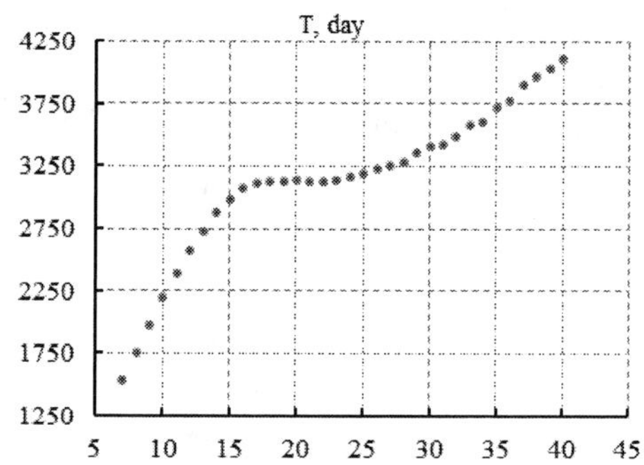

(b)

Figure 2. Neutronic properties of the equivalent system: *(a)* – dependence of $\eta(^{239}Pu)$ on ω_f, *(b)* – dependence of irradiation time T on ω_f

The results of computer simulation show that the increase of ω_f by more than 17–18 % does not lead to any noticeable increase of the reactor fuel campaign T (days) (see Figure 2*b*). The increase in ω_f is also pointless for the spectrum $\varphi_{Vn}(E)$ of the reactor under study [1].

Therefore for further research we chose the fuel pellet of the type 1017 with $\omega_f = 17$ %. According to the results obtained in WIMS-D, the reactor with such fuel pellet is capable of operating for 3110 effective days at the power P = 60 MW$_{th}$. The neutron flux density in the core φ_{Vn} is 8.14×10^{13} n·cm^{-2}·s^{-1}. There are 9.41% of neutrons with the energy E_n up to 4eV, 35.57 % of neutrons with E_n from 4 eV up to 183.2 keV, and 55.02 % of neutrons with E_n from 183.2 keV to 10.5 MeV. In this conditions, k_{eff} is 1.24, $\rho_{initial}$ is $(k_{eff} - 1)/k_{eff} \approx 19.35$ %, $(d\rho/dt) \approx 6.2\times10^{-3}$%/days. The burnup of ^{239}Pu is 85 % ($\eta = N(0) - N(3110))/N(0)$), and the burnup of ^{232}Th is 9.66 %. The computation results of fuel isotopic composition evolution $N(t)$ are presented in Figure 3.

Additionally, concentrations change dependence of isotopes $^{231,233}Pa$ on time $^{231,233}N(t)$ were derived and concentrations of ^{149}Sm и ^{135}Xe were analyzed. The data of $^{231,233}Pa$, ^{233}U, ^{149}Sm и ^{135}Xe concentrations will allow evaluating the required intensity $I_n(t)$ of neutron emission from a thermonuclear reaction in a plasma column. By varying percentage of DT-reactions in comparison with DD-reactions in the plasma column, one can change I_n within the range from 10^{13} n·s^{-1}·cm^{-1} to 2×10^{14} n·s^{-1}·cm^{-1}. That should provide the condition k_{eff} = constant during both the first operation days of the facility and the whole fuel irradiation cycle.

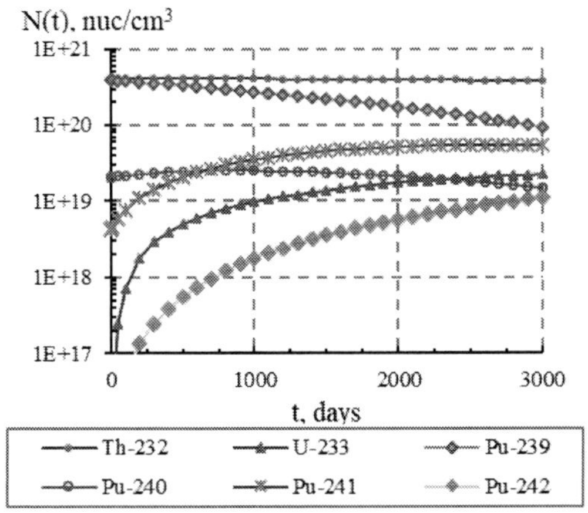

Figure 3. Fuel isotopic composition evolution in HGTRU with the fuel pellet of the type 1017.

The burnable absorber can compensate large initial reactivity excess $\rho_{initial}$ ($\rho_{initial}$=19.35%). For compensation of $\rho_{initial}$ in the LWR and BN-type reactors, for example, Gd_2O_3, Er_2O_3 and B_4C are used. In this case daughter nuclides formed as a result of radiation capture reaction on ^{157}Gd, ^{167}Er and ^{10}B do not have significant impact from the point of view of further neutronic processes in the core.

In solving the task of choosing the appropriate burnable absorber it is of interest to search for nuclides, daughter nuclides of which can have some favourable effect on the development of a self-supporting chain reaction. In the work [8, 9] it is suggested to use ^{240}Pu as a burnable material, a method of enrichment of fuel with various initial content of even-numbered isotopes with proposed by the authors in the work [10]. In our case, the presence of isotope ^{240}Pu in a fuel composition results in significant accumulation of ^{241}Pu, which can be well fissioned by epithermal neutrons, and its nuclear concentration $^{241}N(t)$ exceeds the nuclear concentration of ^{233}U by more than 2.5 times during the whole fuel campaign (see fig. 3). The computation showed that ^{241}Pu and ^{233}U specify the dependence type of $k_{eff}(t)$ and provide steady reactor operation for 8.5 years.

It the loading option discussed above the initial reactivity excess is compensated by the isotope ^{240}Pu: its content (wt. %) in the fuel increased from 5.4 up to 9 %; the content of ^{239}Pu decreased by 3.6%, correspondingly. The use of ^{240}Pu in HTGRU does not lead to any undesired increase of reactivity in the middle of the fuel cycle as it happens in the core of PWR [8, 9]. Using ^{240}Pu $\rho_{initial}$ was reduced from 19.35 to 11.51%, and $(d\rho/dt)$ decreased by 1.51 times. According to the concept suggested by the authors in the works [4], the core of HGTRU, the axial part of which is substituted by a long magnetic trap with high-temperature plasma, starts up from the subcritical state (k_{eff} = 0.95). To transfer the core into the subcritical state it is enough to cover the side surface of fuel pellets with the 13 μm thick layer of ZrB_2.

2.3. Computational model of the axial zone of the core. Specific neutron yield of plasma thermonuclear source

The schematic drawing of the facility intended to study the fuel properties is presented in Figure 4. The axial zone of the core is substituted by the cylindrical vacuum chamber that contains high-temperature plasma generating high-energy neutrons because of D-D or/and D-T thermonuclear reactions. To inject deuterium beam in the plasma, a special chamber is attached to this cylindrical chamber. A magnetic field in these two jointed vacuum chambers containing the plasma provides heat isolation of this plasma from the chamber walls in a radial direction. Heat isolation of plasma along the magnetic field lines is provided by the multimirror sections of the magnetic field, which are adjacent to ends of the two jointed chambers with high-temperature plasma. The total length of the two chambers with high-temperature plasma together with two adjacent regions with the corrugated magnetic field is about 12 meters. The chamber for thermonuclear neutrons generation in the axial zone of the core [1] has the required length that is 3 meters.

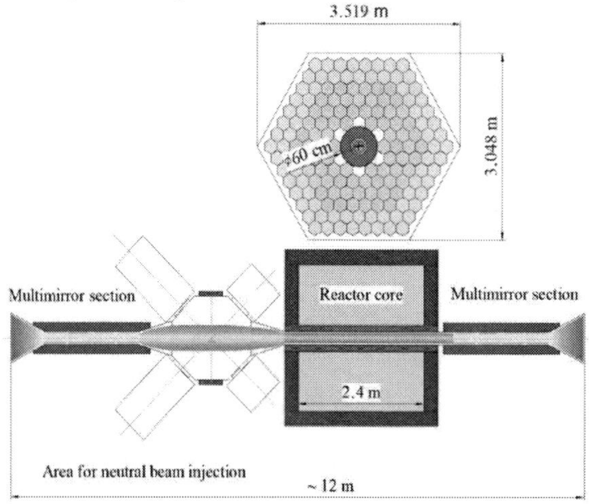

Figure 4. Schematic of the facility to study the fuel properties in long time operation of the reactor core.

Distribution of the magnetic field induction along the axis of the long (12-meters) plasma magnetic trap was chosen from considerations of the maximum neutron yield in the core plasma region upon condition of good homogeneity of radial neutron flow [5].

The distribution of the specific neutron yield per linear centimeter of the plasma column in case of D-D reaction along Z-axis calculated in the work [5] is plotted in Figure 5.

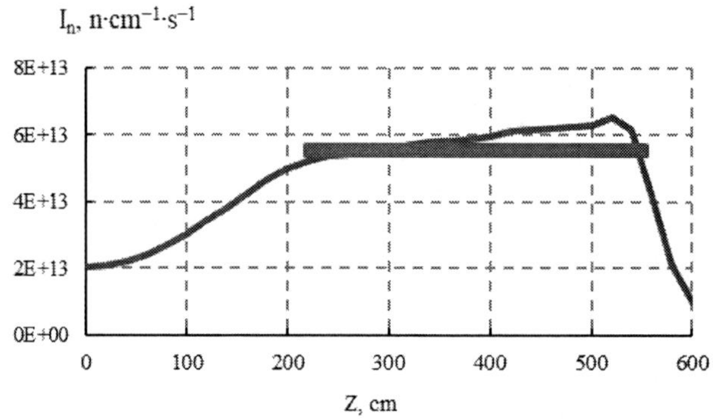

Figure 5. Distribution along Z-axis the specific neutron yield per linear centimeter of the plasma column I_n for the case of D-D fusion reaction. A red bar marks the part of the column that is situated inside of the reactor core.

A red bar located on the interval of z-coordinates from 200 cm to 500 cm marks the part of the column that is situated inside of the reactor core. One can see in Figure 5 that for chosen parameters of the magnetic field and the plasma column, the specific neutron yield changes insignificantly in the plasma column part that is placed in the core. Nevertheless, non-homogeneity of plasma distribution along Z-axis may be taken into consideration in our simulations. As examples of possibility to analyze various plasma distributions along Z-axis, the specific neutron yield of plasma neutron sources with cylindrical and conical shape of the plasma column was performed.

In numerical simulation, it was considered that plasma column is a volume source of monoenergetic neutrons with isotropic velocities distribution in the point of their origination. Neutrons are generated uniformly in the total volume of plasma and the neutrons energy is 2.45 MeV (for D-D reaction) or 14.1 MeV (for D-T reaction).

In accordance to results of the paper [5], when the density of high energy sloshing ions in the column of deuterium plasma 1.5×10^{14} cm^{-3}, the specific neutron yield in DD-reaction is 6.00×10^{13} n·s^{-1}·cm^{-1} (specific emission of neutrons from a volume unit is 1.76×10^{12} n·s^{-1}·cm^{-3}).

We calculated a spatial neutron flux distribution $\varphi_{Zn}(Z)$ on the external surface of the plasma neutron source with respect to axis Z for both two cases: 50 % T with 50 % D and 100 % D at the same specific neutron yield 6.00×10^{13} n·s^{-1}·cm^{-1}. We analyzed the neutron flux distribution for cylindrical and conical shapes of the plasma column. The diameter of the cylindrical plasma column was 17 cm. For conical shape of the column, its diameter was decreased from 17 cm down to 13 cm in the length 300 cm. The external surface of a cooper winding with the thickness 5 cm covering the vacuum chamber for the plasma column was the border between the plasma neutron source and the reactor core. Taking into account the thickness of the winding, the external surface of the plasma neutron source had the diameter varied from 22 cm to 18 cm for the conical shape of the plasma column and no varied 22 cm for the cylindrical one. The temperature of all materials used for the construction of the vacuum chamber and the solenoid was chosen at the level of 0.05 eV. The neutron flux $\varphi_{Sn}(E)$ on the external surface was generated by usage the program in the energy group structure ABBN-78 (of 28 energy neutron groups). Spatial neutron flux distribution $\varphi_n(Z)$ with respect to axis Z of the facility and its radius Y was obtained by placing the computational grid with the size of 1000 to 1000 cells along the height Z ($\varphi_{Zn}(Z)$) and radius Y ($\varphi_{Yn}(Z)$) of the model. The classic method of Monte-Carlo was used to solve the task in the frame of MCNP5 (ENDF/B-VII.0) [11], Serpent 1.1.7 (ENDF/B-VII.0) [12] and PRIZMA (ENDF/B-VII.I) [13] application.

3. Calculation data and conclusion

In accordance calculations, the total escape of D-D neutrons from the facility is 80.4%, neutrons escape from the side and end surface is 66.5 ($\varphi_{Zn} = 6.67\times10^{12}$ n·s^{-1}·cm^{-2}) and 13.9 % ($\varphi_{Yn} = 1.39\times10^{13}$ n·s^{-1}·cm^{-2}), correspondingly.

The total escape of D-T neutrons from the facility is 98.5%, neutron escape from the side and end surface is 83.6 ($\varphi_{Zn} = 8.39\times10^{12}$ n·s^{-1}·cm^{-2}) and 14.9 % ($\varphi_{Yn} = 1.49\times10^{13}$ n·s^{-1}·cm^{-2}), correspondingly. It should also be mentioned that thorium in the first row of the assembly transmutates into ^{231}Pa under the influence by D-T-neutrons form the 26th (that is, for neutrons in the energy range E_n from 6.5 to 10.5 MeV), 27th (for neutrons in the energy range E_n from 10.5 to 14 MeV) and 28th (for neutrons in the energy range E_n from 14 to 14.5 MeV) groups due to (n, xn)-reactions. Therefore, the core provides extra neutrons and fissionable isotope ^{233}U.

In accordance calculations, neutrons escaping from the side surface of the facility can be divided in four main group with the different energy E_n. Percentage of these groups of the neutrons is the following: 2.03% – E_n from zero to 4 eV (thermal neutrons), 30.19% – E_n from 4 eV to 183 keV (epithermal neutrons), 50.11 % – E_n from 183 keV to 10.5 MeV and 17.69 % – E_n from 10.5 to 14.5 MeV (fast neutrons). In the reactor core [1,3] there are 9.41 % of thermal neutrons, 35.57 % of epithermal and 55.02 % of fast neutrons, correspondingly.

To achieve possibility of exactly correct usage of neutron emission from the plasma source for studying neutronic and thermal-physical properties of thorium-plutonium fuel in the core of HGTRU [1,3] we have to achieve the specific neutron yield on the level $I_n = 1.8\times10^{14}$ n·s^{-1}·cm^{-1}. This increase in the neutron yield will be realized in case of replacing some quantity of deuterium by tritium in the plasma column without changing any other parameters of the plasma.

Computer simulation of neutronic processes in the core of the high-temperature gas-cooled thorium reactor for 30 different options of core loadings was performed in the frame of our work. The quantity of fuel compact dispersion phase and the composition of fissionable nuclide were selected. Production of extra neutrons by means of thermonuclear reactions occurring in high-temperature plasma and of reactions of (*n*, *xn*) type was analyzed. Possibility to apply plasma D-T-source of neutrons to modify the near-axial region of HGTRU core was demonstrated. It was shown that the developed models and computer codes for description of the core and the thermonuclear neutrons source allow turning to full-scale study aimed at creation of a thorium subcritical assembly with supply of extra neutrons from thermonuclear plasma at its confinement in the long magnetic trap.

These studies are of undoubted practical interest, since they are necessary to substantiate the safety of operation of such hybrid systems. The research results will allow optimizing the active zone of the hybrid system with leveling the resulting offset radial and axial energy release distributions. Results of our study will be the basis for the development of new and improvement of existing methods of criticality control in related systems such as "pulsed neutron source - subcritical fuel assembly".

Acknowledgment
This work was supported by RFBR, project number 19-29-02005.

References

[1] Shamanin I V, Bedenko S V, Chertkov Yu B et al 2015 *Izvestiya Wysshikh Uchebnykh Zawedeniy, Yadernaya Energetika* **3** 124–134 (in Russia)
[2] Bedenko S V, Ghal-Eh N, Lutsik I O *et al* 2019 *Applied Radiation and Isotopes* **147** 189–196
[3] Shamanin I V, Bedenko S V, Chertkov Yu B et al 2016 *Izvestiya Wysshikh Uchebnykh Zawedeniy, Yadernaya Energetika* **2** 121-132 (in Russia)
[4] Arzhannikov A, Bedenko S, Shmakov V *et al* 2019 *Nuclear Science and Techniques* volume **30** 181 doi:10.1007/s41365-019-0707-y
[5] Yurov D V, Prikhodko V V, Tsidulko Yu A 2016 *Plasma Physics Reports* **42(3)** 210–225
[6] Bagryansky P A, Ivanov A A, Kruglyakov E P *et al* 2004 *Fusion Engineering and Design* **70(1)** 13–33
[7] WIMS-D5B, Deterministic Multigroup Reactor Lattice Calculations. Available at: http://www.oecd-nea.org/tools/abstract/detail/nea-1507/
[8] Hussain A, Xinrong C 2010 *Progress in Nuclear Energy* **52(6)** 531–535
[9] Hussain A, Xinrong C 2009 *Progress in Nuclear Energy* **51(6–7)** 742–745
[10] Smirnov A, Gusev V, Sulaberidze G *et al* 2019 *AIP Conference Proceedings* **2101** 020006
[11] Brown F B 2017 The MCNP5 Random Number Generator *LA-UR-07-7961*
[12] Leppänen J, Pus M, Viitanen T *et al* 2015 *Annals of Nuclear Energy* **82** 142–150
[13] Kandiev Y Z, Kashaeva E A, Khatuntsev K E *et al* 2015 *Annals of Nuclear Energy* **82** 116–120

IOP Conf. Series: Materials Science and Engineering 1019 (2021) 012038 doi:10.1088/1757-899X/1019/1/012038

Time indicators of effective optimization algorithms in group load control modelling

A P Dimitriev[1] and R I Bazhenov[2]
[1]Department of Computer Technologies, Chuvash State University, Cheboksary, Russian Federation
[2]Department of Information Systems and Mathematics, Sholom-Aleichem Priamursky State University, Birobidzhan, Russian Federation
E-mail: dimitrie1@yandex.ru

Abstract. The purpose of the work is to determine the optimal set of parameters of the genetic algorithm for optimization in the control model of the group load of electricity consumers taking into account time parameters. The analysis based on the approximation of empirical results is used as a research method. The results of two efficient algorithms used for optimization in this model are compared. Computational experiments were carried out and their results were processed, the location of efficiency of the parameters of the combined genetic algorithm was determined.

1. Introduction

In shops on smelting of ferrous and nonferrous metals of large foundries often there are several melting electric furnaces, which can be up to 12 installations. Power consumption in them is sometimes controlled by pulse-width modulation (PWM) power control. It consists in periodic switching on and off of these consumers of the electric power, thus the average power consumption of this or that consumer depends on time of stay in the switched-on state.

It is known that simultaneous activation of all consumers of the electric power causes an overload of an electrical transmission network owing to what penalties from system of power supply are exposed. In order to avoid such situations, it is necessary to rationally distribute in time the work of such consumers, that is, to make and optimize the schedule. Optimization of the schedule of switching on of consumers can serve for increase in the power factor in the network. This allows to reduce the permissible power of supply and reduce general electrical losses in the current-carrying sections of the load group. Thus, optimizing the PWM schedule is an urgent task.

When scheduling PWM should take into account the parameters of each power consumer: first, structurally constant parameter, which is expressed by the height of the pulses and physically means the current, and secondly, the width of the pulses. The period is the same for all consumers and can reach 1800 seconds. This allows for long-term calculations in scheduling. Consumers are turned on and off through controllers that can be connected to a microprocessor or microcontroller, in which is necessary to implement some algorithm for scheduling and optimization. Realization of digital PWM on microprocessors described in [1].

In [2] proposed a method for optimizing the schedules of the PWM. According to the condition, it is necessary to get the maximum power factor, which the range of values from 0 to 1. Introduced designations:

IOP Conf. Series: Materials Science and Engineering 1019 (2021) 012038 doi:10.1088/1757-899X/1019/1/012038

- γ_i is the relative pulse duration of the i-th regulator, $i = 1,...,n$, where n is the number of regulators in the group;
- I_i is the load current of the i-th regulator.

The power factor is expressed by the equation:

$$\chi = \left(\sum_{i=1}^{n} I_i \gamma_i\right) \cdot \left(\sum_{i=1}^{n} I_i^2 \gamma_i + 2\sum_{i=1}^{n-1} \sum_{k=i+1}^{n} I_i I_k \gamma_{ik}\right)^{-1/2}, \tag{1}$$

where γ_{ik} is the relative duration of the interval of the joint action of the pulses of the i-th and k-th regulators. When using the above method, the power factor reaches the value of 0.9723 for the four consumers and the value of 0.9906 for the eight ones. All source values are integers, since in the 1990s, when the problem was set, the processing of integers was faster than real ones. Considered practical problem can be considered from an abstract point of view as a mathematical problem of discrete optimization, if the number n increases above the really necessary. When $n = 40$, the power factor for files [3] is from 0.9829 to 0.99941 using this method.

There are other algorithms of scheduling in this problem, sometimes leading to a greater χ, for example, in [3], where the objective function is formulated somewhat differently, which does not change the essence of the problem. Here consumers have parameters: the share of power for the i-th consumer from the maximum value is set by the duration of the corresponding work, p_i. Each consumer has its own constant importance, denoted by t_i. The time period is divided into m equal intervals (discrete moments of time). A regulator may be activated at any such time and only once per period. The set of such moments of time $\{\sigma_i, i = 1,...,n\}$ forms a schedule stable for the vector of input data. It is necessary to minimize overloads according to the following objective function:

$$C = \sum_{h=1}^{m} \sum_{i=1}^{n-1} \sum_{k=i}^{n} t_{h,i} \cdot t_{h,k}, \tag{2}$$

where m is the number of discrete intervals in the period, n is the number of consumers, $t_{h,i}$ is the electricity consumption of the i-th consumer at the h-th moment of the period, it is either t_i or zero, which depends on p_i and the moment of inclusion relative to the beginning of the period.

Equation (2) is included in the denominator of the (1) as an additive component, the other components (1) are constant, so minimization (2) also means minimization (1).

In [3] some other problems of placement for which the (2) is used are given, in particular, a problem of timetabling on high schools, and also a number of the applied algorithms is given. For the timetabling problem, the number n can often reach the above value of 40, so this value is not so abstract.

2. Used optimization algorithms

At such values of n, it is possible to obtain placements with the values of the objective function C exceeding the optimal value by less than $8 \cdot 10^{-5}$ for 600—900 seconds on a personal computer only with the help of two algorithms from the ones given in [3]. This is a selection sequence optimization (SSO) algorithm based on the simulated annealing method [4], and a combined genetic algorithm (GA).

Consider the SSO algorithm. Algorithm parameters:

- Y is the initial number of iterations;
- T is the preset temperature;
- t is the current temperature;
- d is the increasing the number of iterations;
- R is the scatter.

Objects represent consumers. The algorithm is as follows.

1. Sort objects according to some criterion, for example, in ascending order of importance. The sequence of numbers of the sorted list is designated as the starting point S in the n-dimensional space.

Place the objects in the sequence specified by S in a relatively optimal position (by placing them sequentially in the best positions at the time of location, performing an overview of each of the m positions for the newly located object).

Calculate the objective function $C(S)$ by the (2).

2.0. Number of iterations performed $j := 0$.

2.1. In the neighborhood of S, defined by the scatter R, select point, denoted S_1, by pseudo-random permutation from two to four elements in S.

2.2. Place the objects in the sequence specified by S_1 in a relatively optimal position (same to the one in step 1).

2.3. Calculate $C(S_1)$ according to the resulting placement. If $C(S) > C(S_1)$, then $S := S_1$.

2.4. Otherwise, with probability t/T, if $|C(S) - C(S_1)|$ is less than some constant threshold, then $S := S_1$.

2.5. $j := j + 1$.

2.6. If $j > Y$, then $\{t := t - 1; Y := Y + d\}$. Otherwise go to 2.1.

2.7. If $t > 0$, go to 2.0.

3. End.

Using this algorithm it is often possible to get the best value for C than for any other algorithm. However, SSO is little known, unlike various GA [5, 6]. Therefore, the study mainly concerns only GA. The essence of the combined GA is set out below. This algorithm uses some specific parameters. The values of these parameters affect the efficiency of its application. It is necessary to determine the optimal set of GA parameters for optimization in the above model. To do this, requires to:

- compare the results of the GA and SSO;
- determine which parameters should be investigated;
- carrying out computational experiments and process their results.

The analysis based on the approximation of empirical results is used as a research method [7].

First, the approximation of the results of series of three computational experiments using trend lines in Microsoft Excel was performed. The dependence of optimization quality on GA and SSO operation time for f1 [8] and f5 [3] files is shown in the diagram (Figure 1).

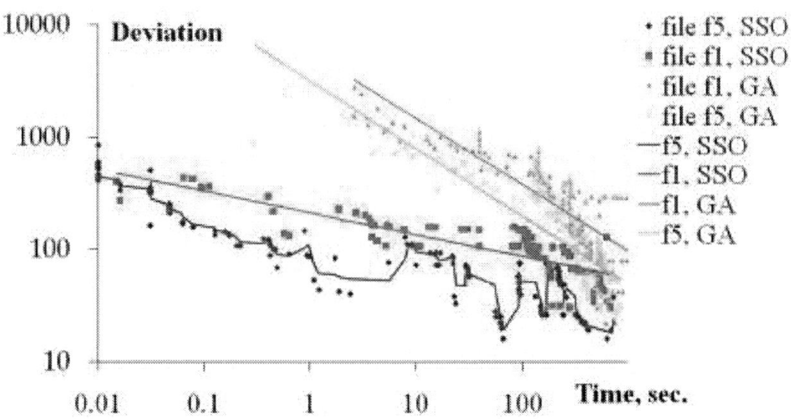

Figure 1. Time dependence of the quality of the optimization for different input data and algorithms.

In Figure 1, on the ordinate axis is the deviation of the value C from the found minimum. The trend line for the lower chart is a linear filter, for the rest it is a power one.

A personal computer with a frequency of 3344 MHz was used for plotting the diagram. For each file, the work of the two above-mentioned algorithms was studied.

As follows from the diagram, usually with the help of SSO it is possible to get closer to the optimal results in less time compared to GA. However, when approaching optimal values, sometimes the opposite is true. At the beginning of the optimization process, as a rule, SSO works more efficiently than GA.

The combined GA largely repeats the Genitor [9], but has some features. The parameters of the combined GA include the size of the population and the level of mutations. In addition, the results are affected by the input data, so it is necessary to examine different sets of input data.

When using a combined GA, it should be noted that if "mutation" less than 50 is selected, it is possible to get values of C, which closer to the optimal one than if this value is more than 50. Unlike Genitor, this algorithm uses a different number of genes from two parents when creating a new chromosome.

Denote by V_1 the proportion of genes from the total number of genes that are copied to a new chromosome from the first ancestor, by V_2—respectively from the second ancestor, by V_3 the proportion of mutating genes, by M the value of the parameter "mutation", and by P the size of the population (the number of chromosomes).

The algorithm uses a random number generator to distribute genes in V_1, V_2, V_3. According to the algorithm, the average $V_3 \approx M$. If $M < 50$, then $V_1 \approx 1/2$, $V_2 \approx 1 - M$. Otherwise $V_1 \approx 1 - M$, $V_2 \equiv 0$ Thus, $V_1 \approx (1.5 - M - |\,0.5 - M\,|)/2$; $V_2 \approx (0.5 - M + |\,0.5 - M\,|)/2$.

For example, if $M = 30$, then about 50% of genes are used from one individual, about 20% from the second, and about 30% of genes mutate.

Next, a series of five computational experiments for the same set of parameters were carried out to reduce random errors.

3. The analysis parameters of genetic algorithm with 40 consumers

On the computer with a frequency of 1500 MHz was investigated the influence of P, to be modified in a wide range, in order to select a more narrow field and the subsequent more precise analysis. This computer is multi-core, so in order to speed up the work in some cases, two equal processes with different parameters were launched simultaneously. The value M was used as a variable within the experimental data acquisition for each size of population. Studied the influence of the following values of P on the optimization result: 3, 10, 30, 100. In Excel, point diagrams of the dependence of the minimum obtained C values from time in the range from 1 to 1000 seconds are plotted in logarithmic scale on both coordinate axes. For each P, trend lines of the power type were displayed. Other types of trend lines significantly distorted the result of the approximation.

From the relative position of the trend lines revealed that more promising for further research area closer to the value of $P = 30$. The worst value listed is 100. This is the highest value of the studied ones, therefore, a further increase of P is not effective.

Further experiments were carried out on a computer with a frequency of 3344 MHz. Similarly, the influence of P, which is changed in a narrower range, was investigated at a fixed M. The following values P: 15, 20, 30, 40 were studied. It is determined that $P = 20$ is more optimal, and the worst of these is 40. Thus, neither decreases below 15 nor increases above 40 are effective.

Similarly, the influence of M from the values of 20, 30, 40 at a fixed P was studied. It was found that $M = 20$ is more optimal, and the worst M is 40. However, the number $M = 20$ in some cases leads to far from optimal results. Thus, increasing M to 40 and above is not effective.

As a continuation of the previous studies, the study of the rectangle area in coordinates (M, P) from $(20, 20)$ to $(30, 30)$ with step 5 was carried out. It is revealed that $M = 30$ and $P = 20$ are more optimal, but there is a significant error when using trend lines.

The use of trend line approximation in Excel results in either straight lines or excessively broken lines, which makes it difficult to compare results and can lead to incorrect conclusions. For example, it is possible to can conclude that $P = 10$ is better, while in reality 30 is better, because it leads to more optimal values in a reasonable time.

When using the "Data Analysis" item in Excel, an attempt is made to express the contents of some cells through formulas, but these formulas are not suitable in this case. The use of the trend function for interpolation is also not suitable, since in this case a linear trend line is needed [10].

In this regard, developed a Windows Forms application (the name of the program Form3), creating the set values of coordinate points on the basis of the other set. To do this, the program interpolates the data as follows. Creates a new set of points with given abscissa. To obtain the ordinate of a point (x, y), two starting points (x_1, y_1) and (x_2, y_2), abscissa of which are located directly to the left and to the right of the abscissa of the point (x, y), respectively, are used. The required ordinate is determined by the equation:

$$y = y_1 + (y_2 - y_1) \cdot (x - x_1)/(x_2 - x_1).$$

For the leftmost point, a large number is used as y_1. For the rightmost point, the y_2 is the ordinate of the previous origin. This is possible due to the shape of the graph in the form of a damped exponent, since at the beginning of minimization there are often points with smaller values of C than the previous and closer to the end rarely.

The results of the diagrams based on the data obtained in this way are shown in Figure 2.

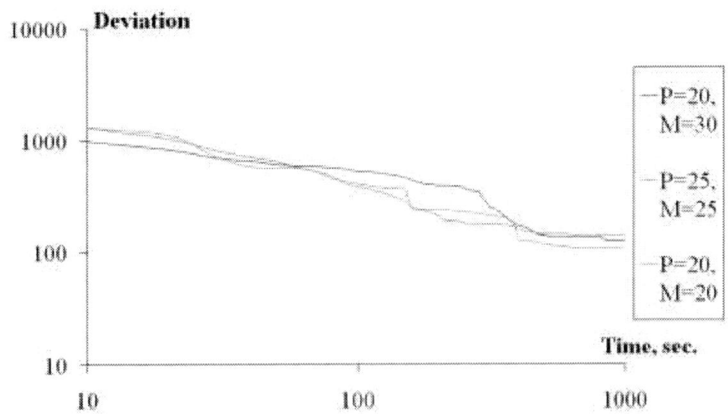

Figure 2. Time depending constructed via interpolation.

As can see, it is more convenient to compare charts here than in Figure 1. From Figure 2 it follows that values $P = 20$ and $M = 20$ are more optimal.

4. The analysis parameters of genetic algorithm with 28 consumers

It is possible that for the source data files containing information about objects in a different quantity, the found optimal values of P and M may be different. Therefore, a file named in28-100-t10-25-p8-50.txt was created. This file contains information about 28 objects instead of 40 objects, as in previous files, with the number of discrete intervals still being used equal to 100.

The data in this file, together with the values of activation moments denoted by o_i at some near-optimal schedule, are presented in table 1.

Table 1. Data in the file with information about 28 objects.

i	1	2	3	4	5	6	7	8	9	10	11	12	13	14	15	16	17	18	19	20	21	22	23	24	25	26	27	28
t_i	30	34	27	30	23	34	24	30	22	29	18	11	16	22	31	14	24	12	26	10	15	23	13	23	26	13	16	14
p_i	58	44	47	41	46	29	40	27	33	21	33	50	32	21	14	29	16	29	13	33	22	13	22	12	9	14	8	8
o_i	48	53	97	7	32	83	43	12	99	78	82	28	50	22	39	100	6	78	37	50	78	53	15	66	44	14	29	99

Similar to the study of the previous files, a series of computational experiments on optimization using GA at constant *M*, *P* with steps of 5 units were carried out. Here, value $P = 2$ was used from the left edge of the study area, not $P = 0$, since there must be at least two chromosomes in the population for crossing.

For each point, a series of 10 experiments was performed. All the experiments lasted from 600 to 900 seconds on a desktop PC or twice as much on a laptop, because the clock frequency on it is half as low. All of these computers have four CPU, so it was often runs four equivalent process with the same initial data. Since the absolute time under such conditions is unpredictable, the achievement of a given number of crossovers is used as a stop criterion. Since time is proportional to the number *M*, then at $M = 20$ it is equal to 150,000, at $M = 30$ it is equal to 100,000, etc. A GA settings file and user interface elements were created to load these settings from a file and save them to a file.

In the study of the file with $n = 28$ to reduce the time, the graph was not built, since the construction of one curve of the graph takes 300 seconds of PC user work. Instead, only the *C* value reached in the specified time was used. Denote by *D* the deviation of the value *C* from the found minimum, rounded to 1247,000. The first point of study in connection with the results of studies for files with 40 objects is $P = 20$, $M = 20$. In it, *D* averaged is equal to 378.6 units with a standard deviation $S_0 = 176.4$ units. In its neighborhood with a step of ± 5 units found point $P = 15$, $M = 20$, where the deviation from 1247011 minimum for four points (excluding the original) and is equal to 362.4 when $S_0 = 166.6$, which is 111.6 less than the next ascending value. This point is taken as the current one, and the study continued in its vicinity with a step of ± 5 units. The results of the study of the parameter values are shown in Figure 3.

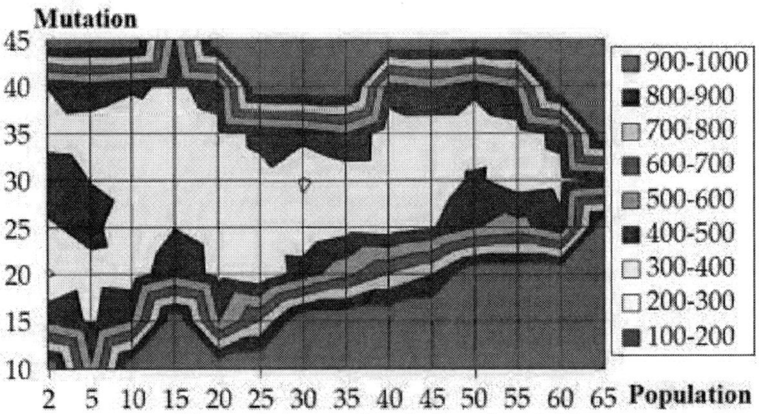

Figure 3. Depending *D* from *P* and *M*.

The criterion of transition to the next point in order to study its neighborhood is the condition that *D* in it was not less than 400 units. The number 400 is chosen because several points where $D < 300$ were found and 400–300 produce the stock of the same order as S_0 is left.

In the process of constructing points, the number 1247011 found in about 7,000 seconds with the help of SSO was first used as a minimum. Then, for about 360,000 seconds of single-processor time using GA were found two numbers smaller than this – 1247007 and 1246962. Placement, presented in table 1, obtained using the GA. The power factor for this location is 0.999945, the value $C = 1246962$

It is expected the nature of the diagrams for the investigated file, the same to nature of file f1, i.e., with increasing P or M any improvement effect is not achieved. Thus, the criterion for the termination of the study is the condition: the study area should be closed by the boundary at which all points should have D at least 400 units.

66 points were investigated in accordance with the termination criterion. Here, S_0 in the whole population is in the range from 79.5 to 279.2. The average value for S_0 is 164.4. The standard deviation for S_0 is 40.2. There were no regularities in the distribution of S_0 in the study area.

For three points: $(2, 35)$, $(10, 20)$ and $(30, 25)$, in which the minimum values of D—259.5, 302.4 and 227, respectively, were found, the series of experiments were repeated. According to the results of 20 experiments at these points S_0 is approximately the same as the average value of S_0 for the whole set of points and is 150.1, 176 and 130, respectively. The averaged values of D for 20 experiments each time became more, namely 288, 334.3 and 291.3. However, the first and third of these points still have two minimum values of D from the whole population.

Two more points with coordinates $(100, 20)$ and $(300, 20)$ were constructed. For the first point $C = 1247384$ that is 422 greater than the minimum value, and for the second one about 1600. This suggests that the increase in P more than 55 is not effective. This is due to the fact that the larger the population, the slower it is possible to find a chromosome, which leading to more effective values.

Thus, if there are 28 objects, the optimal values are in the region $P = 2 \ldots 30$, $M = 20 \ldots 35$.

5. Summary

For the f1 file used in a number of scientific publications authors, 20 chromosomes in the population and the value of the parameter "mutation" 20 are more optimal when using a combined GA.

For source data describing other number of consumers or with other qualities, these values may be different, but calculations show that the differences between these values are within 20 units. That is, neither P nor M should be very large. The SSO algorithm achieves near-optimal values faster than the combined GA.

References

[1] Peterchev A V and Sanders S R 2006 *Digital pulse-width modulation control in power electronic circuits: theory and applications* (Berkeley: University of California) p 163
[2] Il'gachev A N, Abramov A V and Vasil'yev A V 2011 Control method of group of pulse-type voltage regulators (versions) *RF Patent* 2409829 p 16
[3] Dimitriev A P 2013 *Applied Informatics* **6** (48) 045–058
[4] Kirkpatrick S, Gelatt C D and Vecchi M P 1983 *Science* **220** (4598) 671–680 doi: 10.1126/science.220.4598.671
[5] Herrera F, Lozano M and Verdegay J L 1998 *Artif. Intell. Rev.* **12** 265–319 doi: 10.1023/A:1006504901164
[6] Puechmorel S and Delahaye D 2014 Order statistics and region-based evolutionary computation *J. Glob. Optim.* **59** 107–130 doi: 10.1007/s10898-013-0079-5
[7] Zausaev A A and Shurupov A 2016 *Proc. of the 10th all-Russian scientific conf. with int. participation Mathematical modeling and boundary value problems* **2** 177–178
[8] Dimitriev A P 2015 *Modern problems of science and education* **2-1** 185
[9] Kashirina I L 2007 *An introduction to evolutionary simulation: proc. benefit* (Voronezh: Voronezh State University) p 39
[10] Karchevskaya M P, Ramburger O L and Tarkhov S V 2019 *Modern high technologies* **1** 63–68

The dependence between the ignition value of air suspension and the size of the reaction chamber

A I Sechin[1], A A Sechin[1], Yu F Patrakov[2], I L Mezenceva[1] and Yu A Amelkovich[3]
[1] School of Basic Engineering, Tomsk Polytechnic University, Tomsk, Russia
[2] Federal Research Center of Coal and Coal Chemistry Russian Academy of Sciences, Siberian Branch, Kemerovo, Russia
[3] School of Non-Destructive Testing, Tomsk Polytechnic University, Tomsk, Russia
E-mail: amely@tpu.ru

Abstract. The influence of the geometric factor of the reaction chamber on the limit value of the air suspension ignition is studied. It is found that in determining the lower concentration limit of the flame propagation (LCLFP), it is most promising to use an installation with a cylindrical reaction chamber whose height is equal to the diameter. The dependence of the determination of the LCLFP value on the geometric factor of the reaction chamber is proposed, tested on metal powders and organic dust-forming materials. The expression allowing to carry out the comparative analysis of scientific results received on other technical devices with the results received on the installations executed according to the requirements of normative and technical documentation is received that significantly reduces the volume of experimental works.

1. Introduction

The properties of industrial dust determine the level of fire and explosion hazard of many technological processes: crushing and grinding, drying, pneumatic transportation, grinding, grain processing, etc. Annually dust explosions and fires cause significant damage to the national economy [1–3].

It is known [4] that a considerable amount of dust-forming substances and materials in the state of air suspension become explosive. The literature provides a large amount of experimental data [5] on the critical conditions of combustion propagation in air suspensions, which are not always consistent with each other since they are obtained on reaction chambers with different geometric dimensions and at different time periods. The introduction of the normative document [6], led to the ordering of the research results and nevertheless a large layer of experimental data obtained outside the framework of the normative document, is of great scientific interest. The solution of the problem covering of these data, the influence of the geometric factor of the reaction chamber on the limit value of the ignition of the air suspension and is devoted to this work.

Practical observations of the processes of production, storage, transportation and processing of combustible dust-forming materials formed the basis of the principles of ensuring their fire and explosion safety. Development of M.G. Godzhillo formed the basis of the national system of assessment of fire and inflammation of industrial dusts. In the future, this system was improved by V.T. Monakhov, which was included in the state standard regulating the nomenclature of indicators of fire and explosion hazard of oil products and chemical organic products [6–8].

The researchers noted that the value of the lower concentration limit of flame propagation in air suspension is significantly affected by the method of creating an air suspension, the shape of the reaction chamber and its volume, the source of initiation of combustion, etc.

Great difficulties in the comparative analysis of scientific results cause a variety of experimental facilities and techniques to determine the critical conditions of flame propagation.

From the list of factors influencing the result of the experiment, the geometric factor is of interest – the ratio of the reaction chamber volume to its surface. The researchers note that heat losses are crucial in the study of critical conditions of flame propagation in the air suspension.

2. Results and discussion

Let us analyze the case of ignition of an air suspension in the center of a volume of arbitrary shape, the concentration of particles is uniformly distributed. It should be assumed that the ignition of the air suspension is more expedient to make as far as possible from the walls of the reaction vessel in order to minimize the breakage of the reaction chains of combustion on it. Assuming that the flame front has a cellular structure [9], heat transfer in the form of a stream of parallel rays from the moving front is most effective when located in the center of the reaction volume.

Consider the case in which the radiation energy f is completely absorbed by the particles and the flux, which is described by the following equation [10, 11]:

$$\Phi = \Phi_0 e^{-N\sigma x} = \Phi_0 e^{\frac{-x}{L}} \quad (1)$$

where Φ_0 is the total radiation energy of the source, Joule; x is the thickness of the air suspension as an absorption layer, m; σ is the current absorption area, m^2; N is the number of air suspension particles per unit of elementary volume; L is the free path of radiation, m.

For the case when σ and N are constant, the absorbed part of the radiation will be characterized by the value of x, and, consequently, part of the freely transmitted radiation flux. But it should be borne in mind that the value of the absorption layer depends on the geometry and dimensions of the reaction volume. In this case, the value x is the distance from the heat source located in the center of some volume of arbitrary shape to the limiting surface of this volume, then from the expression (1) it follows that the direction of the heat flow with less x will be accompanied by a smaller result of absorption of radiation. Since the reaction volume is characterized by a certain surface, the net losses that do not affect the parameters of the air suspension will be minimal.

Thus, in order to establish the ratio between the amount of heat absorbed by the air suspension and the amount of heat loss by radiation in the reaction chamber of a particular experimental installation, it is necessary to substitute a certain value of the current x_{ef} coordinate in equation (1). As a first approximation, the value of the x_{ef} can be determined on the basis of the following considerations (without taking into account the volume occupied by the particles and the particle shadow area). The geometric place of all distances is the volume of the reaction chamber of the experimental setup taken by us. The geometric location of the points of exit of the rays from any volume is a surface that limits this volume, this will be the surface of the reaction chamber of the experimental setup (or the surface of the technological volume). Therefore:

$$x_{ef} = \frac{V}{S} \quad (2)$$

where V is the volume occupied by the air suspension, m^3; S is the surface limiting the cloud of the air suspension, m^2.

In this case, the equation (1) takes the form:

$$\Phi = \Phi_0 e^{-\sigma N x_{ef}} = \Phi_0 e^{-\frac{V}{SL}} \quad (3)$$

From the equation (3) it follows that the larger the reaction volume of the experimental setup, the greater the probability of absorption of radiation, the less heat loss by radiation. Conversely, the larger the surface, the greater the probability of free release of radiation beyond the cloud of air suspension, the greater the heat loss by radiation.

The solution of the variational problem of finding a geometric figure with a maximum volume at a minimum surface is a sphere. Consequently, the reaction chamber of the experimental setup in the form of a spherical vessel has the lowest probability of free radiation output, that is, has a minimum amount of heat loss. Taking into account the fact that in a spherical chamber the creation of a uniform concentration of the air suspension meets technical difficulties, it is necessary to choose the geometric parameters of the reaction chamber of the experimental setup of another type. The solution of the variation problem leads to the conclusion that of all the figures (cube, cylinder, cone, etc.), the cylinder has a maximum volume-to-surface ratio, and the extremum is reached when its height is approximately equal to the diameter. In this case we have:

$$x_{ef} = \frac{1}{3}R \tag{4}$$

where *R* is the radius of the cylinder base, m.

In an experimental setup with a reaction chamber of cylindrical shape (pipe type), the possibility of creating an air suspension of controlled concentration is much wider (pulsed, gravitational, etc.) than a spherical vessel.

Thus, when determining the lower concentration limit of the flame propagation (LCLFP), an installation with a reaction chamber of cylindrical shape, the height of which is equal to the diameter, should be used.

For this case, the equation of radiation absorption in the volume of the air suspension will be as follows:

$$\Phi = \Phi_0 \exp\left(-\frac{R}{3L}\right) \tag{5}$$

Equation (5) allows us to draw the following conclusions: the increase in the size of the reaction chamber of the experimental setup (*R*) leads to a decrease in heat loss by radiation and, as a consequence, to a decrease in the LCLFP. The latter is not difficult to prove, based on the results of [11, 12]. It is shown that the limiting conditions of ignition of the air suspension are such conditions under which a constant share of heat loss by radiation is realized, determined by L = const. Therefore:

$$\frac{N\sigma V}{S} = const$$

and for a strictly fixed value of the particle size we have:

$$\frac{NV}{S} = const \tag{6}$$

It follows that as the dimensions of the reaction chamber of the installation increase, the concentration of particles *N* can be reduced, that is, the LCLFP decreases. Transformations (6) lead to the dependence of the LCLFP value on the geometric factor of the reaction chamber of the experimental setup:

$$M = \frac{M_0 S V_0}{V S_0} \tag{7}$$

where M – the value of the LCLFP for installation with variable dimensions V and S, g/m^3; M_0 – the value of the LCLFP in the installation with dimensions V_0 and S_0, g/m^3.

Calculation according to the equation (7) finds not only qualitative, but also quantitative confirmation in [8, 11].

In the work [12] on the example of air suspension of titanium powder with dispersion of 50–56 microns, it is shown that with an increase in the volume of the chamber from 0.33 to 21.6 liter LCLFP decreases, and the explosion pressure increases, the most dramatic change in these values occurs in the range of reaction volumes from 1 to 4 liter. In this case, the authors used sealed metal chambers with pulsed powder spraying.

In work [10, 12] influence of diameter of a pipe on a limit of ignition of air suspensions was studied. The researchers used glass pipes with diameters of 0.055, 0.105, 0.14 m and a height of 2 m. At the upper end of the pipes there was a screw dust generator, which almost closed the end of the pipe; at the bottom, at the open end of the pipe, there was a source of ignition – a nichrome spiral.

Particular attention is drawn to the significant methodological differences in the determination of experimental values of the LCLFP in [10, 12]. Comparative data on the calculated and experimental values of the LCLFP are given in tables 1 and 2.

Thus, the expression (7) shows that when comparing various experimental studies on the LCLFP, it is necessary to pay attention to the geometric factor of the reaction chambers of the installations in which it is determined.

Table 1. The dependence of the LEL titanium dispersion 50-56 μm from the geometrical factor of the installation.

The diameter of the chamber [m]	Height [m]	V/S [m]	LCLFP [g/m^3]	
			The experiment	*Calculation*
0.05	0.175	0.011	80	80.0
0.10	0.345	0.021	35	40.0
0.11	0.390	0.024	30	36.2
0.20	0.200	0.033	25	26.4
0.20	0.692	0.044	23	20.0

Note: In table 1 the experimental values of the lower explosive limit of titanium is taken from [10].

Table 2. Experimental [13] and calculated values of LCLFP depending on the diameter of the reaction chamber.

Name of air suspension substance	The diameter of the reaction chambe [mm]					
	55		105		140	
	The results of the determination of the LCLFP [g/m^3]					
	Experiment	*Calculation*	*Experiment*	*Calculation*	*Experiment*	*Calculation*
Alumomagne-sium PAM - 4	123.00	106.40	56.00	56.00	36.00	42.00
Magnesium MPF - 4	68.00	70.60	37.00	37.00	26.00	27.75

Analyzing tables 1 and 2, it should be noted a satisfactory coincidence of the calculated values of the LCLFP according to the equation (7) with the experimental data obtained in various methodological conditions: the pressure, the material of the reaction chambers of the plants, the method of spraying powders. However, it should be borne in mind that the equation (7) is limited due to the approximate value of x_{ef}, the calculation of which did not take into account the volume occupied by the particles and the shadow area of the particles. However, this expression allows for a comparative analysis of the scientific results obtained on other technical devices with the results obtained on the installations performed in accordance with the requirements of normative and technical documentation, which significantly reduces the amount of experimental work.

This regularity (7), obtained on metal powders, is applicable for organic dust-forming materials, and in particular, for fine-crystalline powders of drugs and their intermediates. The uniformity factor of the dust phase distribution in the volume can now be considered as the main one.

Based on experimental data [13], we found $10^3 \geq P \geq 0.1$ kPa. Where it was determined that the optimum diameter of the reaction chamber 0.1–0.11 m. Considering the dependence of the final flow rate on the second air flow rate in the chambers of different diameters as a family of linear equations $y=Kx$, and it was observed that they all have one difference – a dimensionless ratio K (Figure 1). K shows that the increase in the flow rate, structurally means a decrease in the diameter of the chamber. The analysis shows that the diameter of the reactor vessel d is not less than 0.08 m K varies slightly, which indicates the difficulties in creating uniform flows at shorter distances. More than 0.12-0.14 m – ratio K also varies slightly, which indicates the difficulties of creating a uniform velocity field in the flow of small pulses, there is a need to increase the V_C to obtain the most stable distribution. It should be noted that this reduces the area of distribution of the dust cloud, its height decreases, which will introduce a systematic error in the processing of the results.

Based on the above, it follows that the most important is the interval of values of the site from $0.15 > K > 0.08$. In the chambers of this range, it is most easy to adjust the V_C and the speed distribution should be the most uniform.

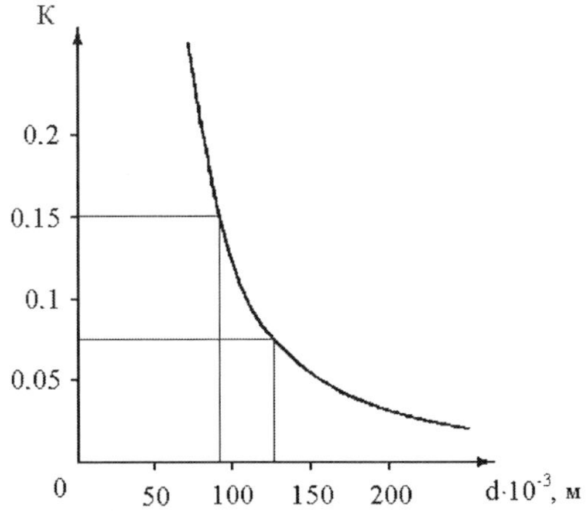

Figure 1. The change in the value of the dimensionless ratio K from the final flow rate V_C in chambers of different diameters.

Thus, it can be stated that the closer the geometric dimensions of the experimental volumes are to the established values, the greater the accuracy of the recalculated value.

After analyzing the design dimensions of the reaction chamber, it is necessary to analyze its experimental capabilities to create a uniformly distributed dust cloud.

Based on the conditions of uniform particle concentration at a flow rate close to the weighing rate, we conclude that the weighing rate for most organic substances does not exceed 10 m/s. Therefore, the height of the reaction chamber from the nozzle to the ignition source should be such as to provide a flow rate of not more than 10 m/s. From the analysis [13] it is clear that the height should be within 0.4–0.8 m. In this case, a minimum velocity difference in the granulometric dispersion of a certain fraction is provided.

3. Conclusion

As a result of the study, the influence of the geometric factor of the reaction chamber on the limit value of the ignition of the air suspension was studied. It is found that in determining the lower concentration limit of the flame propagation, it is most promising to use an installation with a cylindrical reaction chamber whose height is equal to the diameter. The dependence of the determination of the LCLFP value on the geometric factor of the reaction chamber is proposed, tested on metal powders and organic dust-forming materials. The expression allowing to carry out the comparative analysis of scientific results with the results received on the installations executed according to the shown requirements of normative and technical documentation is received.

Acknowledgments

The research is carried out at Tomsk Polytechnic University within the framework of Tomsk Polytechnic University Competitiveness Enhancement Program grant.

References

[1] Beschastnov M V 1991 *Industrial explosions. Assessment and prevention* (Moscow: Chemistry)

[2] Husniyarov M H, Popkov V F, Rudnev N A 2002 *Explosion hazards of oil refining units* (Ufa: USOTU)

[3] Eckhoff R K 2003 *Dust explosions in the process industries* (Boston: Elsevier)

[4] Scholl E W, Reeh D, Wiemann W 2016 *Brenn and Explosions* (Switzerland: ISO/IEC)

[5] GOST 12.1.044-89. Occupational safety standards system. Fire and explosion hazard of substances and materials. Nomenclature of indices and methods of their determination.

[6] Monakhov V T 1979 *Methods of investigation of fire hazard substances* (Moscow: Chemistry)

[7] Lewis B, Elbe G 1987 *Combustion, flames and explosions in gases* (Boston: Academic Press)

[8] Popov B, Medvedeva V 1974 *all-Union chemical Journal* **5** 520–525

[9] Nazarenko O, Amelkovich Y, Sechin A 2014 *Applied Surface Science* **321** 475–480

[10] Sechin A I 2003 *Safety of systems processing dust-forming materials* (Tomsk: Publishing house of scientific and technical literature)

[11] Baratov A N, Korolchenko A Ya, Kravchuk G N 1990 *Pozharovzryvoopasnost substances and materials and means of their extinguishing* (Moscow: Chemistry)

[12] Zeldovich Y B, Barenblatt G I, Librovich V B, Makhviladze G M 1980 *Mathematical theory of combustion and explosion* (Moscow: Nauka)

[13] Patrakov Yu, Sechin A I, Sechin A A 2017 *Gornyi Zhurnal* **12** 87–90

Modification of ANFO detonation parameters by biowaste addition

E Uranchimeg[1], M Narantsetseg[2] and L Purev[3]
[1]Blasting research and training center, Mongolian University of Science and Technology Ulaanbaatar, Mongolia
[2]Chemical engineering department, School of Applied Science, Mongolian University of Science and Technology, Ulaanbaatar, Mongolia
[3]School of geology and mining, Mongolian University of Science and Technology, Ulaanbaatar, Mongolia
E-mail:narantsetseg0427@gmail.com

Abstract. The present study relates to modification of ammonium nitrate fuel oil mixture. For many years, ammonium nitrate fuel oil has been one of the most popular explosives for use in mining operations. Therefore, it is manufactured and used widely in large volume. Although ammonium nitrate is a relatively strong oxidation agent, it is insufficient to detonate by itself. Therefore, it is generally admixed with liquid fuel such as diesel and various modifiers. Trinitrotoluene (TNT) and metal aluminium are typically used as additive. This paper studies ammonium nitrate fuel oil (ANFO) composition modified by biowaste dung as additive to generate efficient explosion and to result low cost blasting.
The experiment results of modified ANFO with biowaste dung as additive illustrate that physical stability and absorption increased by 5% percent compared to regular ANFO. Also was revealed that average detonation brisance increased from 20.5 to 27.5 mm and detonation speed increased from 2300 to 3800 m/s compared to regular ANFO. This modified ANFO is lowered blasting cost from 50 to 150 tugriks per 1kg ANFO compared to aluminium and TNT additives.

1. Introduction

According to statistics, usage of explosives was approximately 69 and 77 million ton in 2014 and 2017 respectively, which reveals evident fact that the market is grown [1]. ANFO compositions are formed of porous ammonium nitrate (AN) prills and diesel oil that is situated within many of the voids of the porous ammonium nitrate prills [2]. In Mining operations, the use of ANFO typically involves drilling a blast hole in the earth that was being mined. Drilling and blasting operations are widely used as rock powdering process in mining industry and road construction projects in Mongolia. In the mining industry, it is vital to define an optimal composition of ANFO which can provide a convenient average detonation brisance [3]. In this study, we are aimed to make an ANFO composition which will be more powerful and cheaper by using a biowaste dung as an additive.

2. Materials and methods

Samples of ANFO with additives and without additives were made in "Monmag" LLC and field tests were conducted on the authorized experimental field of the company "Monmag". Our new powerful ANFO explosive with biowaste dung were used in blasting work of Tumurtolgoi iron ore open-pit mine. Ammonium nitrate (NH_4NO_3) is the major oxidation component of the ANFO. The ammonium nitrate (AN) is typically used in the form of prills (spherical or roughly spherical particles up to a few mm in diameter, for example 0.9 to 3.0mm) [4]. In our practice AN is imported both from China and Russia which were compared to physical characteristics and the results are illustrated in table I.

Table 1. Physical Characteristic Of The Ammonium Nitrate

Parameters	Standart specific method	Technical requirements	NH_4NO_3	
			Russian	China
Adsorption ability, no more, %	MNS 4301:96	8.0	8.6	13.6
Particle size, 1-4mm, no more, %	MNS 6070:2010	95.0	99.9	99.9
Moisture, no more, kg/per granule	MNS 4302:96	0.4	0.4	0.4
Appearance	ТУ 2143-639-00209023-99	No mechanical impurity	-	-
Brittleness, no more, %	MNS 6071:2010	100	100	100
Physical fixity, no more, %	CS 11-0267:2014	5	5	5

The results of physical characteristic are shown that AN imported from both south and north neighbors are met the required standard values. In this study, we used the AN imported from Russia.

3. Results and discussion

The detonation strength of ANFO strongly depends on the characteristics of additives [5]. The technical characteristics of natural biowaste dung are shown in table 2.

Table 2. Technical Characteristic Of Biowaste Dung

	Technical Characteristics				
	Moisture, W^a, %	Ash, A^d %	Volatile compounds, V^{daf}, %	Calorific value Q^{daf}, kcal/kg	Absorption ability, %
Biowaste dung	7.37	16.52	79.27	5001.7	100
	Elemental Analysis, %				
	C^{daf}	H^{daf}	N^{daf}	S^{daf}	O^{daf}
	49.50	6.80	2.80	0.40	40.50

According to table 2, dung characterized as follows: moisture less 7.37%, ash content more than 15.31–16.52%, volatile compounds content 61.31–79.27, calorific value 5001.7 kcal/kg. Calorific value is calculated using elemental analysis by the equation: (Q_s =81C+300H-26(O-S) [6].

Then the ash composition of dung is defined by X-ray fluorescence spectroscopy and is shown in Figure 1.

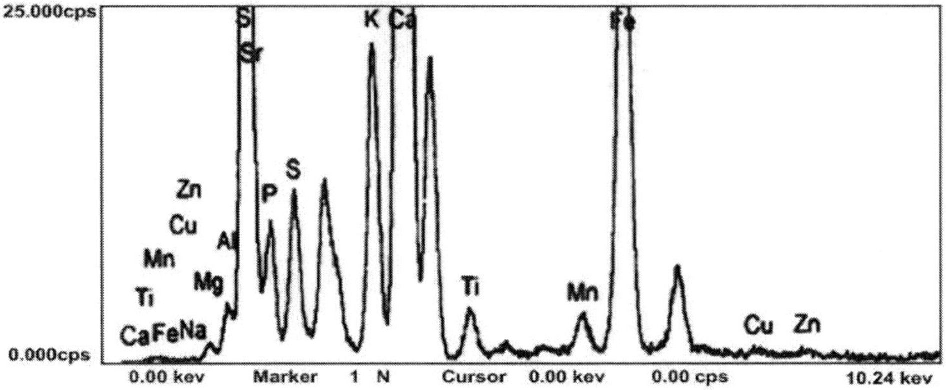

Figure 1. X-ray analysis of biowaste ash.

The ash chemical composition is shown in table 3.

Table 3. Ash composition of dung.

Composition of Ash, %						
SiO_2	Al_2O_3	Fe_2O_3	CaO	MgO	K_2O	Na_2O
49.70	49.70	2.81	19.1	19.10	2.93	4.92
Composition of ash, %						
Mn_2O_3	P_2O_5	TiO_2	CuO		ZnO	SrO
4.92	4.99	0.45	0.01		0.01	0.04

The results of ash chemical composition of dung show that the ash has a comparably high content of silica, aluminum, iron, sulfur, potassium, calcium than content of titan, copper, zinc, zircon, manganese oxides.

Then was studied chemical composition of ANFO. The AN, diesel fuel and modifier dung mixed at specific ratios in order to control oxygen balance and to cause the most efficient explosion possible. Mainly the weight ratio AN: diesel fuel is used 94:6 [7]. However, in our case we used modifier-biowaste dung in composition of ANFO and optimal ratio with combustion additive was determined and it is illustrated in table 4.

Table 4. Optimal weight ratio of the components of ANFO

№	Chemical Composition of modified ANFO (AN: diesel fuel: dung)	Absorption ability, C_{ab}, (%)	Physical stability ability, C_{st}, (%)
1	90:5:5	5	5
2	94:5.5:0.5	4.58	4.91
3	93:5:2	4.87	4.74
4	93:3.5:3.5	4.15	4.29
5	95:2.5:2.5	4.5	5

From the results, can be seen the optimal composition ratio determined to be AN: diesel fuel: dung is 90:5:5. This ratio shows higher absorption and physical stabilities.

Then we compared explosive efficient of ANFO compositions modified with additive and the regular one and explosives characteristics are illustrated in table 5.

Table 5. Explosive Characteristic Of The ANFO With Dung Additive And Regular ANFO

№	Standard Parameters	*Standard* Specific Method	Technical Requirement	Regular ANFO AN:diesel fuel	Modified ANFO AN: diesel fuel: dung
1	Density, (g/cm^3)	MNS 6072:2010	0.8–0.9	0.8	0.85
2	Detonation speed, (m/s)	CS 11-0263:2014	2200–2800	2300	3800
3	Absolute diameter of the blast hole, (mm)	CS 11-0265:2014	120–160	120	120
4	Brisance, (mm)	CS 11-0264:2014	15–20	20.5	27.5
5	Impact sensivity, (%)	CS 11-0266:2014	0	0	0

From the experiment results, it is evident that the modified ANFO shows the increase in detonation speed from 2300 to 3800 m/c, brisance from 20.5 to 27.5 compared to regular ANFO.

Then were done field experiments using modified ANFO with optimal composition ratio of such as AN: diesel fuel: dung (90:5:5) and regular ANFO to compare their detonation characteristics at of Tumurtolgoi's iron ore open pit mine. Produced blast characteristics are illustrated in table 6.

Table 6. Blasting Characteristics of ANFO In Open Pit Mine

№	Main parameters	Regular ANFO		ANFO With Dung Additive	
		Soil	Ore	Soil	Ore
1	Hole diameter (mm)	165	165	165	165
2	Rock hardness (f)	10	13	10	13
3	Drilling depth (m)	999.5	77.5	822.5	47.5
4	Amount of holes (quantum)	230	17	185	10
5	Blasting volume (m^3)	24987.5	1550	20562.5	1187.5
6	Sum of explosive (Kg)	10000	1040	7960	480
7	Specific consumption of explosive (kg/m^3)	0.56	0.68	0.4	0.55
8	Yield of oversized rocks (%)	-	5	-	3

From the results of detonation experiments, it is suitable to use the modified ANFO in ore mines, according, to the following characteristics: specific consumption of explosive declined from 0.68 to 0.55 kg/m^3, yield of oversized rock declined from 5 to 3% compared to regular ANFO.

Lastly, we calculated the prime price of ANFO with dung comparing to ANFO with aluminum and with TNT (2,4,6-trinitrotoluene) additives and results are illustrated in table 7.

Table 7. Calculation of The Prime Cost of 1Kg ANFO.

Explosives	Modified ANFO	ANFO with aluminum additive	ANFO with TNT additive
Prime cost			
Mongolian currency (tugrik)	1300	1350	1450
Difference in prime cost (%)	-	3.7	10.34
Difference in tugrik (tugrik)	-	50	150

The results show that the modified ANFO by biowaste is affordable /cheaper/ than explosives with aluminum additive in 3.7% and with TNT additive in 10.34%, that means cheaper on 50 and 150 tugriks /Mongolian national currency/ per 1 kg ANFO in case of aluminum and TNT additives respectively.

4. Conclusion

1. A complete reference should provide the reader with enough information to locate the article concerned, whether published in print or electronic form, and should, depending on the type of reference, consist of: From the experiments of ANFO composition with biowaste additive was determined the optimal weight ratio NH_4NO_3 : diesel fuel : biowaste to be 90:5:5.
2. Was defined that the uses in ore mines of the modified ANFO by biowaste effect in growth of physical stability, results in detonation speed from 2300 to 3800 m/s, brisance increased from 20.5 to 27.5mm and explosives power increased on 26.31%.
3. Was determined that specific combustion of explosive declined from 0.68 to 0.55 kg/m^3 and yield of oversized rock declined from 5 to 3% when used modified ANFO in iron ore open-pit mine compared to regular ANFO.
4. Was found that the uses of modified ANFO saved 50 to 150 tugriks per 1kg compared to ANFO with aluminum and ANFO with TNT additives.

References

[1] Mongolian mining industry report 2018 *Blasting techniques and explosives circulation* (Ulaanbaatar: Ministry of Mining and Heavy Industry Report)
[2] Laikhansuren B 2018 *Open pit blasting work* (Ulaanbaatar: MUST) pp 5–15
[3] Kutuzov B N 2008 *blasting techniques. Part 2. Blasting in mining and industry Textbook for universities* (Moscow: Mountain Book) pp 512
[4] Kukib B N, Ioffe V B, Zhuchenko E I and Frolov A B 2007 *Mining Information and Analytical Bulletin* 129 – 136
[5] Ilyushin M A, Savenkov G G and Masuria A S 2016 *Industrial explosives* pp 72– 109
[6] Narangerel J 2011 *Coal Chemistry and Technology Foundation* (Ulaanbaatar: Edmon) pp 68–71.
[7] Kutuzov B N 2007 *Methods of blasting part 1. Destruction of rocks by explosion* (Moscow: Mountain Book) pp 41–50.

IOP Conf. Series: Materials Science and Engineering 1019 (2021) 012041 doi:10.1088/1757-899X/1019/1/012041

The mechanoenzymatic method for enhancing the biological value of condensed cream soups

E Bychkova[1], V Pogorova[1], D Gosman[1] and K Dome[2]
[1]Department of Technology and Organization of Food Production, Novosibirsk State Technical University, Novosibirsk, Russia.
[2]Laboratory of Solid State Chemistry, Institute of Solid State Chemistry and Mechanochemistry, Siberian Branch of the Russian Academy of Sciences, Novosibirsk, Russia.
E-mail: bychkova.nstu@gmail.com

Abstract. Condensed creamy pea soups of high biological value have been developed. A stage comprising enzymatic hydrolysis in combination with mechanochemical treatment of raw material was proposed as a method to increase the content of free amino acid and short peptides in the finished product. Evaluation of physicochemical quality parameters demonstrated that the samples had high nutritional and biological value.

1. Introduction

The demand for protein is an evolutionary feature of human nutrition. It is associated with the need to meet the physiological requirements for the intake of essential amino acids, the compounds having a direct impact on vital functions of the human body. Unlike fats and carbohydrates, proteins are neither accumulated in the organism nor synthesized from other nutrients. In other words, adequate intake of proteins is ensured only by nutrition [1].

All proteins in human diet can be divided into two main groups: plant- and animal-derived ones. Animal protein contains all essential amino acids. However, a person eating animal proteins with proper amino acid content also consumes saturated fatty acids and cholesterol, which increase the risk of developing lifestyle diseases such as obesity, diabetes mellitus, and cardiovascular disorders [2]. Plant-based protein foods contain much less saturated fat, while the content of polyunsaturated fatty acids is rather high. Polyunsaturated fatty acids are involved in maintenance of the cell membrane function and reduce the risk of atherosclerosis and cardiovascular disorders. For this very reason, one needs to maintain the balance between consumption of animal and plant proteins [3, 4].

An analysis of the current global trends demonstrates that the percentage of manufactured animal source foods is rather low (80% of all protein produced worldwide is plant protein and only 20% is animal protein). Among the plant-based sources of protein, 50% of protein comes from grains and 25% comes from grain legumes and oil-bearing crops. Therefore, plant-based raw materials are a promising protein source for solving the problem of protein deficiency in the population [5, 6].

All the existing cultivated plants are classified into the following groups: ornamental plants (rose); cereal grains (rice, corn, and wheat); legumes (beans, peas, and soybeans); starch crops (sweet and regular potatoes); sugar crops (sugar beet); oil-bearing crops (sunflower); fiber crops; cucurbit crops (watermelon); vegetable plants (tomatoes, cucumbers, and dill); fruit plants (pineapples and coconuts), and plants exhibiting a stimulant effect (tea, coffee, and poppy). Among these species, the legumes, grains, oil-bearing and nuciferous crops (the latter group belongs to the class of fruits) can be used as a source of plant protein.

This classification infers that there is a wide variety of raw materials based on plant proteins, which can be processed to manufacture functional and specialized food products. However, when searching for sources of plant protein, it is important to assess the economic expediency of using a certain crop as the finished product may be too expensive. Thus, protein content in algae can be as high as 50%, making them superior even to legumes. However, it is infeasible to use this crop in Russia, since it is non-traditional for Russian population and, consequently, economically unviable for systemic use in the diet.

It is a known fact that only 62–80% of consumed plant proteins is absorbed in the human body. Substances known as antinutrients are the reason for this level of protein absorption. The data on the effects of the main inhibiting substances of plant origin are summarized in table 1 [7].

Table 1. Effect of antinutrients on the human body.

Antinutrient name	Effect on the human body
Proteolytic enzyme inhibitors	react with enzymes to form stable compounds, thus impeding protein biosynthesis.
Phytic acid	impairs absorption of iron, zinc, calcium, phosphorus, and other minerals from the food product in which it is contained.
Stachyose Raffinose	are not completely absorbed in the small intestine, resulting in excessive formation of intestinal gas.
Lectins	impair digestion, worsen nutrient absorption, and inhibit maltase and aminopeptidase.

The problem related to insufficient absorption of plant proteins can partially be solved using the chemical method consisting in removal of nonprotein constituents with inhibitory properties from the raw material. In this method, protein concentrates and isolates are obtained, which differ in terms of such parameters as the procedure used for their production, degree of purification, the nature of raw material, and total protein content. Protein concentrate is a purified protein product with 56–66% protein content. In order to produce it, the raw material needs to undergo such stages as defatting, suspension, salt or alkaline extraction, centrifugation, precipitation, and fractionation. Today, two methods are utilized to produce protein concentrates: turbo-separation (the "dry" method) and extraction in a liquid medium (the "wet" method). The protein isolate is a maximally purified product with protein content > 90 %. The conventional flowchart for production of a protein isolate involves protein extraction followed by addition of an acid (in order to ensure isoelectric precipitation of the protein), centrifugation, purification, and drying [8–11].

Protein concentrates and isolates are the purified forms of plant proteins neutral in both taste and aroma, so they can be used even at high doses. It is worth mentioning that plant-based raw materials utilized for protein production are much less expensive and more readily available than animal-derived raw materials. The costs of their storage and transportation are also lower. These factors are especially important for the countries having limited economical resources [9].

Today, protein derived from soybeans and wheat is a leader on the market of concentrates and isolates; however, there is also a gradually growing interest in pea protein. The key drivers of market growth are as follows: spreading vegetarianism, environmental safety of plant protein, as well as the fact that this type of protein is multi-purpose and functional. The key barriers for expansion of the market of protein concentrates and isolates involve allergies to protein and antinutritional properties of the resulting food products, which cannot be completely eliminated using the technologies discussed in this paper [6].

The germination method has proved to be a more efficient technique to enhance absorption of plant proteins. Germination is a transition between seed dormancy and sprout growth, accompanied by

IOP Conf. Series: Materials Science and Engineering 1019 (2021) 012041 doi:10.1088/1757-899X/1019/1/012041

modification of the chemical structure of compounds. Peptide zones become diffuse and eventually break down during imbibition and initiation of seed germination, while new ones are synthesized. As a result, availability of amino acids is enhanced. Enzymes present in the seeds cleave conjugated proteins, fats, and carbohydrates to simpler compounds that are needed for future plant development. For this reason, when a person eats sprouts, the substances supplied to his/her body are already processed and the impact of inhibitory compounds on the human organism is reduced [12]. However, seed germination under natural conditions is a rather labor-intensive and long-lasting process (taking up to 35 hours), which has not yet been brought to a large scale in Russia. The use of artificial enzymatic hydrolysis is an easier attainable technique for improving the nutritional value of legumes.

The objective of this study was to develop novel condensed cream soups using pea protein hydrolysate and to asses their quality.

2. Materials and metods

Pea seeds (grade I according to the State Standard GOST 6201-68) and Fidbest VGPro complex enzyme product manufactured by Sibbiopharm Ltd. were used as study objects.

At the initial stage, peas and the Fidbest VGPro complex enzyme product were sequentially subjected to mechanochemical disintegration on two mills (a rotor mill and a TM-30 centrifugal roller mill; "Novic" ltd, Novosibirsk, Russia) in the conditions described earlier [13, 14]. Enzymatic hydrolysis of pea flour was carried out at the next stage. The hydrolysis was performed at 50°C, following the manufacturer's recommendations. Hydrolysis duration was determined experimentally according to saturation of the solution with water-soluble compounds.

The resulting hydrolysate was brought to a boil in order to inactivate the enzyme and subsequently added to the cream soups. Other ingredients of the soups were as follows: fresh carrots, cauliflower, turkey meat, and 10% dairy cream.

Protein quantification in the ready-to-use food products was performed using the Bradford protein assay. The calibration curve was plotted using 1 mg/ml bovine serum albumin.

Water-soluble compounds were quantified by exhaustive extraction in a Soxhlet extractor for 24 h. The contents of reducing sugars were determined refractometrically; free amino acids were quantified by chromatography/mass-spectroscopy on an Autoflex Speed system (Bruker Daltonics, Germany). The mineral composition of the samples was evaluated by atomic absorption spectroscopy.Other paragraphs are indented (BodytextIndented style).

3. Resultes and discussion

The nutritional value of peas is shown in Figure 1. The diagram demonstrates that this crop is a source of a wide range of valuable biologically active substances.

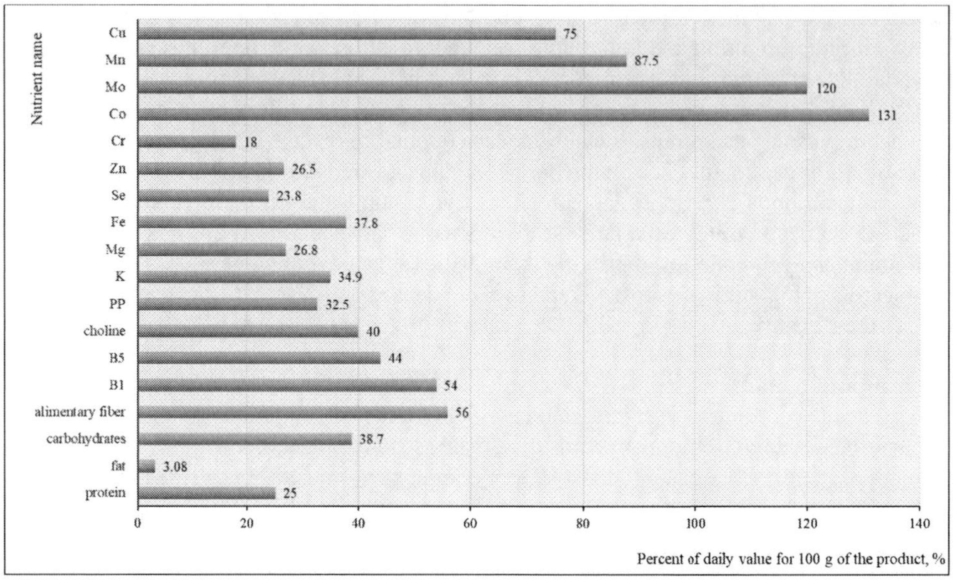

Figure 1. Nutritional value of pea seeds.

It is known that most of water-soluble highly absorbable compounds are contained within the cell. Mechanochemical treatment allows one to disrupt the cell wall and make the biologically active compounds available for the subsequent enzymatic hydrolysis. Addition of an enzyme during mechanochemical treatment is the most efficient way to reach the maximum degree of enzymatic hydrolysis as compared to other techniques. Mechanical activation of pea seed biomass gives rise to a reactive composite. Further enzymatic hydrolysis of this composite and exhaustive extraction allow one to obtain water-soluble compounds at high yields [15].

We determined the capacity of pea protein hydrolysate with respect to enzyme molecules. Test samples of pea flour in the presence of Fidbest VGPro enzyme product at different concentrations (0.5, 1, 2, and 3%) were subjected to hydrolysis. It was found that the amount of water-soluble compounds in the solution remained unchanged at enzyme concentrations higher than 2%. Further addition of the enzyme had no effect on the rate of enzymatic reaction, since the substrate was completely filled with the enzyme at its concentration of ~ 2%. The experimental data are shown in Figures 2 and 3.

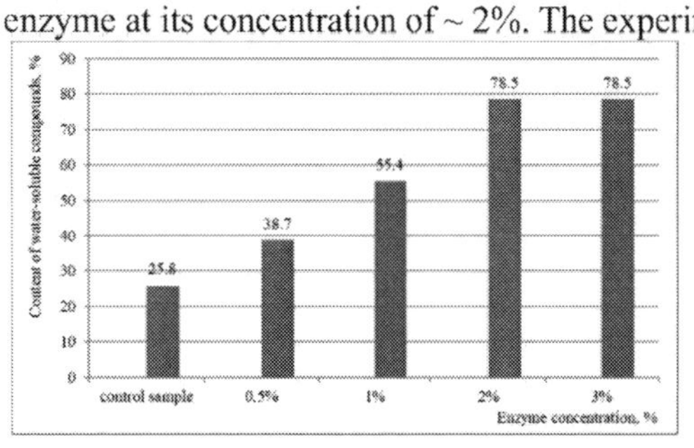

Figure 2. The total content of water-soluble compounds in hydrolysates depending on concentration of the enzyme product.

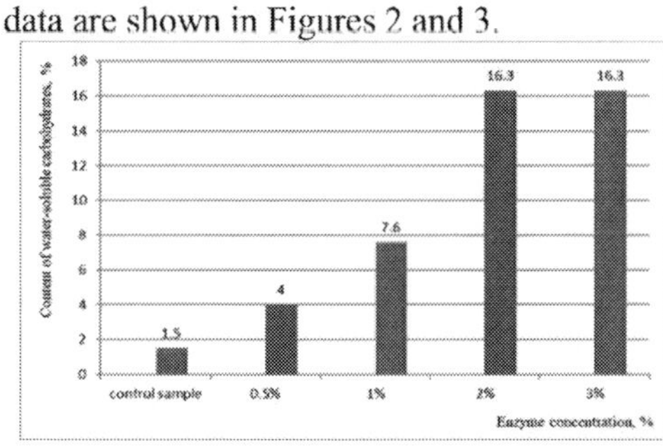

Figure 3. The amount of water-soluble carbohydrates in hydrolysates depending on concentration of the enzyme product.

Enzymatic hydrolysis lasting longer than 5 hours is inefficient, since the amount of water-soluble compounds remains virtually unchanged after this point. It is also worth mentioning that the rate and degree of hydrolysis of mechanically activated pea flour are higher than those for the non-activated sample.

The resulting pea hydrolysate was used to develop cream soups. The highest scores for the sensory quality parameters were assigned to sample 1 (creamy pea soup with vegetables) and sample 2 (creamy pea soup with turkey) (Figure 4).

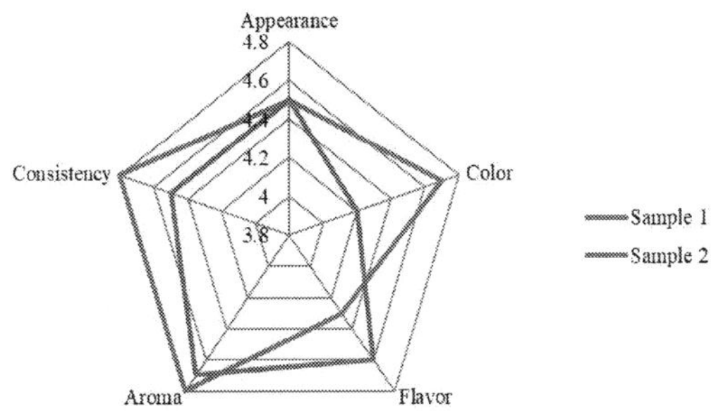

Figure 4 Sensory profile of the samples.

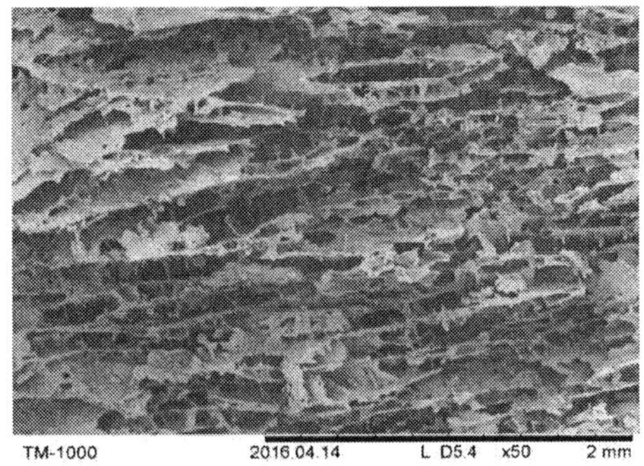

Figure 5. Scanning electron microscopy image of cream soup lyophilizate.

The cream soup samples were freeze-dried. The main benefit of using this method for food products is that nutrients are not destroyed, since the samples are treated at low temperatures. Furthermore, it significantly reduces the risk of oxidation of unstable substances with atmospheric oxygen.

The scanning electron microscopy data (Figure 5) show that the cream soup samples are lightweight, homogeneous and porous, and have a well-developed surface, thus indicating that proper freezing/drying modes have been selected.

Table 2 summarizes the physicochemical quality parameters of the reconstituted cream soups. The finished products are characterized by low acidity and low fat content.

Table 2. Physicochemical quality parameters of cream soups.

Parameter	Sample number			
	1[a]	*2*[b]	*3*[c]	*4*[d]
Dry solids, wt. %	13.70	15.70	20.80	23.50
Protein, wt. %	-	13.60	-	16.80
Acidity, %	0.60	0.60	0.50	0.50
Fat, wt. %	2.10	2.10	2.50	2.30
Ash content, %	0.69	0.78	0.50	0.65

[a]sample 1 – creamy pea soup with vegetables (control sample);
[b]sample 2 – creamy pea soup with vegetables based on pea protein hydrolysate;
[c]sample 3 – creamy pea soup with turkey (control sample);
[d]sample 4 – creamy pea soup with turkey based on pea protein hydrolysate.

Figure 6 shows a diagram indicating changes in the yield of water-soluble compounds in cream soup samples before and after they were subjected to hydrolysis and mechanochemical treatment. It was found that the additional stage of pea flour treatment increases the yield of the water-soluble fraction approximately threefold. This indicates that high-molecular-weight compounds are converted into simpler substances and, consequently, their absorption is enhanced. In a similar fashion, the content of free carbohydrates is increased in soup samples based on pea protein hydrolysate (Figure 7). As a result, the glycemic index of the ready-to-eat meal also increases.

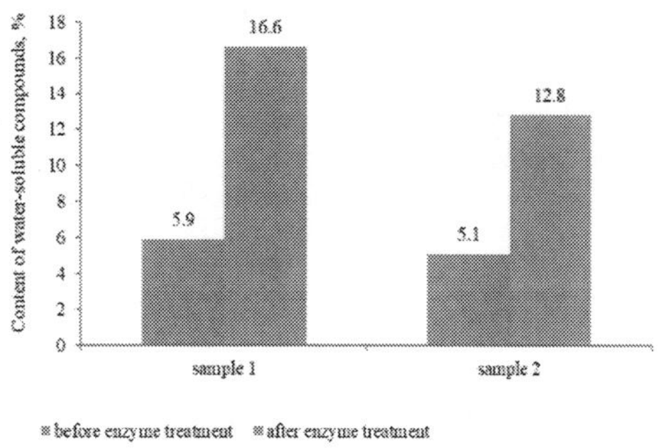

Figure 6. Changes in the contents of water-soluble compounds in the samples.

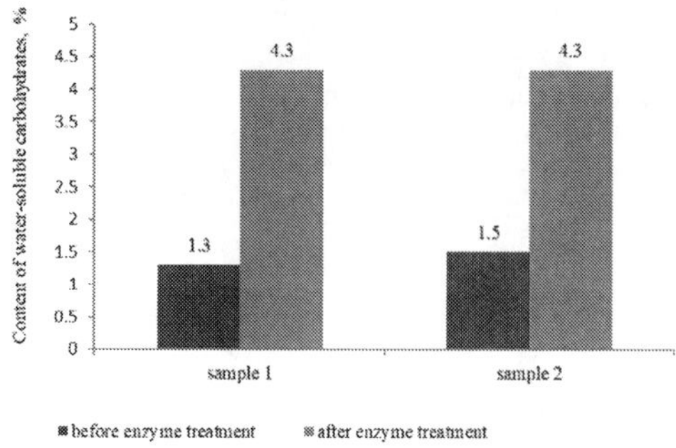

Figure 7. Changes in the contents of water-soluble carbohydrates in the samples.

The increased amount of free amino acid in ready-to-eat soups also indicates that the enzyme product is efficient (Figures 8 and 9).

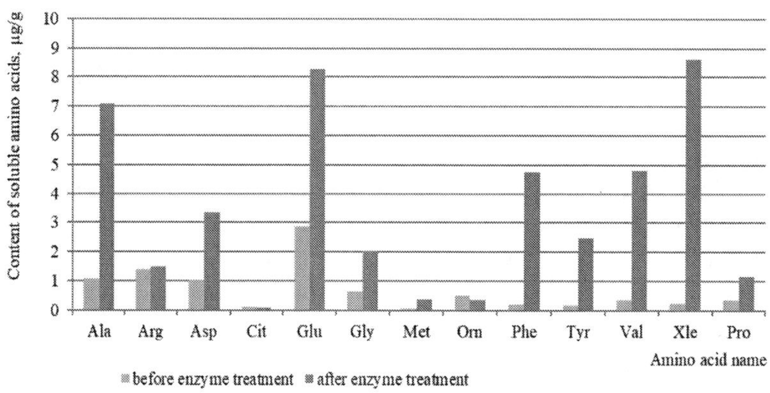

Figure 8. Changes in the contents of free amino acids in the cream soup with vegetables.

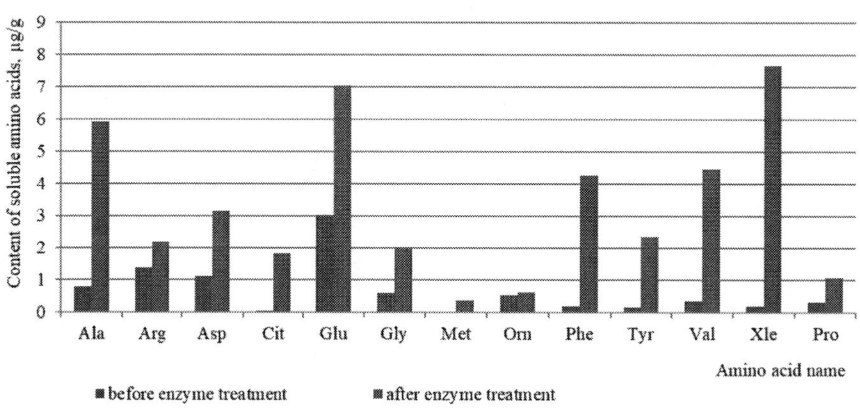

Figure 9. Changes in the contents of free amino acids in the cream soup with turkey.

Figure 10 shows the data on the percent daily value of minerals for one serving of cream soup. The soup samples were found to be a valuable source of potassium and iron.

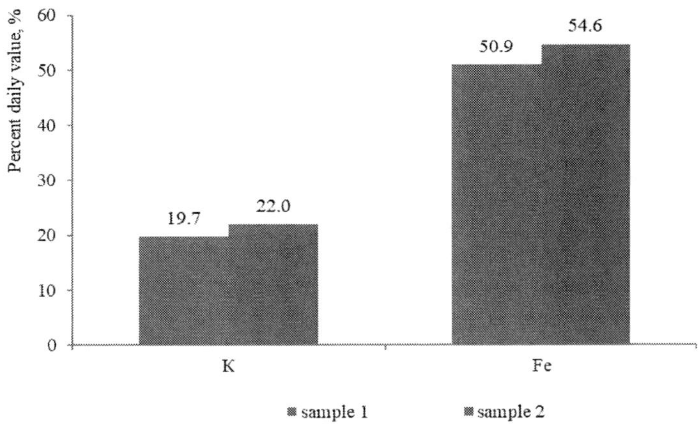

Figure 10. Mineral composition of the cream soups.

The ready-to-use lyophilisates of creamy pea soups were vacuumized and stored. Sensory evaluation showed that the dry soups had neither off-flavor nor off-aroma. They were easily reconstituted into the original state by adding water; the initial quality parameters of the soups were not altered.

The microbiological quality parameters of the reconstituted cream soups fall within the normal range.

4. Conclusions

Specialized cream soups based on pea protein hydrolysate have been developed in this study. The optimal concentration of the enzyme product has been identified. We have achieved the objective of increasing the contents of water-soluble compounds (including free amino acids, peptides, and simple carbohydrates) in the ready-to-serve soup as the mechano-enzymatic hydrolysis stage has been added to the preparation process. A rationale for using the freeze-drying stage to improve shelf life and make soup preparation more convenient has been presented. The recommended daily serving size of the reconstituted cream soup is 200 g. The developed soups consumed on a daily basis provide peptides, free amino acids, simple carbohydrates, and iron for the human body, while preventing allergic reactions and alleviating symptoms of gastrointestinal diseases. They also make sports training more efficient.

References

[1] Hoffman J R and Falvo M J 2004 *J Sports Sci Med* **3** 118–130
[2] Valsta L M, Tapanainen H and Männistö S 2005 *Meat Sci* **70** 525–530
[3] Aberoumand A 2011 *J Food Process Techno* **2** 114–117
[4] Mattila P *el at* 2018 *Foods Hum Nutr* **73** 108–115
[5] Henchion M *el at* 2017 *Foods* **6** 53
[6] Wild F *el at* 2014 *Agro food Industry Hi Tech* **25** 45–49
[7] Gemede H F and Ratta N 2014 *Int. J. Food Sci. Nutr* **3** 284–289
[8] Vioque J, Sánchez-Vioque R, Pedroche J and Yust María del Mar 2001 *Grasas y Aceites* **52** 127–131
[9] Garba U and Kaur S 2014 *Int. J. Curr. Res.* **6** 35–39
[10] Nehete J Y *el at* R 2013 *Pharmacogn Rev* **7** 107–116
[11] Bychkov A *el at* 2019 *Biotechnol. Bioeng* **116** 1231–1244
[12] Hübner F and Arendt E 2013 *Crit. Rev. Food Sci. Nutr* **53** 853–861
[13] Bychkov A L, Buchtoyarov V A and Lomovsky O I 2014 *Cell Chem Technol* **48** 545–551
[14] Podgorbunskikh E M, Bychkov A L, Bulina N V, Lomovskii O I 2018 *J Struct Chem* **59** 201–208
[15] Gavrilova K *el at* 2019 *Raw Mater* **7** 255–2631

Optimal voltage regulator of power supply system for submersible processing equipment

V G Bukreev[1], V M Rulevskiy[2], E B Shandarova[1] and E A Bystrov[1]
[1]Division for Power and Electrical Engineering, National Research Tomsk Polytechnic University, Tomsk, Russia
[2]Research Institute of Automation and Electromechanics, Tomsk State University of Control Systems and Radioelectronics, Tomsk, Russia
E-mail: bukreev@tpu.ru

Abstract. The article considers the synthesis of the optimal voltage regulator for the electric power supply system of the electromechanical complex of the submersible apparatus with the power transmission through the cable line. Based on the reduced linearized model of the multidimensional nonlinear control object the regulator with adjustable parameters is synthesized, providing the DC voltage stabilization at the power supply system load. The variant of building proportional feedback on output state variables is used for the regulator in the system with parametric and external disturbances. Due to the technical impossibility of measuring the voltage on the load, a combined optimal regulator with negative feedback is organized, which makes it possible to stabilize the voltage at the payload within the specified limits when receiving data from the output filter of voltage source inverter.

1. Introduction

The efficiency of submersible processing equipment such as electric submersible pumps for extracting oil and electromechanical complex of the deep-diving submersibles is largely determined by the characteristics of power supply systems (PSS) with the power transmission through the cable line. The main problem in the construction of such special power supply systems is the problem of synthesis of optimal control, providing voltage stabilization at the remote payload of the submersible part [1–3]. At the same time, the impossibility of measuring the regulated voltage on electromechanical load makes it difficult to directly apply the classical structures of closed loop systems with negative feedback on the output variable [4, 5]. In addition, the limitations on the computing resources of the control device that implements the laws and control algorithms, predetermines the allowable simplification of the mathematical model of complex, multidimensional and nonlinear PSS with non-stationary parameters. Therefore, along with the task of synthesis of the optimal regulator, the task of reduction of the initial model of PSS becomes actual, i.e. construction of a lower order model, which adequately reflects the behavior of the power supply system. It is assumed that the basis for the construction of the law providing the required quality of the output voltage stabilization of PSS can be a linearized mathematical model that allows the use of methods of linear control theory. The obtained reduced model of PSS in the form of differential equations allows to further take into account the variable disturbance pattern and nonlinear properties of the control object.

2. Structure of the power part of the PSS and principles of formation of the reduced mathematical model

The power supply system (Figure 1) considered in this paper contains a three-phase voltage source inverter (VSI) with pulse width modulation (PWM), to the input of which a DC voltage U_s is supplied through the L-shaped LC-filter (F1). To smooth out the higher harmonics between the VSI and the step-up transformer (T1), the LC-filter (F2) is included. The increased voltage from the transformer T1 is supplied to the cable line (CL). The reduced voltage from the output of transformer T2 through three-phase bridge rectifier (R) and smoothing LC-filter with parameters L_d, C_d («filter 3» in Figure 1) is supplied to the load of the considered power supply system, which may be active and shown in the circuit by an equivalent resistance R_n. The windings of the step-up transformer T1 are connected according to the Δ/Y connection, and the step-down transformer T2 according to the Δ/Δ connection. The transformers parameters are windings active resistance and leakage inductance, taking into account the leakage flux influence.

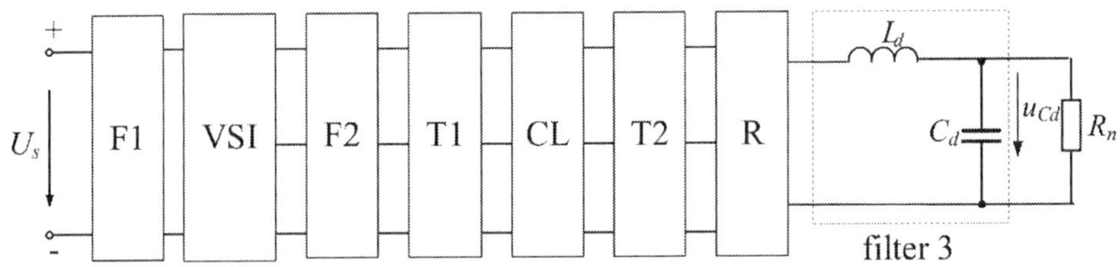

Figure 1. Structure of the power part of the PSS.

Control of VSI in PSS is organized on the basis of PWM with third harmonic injection. For the description of the output phase voltage VSI the switching functions [6] are usually used. In our previous papers the mathematical model of PSS is described in detail. It is presented by the equations in the form of Koshi, which allows to simulate both dynamic and steady-state modes of operation of PSS power part [7, 8]. The complexity of the analysis of processes in PSS lies in the high order of the system (eleven differential equations per phase), as well as the presence of switching functions that describe the operation of VSI and rectifier, which consequently brings the complexity to the synthesis of a voltage regulator. To correctly simplify the mathematical model of PSS the following assumptions were accepted: discrete switching functions on the basis of the analysis of an amplitude spectrum were represented by continuous functions; the part of the circuit including transformers and a cable line, was replaced by an equivalent second order RLC-circuit. Such replacement allowed reducing the number of differential equations from eleven to eight per one phase [9].

The parameters of the equivalent circuit were determined from the condition of conformity of parameters of the second order transfer function on voltage and parameters of the transfer function on the basis of the analysis of the magnitude Bode plot of the PSS. The verification of the adequacy of the reduced mathematical model with the initial circuit of replacement of PSS was carried out by comparison of dependences of flowing currents and voltages in power elements [9]. As researches have shown, the approximation error does not exceed 4% that indicates the adequacy of mathematical model of PSS on the basis of transfer functions.

3. Analytical synthesis of the optimal regulator for PSS

Assuming that the parameters of the power supply system are stationary at the i-th interval of its electromechanical load operation, it is possible to obtain a reduced model of an open loop system (an object under to be controlled):

$$\dot{x}(t) = A_i\, x(t) + b_i\, u_i(t)\,,\ x(t = t_0) = x_0\,,\ t_0 \neq 0\,, \tag{1}$$

IOP Conf. Series: Materials Science and Engineering 1019 (2021) 012042 doi:10.1088/1757-899X/1019/1/012042

where $x(t)$ – n-dimensional vector of control object state variables; A_i – $n \times n$ - matrix of stationary parameters of PSS; b_i – n-dimensional vector stationary control parameters; $u_i(t)$– scalar control input vector, which is the output variable of the voltage regulator (in our case it is the pulse width factor for VSI); t_0 – arbitrary initial moment of time; x_0 – n- dimensional vector of initial values of state variables.

Without loss of commonality for the procedure of synthesis of the optimal regulator it is possible to consider a variant of measurement of the entire vector $x(t)$ of state variables, which takes place at a stage of mathematical modeling of the PSS adequate model. The optimized criterion of quality of regulated processes in PSS can be written in the form of a smooth quadratic function $J_i(x,u)$ of two components:

$$J_i(x,u) = \int_{t_0}^{t_N} \left(x^T(t)Qx(t) + r_i u_i^2(t) \right) dt, \tag{2}$$

where $(*)^T$ – transpose symbol; t_N – moment of time of the final state of the object; Q – positive semidefinite matrix $n \times n$, the components of which are penalty coefficients for the corresponding state variables $x(t)$; r_i – positive coefficient, the value of which reflects the influence of scalar control actions $u_i(t)$ on the state of extreme quality criteria and indirectly reflects the requirements for energy efficiency management.

The most suitable form of weight (penalty) matrix Q in the formation of requirements to the processes of many objects is a diagonal form. This form of matrix Q does not require checking the necessary condition of its positive semidefinition and significantly simplifies the structure of the quality criterion. For optimal regulators, the structure of which is determined by the law of linear feedback on variables of the state of the controlled object of the type (1), the minimum functionality (2) will be provided for the control action:

$$u_i(t) = -r_i^{-1} b_i^T P_i(t) x(t), \tag{3}$$

where $P_i(t)$ – matrix $n \times n$, the components of which are determined by solving the differential (algebraic) Riccati's equation or Lyapunov's equation [10].

In the process of synthesis of an optimum regulator for considered system of an electrical supply the transfer function was determined during its operation in a mode close to a no-load one with a pulse width factor of $k_m = 0.7$. According to the obtained transfer function, the system of differential equations in the form of Cauchy (1) is obtained, which allows to write the matrixes A and b. Regulator coefficients were calculated based on the expression $k^T = r^{-1} b^T P$.

Figure 2 shows a block diagram of the PSS with an optimal regulator (OR), which allows to stabilize the load voltage. The equivalent resistance varies from $R_n = 10R_{nr}$ to $R_n = 3.4R_{nr}$, where R_{nr} – the rated load resistance, L – the load.

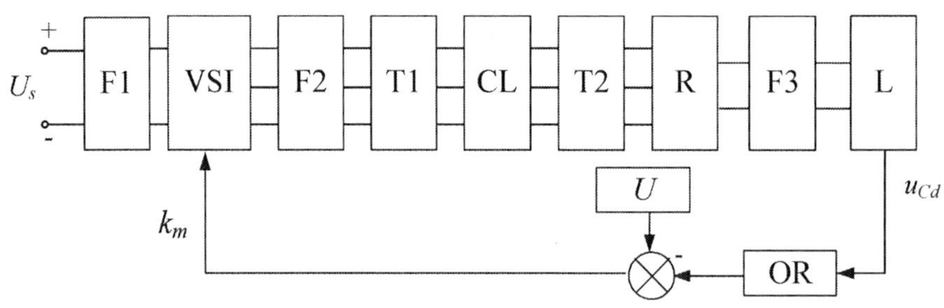

Figure 2. Structure of the PSS model with optimal output regulator.

Figure 3*a* shows the result of the regulator's operation when operation in a no-load mode with the load resistance $R_n = 10R_{nr}$.

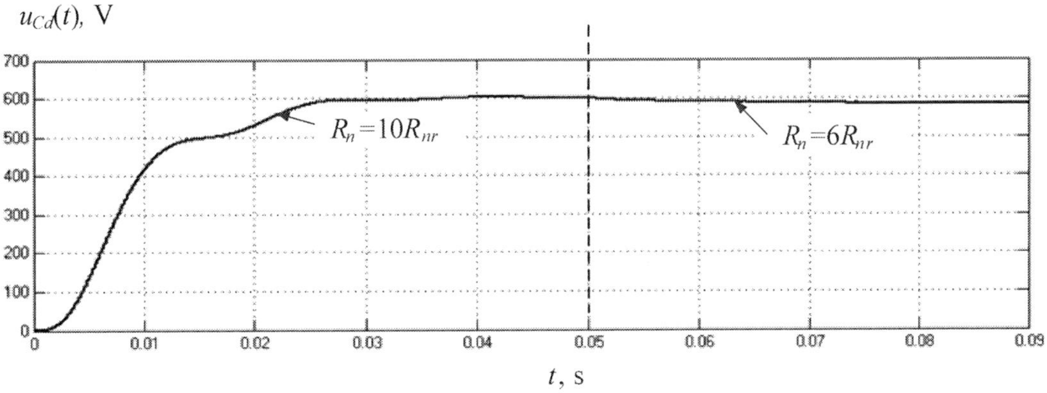

a)

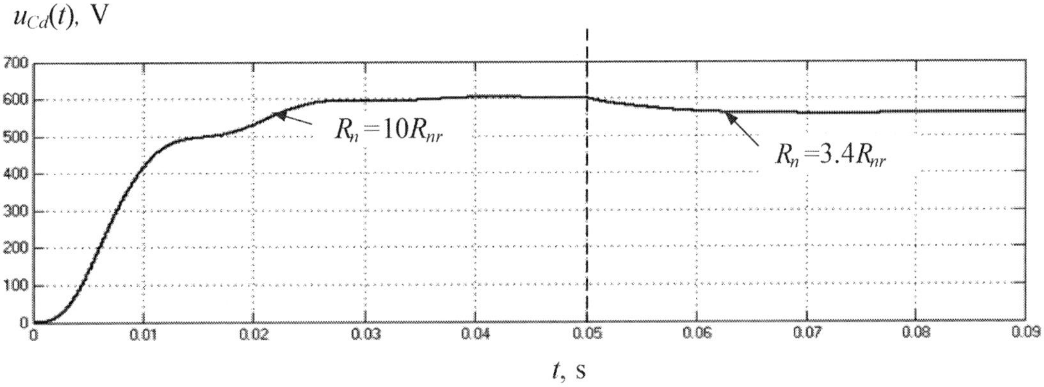

b)

Figure 3. Load voltage in PSS with optimal regulator (switching at a time of 50 ms: a – $R_n = 6R_{nr}$; b – $R_n = 3.4R_{nr}$).

After reaching the steady-state mode, the load increases at the moment of time $t_P = 50$ ms (resistance changes up to $R_n = 6R_{nr}$). Figure 3*b* reflects that at the moment of time 50 ms the load resistance changes from $R_n = 10R_{nr}$ to $R_n = 3.4R_{nr}$, and the error of voltage stabilization increases from 2% (Figure 3*a*) to 10%. If the load continues to increase, the voltage stabilization error increases.

It is known that according to the technical requirements to the existing PSS, the error of voltage stabilization at the load in 20 ms after changing the operating mode should not exceed 10%. To extend the limits of voltage stabilization it was proposed to use a regulator with reconfigurable parameters, which includes three optimal regulators, the coefficients of which were calculated for a certain range of changes in the load of the PSS (close to a no-load mode, mode of reduced load, the nominal mode).

Figure 4*a* shows the results of the combined optimal regulator operation when the system is switched to a no-load mode (the regulator tuned to the mode close to a no-load one is activated) and then the nominal load is connected (the regulator tuned to nominal mode is switched on). Figure 4*b* shows the operation of the combined regulator when connecting a load with resistance $R_n = 2R_{nr}$, the voltage is stabilized by the regulator tuned on a reduced load mode. In all the Figures shown, the load is connected at 50 ms and the stabilization error is less than 10% of the specified voltage *U*.

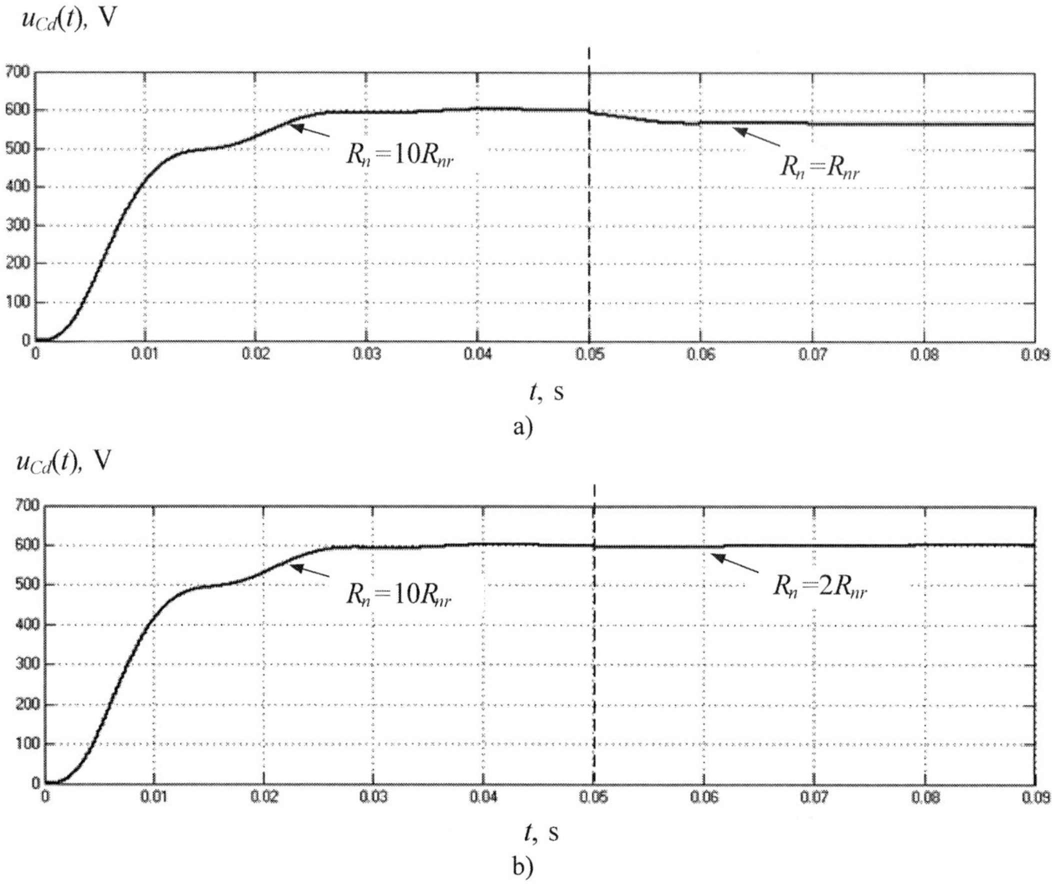

Figure 4. Load voltage with the combined optimum regulator: a – change R_n from $R_n = 10R_{nr}$ to $R_n = R_{nr}$; b – from $R_n = 10R_{nr}$ to $R_n = 2R_{nr}$.

As in power supply system there is no technical possibility to measure voltage on load, it is expedient to use rectified voltage from the output VSI filter for construction of the optimal regulator with negative feedback. Figure 5 shows the combined optimal regulator scheme, which allows to stabilize the voltage at the power supply system load in case of direct measurement at the *LC*-filter output (F2). The voltage received from the filter capacitance F2 is rectified by rectifier R2 and smoothed by capacitive filter F4.

Assuming that the mathematical model of the investigated PSS is linear, we can suggest that in the steady state of operation the load voltage and the rectified voltage from the filter F4 will be linked by a linear ratio. The coefficient of linear coupling k=15.9 was determined experimentally, which was used in the synthesis of the optimal regulator to recalculate the change in load voltage reduced to the measured rectified voltage of the filter F4 (Figure 5). The proposed regulator also consists of three optimal regulators. Each of them is tuned to their own control range (mode close to a no-load one, a reduced load mode, the nominal mode). The regulators are switched according to the signal S coming from the control system CS.

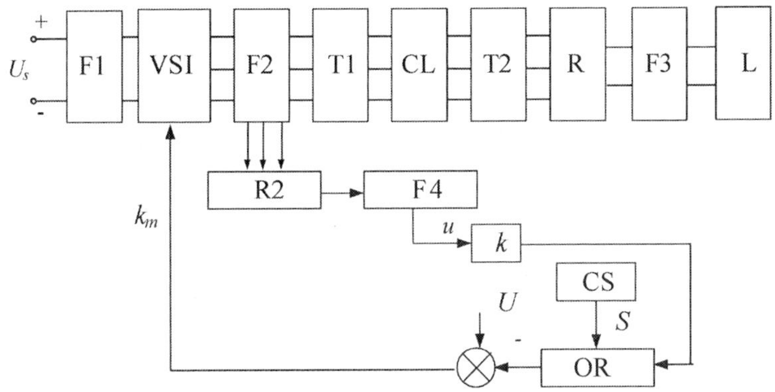

Figure 5. Combined optimal regulator structure for voltage measurement from the output VSI filter.

Figure 6 shows the result of the work of the proposed combined optimum regulator. At the moment of start the PSS, the control system CS acquires a control signal *S*, which causes the activation of the regulator configured to stabilize the voltage in a no-load mode, after 50 ms the nominal load is connected, and the control system sends a signal to disconnect the first regulator and connect the second regulator configured for the operating modes close to the nominal ones (Figure 6*a*).

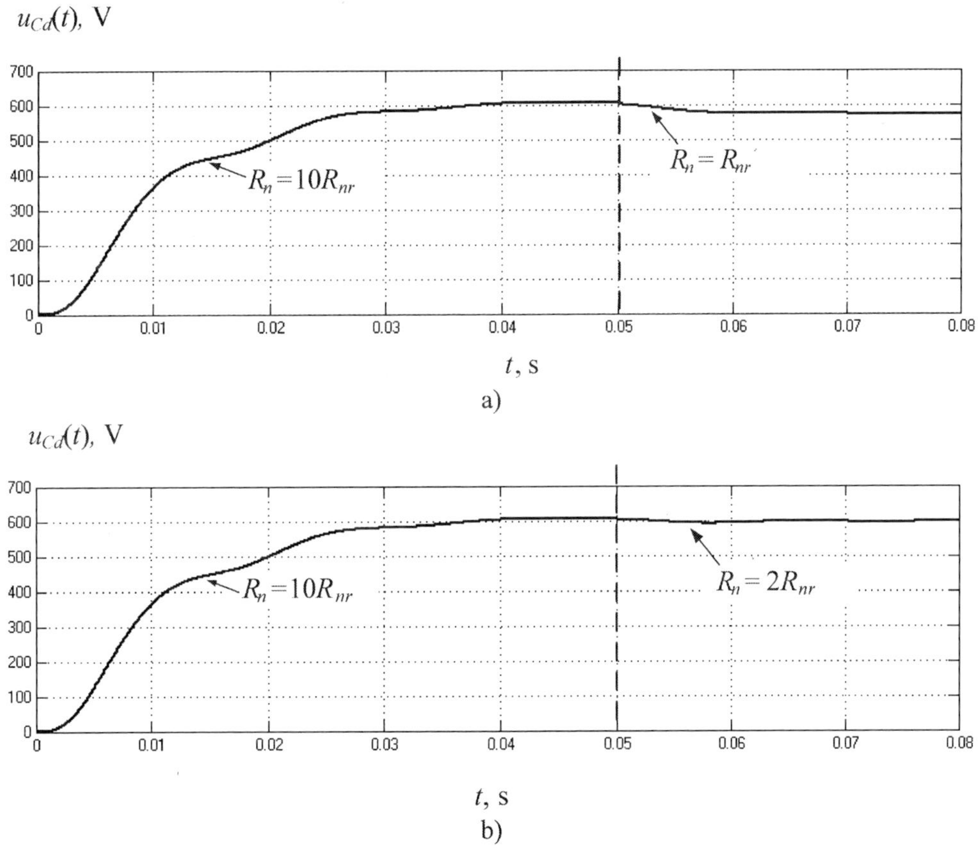

Figure 6. Load voltage with combined optimum regulator and measuring the voltage from the filter VSI: a – change R_n from $R_n = 10R_{nr}$ to $R_n = R_{nr}$; b –from $R_n = 10R_{nr}$ to $R_n = 2R_{nr}$.

In case of further load changes, CS switches regulators according to the current state of the system. So, in Figure 6*b* at the moment of time *t*=50 ms there is an abrupt change of system load to $R_n = 2R_{nr}$, thus the operating regulator is switched off and the regulator adjusted on reduced load mode is included into operation. In all operating modes of the PSS with the proposed regulator the voltage error at the equivalent load does not exceed 10%.

4. Summary

The research showed that the optimum regulator with a variable structure allows to stabilize voltage on PSS payload and to provide the quality indicators established for the voltage measured on the output VSI filter. This is due to the assumption that the characteristics of the PSS are linear. For a more accurate description of the PSS, it is necessary to take into account the non-linearity of the elements, caused in particular by the influence of the hysteresis in transformers. The load conversion factor for the regulator will be a non-linear relationship. Synthesis of the optimal regulator, taking into account nonlinearities of the PSS elements, is the subject for further research.

References

[1] Kolluri S, Thummala P, Sapkota R, Kumar Panda S and Rendusara D 2016 *Proceedings of 17th Workshop on Control and Modeling for Power Electronics* 1–6
[2] Wrinch M C 2009 *Sea Technology* **50** (7) 27–29
[3] Xiao S, Wei T, Xiaojuan K and Ying P 2011 *Proceedings of 4th International Conference on Electric Utility Deregulation and Restructuring and Power Technologies* 1760–1763
[4] Skaanoey T, Kerin U, Van Luijk N and Thibaut E 2017 *Proceedings of the Annual Offshore Technology Conference* **4** 2920–2928
[5] Hussain H A, Anvari B and Toliyat H A 2017 *Proceedings of IEEE International Electric Machines and Drives Conference* 1–7
[6] Zinovev G S 2015 *Power Elecroincs [Silovaya elektronika]* (Moscow: Yurayt Publ) pp 667
[7] Rulevskiy V M, Bukreev V G, Shandarova E B, Kuleshova E O, Shandarov S M and Vasilyeva Y Z 2017 *IOP Conference Series: Materials Science and Engineering* **177** 1–6
[8] Rulevskiy V M, Bukreev V G, Shandarova E B, Kuleshova E O, Shandarov S M and Vasilyeva Y Z 2017 *IOP Conference Series: Materials Science and Engineering* **177** 1–6
[9] Bukreev V G, Shandarova E B and Rulevskiy V M 2018 *Bulletin of the Tomsk Polytechnic University. Geo Assets Engineering* **329** 119–131
[10] Kim D P 2007 *Theory of automatic control. V.1. Linear systems Teoriya avtomaticheskogo upravleniya. T.1. Linejnye sistemy* (Moscow: Fizmatlit) pp 312

Measurement of electron beam transverse flux density distribution

A A Bulavskaya[1], Yu M Cherepennikov[2], S V Chakhlov[3], A A Grigorieva[1], I A Miloichikova[2,4], A V Vukolov[1] and S G Stuchebrov[1]

[1] Research School of High-Energy Physics, Tomsk Polytechnic University, Tomsk, Russia

[2] School of Nuclear Science & Engineering, Tomsk Polytechnic University, Tomsk, Russia

[3]School of Non-Destructive Testing, Tomsk Polytechnic University, Tomsk, Russia

[4] Cancer Research Institute of Tomsk NRMC RAS, Tomsk, Russia

E-mail: bulavskaya@tpu.ru

Abstract. The work presents experimental measurements of electron beam transverse flux density distribution. Experimental data is recorded during the multiple beam scanning in different directions with the particular angle step by the thin scintillation strip. The intensity of the light generated in the scintillator is proportional to the intensity of the radiation going through the strip. Generated photons is guided by the optical fiber to the photomultiplier and registered by the analyzer. The result of the work is the experimental data demonstrated dependence of the radiation intensity on detecting strip position and angle orientation. This dependence is transformed to the radiation intensity dependence on the coordinates in the beam transverse plane using special mathematical processing.

1. Introduction

Modern technology for oncology treatment causes high demands on the dose delivery accuracy especially accounting improvements in external beam radiation therapy. It is necessary to control spatial parameters of the radiation beams before radiation therapy sessions and during ones to provide accuracy of dose delivering and particularly avoid of overexposure of the healthy tissue and insufficient dose values in the tumors [1]. Currently, the control of radiation beam parameters is provided in medical practice accordingly to international protocols for clinical dosimetry [2, 3].

There are few methods to determine the radiation flux density distribution in the transverse plane of the beam. For example, it is possible to measure beam spatial parameters using matrix detectors consisted of ionization chambers [4–6] or diodes [7, 8], that is placed on the plane or inside special phantoms. The same approach is a basis for use of fluorescent screens in conjunction with optical cameras [9–11]. Beam parameters measuring with these devices provides high speed. However, these approaches are invasive and usually the beam is completely absorbed during the measurement. Simultaneously, the arrangement of detecting elements allows estimation only the shape and size of the radiation field but not the exactly intensity distribution.

Another approach to determine beam transverse flux density distribution is based on using materials, that changing properties when interact with the radiation, in a form of plates or films [12–14]. These methods have high resolution, but require consumables that leads to increases the operational cost of these devices and necessity of their parameter monitoring, for example, to perform additional calibration for particular batch of detectors.

Previously proposed method to measure electron beam transverse flux density distribution, which is devoid of the above disadvantages, is tested in a frame of this work.

2. Materials and methods

2.1. Measurement method.

The method to measure electron beam transverse flux density distribution is based on multi-angle measurements. This method means multiple strip scanning of the beam under different angles with fixed angle step. The set of measured intensity dependences on scanning angle and strip position can be transformed to the flux density distribution of the particles in beam transverse plane using mathematical reconstruction. It is previously demonstrated [15] that the minimal scan number to perform this method with reliable results is not less than 18.

2.2. Radiation source.

In the experiment TPU microtron [15] is used like a radiation source. The energy electron energy in output beam is 6.0 MeV with 1.0 mA current.

2.3. Registration system.

Scanning element is a thin scintillation strip 2×6×150 mm^3 sized. In the experiment, we use scintillator BC-408 by Saint-Gobain [16]. Generated in the scintillator light photons is guided by the optical fiber to the photomultiplier with the 6 mm cross-section diameter to the silicon photomultiplier [17]. The parameters of the photomultiplier is following: active area is 6×6 mm2, spectral range is 300–800 nm, and registration efficiency for photons is up to 47% at 420 nm wavelength. The signal of the photomultiplier is amplified and hardware processed, and registered by the computer.

Ad-hoc software is used to control scanning system and to record experimental data. In the recording intensity data and scintillation strip position is synchronized between each other.

To verify experimental results electron beam transverse flux density distribution is measured using Gafchromic EBT2 dosimetry film [18] located perpendicularly to beam propagation axis. These films are digitized with Epson Perfection V750 Pro scanner [19] and mathematically processed with MATLAB package [20].

2.4. Experimental scheme.

Output beam is going through a circular shaped collimator with pinhole 10 mm diameter and 50 mm length. Electron beam transverse flux density distribution is measured at 50 mm from the collimator.

Scintillation strip is shifting linearly in the plane perpendicular to the beam axis by 25 mm; the angular step equals 10°. For the experiment, we use controller, rotation and linear stages by Standa [21–23].

Scintillation strip moves continuously, with speed chosen to provide 1 mm movement during 2 seconds. Measured for 2 seconds signals are averaged for higher accuracy.

Figure 1 shows experimental layout for measurement of electron beam transverse flux density distribution.

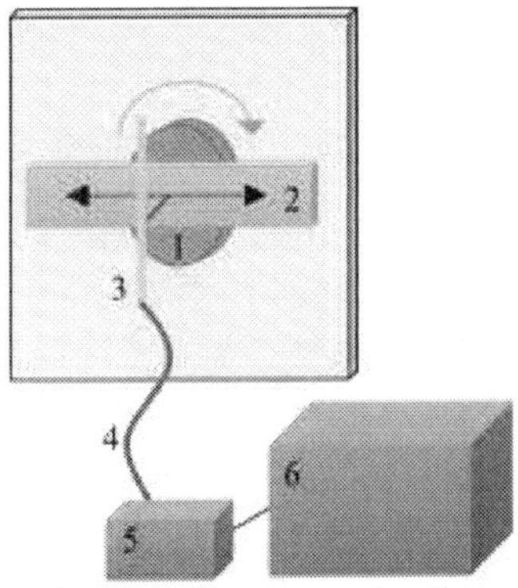

Figure 1. Experimental layout: 1 – rotation stage; 2 – linear stage; 3 – scintillation strip; 4 – optical fiber; 5 – photomultiplier; 6 – PC.

3. Results and discussions

The set included 19 dependences of radiation intensity on scintillation strip location and angular orientation is measured in the experiment. The first measurement was performed for the beam went through the circular shaped collimator. Measured data is used for synogram formation, shown in Figure 2.

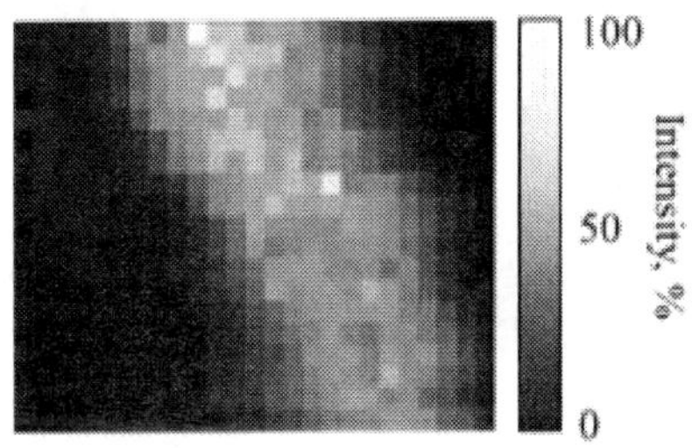

Figure 2. Scan results of the electron beam by the scintillation strip in synogram form.

Figure 3 shows electron beam shape, measured by dosimetry film and by the proposed method.

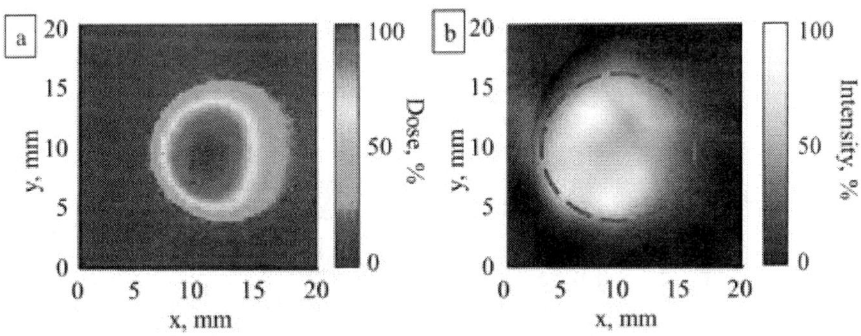

Figure 3. Results of collimated beam profile measured by: a – dosimetry film; b – scintillation strip.

As can be seen from Figure 3 the proposed method allows measurement of electron beam transverse flux density distribution. The difference between results measured by different system is caused by the resolution of the detector based on scintillator that depends on thicknesses of the strip. Besides, high quantity of the secondary photons generated in the collimator, that is not registered by the film due to low thickness. In order to avoid errors, connected with Cherenkov radiation generation in optical fiber, it is shielded with lead foil. However, it is impossible to shield it in a region of optical fiber and scintillator connection. Therefore, in a case of sensitive detector position close to Cherenkov radiation critical angle electrons can penetrate through the optical fiber and generate Cherenkov photons in there, that cause edges blurring (Figure 3*b*).

The second experiment with the deformed electron beam is also performed to estimate method applicability. Metal target, which is partly cover the beam, is placed in the beam propagation axis in the second experiment in order to deform it. The thicknesses of the target provide total absorption of the beam. Figure 4 shows registered data in synogram form.

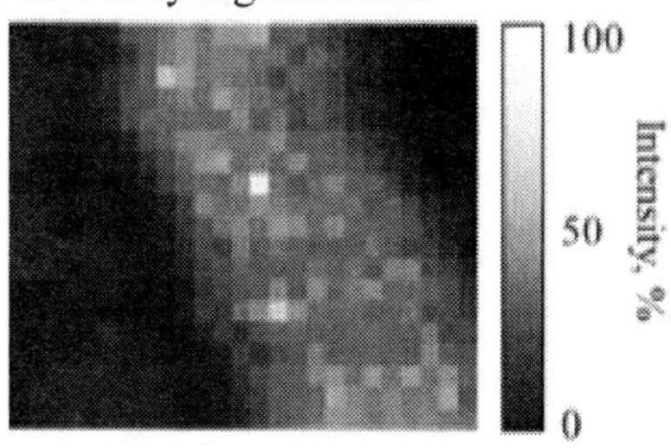

Figure 4. Deformed beam scan results registered by the scintillation strip in a synogram form.

Figure 5 shows shape of electron beam deformed by the additional metal target, registered by dosimetry film and by the investigated method.

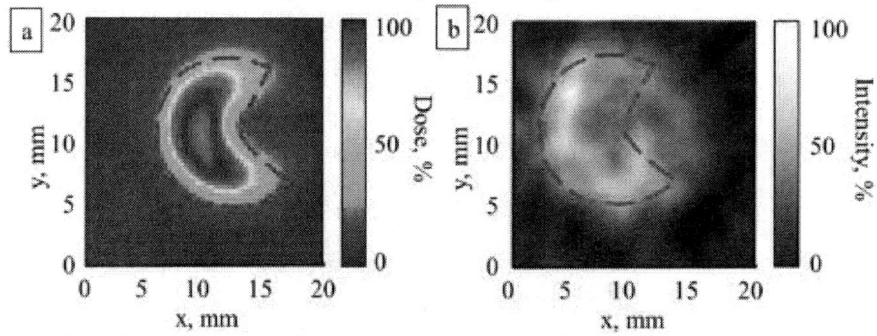

Figure 5. Deformed beam profiles measure by: a – dosimetry film; b – scintillation strip.

As can be seen in Figure 5*b* there is hot spot in a region where the beam is not covered. Besides, the sizes of the beam can be estimated with this method, however, the level of the results is not high, because of previously described errors. Besides, the secondary gamma-radiation is generated in a metal target covering the beam and provide contribution to the registered signal. Therefore, in Figure 5*b* one can see only partial decrease of the intensity after the target (about 40%) instead of total absence.

In order to estimate electron beam transverse flux density distribution accounting described factors that influence on the reconstruction results, the data is renormalized to equate 40% intensity level to 0%, while keeping the maximum at 100% level. Figure 6 shows results after renormalizing.

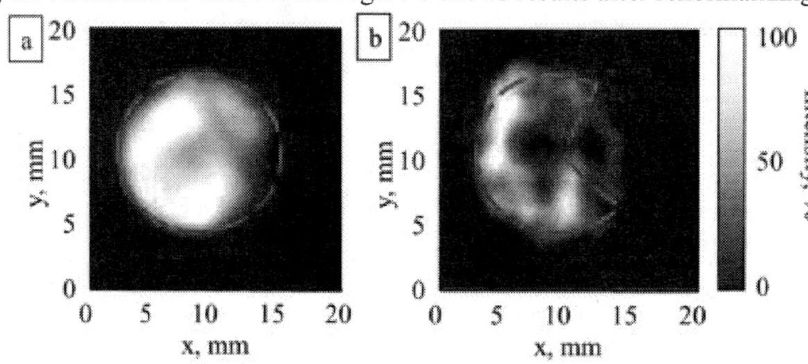

Figure 6. Beams profiles after final processing: a – beam after circular shaped collimator only; b – after circular shaped collimator and metal target.

Figure 6 demonstrates possibility of electron beam transverse profiles estimation with proposed method.

4. Conclusion

In this work, we test the method for measurement of electron beam transverse flux density distribution based on multi-angular scan. It is shown that obtained data allow us to estimate the transverse profiles of electron beams, despite of the presence of many factors that adversely affect the result of the measurement. Approach of background subtraction allows significant improvement of the results quality. Further work is aimed at method accuracy improvement by increasing the speed of the registration system, and accordingly, the statistical sample.

Acknowledgment
This work is supported by the Russian Science Foundation, project No. 19-79-10014.

References

[1] Van Elmpt W *et al* 2008 *Radiother. Oncol.* **88 (3)** 289–309
[2] Almond P R *et al* 1999 *Med Phys.* **26**(9) 1847–1870
[3] Andreo P 2000 *IAEA technical report series* **398** 242
[4] Amerio S 2004 *Med. Phys.* **31**(2) 414–420
[5] Spezi E, Angelini A L, Romani F and Ferri A 2005 *Phys.Med. Biol.* **50**(14) 3361
[6] StarTrack with OmniPro-Advance The universal QA solution. Available at: http://www.el-se.com/wp-content/uploads/2015/07/StarTrack.pdf
[7] Létourneau D *et al* 2004 *Radiother. Oncol.* **70**(2) 199–206
[8] Jursinic P A and Nelms B E A 2003 *Med. Phys.* **30**(5) 870–879
[9] Monti A F and Frigerio G 2006) *Radiother. Oncol.* **81**(1) 88–96
[10] Hesse B M, Spies L and Groh B A 1998 *Phys.Med. Biol.* **43**(12) 3607
[11] Cremers F *et al* 2004 *Med Phys.* **31**(5) 985–996
[12] Borca C *et al* 2013 *J. Appl. Clin. Med. Phys.* **14**(2) 158–171
[13] Zhu X R *et al* 2003 *Med Phys.* **30(5)** 912–919
[14] Childress N L and Rosen I I 2004 *Med Phys.* **31**(8) 2284–2288
[15] Stuchebrov S *et al* 2016 *AIP Conf. Proc.* **1772** 060016
[16] BC-400, BC-404, BC-408, BC-412, BC-416 Premium Plastic Scintillators. Available at: https://www.crystals.saint-gobain.com/sites/imdf.crystals.com/files/documents/bc400-404-408-412-416-data-sheet.pdf
[17] Sensl: J-Series High PDE and Timing Resolution, TSV Package. Available at: http://sensl.com/downloads/ds/DS-MicroJseries.pdf
[18] GAFCHROMIC EBT2. Available at: http://www.elimpex.com/new/products/radiation_therapy/Gafchromic/content/GAFCHROMIC%20EBT2.pdf
[19] EPSON Perfection V700 Photo and V750 Pro: User's Guide. Available at: https://files.support.epson.com/pdf/prv7ph/prv7phug.pdf
[20] Matlab. Available at: https://uk.mathworks.com/products/matlab.html
[21] Standa. 8SMC5-USB-Stepper & DC Motor Controller: Available at: http://www.standa.lt/products/catalog/motorised_positioners?item=525&prod=stepper-dc-motor-controller
[22] Standa. 8MR190-90-Large Motorized Rotation Stage. Available at: http://www.standa.lt/products/catalog/motorised_positioners?item=325&prod=motorized_rotation_stage
[23] Standa. 8MT177-100-Motorized Stage. Available at: http://www.standa.lt/products/catalog/motorised_positioners?item=242&prod=motorized_stage

IOP Conf. Series: Materials Science and Engineering 1019 (2021) 012044 doi:10.1088/1757-899X/1019/1/012044

Computing unite of a mobile computer vision system

I V Zoev, N G Markov and T A Yagunov
Department of Information Technologies, Tomsk Polytechnic University, Tomsk, Russia
E-mail: ivz3@tpu.ru

Abstract. The work is devoted to the computing unite of a mobile computer vision system and developing his algorithmic software. We developed hardware-implemented the convolutional neural networks on a field programmable gate array. A study of the performance and power consumption of variants of the computing unite.

1. Introduction
Nowadays, the scientific direction moving to using artificial neural networks for development of mobile intelligent computer vision systems (CVS). These CVSs should automatically analyze the images with the help of neural networks trained by modern deep-learning methods. The most promising mobile CVSs are developed using convolutional neural networks (CNN) [1].

The paper proposes various ways for organization of computations in a computing unite (CU) of a mobile CVS, which consists a hardware-implemented CNN on a modern system on a chip (SoC). The results of a study are presented for performance and power consumption of three variants of this CU for various ways of organizing computations.

2. Organization of computations in CU based on SoC

2.1. General statement on the organization of computations
A modern intellectual CVSs are systems which allow you to automate the process of analyzing visual information using artificial neural networks trained using deep learning methods. An important class of these CVSs are mobile CVSs have been installing on autonomous vehicles, for example, on unmanned aerial vehicles, on unmanned vehicles, etc. These CVSs should have low power consumption, for example, in the case of unmanned aerial vehicles not more than 10 watts. Today, mobile CVSs are developed using hardware-implemented CNN on modern SoCs, which have field programmable gate array (FPGA) for high-performance CNN computations. The following arguments shown advantages of high performance of the hardware implementation of the CNN on the FPGA:

• comparison with other implementations of the CNN, for example, on graphics processors, gives that the CU on SoC on modern FPGAs consumes significantly less power;

• flexibility, it is mean possibility to implement on the FPGA, different and even simultaneously operating several CNN architectures.

Therefore, having these advantages of the hardware implementation of the CNN on the FPGA, it is possible to development mobile and high-performance CVS based of modern SoC, which is intellectual in nature because of using CNN.

In [2], it was proposed to use the computational resources of not only the FPGA, but also other components of modern SoCs for the hardware-based CNN. The architecture of most SoCs allows you to organize direct access of a hardware-implemented CNN to external memory. The implementation of this method of interaction of the FPGA with external memory will allows you to perform some operations that differ from the convolution and subsample procedures of the CNN on the SoC processor.

Also in [2] an original method for performing computations in a hardware CNN on a FPGA was proposed. It differs from the known methods by using unified computational convolution and subsampling units. The unification of the convolution / subsample blocks is achieved by extracting a parameters of this blocks, usually specified at the stage of their synthesis, and placing them in a separate variable FPGA memory area, called the configuration memory region. Thereby it allows to use the blocks of the suitable type in the layers of the CNN with different architectural parameters. The implementation of the method assumes that the number of computing blocks involved in the hardware CNN can be variable. And the number of computing blocks is determined only by the FPGA resources. Scaling by the number of using blocks carried out both as a whole for the CNN and for individual layers. It will significantly reduce the required computational resources of the FPGA and will help implement various CNN architectures without FPGA reconfiguring.

When implementing the unification method, in addition to allocate the configuration memory area, it is necessary to organize access of the SoC processor to this memory area and to develop the appropriate software, primarily in the form of a multifunctional driver of hardware CNN.

2.2. Ways of organizing computations in the CU and their implementation

The unification of computing blocks of the CNN hardware and the way proposed in [2] (below – Way 1) of organizing computations in SoC, taking into account the interaction of the FPGA with external memory, made it possible to implement a mobile CU of CVS based on the Terasic SoCKit developing board with the SoC Cyclone V SX [3]. The enlarged functional diagram of this CU is shown in Figure 1.

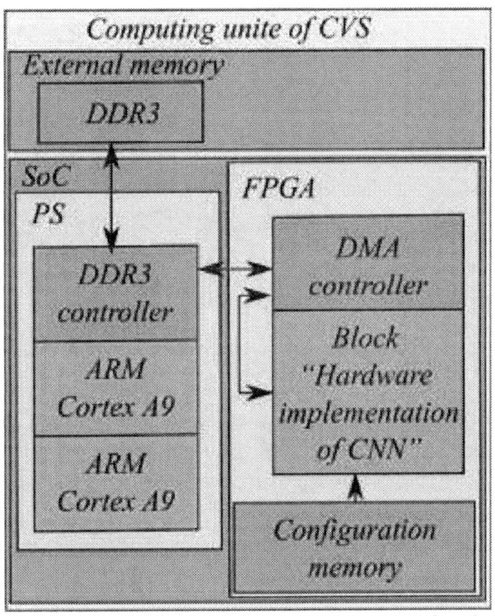

Figure 1. The enlarged functional diagram of the CU for the mobile CVS.

The diagram shows two main functional units. The SoC is the first one. It contains a processor system (PS) and FPGA. The PS in turn includes a DDR3 external memory controller and two ARM Cortex A9 processor cores. On FPGA we developed a controller of direct access to external memory (DMA controller), a block "Hardware implementation of CNN" and configuration memory. The block "Hardware implementation of CNN" consists of a neural computing unit (NCU). The NCU performs the computations of the convolution and subsampling procedures for different CNN architectures of the LeNet5 subclass. The second functional unit is external DDR3 memory.

An executing the convolution procedure for each convolutional layer of the CNN was required two types of data independent of each other. The first is the input and output data (the analyzed image and feature maps) of the CNN convolution layer. The second type is the weights of the convolution layer. This statement allows us to propose a second way (below – Way 2) of organizing computations in a CU, which includes using two channels of external memory in SoC and two different external memory modules. One of these channels will be reserved only for the transfer of the first type of data, and the second – for the transfer of data of another type. The Figure 2 shows the enlarged functional diagram of the CU with that way of implementation.

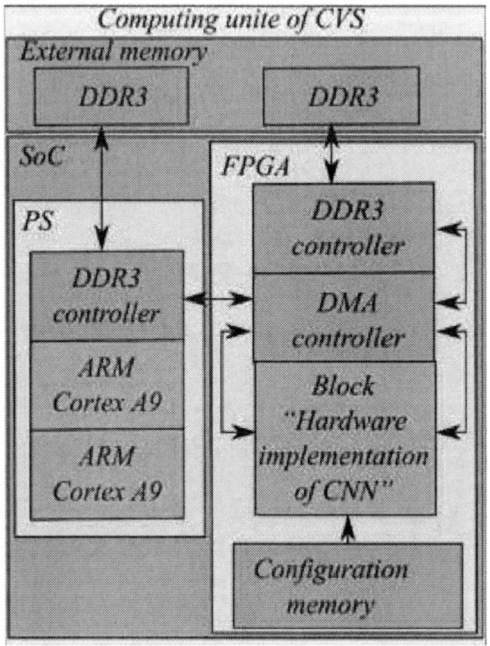

Figure 2. The enlarged functional diagram of the CU for implement Way 2 computations of organization.

The difference between the new functional diagram of the CU (new version of the CU) and the diagram of the CU in Figure 1 (the first version of the CU) is that the controller of direct access to external memory (DMA controller) has access in two physically separated memory modules. It provides data transfer between these two external memory modules and the hardware implementation of the CNN. Considering this, it make possible the simultaneous loading input data and weights into the internal buffers of the NCU module.

The best option for this variant of CU is the placement of the CNN weights in the external memory module connected directly to the FPGA through its own DMA controller. Because when perform calculating the CNN, the weights do not change, and the necessity of their unloading from the PS memory appears only during the initialization of the CU. The input and output data of the CNN layers should be stored in the PS external memory module. Because it gives the possibility of their operational pre- and post-processing.

It is possible to propose a third way (below – Way 3) of organizing computations in a CU, which includes using several modules of the NCU in the "Hardware implementation of the CNN" block. In fact, this method is an extension of the capabilities of Way 1, which have only one NCU module for use. Figure 3 shows the enlarged functional diagram of the CU (the third version of the CU) with the implementation of Way 3 of the organization of computations. This diagram shows the usage of n NCU modules.

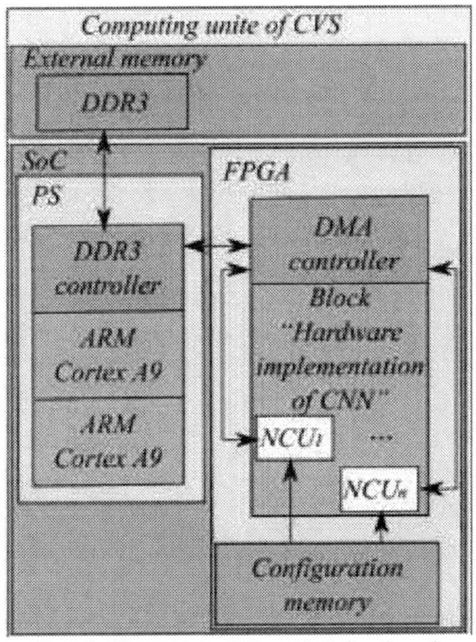

Figure 3. The enlarged functional diagram of the CU for implement Way 3 organization of computations.

The number n of NCU modules can be variable and their number depends only on the amount of computational resources of SoC. Also, the each NCU is a completely independent and self-sufficient module. The internal architecture of individual NCU in the "Hardware implementation of CNN" block also may be differ.

3. Algorithmic software of CU

We implemented the hardware components (diagrams in Figures 1–3) and also developed the corresponding algorithmic software executing on PS. The algorithmic software have difficult original algorithm of locating and marking external memory (memory mapping) for hardware-implemented CNN on FPGA with various architectures. The point of this algorithm is to reduce the amount of external memory required for CNN computations by providing direct access for NCU to external memory areas that are common to different layers of the CNN. Also we have developed the algorithms for the reconfiguration of the NCU in the FPGA.

A multifunctional software driver for the Linux kernel OS was developed for ensure interaction between the FPGA, PS and external memory. This driver was written by using C language. The main functions of the driver is following:

- functions for interaction with FPGA registers;
- functions that work with data buffers in external memory modules;
- functions that mapping the external memory;
- reconfiguration of the NCU in the FPGA;
- interrupt handling, which come from FPGA.

This driver easily adapts to the peculiarities of different the CU versions and the CNN architectures.

In C++ language, we have developed a library for the interaction of user with the CU settings. Library functions allow him to interact with the CU through the character devices registered by multifunctional driver. Each operation in this driver is initiated by a user's command transferred to the character device through a specific library function.

4. The task of studying the effectiveness of CU

In each of the three variants of CU, was implemented one of the methods of computations organization discussed above. The original CNN architecture of the LeNet5 subclass from [4] was used as a hardware-implemented CNN in CU. This CNN architecture has the following parameters. The first layer is a convolutional, the number of input feature maps 3, output - 6, the kernel size of convolution is 7×7, stride is 1. The second layer is subsample, the number of input feature maps is 6, the output – 6, kernel size of the subsample is 2×2, stride is 2. The third layer is convolutional, the number of input maps is 6, output – 32, kernel size of convolution is 5×5, stride is 1. The fourth layer is subsample, the number of input feature maps is 32, output 32, kernel size of subsample is 2×2, stride is 2. The fifth layer is convolutional, the number of input feature maps is 32, output - 100, kernel size of convolution is 5x5, stride is 1. The sixth (fully connected) layer is the convolution, the number of input feature maps 100, output - 10, kernel size of convolution is 1×1, stride is 1.

Initially, training was carried out on the software-implemented CNN of this architecture on a sample of images of hand-written numbers MNIST [5]. It contains 60,000 training images and 10,000 test pairs (tag-image). The CNN training was carried out using the Caffe library [6], used the method of back propagation of error and the method of stochastic gradient descent. Optimization parameters were taken from the library's training examples. The accuracy of the classification of handwritten numbers for the software-implemented CNN of this architecture, using the 32-bit floating-point format, is 98.7%. The weight coefficients, obtained during training, then were transferred to each of the CU variants.

In all experiments, was solved the problem of classifying each image from 10,000 MNIST test images. The classification of images was carried out using each of the three variants of the CU. For classifying images, we used configurations K1, K2, K3 and K4 of the NCU module presented in table 1. The operating frequency of the NCU in each of the configurations is 50 MHz.

Table 1. Configurations of NCU module.

Configuration Name of NCU	K1	K2	K3	K4
Number of computational convolution blocks	6	10	32	100
Number of computational subsample blocks	6	10	32	32

The performance of each CU version was estimated as the difference between the time getting the result from the SoC as a handwritten digit class and the launch time of this CU for classification. The moments of time were determined using the standard chrono library of the C++ programming language. The average time for classifying one image was calculated as the average of 10,000 times the classification of test images.

There are we also researched the energy consumption of each variants of CU and their efficiency.

5. The results of the research and their discussion

Table 2 shows the results of studying a performance for two variants of the CU that use Way 1 and Way 2 for organization of computations.

Table 2. CU performance with Way 1 and Way 2 computing organization.

Configuration Name of NCU	Runtime Way 1, ms	Runtime Way 2, ms
K1	16.464	15.487
K2	13.985	11.27
K3	6.933	6.579
K4	7.202	5.986

The results from table 2 show that in the case of the variant CU implementing Way 2, for all configurations of NCU, there is a reduction in the average time for classifying one image with comparing to the variant CU implementing Way 1. At the same time, optimal performance according to the CU performance criteria, related to the amount of computing resources, has a CU with NCU K3 configuration (32 convolution blocks and 32 subsample blocks).

The following series of experiments was carried out for a variant of CU implemented Way 3 computing organization, which containing two or three NCU modules with a K3 configuration. The maximum possible number of NCU modules is three. It is the limit amount of FPGA resources in the Cyclone V SX SoC. Table 3 shows the results of these experiments. For comparison, table 3 includes the performance of CU variant that implements Way 1. This way, like Way 3, takes into account the interaction of the FPGA with only one external memory module. In experiments when computations of computations in CU Way 3 for each analyzed image was performed synchronous launch of NCU modules.

Table 3. CU performance with Way 1 and Way 3 computing organization.

Performance variants CU	Way 1	Way 3 (2 modules NCU)	Way 3 (3 modules NCU)
Number of simultaneously classified images, pcs	1	2	3
Average image classification time on one module of the NCU, ms	6.933	7.811	12.763
Average classification time per image, ms	6.933	3.905	4.161

The best performance has the version of the CU, which implements Way 3 of the organization of computational processes with two NCU modules. So the preference should be given this variant of CU.

The measuring of power consumption in this experimental shows that the CU variants in organization of computations for Way 1 and for Way 3 is 5.1 watts. The power consumption of the variant of the CU implementing the Way 2 of the organization of computations increases to 7.4 watts. The reason of such increase covered in connection of another DDR3 external memory module.

All results obtained in the experiments indicate that taking into account modern requirements for CVSs in terms of performance and power consumption, in principle any of the three CU variants can be used as part of a mobile and high-performance CVS with low power consumption. However, the developer must choose the variant of the CU in accordance with the scale of the real-time work for the CVS.

6. Conclusion

Today, the current research area is focused on the development of mobile and high-performance CVSs using modern SoCs. This paper is devoted to the development of a CU these CVSs. In this CU the hardware is implemented on the FPGA SoC Cyclone V SX CNN allows to solve the tasks of classifying objects in images.

Two new ways for organization of computations were proposed. And different CU variants were implemented based on proposed functional diagrams. Also algorithmic software for these CUs was developed.

The task of the researching of the performance was solved for all proposed variants of the CU with the hardware implementation of the original architecture of the CNN of the LeNet5 subclass. The results of carried out research show that the best performance has a CU version that implements Way 3 of the organization of computations, and preference should be given to CU with two hardware-implemented neural computing units. Also we found that the lowest power consumption have devices that implement Way 1 and Way 3 of the organization of computations.

Acknowledgment
The reported study was funded by Russian foundation for basic research, project no. 18-47-700010 p_a.

References

[1] Russakovsky O *et al* 2015 *International J of Computer Vision* **115** (3) 211–252
[2] Beresnev A P, Zoev I V and Markov N G 2018 *Proc. of 28th Int. Conf. on Computer Graphics and Vision* 184–187
[3] SoCKit – the Development Kit for New SoC Device Terasic. Available at: http://www.terasic.com.tw/cgi-bin/page/archive.pl?CategoryNo=167&No=816
[4] Zoev I V *et al* 2017 *Computer Optics* **41** (6) 938–949
[5] LeCun Y, Cortes C and Burges C 2018 The MNIST database – MNIST handwritten digit database. Available at: http://yann.lecun.com/exdb/mnist/
[6] Caffe - Deep Learning Framework 2019. Available at: https://caffe.berkeleyvision.org/

IOP Conf. Series: Materials Science and Engineering 1019 (2021) 012045 doi:10.1088/1757-899X/1019/1/012045

Knowledge models in computer-aided manufacturing systems

A A Zakharova[1,2] **and Y V Grebenyuk** [1]
[1]Digital Technology Department, Yurga Institute of Technology, Tomsk Polytechnic University, Yurga, Russia
[2]Automated Control Systems Department, Tomsk State University of Control Systems and Radioelectronics, Tomsk, Russia
E-mail: zacharovaa@mail.ru

Abstract. Despite of vast amounts of information and considerable opportunities to process it with computer-aided manufacturing systems, personal experience and lore of experts become a role-defining category in the issues of creating and making technical decisions. This should be factored into the process of creating instruments for decision making, which are based on computer-aided manufacturing systems. Efficient decision-making models, as well as models of factors (determinants) evaluation, should include expert knowledge data. In this connection, it is of burning importance to find ways to formalize expert knowledge into a code readable by computer-aided manufacturing systems. The research paper presents three determinant (criteria) evaluation models which help make technical decisions. The models include expert knowledge data formalized by means of linguistic variables. The research paper also presents recommendations on how to opt for a proper model, including requirements, and peculiarities of expert information collection matters.

1. Introduction

Each stage of manufacturing engineering products includes decision making in structure, material, operation factors, technology and other matters [1, 2]. Very often, these specifications reveal lack of formalization, being multivariate and holding qualitative description of decision-making determinants (criteria), and their relations. In this situation decision making is carried out in the conditions of information incompleteness and ambiguity, poor simulation and prognostication of processes and events. This brings personal experience and lore of experts to the forefront, when choosing between alternatives, as soon experts are true bearers of deep and unformalized understanding of issues [3]. To retrieve, store and process expertise, special classes of Automated Information Systems (AIS) such as decision support systems (DSS) and expert systems (ES) have been recently developed. They are computer applications to Computer-Aided Manufacturing systems (CAM), and use data collected and stored in CAM. At the same time, recommendations generated by AIS and ES are then transferred to CAM, and are viewed as a background for technical solutions. AIS and ES are specific as soon as they contain decision making models and knowledge databases, which include instructions on how to create new knowledge, as well as information about human experience and knowledge in a definite subject. To develop these systems, it is necessary to create new methods which help elicit and formalize expert knowledge into a code readable by CAM.

Fuzzy sets and linguistic variables, formalisms most frequently used in expert systems [2–5], allow:

- modelling smooth shift in an analyzed determinant characteristics;
- processing qualitative value of a determinant development;

- formalizing expert knowledge in case of determinants which are multiple-valued in their measurement specifications;
- collecting quantitative value of determinants in a form of a comparable scale;
- finding and describing relations between determinants;
- utilizing obtained linguistic variables and fuzzy sets as input determinants to solve different tasks when making technical decisions.

It is worth mentioning that it is important to choose methods and plot membership functions for linguistic variables, when making decisions based on fuzzy sets. Authors usually do not ground, or give details to this aspect. Consequently, it is of burning importance to develop a set of universally valid models which allow formalizing expert knowledge in technical decision making.

2. Requirements, particular qualities, and methods used to collect and process expert information

In this paper a determinant is any value specification (technical, economic, ecological, etc) of an object, process or event in engineering production, which influences the option, decision makers do in preference to one or more alternatives, like material strength, production labor intensity etc. Alternatives are possible technical decisions like different kinds of material, production techniques etc, and are specified by a set of criteria decision makers define.

A linguistic variable of a determinant is characterized by a triple [6]:

$$<\beta, T, X>, \tag{1}$$

where β is the name of a linguistic variable; T is the collection of its linguistic values, where $T = \{T_1, T_2 ... T_s\}$, and which shows a desirable (allowed, demanded etc) level of a given determinant development and/or its importance when choosing between alternatives [7]. Each value of any linguistic variable is the name of a fuzzy set $\alpha_s, s = \overline{1,h}$, which formalizes the s-level of a determinant; X is the domain of a linguistic variable definition.

Fuzzy sets, which specify the level of a determinant development, are characterized as follows:

$$<\alpha_s, X, C_{\alpha_s}> \tag{2}$$

where α is the name of a fuzzy set;

$C_{\alpha_s} = \{\mu_{\alpha_s}(x)/x\}$ is a fuzzy set, which specifies the value of a fuzzy set α_s;

$\mu_{\alpha_s}(x)$ is a membership function of a fuzzy set C_{α_s}. Each value $x \in X$ can obtain a degree of membership of a determinant in a fuzzy set.

Membership function of a fuzzy set should allow indicating requirements, particular qualities, as well as methods which help collect and process expert information to evaluate each definite determinant, such as:

- type of a determinant (qualitative or quantitative);
- universal measurement specifications. Quantitative determinants are essentially universal unlike qualitative determinants, which are not always universal and might exist in a form of general indices (calculated via classical methods), relative ratings of objects, etc;
- source of information (an expert, any other individual engaged in decision making, or both);
- decision makers (individuals or a group of experts);
- number of experts;
- type of a scale used to evaluate determinants;
- measurement patterns, etc.

To factor listed above matters into practice, authors opted for three basic methods which help formalize expert knowledge while evaluating determinants. They are methods based on a pair wise comparison, statistical information processing and on a base of standard functions. Each method helps create definite determinant evaluation models. Expert knowledge formalization models based on standard (exponential) functions have already been characterized in details [5]. The model is due when a linguistic variable denotes a determinant with one or several specifications, which follow: there are universal measurement specification, which describe the notion; it is not necessary to give highly precise definitions to some separate meanings (values) of a determinant; experts can give some approximate evaluations of determinants levels through direct methods like (call membership degree or membership function parameters, which help determine its value, directly) [5]. The research also deals with two other determinant evaluation models.

3. Determinant evaluation model, based on pair wise comparison

Let linguistic variable $<\beta, T, X>$ describe a determinant which:

- cannot be described through universal measurement specifications, as soon as there is no any to describe the notion;
- or, it is only possible to evaluate the determiner level for a restricted number of alternatives (objects, processes, events);
- or, it might be difficult to evaluate alternatives directly, based on the level of the determinant development;
- or, it is necessary to alleviate expert's subjectivism.

Let us specify some elements of formulae (1) and (2) in this model. The image of T linguistic variable includes 3-4 terms. Domain of X linguistic variable includes comparison alternatives finite set $x_i, i=\overline{1,n}$. As a rule, the number of these alternatives does not exceed 9 [8]. A fuzzy set, which specifies a fuzzy set α_s can be shown as this $C_{\alpha_s} = \{\mu_{\alpha_s}(x_i)/x_i\}$. Herewith, each definite determiner can obtain its s-level membership degree value $\mu_{\alpha_s}(x_i)$.

1. Indices with numerical interpretation have X alternative domain with definite indicator values, typical to each enterprise. For example, it is necessary to evaluate a total production determiner. To denote the indicator we use a linguistic variable β for total production, set of basic values T = ("low", "middle", "high"), definition domain $X = [80, 100]$ (thousand rubles). Then, we can take X = {80, 90, 100, 110, 120} (thousand rubles) as comparison alternatives;

2. When it is difficult to evaluate a determinant numerically, its development in variable conditions might be evaluated (e.g., different techniques applied, materials used, segments, items of production, etc.). Let us evaluate the competitiveness of goods produced by an enterprise. In this case, analogous goods, produced by other enterprises, and available on a market, might be taken as comparison alternatives.

Pair wise comparison technique helps divide a determiner development level evaluation, applicable to some alternatives, into several less complicated steps. Investigators carry out alternative comparison by pairs. This, in its turn, helps lower subjectivism experts evince when they produce evaluation directly [9].

To plot a membership function $\mu_{\alpha_s}(x)$ for each term of a linguistic variable, it is necessary to form a matrix of alternatives $M_s = \|m_{ij_s}\|$ pair wise comparison. The matrix elements m_{ij_s} $(i, j = 1,2,..., n)$ are comparative evaluations of an investigated determinant development degree upon the given alternatives. An expert should evaluate the $x_i \in X$ alternative proximity to the notion, described via a C_{α_s} fuzzy set in comparison with an $x_j \in X$ alternative. In this regard, investigators use an interpretation scale for m_{ij} indices shown in table 1 [10]. If alternative x_i exceeds alternative x_j, index $m_{ij} \in [1,2,...9]$. If x_j element exceeds x_i element, $m_{ji} = \dfrac{1}{m_{ij}}$.

Table 1. m_{ij} indices interpretation.

Meaning	m_{ij}
$\mu(x_i)$ approximately equal $\mu(x_j)$	1
$\mu(x_i)$ a little bit bigger $\mu(x_j)$	3
$\mu(x_i)$ bigger $\mu(x_j)$	5
$\mu(x_i)$ visibly bigger $\mu(x_j)$	7
$\mu(x_i)$ much bigger $\mu(x_j)$	9
Intermediate value	2, 4, 6, 8
Reciprocal value – when $m_{ij} \neq 0$	$m_{ji} = \frac{1}{m_{ij}}$

Membership function value $\mu_{\alpha_s}(x_1), \mu_{\alpha_s}(x_2) ... \mu_{\alpha_s}(x_n)$ for $x_1, x_2, ... x_n$ is calculated through the equation $M \cdot r = v_{\max} \cdot r$, where M is a pair wise comparison matrix, $r = (r_1, r_2 ... r_n)$ is a proper vector; $v_{\max}$ is a maximal proper number M. In accordance with a simplified proper vector calculation procedure [9], values of the elements of vector $r_s = (r_1, ..., r_j ..., r_n)$ and corresponding unnormalized value of element $\mu_{\alpha_s}(x_i)_{HH}$ membership degree might be calculated through the equation (3).

$$r_{s_j} = 1 / \sum_{i=1}^{n} m_{ij_s} \; ; \; \mu_{\alpha_s}(x_i)_{HH} = 1 / \sum_{i=1}^{n} m_{ij_s} \tag{3}$$

Vector r can be used to assess consistency of an expert's opinion (pair wise comparison matrix). The equation is solved through $M \cdot r = v_{\max} \cdot r$ at this point the degree $v_{\max}$ deviates from n is used to evaluate accuracy [9]. Index and homogeneity ratio, calculated through the equation (4) are used.

$$HI = (v_{\max} - n)/(n-1); \; HR = HI / M(HI), \tag{4}$$

where HI is a homogeneity index; HR is a homogeneity ratio ($HR \leq 0{,}10$ is admissible); $M(HI)$ is a mean value of homogeneity index, defined in accordance with [9];

$$v_{\max} = \frac{1}{n} \sum_{i=1}^{n} v_i \; ;$$

n is a number of alternatives;

v_i are values obtained as a result of element-wise division of vector ρ_s elements by elements of vector r_s, e.g., $v_1 = r_{s_1} / \rho_{s_1}$;

$\rho_s = M_s \cdot r_s$ is a vector.

In case of group expertise, individual matrices of each expert's comparisons lie in a base of an aggregated matrix of a pair wise comparison. An aggregated evaluation is calculated through the following equation (5).

$$(m_{ij_s})_A = (m_{ij_s})_1^{w_1} (m_{ij_s})_2^{w_2} ... (m_{ij_s})_k^{w_k} \tag{5}$$

where $(m_{ij})_A$ is an aggregated evaluation of the element of matrix;

k is a number of individual matrices of pair wise comparisons (number of experts);

w_k is the weight of an expert, with $w_1 + w_2 + ... + w_k = 1$.

Next, it is necessary to normalize the membership function. For this purpose it is important to divide each r_{s_j} element of r_s vector (or coincident unnormalized value of element membership degree) $\mu_{\alpha_s}(x_i)_{HH}$) by its maximal value (equation 6).

$$r_{s_j}' = \frac{r_{s_j}}{r_{s_j \max}}; \tag{6}$$

The suggested model of a determinant evaluation, based on a pair wise comparison method, allows obtaining membership function values of comparison alternatives for each basic value of a linguistic variable (determinant level) within a specific domain

4. Determinant valuation model, which uses statistical information

Let a linguistic variable $< \beta, T, X >$ describe a determinant which:

- can be described through universal measurement specifications, which characterize the notion;
- to be evaluated, needs processing a big amount of information received from many respondents.

Let us specify some elements of equation (1) and (2) of the model. The image of a T linguistic variable includes 3–4 terms. The domain of an X linguistic variable is divided into l number of intervals of equal $j = \overline{1, l}$ length. A fuzzy set, which describes α_s values, can be described as $C_{\alpha_s} = \{\mu_{\alpha_s}(x_j)/x_j\}$; where x_j is a subset X which enters a j interval; $\mu_{\alpha_s}(x_j)$ is a C_{α_s} fuzzy set membership function. Each $x_j \in X$ interval can obtain its s-level membership degree value.

Polled data are presented in a form of an empirical table, where lines are values (terms) of a linguistic variable, columns are intervals of a linguistic variable domain. Cells contain polled data –n number of respondents' replies where they used some definite value of a linguistic variable (b_{sj}) to the definite determinant value interval (table 2).

Table 2. Statistical observation data.

Linguistic variable value	**Linguistic variable domain intervals**					
	x_1	x_2	...	x_j	...	x_l
α_1	b_{11}	b_{12}	...	b_{1j}	...	b_{1l}
...	...	...	...	...	...	...
α_s	b_{s1}	b_{s2}	...	b_{sj}	...	b_{sl}
...	...	...	...	...	...	...
α_h	b_{h1}	b_{h2}	...	b_{hj}	...	b_{hl}

To find membership degree of each linguistic variable value interval, number of replies, given by respondents who used the setting linguistic variable value in the ratio of the determinant value interval, should be divided by the maximal number of replies. It is worth noticing that the number of observations within each interval can vary. That is why polled data need further processing.

Membership function can be determined through the equation (7).

$$\mu_{sj} = c_{sj} / c_{s\max}, \tag{7}$$

where μ_{sj} is the membership function value of an s-term of a linguistic variable within a j interval;

c_{sj} are transformed elements of b_{sj} from the Statistical Observation Data Table, given above;

$c_{s\max}$ is the maximal c_{sj} element if j is taken.

b_{sj} elements transformation is carried out through the equation (8):

$$c_{sj} = \frac{b_{sj} k_{\max}}{k_j}, \tag{8}$$

where $k_{\max} = \max k_j$;

k_j are the prompts matrix elements, calculated to smooth the function through the equation (9).

$$k_j = \sum_{s=1}^{h} b_{sj}, \tag{9}$$

If $k_j = 0$, transformation of elements from the j-column is carried out through the approximation by linearization (equation 10).

$$c_{sj} = \frac{c_{s(j-1)} + c_{s(j+1)}}{2}. \tag{10}$$

5. Recommendations on how to choose a determinant evaluation

Determinant evaluation models described in the paper play two important roles:

1. First of all, determinant evaluation models are efficient instruments which help formalize an expert's perception of a desirable (available, demanded, etc) level of a determinant development. The models allow collecting linguistic evaluations, as well as clear and accurate quantitative analyses of determinants with prescribed initial values. They also allow setting smooth shifts in determinants values membership intensity, and formalize experts' confidence in this or that value of a determinant.
2. Secondly, determinant evaluation models are setting determinants for other decision-making models. Membership function values are considered to be initial when evaluating alternatives for technical decision through multi criteria filter.

The research proceeds with recommendations on how to choose the most suitable and efficient model which help evaluate determinants (table 3).

6. Summary

The paper presents a final kit of models which help plot the membership functions for linguistic variables terms, which allow creating new expert knowledge-based models of evaluation determinants that influence the choice between alternatives when making technical decisions. All the possible requirements, methods of collecting and processing expert information, which help evaluate determinants, have been observed. Elaborated recommendations on how to choose a determinant evaluation model allow choosing the model, which can formalize expert knowledge in any type of a determinant into a code readable by computer-aided manufacturing systems.

Table 3. Recommendations to choose the model.

Determinant specification (determinant evaluation process specification)	**Determinant evaluation model (three methods)**		
	Pair wise comparison	Based on statistical information	Based on standard functions
Source of information, which is used to evaluate determinants	Experts	Polled data/ Experts	Experts
Recommended number of experts-group participants	No more than 9	9 and more	Unrestricted
Type of a determinant	Qualitative/ Quantitative	Quantitative	Quantitative
Universal measurement specifications of a determinant	No/Yes	Yes	Yes
Number of comparison alternatives (intervals)	No more than 9	10 and more	Not used
Required degree of detailed elaboration and accuracy when describing a determinant	High	High and medium	Medium and low

Acknowledgments
The reported study was funded by RFBR according to the research project № 19-013-00486A.

References

[1] Szafranko E 2018 *Proceedings of Materials Today* 2002–2009
[2] Gorbatova E, Emelianenko E, Zaretckii M 2017 *IOP Conference Series: Earth and Environmental Science* **87** 072001 doi:10.1088/1755-1315/87/7/072001
[3] Malyshev N, Berstein L, Bozhenuk 1998 *Proceedings of Int. Conf. Fuzzy Sets in Informatics* 44
[4] Yadav O, Singh Chinnam N, Goel P 2003 *Reliability Engineering and System Safety* **80** 63–74
[5] Zakharova A, Ostanin V 2015 *IOP Conf. Ser.: Mater. Sci. Eng.* **91** 012070 doi:10.1088/1757-899X/91/1/012070
[6] Zadeh L 1965 *Proceedings of Inf. Control* **8** 338–353
[7] Zadeh L 1978 *Fuzzy Sets and Systems* **1** 3–28
[8] Saaty T, Ozdemir M 2003 *Mathematical and Computer Modelling* **38** 233–244
[9] Saaty T 2002 *Scientia Iranica* **9** 215–229
[10] Saaty T 1978 *Fuzzy Sets and Systems* **1** 57–68

Intelligent analysis of digital economy competencies in agriculture labor

A A Aletdinova, M A Bakaev and V A Astapchuk
Automated Control Systems department, Novosibirsk State Technical University, Novosibirsk, Russia
E-mail: aletdinova@corp.nstu.ru

Abstract. As information becomes the main resource and production factor, there is widely declared need for digital competencies and skills in employees. This demand is uneven between industries and regions and is often biased due to companies' and organizations' desire to "keep up with the times". In our paper we focus on employees' digital economy competencies in agriculture, which is traditionally considered a rather low-tech sector, and compare them to several other industries with varying degrees of informatization. To negate the possible biases, in our analysis we rely on online data – vacancy ads published by companies at job-related portals – instead of traditional statistics. The online data was collected by our dedicated labor market monitoring intelligent system that has been in operation since 2011. The digital economy competencies were divided into 6 types: 1) intellectual and educational, 2) communicational, 3) motivational, 4) managerial, 5) ICT, and 6) ecological. We further use the groups and the wordings of the competences in online job-related ads to fill-in the "competencies specifications" thesaurus. Finally, we calculate frequencies for the digital economy competencies in vacancies for the different industries. We found that in agriculture the most demanded were the intellectual and educational, managerial and ICT-related ones.

1. Introduction

As the developed countries undergo transition towards digital economy, the demand for the related competences in workers is increasing in the labor market. There is now a term "critical competences" in use in the USA, similarly to the "critical materials" and "critical technologies" [1–3]. In our work, however, we employ a wider concept of digital economy-related competences (DECs), which correspond to the set of knowledge and skills most needed in the coming shift of the technological paradigm. Indeed, many researchers believe that "digital literacy" is the most essential competence in the XXI century [4, 5], while others register the corresponding synthesis of science, arts, engineering and design [6].

The American Occupational Therapy Association (AOTA) has developed standards for the "continuing competence", which include such groups as knowledge, performance skills, interpersonal abilities, critical reasoning and ethical reasoning skills [7]. In [8], the authors identify computer literacy, media literacy, and media education as the key competences for the digital economy – and indeed, most research works related to DECs focus on ICT. However, we believe (somehow similarly e.g. to [4] who adds critical thinking to the set) that the set of skills essential for the new economy needs to be extended, based on the actual requirements by employers. Indeed, the World Economic Forum stated that by 2020 such competences should include complex problem-solving, critical thinking, creativity, managing and coordinating people, service orientation, evaluation and decision-making, negotiations, and cognitive flexibility. The Boston Consulting Group, based on their survey, identified such groups of DECs as cognitive, social-behavioural, and digital skills [9].

Understanding which competences will become truly critical should be based on expert evaluations, surveys, and monitoring and analysis of the actual labor market data. The latter approach has advantages in its natural capability to both register demand for DEC at a particular moment of time, and to study its changes with relation to the economic situation and the market conditions, even in retrospect. Since 2011, we have been collecting online data on job offers and ads posted by job-seekers (who mostly put online their resumes) for several regions of Russia. For this end, we employed the dedicated intelligent labor market analysis system that we developed and maintained [10]. The online data sources are popular Russian job-related websites, including the renowned HeadHunter.ru (http://headhunter.ru), which nowadays offers API for interested external parties and robust capabilities for analysis. The software's database currently has more than 10 million unique records, structured per time periods, regions, industries and professions, etc. Previously, our analysis of the collected data was mostly focused on proposed and required wages and their relations. However, the data on job offers that we collect in the intelligent system also includes descriptions of job functions and skills, which call for semantic analysis with respect to the requirements towards DECs.

So, in our current work, we study the online data that is being collected by our software with respect to employers' demand for DECs in such digitally "conservative" industry as Agriculture. In Section 2, we outline some particulars of online data mining, based on the experience from our labor market intelligent system project. We also list the DECs that were refined for the thesaurus subsequently used for the analysis of larger sample of job offers, collected by our software system. In Section 3, we present the results of the analysis of demand for DECs per different industries, with particular focus on Agriculture. In Conclusions we summarize our findings and outline perspectives for further research.

2. Methods and tools

2.1. Particulars of labor market online data mining.
The collection of online data can be performed through two principal approaches: either via API provided by the data source website (if available), or through web scraping of unstructured data from web pages. In the latter case, generally a virtual browser sends request to the website and returns the web page HTML code for processing.

For the web scraping-based collection, the challenges are changes frequently introduced in the source website's web pages structure, and the web technologies that complicate the pure HTML-based data (such as AJAX, Web Components, etc.).

The necessity to monitor the rapidly changing labor market, with relatively short live span of an online job offer, calls for at least daily data collection. However, there's the need to deal with repeated posting of the ads, which can be slightly modified by the owners.

The data gathered from online sources is often incomplete, both due to problems in its collection and to fundamental omissions. E.g. wages in job offers are not always stated, and if they are, only one border of the range is often supplied.

The target content, job-related ads and resumes, has weak formalism and possible spelling mistakes, which further complicates the data processing. Special methods of syntax and semantic analysis of the textual data need to be employed, often with increased computational costs.

This consideration together with generally large volumes of the collected data imposes requirements upon the storage system, involving not just relational databases, but also the modern NoSQL approaches.

Thesaurus is essential for structuring of collected data, so its population is the pre-requisite for any semantic data mining. With respect to our goals, it must include the concepts related to DECs, de-facto representing the conceptual model of the field involving job functions and requirements. The initial population of the thesaurus necessitates both study of literature and the state-of-art data sources, and we describe the results of the respective analysis in the next sub-chapter.

2.2. Groups of digital economy competences.
From the analysis of existing sources [11–13] and based on the actual job offers published at HeadHunter.ru, we identified several major groups of competences related to digital economy:

1. **Intellectual and education-related DECs**, such as: Ability and readiness for life-long learning, Disposition towards improvement, Willingness to retrain, Professional mobility, Striving for new things, Critical thinking ability, Creativity, Self-development, self-improvement, Tendency towards non-standard solutions, Savviness, Resourcefulness, Systems thinking, Artistic creativity skills.
2. **Communicational DECs**, such as: Multi-language and multi-cultural understanding, Interdisciplinary communication, Customer focus.
3. **Motivational and will-related DECs**, such as: Disposition towards reasonable risk, Enterprise, Self-dependence, Self-realization, Striving for success, Self-respect.
4. **Managerial DECs**, such as: Project Management, Working with people, Willingness to work in a team and in a highly competitive environment, Working in conditions of uncertainty, Willingness to work both independently and in a group, Organized, Emotional intelligence.
5. **ICT-related DECs**, such as: Mastery of office programs, Data warehouses, Cybersecurity, Global data transmission networks and communication channels, Technical writing, Web development, Programming fundamentals, System administration and network configuration.
6. **Ecology-related DECs**, such as: Knowledge of requirements for waste disposal, Pollution control, Compliance with industrial hygiene, Safety and labor protection, Participation in resource saving.
7. Correspondingly, our thesaurus was populated with the above DECs and was used in the analysis of their frequencies in the published job offers.

3. Results

3.1. The establishing analysis (2018).
The employers' demand for the DEC competencies in employees is specific per industries, so in our analysis sample we chose 5 diverse ones: 1) Agriculture, 2) ICT, 3) Health care and pharmaceuticals, 4) Mining, and 5) Financial sector. To characterize the demand, we measured "frequencies" – the share of job offers that mentioned at least one DEC from our thesaurus. The resulting frequencies extracted from the job offers we selected from HeadHunter.ru for December 2018 (106342 in total), are presented in table 1.

Table 1. Job offers and DECs frequencies per the selected industries (2018).

Industry	Total job offers in the sample	Share of job offers with DECs, %
Agriculture	4333	68.2
ICT	44358	93.6
Health care and pharmaceuticals	19682	85.7
Mining	4321	51.6
Financial sector	33648	88.9
Total/avg.	106342	77.6

Since all the frequencies are above the 50%, it suggests that employers seek for employees with not just education and experience, as it was the case in "traditional" economy, but with DECs. The demand varies per industries, and our more detailed analysis discovered the following (Figure 1):

1. The intellectual and education-related DECs are in most demand in Agriculture and Health care and pharmaceuticals.

2. The communicational DECs are most asked for in Health care and pharmaceuticals.
3. The motivational DECs are most needed in Mining industry.
4. The managerial DECs are in most demand in Agriculture.
5. The ICT DECs are ubiquitous, and they are in nearly equally high demand in Financial sector, Mining, ICT, and Agriculture.
6. The demand for the ecology-related DECs are relatively low and was only found in Agriculture and Health care and pharmaceuticals, which suggests that this group is underestimated.

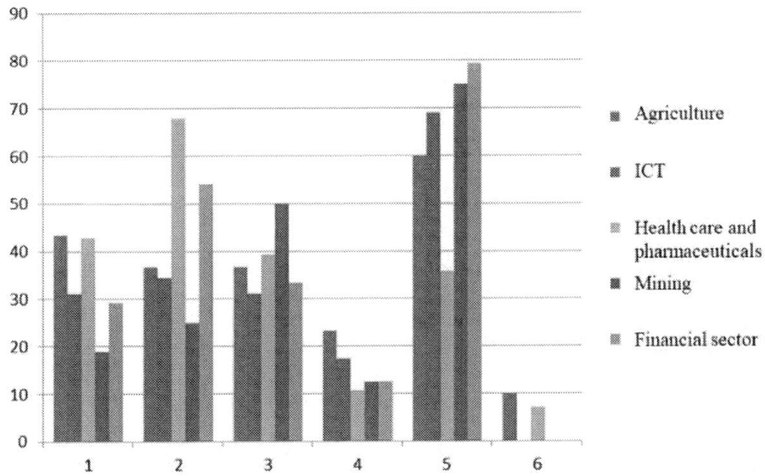

Figure 1. Comparative analysis of the demand for DECs per industries and the groups (1–6).

Among the considered industries, Agriculture is second to Mining in lowest demand for DECs (Mining actually had the least detailed job offers, 1/3 of them didn't even have any requirements towards education or work experience). However, at the same time more than 2/3 of job offers in Agriculture require at least one DEC, and the proposed wages are not particularly low: e.g. 15000–140000 Rub for tractor drivers, 17000–45000 Rub for milking machine operators, 13000–250000 Rub for zoo technician, 12000–250000 Rub for agronomists (1 USD equaling approximately 65 Rub). Another particular feature is relatively fair demand for ecology-related DECs (in 10.02 % of the considered job offers), while for ICT, Mining or Financial sector this group is absent whatsoever.

3.2. The Additional Analysis (2019)

The vacancies data was repeatedly analyzed in May 2019, i.e. nearly 6 months after the first analysis. For each of the 5 selected industries, 100 job offers were randomly selected from HeadHunter.ru, with the scope set as all Russia (without specification of any particular regions). The work was performed by student workers who were given instructions replicating the establishing analysis. The results of the sampling analysis and the comparison with the 2018 are presented in table 2.

Table 2. Job offers and DECs frequencies per the selected industries (2019)

Industry	Total job offers in search results	Share of job offers with DECs, %	Change vs. the 2018 data, %
Agriculture	5712	81	+18.8
ICT	60298	84	-10.3
Health care and pharmaceuticals	22083	74	-13.7
Mining	5384	55	+6.6
Financial sector	33494	85	-4.4
Total/avg.	**126971**	**75.8**	**-2.3**

IOP Conf. Series: Materials Science and Engineering 1019 (2021) 012046 doi:10.1088/1757-899X/1019/1/012046

As the processed job offers were processed, the encountered DECs, as worded by the job posters (companies' representatives) were included in the thesaurus that we are developing for further automation of the analysis. Each of the encountered *DEC formulations* was linked to a Core DEC belonging to ones listed in Section 2 (see *B. Groups of Digital Economy Competences*). In total, we selected 654 formulations – there was virtually no aggregation, since the goal was to have as many different workings as possible, for the sake of completeness of the planned automated analysis.

The thesaurus was implemented using CMS Drupal's Content Construction Kit functionality. Examples of the Core DECs and of DEC formulations (for the popular Web Development competence) are presented in Figures 2 and 3 respectively (in Russian).

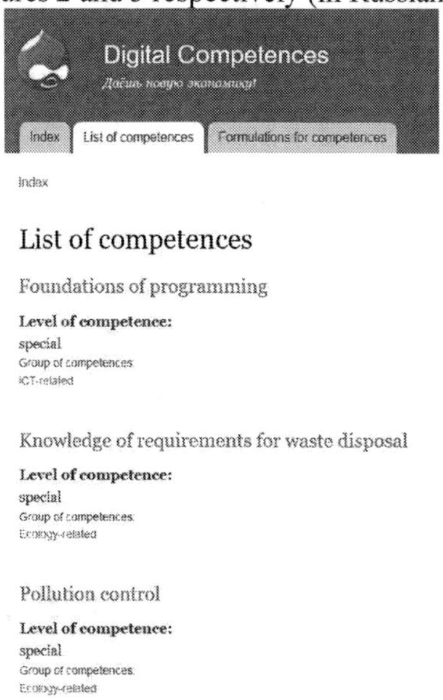

Figure 2. Core DECs composed according to the identified groups.

The current version of the thesaurus is available at http://nstu.khvorostov.ru/testcorolla/.

Figure 3. DEC formulations extracted from job offers (for the Web Development competence).

4. Conclusions

The job-related websites providing online data can be used as indexes of labor market condition and the economic situation in general [11]. Particularly, in our work we used data collected from the popular HeadHunter.ru website to study the demand for DECs in several different industries, with particular focus on Agriculture. We devised the groups for the DECs, with the corresponding wordings of particular competences as specified in research works and found in the actual job-related postings. The thesaurus was used for semantic analysis of the large sample of data – more than 100 thousands of records for December 2018. After about 6 months we performed additional manual analysis of the data with respect to the DECs, using the same algorithm.

We found generally high demand for DECs (measured as share of job offers mentioning at least one DEC), as they were required in more than 50% of employers' ads for all the considered industries. As one would expect, ICT and Financial sectors were the leaders, although in 2019 the former somehow lost the lead. The lowest demand was discovered for the Mining industry (51.61% in 2018 and 55% in 2019), while Agriculture considerably improved (from 68.18% in 2018 to already 81% in 2019). At the same time we should note that the two trailing industries gained the most during the 6-months period, while all the other industries lost in DEC share. It may suggest that there is the process of "regression to the mean": after a while the degree of demand for DEC may become independent of a particular industry. It is likely that each employee would be expected to be in possession of up-to-date digital skills, just like in industrial society it has become customary to be able to read and write.

Further, we found that in Agriculture industry the employers did ask for DECs from the Intellectual, Managerial, ICT, and even the underestimated Ecological groups. Thus, even such conservative industry as Agriculture nowadays pushes employees towards self-development and self-learning, mastering systematic thinking, communication skills, project management, etc. Particularly impressive are the requirements towards computer literacy and computer professional skills – from competence in office software to network administration, programming, information systems development, and so on.

Our results suggest that the Agricultural industry is not left outside of the digital transformation and that the demand for DEC-skilled workers is becoming critical. Clearly, hiring of such employees can be possible only if the social infrastructure and the working conditions in rural areas improve considerably. We also expect that the currently underrated Ecology-related DECs may become crucial competences for the Agriculture in the near future, as the need for intensive and sustainable development of farming and animal husbandry becomes more prominent.

Our further research plans involve quantitative analysis of effects of particular DECs on wages in different industries and regions – e.g. how much more or possibly less money an employee possessing Technical writing competence would make in Agriculture. The millions of data records that have been accumulated in our labor market monitoring and analysis software system over the years should allow rather reliable estimations. They could be subsequently used by specialists in their career planning, as well as by HR specialists in planning job offers and seeking personnel with competencies that are critical for their companies' businesses. Finally, educational institutions would be able to devise their study plans according to the current and prospective needs of the modernizing digital economy.

Acknowledgment

The reported study was funded by RFBR according to the research project No. 17-32-01087 a2 OGN.

We would also like to thank the Bachelor and Master students of NSTU (Applied Informatics major) who contributed to the data collection: Daria Savchuk, Kirill Pyzhlakov, and Olga Tarsukova.

References

[1] Sousa M J 2014 *International Journal of Economics, Commerce and Management* **II**(9) 1–10
[2] Ala-Mutka K, Punie Y and Redecker C 2008 *Digital competence for lifelong learning* (Luxemburg: Office for Official Publications of the European Communities) pp 271–282
[3] Gallardo-Echenique E E, de Oliveira J M, Marqués-Molias L 2015 *MERLOT Journal of Online*

Learning and Teaching **11 (1)** 1–16
[4] Passey D, Shonfeld M, Appleby L 2018 *Technology, Knowledge and Learning* **23 (3)** 425–439
[5] Hatlevik O E, Guðmundsdóttir G B and Loi M 2015 *Computers & Education* **81** 345–353
[6] Walker I, Chan D K G, Nagami M and Bourguignon C (Eds) 2018 *New Perspectives on the Development of Communicative and Related Competence in Foreign Language Education* (Berlin:Walter de Gruyter GmbH & Co KG)
[7] Myers C T 2019 *American Journal of Occupational Therapy* **73**(2) 8
[8] Normore A T, Brooks J S and Silva SA 2016 Global Core Leadership Competencies: A Response to "Institutional Culture" and (In)-Competence in Higher Education *The dark side of leadership: Identifying and overcoming unethical practice in organizations* (UK:Emerald Group Publishing Limited) pp 155–176
[9] Russia 2025: from human resources to talents: the Boston Consulting Group 2017 report. Available at:http://image-src.bcg.com/Images/Skills_Outline_web_tcm27-175469.pdf
[10] Bakaev M, Avdeenko T 2016 *Lecture Notes in Computer Science* **9714** 15–23
[11] Louvet J P 2000 *Les principaux résultats de l'étude «Technologies clés 2005* (Ministère de l'Économie, des Finances et de l'Industrie)
[12] Alam K, Erdiaw-Kwasie M O, Shahiduzzaman M 2018 *Journal of rural studies* **60** 60–69
[13] Voogt J, Erstad O, Dede C and Mishra P 2013 *Journal of computer assisted learning* **29** 403–413

Crowd intelligence methods for identifying corruption-prone legislative acts

D Rakhvalova[1], G Kurcheeva[2], M Rakhvalova[3] and M Bakaev[2]
[1] Dept. of Audit, Accounting and Finance, Novosibirsk State Technical University, Novosibirsk, Russia
[2] Dept. of Automated Control Systems, Novosibirsk State Technical University, Novosibirsk, Russia
[3] Dept. of Civil Law, Novosibirsk military Institute of National Guard Troops, Novosibirsk, Russia
E-mail: bakaev@corp.nstu.ru

Abstract. The article is dedicated to the problems in automation of anti-corruption expertise of legislative acts, which gets increased attention within today's Computational Law field. We carried out the analysis of the corruption-related factors frequency and outlined approaches for assessing the risks involved with their presence in a legal document. The frequency was classified per the identified attributes of regulatory legal acts: the year of adoption, the social field of the legal regulation, the level. The analysis of legal documents was carried out using the "crowd-intelligence" approach – performing relatively elementary tasks of annotating data for machine learning by people who are not expert lawyers. The results of our work can contribute to improving the effectiveness of anti-corruption expertise and support decision-making by the experts in the field.

1. Introduction

In the *Strategy of Scientific and Technological Development of the Russian Federation* [1], top priority is given to the development of Computational Law. Its viable component is anti-corruption inspection (ACI) of existing and drafted legal acts, which is to be carried out on federal, regional and municipal levels. Particular attention is to be paid to legal acts that relate to human rights, freedoms and duties of citizens, legal statuses of organizations, etc. [2]. Several hundred of only Federal Laws were enacted in Russian Federation in 2018, not even counting the by-laws, so the scope of the ACI is truly remarkable.

Information available for the Ministry of Justice (table 1) [3] and other government bodies (table 2) [4] in the region of Novosibirsk, Russia, suggests that the number of discovered corruption-prone factors (CPFs) has halved during the last reported years. One might wonder if this means that the legal act drafts are improving, or that the share of the undiscovered factors grows.

Although no mathematical model or software tool can completely replace a human expert with adequate qualification, the ever-growing volume of legislation calls for automation and support of the expert in ACI [5]. In our research work we perform preliminary study of Artificial Intelligence methods' feasibility in computerization of ACI. Particularly, in Section 2 we investigate different types of CPFs with regard to feasibility of their automated detection, while in Section 3 we use crowd workers to extract meta-data in several dimensions and provide related assessment of CFP risks.

Table 1. ACI for regional legislation acts and drafts (the Ministry of Justice for the Novosibirsk region) [3].

Year	Number of legal acts in ACI (by the Ministry of Justice)	Number of CPFs discovered
2013	3294	511
2014	5354	345
2015	3276	355
2016	2878	298
2017	3215	252
2018	3291	358

Table 2. ACI for regional legislation acts and drafts (the Novosibirsk region government) [4].

Year	The total number of regional legal acts drafts	Number of legal acts in ACI (by the regional Government)	Number of CPFs discovered
2013	No data	3294	511
2014	5918	5354	345
2015	6207	3276	355
2016	4951	2878	298
2017	4929	3215	252

2. Methods and approaches towards automation

2.1. The ACI Methodology

The significance of ACI integration into the Russian legal system is highlighted by the registry of independent experts who got accreditation to carry out ACI. As of fall 2018, the registry includes nearly 2500 persons and 400 companies [6]. However, challenges in ACI remain and surge, due to:

1. The work effort per each expert and company keeps increasing.
2. The need for careful selection of experts for every legal act, so that the level of related qualification is warranted.
3. Poor availability of legal acts drafts and the previous ACI results. They often lack structuring or miss the final legal opinions, as of 2019[1].
4. Ambiguity in formulations for some of the CPFs and lack of consistency in approaches from law enforcement bodies.

To illustrate the latter point, we composed a sample of 350 legal acts and reflected on the approaches and legal decisions (table 3).

Meanwhile, the Constitutional Court of the Russian Federation has repeatedly highlighted the need for certainty, clarity and unambiguity of legislative regulation, "which can be ensured only through uniform understanding and interpretation of legal norms by all law enforcers" (e.g. in [7]). To ensure at least some consistency in the domain, we propose relying on ACI methodology specially devised by the Russian Ministry of Justice [8]. The CPFs in the ACI methodology are formulations in legal acts that establish ungrounded wide limits or the rights to make exceptions from common rules. Further we

[1] See e.g. https://novo-sibirsk.ru/dep/expertise/docs/ or
http://mec.omskportal.ru/ru/RegionalPublicAuthorities/executivelist/MEC/corruption.html

explore the feasibility of automating the ACI with respect to automated discovery of the identified CPF.

Table 3. Ambiguity in approaches towards CPFs in legal acts.

Discovery / permission of CPF	Number of legal acts
ACI discovered CPF(s); the court recognized the legal act as inappropriate.	0[a]
ACI discovered CPF(s); the court refused to recognize the legal act as inappropriate.	9 (2.57%)
ACI didn't discover CPF(s); the court recognized the legal act as inappropriate.	14 (4%)
ACI didn't discover CPF(s); the court refused to recognize the legal act as inappropriate.	169 (49.28%)
No data on ACI available; the court recognized the legal act as inappropriate.	8 (6.96%)
No data on ACI available; the court refused to recognize the legal act as inappropriate.	62 (17.71%)
ACI wasn't performed; the court recognized the legal act as inappropriate.	17 (4.85%)
ACI wasn't performed; the court recognized the legal act as inappropriate.	52 (14.85%)

[a] Which makes one wonder why bother to perform ACI at all.

2.2. The ACI Automation

Full translation of legal acts into machine readable form seem problematic, but the combination of partial semantic interpretation and statistical approach that has proved itself very worthy in computer linguistics should be feasible in identification of at least some of the CPFs [9]. Further, we should enquire if these discoverable factors are frequent enough in legal acts.

For instance, identifying lack of clearly stated timeframes for a government or municipal service – probably the most frequent breach – doesn't require deep analysis of a legal act text. Another popular ambiguity is "entitled" formulation for actions that can be performed by a government or municipal body towards citizens and companies. All similar formulations, such as "can", "defines", "sets", "decides", "makes decision", "authorized", "have the authority", "have the right" are explicitly noted in the ACI Methodology as corruption-prone "discretionary power" factor [8]. Thus, it can be reasonably well identified using linguistic analysis methods [10], particularly gramathematic and morphologic analyses [11]. The sub-factors and the corresponding linguistic approaches are outlined in table 4.

Another frequent CPF is "legal language ambiguity" – the use of not well-established, uncertain terms and categories of evaluative nature. In some cases this may be unavoidable and should not be considered a CPF: e.g. civil law contains plenty of dispositive norms, such as "reasonable", "commensurate" and so on. In the long terms, however, legal language ambiguity is disadvantageous for legislation, as the number and complexity of legal acts increase. So, consistency in using all kinds of terms, including the ones from external fields, is highly preferable, even without relation to corruption. The approaches that we propose for identification of the "legal language ambiguity" factor is based on linguistic analysis methods [10] and are outlined in table 5.

Table 4. Approaches for identification of the "Discretionary Power" CPF.

Sub-factor	Approach for identification
Lack or ambiguity of timeframe for making a decision	Search for: "immediately", "reasonable period", "recommended period", "period established/stipulated in legislation", "period may be extended/postponed/changed", "within the shortest time", "the respected/corresponding period".
Lack or ambiguity of conditions for making a decision	Search for: "if good reasons exist", "sufficient conditions".
Lack or ambiguity of justification for making a decision	Search for: "the decision can be made", "the decision is made in the cases stipulated in the law", "sufficient grounds", "at its discretion", "entitled", "authorized".
Duplicate authority in responsible bodies or decision-makers	Search for matches between the analyzed text and the other legal acts.

Table 5. Approaches for identification of the "legal language ambiguity" CPF.

Sub-factor	Approach for identification
Using not well-established terms Using of uncertain terms	Search for: • Verbs with enantiosemic prefixes (e.g. re-elect can mean both "elect again" and "elect someone else". In Russia, such prefixes are "pere-" and "re-". • Nouns with enantiosemic prefixes (e.g. "re-statement" can mean both stating something again or making alternations to something). • Nouns, for which legal definitions of terms exist ("… in essence is…", "… should be understood as …", "… means …").
Using categories of evaluative nature	Search for: "the decision can be made", "the decision is made in the cases stipulated in the law", "sufficient grounds", "at its discretion", "entitled", "authorized".

Having considered different types of CPFs with respect to the automated text analysis feasibility, we subsequently summarize our thoughts, per the factors listed in the ACI Methodology:

a) **Discretionary power.** Automated search for textual constructs that identify lacking or ambiguous terms, conditions or justifications for decisions, as well as duplicate authorities at local or federal levels, as we detailed above. Detection of a construct (irrespective of their frequency in the text) implies increased risk of CRF presence, so expert's attention should be drawn to it.

b) **Authority based on "entitled" formulation.** Automated search for textual constructs that identify dispositive establishment of authority for local or federal bodies to perform actions in regard to citizens and companies. Again, expert's involvement is needed for the final assessment of the CPF risks.

c) **Selective changes in the amount of rights.** Seemingly no automation is possible.

d) **Excessive freedom of bylaw rulemaking.** Seemingly no automation is possible.

e) **Adoption of regulatory legal acts outside of one's authority.** Seemingly no automation is possible.

f) **Filling legislative gaps with the help of by-laws in the absence of a legislative delegation of relevant powers.** Seemingly no automation is possible.

g) **Absence or incompleteness of administrative procedures.** Automated analysis is problematic, since it's hard to find common regularities in the diverse administrative procedures, whose content is not consistent. Still, the analysis could be feasible for the most standardized regulation documents, such as provision of state services.

h) **Refusal of due competitive (auction) procedures.** Seemingly no automation is possible.

i) **Regulatory collisions.** Automated analysis is problematic, as involvement of expert knowledge is required. Conflict in law is a phenomenon when two or more norms contain different rules of behavior applied in the same conditions. However, opposing formulations do not necessarily use similar language constructs, so a human's involvement would be generally required.

j) **Overstated requirements towards a person to exercise the belonging rights.** Seemingly no automation is possible.

k) **Abuse of the applicant's right by state bodies, local self-government bodies or organizations (their officials).** Seemingly no automation is possible.

l) **Legal language ambiguity.** Automated search for uncertain, ambiguous terms and categories of evaluation character (as we detailed above). It would need to rely on pre-defined thesauri of legal definitions and evaluative terms. A CPF risk would emerge, e.g., if the legal act text contains a definition of a term already defined elsewhere. Still, expert's involvement would be needed for the final assessment of the risk.

In the next Section of our paper, we employ a group of non-experts (crowd-workers, see review e.g. in [12]) for labeling of legal documents, thus starting collection of training data for the future identification of the above CPFs based on machine learning techniques.

3. Crowd-intelligence analysis of legal acts

Our goal was to perform analysis of available legal acts with respect to potentially identifiable CPFs. For this end, we involved 30 student members of crowd-intelligence lab of Novosibirsk State Technical University. In total, they processed (see the output form in table 6) 4763 legal documents obtained from websites of the High Court of the Russian Federation (http://www.vsrf.ru/) and "Pravosudie" State Automated System (https://sudrf.ru/). The legal documents were of two types: 1) the actual legal acts/regulations, and 2) judicial acts after consideration of court applications dedicated to challenging a legal act. Particularly, the crowd workers processed this number of the documents:

- 3050 ACI documents;
- 429 documents related to CPFs;
- 247 documents with the legal language ambiguity factor;
- 65 documents with the authority based on "entitled" formulation factor;
- 669 documents with regulatory collisions;
- 225 documents related to point 3 of the ACI Methodology;
- 48 documents related to the point 4 of the ACI Methodology.

Considering rather inconsistent approaches of various Russian regions towards ACI and controversial law enforcement practice, we have devised the following scale for assessing presence or absence of CPF in legal acts drafts:

1. identified CPF (name) involves the necessity to develop the new legal act draft;
2. identified CPF (name) involves the necessity to make corrections to the current legal act draft;
3. no identified CPFs.

Table 6. The output form for the crowd workers processing the legal documents.

№	**Judicial act**					**Legal act**			
	Description			Analysis results		Description			CPF
	title	adoption date	case number	ruling	formulation	title	date	number	
1	2	3	4	5	6	7	8	9	10

For the effective legal acts, the scale was adjusted as follows:

1. identified CPF (name) involves the necessity to develop draft of the new legal modifying the current legal act;
2. no identified CPFs.

On the first stage, the analysis of meta-data, we made random selection of judicial acts from cases dedicated to challenging legal acts. We present the distribution of legal acts that were recognized as inappropriate:

1) By the year of adoption (table 7).
2) By the social field that is regulated by the legal act (table 8).
3) by the legal act level: federal, regional, local (table 9).

Table 7. Meta-data analysis: inappropriate legal acts by years (selected from 1993–2018).

Year	**Number of the cases challenging the legal acts**	**Number of legal acts recognized as inappropriate**
1993	1	1 (100%)
1994	1	0 (0%)
1995–2003	0	0 (-)
2004	1	0 (0%)
2005	1	1 (100%)
2006	5	4 (80%)
2007	5	4 (80%)
2008	26	17 (65%)
2009	11	6 (55%)
2010	33	28 (85%)
2011	9	6 (67%)
2012	34	31 (91%)
2013	20	8 (40%)
2014	14	9 (64%)
2015	22	18 (82%)
2016	17	8 (46%)
2017	8	2 (25%)
2018	2	1 (50%)

Table 8. Meta-data analysis: inappropriate legal acts by social fields (public and private law).

Year	Number of the cases challenging the legal acts	Number of legal acts recognized as inappropriate
Tax	1	1 (100%)
Finance	5	5 (100%)
Housing	6	4 (67%)
Social security	13	9 (69%)
Land and urban	20	17 (85%)
Administrative	56	36 (64%)
Other public law	99	63 (64%)
Labor	1	0 (0%)
Civil	8	6 (75%)
Other private law	1	0 (0%)

Table 9. Meta-data analysis: inappropriate legal acts by levels (federal, regional, municipal).

Year	Number of the cases challenging the legal acts	Number of legal acts recognized as inappropriate
Federal laws	0	0 (-)
Federal by-laws	17	2 (12%)
Regional laws	15	10 (67%)
Regional by-laws	30	14 (47%)
Municipal legal acts	148	114 (77%)

Based on the meta-analysis (tables 7–9), we calculated the CPF risk coefficients, as the share of the particular identified factor per all the identified factors (see in table 10).

4. Conclusions

Existing literature in legal domain includes plenty of works dedicated to legal aspects of anti-corruption inspection (ACI). However, the issues of its automation have drawn attention rather recently, with the development of Computational Law (see in [13]). This field is multi-disciplinary and merges the traditional law studies with the methods and approaches provided by modern information and communication technologies. So, seeking to support experts in ACI, in our current research we explored directions for its automation through employment of crowd-intelligence methods.

After considering the CPFs listed in the ACI Methodology prescribed by the Russian Ministry of Justice, we can conclude that automated analysis is feasible for at least several of the most widely encountered factors. Further, we performed pilot crowd-intelligence processing of legal documents and extracted meta-data from a sample of legal acts in several categories: adoption year, social field, and level. Based on the legal acts inappropriateness frequencies as recognized by courts, we assessed CPF risk values per the selected dimensions of meta-data.

Table 10. CPF risks based on the meta-data analysis (for a legal act in ACI).

The meta-data dimension	The meta-data parameter	The CPF risk assessment value
The considered legal act's year of adoption	1993	1.1
	1994	0
	1995–2003	0
	2004	0
	2005	1.1
	2006	4.4
	2007	3.3
	2008	12.1
	2009	4.4
	2010	18.7
	2011	6.6
	2012	22
	2013	8.8
	2014	9.9
	2015	16.5
	2016	8.8
	2017	2.2
	2018	1.1
The considered legal act's social field	Tax	1.04
	Finance	5.2
	Housing	3.12
	Social security	7.28
	Land and urban	15.6
	Administrative	37.44
	Other public law	37.44
The considered legal act's level	Labor	0
	Civil	1.04
	Other private law	0
	Federal laws	0
	Federal by-laws	2.04
	Regional laws	10.2
	Regional by-laws	14.28
	Municipal legal acts	77.52

More generally, we envision the following stages in the ACI with respect to the automation:

1. Automated processing of legal acts for assessment of CPFs risks.
2. Preliminary classification of legal acts into the ones that can and cannot undergo the inspection.

3. Division of the inspected legal acts into groups based on the CPFs risk levels.
4. Cross-inspection of legal acts with the highest CPFs risks by several experts independently.

We see limitations of our current paper mostly in the absence of the textual analysis of legal acts with regard to the identified automatable CPFs, which is the main direction of our further research work. So, we plan to use the crowd-intelligence approach for more detailed labeling of the legal acts, which could further be used to train models for detection of the CPFs in the texts. Still, we believe that already our current results might be of interest to researchers and developers working in Computational Law, as well is to ACI experts and organizers.

Acknowledgments
The reported study was funded by RFBR according to the research project No. 17-32-01087 a2 OGN.

References

[1] Decree of the President of the Russian Federation of 01.12.2016 #642 "On the Strategy of Scientific and Technological Development of the Russian Federation," The Russian Legislation Collection. 05.12.2016. #49. Article 6887. (in Russian) Available at: http://kremlin.ru/acts/bank/41449
[2] Order of the Russian Ministry of Science and Education of 26.07.2018 #9n "On the approval of the Order of the anti-corruption inspection for legal acts and legal acts drafts for the Ministry of Science and Education of the Russian Federation," the Official Legal Information Portal. Available at: http://www.pravo.gov.ru
[3] Statistical Data. The Ministry of Justice of the Novosibirsk Region of Russia. Available at: http://minjust.nso.ru/page/2303
[4] Statistical Data. The Government of the Novosibirsk Region of Russia. Available at: http://www.nso.ru/page/26841
[5] Boella G, Di Caro L, Humphreys L, Robaldo L, Rossi P and van der Torre L 2016 *Artificial Intelligence and Law* **24** (3) 245–283
[6] Statistical Data. Ministry of Justice of the Russian Federation. Available at: http://minjust.ru/ru/activity/legislative/anticorrekspert/ accredited_persons_as_anticorruption_experts on 15.02.2019.
[7] Constitutional Court of the Russian Federation. Ruling of 16.03.2018 #11-P "On the case of validating the correspondence to the Constitution of the 4th part of the 13th article of the Law of the Russian Federation "On the social security of citizens who suffered radioactive impact in Chernobyl Atomic Power Plant disaster", in regard with complaint by V.N. Fomina," Rossiyskaya Gazeta, № 61, 24.03.2018. (in Russian) Available at https://rg.ru/2018/03/24/chernobyl-dok.html
[8] Astanin V V et al. 2017 Methodological recommendations for performing the legal and anti-corruption inspection of legal acts of Russian Federation subjects and municipalities (with concrete illustrative examples). (Moscow: FBU NSPI of the Ministry of Justice of Russia) 56 p (in Russian)
[9] Capuano N, De Maio C, Salerno S and Toti D A 2014 *Proc. of the 4th International Conference on Web Intelligence Mining and Semantics* (WIMS'14) 1–6
[10] Barabash O V 2016 *Proceedings of Penza State University* **1** 17–21 (in Russian)
[11] Batura T V 2016 Mathematical linguistics and automated processing of texts (textbook Novosibirsk State University) 166 p. (in Russian)
[12] Daniel, F, Kucherbaev P, Cappiello C, Benatallah, B and Allahbakhsh M 2018 *ACM Computing Surveys (CSUR)* **51** (1) 7 doi:10.1145/3148148
[13] Kurcheeva G, Rakhvalova M, Rakhvalova D, Bakaev M 2019 *Communications in Computer and Information Science* **947** 123–137

IOP Conf. Series: Materials Science and Engineering 1019 (2021) 012048 doi:10.1088/1757-899X/1019/1/012048

Registration and processing system of digital speckle images

E A Korotina[1], F A Gubarev[2], T V Gandzha[3], I D Liushnevskaya[2]
[1]Engineering School of Non-Destructive Testing, Tomsk Polytechnic University, Tomsk, Russia
[2]Research School of Chemistry & Applied Biomedical Sciences, Tomsk Polytechnic University, Tomsk, Russia
[3]Department of Computer Control and Design Systems, Tomsk State University of Control Systems and Radioelectronics, Tomsk, Russia
E-mail: k.korotina28@gmail.com

Abstract. In this paper, we discuss an application of the method of laser-speckle correlation for studying the blood coagulation process. A module for recording and processing speckle images based on a debugging kit for the Atxmega128A1 microcontroller has been developed. The results of testing the developed module as part of a laboratory setup based on a semiconductor laser using standard reagents are presented.

1. Introduction

Blood coagulation (hemocoagulation) is the most important step in the work of the hemostasis system, which is responsible for stopping bleeding when the vascular system of the body is damaged. Disruption of the blood clotting process can be caused by various reasons and entail undesirable consequences. Estimation of hemostasis parameters in real time is an essential task, despite the large number of existing methods and means of measurement. Nowadays several methods have been developed to observe and measure the physical characteristics which accompanied the coagulation process, for example, blood viscosity [1–4], surface tension [5] or electrical conductivity [6]. The main disadvantages of the majority of known methods are invasiveness, the inability to carry out clotting monitoring in real time and performing point-of-care testing.

Currently, diagnostic methods are being developed based on analysis of laser speckle images [7, 8]. Optical methods based on laser speckles make it possible to analyze blood clotting non-invasively. In particular, our team is developing a device for estimating the clotting time of native blood based on the digital speckle image correlation method [8], and an electronic module is required, which is used for an automated processing of speckle images.

In use, the device, which will be developed based on the hardware-software complex presented in the work, makes it possible to carry out actions inaccessible to competitors, namely: carrying out express analysis and continuous monitoring. The first feature will be important for people who need constant monitoring of blood clotting factor. Too fast or too slow blood clotting time may indicate the presence of disorders of the cardiovascular system. The second feature will make a great contribution to surgery. Continuous monitoring of this factor during surgery significantly simplifies the work of the surgeon.

The purpose of this work is to develop an automated microprocessor system for recording and processing digital speckle images.

2. Technical means

The experiments on plasma clotting were conducted using the existing experimental setup with the designed microcontroller module. The flow chart of the experiment and developing module are shown in Figure 1. The view of the experimental part is shown in Figure 2. Before the experiment, the camera with microcontroller was fixed in the experimental setup, available in the laboratory of the department. For the experiment, a continuous diode laser with the radiation power of 10 mW was used, a control blood plasma and a Techplastin reagent (Technologiya Standard, Russia) with a given coagulation time [9]. A cuvette for the plasma had 0.1 ml volume and was made from plastic using 3D printing.

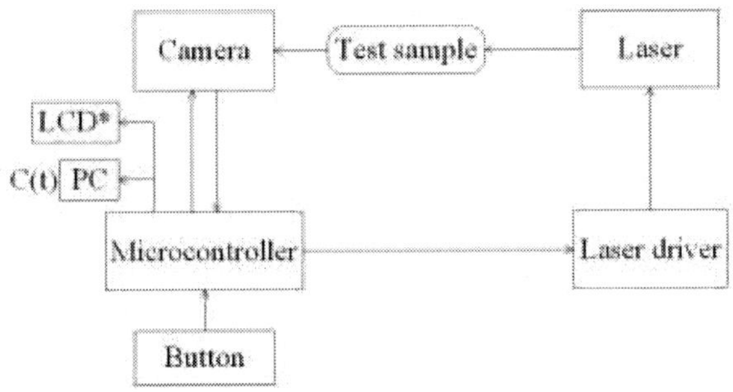

Figure 1. The flow chart of developing module.

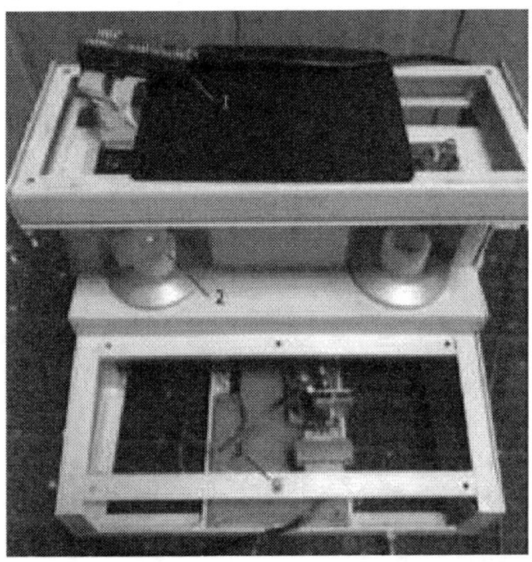

Figure 2. Experimental setup: 1 – developed module, which consist of camera and microcontroller, 2 – control plasma in a cuvette on the sample table and a solid-state laser, 3 – laser driver.

During the process of coagulation, the movement of microparticles occurs, and clots form in the sample under study. The illuminating laser beam, due to scattering, forms a speckle images which are presented in Figure 3. We used a monochrome camera mode for recording speckle images because the colour data is excess for the correlation analyses.

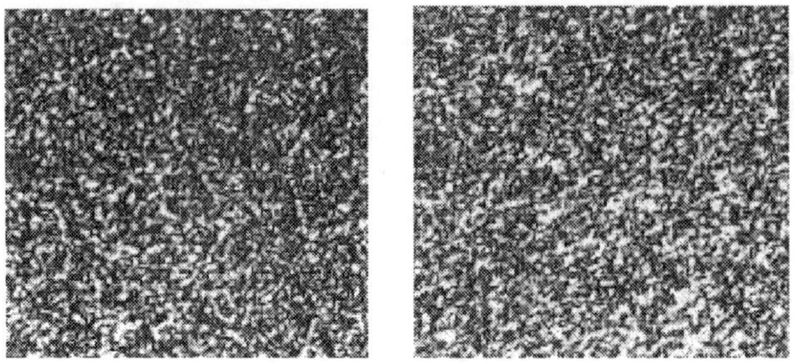

Figure 3. Laser speckle images during the process of control plasma clotting. Moments of time are random.

The hardware for the microcontroller module was selected according to the requirements specification. The requirements were: the speed of registration of speckle images – 30 frames per second, the frame resolution – 100×100 pixels and the time of continuous recording – 10 minutes.

We used the speed of registration of speckle images 30 frames per second, because it's an average speed that enough to record speckles and in order to optimize the performance of selected resources. The frame resolution is also average, because minimum is 50×50, when it is possible to recognize speckles. The time of continuous recording is 10 minutes because in future we are planning to work with the whole blood and the time of the whole blood coagulation is 3–5 minutes, we take 10 minutes to record pathologies or disorders if they exist.

To develop an automated microprocessor system, microcontroller ATxmega128A1 (Atmel) and video camera module OV7670 with the with parallel connection interface (Omni Vision) were chosen. The selected hardware in connection is shown in Figure 4. The chosen hardware meets the requirements and sufficiently productive. Important factor is that it is possible to write a script for every register either of camera or of microcontroller.

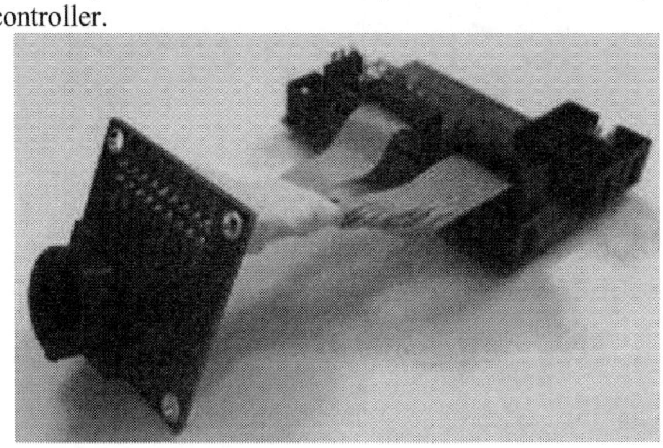

Figure 4. The selected hardware in connection.

To write scripts for the microcontroller and the video camera, the XRobot (developed in TUSUR, Russia) programming language was used to simplify the writing of scripts. The implemented script was loaded into the LabView software product, with the help of this modeling environment a virtual device – programming unit for the microcontroller ATxmega128A1 was developed. To process the speckle images, the multi-level computer simulation environment MARS, also developed in TUSUR, was used.

In computer studies of complex technical objects and their virtual analogs, the actual question is the visualization of measurement and simulation results in their natural form, as well as the implementation of the ability to control the topology and parameters of the object model. MARS has a multi-layer editor that makes it possible to visually separate the computer model of the object under investigation, the system for processing the simulation results and the means for visualizing them and controlling the parameters of the technical object [10].

The camera registers the speckle image, then, with the help of the microcontroller and the multi-level computer simulation environment MARS [10], the obtained frames are processed. The developed software script allows to calculate the correlation coefficient automatically by the formula given in [11], which compares the subsequent image with the previous one. The graph of correlation coefficient versus fibrin formation time is plotted in real time.

In the process of working on the hardware and software part, a device was created for recording and processing digital speckle images. The performance of the device was confirmed by experiments.

3. Results

Using the developed module of images registration it was possible to record image sequences with the length of about 1500 frames. The primary calculation of the correlation coefficient is a curve with quite intense noise (black curves in Figures 5, 6). Therefore, the data were smoothed using the sliding averaging method (red curves in Figures 5, 6). For smoothing the Mathlab software was used. According to the data obtained, it is possible to determine the time of fibrin formation. The obtained time coincides with the visually observed change of speckles during the coagulation recorded on a video camera.

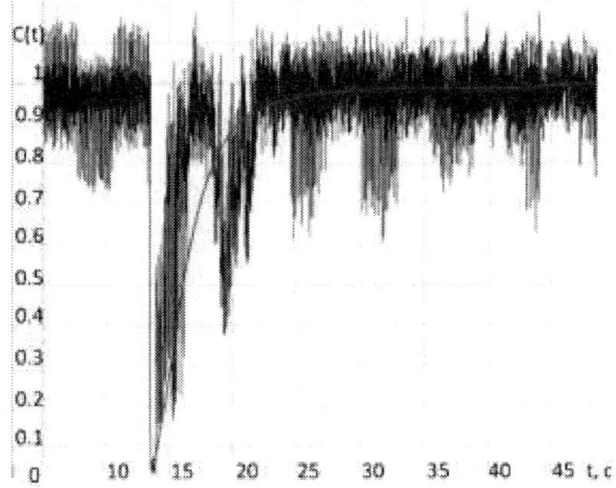

Figure 5. Real and smoothed correlation coefficient versus time of fibrin formation.

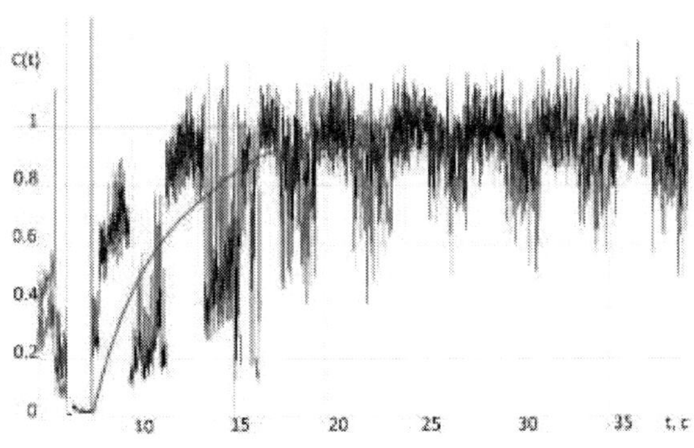

Figure 6. Real and smoothed correlation coefficient versus time of fibrin formation.

We conducted eight experiments with the same reagents. The resulting time was 11–13 seconds and it is coinciding with the known coagulation time of the reagent. The results of experiment are shown in Figure 7.

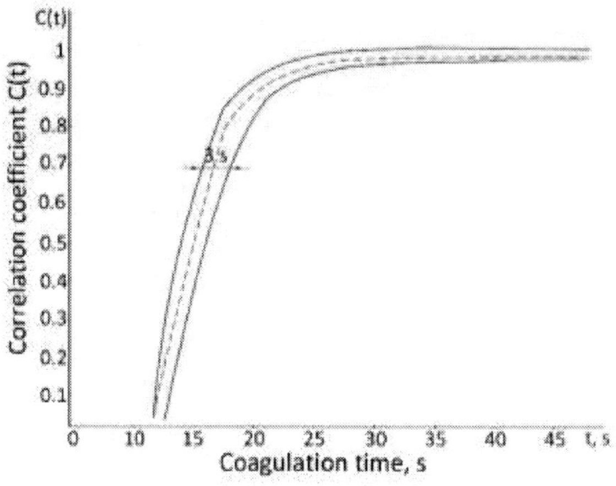

Figure 7. Range of curves of correlation coefficient versus fibrin formation time.

4. Conclusion

In this work, an automated module based on ATxmega128A1 microcontroller was developed. The module allows to record and process digital speckle images. Image processing is carried out in real time with a slight delay. The module was tested using the standard reagents for blood clotting time measurement.

In the future, it is planned to improve the algorithms of the system, to work out the method of analysis of blood coagulation, to install visualization tools on-board to display the dependence of the correlation coefficient on the fibrin formation time.

Acknowledgment

The authors are thankful to the company "Mednorth-Tehnics" (Tomsk, Russia) for the technical support.

References

[1] Hansson K M, Vikinge T P *et al* 1999 *Biosens. Bioelectron.* **14** 671–682
[2] Puckett L G, Barrett G *et al* 2003 *Biosens. Bioelectron.* **18** 675-681
[3] Huang C C, Wang S H *et al* 2005 *Ultrasound Med. Biol.* **31** 1567–1573
[4] LibgotCallé R 2008 *Ultrasound Med. Biol.* **34** 252–264
[5] Mintz M 1989 Method and Apparatus for Detecting a Blood Clot: US Patent No 4787369
[6] Tyutrin I I, Sorokozherdiev V O *et al* 2006 A Method for Evaluating the Functional State of the Hemostatic System: RF Patent No 2282855
[7] Li L, Sytnik I D *et al* 2018 *Biomedical Engineering* **52** 177–180
[8] Li L, Sytnik I D *et al* 2019 *Proc. SPIE* **11065** 110650E
[9] Kits and reagents for evaluating the hemostatic system. Available at: https://tehnologia-standart.ru/
[10] Dmitriev V M, Shutenkov A V *et al* 2011 *MARS – environment for modelling technical devices and systems* (Tomsk: In-Spectrum) p 278
[11] Gubarev F A, Li L *et al* 2016 *MATEC Web of Conferences* **48** 04003

IOP Conf. Series: Materials Science and Engineering 1019 (2021) 012049 doi:10.1088/1757-899X/1019/1/012049

New technology for assessing platelet aggregation activity

I I Tyutrin[1], L Li[2], F A Gubarev[2], E L Zhukov[3], D S Slizevich[3], I D Liushnevskaya[2] and I S Kalganov[4]
[1] Department of Professional Development and Retraining, Siberian State Medical University, Tomsk, Russia
[2] Research School of Chemistry & Applied Biomedical Sciences, Tomsk Polytechnic University, Tomsk, Russia
[3] Mednord-T LLC, Tomsk, Russia
[4] Center Orbita LLC, Moscow, Russia
E-mail:info@mednord-t.ru

Abstract. In this paper, we discuss an application of the method of the Digital Laser Speckle Image Correlation method (DIC) for studying the aggregation function of platelets in whole blood with the maximum standardization of the pre-analytical and analytical stages of the study. Materials and images are provided, methods are explained.

1. Introduction

Currently, antiplatelet therapy is of high importance to prevent thrombotic complications [1,2]. Despite the different mechanism of action of antiplatelet agents (inhibitors of COX-1 platelet receptors, P2Y12 receptor blockers, GPIIb/IIIa receptor antagonists, phosphodiesterase inhibitors, etc.), ultimately, their task is to inhibit platelet aggregation. Due to the increase in cases of resistance and "perverted responses", the need for a personalized assessment of platelet aggregation activity is of great clinical significance.

However, the problem of individual selection of antiplatelet therapy is currently at the stage of clinical research. The main obstacle to the widespread introduction of such an approach is the presence of a variety of methods for determining platelet function with the often-encountered inconsistency of their results with each other. The variability of the results is due not only to the difference in the test methodology, but also to the clinical heterogeneity of patient groups, different antithrombotic therapy regimens, the unequal determination of the threshold value of the method for detecting drug sensitivity, and the difference in test execution techniques by different researchers [1–5].

To date, there are many methods for assessing platelet aggregation activity: assessment of changes in plasma optical density ("Gold Standard" – Born-O'Brien's turbidimetric method) [3], assessment in whole blood - by the method of impedance aggregometry (Multiplate® Analyzer) [1], classical thromboelastography (test platelet mapping), determination of thromboxane B2 in serum or plasma, determination of 11-dehydrothromboxane B2 in the urine, etc. [2].

The main problems in assessing platelet aggregation activity are a lack of correlation between different methods, variability of results, sample preparation time (40–50 minutes) and a lack of standardization of the study.

At the same time, a new well-proven method of measuring platelet aggregation activity in whole blood has been established – Low-Frequency Piezoelectric Tromboelastography (LPTEG Method) with the usage of the ARP-01M "Mednord" hardware-software complex produced in Russia (Registration certificate 2010/09767) [6,7]. The method allows to make an assessment in the Point-of-Care test mode and does not require complicated sample preparation. The disadvantages of the method include the need for highly qualified personnel to correctly conduct the test and interpret the measurement results [8].

The aim of this work is to develop a method for evaluating the aggregation function of platelets in whole blood with the maximum standardization of the pre-analytical and analytical stages of the study.

2. Materials and methods

The technology proposed in this paper is based on the use of the Digital Laser Speckle Image Correlation method (DIC) [9–12]. The optical method is non-invasive and allows for measuring the temporal parameters of the processes by measuring the changes in the correlation coefficient of laser speckle images recorded on a digital camera. Previously, the team showed the possibility of using the DIC method for estimating the fibrin formation time using citrate blood plasma [11]. The results obtained [11] have provided an evidence to believe that this optical method can also be used to evaluate the blood cells aggregation activity in whole blood.

Figure 1 shows the appearance of an experimental setup for the study of blood clotting using the DIC method. The installation had two channels differing only in the radiation source. A laser diode with a power of 5 mW and a wavelength of 650 nm (left channel) and a He-Ne laser with a power of 5 mW and a wavelength of 632 nm (right channel) were used. Registration was carried out by the same monochrome VAA-136-USB USB-cameras. The frequency of registration of a digital camera is not a defining parameter, so the registration was carried out at a standard shooting frequency of 30 frames per second.

The cuvettes were made by 3D printing from transparent Watson-branded plastic manufactured by Best Filament. According to the results of a previous study [11], the optimal amount of analyte for the DIC method is 50-100 μl. Therefore, the cuvettes had a volume of ~100 μl with a diameter of 12 mm.

A typical view of speckle images is shown in Figure 2.

The calculation of the correlation coefficient was carried out using the correlation formula and the calculation algorithm described in [12], using the MatLab® software package.

As a control method for assessing platelet aggregation, the highly sensitive method of LPTEG was also applied, that also utilizes whole blood and has a high correlation with the gold standard by the Born-O'Brien turbidimetric method.

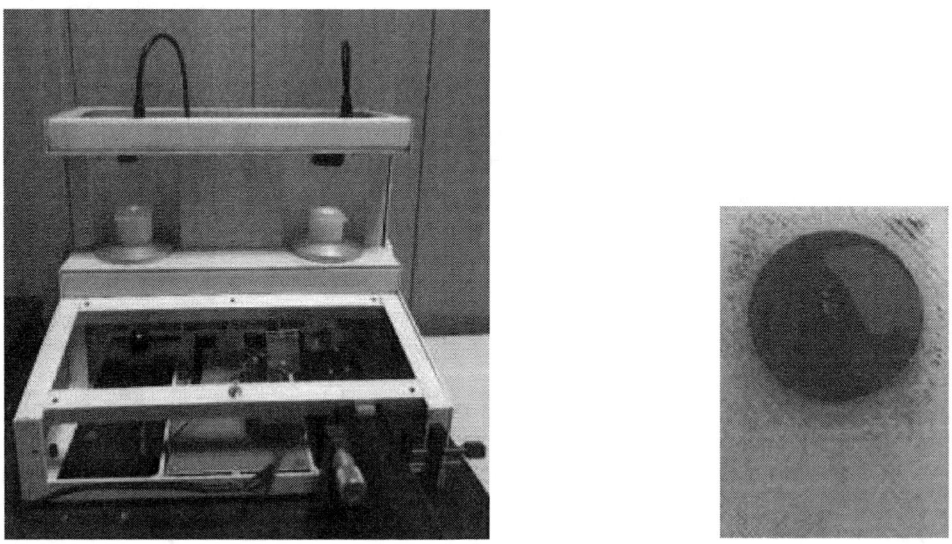

(a) (b)

Figure 1. Laboratory setup for a comparative study of the aggregation activity of the blood cells (a) and the experimental cuvette (b).

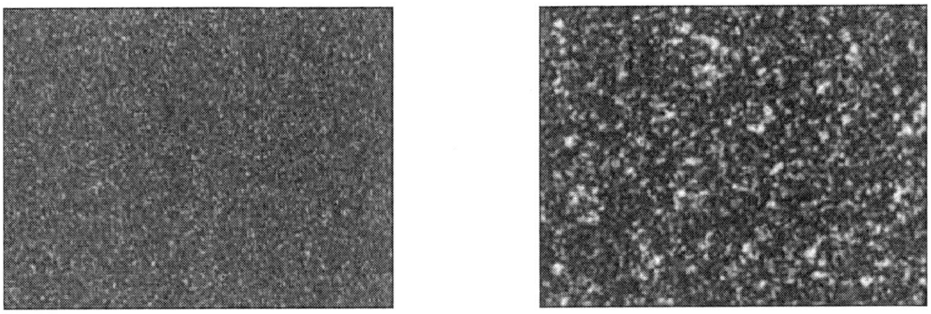

(a) (b)

Figure 2. Image view of speckle images obtained using a laser diode (a) and a He-Ne laser (b).

The study involved 4 volunteers of the Siberian population with different states of the hemostasis system who had previously signed a voluntary informed consent. The blood from the cubital vein was obtained into a three-component siliconized syringe ($V = 1$ ml) without applying a tow.

The blood was immediately (10–12 seconds) placed in a cuvette ($V = 0.45$ ml) made of the medical plastic (Mednord) and in two cuvettes printed on a 3D printer, and at the same time the registration of the platelet aggregation process was started.

The data from ARP-01M "Mednord" was processed in the specialized software X-Gemo-3M. Figure 3 shows the appearance of the device ARP-01M. Detailed information about the device is given in [13]. LPTEG curves were recorded to the T1 point - a point characterizing the aggregation activity of blood cells.

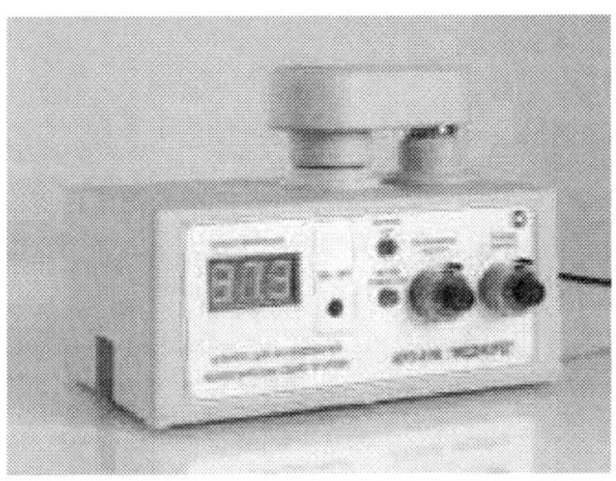

Figure 3. Low-frequency piezoelectric thromboelastograph ARP-01M "Mednord" (Russia).

The comparison of the obtained data was made in the SPSS Statistics 15.0 package.

3. Results and discussion

Figures 4 and 5 present the results obtained using two different methods: low-frequency piezoelectric tromboelastography (LPTEG) and laser speckle imaging (He-Ne laser and laser diode).

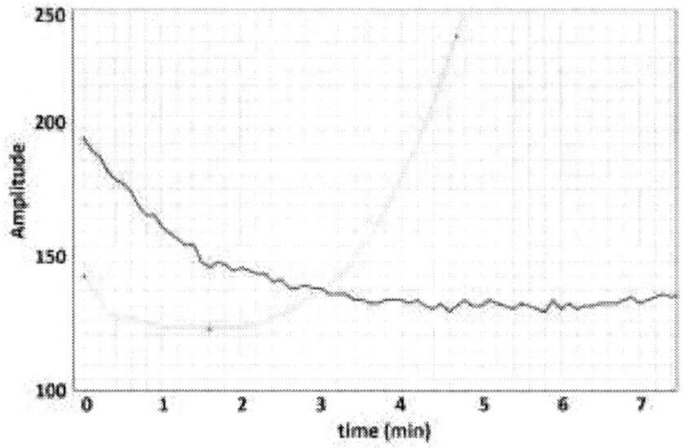

Figure 4. An example of data obtained by the LPTEG method, volunteers of the Siberian population: Patient 1 — yellow graph, Patient 3 — blue graph.

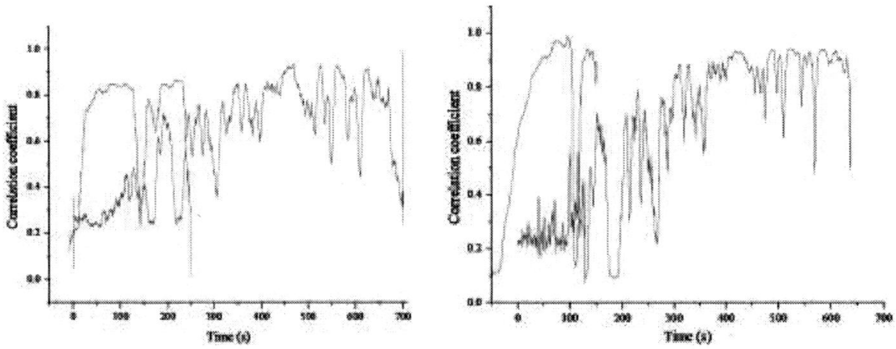

(a) (b)

Figure 5. An example of data obtained by laser speckle imaging, in Siberian volunteer s(Patient 1 – red graph, Patient 3 – blue graph), using a laser diode (a) and a He-Ne laser (b).

After conducting research, the data obtained were summarized in table 1.

Table 1. Obtained data.

Patient №	T1, min	He-Ne, sec	Diode, sec
1	1.6	61	92
2	2.1	103	-
3	5.8	354	418
4	2.9	146	-

Correlation analysis of the obtained results showed a high correlation of the proposed technique (both when using a laser diode and a He-Ne laser) with the method of LPTEG. The results of the correlation analysis is summarized in the table 2.

Table 2. The results of the correlation analysis.

		T1	He-Ne	Diode
T1	Pearson Correlation		0.989(*)	0.994(**)
	Sig. (2-tailed)		0.011	0.006
	N		4	4
HeNe	Pearson Correlation	0.989(*)		0.968(*)

	Sig. (2-tailed)	0.011		0.032
	N	4		4
Diode	Pearson Correlation	0.994(**)	0.968(*)	
	Sig. (2-tailed)	0.006	0.032	
	N	4	4	

*Correlation is significant at the 0.05 level (2-tailed).
**Correlation is significant at the 0.01 level (2-tailed).

4. Conclusion

The proposed method for assessing the functional state of the aggregation activity of blood cells has a strong correlation with the existing low-frequency piezothromboelastography method (> 0.9).

The method minimizes the errors of the preanalytical and analytical stages of the study, due to standardization of the intake (volume of the analyte, speed of the intake, thickness of the needle) and the use of native blood.

The results of tests demonstrates the perspective of development of the method of rapid assessment of the aggregation activity of blood cells.

References

[1] Capodanno D, Angiolillo D J 2009 *Hamostaseologie* **29** 376–380
[2] Solovyov M A, Udut V V 2015 *Medico-Biological and Socio-Psychological Problems of Safety in Emergency Situations* **1** 96–102
[3] Born G V R 1962 *J. Physiol. (Lond)* **162** 67-68
[4] Gabbasov Z A, Popov E G 1989 *Laboratornoye delo* 10 15–18
[5] Vorobyova I I 2012 *Kreativnaya Kardiologiya* **1** 50–63
[6] Udut V V and Kinzersky A A Method for assessing platelet aggregation activity *Patent RU 2666945* **2** 26
[7] Tuchin V V 1999 *J. Biomed. Opt* **4** 106–124
[8] Udut V V, Tyutrin I I 2017 *Experimental and Clinical Pharmacology* **80** 10–17
[9] Vilensky M A, Agafonov D N 2011 *Quantum Electron* **41** 324–328
[10] Golberg M, Fixler M 2013 *J. Biomed. Opt* **18** 101310-1–101310-5
[11] Li L, Sytnik I D 2018 *Biomedical Engineering* **52** 177–180
[12] Gubarev F S, Li L 2016 *MATEC Web Conf* **79** 01053
[13] Official web-site of the “Mednord-T” company. Available at: http://mednord-t.ru/

Numerical 3D simulation of enhanced oil recovery methods for high-viscosity oil field

M G Persova[1], Y G Soloveichik[1], A S Ovchinnikova[1], I I Patrushev[1], A V Nasybullin[2] and E V Orekhov[2]
[1] Research Laboratory "Modeling and data processing of high technologies", Novosibirsk state technical university, Novosibirsk, Russia
[2] Department of development and exploitation of oil and gas fields, Almetyevsk state oil institute, Almetyevsk, Russia
E-mail: mpersova@mail.ru

Abstract. The paper is devoted to the numerical simulation of oil recovery process using the technologies of polymer flooding and surfactants. The simulation is performed on the basis of the finite element method as well as the special scheme of flow balancing and phase flows between the cells, which along with high accuracy of pressure approximation, ensures the conservation of mass of filtering mixture components. The possibilities of simulation of different mechanisms of interaction of injected compositions with a medium are presented. The results of numerical simulation of different reservoir stimulations are shown using the digital model of the high-viscosity oil field in Tatarstan. It is shown that the application of polymer flooding and surfactant-polymer solutions makes it possible to increase the oil fraction in the produced fluid and, consequently, enhance the oil recovery.

1. Introduction

In order to enhance high-viscosity oil recovery, the technologies based on the application of high viscosity compositions are used [1]. In this connection, it is important to design the field development. For this purpose, the numerical simulation of the multiphase filtration is widely used [2–10].

Commercial software packages (most of them are mentioned in [10]) are based on the mathematical models [3] and various modifications of the finite difference method (FDM). Unlike the FDM, the finite element method (FEM) allows reasonably accurate description of wells and curvilinear surfaces between the layers, the local increase of the basic functions order (for example, in wells vicinities), and application of the non-conforming finite element mesh [11] (for example, in order to include local inhomogeneities into the computational domain). However, due to the fact that the formal application of the FEM does not guarantee the conservation of mass of filtering mixture components, some modern approaches are based on the hybrid finite element – finite volume method [5–7]. The approach considered in this paper is based on the FEM and special scheme of flow balancing [12], which ensures the conservation of mass of filtering mixture components.

In the technologies of the enhanced oil recovery, different compositions can be applied. These compositions can be characterized by different mechanisms of interaction with a medium (geological formation and fluid). These mechanisms are studied, as a rule, in laboratory conditions. However, the effectiveness of their application can be estimated using the digital model of the specific oil field for which further development is designed. Naturally, this digital model must be adequate, and numerical methods should allow taking into account different mechanisms of interaction with a medium, i.e. makes it possible to simulate multiphase filtration taking into account the mass fraction of the components of filtering phases, rules of their interaction, possible transition of components from one phase to the other as well as possible formation of new phases, which can have their own values of viscosity and/or relative permeability different from other phases. Therefore, water, oil, emulsion, immiscible polymer solutions, etc. can be a separate phase.

In this paper, we present the results of simulation of reservoir stimulation by high-viscosity agents using the digital model of the high-viscosity oil field in Tatarstan.

2. The mathematical models used for multiphase filtration calculation

Taking into account the specifics of the considered oil field (very low gas concentration), we use the multiphase filtration model for the case of incompressible and isotherm fluid [8, 9].

The field of the pressure P at each time instant can be obtained from the solution of the partial differential equation

$$-\operatorname{div}\left(\sum_{m=1}^{M}\mathbf{K}\frac{\kappa^{m}}{\eta^{m}}\operatorname{grad}\left(P+P_{c}^{m}\right)\right)=f \quad (1)$$

with boundary conditions

$$P\big|_{\Gamma_1}=P_g, \quad (2)$$

$$\left.\sum_{m=1}^{M}\mathbf{K}\frac{\kappa^{m}}{\eta^{m}}\frac{\partial P}{\partial n}\right|_{\Gamma_2}=0. \quad (3)$$

Equations (1)–(3) use the following notifications: f is the function describing the withdrawal or injection of mixture as well as the change of the mixture volume caused by chemical reactions, $\mathbf{K}$ is the tensor of absolute permeability, m is the number of the phase, which is characterized by the relative permeability κ^m, the viscosity η^m, and capillary pressure P_c^m, and Γ_1, Γ_2 are the sets of remote and isolated boundaries of the computational domain Ω, respectively.

We assume that the porosity Φ and the tensor of absolute permeability $\mathbf{K}$ are time-independent piecewise constant functions in space, and at any time instant the pores of a medium are fully filled with a multi-phase mixture.

Each m-th phase can consist of L^m component and, besides the parameters mentioned above, is characterized by the saturation S^m, density ρ^m and set of values χ^{lm} defining the mass fraction of the l-th component in the m-th phase.

The boundary problem (1)–(3) is solved by the finite element method on the non-conforming meshes with hexahedral cells.

Making use of the pressure field distribution, we calculate the flowing mixture volumes as

$$V=\left|\int_{\Gamma}\left[\sum_{m=1}^{M}\mathbf{K}\frac{\kappa^{m}}{\eta^{m}}\operatorname{grad}\left(P+P_{c}^{m}\right)\right]\cdot\vec{n}\,d\Gamma\right|\Delta t \quad (4)$$

where Δt is a time step, $\vec{n}$ is the outward normal to the finite element boundary. In order to ensure the conservation of mass of filtering mixture components, we use the special method of the flow balancing proposed in [12].

On the basis of the mixture volumes V_{Γ_i,Ω_e} flowing from the cell Ω_{k_i} to the cell Ω_e through the face Γ_i, new values of saturations are calculated as

$$S_{new}^m = \frac{\mathrm{mes}(\Omega_e)\Phi S^m + V_{in}^m - V_{out}^m}{\mathrm{mes}(\Omega_e)\Phi} \tag{5}$$

where $\mathrm{mes}(\Omega_e)$ is the cell volume, and V_{in}^m and V_{out}^m are the volumes flowing into and out of Ω_e. These volumes, in their turn, for each face Γ_i of the cell Ω_e are defined by the ratio

$$V_{\Gamma_i,\Omega_e}^m = V_{\Gamma_i,\Omega_e} \cdot \frac{\kappa^m}{\eta^m \sum_{i=1}^{M} \kappa^i / \eta^i}. \tag{6}$$

New values of the mass fraction χ_{new}^{lm} are calculated by the equation

$$\chi^{lm} = \left(n_{\Omega_e}^{lm} \cdot \mathrm{M}_l\right) \Big/ \left(\sum_{k=1}^{L^m} n_{\Omega_e}^{km} \cdot \mathrm{M}_k\right). \tag{7}$$

In equation (7), M_l is the molar mass of the component l, and $n_{\Omega_e}^{lm}$ is the number of moles of the substance in the cell Ω_e, which can be found by the ratio:

$$n_{\Omega_e}^{lm} = \sum_{i \in I_{in,\Omega_e}} \rho_{\Omega_{k_i}}^m \chi_{\Omega_{k_i}}^{lm} V_{\Gamma_i,\Omega_e}^m \Big/ \mathrm{M}_l + \rho_{\Omega_e}^m \chi_{\Omega_e}^{lm} \left(V_{\Omega_e}^m - \sum_{j \in I_{out,\Omega_e}} V_{\Gamma_j,\Omega_e}^m \right) \Big/ \mathrm{M}_l, \tag{8}$$

where Ω_{k_i} is the finite element from which the mixture volume V_{Γ_i,Ω_e}^m flows into the cell Ω_e through the face Γ_i, I_{in,Ω_e} and I_{out,Ω_e} are the sets of the global face numbers corresponding to the phase volumes flowing into and out of Ω_e, and $V_{\Omega_e}^m$ is the volume of the phase m in the cell.

In order to define the chemical reaction, we use two sets of number pairs $(l,m) \in I_{ch,1}^2$ and $(i,k) \in I_{ch,2}^2$, which designate the numbers of the components and phases of the reacting substances and substances obtained as a result of the reaction, respectively. For each number pair, we preset the value n_{rs} corresponding to the quantity of reacting moles of the component r in the phase s. Besides, we preset the reaction rate defined in terms of the amount of one of the reacting substances per unit time.

In order to simulate a chemical reaction in the cell Ω_e, the current time interval Δt is divided into the subintervals τ, and the quantity of moles $\widehat{n}_{\Omega_e}^{lm}$ reacting in a time τ is calculated

$$\widehat{n}_{\Omega_e}^{lm} = \rho_{\Omega_e}^m \chi_{\Omega_e}^{lm} V_{\Omega_e}^m \cdot \nu \cdot \tau, \tag{9}$$

where the rate ν is defined in terms of the substance corresponding to the component l in the phase m.

On the basis of the values $\widehat{n}_{\Omega_e}^{lm}$ and n_{lm}, the molar quantity of other substances reacting and obtained as a result of the reaction in a time τ in the cell Ω_e are defined.

Then, new mass fractions of reacting and obtained substances are calculated in the form

$$\chi_{\Omega_e}^{lm} = \frac{\left(n_{\Omega_e}^{lm} - \widehat{n}_{\Omega_e}^{lm}\right)\cdot \mathrm{M}_l}{\sum\limits_{j\in I_{\chi m}} \left(n_{\Omega_e}^{jm}\cdot \mathrm{M}_j\right) - \widehat{n}_{\Omega_e}^{lm}\cdot \mathrm{M}_l}, \quad (l,m)\in I_{ch,1}^2, \tag{10}$$

$$\chi_{\Omega_e}^{ik} = \frac{\left(n_{\Omega_e}^{ik} + \widehat{n}_{\Omega_e}^{ik}\right)\cdot \mathrm{M}_i}{\sum\limits_{j\in I_{\chi k}} \left(n_{\Omega_e}^{jk}\cdot \mathrm{M}_j\right) + \widehat{n}_{\Omega_e}^{ik}\cdot \mathrm{M}_i}, \quad (i,k)\in I_{ch,2}^2. \tag{11}$$

New values of the molar quantities are calculated as $n_{\Omega_e}^{lm} = n_{\Omega_e}^{lm} - \widehat{n}_{\Omega_e}^{lm}$ and $n_{\Omega_e}^{ik} = n_{\Omega_e}^{ik} + \widehat{n}_{\Omega_e}^{ik}$.

Using new values of mass fractions, we calculate new values of densities on the basis of which, in their turn, new volumes of phases are defined.

Before the pressure calculation at the next global time step, new values of saturations, relative permeabilities, and viscosities for all phases as well as the change of the mixture volume appearing as a result of chemical reactions are calculated.

The numerical method presented is implemented in the software with a graphical user interface (GUI).

3. Numerical simulation results

3.1. Verification

In order to verify the software, we use the results of the 10th SPE Comparative Solution Project [10].

The model dimensions are 1,200×2,200×170 ft^3. The distributions of porosity and tensor components of absolute permeability are set on the grid containing 60×220×85 cells. The porosity distribution is shown in Figure 1 (the screenshot of the GUI software developed). The injection well (I1) is placed in the domain center and has an injection rate of 5,000 B/D. The four producing wells (P1-P4) are placed in the corners and produce at 4,000 psi bottomhole pressure.

Figure 2 shows the water cut curves obtained with the use of our software (the black curves with labels) in comparison with Landmark (red curves) and Roxar (green curves) solutions.

Figure 3 shows the oil rate curves obtained with the use of our software (the black curves with labels) in comparison with Landmark (red curves) and Roxar (green curves) solutions.

Figure 1. The distribution of porosity (the screenshot of the GUI software developed).

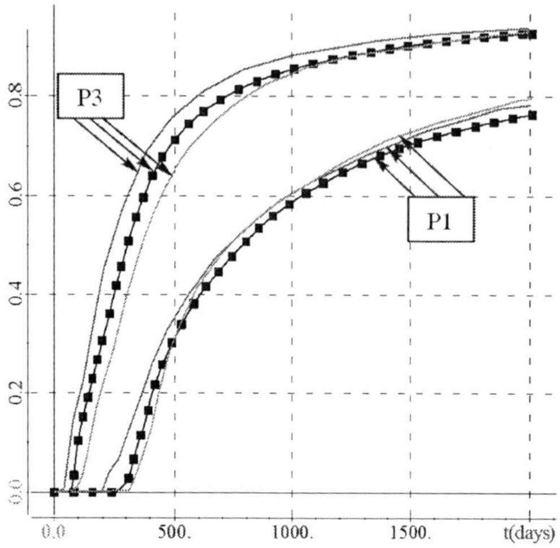

Figure 2. The water cut curves: our (the black curves with labels), Landmark (red curves), and Roxar (green curves) solutions.

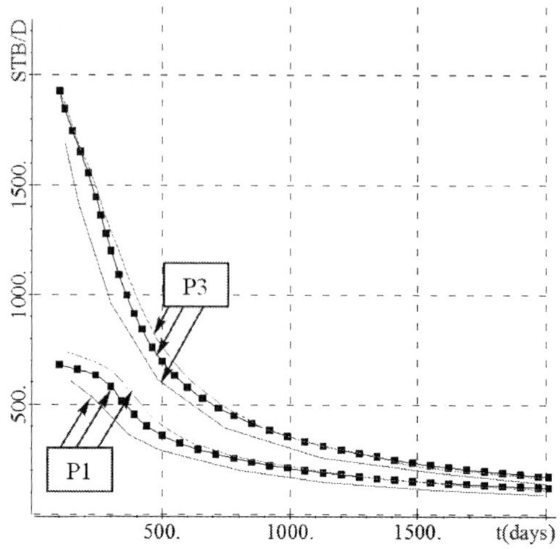

Figure 3. The oil rate curves: our (the black curves with labels), Landmark (red curves), and Roxar (green curves) solutions.

As seen from these Figures, our curves are either between the curves obtained by the use of other codes or quite close to one of them (or to both if they are close to one another).

3.2. Numerical simulation of reservoir stimulations

We simulate the injection of displacement agents using the geological model which is constructed on the basis of the historical well data observed on one of the high-viscosity oil field in Tatarstan. The initial distributions of the oil saturation and absolute permeability are shown in Figure 4. The total number of wells used in simulation is 61. Of them three wells were used only for injection, and fifteen wells were firstly used for exploitation and then, at different time instants, they became injection wells. Other wells were used for exploitation within the whole time interval. The whole time interval of simulation is 12000 days. Most of the wells contain several perforations which are made and abandoned at different time instants, i.e. injection or production for these wells is performed in different layers. The data on injection and production flow rates as well as the time instants of making and abandoning perforations are taken from the historical data. The injection of displacement agents is carried out in two wells (see wells designated by 1 and 2 in Figure 5).

We consider three variants. In the first variant, we assume that only water is injected. In the second variant, the polymer solution is injected into the wells during a year, after that water is injected again. The third variant differs from the second one in that instead of the polymer solution, the surfactant-polymer solution is injected.

The polymer solution moving in the reservoir mixes with the reservoir and injected water and, as a result, the solution viscosity reduces proportionally to the decrease of the polymer concentration.

In order to simulate the injection of the surfactant, different mechanisms of its interaction with a medium are considered. The first mechanism is as follows. When the surfactant enters the cell, it changes the property of the formation to retain oil (i.e. residual oil saturation).

The results of simulation for the variant in which the surfactant-polymer solution is injected into the wells 1 and 2 are shown in Figure 5. They show that the surfactant is gradually transferred (figures in the centre) with continuous reduction of its concentration (nonaxisymmetric distribution of the surfactant density is caused by the fact that during its injection, all wells of the oil field produce or inject; this leads to a complex distribution of flows). The right figures show the change of the property of the formation to retain oil. Before the formation interacts with the surfactant, its retention level is 0.3 (red color). If the surfactant concentration in the cell is sufficient, this level reduces to 0.1 (white color). The left figures show the "washing" residual oil with its displacement by the high-viscosity composition as evidenced by the increase in oil saturation (red rings) around light ("washed") zones.

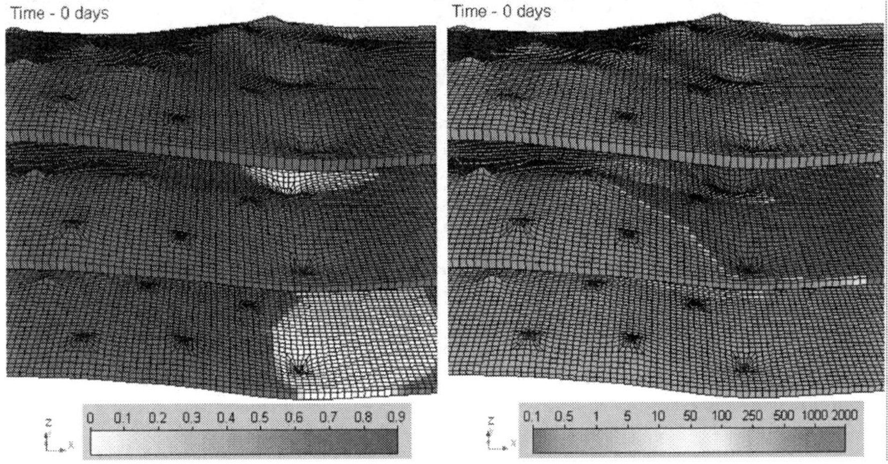

Figure 4. In-layer distributions of the initial oil saturation (on the left) and absolute permeability (on the right).

Special attention should be paid to the following. As seen from the distributions presented in Figure 5, approximately the same situation is in all three layers in the vicinity of well 1. It is related to the fact that all three layers are almost homogeneously oil saturated in this zone. The injection into well 2 is performed only in the top layer. The layer which follows the top layer has low permeability and contacts with the water-bearing layer. Therefore, the situation in layers in this zone differs greatly. Besides, the central figures show that near well 1, the surfactant is weakly transferred to the north. It is related to the fact that low permeability "wall" in the digital model lies near well 1.

Figure 6 shows oil saturation distributions for all three variants considered above.

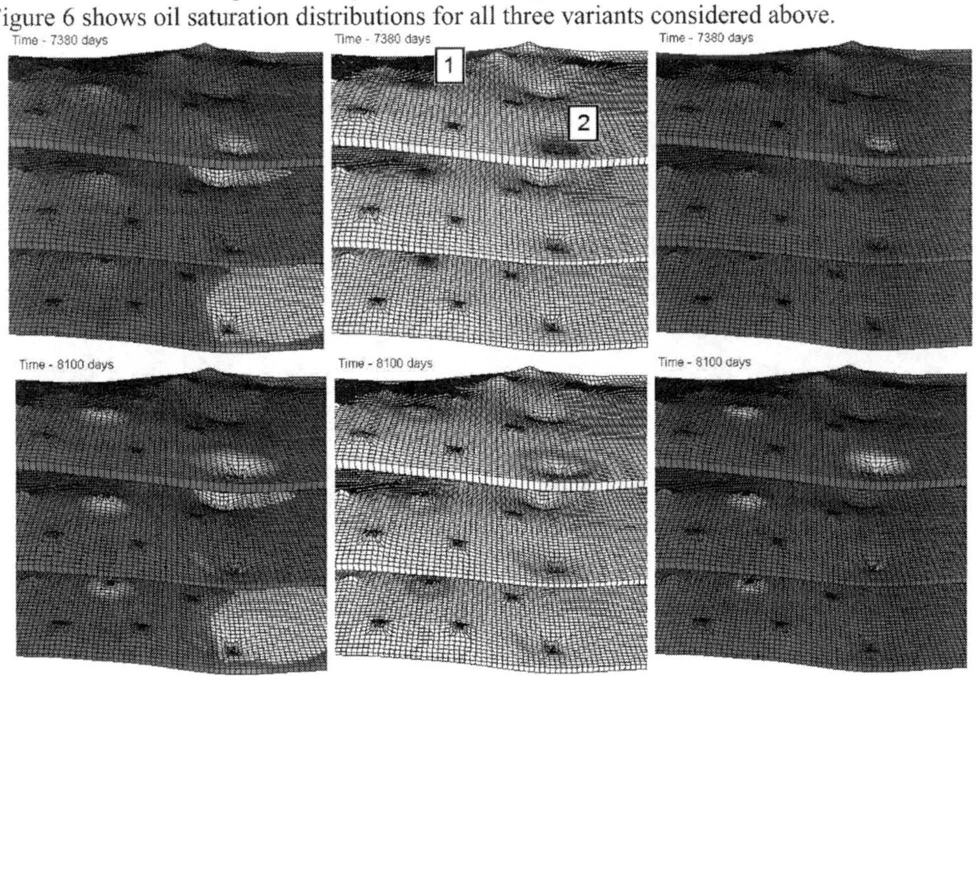

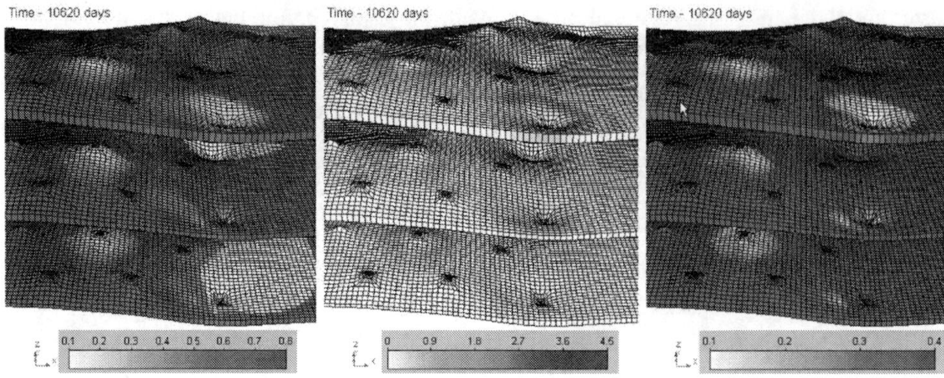

Figure 5. In-layer distributions of oil saturation (on the left), surfactant density (in the centre), and property of the formation to retain oil – the retention level (on the right) at different time instants.

Results presented in the figure show that when the polymer solution is injected, the portion of oil is concentrated at some distance from the injection wells (red rings surround zones around these wells), but these zones around wells contain the residual oil. If the surfactant-polymer solution is injected, the oil is also concentrated at some distance, and the residual oil is "washed" from the zones near the wells (the corresponding zones around the injection wells are lighter).

The second variant of the interaction mechanism of the surfactant with the medium is as follows. The surfactant entering a cell is combined with oil in the following proportion: 0.024 moles of surfactant and 1 moles of oil. As a result, a new "combined oil" phase in which surfactant and oil are included in the same proportion is formed. The residual saturation of this phase either significantly decreases or equals the zero.

For this variant of simulation, Figure 7 shows in-layer distributions of the surfactant density and the saturation of the "combined oil" with a different value of residual saturation. The distributions are presented at two time instants. As seen from the distribution presented in Figures 7*b,c* the "combined oil" phase is transferred from injecting to producing wells. The analysis of saturation distributions of oil in the "oil" phase (this is the oil without surfactant) shows that near the injecting wells, white sub-domains appear (similar to Figure 6*c*). This indicates that at least the portion of the "combined oil" is obtained from the residual oil (and, therefore, as in the previous case, the residual oil is "washed").

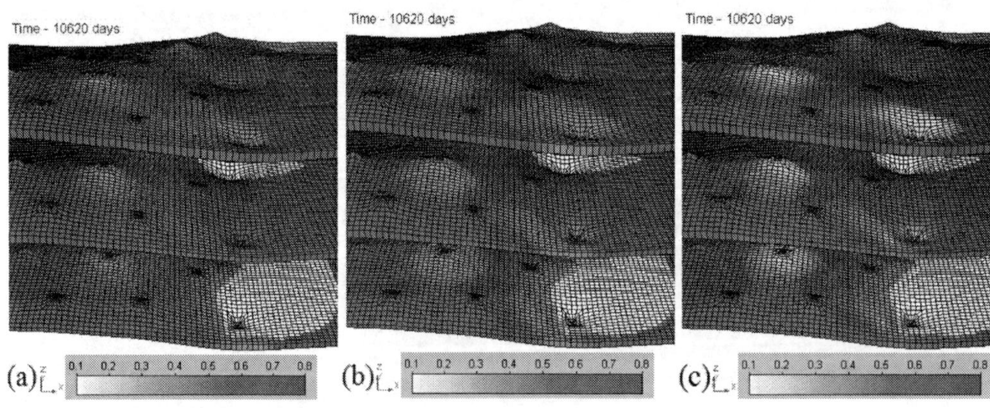

Figure 6. In-layer distributions of oil saturation for three variants: water injection (a), polymer solution injection followed by water injection (b), and injection of surfactant-polymer solution followed by water injection (c).

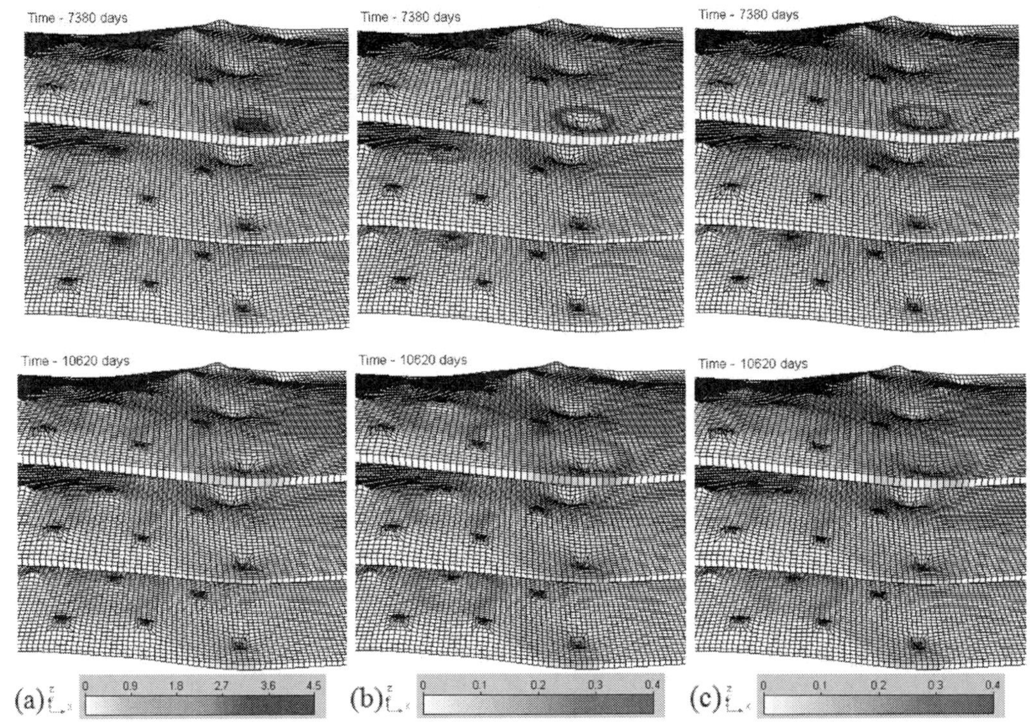

Figure 7. In-layer distributions of the surfactant density (a) and the saturations of the "combined oil" with the residual saturation equal to the zero (b) and 0.1 (c) at two time instants.

Figure 8 shows the oil fraction in the produced fluid for all three variants: water injection (green color), polymer solution injection followed by water injection (dark blue color), and injection of surfactant-polymer solution followed by water injection. The latter variant, in its turn, is presented for two interaction mechanisms of surfactant with the medium: the change of the property of the formation to retain oil (red color) and formation of a new "combined oil" phase (black color curve with labels). The portion of oil, which "comes to" the producing well in the "combined oil" phase, is shown by black color without labels.

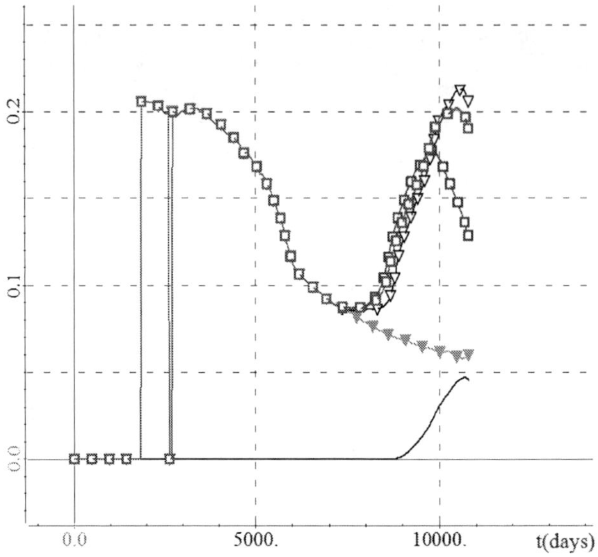

Figure 8. The oil fraction in the produced fluid for three variants: water injection (green color), polymer solution injection followed by water injection (dark blue color), and injection of surfactant-polymer solution followed by water injection (red and black colors, see text for more details).

On the basis of results obtained we can conclude the following. The application of the polymer and surfactant-polymer solutions at the later stage of oil field development makes it possible to increase the oil fraction in the produced fluid. Moreover, additional investigations show that the polymer solution application makes sense at more early stages of oil field development. This allows significant enhancement of high-viscosity oil recovery. The simulation results show that the considered interaction mechanisms of surfactants with a medium give similar effects, but the time delay for these effects is still more than for the effects of polymer solution application.

4. Conclusion

In general, the simulation methods presented enable us to study the possibilities of the application of various displacement agents and their compositions in order to enhance high-viscosity oil recovery and design further development of the oil field on the basis of its digital hydrodynamic model. The specifics of interaction mechanisms of compositions with a medium should be preliminary studied in laboratory conditions.

Acknowledgment

The reported study was funded by RFBR, project number 20-31-90049 (the results of simulating the polymer solution injection) and the Ministry of Science and High Education of the Russian Federation under the state assignment (Research Laboratory "Modeling and data processing of high technologies", project № 2019-0851, research topic № FSUN-2020-0012).

References

[1] Green D W and Willhite G P 1998 *Enhanced Oil Recovery* (Richardson, Texas: Henry L. Doherty Memorial Fund of AIME, Society of Petroleum Engineers)
[2] Aziz K and Settari A 1979 *Petroleum Reservoir Simulation* (London: Applied Science Publishers)
[3] Cao H, Crumpton P I and Schrader M L 2009 *Proc. of SPE Reservoir Simulation Symp. Proc.* **2** 1075-1086
[4] Zhang R han, Zhang L hui, Luo J xin, Yang Z dong and Xu M yang 2016 *J. Pet. Sci. Eng.* **146**

IOP Conf. Series: Materials Science and Engineering 1019 (2021) 012050 doi:10.1088/1757-899X/1019/1/012050

1211–1225
[5] Nick H M and Matthäi S K 2011 *Vadose Zo. J.* **10** 299-312
[6] Nick H M and Matthäi S K 2011 *Transp. Porous Med.* **90** 421–444
[7] Schmid K S, Geiger S and Sorbie K S 2013 *J. Comput. Phys.* **241** 416–444
[8] Persova M G, Soloveichik Y G, Patrushev I I and Grif A M 2018 *Proc. of 14th Int. Scientific-Technical Conf. on Actual Problems of Electronic Instrument Engineering (APEIE 2018)* 212–215
[9] Soloveichik Y G, Persova M G, Patrushev I I and Glushkov S A 2018 *Proc. of 14th Int. Scientific-Technical Conf. on Actual Problems of Electronic Instrument Engineering (APEIE 2018)* 301–306
[10] Christie M A and Blunt M J 2001 *SPE Reserv. Eval. Eng.* **4** 308–316
[11] Soloveichik Y G, Persova M G, Domnikov P A, Koshkina Y I and Vagin D V 2018 *Geophys. J. Int.* **212** 2159–2193
[12] Persova M G, Soloveichik Y G, Grif A M and Patrushev I I 2018 *Proc. of XIV Int. Scientific-Technical Conf. on Actual Problems of Electronics Instrument Engineering (APEIE 2018)* 205–211

IOP Conf. Series: Materials Science and Engineering 1019 (2021) 012051 doi:10.1088/1757-899X/1019/1/012051

Feasibility demonstration of back-to-back method for induction traction motors testing

E V Beyerleyn, P V Tyuteva and B S Dorzieva
School of Energy and Power Engineering, Tomsk Polytechnic University, Tomsk, Russia
E-mail: Tyutevapv@tpu.ru

Abstract. The authors present the feasibility demonstration of energy-saving back-to-back test method implementation for induction traction motors testing. The results of theoretical and physical modeling that support the energy effectiveness and usefulness of the proposed back-to-back method are presented in the article. The effectivity ratio of the test circuit has been calculated due to confirm the energy effectiveness and feasibility of the proposed method, and also authors identified a number of advantages of the back-to-back method for high power induction traction motors.

1. Introduction

In the Russian Federation, the fleet of traction electric machines (brushed and brushless) involved in various industries totals hundreds of thousands, while the average power of the traction electric motor is about 1000 kW, which have found their application as traction or auxiliary machines (for example, direct current traction motor EDP810 mounted on a 2ES6 locomotive has a rated power 810 kW, the NTA-1200 asynchronous traction motor set on an EP10 electric locomotive has a rated power 1200 kW). The operational life of these machines enshrined at the design, manufacture and maintenance stages and is about decades, it is economically feasible to carry out scheduled and preventive repairs of these machines, if the machine has not exhausted its lifetime potential. As the cost of such an electric motor is high, therefore, enterprises operating these machines are rarely interested in timely repair based on their actual condition. Often the lifetime resource of these machines is either worked out or close to this.

For the reasons stated above, the replacement of obsolete traction motors with their new modifications is not carried out at a sufficient pace. Which leads to an increase in the number of electric machines failures, as well as to an increase in the repair work and, accordingly, to rise of unbudgeted expenses. According to the global trend, the changeover from the brushed electric drive of railway rolling stock to asynchronous is inevitable, due to the undeniable advantages of these machines. However, in the Russian Federation, according to experts, direct current motors as traction motors will be used for about another 30 years, despite the fact that, the use of asynchronous machines as traction motors has a number of advantages compared to brushed machines [1, 2].

An increase in the share of asynchronous traction electric motors makes it necessary to solve a number of urgent issues related to their operation. Among these issues, we can single out the need to create new or re-equip existing repair bases and test stations, while the test stations must meet modern requirements for reliability and energy efficiency. This can be achieved by introducing modern energy efficient methods and techniques of asynchronous machines diagnosing and testing [3].

While testing traction machines, it is necessary to provide long-term test conditions, which leads to high energy costs. As a solution to this problem, a methodology and test circuit for the traction asynchronous machines according to the back-to-back method can be proposed; this method allows obtaining the largest percentage of energy savings during the test time. The practical application of this method up to the present moment has been limited only to direct current machines, research team of the School of Energy and Power Engineering of Tomsk Polytechnic University have proposed a circuit that implements the back-to-back method for traction induction motors testing [4]. The aim of this work is a theoretical study and practical application of the proposed method for induction traction motors testing.

2. Theoretical study

The selection of circuit for testing of electric machines is determined by parameters of electrical equipment used on test station, all of them can be characterized by range of quantitative and qualitative parameters which include the rated power of auxiliary equipment and the amount of consumed electrical power during the test procedure. The interrelation of these factors can be expressed in terms of effectivity ratio that could be determined as [5]:

$$k_E = \frac{P_1 - P_{test.cir}}{P_1} \tag{1}$$

where P_1 – power consumed by the tested electric machine; $P_{test.cir}$ – power consumed by the circuit as a whole. The effectivity ratio shows the fraction of electricity required for the motor compared with its rated power.

There are several circuits and methods used for high rated power alternative current machines testing [5], some of them are presented in table 1. All test circuits have sufficiently high effectivity ratio values, thus differ in the number of auxiliary machines included in the test station.

Table 1. Effectivity ratio.

Test Circuit	k_E	Number of Auxiliary Machines
Direct load of traction motor	0.25 ÷ 0.60	5
Circuit with power regeneration to main supply	0.65 ÷ 0.70	6
Circuit with load reversal, power losses covered by booster generator	0.70 ÷ 0.60	5
Circuit with load reversal, power losses covered by direct current generator	0.75 ÷ 0.65	4

As the table 1 shows test circuits have relatively high effectivity ratio values, however taking into account the ratings of modern induction traction motors (power about 1000 kW, efficiency 95 %) the test station must be capable of loading the tested motor with the load that is equal to or greater than the rated value, after that the number of energy conversion levels must be the smallest. The back-to-back test method complies with all of those requirements.

To obtain the numerical value of the effectivity ratio let's consider in detail the power flow of the back-to-back test method for induction traction motors, as shown in Figure 1, where P_{IG} – power generated by the induction machine in the generator mode, P_2 – power output of the tested motor.

The power flow diagrams of the machines being tested, which worked in a generator and mo-tor mode, will be similar to each other, apart from the overall mechanical losses, which will be compensated in the machine under test that works in a motor mode.

The motor and the generator work jointly during test, so they are connected mechanically and electrically, hence, their power flow diagrams represent sequential combination of the machines' diagrams that work in motor and generator modes, as shown in Figure 2.

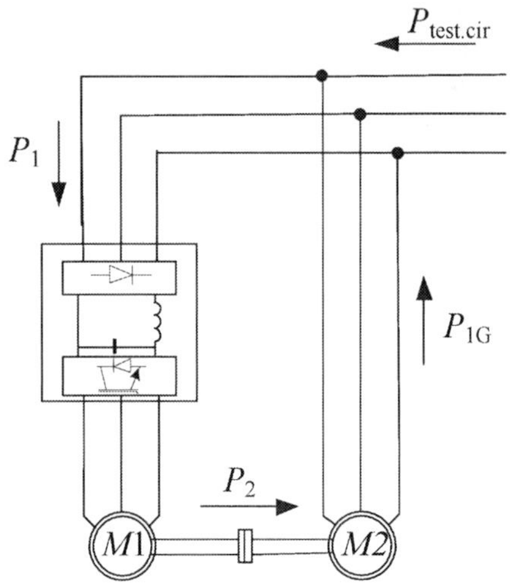

Figure 1. Energy flow in back-to-back test circuit.

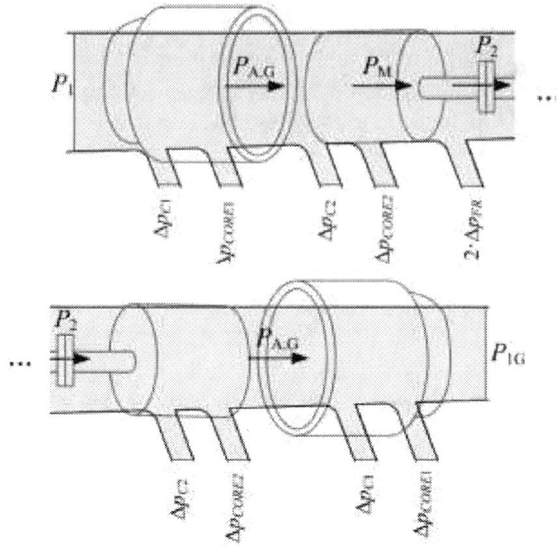

Figure 2. Power flow diagram of the induction motor – the induction generator system.

To determine the power losses during the test with a standard technique was used [6], in the (2) are the power losses taken into account during the experimental studies of back-to-back test method:

$$P_2 = P_1 - \Delta p_{C1} - \Delta p_{CORE1} - \Delta p_{C2} - \Delta p_{C1} - \Delta p_{CORE2} - \Delta p_{FR} - \Delta p_{STR} \quad (2)$$

where P_2 – output power of the tested motor; P_1 – input power of the tested motor; Δp_{C1} – copper losses in the stator winding; Δp_{CORE1} – stator core losses; Δp_{C2} – copper losses in the rotor winding; Δp_{CORE2} – rotor core losses; Δp_{FR} – summative friction losses; Δp_{STR} – stray load losses. Also Figure 2 contains the following legend $P_{A.G}$ – – air-gap power, P_M – converted power or mechanical power of the machine.

For determination of the effectivity ratio we use the expression (1), where power consumed by the tested electric machine could be determined from the expressions for the direct determination of efficiency:

$$P_1 = \frac{P_2}{\eta_M} \quad (3)$$

where η_M – efficiency of the tested motor.

The power produced by the machine that works in generator mode could be expressed as:

$$P_{1G} = P_2 \cdot \eta_G \quad (4)$$

where η_G – efficiency of the machine in generator mode.

Then taking into account the power losses occurred in machines the summative power consumption of the test circuit:

$$P_{test.cir} = P_1 - P_{1G} \quad (5)$$

Substituting (2) and (4) in the expression (5) we obtain the effectivity ratio of the back-to-back test method:

$$k_E = \frac{\frac{P_2}{\eta_M} - P_2 \cdot \left(\frac{1-\eta_M \cdot \eta_G}{\eta_M}\right)}{\frac{P_2}{\eta_M}} = \eta_M \cdot \eta_G \quad (6)$$

As the expression shows, the effectivity ratio of the back-to-back test method depends on the efficiencies of the tested machines. The power losses of both machines are covered in the test station due to a main source supply.

3. Experimental study

According to the back-to-back test circuit (Figure 1), a test bench was created that allows testing using the back-to-back method, the data obtained during the test make it possible to evaluate the energy efficiency of the test circuit as a whole [3,7]. A general purpose industrial asynchronous electric motor (6A series, rated power P_r = 0.55 kW) is used as the tested electric motor and load machine. Since in laboratory conditions it was not possible to use a real traction machine, the physical test model was implemented using low-power machines, while the similarity theory allows the results to be applied to larger machines with high accuracy. A static frequency converter with a braking resistor (Danfoss VLT 2800 series) was used to provide control and source to the motor under test. Data collection during the tests was carried out using an analog-to-digital converter connected to a personal computer via USB bus (National Instruments NI USB 6009 model). Current and voltage sensors working on the Hall effect (current sensor LEM LA25-P, voltage sensor LEM LV25-P) were connected to the phases of the tested electric motor, as well as to the phases of the electric motor used as a load, these sensors were used as data sources.

According to the law of change of main supply voltage and frequency, the experimental study was performed with increased frequency and amplitude of the main supply voltage, the description can be found in earlier publications [8]. Experimental data for effectivity ratio determination of the back-to-back test method are shown in Figure 3. The principal interest is of dependences that show the energy flow in the system, i.e. the power consumption of the tested motor, power produced by the machine in generator mode and the amount of power taken from the main supply for covering the losses [9].

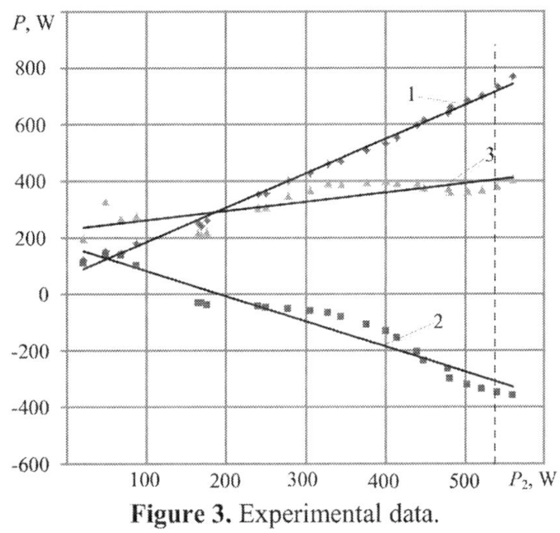

Figure 3. Experimental data.

Figure 3 shows the following experimental data: 1 – power consumption versus power output of the tested motor, 2 – power produced by the machine in generator mode versus power output, 3 – power taken from the main supply for covering the losses versus power output. According to the results obtained by experimental study, it is obvious that in order to achieve the rated mode the tested induction motor needs to consume about 730 W, while the load machine generates 320 W. To cover the losses occurred during the test circuit consumes 400W. Taking into account that efficiency of the tested machines is equal to 0.705, thus for covering the losses circuit consumes about 50% of the power, this theoretically calculated value is confirmed by the value of effectivity ratio, whose value in this particular case is equal to 0.5.

The experimental data confirm the interdependence between efficiencies of tested motors and effectivity ratio. This brings us to the conclusion that the implementation of back-to-back text method makes sense for high rated power induction traction motors testing. As high rated power induction motors have higher energy characteristics, the effectivity ratio in this particular case is equal to 0.8÷0.9.

4. Summary

Experimental studies performed on general purpose induction motors have shown that the proposed back-to-back test method is efficient. Performed tests confirm that the summative efficiency of the circuit depends on the efficiency of the tested machines and can be used for induction traction motors testing. Energy saving in this context will be up to 90 % of rated power of tested machines.

The analysis shows that the back-to-back circuit possesses a number of advantages:

- the experimental circuit consist of the smallest number of auxiliary machines in comparison with alternatives;
- the test circuit allow testing two motors simultaneously that could not be realized in alternatives;
- due to smaller levels of energy conversion the back-to-back test circuit has the highest effectivity ratio.

However, the final decision on usefulness of back-to-back test method implementation should be based on the number of studies performed in association with enterprises involved in designing, production and repair of induction traction motors, and in the following develop the test procedures, standards and guidelines.

References

[1] Kachin O S 2009 *J. Experimental Techniques* **33** 47–50 doi:10.1111/j.1747-1567.2008.00446.x
[2] Kotelnikov A V, Nestrahov A S 2000 *Bulletin of VNIIZhT* **5** 3–15
[3] Goldovskaya A A, Dorokhina E S, Rapport O L, Aslanyan R O 2014 *Proceedings of TUSUR University* **2** 315–318
[4] Beyerleyn E V, Cukublin A B, Rapoport O L 2009 The device for traction motor testing Useful model patent 800018
[5] Sherbakov V G 1998 *Electric traction motors* (Novocherkassk: Nautilus Agency) p 672
[6] Kopilov I P 2002 *Electrical machines* (Moscow: Higher education) p 757
[7] Beyerleyn E V, Tyuteva P V 2014 *Proceedings of Conf. of Young Specialists on Micro/Nanotechnologies and Electron Devices* 359–361 doi:10.1109 / EDM.2014.6882547
[8] Beyerleyn E V, Cukublin A B, Rapoport O L 2006 *Bulletin of higher education Electromechanics* **3** 46–48
[9] Beyerleyn E V, Tyuteva P V 2013 *Naukovedenie* **3** 1–5

IOP Conf. Series: Materials Science and Engineering 1019 (2021) 012052 doi:10.1088/1757-899X/1019/1/012052

Megawatt PV-diesel power stations in the Arctic

V N Dmitrienko, B V Lukutin, M M Popov
Department of electric power and electrical engineering, National Research, Tomsk Polytechnic University,Tomsk, Russia
E-mail: vnd2@tpu.ru

Abstract. The work of hybrid energy complexes of significant installed capacity based on photovoltaic and diesel generation in the northern regions of Yakutia is considered. The considerable remoteness of settlements from the country's central energy system, the cargo delivery inaccessibility, as well as extreme climatic conditions determines the use of a hybrid energy complex with continuous diesel generation (without energy storage use).

1. Introduction

There are many areas without access to centralized electricity supply on the territory of the Russian Federation. There are remote areas of Siberia and the Far East. For example, there are about 130 isolated power supply systems situated in Sakha Republic, their total installed capacity is of 210 MW. Decentralized power supply systems of settlements are built, as a rule, using diesel power plants (DPP) of different installed capacity. Delivering diesel fuel to remote, hard-to-reach areas with undeveloped infrastructure significantly increases the cost of electricity produced. The fuel component in the tariff for electric energy can reach 50 ÷ 80% of the total cost of diesel power plant operation [1, 2].

It is possible to improve the technical and economic characteristics of diesel power supply systems by including in their composition generating plants of renewable energy, in particular photovoltaic power plants (PV). This is achieved by saving diesel fuel and the lifespan of DPP [3].

The most common variants for the schematic construction of photovoltaic diesel hybrid power plants (PVD) are: a) separate work of diesel power station and PV power station; b) PV power plant operation in parallel with the local electrical grid formed by DPP; c) hybrid variant, providing for the possibility of separate and joint work PV and DPP. Variant of parallel work PV and DPP provides for the DPP continuous operation with the part replacement of the diesel generation with the energy of the photovoltaic power station. The advantage of this option is the ability to maximize the use PV installed capacity without expensive storage batteries of electricity [4]. The disadvantage of this option is the need to limit the network inverters power at the level of less than half of the current generation power of the diesel power complex. This limits the contribution of "green energy" to the PVD power station energy balance.

2. Experimental

2.1 Goal of work

The photovoltaic component installed power optimization of an autonomous PVD with continuous diesel generation in the North according to technical and economic criteria.

2.2 Subject of research

The PVD power supply system of the northern village, in particular – Batagay, Yakutia. Batagay village is a large settlement located beyond the Arctic Circle with geographic coordinates 67.65^0 N,

134.64⁰ E and population of 3676 people (for 2017), characterized by the significant electrical load level: $P_{load.max}$ = 5186 kW (winter), $P_{load.min}$ = 907 kW (summer). The installed capacity of the existing diesel power plant is 11 MW. Insolation during the year varies from the polar day to the polar night [5].

2.3 Research methodology

Optimization objective functions are: maximization of installed capacity utilization factor (ICUF) and minimization of produced energy cost price. ICUF – it is the ratio between actual energy production to theoretical, determined according to expression:

$$ICUF = W/P_{INS}\cdot T \tag{1}$$

Where: W – the energy volume produced by part of the hybrid power plant (PV or DPP), for the reading period, kWh; P_{INS} – installed capacity of hybrid power plant part, kW; T – reference period, h.

kWh electric energy cost price – is one of the most important parameters reflecting the economic feasibility of options for constructing generating facilities. For PVD power plants determined according to expression:

$$C_{total} = (p_{N.DPP}\cdot K_{DPP} + C_{DPP} + p_{N.PV}\cdot K_{PV} + p_{N.BS}\cdot K_{BS})/W \tag{2}$$

Where: K_{DPP}, K_{PV}, K_{BS} – investments for hybrid complex construction consisting of diesel, PV power plants and batteries; In the case of building a system operating in parallel with a grid formed by DPP (without batteries) K_{BS} = 0. C_{DPP} – operating costs for the diesel station functioning.

Operating costs for servicing the PV part of the PVD station functioning are not taken into account, due to their relatively small size.

Operating costs for maintenance of battery storage for the considered PVD option are missing.

$p_{N.DPP}$, $p_{N.PV}$, $p_{N.BS}$ – diesel, photovoltaic power plants and batteries profitability coefficient.

Target functions optimization is carried out using a computer model of the power supply system with PVD built on the basis of a specialized software product for designing photovoltaic systems PVsyst and software complex MS Excel. The optimization algorithm for the scheme with continuous diesel generation is shown in Figure 1.

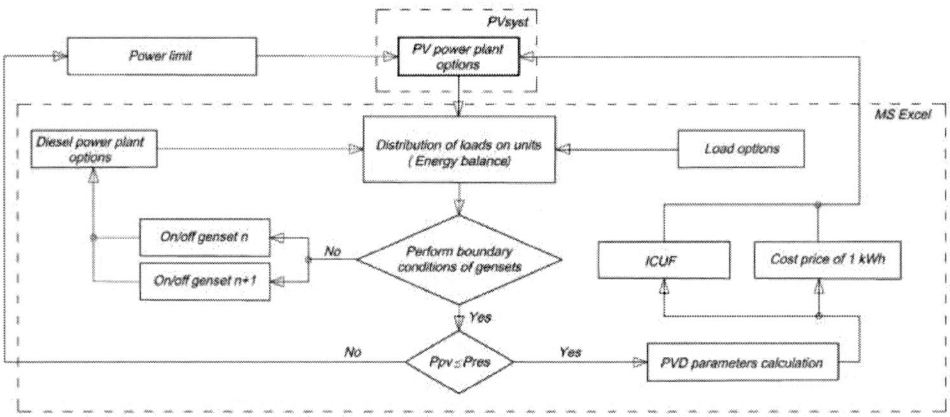

Figure 1. The optimization algorithm of installed capacitance PV power plant working parallel with local diesel power supply system.

According to the presented algorithm, the PV part initial data of the hybrid power plant are entered into PVsyst: current daily insolation with hourly sampling, PV panels quantity taking into account their orientation to the terrain, PV panel's type, spacing between rows, etc. At the output, the operator gets hourly data on the electrical energy production from PV power plant which are further exported to MS Excel, and averaged to the characteristic days of each month by insolation.

In the MS Excel software environment, the hourly data of the village load schedule are recorded, which are also reduced to the characteristic days of the month for the electricity consumed. As a result, characteristic days of each month are formed, taking into account electricity consumption and the diesel and photovoltaic parts of PVD generation.

Information on the DES composition and diesel units parameters (installed capacity, load limits) is entered. The diesel units loading optimized between operating and backup generators by redistribution taking into account the stability conditions of the PV network inverters – limiting its current capacity at the 30-40% level of PVD generation [6, 7].

If the conditions are met, the values of the objective functions are calculated: the ICUF and the electricity cost. The obtained values are analyzed by the operator, and, if necessary, the study is repeated with the modified PV installed capacity to obtain optimal results.

It should be noted that for the PVD with continuous diesel generation, the required power of the diesel power plant is not static, it is constantly changing and depends on many factors: varying seasonal and daily insolation, corresponding changes in the electricity consumption schedules of the electrification facility, etc.

Daily change graphic dependencies in the generated power of the PVD are shown in Figure 2. In the winter period, the decrease in the DPP power in the daytime is determined by the PV generation, corresponding to the change in insolation. With the PV installed capacity, PPV =1.5 MW all the energy in February is consumed by the load, replacing the corresponding part of the DPP energy. The current values of the PV and DPP generated capacity in this month correspond to the conditions for the stability of their parallel operation. Together with the increase in the length of daylight in April, the intensity of solar radiation goes up too, growing the photovoltaic part contribution of the energy complex to the PVD energy balance. For the considered installed capacities, the conditions for the parallel operation stability of the power plant photovoltaic and diesel parts at noontime of the April characteristic day are violated (see Figure 2).

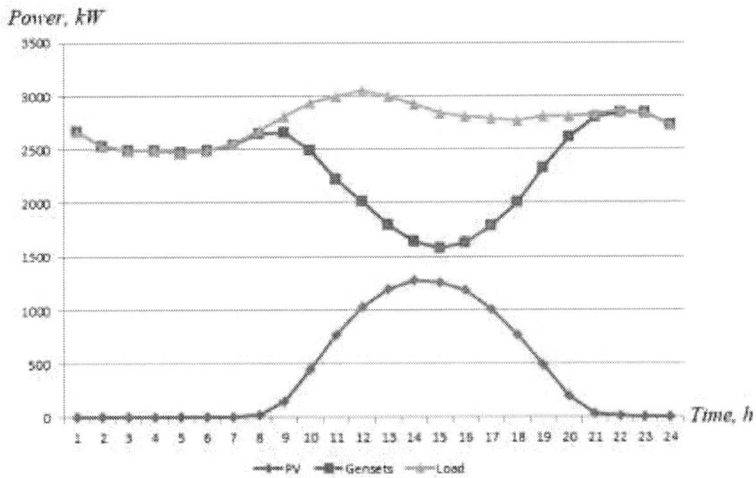

Figure 2. Hybrid complex power change, PPV =1.5 MW.

Using the developed PVD mathematical and computer models in accordance with the algorithm shown in Figure 1, the main energy indicators of the hybrid power supply system photovoltaic part in the Batagay settlement were optimized.

3. Research results

In the research process, optimal values of target functions and their monthly changes during the year were determined (see table 1). The meteorological station Verkhoyansk, nearest to the Batagay settlement, was used as a source of initial meteorological data.

Table 1. Main characteristics of hybrid power plants.

	PV power capacitance					
	1 MW		*1.5 MW*		*2 MW*	
Month	ICUF PV, %	En. cost, rub/kWh	ICUF PV, %	En. cost, rub/kWh	ICUF PV, %	En. cost, rub/kWh
Jan.	0.70	88.46	0.70	89.98	0.70	91.18
Feb.	8.82	51.19	8.82	51.94	8.82	52.53
Mar.	22.49	37.74	22.49	38.15	22.19	38.46
April	27.22	32.01	27.22	32.25	25.01	32.43
May	23.34	28.79	23.35	28.92	21.32	29.03
June	21.53	27.23	18.57	27.32	15.25	27.41
July	20.00	26.04	16.33	26.10	13.17	26.17
Aug.	16.53	24.77	15.45	24.79	13.51	24.84
Sept.	12.92	23.09	12.92	23.09	12.92	23.10
Oct.	9.29	21.51	9.29	21.48	9.29	21.48
Nov.	2.86	20.16	2.86	20.13	2.86	20.12
Dec.	0.00	18.93	0.00	18.91	0.00	18.90
Yearly	13.81	18.93	13.17	18.91	12.09	18.90

The diesel generators installed capacity currently participating in joint work with PV was determined in accordance with the optimization algorithm shown in Figure 1.

Graphic dependences, built according to the table 1 data, are shown in Figures 3 and 4 shows averaged annual ICUF values as a function of the PV power plant installed capacity.

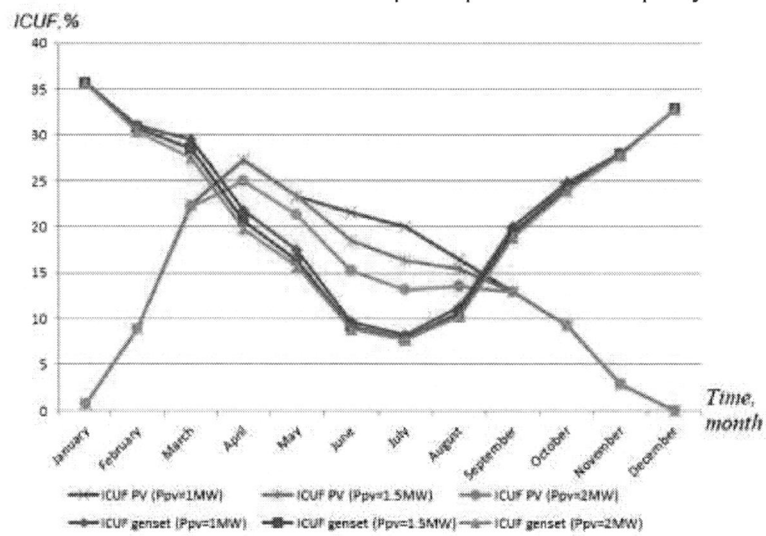

Figure 3. Hybrid power plant ICUF change during the year depending from installed PV plant capacitance.

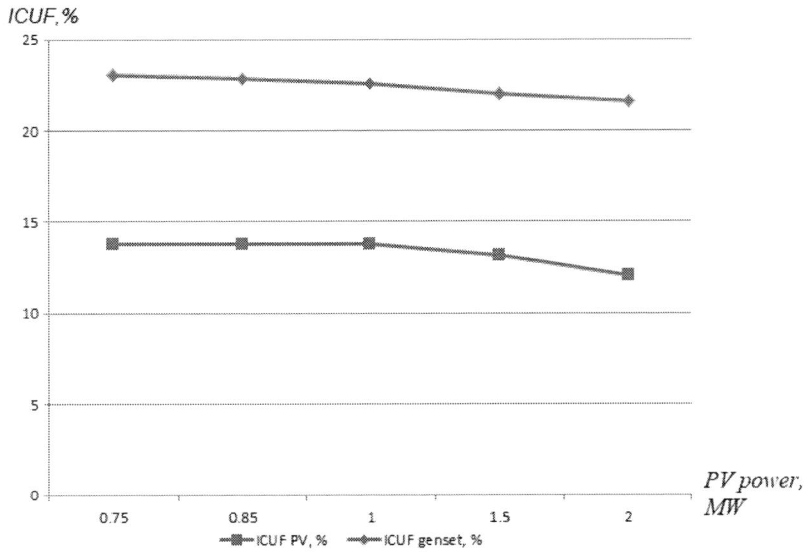

Figure 4. Hybrid power plant ICUF annual change depending from installed PV plant capacitance.

Analysing the table data, as well as the graphical dependencies given, it can be noted that with an increase in the PV part installed power of the hybrid power plant, the photovoltaic generation volume is grown too. Hence, the amount of saved diesel fuel for the electrical energy production increases, which leads to a decrease in the ICUF of the PVD diesel power plant (Figures 3 and 4).

With the growth of the photovoltaic part installed power of the hybrid energy complex, in the sunniest seasons of the year, increase in energy is observed, exceeding the requirements of the grid inverters stable operation for the local diesel power supply. As a result, the installed capacity utilization factor of the PV installed power is reduced. With the PV installed power of 1 MW, the PV ICUF has the maximum value throughout the year. With an increase in the PV power to 1.5 MW, in summertime during the hours of the sun greatest activity and the reduction of the PVD power load, an excess of electricity appears. With further increase in the PV power, the possible electricity "overproduction" begins to manifest itself in the offseason: autumn, spring. Consequently, for the considered example, an increase in the PV power above 1 MW is accompanied by a drop in the PV ICUF value, in the context of day, month and year.

Reducing the PV installed capacity below 1 MW is also not advisable for the reasons of fuel economy - the fuel saved amount will be less, which will lead to an increase in the cost of energy produced by the hybrid power plant.

For the conditions under consideration, the installed power of a 1 MW FES is the optimum point for the indicated objective functions.

Based on the calculated values of the electricity cost, shown in table 1, a graphical dependence (see Figure 5) of the PVD electricity cost change for the changes range in the PVD installed power from 1 to 2 MW was constructed.

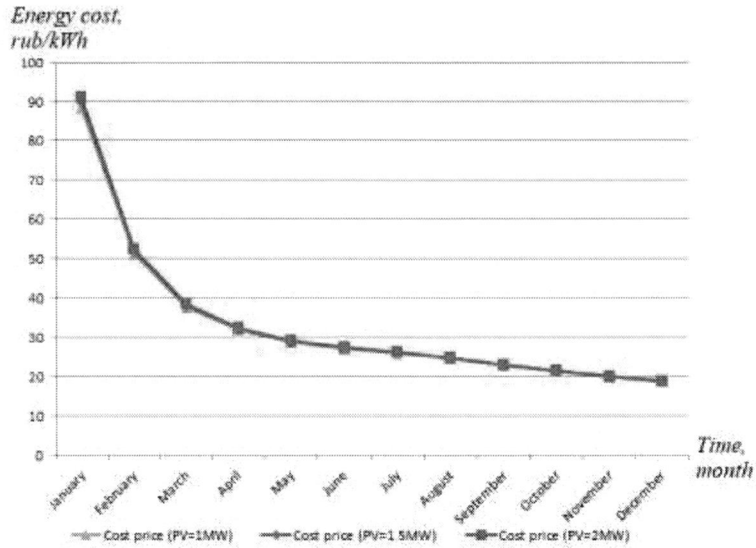

Figure 5. Annual changing of Hybrid power plant kWh depending from PV installed capacitance.

The kWh production cost of electric power of the hybrid power plant with a change in the PV installed power from 1 to 2 MW practically does not change, therefore, an increase in the installed power of the photovoltaic station by more than 1 MW for PVD Batagay village does not provide economic benefits (difference 1 ÷ 3 kopecks). The large PV installed capacity is accompanied by an increase in investments, approximately compensated by the resulting fuel savings.

In the southern Yakutia regions, the annual solar radiation arrival to the earth surface has great importance relative to the northern ones. In this regard, an additional study was conducted on the operation of a hybrid power plant with continuous diesel generation in the south of Yakutia. Studies have been conducted for a PVD power plant in the area of the Aldan city with coordinates 58.60 N, 125.40 E.

Based on the study results, we can conclude - the displacement of the PVD installation site from the northern to the southern territories gives an increase in the average annual PV utilization factor. Due to this, an increase in the optimum installed power of the photovoltaic component of the hybrid energy complex is observed. For Batagay village the optimal capacity is 1 MW, if this complex is shifted to the latitude of the city of Aldan, the optimal PV installed capacity can be increased to almost 1.5 MW. This has a positive effect on the change in the cost of energy produced by PVD.

The results are valid for the existing DPP equipment of the Batagay village. It can be expected that modern diesel generators will improve the achieved technical and economic characteristics of the PVD in Batagay village.

Consider the variant of diesel power plant modernization, providing for the replacement of the existing diesel generators fleet with more modern, automated ones. Modern automated gensets of a large manufacturers number allow their long-term loading in the range from 30% to 100%, without significant negative consequences. The operating range expansion of the generated power leads to an increase in the total reserve power (P_{res}), which diesel generators can provide at any one time and, accordingly, to better use of the photovoltaic component potential of the PVD, while maintaining the stability of its operation. By increasing P_{res}, it becomes possible to reduce the required value of the power produced by the diesel part of the hybrid power plant, which contributes to a certain increase in the PV ICUF, reducing the diesel fuel consumption and, as a result, the electricity production cost. All this improves the characteristics of the hybrid energy complex as a whole.

The DPP provision redundancy is achieved by meeting the conditions for ensuring the maximum load (P_{load}) when two of the most powerful generators exit the operation, it is determined by the expression:

$$\Sigma P_i - 2P_{max} > P_{load.max} \quad (3)$$

Where: P_{max} – units with the highest installed capacity; P_i – unit with i-th installed capacity; $P_{load.max}$ – maximum load power.

To modernize the diesel power plant of the Batagay settlement, it is necessary to install: 1000 kW diesel units – 6 pcs., 750 kW diesel units – 3 pcs. The total installed capacity of diesel power plants will be 8250 kW, which is less than the existing one. For a simulated emergency, the installed capacity of diesel power plants will be 6250 kW, which fully meets the requirements of a reliable power supply of the village.

Units installation with different installed capacity is advisable to ensure greater units maneuverability f a diesel station and to create the possibility of maintaining the required power level at different load levels with the best technical and economic indicators of a hybrid power plant.

The study results before and after the diesel composition replacement of the hybrid power plant reflect a significant increase in the utilization rate of the PV installed power due to the expansion of the gensets operating range.

There is also significant fuel savings due to a decrease in specific fuel consumption for the production of electrical energy from 250 g/kWh to 222 g/kWh.

Some comparative characteristics of the PVD with the existing and modernized DPP, with different installed PV capacities, are shown in table 2.

Table 2. Hybrid power plants characteristics before and after modernization.

	PV power plant capacitance, MW	***0***	***1***	***1.5***	***2***	***2.5***
Fuel consumption (yearly, tn)	Before gensets modernization	5881	5544	5432	5349	5301
	After gensets modernization	5176	4870	4754	4654	4580
PV utilization factor (yearly,%)	Before gensets modernization	0	1	0.89	0.8	0.72
	After gensets modernization	0	1	0.96	0.9	0.84

Based on the values of table 2, graphical dependencies are plotted, shown in Figures 6 and 7, respectively.

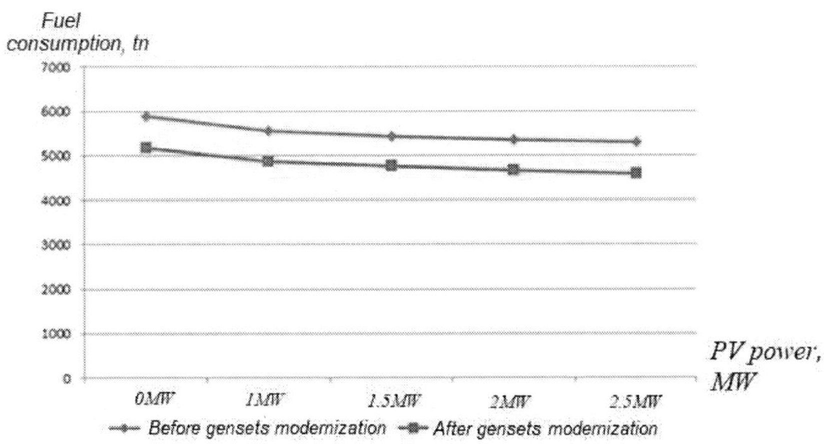

Figure 6. Diesel consumption yearly schedule depending from PV plant installed capacitance for existing and modernized gensets.

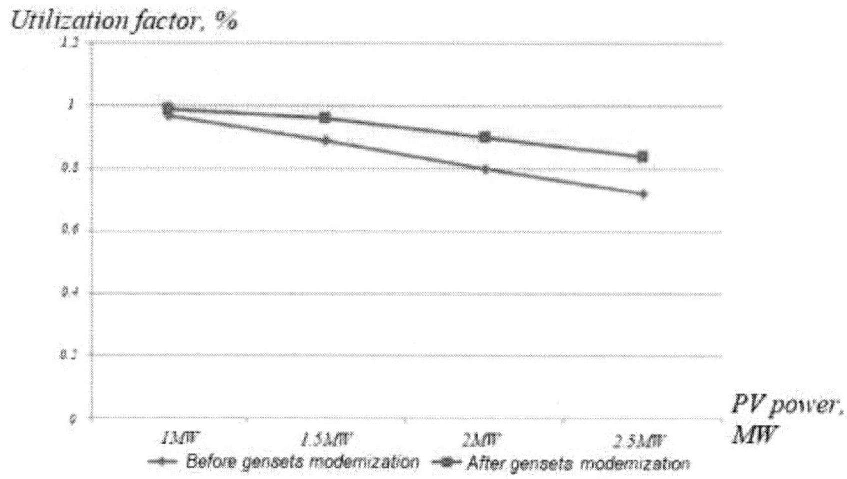

Figure 7. Utilization factor yearly change depending from PV plant installed capacitance for existing and modernized gensets.

When creating a PVD with a modernized diesel power plant, it is advisable to increase the PV installed power to 2 MW, despite the under-utilization of part of the energy (the station loses no more than 25% of solar energy in the summer months), because in addition to the greater amount of fuel saved, the hybrid station gets the best performance at the cost of electricity produced. A further increase in the PV installed capacity is not advisable because of the need to significantly reduce the beneficial use of the energy generated by it in the spring and summer months.

When constructing high-power hybrid PVD power stations based on a continuous diesel generation scheme, it is recommended to upgrade local diesel power stations to ensure the best technical and economic characteristics. This will reduce the dependence degree of the settlement on imported diesel fuel by reducing its specific consumption for the kWh electricity production by new diesel units, as well as expanding their load range. This will increase the PVD sustainability and make it possible to more fully utilize the PV power station capabilities. Also, the upgrade will eliminate the error of personnel when switching by automating the work of the hybrid power plant components.

4. Conclusions

Autonomous PV-diesel power plants with an installed capacity of 1 MW and more, in remote areas conditions with harsh climate, for technical and economic reasons require the use of PVD with continuous diesel generation, without using power storage devices.

The optimization of the PVD technical and economic characteristics with continuous diesel generation was carried out on the basis of a particular example of the Batagay settlement.

It is shown that the PV power plants installed capacity with continuous diesel generation of significant power (several MW) at latitudes similar Batagay village should be about 20% of the maximum electrical load value.

The study was conducted of the terrain latitude influence on the optimal composition of the PVD power plants equipment. With the displacement of a hybrid PVD with continuous diesel generation similar Batagay in the southern territories of Yakutia, the PVD installed power can be increased to 30% of the maximum load value due to the higher solar radiation intensity in the spring and autumn periods when the amount of power consumed by the load is large enough.

References

[1] Dmitrienko V N, Lukutin B V 2014 *Proceeding of XX All-Russian Scientific and Technical Conference Energy: efficiency, reliability, safety* **1** 69–73

[2] Dmitrienko V N, Lukutin B V 2017 *Bulletin of the Tomsk Polytechnic University Geo Assets Engineering* **5** 49–55

[3] Hina A, Palanisamy K 2015 *Renewable and Sustainable Energy Reviews* **45** 431–446

[4] Lim P Y, Nayar C V 2010 *IEEE International Energy Conference* 223–227 doi: 10.1109/ENERGYCON.2010.5771680.

[5] Dmitrienko V N, Lukutin B V 2014 *Modern problems of science and education 25–32*

[6] Celik A N 2002 *Energy Conversion and Management* **44** 1951–1968 doi: 10.1016/S0196-8904(01)00223-6

[7] Gupta A, Saini R P, Sharma M P 2007 *Proceeding of International Conference on Electrical Engineering*1–6 doi: 10.1109/ICEE.2007.4287310

The efficiency of combined electrothermal and electrochemical accumulation of electricity of a photovoltaic power plant

B V Lukutin, D I Muravyev and A V Ryzhkova
School of Energy and Power Engineering, Tomsk Polytechnic University, Tomsk, Russia
E-mail:dim15@tpu.ru

Abstract. The relevance of the research is caused by the fact that renewable energy, in particular, photovoltaic generation is becoming essential support in the decentralized systems in Russia. However, the high cost of the power equipment of photovoltaic power plants is a deterrent to their wide practical application. This paper presents the method for reducing the cost of photovoltaic power plants by optimizing energy conversion processes in isolated power supply systems. The characteristics of equipment for photovoltaic generation and the subsequent conversion of parameters and power storage is an urgent task are presented.

1. Introduction

Currently, there is a problem of a significant discrepancy between daily insolation schedules and electricity consumption. To harmonize the energy balance of a stand-alone photovoltaic system, electrochemical batteries are widely used for that purpose. They are providing storage and exchange of electricity for the target of guaranteed power supply to consumers. Unfortunately, batteries have a limited operating lifetime, they are sensitive to operating temperature and have a high cost.

Typical stand-alone electrification examples are small villages and farm facilities. Among the electrical receivers of such consumers, a significant proportion is occupied by electric heaters. According to [1], the minimum daily consumption rate of hot water per human in a private house is 20~30 liters. Technological needs for milking one cow on dairy farms require an average of 24 to 28 liters of hot water per day [2].

For the consumers of the power supply under consideration, the accumulation of electricity of photovoltaic cells can be carried out not only in electrochemical batteries but also in water-heating installations. These installations will make it possible to reduce the cost of a photovoltaic power plant by reducing the capacity of batteries.

There are a large number of solar water heaters, characterized by simplicity of design and low cost. The specific cost of collectors of various designs differs by an order of magnitude. The simplest solar collectors cost from 2.3 thousand rubles/m2. Such collectors have an efficiency of about 19%. The most advanced evacuated tubular collectors cost already 25 thousand rubles/m^2, which far exceeds the specific cost of photovoltaic panels. The efficiency of such collectors reaches 51~57% [3].

The level of insolation and weather conditions significantly affect the efficiency of solar collectors. Practically, in the conditions of central Russia, only seasonal use of solar water heating installations is possible from March to October with a probability of 70~80% of water heating up to 45~50 Celcius degrees [4]. The efficiency of photovoltaic conversion in a wide range of changes in insolation is almost unchanged, which expands the possibilities of photovoltaic power plants, including for hot water.

In this paper, the authors present investigations of exploring the possibility of reducing the cost of the system for accumulating electrical energy of a photovoltaic power plant through the use of combined electrochemical and electrothermal batteries.

2. Methods and methodology

2.1. Object of the research.
Potential objects of electrifications from power plants of renewable energy are the following: farms facilities, small agriculture enterprises, which are remote from centralized electrical network access. Typical examples are private houses, dairy farms, the power consumption of which includes a significant proportion of electric heaters.

2.2. The characteristics of the combined electricity storage.
The authors show the characteristics of the combined electricity storage of a stand-alone photovoltaic power plant that supplies a private house. Energy needs at home can be evaluated as the sum of electricity consumption determined by social requires. They are the following: an average of 70 kWh per human per month [5], which is 2.3 kWh per day, or 7~9 kWh per family of 3~4 people including additional electricity consumption on hot water. Energy costs for heating water can be determined by the formula [6]:

$$P_t = m \cdot C_w \cdot (Q_2 - Q_1), \tag{1}$$

where t – a time of heating water [s], P is the power of the heating element [kW], m is the mass of water [kg], C_w (4.2 kJ/kg·deg) – specific heat capacity of water, Q_2 and Q_1 – the final and initial temperature of the water.

With an initial water temperature of 10 Celcius degrees, the daily energy demand of a water heater for heating 60~80 liters of water to 50 Celcius degrees will be about 10~13 kWh.

The total daily power consumption of the house for the considered example will be:

$$W = W_e + W_{wh} = 17 : \ 20 \ \text{kWh}, \tag{2}$$

where W_e – is the electricity consumed by electrical loads, W_{wh} – the electric power consumed by the water heater.

This amount of electricity must be generated daily by a photovoltaic power plant, and the necessary amount of this energy must be supplied to the electric receivers through the storage systems.

A typical graph of the daily summer consumption of a rural home is shown in Figure 1 [7,8] together with a photovoltaic generation schedule.

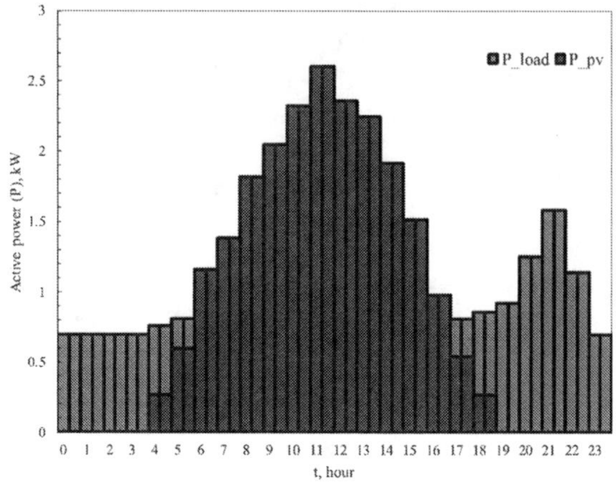

Figure 1. Daily dependencies of the generation of a photovoltaic plant in Russia and the total load of a rural house.

Electricity consumption for water heater of accumulative type can be assumed uniform throughout the day. The average power consumption of the water heater, with a daily energy consumption of 11.5 kWh, will be 0.48 kW. Maximum total power required 1.58 kW. The average power of night load is 0.9 kW (at 21:00 p.m.), daytime – 0.84 kW (at 15:00 p.m.).

The nominal power density of a photovoltaic solar battery is determined by the formula:

$$P_{sb} = P_i \cdot K = 180 \ W/m^2, \tag{3}$$

where P_i – is a benchmark of insolation (1000 W/m^2), K – is the efficiency of monocrystalline silicon panels (18%).

The maximum power density of the solar battery at a latitude of 50 degrees, for example, the village of Kosh-Agach of the Altai Region, taking into account the maximum summer insolation of 750 W/m^2 [8], will be 135 W/m^2. The average specific capacitance of a solar battery per day in these conditions will be determined through the operating load factor of the daily insolation schedule [9] as:

$$P_{avg} = P_{sb} \cdot K_i = 68.4 \ W/m^2, \tag{4}$$

where K_i (0.38) – the operating load factor for the daily insolation schedule in the area of the village of Kosh-Agach. The specific daily energy of the photovoltaic power plant will be 1.23 kWh (W_{spec}).

The area of the solar battery is determined by the ratio of the required daily energy of the load and the specific daily capacitance of the photovoltaic conversion:

$$S_{sb} = W/W_{spec} = 16.3 \ m^3 \tag{5}$$

To design a photovoltaic power plant, single-crystal panels can be chosen, such as, for example, type HH-MONO-200 W. The nominal voltage and current of this panel are 36.5 V and 5.5 A, respectively. To generate the required amount of electricity should be used not less than N panels:

$$N = S_{sb}/S_p = 13 \tag{6}$$

Taking into account the efficiency of the batteries as 0.8, we take $N = 16$. The average daily power generation of a photovoltaic power plant will be $P_s = 1$ kW, respectively, the daily energy is 24 kWh. The maximum noon power of photovoltaic generation is determined taking into account the specific capacitance of the daily insolation graph K_i at the considered latitude [9], as the following:

$$P_{max} = P_s / K_i = 2.6\ kW \tag{7}$$

Graphs of the total daily power consumption and generation of photovoltaic power plants for the considered example are shown in Figure1. According to the nature of the distribution of these quantities during the day, there is a noticeable discrepancy between the maximum of insolation and maximum loads. To eliminate this contradiction, it is necessary to use the energy storage.

3. Research results

Universal storage of electricity is an electrochemical battery. The daily energy balance of power consumption of a photovoltaic power supply system of a house with a battery storage for the example in question can be obtained by the following formula:

$$(P_{si} - P_{ei} - P_{hwi} + P_{ai}) \cdot t_i = \min + P_{ai}, \tag{8}$$

where P_{si}, P_{ei}, P_{hwi}, P_{ai} – is, respectively, the power of photovoltaic generation, electricity consumption, electricity consumption for hot water supply, charge or discharge of a battery in the i-th time interval, taken to be as 1 hour.

An optimum energy balance, taking into account the discreteness of the equipment, will correspond to a small excess of photovoltaic generation (8), which will be unclaimed from the moment the batteries are fully charged. For the considered example, using the data of table 1, the authors found that the energy of charging batteries during the day is 11.75 kWh. The discharge energy is 9.95 kWh.

Table 1. Aggregated data of calculations

t, hour	Power balance parameters		
	P_s, kW	P_e, kW	P_{sb}, kW
1	0	0.7	-0.7
2	0	0.7	-0.7
3	0	0.7	-0.7
4	0	0.7	-0.7
5	0.27	0.76	-0.49
6	0.6	0.81	-0.21
7	1.16	0.92	-0.24
8	1.38	0.97	0.41
9	1.82	0.92	0.9
10	2.05	0.81	1.24
11	2.32	0.81	2.22
12	2.6	0.81	1.79
13	2.36	0.86	1.5
14	2.25	0.81	1.44
15	1.92	0.81	1.11
16	1.51	0.81	0.7
17	0.98	0.81	0.17
18	0.54	0.81	-0.27
19	0.27	0.86	-0.59
20	0	0.92	-0.92
21	0	1.25	-1.25
22	0	1.58	-1.58
23	0	1.14	-1.14
24	0	0.7	-0.7

Thus, the daily exchange rate of electricity will be 10 kWh. Taking into account the permissible (for reasons of service life of batteries), the discharge depth of 50% relative to the nominal capacity of the battery energy reserve will be 20 kWh. A sequence diagram of the battery's energy reserve is shown in Figure 2. *Curve 1* shows the change in charge per day, starting with the original 100% charge. *Curve 2* corresponds to the established periodic daily operation mode of the storage device

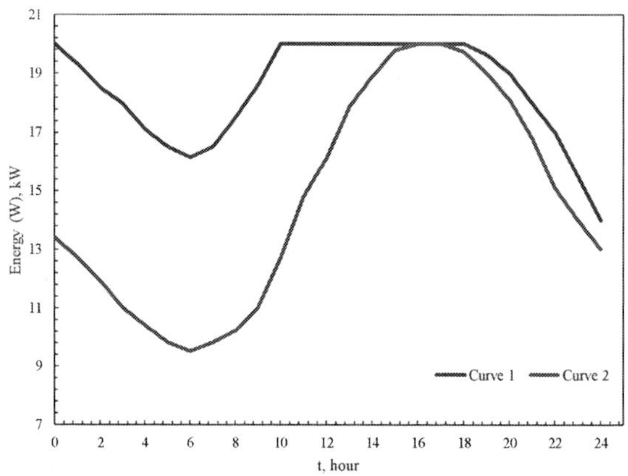

Figure 2. The cyclogram of the battery power reserve.

The cost of a rechargeable battery, with a unit cost of lead-acid batteries of 0.06 $/Wh [10], is $ 1200. Another variant of the power supply system with the release of hot water supply and accumulation of the necessary electric power for this purpose as a supply of hot water in a thermally insulated tank will reduce the capacity of electrochemical batteries by the amount of energy required for hot water supply at home. For this example, this value will be 11.5 kWh. The condition for this is to heat the daily volume of water to the required temperature during the active operation of the photovoltaic converters with a sufficient level of insolation. For this example, this time from 8 to 16 hours. During this time, a 1.5 kW electric water heater will provide the necessary heating of water with a power consumption of about 12 kWh.

The price of accumulative electric water heaters with a volume of 100 liters for most manufacturers ranges from $ 120 to $ 220 [11]. In this case, to provide electricity for electrical loads will require batteries capable of exchanging energy in the amount of 3.5~4.5 kWh per day. The total capacity of electrochemical batteries of a photovoltaic power plant will decrease to 7~9 kWh. The cost of storage devices will be $ 420~530. The rechargeable batteries will be charged from 6 to 18 hours, consuming the charging power in accordance with the current energy balance of the stand-alone photovoltaic power plant.

The cost of a combined solar energy storage system will not exceed $ 750, which is almost 2 times cheaper than electrochemical storage for this example. Figure 3 shows a block diagram of a photovoltaic power plant with a combined storage of solar energy.

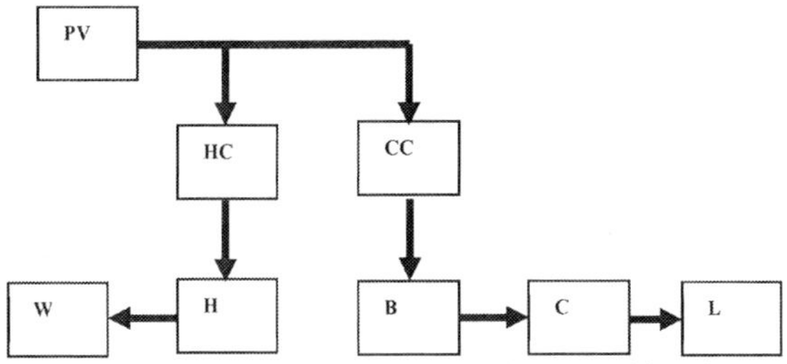

Figure 3. Block diagram of a photovoltaic power plant with a combined accumulation of solar energy.

The electricity from the photovoltaic panels (PV) is fed to the intelligent controllers: the charge (CC) of the battery (B) and the power to the water heater (HC). Battery energy through an autonomous inverter (C) feeds electrical loads (L). A water heater (H) receives electricity from photovoltaic panels that are not currently required by the storage battery. The water heater of the accumulative principle of action ensures the accumulation of solar energy during periods of maximum insolation into the thermal energy of water and ensures its storage during the day due to thermal insulation. When the water heater is on, the power of the photovoltaic power plant exceeds the electric consumption of the water heater – 1.5 kW.

Ensuring the maximum use of insolation is achieved by using MPPT technologies (Maximum Power Point Tracking) in the design of the considered controllers. These intelligent technologies ensure the operation of panels in the mode of generating maximum power corresponding to the current level of insolation.

The algorithm of the functioning of the circuit shown in Figure 3 can be formulated as the following conditions:

a) With a high level of insolation, exceeding the optimal charge energy of electrochemical batteries, solar energy is converted into electricity and is distributed to the battery charge and water heating in the water heater. At the same time, the electric energy of a photovoltaic power plant is accumulated in electrochemical batteries and in the form of thermal energy of heated water. With charged batteries the energy is spent on water heating.

b) With an insufficient level of insolation, the electrical energy of the panels is directed only to the charge of the batteries.

c) In the absence of insolation, electrical and thermal characteristics are fed from the corresponding batteries of electrical and thermal energy.

4. Conclusion

A. The reduction of energy intensity

The use of combined solar energy storage systems with electrochemical and heat accumulators in stand-alone photovoltaic power plants makes it possible to reduce the energy intensity of electrochemical accumulators by the amount of energy consumption for thermal characteristic.

B. The efficiency growth

The efficiency of direct conversion of photoelectricity into heat is significantly higher that its conversion through an intermediate in the form of an electrochemical battery.

C. The reduction of unitary cost

The unitary cost of the thermal accumulator is several times less than the electrochemical one. The work shows almost a double reduction in the cost of a combined solar energy storage system of a photovoltaic power plant for a private house.

D. The reduction of installed power of inverter

It is advisable to supply the electric water heater with DC from photovoltaic panels, which makes it possible to reduce the installed power of the inverter to a value determined by the peak value of electrical loads of AC. Considering the lower cost of the water heater controller compared to the inverter, it also reduces the cost of this photovoltaic power plant.

References

[1] DomEkonom.su: Goryachee vodosnabzheniye chastnogo doma [Hot water supply of a private house]. Informatsionnii portal, 2014. [Online]. Available at: https://domekonom.su/gvs-chastnogo-doma.html
[2] *Water for Sustainable Food and Agriculture Food and Agriculture Organization of the United Nations 2017 A report produced for the G20 Presidency of Germany.* Available at: http://www.fao.org/3/a-i7959e.pdf
[3] Shishkin N D 2015 *Bulletin of Astrakhan State. tech. University* 51–59
[4] Belgorodskiy insitut alternativnoy energetiki: Energiya solntsa [Belgorod Institute of Alternative Energy: Energy of Sun]. Informatsionnii portal, 2019. [Online]. Available at: http://altenergo-nii.ru/renewable/solar/
[5] Allcott H 2011 Jornal of Public Economics **95** 1082–1095
[6] Retseck G 2019 How It Works: Water Heater. Available at: https://web.archive.org/web/20061123014507/http://www.popularmechanics.com/home_journal/how_your_house_works/1275141.html
[7] Alunov A N Onlayn electric: Interaktivniye raschety sistem elektrosnabzheniya [Online Electric: Interactive calculations of power supply systems]. Available at: https://online-electric.ru
[8] Scientific and Applied Climate Handbook. Series 3. Long-term data. Parts 1–6. Vol. 20. Tomsk, Novosibirsk, Kemerovo Regions, Altai Territory. St-Petersburg, Gidrometeoizdat Publ., 1993. 718 p.
[9] Lukutin B V, Shandarova E B, Fuks I L 2016 *Proceeding of 2nd International Conference on Industrial Engineering, Applications and Manufacturing* 7911430 doi: 10.1109/ICIEAM.2016.7911430
[10] Ganzha D Sovremenniye inzhenerniye resheniya dlya COD [Modern engineering solutions for data centers]. Network Solutions Log [Online]. Available at: https://www.osp.ru/lan/2018/04/13054571/
[11] Butuzov V A, Lychagin A A 2002 *Electronic journal «Ecological systems» 1–9*

Automated filtering and polygonal reconstruction of virtual skull model based on computed tomography

M G Grif, A M Grif
Automated Control Systems Department, Novosibirsk State Technical University, Novosibirsk, Russia
E-mail: grifmg@mail.ru

Abstract. A method of automated filtering and polygonal reconstruction of the skull virtual model using the methods of "Marching cubes" and "Connected component labeling" is considered. An algorithm for effective software implementation is described. A computational experiment on computed tomography data consisting of 244 DICOM images is presented. The experimental results confirm the correctness and computational efficiency of the algorithm.

1. Introduction

According to statistics, about 30% of people with traumatic brain injury have a trepanation defect [1]. In modern medical practice, the method of 3D printed implants, which are designed for each patient individually by using a virtual skull model, is recognized as an advanced solution [2]. In the vast majority of cases, the results of computed tomography (CT) examination conducted in a medical institution are used to obtain this model. Moreover, these results are used to create skull models both voxel-based and polygonal which are user-friendly [3].

However, the created polygonal model of the skull may require significant improvement, because it may contain many small three-dimensional objects the so-called artifacts, that do not correspond to the patient's skull. The fact is that the CT scans may show an imaginary intensity of x-ray radiation that does not correspond to the tissue parameters of the study area, which occurs due to physiological (for example the presence of dental crowns), systemic or hardware nature.

This paper is devoted to the description of the artifacts filtering method at the stage of the polygonal model reconstruction.

2. Theory

3D polygonal model of the patient's skull. The virtual model of the patient's skull is reconstructed from a set of CT scans taken in a medical institution. Most modern x-ray equipment, including computed tomography (CT) and magnetic resonance imaging (MRI), supports a single standard for storing, transmitting and visualizing medical data. An example of such international standard is DICOM [4], developed by the International Organization for Standardization (ISO).

A quantitative evaluation of the density of the study structures is the scale of the x-rays attenuation the so-called Hounsfield unit scale (HU). To create a 3D voxel-based human body model, the linear interpolation of pixel values of neighboring images is used. According to this model and the correspondence between the HU scale and human organs and tissues (Figure 1), it is possible to segment the study structures. In this case, the area of the skull bone tissue corresponds to the value of +700 HU.

Substance	HU
Air	−1000
Lung	−700
Soft Tissue	−300 to -100
Fat	−84
Water	0
CSF	15
Blood	+30 to +45
Muscle	+40
Bone	+700(cancellous bone)to +3000 (dense bone)

Figure 1. Correspondence between human anatomy and the HU scale.

The voxel-based model is transformed into a polygonal one, and the most effective way to obtain such triangulation of a surface is the "Marching Cubes" algorithm [5]. The substance of this algorithm is to create triangular isosurfaces of a scalar 3D field by evaluating the values of the voxel function. Figure 2 shows an example of using this algorithm [6, 7] to construct a volumetric digital model of the patient's skull.

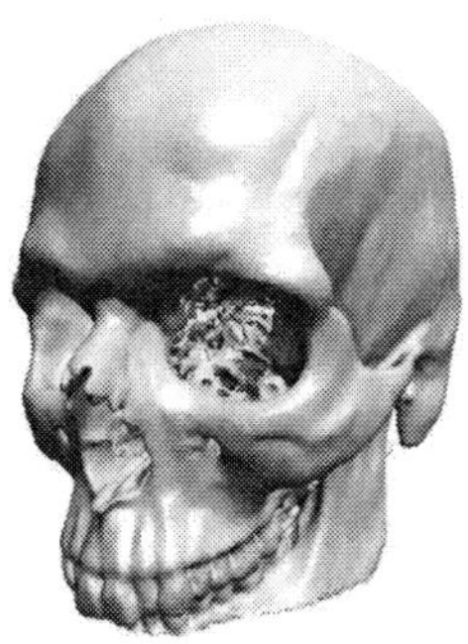

Figure 2. Reconstruction of the 3D skull model.

Artifacts filtering of topical diagnosis. In DICOM images, the imaginary signal intensity may not correspond to the tissue parameters of the study area the so-called artifacts. Artifacts can be physiological, system or hardware in nature. For example, several metal dental crowns which are adjacent can light up the image due to the non-falling of the metal parameter in the coverage area of the HU scale. For such cases, it is advisable to use image filtering. Figure 3 shows an example of the image reconstruction with artifacts in the research area.

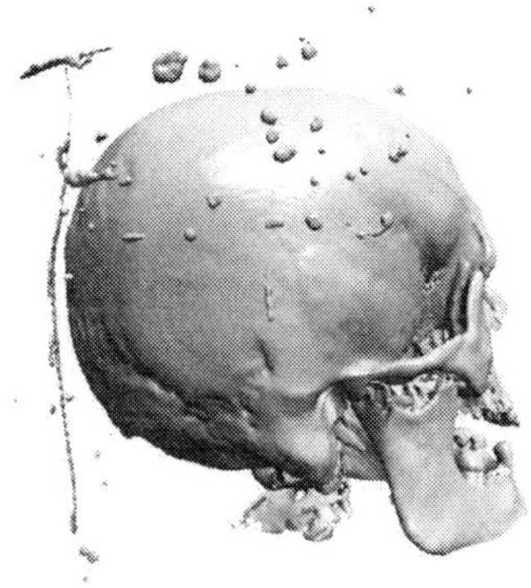

Figure 3. The skull model with the artifacts.

Modern soft and algorithmic filtering tools are mainly represented by "manual" functional. The application allows you to "cut off" the necessary study area with the help of a 3D parallelepiped. This process is simple, but it requires a large number of similar actions to select a new artifact object and its cutting. For example, for the model considered earlier in Fig. 3, it is necessary to make about 20 of such manipulations.

The natural idea of an automatic filtering is to remove objects that are not related to the skull from the research area. It requires the knowledge of the morphology of the study image, i.e. segmenting it into related three-dimensional objects with their identification and subsequent removal of those objects that are not included in the structure of the patient's skull. 3D segmentation of the image can be performed using the "Connected Component Labeling" algorithm [8], which is one of the most effective ways of matrix selection of related components. However, preliminary studies have shown that the "direct" application of a segmentation model of about 5 million polygons (Figure 3) is a computationally expensive task even with the use of preliminary indexing of polygons by an octree.

A method of image filtering combining the process of image reconstruction and filtering is proposed. Its essence is as follows:

At the first stage, a three-dimensional graph of the model described by the function is constructed

$$\chi(i,j,k) = \begin{cases} 1, \textit{if voxel}\ (\mathrm{i, j, k}) \\ \quad \textit{contains polygons}, \\ 0, \textit{else}, \end{cases}$$

moreover, the correspondence of polygons to a given voxel fabric is determined by the "Marching Cubes" algorithm, and for all values of the function $\chi(i,j,k)$ equal to one. Polygons for this voxel are stored in a separate hash table.

At the second stage, the values of the function $\chi(i,j,k)$ are equated so that a certain value $c = 1 \ldots n$ corresponds to a set of numbers $\{(i,j,k)\}$ that describe a geometrically related component. To do this, the "Connected Component Labeling" algorithm is used.

At the third stage there is such a number $\xi \in \{1 \ldots n\}$ that the power (number) of connected voxels of the formed ξ component is the maximum among the whole set of components. It is natural to assume that many voxels will correspond to the geometry of the skull.

At the fourth and final stage, only those polygons that correspond to the voxels found at the third stage are extracted from the hash table built at the first stage.

IOP Conf. Series: Materials Science and Engineering 1019 (2021) 012054 doi:10.1088/1757-899X/1019/1/012054

3. Experimental results

We can consider the application of the proposed approach by using the example: patient's MRI data with a traumatic brain injury obtained in Novosibirsk Research Institute of Traumatology and Orthopedics of Ya.L.Tsivyan.

Tomographic data is represented by a set of 244 DICOM files. Figure 4 shows the 10th, 80th, 140th and 180th images of the series.

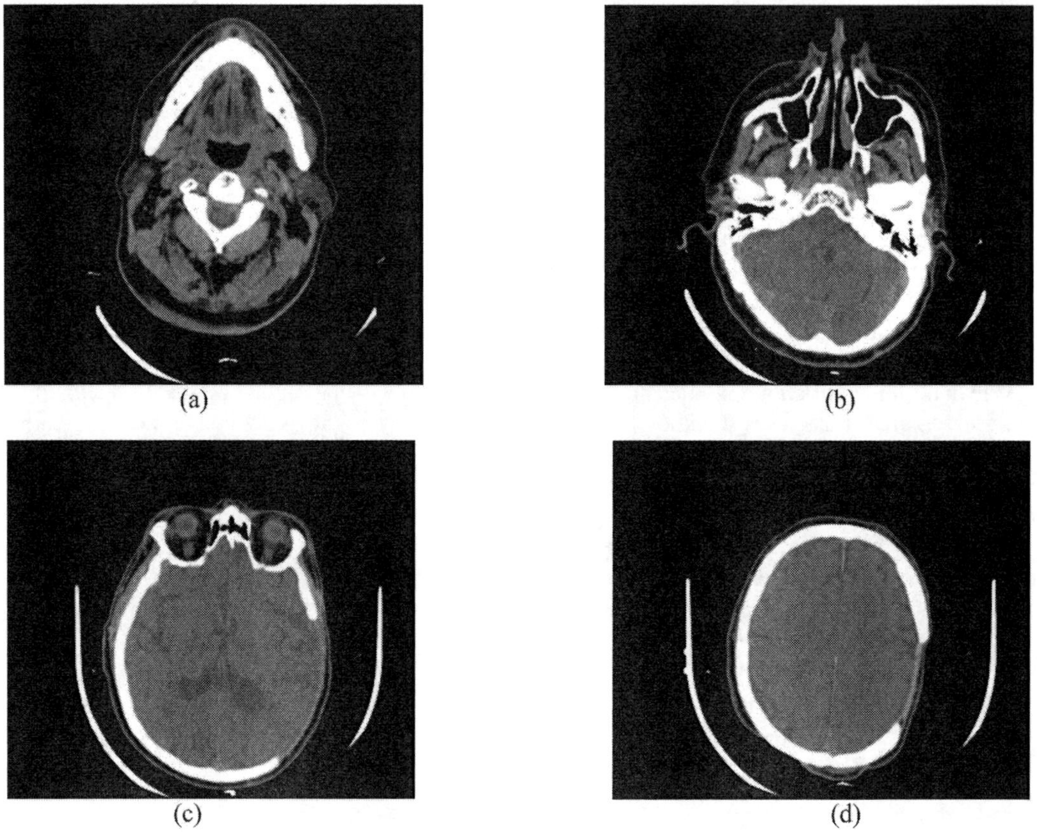

Figure 4. A series of research images are 10 (a), 80 (b), 140 (c) and 180 (d).

The digital three-dimensional skull model is reconstructed from a series of images. Figure 5 shows the obtained model.

It can be seen that in the skull there are many artifacts. Moreover, some of them are formed due to the semicircular design of the tomographic table (it can be seen at the bottom of all the images in Fig. 4), and some reflect the presence of the dental crowns.

Moving to the simultaneous reconstruction and filtering of the practice model. Figure 6 shows the obtained model.

(a) (b)

Figure 5. Skull model with the artifacts: front view (a) and side view (b).

(a) (b)

Figure 6. Skull model after filtration: front view (a) and side view (b).

It is seen that the proposed algorithm automatically filters the image in a proper way. It is worth noting that the reconstruction and filtering of this model using an Intel Core i7-2820QM processor and a frequency of 2.30 GHz takes less than three minutes.

4. Conclusions

The method and algorithm of automated filtering and polygonal reconstruction of the virtual skull model are considered.

The results of the reconstruction and simultaneous filtering of tomographic data, presented by Novosibirsk Research Institute of Traumatology and Orthopedics of Ya.L.Tsivyan, confirmed the effectiveness of the proposed method. The software implementation of the algorithm does not require preliminary indexing of model's polygons and uses a small amount of memory as well as it is computationally low-cost.

References

[1] Likhterman L B 2014 *Traumatic brain injury. Diagnosis and treatment* (Moscow: GeotarMedia) p. 488 (In Rissian)
[2] Eolchiyan S A 2014 *Burdenko's Journal of Neurosurgery* 78(4) 3–13 (In Russian)
[3] Grif A M, Mishinov S V 2016 *Proceedings of the VI Russian scientific and practical conference of young scientists with international participation* 939–942 (In Russian)
[4] *Digital imaging and communication in medicine* Available at:http://dicom.nema.org/
[5] Harvey E C, William E 1985 Patent US 4710876 A. *System and method for the display of surface structure contained within the interior regions of a solid body*

[6] Grif M G, Grif A M 2017 *Proceedings of International multi-conference on engineering, computer and information sciences (SIBIRCON)* 544–549
[7] Gavrilova L O *et al* 2017 *Proceedings of International journal of applied and fundamental research* 209–213 (In Russian)
[8] Dillencourt M B, Samet H and Tamminen M 1992 *J. ACM* **39** 253–280 doi: 10.1145/128749.128750

IOP Conf. Series: Materials Science and Engineering 1019 (2021) 012055 doi:10.1088/1757-899X/1019/1/012055

Prospects for the development of functional expansion of individual low-cost prosthetic limbs for people with disabilities with intelligent sensors

D Gryaznov[1], V Zhmud[1], V Avrmachuk[2], D Malakhov[1], J Nosek[3], L Dimitrov[4]
[1] Department of Automation, Novosibirsk State Technical University, Novosibirsk, Russia
[2] Department of Complex Information Security of Computer Systems, Tomsk State University of Control Systems and Radioelectronics, Tomsk, Russia
[3] Department of Mechatronics and Computer Engineering (MTI), Faculty of Mechatronics, Informatics and Interdisciplinary Studies Technical University of Liberec Liberec, Czech Republic
[5] Vice rector of Technical University of Sofia, Sofia, Bulgaria
E-mail: oao_nips@bk.ru

Abstract. Despite the development of medicine, as well as robotics aimed at helping people with disabilities, most disabled people with missing limbs cannot hope for a full life with the help of bionic prostheses, since these products are made strictly individually, and their control requires expensive surgeries to implant sensors. In addition, such prostheses themselves are very expensive. An alternative solution is to reduce the cost of such prostheses as much as possible and abandon the implantable sensors of signals of the human nervous system. The paper proposes a method for solving such a problem using a variety of sensors that remove signals from the skin surface without invasive surgery.

1. Introduction
Humanitarian goals sometimes force us to think not only about how to solve some difficult problems to restore the motor capabilities of disabled people, but also take into account the price ranges of such products. The incomes of the vast majority of disabled people from all sources are such that they cannot afford to purchase unique prostheses that almost completely replace the lost limbs. Despite the fact that there are many examples of such unique prostheses, the movement capabilities of which are completely identical to the movements of real human limbs, their cost makes them inaccessible to most disabled people. In addition, implantable sensors are needed to control such prostheses, which receive signals directly from human living nerves. Operations to implant such sensors also cannot be widely used, firstly, because of their high cost, secondly, such operations are contraindicated for some patients, and thirdly, not every patient is ready to agree to such additional surgical intervention. This paper offers an alternative method for solving this problem without the need for expensive technologies, without implanting sensors.

2. Statement of the problem
Biomechanical prostheses using invasive transducers are developing rather quickly, but at present these are piece goods of high cost [1–4]. But these items are very expensive. The development of robotics has led to the fact that individual MEMS nodes [5], which in very small dimensions implement complex

electromechanical functions, have become widely available. Also, three-dimensional printing methods make it easy and relatively inexpensive to manufacture plastic parts of any arbitrarily complex shape [6], and computer-aided design methods greatly facilitate the design of such parts. Mathematical packages Solid Works and similar ones [7–9] allow not only to design such parts, but also to calculate their strength, and, if necessary, to correct the structure, for example, by strengthening it with stiffeners.

In order for a disabled person to benefit from the use of an artificial limb, in its simplest form, it is sufficient that he can carry out grasping movements with his hand, as well as turning the hand. In the simplest version, it is enough to bend separately on command the thumb prosthesis and jointly the prostheses of the other four fingers. A more advanced prosthesis may offer separate flexion and extension of the index finger. Finally, if each finger is equipped with a separate actuator at each joint and a separate command input that distinguishes the signal level, then each individual finger can bend and unbend on a separate command. In addition, clockwise and counterclockwise rotation of the wrist can be controlled by another signal, in which the level can indicate the angle of rotation or strength. Ideally, the steering force and steering angle are controlled separately. To improve the control technique, it is necessary to work in more detail with a specific disabled person, and this work can be performed by a computer simulator using neural computation methods.

At the same time, methods of reading neurosignals without implanting electrodes are actively developing [10]. The developers claim that the signals read are directly signals from the human brain. They are probably wrong, in fact, various rather sensitive sensors can read the external manifestations of changes on the skin, which include changes in temperature, humidity, and color changes in the skin that occur under the influence of changes in blood supply can also be read. Individual muscle movements can also be detected using optical sensors or touch sensors. Blood speed and pulse can also be read. Most of these sensors can be quite cheap, since they can be made using MEMS technologies or using hybrid optics and microelectronics (as well as nanoelectronics).

Thus, a lot of information about a person's feelings and intentions can be obtained using sensors that are simply placed on the human body without invasion. It remains only to discuss which sensors to choose, how exactly the sensors should be positioned, and how to establish a connection between the signals from these sensors and the necessary actions of the prostheses.

3. The proposed method for solving the problem

It is known that neural networks are able to reveal characteristic connections between various events by the training method. Mapping can be done by training such neural networks. For example, the patient cannot use those remaining limbs in order to operate the prosthesis, since this makes no sense. The purpose of using prostheses is to provide the patient with additional opportunities that he is deprived of due to the absence of a limb. Therefore, the sensors should pick up signals not from finger movements, but from those areas of the skin that are most convenient for the patient to control prostheses. Presumably, these are the areas of the stump that are closest to the place where a healthy limb should be. But it can also be the scalp, by analogy with how a device for detecting brain activity works [10].

The proposed procedure for establishing the relationship between control signals and control is as follows.

The patient is supplied with a large number of sensors for various skin conditions, including sensory sensors, for example, BPSDual v1.0 [11]. Optocouplers and sensors used to determine the pulse, temperature sensors can also be used. The patient is asked to mentally move an imaginary limb. For a more visual procedure, it is advisable to depict a mechanical limb on a computer screen, which is controlled by signals received from these sensors. The learning process will take place in the same way as the training of an infant to use its limbs. At first, random attempts to carry out the necessary movement lead to failures, but gradually an increasingly stable and successful unambiguous connection is established between the desired movement and the actual movement performed. If signals are not used, these sensors can be further excluded, if there are not enough available signals for control, you can try to find additional points for picking up commands.

4. Additional possibilities

Additional possibilities to this method are created by the use of sensors to control a virtual image formed on a computer. Even if the patient does not yet have a prosthesis, and also in the case when he does have one, but he needs to work with a computer, establishing unambiguous correspondences between the patient's desire and the direction of movement of at least the mouse cursor allows any patient to control actions on the computer without using fingers and without using voice commands.

For example, if with the help of any efforts formed mentally, the patient has learned to stably change some certain signals on the sensors placed on his body, then he can learn to move the cursor up and down, right and left, and also make a "click" ... This allows you to select the desired icon on the screen and click on it, that is, launch any program the user needs. If the keyboard is called up on the screen, the patient can type text in this way. If necessary, and after some training, the patient can also draw structures using such software tools as, for example, Word and Paint programs, can create and fill in spreadsheets, and carry out fairly complex calculations. This opportunity can be especially valuable for people who lack both arms or are paralyzed. A person in this state can write letters, memoirs, even scientific articles (for example, it is argued that Stephen Hawking participated in the writing of scientific articles even in the absence of the possibility of any movements and without the ability to speak) [12].

It may seem that the proposed technical solution requires placing too many sensors on the patient's body. This may require an unnecessarily large amount of wires that will wrap the patient like a spider web. Firstly, these wires can be placed in a small number of fabric belts with Velcro. In addition, individual sensors can be linked to the prosthesis using low-power Bluetooth channels. A large number of such channels can lead to an unnecessary increase in the cost of the device, therefore we do not insist on such a solution, however, in some cases, the connection between the place of pickup of signals and the place of their application may turn out to be inconvenient for using wires, in this case such a solution is also not excluded, the cost this decision is not prohibitively high. According to our estimates, a prosthetic hand with a sufficient number of sensors can be manufactured for 250-350 US dollars, and in case of mass production, this cost may be even lower. The training stand and software costs are negligible compared to these costs, so when these prostheses are mass-produced, they can be supplied for up to USD 100 or less. This can dramatically improve the lives of many patients who do not have enough funds to purchase expensive modern multifunctional limb prostheses.

5. Conclusion

This paper draws the readers' attention to the problem of providing disabled people who do not have sufficient material support with bionic prostheses that could be effectively controlled without the use of invasive sensors implanted into the patient's body. These studies are being developed in cooperation with the Technical University of Liberec and the Technical University of Sofia [13].

References

[1] Biomical arms 2016. Available at: https://habr.com/ru/post/394579/
[2] Morais G D, Neves L C, Castro M C 2016 *Proceedings of the 9th International Joint Conference on Biomedical Engineering Systems and Technologies* **4** 227–231 doi: 10.5220/0005706302270231
[3] Wakita Z, Nagata K, Yamanobe N, Clark M 2019 *The Proceedings of Conference of Kanto Branch* 519—520 doi: 10.1299/jsmekanto.2009.15.519
[4] Park S, Jung Y, Bae J 2018 *Mechatronics* **55** 54–62 doi: 10.1016/j.mechatronics.2018.08.011
[5] SUSS MicroTec - Semiconductor Equipment Manufacturer. Available at: https://www.suss.com/en/mems?gclid=EAIaIQobChMIy5eXpMTK7AIVFPuyCh3IGA3LEAAYAiAAEgKY2_D_BwE
[6] Siemens. Mobility solution. 3D printing for rail industry. Available at: https://www.mobility.siemens.com/global/en/portfolio/rail/services/spare-part-services/easy-sparovation-part.html
[7] Solid Works. Available at: https://www.solidworks.com/ru

IOP Conf. Series: Materials Science and Engineering 1019 (2021) 012055 doi:10.1088/1757-899X/1019/1/012055

[8] Katie Kablack – Industrial Design. Available at: https://www.katiekablack.com/solidworks-library
[9] Solid Works. Products and services. Available at: https://www.3ds.com/ru/produkty-i-uslugi/solidworks/
[10] Neuro Angel. Better me in the better Team. Available at: https://neuroangel.ru/eng
[11] ELEMYO. MYO sensor BPSDual v1.0. Available at: https://elemyo.com/products/sensor_bpsdual_1_0
[12] Physics World. Stephen Hawking' Sientific Legacy. Available at: https://physicsworld.com/a/stephen-hawkings-scientific-legacy/
[13] Zhmud V A, Frantsuzova G A, Dimitrov LV, Nosek J 2018 *Journal of Physics: Conference Series* **1015**(3) 032190.

Research on the effect of test parameters on signal contrast and accuracy of determining the size of a defect

V Radchenko[1], D Palumbo[2], N Ryngach[1], K Bobin[1] and N Kurlaev[1]
[1]Novosibirsk State Technical University, Aircraft Department, Novosibirsk, Russia
[2] Polytechnic University of Bari, Department of Mechanics, Mathematics and Management, Bari, Italy
E-mail: ryngach@corp.nstu.ru

Abstract. Currently, is used several non-destructive testing methods for detecting defects in the composite materials. The most potential is methods non-contact scanning for a large surface area with relative insensitivity to environmental disturbances. In this regard, are developed methods for defect detection in composites, all of which have own peculiarity. A fundamental difference concerns the overall length of time is required for non-destructive testing and analysis of thermographic data from the infrared camera, first of all, for large-scale structure. This research work has been conducted a quantitative way of analyzing data to optimizing the testing parameters in terms of testing time and signal-to-noise ratio. In particular, were carried a few tests out on stringers of carbon fiber reinforced panels. These tests included lock-in thermography technique with simulated defects, which has been used for finding parameters to correct detect dimensions of defects.

1. Introduction

The concept to increase the proportion of composite materials for constructions and components is a normal practice in a lot of areas modern production, aviation, aerospace and shipbuilding industry.
But in spite of the high strength characteristics, the presence of the defect on the surface of composite material may be collapse changed robustness of the structure, so timely non-destructive testing is important on assessing residual strength for a structural component or construction. This's demonstrated in the literature [1].

Currently, is used several non-destructive testing methods for detecting defects in the composite materials: ultrasound [2–4], sherography [5], x-ray [6], electrical potential technique [7], thermography [7–12] and also their combination if it is necessary for detection different kinds of defects quantify. Methods of active thermography are of particular interest, cause method's a really quick and easy for inspection of large-sized constructions. This has an impact on the reduction in the time, alleviation of the financial burden of maintenance and repair organizations aviation technics. Also, possibly the automatization of procedure for inspection and analyze data.

The best time setting for testing/analyze data relationship offered in thermography is techniques as Lock-in (LT), Pulsed (PT) and Stepped (ST/SH). These methods belong to active thermography, it's mean to using a heat source for periodically exciting the specimen surface of heat waves. The presence of the defect in the composite panel may be detected by monitoring the surface temperature decay of the specimen. The defects appearance is as an area of different temperature to a surrounding sound area and it produces an abnormal behavior of the temperature decay curve [11].

The above-mentioned thermography techniques have been used in number of works for the demonstrating propose of the capacity of thermography to detect defects in carbon fiber reinforced panels (CFRP) and carbon fiber reinforced panels (GFRP) composite materials. In particular, those techniques have got especially efficient in the evaluation of interlaminar delamination or debond.

For analyzing thermography data and for visualization defects was developed different approaches, such as a modulated heat source. This aproach have been used for periodically exciting the sample in a desired range of frequencies and finally, accumulation of data about defects in the sample is described theoretically in the work of Mulaveesala [13], and is demonstrated in the works of Palumbo [14–16], where the analyzed data for lock-in techniques as possible with high-level noise for two parameters: amplitude (A) and phase (φ). Lock-in is effective, exact and to a lesser extent susceptible to external influence than Pulsed and Stepped techniques, but requires more inspection time for sample depending on the size surface and depth material because of a few cycles of testing. However, the results of the lock-in approach have not optimized in terms of the inspection parameters and the signal/noise data relationship offered. In this regard in this work is focused on the phase parameter in the processing thermography data, because of the amplitude parameter is less effective for detection defect area. Conducted several testings on the longitudinal stringers of CFRP in order to consider the impact time to heat up and cool the samples on the contrast signal. In addition, quality analyze data is completed with supporting the semi-automatic algorithms.

2.Theory lock-in thermography

Lock-in thermography is based on the generation heating waves inside to inspection specimen for instance by periodically exciting the specimen surface as shown in Figure 1 [7, 10]. The resulting oscillating temperature field in the stationary regime may be recorded on the distance through its thermal infrared emission by an infrared camera. A thermal wave is reconstructed by measuring temperature evolution over the sample surface: by the algorithm might be obtained information about parameters amplitude (A) and phase (φ) of the thermal wave. Phase data are insignificantly dependent of local optical (surface roughness and etc.) and infrared surface features (as an emission), and the phase signal deeper penetration into the material than of the amplitude signal [7, 10], and this impacts on the analyze results.

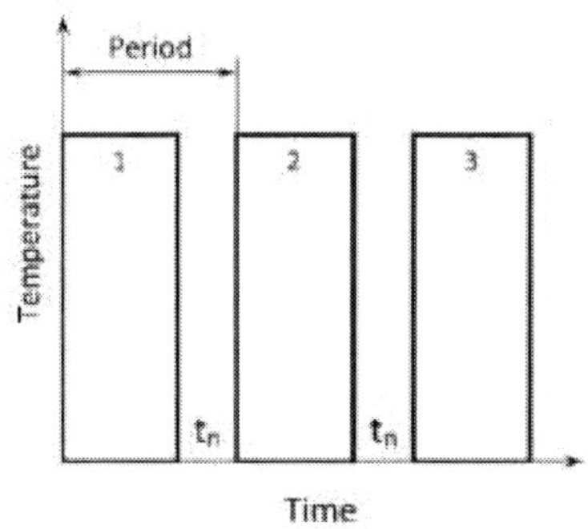

Figure 1. Cycle heating up and cooling time of halogen lamps for Lock-in approach.

The phase parameter has had a close link between depth z and thermal wave:

$$\varphi(z) = \frac{2\pi z}{\lambda} = \frac{z}{\mu} \tag{1}$$

where λ is the length of thermal wave and μ is length of the thermal diffusion:

$$\mu = \sqrt{\frac{2k}{\omega \rho C p}} = \sqrt{\frac{2\alpha}{\omega}} \tag{2}$$

where k is the thermal conductivity, W/(m×K); ρ is the density, kg/m3; Cp is the specific heat at constant pressure, J/(kg×K); ω is the modulation frequency, Hz; α is the thermal diffusivity, m^2/s.
Equation (1) might be extended to composite laminates and this displays a limitation the analysis in a near-surface region in high-frequency thermal ways [11, 17], while lower modulation frequencies propagate deeper and slowly. A serious weakness method remains that the operator has to change and to perform different parameters for inspection, changing the frequency of thermal excitation for depth evaluation. In Lock-in thermography, a modulated square wave has been shown in Figure 1 with using periodical heating up and cooling the surface specimen by two halogen lamps as a heat source with a power 1.5 kW each. It should be noted that the thermal response of composite material contains data about higher frequencies proportional to the main frequency, as shown in other works [16, 17]. And by way of decomposing the thermal response in the time domain as the sum of a singular sinusoidal wave, getting:

$$T_m(t) = a + bt + \sum_{n=1}^{\infty} \Delta T_n \, sin(n\omega t + \varphi_n) \quad (3)$$

where *t* is time, *s*; *a* and *b* constants of the model of the material average temperature growth; ω is the modulated frequency of the main harmonic, in other words, the first Fourier component, Hz; ΔTn is the amplitude of n – th Fourier components; φ_n is phase of n – th Fourier components, n =1,3,4,7.

Values of n-th components invent odd numbers, because of the modulated square wave has had heating up and cooling steps, if the experiment has got the three cycles (Figure 1) then the value n = 5.

3.Proposed method and set-up

In this work has been conducted non-destructive testing on the 6th stringers of CFRP with delamination (Figure 2).

Table 1 is demonstrated stringer in which b1 is the dimension of the defect in the front side sample, and b2 is the dimension defect of backside sample because of the delamination is cross-cutting nature. The D and L are the width and length of the sample. The S is length connective part in which is the defect.

Table 1. Size dimension of stringers

Type of stringer	Size Dimension				
	b1, mm	b2, mm	S, mm	D, mm	L, mm
1D	30	37			
2D	6	11			
3D	1	7	66	55	290
1N	27	3.1			
2N	14	24			
3N	40	39			

In the NDT standard ISO9712 thermography is considered as a superficial technique, for the composite material, it is more a volumetric technique and it being able to analyze with up to 6–8mm thickness of the structure, i.e. the vast majority of composite components [12]. The experimental set up in working order is shown in Figure 2.

Figure 2. Experimental set-up with CFRP stringer

The parameters of non-destructive testing for the detection of delamination in the specimen using Lock-in thermography taking into account different interval of heating up and cooling time in table 2.

Table 2. Parameters for the lock-in technique

Modulation Period, s	Frame rate, Hz	Total frames	Number of cycles
90	2	550	3
60	2	370	3
40	4	500	3

For analyze the dimension of delamination or rather to obtain estimated defect area (AER) is used the semiautomatic algorithm in MATLAB (Figure 3).

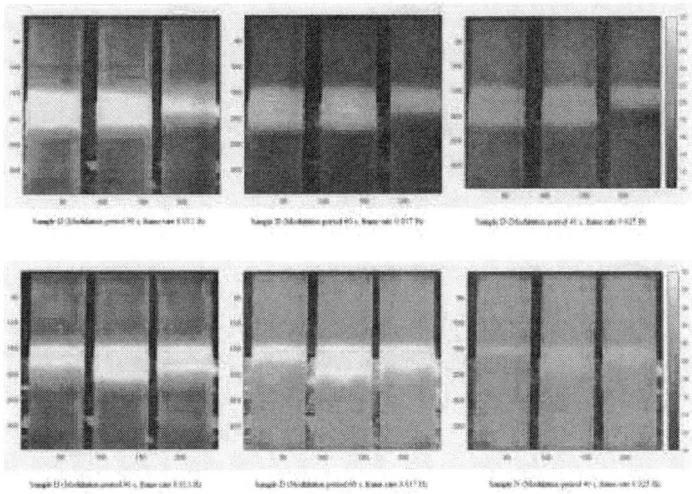

Figure 3. Samples D and N with different parameters of testing.

As can be seen, changing parameters non-destructive testing impact on quality results of thermography data. It is necessary to define the optimal balance between the better values estimated defect area and inspection time. Increasing the inspection time depends on the size of the testing area and depth of composite material.

The existence of only the phase signal along the profile doesn't give in the practical context find out the exact dimension of the estimated delamination area. Available methods are labour-intensive, difficult to achieve and cannot be automaty (Figures 4–6). So, in this work is used a quantitative data analysis. This approach has been widely used for testing time reducing. The theoretical basis of this approach in the work of Palumbo [16] was mentioned. This approach was used in a semi-automatic algorithm in quantitative measured data analysis.

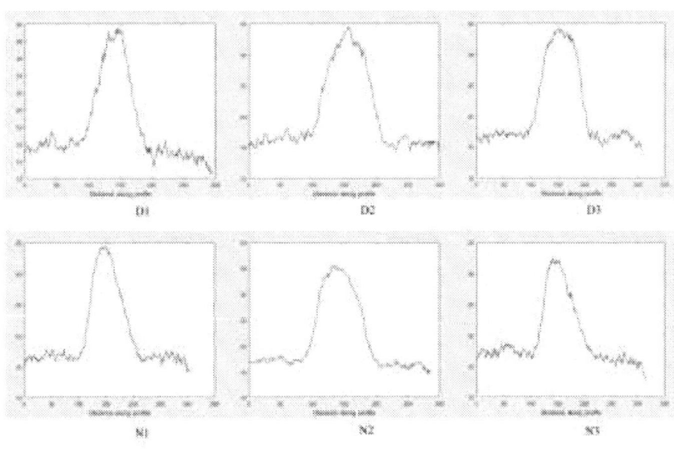

Figure 4. Phase signal along the profile in samples D and N (sinusoidal wave, modulation period 90s, modulation frequency 0.011 Hz).

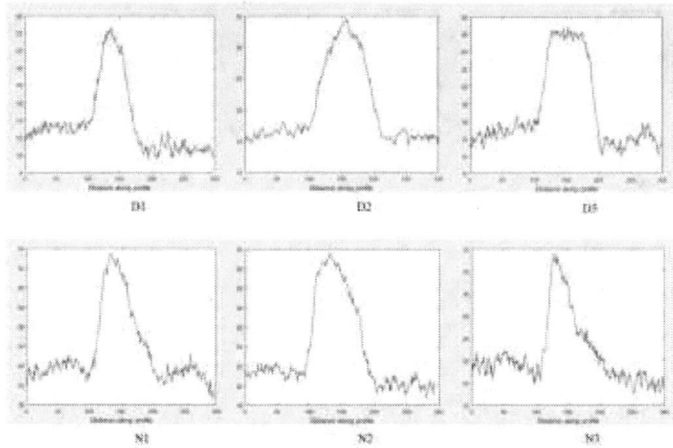

Figure 5. Phase signal along the profile in samples D and N (sinusoidal wave, modulation period 60s, modulation frequency 0.017 Hz).

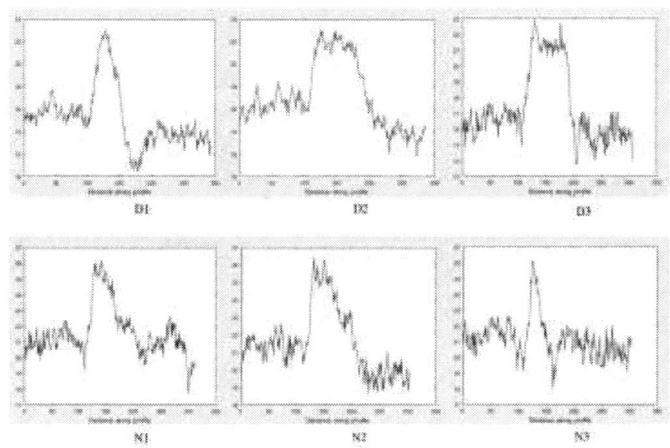

Figure 6. Phase signal along the profile in samples D and N (sinusoidal wave, modulation period 60s, modulation frequency 0.017 Hz).

The main idea the quantitative data analysis lies in the fact to the values in each pixel which is more than the threshold value is changed in the value 0, and values less are changed in the values 1 this's the delamination area. The threshold value (Th) has been defined in the equation (Eq. 4):

$$Th = \alpha[mean(A_d) - mean(A_s)] + mean(A_d) \quad (4)$$

where A_d representing the area with defect, determined by n x n pixels image, centered on the peak value of the signal; A_s is the sound is obtained as previously exposed.

4.Data analysis

The application of the quantitative data analysis approach for lock-in technique allow obtaing binary images (Figures 7–9).

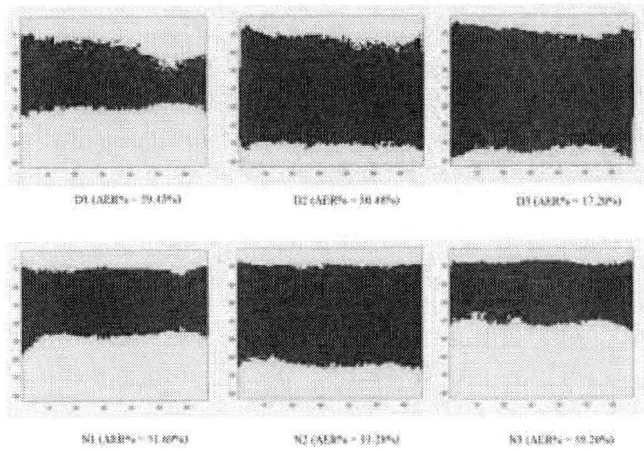

Figure 7. Binary images for detection of delamination in samples D and N (sinusoidal wave, modulation period 90s, modulation frequency 0.011 Hz).

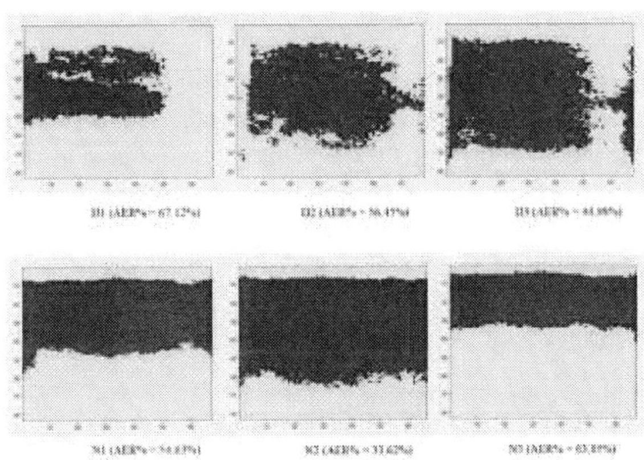

Figure 8. Binary images for detection of delamination in samples D and N (sinusoidal wave, modulation period 60s, modulation frequency 0.017 Hz).

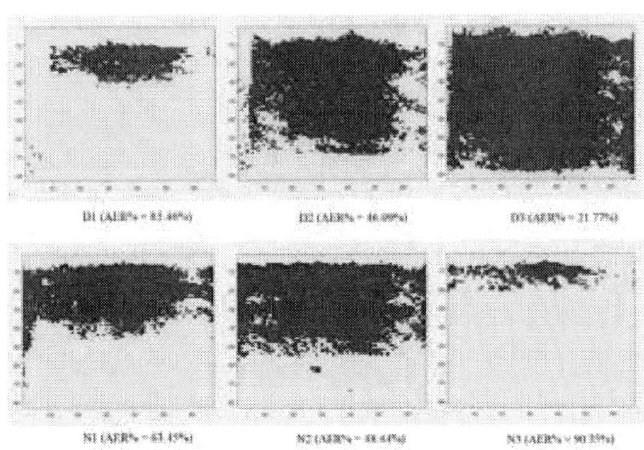

Figure 9. Binary images for detection of delamination in samples D and N (sinusoidal wave, modulation period 60s, modulation frequency 0,017 Hz).

IOP Conf. Series: Materials Science and Engineering 1019 (2021) 012056 doi:10.1088/1757-899X/1019/1/012056

5.Conclusion

As was expected, deeper defects appear in correspondence with a higher modulation period (lower excitation frequencies). The main results may be summarized as follow:

- The modulation period of the signal has been given in 60 second and frequency 0.017 Hz is shown even satisfactory results if the estimated defect area is an interval between 33.28 %≤AER%≤59.43 %. And this value AER% is enough to conclude about scrapping status of the component.
- The modulation period of the signal has been given in 40 second and frequency 0,025 Hz can be shown sometimes interest results if the estimated defect area is AER%≤21.77 %. However, the feasibility of this proposal needs to be further explored.
- The modulation period of the signal has been given in 90 second and frequency 0.011 Hz is the best parameters for non-destructive testing, showing enough exact results, but the feasibility using.
- The modulation period of signal has been given in 90 second and frequency 0.011 Hz is the best parameters for non-destructive testing, showing the enough exact results, but the usefulness using is achieved for surface of small-scale with the values of estimated defect area in a low interval or multilayer composites with the depth of the defect 6–8 centimetres, since for it necessary a long process of scanning and analyze data.
- In this work lock-in thermography set-up proposed to choose optimal testing times by getting the highest values of signal contrast. A directly dependence of the signal contrast on the time of heating and cooling the surface of the sample has been established, as well as the dependence of the signal contrast on the delamination dimension and depth.

References

[1] Baker AA, Dutton S and Kelly D 2004 *Composite materials for aircraft structures* (Reston: American Institute of Aeronautics and Astronautics)
[2] Castellano A, Fraddosio A, Marzano S and Daniele Piccioni M 2017 *Procedia Engineering* **199**1519–1526
[3] Castellano A, Fraddosio A and Piccioni M D 2017 *Composites Part B: Engineering* **116** 122–136
[4] Hosur M V, Murthy C R L, Ramamurthy T S and Shet A 1998 *NDT & E International* **31** 359–374
[5] De Angelis G, Meo M, Almond D P, Pickering S G and Angioni S L 2012 *NDT & E International***45** 91–96
[6] Nikishkov Y, Airoldi L and Makeev A 2013 *Composites Science and Technology* **89** 89–97
[7] Angelidis N and Irving P E 2007 *Composites Science and Technology* **67** 594–604
[8] Usamentiaga R, Venegas P, Guerediaga J, Vega L and López I 2013 *NDT & E International* **54** 123–132
[9] Yang R and He Y 2016 *Infrared Physics & Technology* **75** 26–50
[10] Badghaish A A and Fleming D C 2008 *Journal of Composite Materials* **42** 1337–1357
[11] Maldague X 2001 *Theory and practice of infrared technology for nondestructive testing* (New York: Wiley)
[12] Palumbo D, Ancona F and Galietti U 2015 *Meccanica* **50** 443–459
[13] Mulaveesala R and Tuli S 2006 *Appl. Phys. Lett.* **89** 191913
[14] Palumbo D, De Finis R, Demelio P G and Galietti U 2016 *Composites Part B: Engineering* **103** 60–7
[15] Palumbo D, De Finis R, Demelio G P and Galietti U 2017 St *Composites Part B: Engineering* **117** 49–60
[16] Palumbo D, Cavallo P and Galietti U 2019 *NDT & E International* **102** 254–263
[17] Pitarresi G 2015 *ExpMech* **55** 667–680

IOP Conf. Series: Materials Science and Engineering 1019 (2021) 012057 doi:10.1088/1757-899X/1019/1/012057

The calculation of the campaign of reactor RITM-200

D E Savitsky and A V Kuzmin
The Butakov Research Center, Tomsk Polytechnic University, Tomsk, Russia
E-mail: denis_savickii96@mail.ru

Abstract. In this paper, the campaign of RITM-200 reactor was calculated. The duration of the campaign was determined taking the net capacity factor into consideration. The calculated duration concurred with the known data. The neutron parameters were calculated using the effective temperature method. The presence of burnable absorber rods was taken into account. Their effect was considered using the diffusional approach. The iterative computations were used to finally determine the temperature of the neutron gas. At the end, the reactivity curve displaying different effects inside fuel, namely fuel and gadolinium burn-out, the poisoning and slagging was drawn.

1. Introduction

RITM-200 is the newest reactor for the icebreaking fleet developed by JSC OKBM Afrikantov. Its appearance is presented in Figure 1. It is designed to produce 175 MWt (55 MWe) to power Project 22220 icebreakers, the first of which entered service in the May of 2019. The RITM-200 is an integrated PWR type, i.e. it contains the equipment of the first circuit inside the vessel of the reactor. One of the features of RITM-200 is the refuelling period, which is up to 7 years. In this paper, the calculation of the refuelling period was carried out using the effective temperature method.

Figure 1. View of RITM-200.

2. The core description

The core contains 199 fuel assemblies. According to [1] the structure of the fuel assembly remained the same as in KLT-40S reactor. Each assembly has 69 fuel rods, 9 burnable type-one absorber rods and 6 type-two absorber rods, and 7 control rods in the center of the fuel assembly (Figure 2). Burnable type-one and type-two absorber rods are distinguished by the density of burnable absorber [1]: in type-one rod gadolinium density equals 2.7 g/cm^3, whereas in type-two rod gadolinium density equals 1.4 g/cm^3.

As the structure of the fuel assembly of RITM-200 is the same with the structure of the fuel assembly of KLT-40S, the mean uranium enrichment can be inferred from the data in [1] of the mass of Uranium-235 and the whole uranium load in KLT-40S. Hence the average enrichment is 14.06%.

Structurally, the fuel rod is a smooth rod with a diameter of 6.8 mm with a length of the fuel containing part of 1200 mm. Chromium-nickel alloy 42HNM (41% Cr, 0.2% Mn, 0.2% Mo, 56.1% Ni, etc.) is used as fuel cladding, because it has a higher corrosion resistance and resistance to disruption of the water chemistry regime than the traditional zirconium alloy E110. The thickness of the cladding is 0.5 mm. Inside the cladding, there is a dispersed fuel composition consisting of uranium dioxide granules with low porosity and a silumin (87% Al, 11% Si, 2% Ni). The volume fraction of uranium-containing material in the fuel pellet is 66%.

A pure gadolinium oxide is not used as a burnable absorber. Instead of it a burnable absorber rods contain composition $3Gd_2O_3 \cdot Nb_2O_5ZrO_2$ which has less swelling as a result of neutrons absorbing.

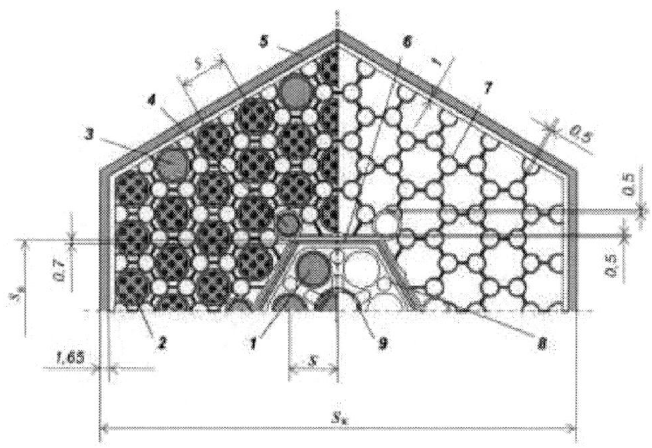

Figure 2. RITM-200 fuel assembly structure: 1 – control rod, 2 – fuel rod, 3 – burnable absorber type-one rod, 4 – burnable absorber type-two rod, 5 – casing, 6 – displacer, 7 – spacing grid, 8 – control rods spacing plates, 9 – central tube.

3. The effective temperature method

In calculations of neutron behavior, all cross sections should be related to the average neutron speed. It should be noted that the Maxwell thermal neutron spectrum gradually passes into the spectrum of slowing down neutrons at a temperature of 293 K at an energy that approximately equals 0.2 eV which is called the "sewing energy". The "sewing energy" is illustrated in Figure 3.

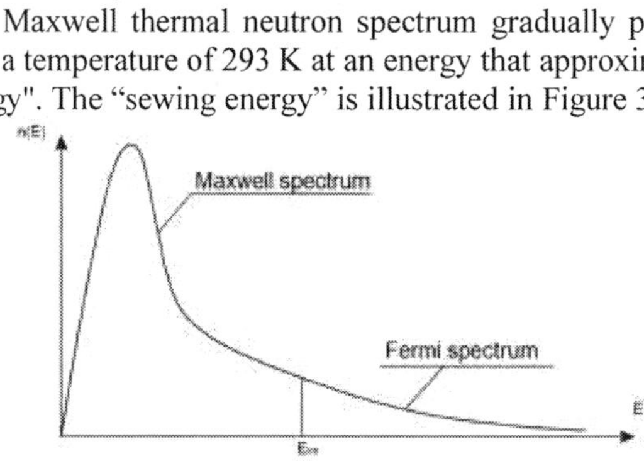

Figure 3. Illustration of the "sewing energy".

In reality, the thermal neutrons distribution does not exactly coincide with the Maxwell distribution, since the absorption of thermal neutrons takes place, the spectrum is shifted to higher energies.

For convenience of calculations in the theory of reactors, it is assumed that thermal neutrons are distributed over the Maxwell spectrum, but have a higher effective temperature (the temperature of the neutron gas — T_{ng}), which exceeds the moderator temperature.

The gadolinium cross sections cannot be used for finding the temperature of the neutron gas directly, as due to its big cross-section of absorption (σ_a = 49000 barns [2]) the strong degradation of the neutron flux occurs in the burnable absorber rod and only the part of the gadolinium volume participates in the neutron absorption. This part is described by the shielding coefficient k_s.

$$k_s = \frac{F_{ba}}{F_c} \tag{1}$$

Where F_{ba} is the thermal neutron flux in burnable absorber rods, F_c is the average thermal neutron flux in the core. As is it shown in the Figure 4, the thermal neutron flux in burnable absorber rods is lower than the average thermal neutron flux due to high absorption. The fall in the F_{ba} decreases throughout the campaign as the density of gadolinium declines.

Therefore, to find T_{ng} it is essential to know k_s.

$$T_{ng} = T_0 \left[1 + C \left(\frac{\sum_j (VN\sigma_a(0.025\ \text{eV}))_j}{\sum_j (VN\zeta\sigma_s(1\ \text{eV}))_j} \right) \cdot \sqrt{\frac{293.6}{T_0}} \right] \tag{2}$$

Where T_0 is the moderator temperature, V and N are volumes per one fuel rod and densities of components of the core, $\sigma_a(0.025\ eV)_j$ and $\sigma_s(1\ eV)_j$ are thermal cross-sections of absorption and scattering, respectively, of materials of the core, C is the empirical constant which equals 1.70 for PWR type reactors. ζ is a mean logarithmic loss of energy in a collision of a neutron and a nucleus.

The parameters of core components necessary to calculate T_{ng} by (2) are presented in table 1.

Table 1. Parameters of core components except burnable absorbers

Materials	$VN\sigma_a(0.025$ eV) (cm)	$VN\zeta\sigma_s(1$ eV) (cm)
U-235	0.310	$5.26\cdot10^{-5}$
U-238	$7.55\cdot10^{-3}$	$2.10\cdot10^{-4}$
O-16	$1.75\cdot10^{-6}$	$2.93\cdot10^{-3}$
H2O	0.0109	0.71
Al-27	$8.39\cdot10^{-4}$	$3.93\cdot10^{-4}$
Si-28	$7.09\cdot10^{-5}$	$6.78\cdot10^{-5}$
Zr-91	$6.75\cdot10^{-4}$	$5.00\cdot10^{-4}$
Cr-52	0.0128	$5.94\cdot10^{-4}$
Mo-96	$1.4\cdot10^{-4}$	$6.53\cdot10^{-6}$
Ni-59	0.0220	$2.88\cdot10^{-3}$
Mn-55	$2.53\cdot10^{-4}$	$1.43\cdot10^{-6}$

In RITM-200 the mean moderator temperature is 296.5°C (569.5 K).

4. Burnable absorber rod calculation

To calculate the shielding coefficient, it is necessary to represent the core as a block with the burnable absorber rod in the center of the block and everything else around it, as it is drawn on the Figure 4.

Other elements except gadolinium may be dismissed as their yield is little.

The calculation is iterative, thus the initial approximation of the temperature of the neutron gas should be assumed. After the assumption the average cross-sections of absorption and fission in the Maxwell spectrum should be calculated.

- For 1/v absorbers:

$$\overline{\sigma_a} = \sigma_{a0} \cdot 0.884 \sqrt{\frac{293}{T_{ng}}} \tag{3}$$

- For non-1/v absorbers:

$$\overline{\sigma_a} = \sigma_{a0} \cdot 0.884 \cdot G_a \sqrt{\frac{293}{T_{ng}}} \tag{4}$$

$$\overline{\sigma_f} = \sigma_{f0} \cdot 0.884 \cdot G_f \sqrt{\frac{293}{T_{ng}}} \tag{5}$$

Transport cross-section:

$$\overline{\sigma_{tr}} = \overline{\sigma_a} + \overline{\sigma_s} \cdot (1 - \mu) \tag{6}$$

Where $\mu = \frac{2}{3A}$, A is the atomic number of the nuclide.

Where G_a and G_f are the thermal correction factors for non-1/v absorbers for absorption and fission cross-sections, respectively.

$$r_1 = \sqrt{4 \cdot V_{Gd} / \pi} \tag{7}$$

Then the average macroscopic cross-section of absorption (Σ_a) and transport cross-section (Σ_{tr}) are calculated in both zones: burnable absorber zone and multiplying zone. Index 2 refers to the multiplying zone.

$$\Sigma_a^{Gd} = \sum_i (VN\sigma_a)_{Gd} / V_{Gd} \tag{8}$$

$$\Sigma_{tr}^{Gd} = \sum_i (VN\sigma_{tr})_{Gd} / V_{Gd} \tag{9}$$

$$\Sigma_a^2 = \sum_i (VN\sigma_a)_2 / V_2 \tag{10}$$

$$\Sigma_{tr}^2 = \sum_i (VN\sigma_{tr})_2 / V_2 \tag{11}$$

Then the material parameters of the media are determined:

$$\chi_{Gd} = \sqrt{3\Sigma_a^{Gd}\Sigma_{tr}^{Gd}} \tag{12}$$

$$\chi_2 = \sqrt{3\Sigma_a^2\Sigma_{tr}^2} \tag{13}$$

IOP Conf. Series: Materials Science and Engineering 1019 (2021) 012057 doi:10.1088/1757-899X/1019/1/012057

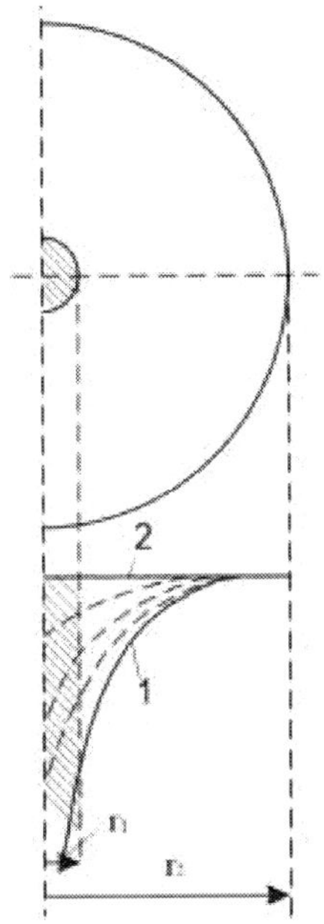

Figure 4. Thermal neutron flux density distribution near burnable absorber rod: 1 – In the beginning of the campaign with initial burnable absorber density; 2 – at the end of campaign when the burnable absorber is completely burned out, r_1 – the radius of burnable absorber rod, r_2 – the radius of volume of water per one burnable absorber rod.

Effective macroscopic cross-section of absorption of gadolinium:

$$\Sigma_{Gd}^{ef} = \frac{\Sigma_{Gd}}{1+(3.593\cdot\frac{r_1}{2}-\gamma_d)\cdot\Sigma_{Gd}+0.426\cdot(\frac{r_1}{2}\Sigma_{Gd})^2} \tag{14}$$

Where

$$\gamma_d = \frac{\chi_2 r_1}{2\Sigma_a^2}\cdot\frac{K_0(\chi_2 r_1)}{K_1(\chi_2 r_1)}+\frac{1}{\Sigma_a^{Gd}}\left[\frac{\chi_2 r}{2}\cdot\frac{I_0(\chi_{Gd} r_1)}{K_1(\chi_{Gd} r_1)}-1\right] \tag{15}$$

The shielding coefficient:

$$k_s = \frac{1}{1+\gamma_d\Sigma_{Gd}^{ef}} \tag{16}$$

5. Computation of the effective multiplication factor

As the shielding coefficient is determined, the temperature of the neutron gas is to be calculated by (2), where the densities of absorber rod components ought to be multiplied by k_s to take into account the part of the rod that does not participate in the process of absorbing neutrons.

If calculated T_{ng} differs by more than 3 percent from the initial assumption, the calculation is repeated from the beginning using T_{ng} calculated in the previous iteration by (16). Equation (2) is omitted as T_{ng} is known from the previous iteration.

The iterative calculation is stopped when found temperature of the neutron gas by (16) concurs with one determined in the previous iteration.

As a result of the calculation, T_{ng} = 1050 K.

The average cross-sections of fission for U-235: $\overline{\sigma_f}$ = 258.91 barn; Σ_f = 0.5516 cm^{-1}.

The average cross-sections of all core materials are presented in table 2.

The next step is the calculation of the effective multiplication factor k_{eff} by the six-factor formula [3]:

$$k_{eff} = \eta f p \varepsilon P_{FNL} P_{TNL}, \tag{17}$$

Where η is Thermal Fission Factor; f is the thermal utilization factor, p is the resonance escape probability, ε is the fast fission factor, P_{FNL} is the fast non-leakage probability, and P_{TNL} is the thermal non-leakage probability. Calculation formulae for each factor are given in [4].

Table 2. Average microscopic and macroscopic cross-sections of absorption of main core materials.

Materials	$VN\sigma_a(0.025\ eV)$ (cm)	$VN\zeta\sigma_s(1\ eV)$ (cm)
U-235	306.64	0.6822
U-238	1.33	0.1342
O-16	$1.3 \cdot 10^{-4}$	0.1092
H2O	0.33	0.8972
Al-27	0.11	0.0268
Si-28	0.08	0.0046
Zr-91	0.09	0.2661
Cr-52	1.52	0.1973
Mo-96	1.30	0.0035
Ni-59	2.17	0.8702
Mn-55	6.52	0.0015

The found k_{eff} is the effective multiplication factor in the beginning of the campaign. The next stage is to calculate k_{eff} and ρ throughout the campaign. The result of the calculation of k_{eff} and ρ in the beginning of the campaign are given in table 3.

Table 3. The result of calculation of keff and reactivity

η	f	P	ε	P_{FNL}	P_{TNL}	k_{eff}	ρ
2.04	0.5984	0.8904	1.02	0.944	0.994	1.040	0.038

6. Computation of reactivity change during the campaign

During burnout of the fuel in the reactor, various nuclides are formed in it. Some of them are fissile, while others are poisonous elements. Fissile nuclides include: Pu239, Pu240, and Pu241. As the density of Pu240 is relatively low in comparison to the density of U238, then the fission of Pu240 by fast neutrons is neglected.

Average cross-sections of fission and absorption in the Maxwell spectrum of aforementioned elements: $\sigma_a^9 = 1588.9$ barns, $\sigma_f^9 = 940.7$ barns, $\sigma_a^0 = 169.1$ barns, $\sigma_a^1 = 1188.5$ barns, $\sigma_f^1 = 886.5$ barns.

The average neutron flux density:

$$\overline{\Phi} = \frac{3{,}1 \cdot 10^{10} \cdot N_{th}}{\sigma_f^5 \cdot N_5 \cdot V_{fuel}} \tag{18}$$

Where $N_{th} = 175$ MWth is the thermal power of the reactor, $N_5 = 0.213 \cdot 10^{22}$ cm^{-3} is the initial density of U^{235}. $V_{fuel} = 366606$ cm^3 is the volume of the fuel in the core. $\overline{\Phi} = 2.68 \cdot 10^{13}$ n/(cm^{2}s).

Effective time:

$$z = \overline{\sigma_a^5}\overline{\Phi} t \cdot CF, \tag{19}$$

Where $CF = 0.65$ is the net capacity factor according to the data of JSC OKBM Afrikantov [5].

According to [6] the total thermal energy yield in the reactor campaign is 4.5·106 TW·h. Therefore, reactor campaign lasts for 1074 effective days (2.93 effective years), which equals 4.52 years taking CF into consideration.

By the formulae given in [7] the densities of all fissile materials are calculated, assuming that density of U^{238} and the average neutron flux density are constant (Figure 5).

Also, the appearance of different non-fissile nuclides with a large neutron absorption cross-section should be taken into account. These nuclides are classified to poisons, which have a short half-life period, and slags, which are long-lived or stable fission products.

To take into account the increase of reactivity due to the burnout of burnable absorber rods it is necessary to calculate the dependence of gadolinium density on time by following equation:

$$y = N_{Gd}(t) / N_{Gd}(0) \tag{20}$$

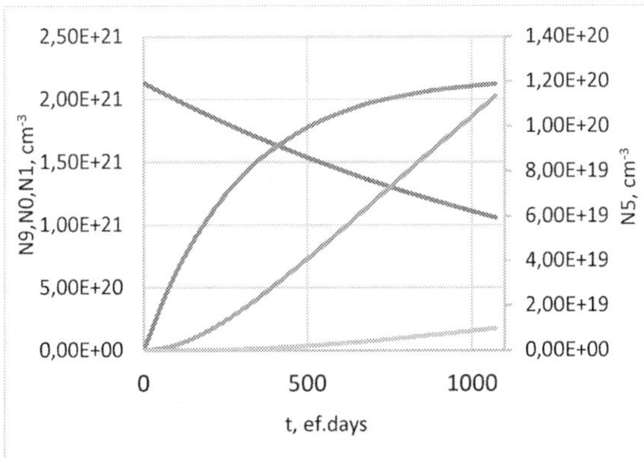

Figure 5. The change of densities of fissile nuclides. The blue line is the density of U235, the orange line is the density of Pu239, the grey line is the density of Pu240, and the yellow line is the density of Pu241.

$$x = N_5(t) / N_5(0) \tag{21}$$

Then:

$$y = x^{\sigma_{Gd}/\sigma_5} \exp[\gamma_d \Sigma^{ef}_{Gd,0}(1-y)] \tag{22}$$

After the dependence of Gadolinium density is known, it is feasible to determine the shielding coefficient throughout the campaign:

$$k_s(t) = \frac{1}{1 + \gamma_d \Sigma^{ef}_{Gd,0} N_{Gd}(t) / N_{Gd}(0)} \tag{23}$$

Having calculated effective multiplication factor using calculated data for core materials it is possible to draw a reactivity change curve, which reveals the change of reactivity during campaign.

The curve in Figure 6, or rather the peak of the curve defines the total weight of control rods that are used to compensate the excessive reactivity. The total weight of the rods should be at least $\rho_{cr} = 0.209$.

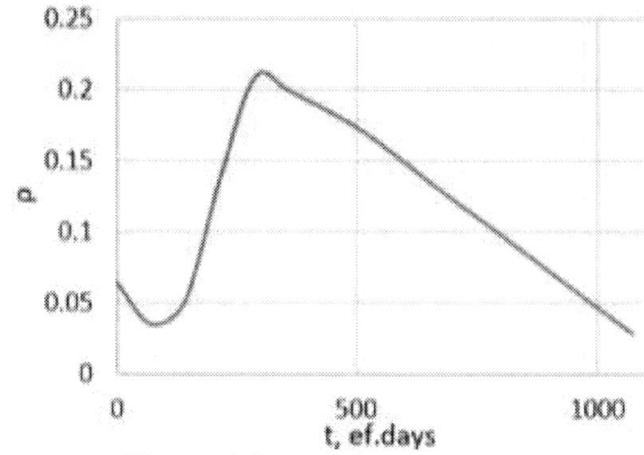

Figure 6. Reactivity change curve.

7. Conclusion

As it was deduced from the total thermal energy yield, the reactor may operate for 1074 effective days, which concurs with the results of the calculation shown in Figure 6. Also, it can be inferred from Figure 6 that at the end of campaign $\rho = 0.028$, which is enough to supply engines of the icebreaker to convey it to the refuelling point.

In addition, Figure 6 reveals that gadolinium burns out too quickly – in the first third of the campaign. The possible reason is the usage of the diffusional approach for calculating the shielding coefficient. Another approach that can be utilized to determine the shielding coefficient is the Wigner approach.

References

[1] Deyev V I, Schukin N V, and Cherezov A L 2012 *Principles of Calculation of Nuclear Marine Propulsion* (Moscow: NRNU MEPhI Press) p 256

[2] Masterson R E 2017 *Nuclear Engineering Fundamentals: a Practical Perspective* (Boca Raton : CRC Press) p 988

[3] Masterson R E 2017 *An Introduction to Nuclear Reactor Physics* (Boca Raton : CRC Press) p 1107

[4] Merzlikin G Y 2011 *Nuclear Reactors Theory* (Sevastopol: Sevastopol State University Press) p 341

[5] *RITM-200 Nuclear Marine Propulsion for New-generation Ice-breakers* (Nizhny Novgorod: JSC OKBM Afrikantov) p 20

[6] Kudinovich I V 2016 *Nuclear Propulsions of Promising Civil Marine Objects and the Safety Justification* (Saint-Petesburg: The Krylov State Research Centre) p 292

[7] Dementyev B A 1986 *Nuclear Reactors Control and Kynetics* (Moscow: Energoatomizdat) p 272

IOP Conf. Series: Materials Science and Engineering 1019 (2021) 012058 doi:10.1088/1757-899X/1019/1/012058

Method of multispectral image fusion using wavelet spectra

I Borisova and V Legkiy
Department of Autonomous Information and Control Systems, Novosibirsk State Technical University, Novosibirsk, Russia
E-mail: i.v.borisova@corp.nstu.ru

Abstract. The method of image fusion for images of the same scene obtained in different spectral ranges with the formation of an integrated monochrome image is considered. The fusion method is based on the wavelet decomposition of the original images. A strategy for fusing detail coefficients by comparing the proportions of their magnitudes for all original images is proposed. The fusion procedure does not require any thresholds. The algorithm does not introduce additional distortions and collects all the necessary information from the original images. The proposed algorithm can be used in optoelectronic systems for automatic image processing.

1. Introduction

Multisensor systems for target detection and tracking are now widely used primarily because of their ability to work in conditions of strong hindrances, which can be natural or artificial. Each individual channel is not able to meet technical requirements in conditions of poor visibility, masking targets and the use of electronic countermeasures. However, the efficiency of the system is increased by common use of information from different channels. The composition of optical-electronic equipment usually consists of television, thermal (multi-band), laser and other channels [1, 2].

The information fusion is divided into three levels: pixel fusion, feature fusion, and decision fusion. The pixel fusion is low level integration, which is applicable for the sensors with compatible receiving rate, dimension and format of data. The algorithms for image fusion can be divided into two groups: pseudo-color image generation algorithms and true-color image generation algorithms. Pseudo-color images are mainly used for remote sensing data processing and in medical applications [3, 4].

In this paper we consider the fusion of images of the same scene obtained in different spectral ranges to form an integrated monochrome image. The main problem of compiling images for visual display is «saving a picture» [5]. The important details from all compiled images should be presented in the resulting image, and the compilation procedure should not introduce distortions. The visual quality of the compiled image should be at least as good as the original image when other channels of the system are blocked. The compilation method should work with any number of observation channels, regardless of their spectral range and combination.

Earlier methods for compiling monochrome images were based on a weighted summation of the original images with some weights [6, 7]. The fusion of essential details was provided by adding of maximum brightness contrasts from all channels to this linear combination.

In the later algorithms, significant brightness differences in each of the original images are detected and combined in the integrated image. In recent years, many methods based on multi-scale transforms have been proposed for image fusion. These methods include contourlet transform [8], curvelet transform, the discrete wavelet transform [3, 6, 9–11], and various pyramid algorithms. The wavelet transform differs from the Fourier transform in that it allows local spectral analysis. Spectral wavelet coefficients correspond both to the amplitudes of different frequencies and to different spatial regions of the image. The advantage of wavelet analysis for image fusion is the ability to use images of various spatial resolutions to reduce the amount of information processed.

2. Image fusion algorithms based on the wavelet transform

We consider the image fusion methods based on the wavelet transform described in [3] and [6, 10]. The wavelet system Haar, which requires a minimum of calculations, is chosen as the basis in these works. A separable two-dimensional basis, obtained by integer translations and binary scaling of basic functions, is used.

Two-dimensional scaling function, $\varphi(x, y)$, and three two-dimensional wavelets, $\psi^{H}(x, y)$, $\psi^{V}(x, y)$, and $\psi^{D}(x, y)$, are formed as the products of one-dimensional scaling functions, $\varphi(x)$ and $\varphi(y)$, and wavelets, $\psi(x)$ and $\psi(y)$:

$$\varphi(x,y)=\varphi(x)\varphi(y), \psi^{H}(x,y)=\psi(x)\varphi(y),$$
$$\psi^{V}(x,y)=\varphi(x)\psi(y), \psi^{D}(x,y)=\psi(x)\psi(y)$$

Two-dimensional wavelets measure intensity variations along different directions: $\psi^{H}(x, y)$ measures intensity variations along columns (separates horizontal edges), $\psi^{V}(x, y)$ responds to variations along rows, and $\psi^{D}(x, y)$ corresponds to variations along diagonals.

All original images are processed by two-dimensional wavelet decomposition: each row is divided into low-frequency and high-frequency halves, and then each column is processed similarly. The result is four half-sized arrays from the original: W_{φ} are approximation or scaling coefficients, W^{H}, W^{V}, and W^{D} are detail or wavelet coefficients.

When the arrays of coefficients are formed, there are several ways to restore the resulting image. The approximation coefficients for the fused image can be taken from one (pre-selected) original image. They can also be obtained by averaging or weighted averaging the approximation coefficients from all original images. The detail coefficients can be formed from the maximum magnitudes of the corresponding coefficients for all original images. They can also be selected based on some feature. In any case, the brightness range of the original images should be normalized before processing.

References [6, 10] proposed that the approximation coefficients were formed as an average of the coefficients of the two input images, and the detail coefficients were composed of maximum magnitudes. In reference [3] the correlation of the approximation coefficients for input images were tested. The array of approximation coefficients is formed from the maximum magnitudes, if the correlation is low, and, otherwise, is a linear combination of coefficients from two input images. High frequency component of transformation changes more intensely, and the authors [3] used the fusion strategy based on variance for the detail coefficients. The detail coefficients are selected from the image, where the variance of the coefficients calculated over the local neighborhood is greater.

Fusion weighting coefficient can be calculated as

$$T_{\text{var}}=\frac{2\sigma_1^2\sigma_2^2}{\sigma_1^4+\sigma_2^4} \quad (1)$$

where σ_1 and σ_2 are respectively standard deviations within the field (from 3×3 to 7×7) from detail coefficients for the first and second images.

If T_{var} is less than the specified threshold t, then the detail coefficient is taken from the image where the standard deviation value is maximum. If T_{var} is greater than the threshold value, then the detail coefficient is a linear combination of coefficients from two input images:

$$W^{i}=w_1W_1^{i}+w_2W_2^{i}, \quad w_1=\sigma_1/(\sigma_1+\sigma_2), \quad w_1+w_2=1, \quad (2)$$

where W^{i} are the detail coefficients for the fused image, W_1 and W_2 are the detail coefficients for the first and second original images, i = H, V, D.

The described algorithms were simulated. The simulation results showed that when the detail coefficients are composed from the maximum magnitudes [6, 10], the synthesized image contains too many small details. This complicates the analysis of the fused image and makes the method unstable with respect to noise. On the other hand, the formation of approximation coefficients [3] from the maximum magnitudes is possible only for images with similar brightness. When fusing TV and thermal images, artifacts in the form of light spots appeared in the resulting image.

The fusion strategy for high-frequency component of wavelet transform according to formulas (1), (2) gives good results, but it requires thresholds setting for comparing fusion weighting coefficients with them. During the simulation, the thresholds were calculated by preprocessing the image. Thus, the fusion procedure became two-pass.

We propose a new fusing method for detail coefficients, which does not require threshold values. The method is based on comparing the proportions of the detail coefficients. If the considered point belongs to the edge, the magnitude of one of the detail coefficients will be greater than the others. The greater the difference, the more visually significant the boundary will be. On the other hand, if there is a point from the neighborhood with constant brightness, the magnitudes of the detail coefficients are almost equal. We introduce the fusing rule as

$$\begin{cases} W^i = W_1^i, & \text{if } \dfrac{\max\limits_{\substack{j=\mathrm{H,V,D} \\ j\neq i}} \left|W_1^j\right|}{\left|W_1^i\right|} < \dfrac{\max\limits_{\substack{j=\mathrm{H,V,D} \\ j\neq i}} \left|W_2^j\right|}{\left|W_2^i\right|} \\ W^i = W_2^i, & \text{otherwise} \end{cases} \quad (3)$$

The original image and the image with added Gaussian noise are shown in Figure 1.

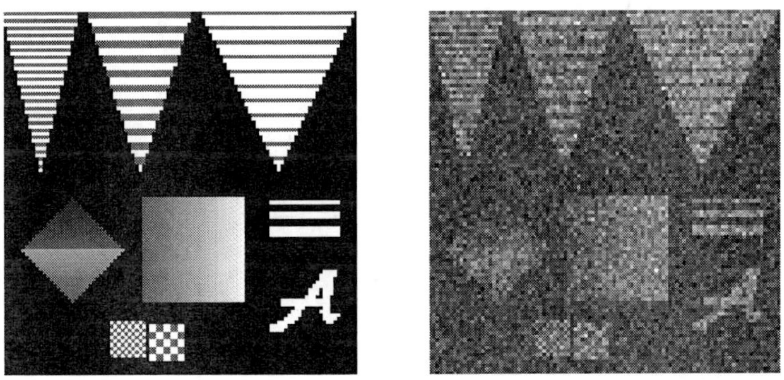

Figure 1. Original image (left) and image with added noise (right).

The results of merging these images are shown in Figure 2. Approximation coefficients were taken from the first original image. The detail coefficients were formed from the maximum magnitudes – the result is shown in Figure 2*a*, according to equationes (1) and (2) – the result is shown in Figure 2*b*. The fusion result image by the proposed method according to equation (3) is shown in Figure 2*c*.

The correlation coefficients are calculated for the first original image and all fusion results. They were 0.7842, 0.7844, and 0.9462 for the images in Figure 2 *a* – *c*. Thus, the considered method is more resistant to noise than the compared methods.

(a) (b) (c)

Figure 2. Fusion result image: a – by the method of maximum magnitudes; b – by calculating the coefficients T_{var}; c – by the proposed method.

3. Experiment and result analysis

The original multispectral images for the fusion by the proposed algorithm are shown in Figure 3. Image fusion can be performed at any level of the wavelet decomposition. The fusion result image according to the considered algorithm using a one level of the wavelet decomposition is presented in Figure 4. The fusion result using two levels of decomposition is shown in Figure 5. Note that only the detail coefficients were combined. The approximation coefficients were taken from the first original image.

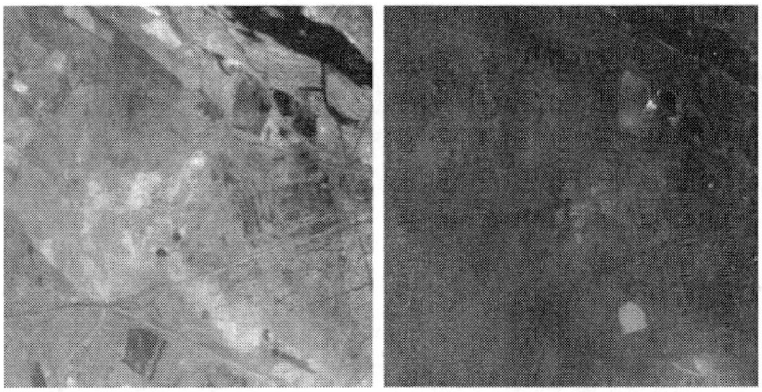

Figure 3. Original images.

Figure 4. The fusion result image (one level of the wavelet decomposition).

Figure 5. The fusion result image (two levels of the wavelet decomposition).

The resulting image contains details that are indistinguishable in one of the ranges. Using two levels of decomposition can significantly reduce processing time. However, increasing the number of decomposition levels can be used to compile images with large objects. The mosaic effect appears on small objects, and smooth curved lines become jagged.

4. Conclusion

The result fusion images according to the considered algorithm are convenient for the operator's perception. The algorithm does not introduce additional distortions and collects all the necessary information from the original images. The fusion procedure does not require thresholds. The proposed algorithm can be used in optoelectronic systems for automatic image processing.

This paper does not consider the possibility of further improving the quality of the generated images, in particular, due to directional filtering during wavelet decomposition or filtering of wavelet coefficients [12]. Further studies are aimed at adapting the algorithm by combining approximation coefficients to improve the quality of the result.

References

[1] Yang X and Pei J 2007 *Chinese optics letters* **5**(7) 400–402
[2] Kaderova G N *et al* 2007 *Journal of Optical Technology* **74**(1) 20–23
[3] Xue X, Xiang F, Wang H and Peng J 2016 *Naukovyi Visnyk Natsionalnoho Hirnychoho Universytetu* **3** 122–128
[4] Gomathi P S and Kalaavathi B 2016 *Circuits and Systems* **7**(8) 1598–1610
[5] Borisova I V, Gorenok V N, Oparin A N and Popov P G 2003 *Proceeding of SPIE* **5126** 255–257
[6] Frolov V N, Tupikov V A, Pavlova V A and Alexandrov V A 2016 *Izvestiya Tula State University. Technical sciences* **11**(3) 95–104
[7] Vetrov A N, Osipova A A, Gakhzar M A and Artyukhin I Yu 2013 *Modern problems of science and education* **6** 173
[8] Xin G, Zou B, Li J and Liang Y 2011 *Information Technology Journal* **10**(6) 1138–1149
[9] Li H, Manjunath B S and Mitra S K 1995 *Graphical Models and Image Processing* **57**(3) 235–245
[10] Teterin V V, Pavlova V A and Aleksandrov V A 2006 *Journal of Optical Technology* **73**(10) 698–701
[11] Min L 2012 *Computer Engineering* **38**(23) 211–214
[12] Voskoboynikov Yu E 2017 *Fundamental research* **8**(2) 246–250

Perspective direction of shaped charges perfecting

E M Grif, A V Guskov, K E Milevskij
Aircraft Faculty, Novosibirsk State Technical University, Novosibirsk, Russia
E-mail: a.guskov@corp.nstu.ru

Abstract. The paper analyzes the achievements of modern research in the area of increasing the penetration diameter in the operation of shaped charges. The functioning of cumulative charges with various forms of cumulative lining is simulated numerically in Ansys Autodyn software. Ways of further development of the issue and explores the possibilities of functioning of shaped charges with a lining complex conical-ring type are discussed. Proposals for improving the cumulative lining of a complex conical-ring type are advanced, substantiated and numerically modelled.

1. Introduction

The shaped charges are widely used in explosive technologies of cutting, perforation, destruction of materials and structures. Industrial shaped charges can be divided into two groups: cutting and punching. Cutting shaped charges are elongated explosive charges with an elongated wedge-shaped or semi-cylindrical cumulative notch covered with a metal or metal-polymer facing. With the explosion of such charges as a result of the deformating of the lining, a flat cumulative jet (cumulative knife) is formed, which "cuts" the barrier. Punching charges are compact axisymmetric shaped charges with shaped notch in the form of a cone or hemisphere, covered with a metal lining. When functioning shaped charges of this type, a high-speed cumulative jet is formed [1].

2. Theory

2.1. Ways for improving shaped charges

Improving the efficiency of shaped charges is carried out in the following ways: by optimizing the size and shape of the shaped charge, shape, geometry and material of the cumulative lining, quality, energy content, density and speed of detonation of the explosive, detonation lens, proper selection of the focal length for charge and accuracy in manufacturing charge parts and assembling them. The most progressive, with broader potential is the direction of increasing the characteristics of the charge due to changes in the shape, profile and choice of material cumulative linings [2].

The improvement the physical and mechanical properties of the material cumulative linings is traditionally done through optimization of technological processes (rotary drawing, stamping, powder technology). Cumulative linings manufactured by the rotary drawing method have technological heredity in the form of heterogeneity of deformation of the liner material, which leads to an additional rotation of the jet during its formation and, as a result, the jet energy dissipation. In order to avoid the described effect, a method was proposed for the manufacturing of cumulative linings, based on the phased bidirectional deformation of the workpiece, alternating with recallocation annealing. The use of this technology allows to obtain blanks with plastic properties similar to those that were before processing, and ensures the absence of shear distortions in the material, since phased bidirectional drawing ensures their full mutual compensation. This method provides a microcrystalline structure with

the smallest and most uniform grain, and its use allows, depending on the specific conditions of application of the distance to the barrier, to increase the penetration of shaped charges by 15-30% [3]. The use of powder technologies also allows to significantly increase the effectiveness of shaped charges. The authors of [4] propose a fundamentally new way to obtain multilayer cumulative linings based on applying a dense powder coating on the facing by detonation method. This technology allows to increase the efficiency of charge operation by 15% compared with the standard.
The way to improve the forms of cumulative linings can also be divided into two ways: the traditional, based on changes and improving the accuracy of geometric parameters, and fundamentally new, based on the creation of new profiles and forms of cumulative lining. For example, the author [5] proposes a shaped charge with a star-shaped profile of a cumulative lining, the principle of which is based on the process of double collapse of the metal of a cumulative lining. Cumulative lining has a conical shape, complemented by V-shaped elements along a generatrix. Due to the energy of cumulative "knives" formed as a result of the deformation of V-shaped "petals", this approach contributes to the increase in the volume of metal involved in the formation of a cumulative jet, obtained by deformation the conical part of the lining. The solution allows to significantly increase the mass-energy characteristics of jet, which allows to achieve high performance in the state and, accordingly, increase the efficiency and penetration depth.

2.2. Methods for increasing the penetration diameter by shaped charges
With the expansion of the spectrum of the use of shaped charges in civilian use, new problems arise in relation to another effective characteristic of cumulative jets – the penetration diameter. Research activities follow the traditional way of achieving the physicomechanical properties of a material (combination of materials) and optimal geometric characteristics of cumulative conical and hemispherical cumulative facings. The authors of [6] propose a shaped charge, the lining of which contains a special liner attached to the inner surface of the lining with a special plastic material. In the process of compression cumulative lining due to the presence of the liner and plastic material between the cladding and the liner decreases the speed of compression and forms a cumulative jet of large diameter. The diameter of the cumulative jet is directly proportional to the diameter of the hole being punched in the barrier.

Such methods for solving the problem of increasing the penetration diameter in advance face a serious problem based on the physic of cumulation process: the cumulative jet in jet formation theory is described by the deformation of a conical lining that forms a long thin metal wire moving longitudinally at very high speed [7]. Therefore, the methods for increasing the penetration diameter through the use of cumulative linings of new forms and profiles are of particular interest. Researchers [8] offer a shaped charge with the presence of longitudinal cumulative grooves in the spherical (elliptical) part of the lining allows you to create additional cumulative "knives" that, when penetrating into the barrier, do additional work to increase the area of the perfringoration channel. The use of charges of the proposed design allows to increase the diameter of the inlet of the perforation channels compared with the use of shaped charges with a conical cumulative lining, and to increase the depth of penetration of obstacles compared to using charges with a hemispherical or semi-ellipsis cumulative facing.

However, at this stage of development of science and technology, the possibilities of these areas are limited: due to the design features of cumulative linings of traditional forms, the metal participating in the jet formation process and (or) pestoing, regardless of its physical characteristics, forms the cumulative element at the center of the charge.

Consequently, when solving the problem of increasing the penetration diameter, it is advisable to develop the direction of creating and improving shaped charges with a complex shape of shaped linings [9], during operation of which, a cumulative element is formed in the central and peripheral areas of the housing.

An example of such a charge is the shaped charge of the perforator [10] (Figure 1).

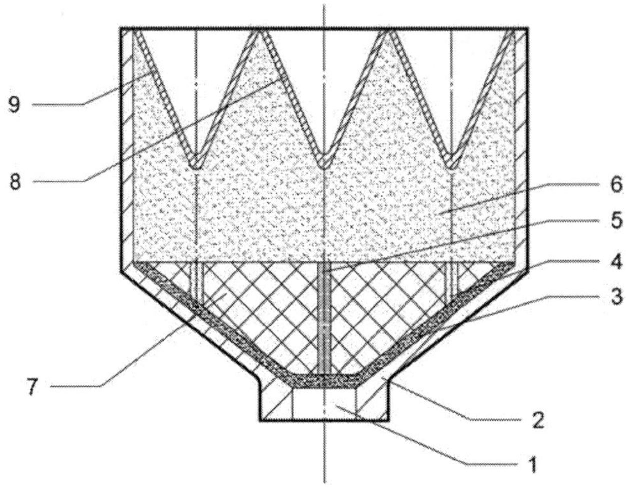

Figure 1. Shaped charge of perforator.

The shaped charge housing is one-piece, has a cylindrical-conical shape, the explosive charge is cylindrical, in the central part of which is a conical cumulative lining element, the charge includes a point initiating device and a detonation explosive lens assembly consisting of an inert material lens and a combination of additional explosive charges substances, while the formation of the ring cumulative element occurs along the axis of the V-shaped part of the cumulative lining. Effective compression of the annular and conical cumulative linings is achieved by installing an explosive lens assembly with a combination of charges of additional explosives, thereby ensuring the formation of two cumulative jets: cylindrical and ring.

Shaped charge works as follows. The initiation means (1), enclosed in the lower part of the charge case (2), transmits a pulse to an additional explosive charge (3). An additional explosive charge (3), in turn, transmits impulses to explosive (6), as well as additional explosive charges (4) and (5), placed in an explosive lens assembly of inert material (7) and selected according to their density and detonation speeds so that the time of transmission of pulses of additional explosive charges (3), (4) and (5) to the main blasting explosive (6) coincides. The initiation of explosives (6) in the surfaces of contact with explosives (4) and (5) ensures the appearance of spherical shock waves, the front of which approaches normal to cumulative linings, which implies the most effective deformation of the cumulative lining (8) and (9) followed by the formation of cumulative jets of cylindrical and ring forms. Initiation of explosives (6) in the contact surface with explosives (3) provides compensation for the shock wave generated by the reflection of the detonation wave from the wall of the housing (2).

The design of the device may vary depending on the purpose and conditions of work performed. For example, if necessary, by changing the geometrical parameters of a cumulative lining, you can perform a shaped charge as a shaper compact punche elements with a parabolic or conical with a wide angle of forms of cumulative lining.

3. Experimental Results

Designing the described shaped charge is a difficult task, primarily due to the optimization of the explosive lens assembly. Therefore, it was decided that it would be expedient to carry out numerical calculations of the functioning of the shaped charge of the simplified design (Figure 2).

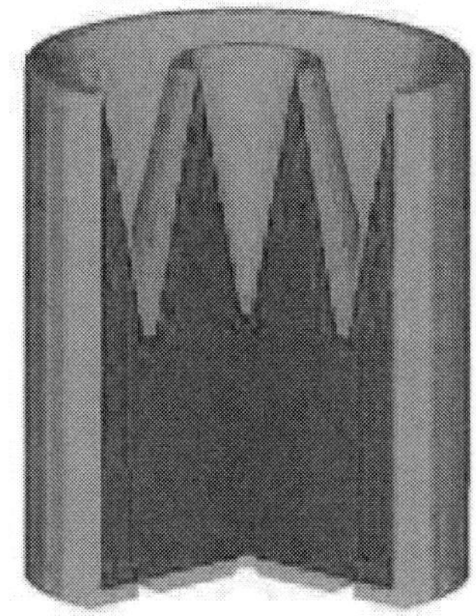

Figure 2. Shaped charge of perforator with complex lining.

To assess the work performance of the proposed design, three functional schemes of shaped charges are calculated: traditional (shaped charge with traditional cumulative conical lining), ring (shaped charge with ring cumulative lining) and complex (shaped charge with combination of conical-ring and conical parts of lining). When performing calculations, the tool of modern numerical modeling Ansys Autodyn was used. The diameter of the explosive checker (explosive) was 50 mm, the copper cumulative lining had a thickness of 1 mm, the barrier (steel plate) - 8 mm. To describe the behavior of explosives, the Jones-Wilkins-Lee (JWL) state equation was used, the ideal gas state equation was used to describe the air behavior, the Johnson-Cook model and the Gruneisen state equation were used to describe the behavior of cumulative lining, charge hull and obstacle.

To evaluate the results of the calculations, the parameters of the cross-section of penetration D, the cross-sectional area of penetration S, were singled out, the uniformity of penetration in the longitudinal direction was estimated.

The calculation of the functioning of the traditional shaped charge scheme is shown (Figure 3).

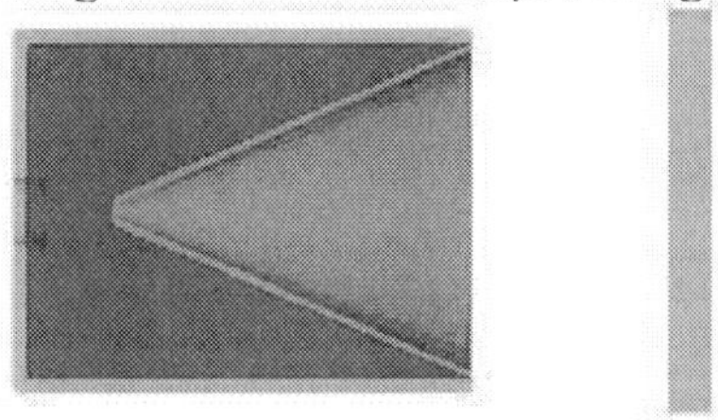

(a)

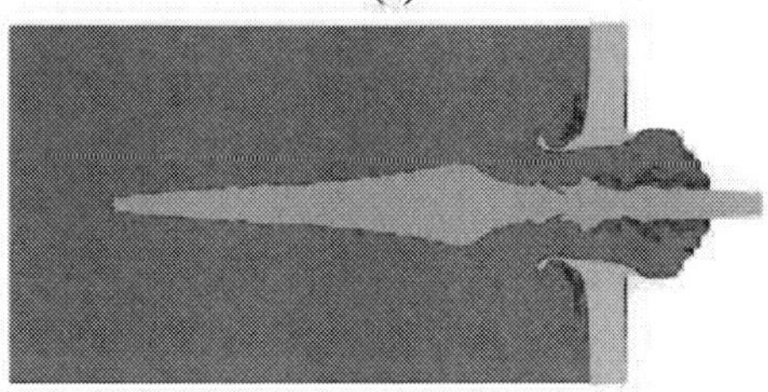

(b)

Figure 3. The calculation of the functioning of the traditional scheme of shaped charge at time points – 0 μs (a), 32 μs (b).

The breakdown cross section for a traditional charge is D = 1.8 cm, the penetration area is S = 2.54 cm^2, the walls of the hole punched are straight.

Also, to evaluate perfomance of the functioning of a charge with a composite form of a cumulative lining, it is necessary to calculate the operation of a charge containing an ring cumulative lining. There are special devices for producing ring cumulative jets, for example, containing W-shaped cumulative linings [11], as well as a device whose design feature is the presence of a special device [12]. However, for the purity of the experiment, interesting was the design, not equipped with special devices. The calculation of the shaped charge scheme described above with a conic-ring cumulative facing is shown (Figure 4).

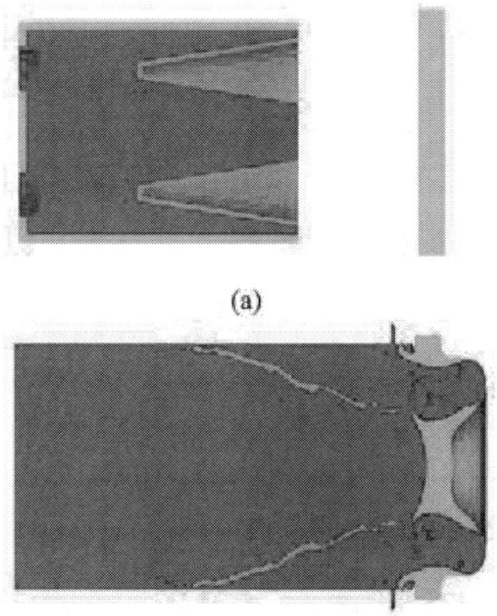

(a)

(b)

Figure 4. The calculation of the functioning of the conic-ring scheme of shaped charge at time points – 0 μs (a), 32 μs (b).

Such a lining design lays down a rigid connection between the angle of the solution and the length of the lining that forms, which influences the jet forming process, the jet diameter and the velocity distribution in the jet. As noted above, the design in Figure 4 is not equipped with a special device (for example the explosive lens assembly) that would ensure the symmetry of the propagation of detonation in the explosive relative to the ring axis of the lining, therefore, during deformation ring low-speed cumulative pest was formed and not ring cumulative jet with a traditional "proper" profile. The obtained results can be analyzed for two cases, depending on the properties of the environment material behind the barrier: if the obstacle is rigidly connected with the material, a through-hole channel is formed in it, and if the environment material behind the barrier does not prevent the introduction of the "plug" knocked out of the barrier, a hole is formed, whose diameter is equal to the penetration diameter of the ring cumulative element.

For the first case, this approach, compared to the traditional charge, made it possible to increase the cross-section of the penetration by 33%, to $D = 2.4$ cm, the cross-sectional area of the penetration - by 344%, to $S = 11.3$ cm^2. In the second case, accompanied by knocking out the "plug", the penetration diameter increased by 133% and amounted to $D = 4.2$ cm, while the area of penetration increased by 445% and amounted to $S = 13.84$ cm^2. Despite the large increment in the quantitative characteristics of penetration, it is worth noting its unevenness in the longitudinal direction. If such a defect is insignificant when knocking out a "plug", in the opposite case, when ring-cilindric penetration is formed, the irregularity of penetration significantly affects the decrease in the size of the outlet opening in the barrier, reducing the bandwidth of perforation channel.

The increase in the volume of the metal involved in the process of jet formation and penetration of the cumulative jet into the barrier, theoretically, increases the volume of the material punched. A new typical shaped charge scheme based on this principle, is obtained as a result of the inclusion in the design of the typical ring cumulative charge of the conical element of the cumulative lining. The calculation of the functioning of the shaped charge with a complex cumulative lining, which is a combination of conical-ring and conical elements is shown (Figure 5).

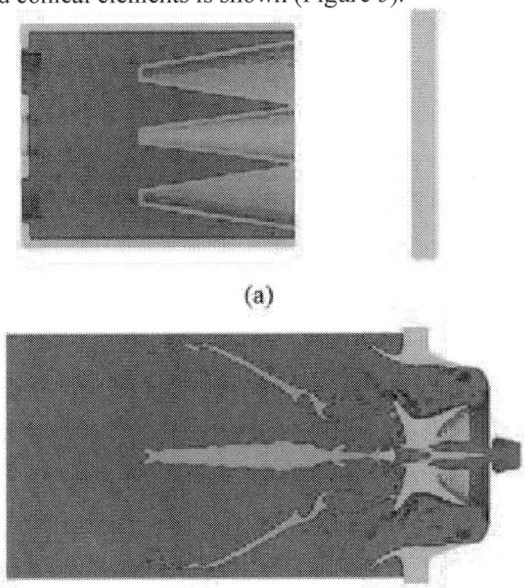

(a)

(b)

Figure 5. The calculation of the functioning of scheme of shaped charge with a complex cumulative lining at time points - 0 μs (a), 32 μs (b).

When shaped charges of this type are functioning, the character of penetration, as in the previous scheme, depends on the properties of the environment material behind the barrier. Depending on whether the emergence of the formed "plug" is possible, the final result of the penetration will be either a set of ring and cylindrical channels, or a hole of large diameter. In the first case, the penetration cross section increased by 44% compared with the traditional scheme, by 8% compared to the ring one and was D = 2.6 cm, the penetration cross section area increased by 360% and 4%, respectively, and amounted to S = 11.78 cm^2. In the case of knocking out the plug, the penetration diameter increases by 122% compared to the traditional scheme, decreases by 5% compared to the ring one and is D = 4.0 cm, the area of penetration increases by 394% compared to the traditional scheme, decreases by 10% in compared with the ring and is S = 12.56 cm^2.

If, when knocking out a "plug", the nature of breaking through an obstacle is similar to that described for a ring pattern, then when forming a complex type of penetration, along with the previously described unevenness of penetration in the form of ring channels in the longitudinal direction, a narrowing of the primary penetration hole occurs. This is explained by the fact that the penetration of the cumulative jet formed by the collapse of the conical part of the lining occurs before the penetration of the low-speed cumulative ring pest, and the latter, deforming the obstacle, causes a narrowing of the primary penetration. To compensate for the defects described above, the operation of the structure with a shortened conical part of the lining was considered, the calculation of which is reflected (Figure 6).

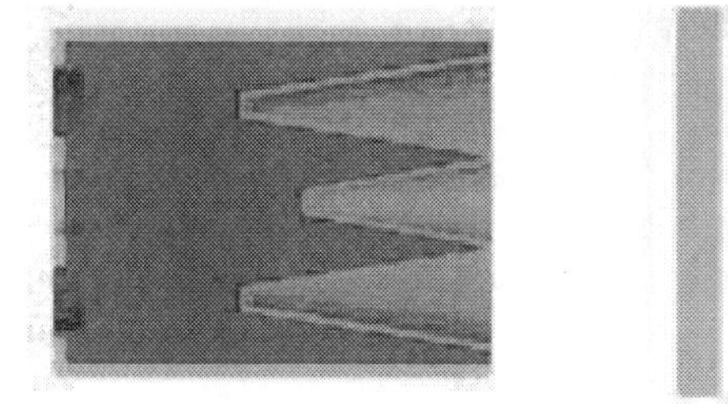

(a)

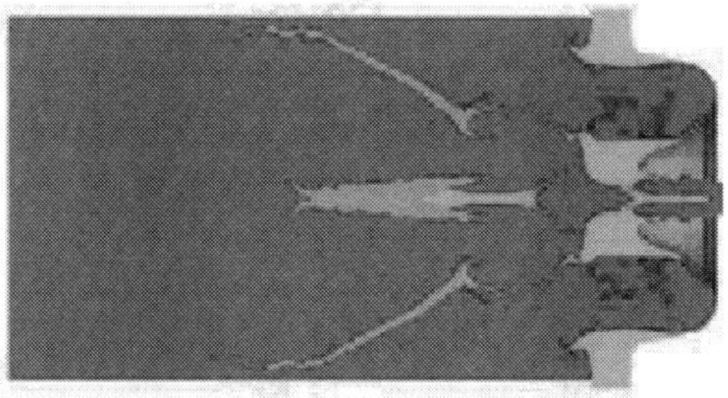

(b)

Figure 6. The calculation of the functioning of scheme of shaped charge with a shortened conical part of the lining at time points – 0 μs (a), 32 μs (b).

Since the length of the lining and the surface area of the conical part of the lining are reduced, the pressure distribution in the detonation products in the zone between the parts of the complex lining, i.e. there is a decrease in the influence of the compression process of the conical lining on the process of compression of the ring part of the lining. There is a slight decrease in the breakdown cross section (by 4%) and the breakdown cross section area (by 5%) compared to the previous calculation, but the penetration uniformity in the transverse direction does increase, which confirms the assumption of a more orderly distribution of energy in the cumulative pest during the operation of the optimized structure.

4. Conclusion

Studies of composite cumulative linings, including elements of conical-ring and conical shape, open up broad prospects for obtaining penetration of large diameters. This combination, made in the dimensions of perforation charges, allows to increase the transverse penetration of the barrier 1.4-1.6 times and increase its area 2.4–2.5 times (2.2–2.9 and 2, 4–2.7 times, respectively, when knocking out a "plug") compared with penetration through, provided by the functioning of a traditional shaped charge with a conical lining. The proposed design is multipurpose, as it allows adjusting the geometrical parameters of charge in a wide range depending on the research needs. The results of the study determined the possibilities and prospects for the use of a complex form of lining of shaped charges, and in the future it is planned to optimize their parameters, including to determine the influence of the characteristics of the explosive, the material and the geometry of the shell [13], in order to study more deeply the processes occurring during charge operation and to obtain the best penetration characteristics both in depth and in diameter of the holes.

References

[1] Kanel G I, Razorenov S V, Utkin A V and Fortov V E 1996 *Shock wave phenomena in condensed matter* (Moscow: Janus)
[2] Andreev S G *et al* 2002 *Explosion Physics ed L P Orlenko* (Moscow: Fizmatlit)
[3] Vladykin E I, Kurepin A E and Semin V A Patent 2180723 Russia IPC F42B1/036, B21D22/14
[4] Kalashnikov V V *et al* 2011 *Bull. of the Samara Sc. Center of the Rus. Ac. of Sc.* **1** 373–376

[5] Vorotilin M S 2011 *Izv. TSU. Tech. sc.* **2** 96–101
[6] Antipinsky S P, Neskin A G, Sokolov M L, Pantyukhin B S and Zelenov A N Patent 2495360 Russia IPC F42B3/08, F42B1/028, E21B43/117
[7] Kalashnikov V V *et al* 2010 *News of the Samara Sc. Center of the Rus. Academy of Sc.* **1** 370–373
[8] Mishnev V I, Guskov A V, Milevsky K E and Trishin Yu A 2003 *Cumulative stream formation: a laboratory workshop for the 4th year of FLA (specialty 171400, 330500, 120400) full-time education* (Novosibirsk: Publishing house of NSTU) p 40
[9] Grif E M, Guskov A V and Milevsky K E 2018 *Science. Industry. Defense: Tr. 19 All-Russia. scientific tech. conf., Novosibirsk, April 18–20. 2018* vol 2 (Novosibirsk: Publishing house of NSTU)
[10] Grif E M, Guskov A V and Milevsky K E Patent 2717853 Russia E21B43/117, F42B1/028
[11] Sitnikov V I, Vedernikov Yu A, Seleznev A I 2001 *Chem. Phys.* **10** 35–37
[12] Malyarov D V, Zhdanov I V and Tarasov I B Patent 2478904C1 Russia IPC F42B1/02
[13] Andreev V V, Guskov A V, Milevsky K E and Slesareva E Yu 2013 *High-energy materials: studies manual* (Novosibirsk: Publishing house of NSTU)

IOP Conf. Series: Materials Science and Engineering 1019 (2021) 012060 doi:10.1088/1757-899X/1019/1/012060

Algorithm for markers detection on fringe images

A I Ubert
Department of Automation, Novosibirsk State Technical University, Novosibirsk, Russia
E-mail: aleksey.ubert@gmail.com

Abstract. Thanks to non-contact scanning technology and its harmlessness to human health, the usage of numerous 3D scanners in various fields of medicine has considerably increased. For example, in orthopaedics, medical topographic system TODP uses structured light with fringe projection to determine the 3D relief of the dorsal surface of the trunk for the urgent task of diagnosing spinal deformities. To improve the accuracy of postural states analysis, anatomical landmarks of the skeletal bone structures are marked with high-contrast small-sized markers of the reflective film. Thus, the task of the marker's detection in the image appears. The article discusses the algorithms for detection of small-sized bright markers in fringe images, obtained by the computer optical topography method. The experimental results have shown that the proposed algorithm can detect 99.94% of the reflective markers in a sample of 1000 images with 8350 markers in total.

1. Introduction

Numerous 3D scanners work by using the optical triangulation principle [1]. Projection of structured light onto a shaped object produces distortions of pattern in proportion to an object shape [2]. These distortions register from other perspectives than that of the projector. The resulting image can be used for the geometric reconstruction of the surface shape with high accuracy and optical spatial resolution [3].

The most common and simple structured light is the fringe projection. Phasemetry methods [4, 5] allow processing a fringe image as a two-dimensional sinusoidal signal and reconstructing the surface shape in each point of the image. To assure safety for human health, 3D scanners with structured light are used in medicine, for example, in the TODP system [6] for scoliosis diagnostic and body posture analysis.

2. Statement of the problem

To improve the accuracy of the analysis of the postural state, the anatomical landmarks of skeletal bone structures are marked with high-contrast small-sized markers of the reflective film. The high contrast of the markers makes it possible to distinguish them on the fringe background as bright dots of 3-4 pixels in size. During the detection stage, the bright markers on the image are pulsed noises, which can significantly distort the local shape of the reconstructed human body surface. Therefore, all reflective markers must be identified and “erased” from the fringe image.

The article discusses the algorithms for the detection of the reflective markers on the fringe images, used in TODP system.

2.1. *Obtaining an image for the reflective marker detection*

Reflective markers are usually clearly visible on a grayscale fringe image, which illustrates Figure 1. However, using amplitude detection often is not enough for its automatic identification. In practice, the individual markers may be inclined at a large angle to the light source. It leads to a significant drop of marker's brightness, making it comparable to the fringe's brightness. To ensure the detection of poor-quality markers, a special image is prepared. Firstly, an image with no fringe is created row by row via low-pass filter "triangular window" with impulse response:

$$h(n) = \frac{2M - |n|}{4M^2}, \tag{1}$$

where is M – window size, n – discrete image's coordinate. Second, using (1) with $M_2 > M$ another image with smoothed markers is generated.

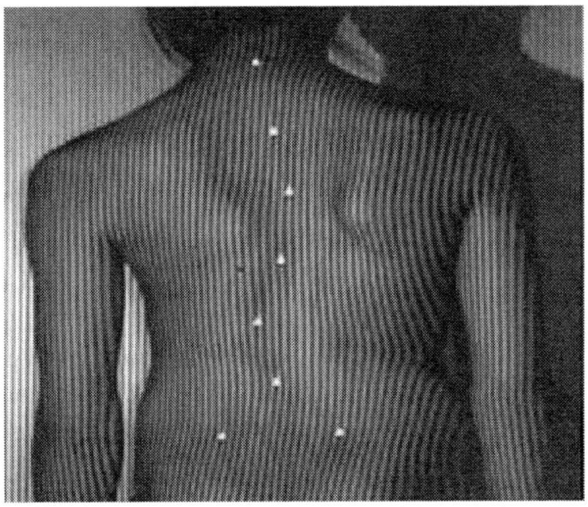

Figure 1. Image with fringe and reflective markers.

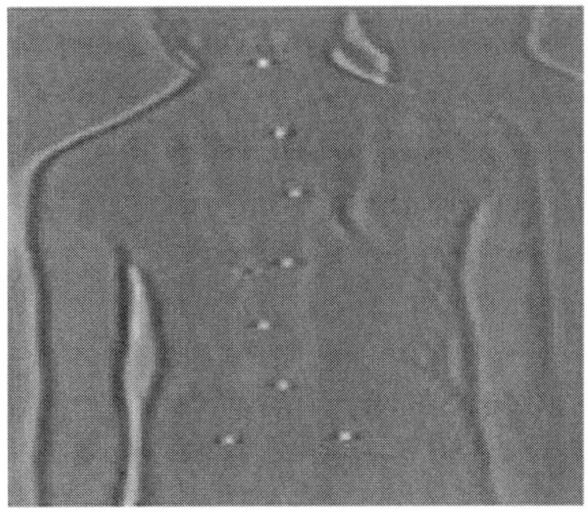

Figure 2. Image for markers detection.

When subtracting from the first image, the second image and addition of brightness code 128, a special grayscale image for markers detection is obtained as shown in Figure 2. Zero values of the difference are represented by gray background (code 128), the positive ones are in the form of light ones, and the negative ones are in the form of the dark pixels. There are no fringes on this image and markers are still visible. However, some pseudo markers can be detected on the external contour of a human's body.

2.2. *Specific marker's shape on an image for markers detection*

All source image (Figure 1) transformations for preparing an image for markers detection (Figure 2) are linear and correspond to a smoothing differentiator (1). It removes the energy of the fringes and leaves reflective markers on the image. The markers have the form of the impulse response of a linear filter, due to small size they can be approximated as a Dirac delta function. The brightness characteristic of the markers along the horizontal axis is presented in Figure 3.

The shape of the marker (excluding background brightness code 128) is the main positive impulse at the marker's center and two negative side lobes are oriented along the smoothing direction. The amplitude ratio of the main pulse and side lobes and their lengths are determined by the characteristics of the smoothing differentiator (1). The specific shape of the markers allows using recognition algorithms in accordance with their characteristics.

2.3. The structure of the algorithm for markers recognition

The algorithms for recognition of markers are developed using three principles: phasing, hierarchy, and adaptability. In accordance with these features, the process of identification of the markers is divided into stages arranged in a hierarchy from simple to complicated.

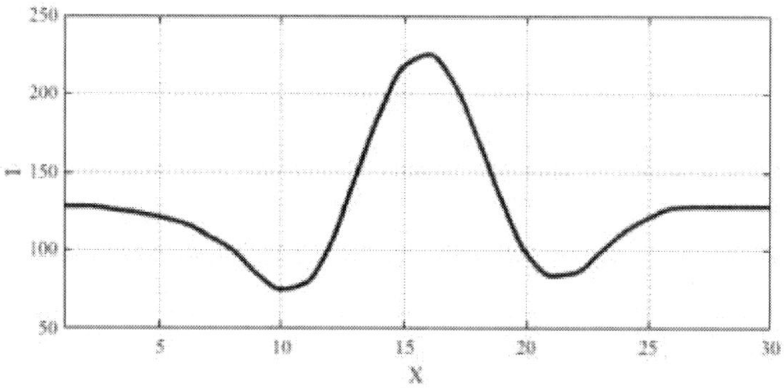

Figure 3. Brightness characteristic of the markers.

In the beginning, the simplest amplitude criteria are used for all pixels in an image, and it gives many pseudo markers. At the next stages, the complexity and amount of calculations increases, but an insignificant part of the pixels of the entire image is being processed. Due to different filming conditions, the quality of the source images varies greatly, so the algorithms are adaptive to processing of poor-quality images.

3. The marker's detection algorithm

The flowchart of the marker's detection algorithm is shown in Figure 4. The algorithm can be divided into three main stages. At the first stage (*Blocks 2–4*) the potential markers (PM) are selected using the amplitude threshold and the specific shape of the markers. These simple and fast algorithms can process a large amount of data in an acceptable time (less than 0.15 sec). At the second stage (*Block 5*) the local area around PM is analyzed by applying more complicated criteria. Then, due to the rejection of all PM with "bad" surroundings, the number of PM significantly decreases. At the third stage (*Blocks* 6–8), all characteristics of PM and its surroundings are compared with the mathematical marker's model characteristics. The marker's model is obtained by the statistical methods via processing a large amount of the training sample data.

Image processing is carried out sequentially in columns. In the first cycle, pixel-by-pixel usage of simple criteria detects potential markers in the columns of the image. The second cycle analyzes PM, rejects the false markers and indicates the true ones. If no true markers are found (due to the poor-quality image), the cycles are repeated with other threshold parameters. Consider the algorithm logic block by block next.

3.1. Block 1. The initial condition's setup

In the initial condition's setup block, the global thresholds and other parameters are to be initialized. These parameters represent the amplitude detection threshold, expected value of the marker's amplitude, shape's characteristics of the marker, asymmetry of lobes, expected value of the marker's surrounding, characteristics of the marker's statistical model, etc. Some parameters, such as the amplitude threshold, depend on the quality of the source fringe image, that is why they are selected automatically and adaptively.

3.2. Block 2. The amplitude detector

In this unit, the brightness amplitude for all pixels in the image for detection markers is calculated via the equation of the second finite-difference for vertical direction:

$$A(y)=\frac{2I(y)-I(y-3)-I(y+3)}{4}, \quad (2)$$

where is $A(y)$ – calculated amplitude of pixel, $I(y)$ – pixel's brightness. Further $A(y)$ is compared with a threshold value A_T. If $A(y) > A_T$, the current pixel is indicated as a potential marker, otherwise, the next pixel in the column is to be checked. All columns are processed sequentially in the same way.

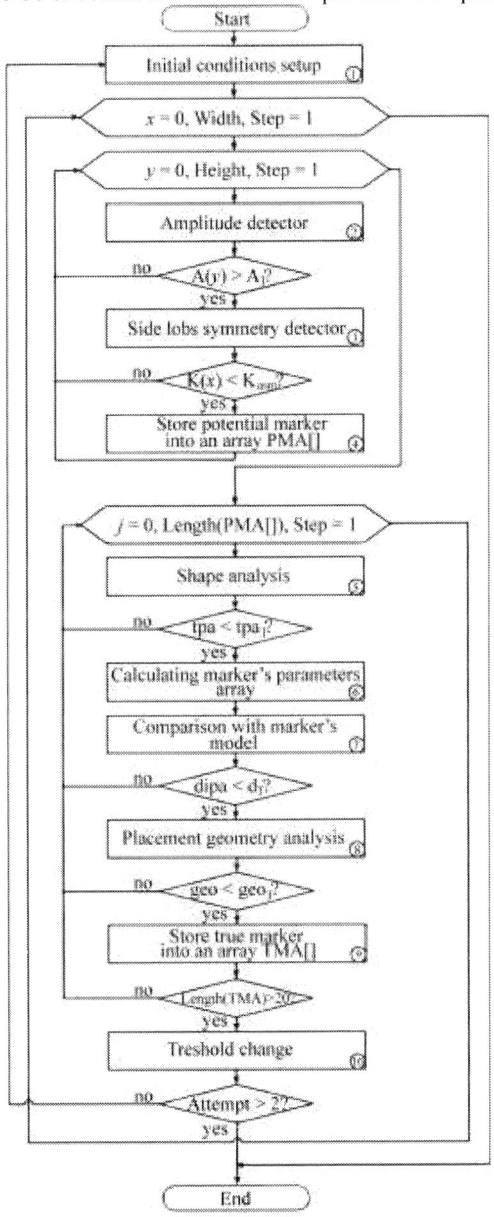

Figure 4. Flowchart of the marker's detection algorithm.

3.3. *Block 3. The side lobs symmetry detector*

As mentioned, the true markers have a specific shape with the main positive impulse and two symmetric negative side lobes stretched in the horizontal direction (Figure 3). In the vertical direction, it has a few bright pixels in a uniform background. For most of PM, the shape is different from the true marker, including the lack of symmetry of the side lobe shape. To detect the true markers asymmetry coefficient of the left and right lobe shapes in the horizontal direction is calculated using the following equation:

$$K(x) = 50\left|\frac{I(x-5) - I(x+5)}{2I(x) - I(x-5) - I(x+5)}\right|. \quad (3)$$

The value $K(x)$ is compared with the threshold K_{asm}. If it is exceeded, PM is to be discarded.

3.4. *Block 4. The storage of the potential markers*

All potential markers, which have passed through *Block 2* and *Block 3*, are to be stored in array PMA for their further selection by more complex criteria.

3.5. *Block 5. The potential marker's surroundings analysis*

In this block, the parameter, showing the heterogeneity of the surrounding pixels, is calculated. Area 9×9 surrounding pixels near PM is considered and determines 8 estimates of the brightness amplitude of the central pixel PM via the second finite difference in eight directions – horizontal, vertical, and diagonals:

$$S_k = 2I(m) - I(m+4) - I(m-4), k = \overline{1,8},$$

where is m – coordinates x or y of the surrounding pixels which depends on direction k. Using $S_1,\ldots,S_8$ calculates a parameter showing the heterogeneity:

$$tpa = 0.125\left(\sum_{i=1}^{4} S_i^2 S_{i+4}^{-2} + \sum_{i=1}^{4} S_{i+4}^2 S_i^{-2}\right) - 1. \quad (4)$$

In (4) parameters S_i and S_{i+4} are calculated in opposite directions. For true marker point $S_1 \approx S_2 \approx \ldots \approx S_8$, hence *tpa* value trends to zero. For the pseudo markers located on the torso edge, values S_i in opposite directions significantly differ due to various brightness: on the one side it is a human body, on the other side it is background, so, *tpa* is compared with some threshold tpa_T. If $tpa > tpa_T$, the current PM is rejected. The given procedure allows to reject most of the pseudo markers onto a human's torso edge.

3.6. *Block 6. Calculating of the marker's parameters*

For the next comparison of PM with the statistical marker's model some additional parameters are to be calculated:

- Marker's brightness *B*. It is taken from the source image (Figure 1).
- The amplitude of the pulse I_P – brightness value in the extreme pulse point.
- Amplitudes of the left and right-side lobes KI_L and KI_R – the value of brightness in lobe's extreme points.
- Asymmetry of the side lobes K_L. It is calculated using (2), but instead of values $I(x \pm 5)$ and $I(x)$ the KI_L, KI_R I_P values are applied.
- The amplitude ratio of the main pulse and side lobes K_{LR}:

$$K_{LR} = \frac{|256 - KI_L - KI_R|}{|I_P - 128|};$$

- PM area size SZ_S on the source image and SZ_P – PM area size on the image for detection of the markers.

These parameters, as well as the parameters previously calculated, are stored in an array PML:

$$PML = [A, K, tpa, B, I_P, KI_L, KI_R, K_L, K_{LR}, SZ_S, SZ_P] \quad (5)$$

Finally, the array *PML* is transferred into *Block 7* for the next calculations.

3.7. Block 7. The comparison with marker's model

By processing a large amount of training sample data, the mathematical model of the reflective maker as an average value of data has been created. The parameters *IML* as in (5) and standard deviation σ_i for each parameter in the marker's model have been calculated. To estimate the proximity of the marker parameters to the model one uses the following equation:

$$dipa = \left(\sum_{i=1}^{11} \left(\frac{PML_i - IML_i}{\sigma_i} \right)^2 \right)^{1/2} . \qquad (6)$$

In (6), every parameter has a different contribution to *dipa*'s estimate. It has been discovered that the maximum weight has *tpa* (up to 40% on false PM) and SZ_S (12%) values. Compared to the marker's model and real reflective marker, at least one of the false marker's parameters does not fit in the standard deviation σ_i. Thus, using an empirical value d_T as *dipa*'s threshold almost each of the false PM can be discarded.

3.8. Block 8. Placement geometry analysis

The algorithm for rejecting markers by geometric criteria is used as the final step for removing the false markers. It is based on the analysis of the relative position of the found markers. To label patients for computer optical topography, the standard scheme is used, which includes marking the spine line and two points of the pelvis. It allows to identify and remove those markers which location does not match the geometry of the standard marking, including the false-positive markers that cannot be rejected by the previous algorithms.

3.9. Block 9. The storage of true markers in the output array

If a potential marker passes all the previous blocks it is considered a true reflective marker. The marker and its characteristics are stored in the output array of the algorithm (*TMA*) for the next proceedings.

3.10. Block 10. New threshold's setup

Fifteen markers are quite enough to compute the optical topography. The algorithm is designed to detect up to 20 markers. If the output array *TMA* counts over 20 markers, some markers in *TMA* can be false-positive. The presented situation may arise due to improper lighting conditions during the filming of the image or presence of any undesired objects in the frame. Facing the problem mentioned above, the new threshold values for *Blocks 2–8* are to be calculated and the algorithm is to be repeated starting with *Block 1*. The number of the possible transitions to *Block 1* is limited to two for the preventing of algorithm looping. With the execution of instructions of this block, the algorithm is finished.

4. The experimental results

The process of marker's detection algorithm and final result are presented in Figure 5.

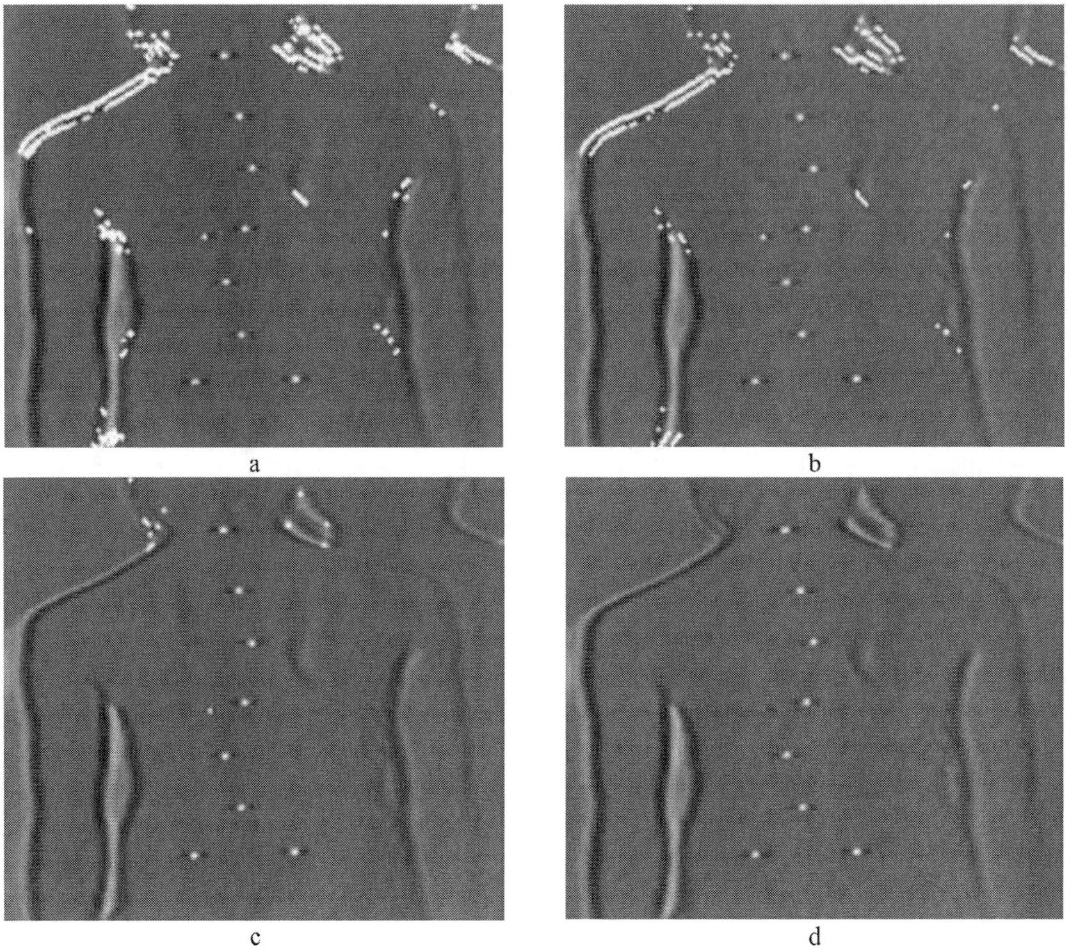

Figure 5. The stages of the reflective markers detection process: amplitude detector (a), side lobs symmetry detector (b), analysis of surroundings (c), comparison with the model (d).

The whole test is run on Intel Core i3-4350 PC with 2 GB RAM. Source fringes image size (Figure 1) is 256×256 pixels (65536 pixels in total) and all reflective markers are clearly visible. An amplitude filter (2) with threshold, as expected, has found many potential markers: 545 PM have been detected as shown in Figure 5*a*. A lot of false PM are located at the edge of the background and human body due to fast brightness drop at that place on the source fringe image. After applying side lobs symmetry detector (3) the number of the PM has decreased to 303 units (Figure 5*b*). The primary image processing cycle takes 0.08 sec to complete and store potential markers in *PMA* array. Analysis of surroundings (4) has allowed to remove most of the pseudo markers from human's body boundary and only 39 PM last including the true reflective markers as shown in Figure 5*c*. All false PM have been rejected while compared with the marker's mathematical model (6). The geometry analysis has confirmed that the remaining 8 markers shown in Figure 5*d* are real ones.

Notice that the source fringe image has a good quality and the detection of all reflective markers does not present any difficulties. The experiment with images of random quality has shown that the algorithms allow to detect 99.94% of the true markers in a sample of 1000 images with 8350 reflective markers in total, and no false-positive markers have been detected. In the sample of 5000 real markers on different quality images, the first type probability error value is about 0.22% (algorithm can't detect 11 markers), and the second type error probability is 0.06% (3 false markers was found).

IOP Conf. Series: Materials Science and Engineering 1019 (2021) 012060 doi:10.1088/1757-899X/1019/1/012060

5. Conclusion

The proposed algorithms allow detecting and recognizing the small-sized reflective markers on the fringe image used for medical topography tasks. The algorithms provide a high accuracy of the detection of the true markers with full automation of processing at high speed due to the principles of phasing and hierarchy. The principles of the adaptability of the basic algorithm and usage of the statistical model of the markers make it possible to detect the reflective markers even on images obtained under incorrect lighting conditions.

References

[1] Chen F, Brown G M and Song M 2000 *Opical Engineering* **39** 10–22
[2] Salvi J, Pages J and Batlle J 2004 *Pattern Recognition* **37** 827–849
[3] Chen H J, Lu J D and Zhang J 2007 *Optics Express* **15** 12318–12330
[4] Abdul-Rahman H 2007 Three-Dimensional Fourier Fringe Analysis and Phase Unwrapping *Ph. D thesis Liverpool John Moores University, UK*
[5] Takeda M and Mutoh K 1982 *Applied Optics* **22** 3977–3982
[6] Ubert A I and Sarnadsky V N 2012 *Procceding of IEEE 11th International Conference on Actual Problems of Electronics Instrument Engineering (APEIE)* 81–87

IOP Conf. Series: Materials Science and Engineering 1019 (2021) 012061 doi:10.1088/1757-899X/1019/1/012061

Study of aluminium nanopowder combustion by method of laser-speckle correlation

L Li[1,2], A V Mostovshchikov[3], A P Ilyin[4] and F A Gubarev[2]
[1]Faculty of Electrical and Control Engineering, Liaoning Technical University, Huludao, China
[2]Research School of Chemistry & Applied Biomedical Sciences, Tomsk Polytechnic University, Tomsk, Russia
[3]School of Nuclear Science & Engineering, Tomsk Polytechnic University, Tomsk, Russia
[4]Research School of Core Engineering Education, Tomsk Polytechnic University, Tomsk, Russia
E-mail:gubarevfa@tpu.ru

Abstract. In this paper, we discuss an application of the method of laser-speckle correlation for studying the combustion process of aluminium nanopowder in air and its mixture of iron nanopowder. It is suggested to characterize the combustion process using the statistic parameter – correlation coefficient. The principle of observation is explained and the experimental schema for investigating is provided. The differences between aluminium nanopowder and its mixture with iron nanopowder are also considered.

1. Introduction

The innovation of modern advanced equipment and instruments is inseparable from the advancement of materials science. Because of the unique properties of new materials, their discovery can provide new development directions and performance enhancements for traditional technologies. Some of these materials are manufactured by one or several high-temperature processing procedures. Therefore, the effective monitoring and observing of these processing procedures is a key technique to guarantee the quality of materials.

The manufacture of aluminium nitride by combustion of aluminium nanopowder in air is classified as a high-temperature process [1–3]. The temperature of the combustion reaches 2500 °C. Moreover, the aluminium nanopowder is wildly used as an additive in solid propellants, thermites, and explosives to improve the energy effect of combustion. Besides, aluminium-based materials provide some potentially high impact commercial opportunities in the automotive industry and microelectronics [4, 5].

Laser speckle is a special coherence phenomenon, which is produced by coherent source of light, when it illuminates rough objects. Analysing the laser speckles is possible to observe the changing of objects and characterize these changes by some statistical parameters. In practice, many sorts of changes is possible to investigate by the laser speckle techniques. Therefore, the laser speckle technique as a new optical detection method in decades, which is currently under investigation and widely used in industry, biomedical and other fields [6–8].

2. Problem statement

The mechanisms of AlN formation by combustion of aluminum nanopowder and its mixtures in air is currently under investigation to understand the nature of the synthesis process and to control the production of ceramic powders with a specified phase composition. High temperatures, intensive background lighting and high reaction rates during aluminum nanopowder combustion make the observation of the combustion process more complicated.

Because of the superior properties of aluminium nanopowder, researchers have developed a number of methods to investigate them [9]. Using the shock tube to study of aluminium combustion is a general method [10, 11]. The shock tube can generate highly controlled elevated temperature and pressure conditions with various oxidizers that provide more possibility to understand the properties of aluminium combustion. In the work [12], the authors added several optical fibers on a shock tube to register the gas phase emission during aluminium nanopowder combustion using photodiodes connected to optical fibers. In the work [13], authors suggested to use an optical system with brightness amplification for visualization of aluminum nanopowder combustion.ther paragraphs are indented (BodytextIndented style).

However, most of these studies depend on the particular reaction properties of aluminium nanopowders. Therefore, finding a simple way to study of nanopowders combustion is a perspective task. All nanosized metals are of dark grey or black color [1–3] and they are interesting objects for optical properties research. Therefore, using an optical method to observe the changes of aluminium nanopowder during the combustion process in air is a relatively easy and effective solution. Based on this, the method of laser speckle correlation seems to be a suitable one for aluminium nanopowder studies.

In this work, the method of laser speckle correlation is suggested to study the temporal characteristics of combustion of aluminium nanopowder and its mixture with iron nanopowder.

3. Experimental technique

3.1. Method of Laser Speckle Correlation

A laser-speckle pattern is a special combination of spots with varying brightness and irregular shape. These spots are produced by artificial spray or interference phenomenon, which is caused by coherent light. The coherent light does not require high brightness and does not damage the detected object. Therefore, the speckle pattern produced by the interference phenomenon has many application fields.

The speckle technique for observing the combustion process in this work based on the method of laser speckle correlation. The essence of this method is characterizing the changing parameter by the correlation coefficient. The correlation coefficient is calculated by the correlation formula, which is usually used for determining differences between two one or two-dimensional arrays. In this case, we used this formula to determine the difference between two adjacent speckle images, which captured using a high-speed camera.

The correlation formulas used in laser-speckle analyses were previously discussed in [14]. In this paper, we used the appropriate correlation formula (1) to calculate the correlation coefficient:

$$C = \frac{\sum\sum[f-\bar{f}]\cdot[g-\bar{g}]}{\sqrt{\sum\sum(f-\bar{f})^2)\cdot\sum\sum(g-\bar{g})^2}} \tag{1}$$

where $\bar{f}$ is the average intensity of the calculating area in the current speckle image; $\bar{g}$ is the average intensity of the previous speckle image; f is the intensity of every pixel in the calculating area; g is the intensity of every pixel in previous speckle image; C is the correlation coefficient of the current and previous speckle images. This formula was used in our previous work [14].

3.2. Experimental Setup

The scheme of the experimental setup is presented in Figure 1. The light source in our experiments was an Nd:YAG laser with the variable power from 5 mW to 200 mW. It generated the coherent light

IOP Conf. Series: Materials Science and Engineering 1019 (2021) 012061 doi:10.1088/1757-899X/1019/1/012061

of 532 nm wavelength for achieving a prerequisite of producing the laser-speckle patterns. The diameter of the laser beam was 0.9 mm. The objects under investigation were the aluminium nanopowder sample and the mixture of aluminium and iron nanopowders. In this work, these samples were shaped into a 2 cm long aluminum nanopowder cuboid (Figure 2), and were ignited by a gas burner. The rough surface can extremely scatter the illuminating laser beam. The filter with a narrow bandpass of 532 nm wavelength was installed to pass the reflected laser speckles and isolate the background radiation light emitted by the burning sample. The high-speed monochrome camera Phantom Fastcam SA1.1 with Canon Macro Lens EF 100 mm objective lens set on infinity focus was used for capturing the speckle images.

The light path in this experimental setup is quite simple. The laser irradiated the center of the nanopowder sample. The reflected scattered light produced the speckle patterns, captured by the camera through the bandpass filter.

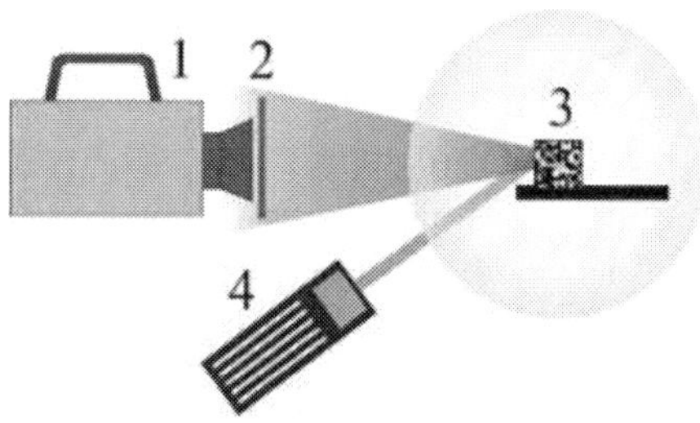

Figure 1. Experimental scheme: 1 – high-speed monochrome camera with objective, 2 – 532 nm narrow-band filter, 3 – cuboid nanopowder, 4 – Nd:YAG laser.

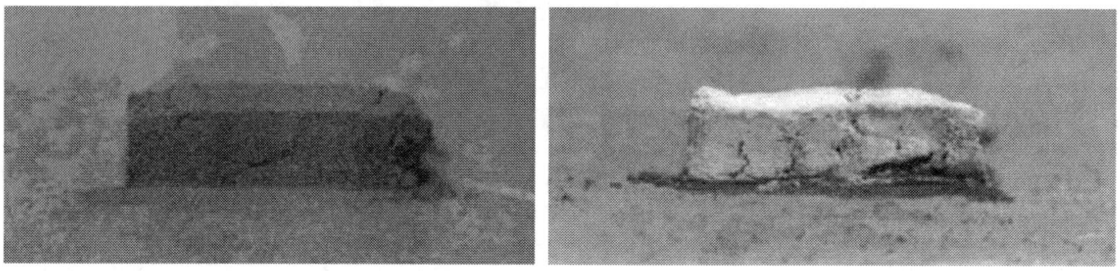

Figure 2. The typical views of aluminum nanopowder cuboid samples before combustion and after.

3.3. Principle of Imaging

Figure 3 provides two different laser-speckle images captured during combustion. These two speckle images show a total different speckle patterns. The difference indicates that the surface shape of the sample is changing during combustion. The velocity of the combustion process is inversely related to the similarity of adjacent laser-speckle images.

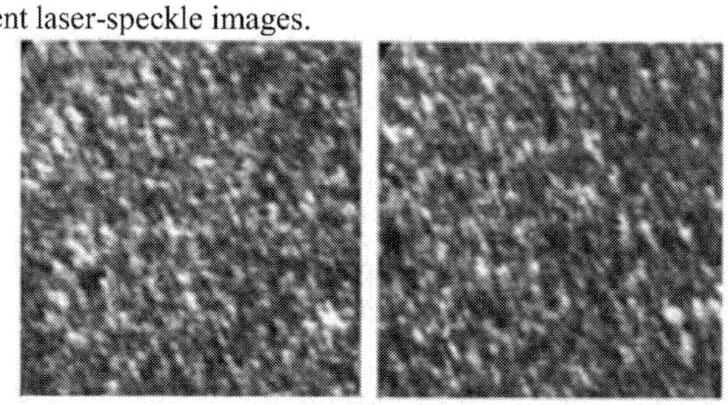

Figure 3. Laser speckle images during the aluminium nanopowder combustion.

The aluminium nanopowder combustion in air propagates via two heat waves [13, 15]. After ignition, the first heat wave propagates through the sample with low temperature 600–800°C. During the second heat wave, the combustion process turns into a thermal explosion mode with a high combustion temperature 2000–2400°C [15]. Figure 4 shows the time dependence of the correlation coefficient for the combustion. The curve can be divided into three parts corresponding to the heat waves propagation. The first part corresponds to the time interval from ignition to the start of the first heat wave. In this part, the first wave propagates from the ignited area to the observing area. The second part record is the duration of the first heat wave, and the third part shows the duration of the second heat wave.

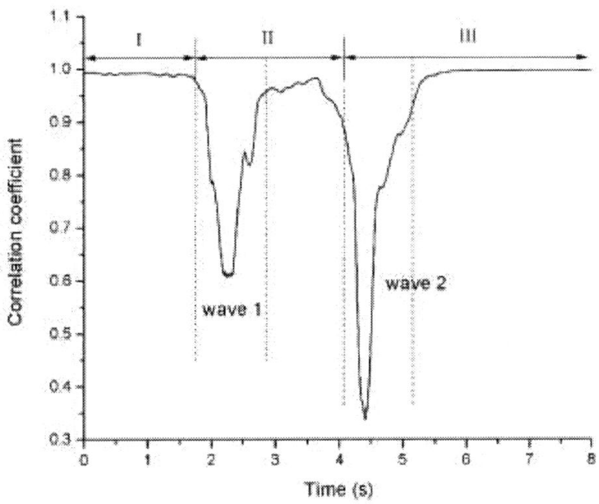

Figure 4. The schematic time dependence of the correlation coefficient during the aluminium nanopowder combustion.

4. Results and discussion

The combustion of four aluminium nanopowder samples are characterized by the correlation coefficients shown in Figure 5. All these curves, following the principle of observation, demonstrate two characteristic parts corresponding to two heat waves. The propagation time of both heat waves at the observing area can be measured using these curves. The time parameters of combustion measured using the data in Figure 5 are summarized in table 1.

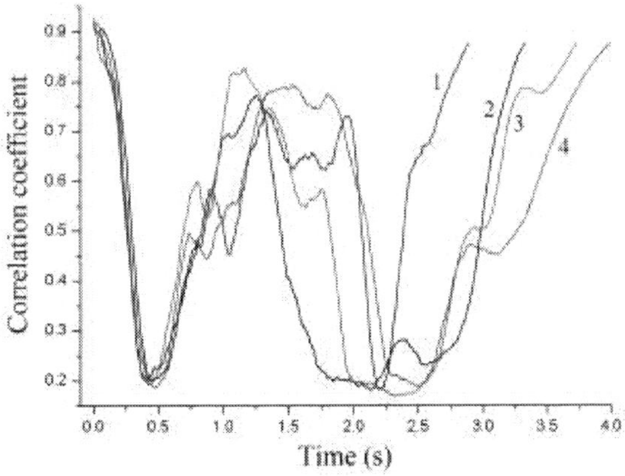

Figure 5. Time dependences of the correlation coefficient during the combustion of four aluminium nanopowder samples.

The combustion process of aluminium and iron nanopowder mixture is characterized by the correlation coefficient shown in Figure 6. These four curves demonstrate three parts of combustion, which correspond to two heat waves. Compared with the curves in Figure 5, the second heat wave in Figure 6 starts earlier, which means that the time difference is smaller. In addition, the duration of both two heat waves in Figure 6 is smaller than for the aluminium nanopowder curves in Figure 5. These time parameters are summarized also in table 1.

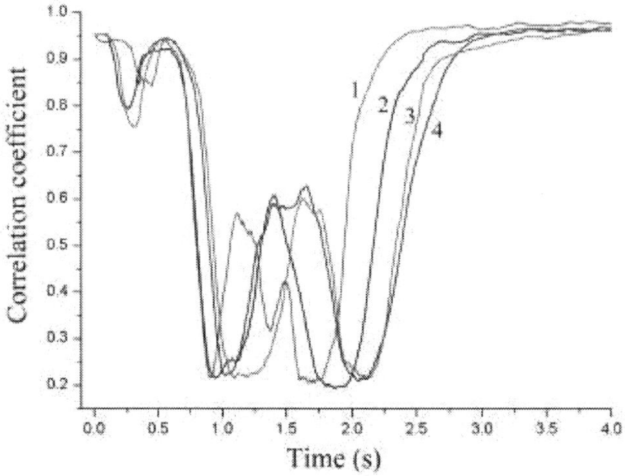

Figure 6. Time dependences of the correlation coefficient during the combustion of the mixture of aluminium and iron nanopowders.

The combustion parameters of two types of nanopowders given in table 1 demonstrate the perspectives of using the laser-speckle correlation method for determining the properties of different kinds of nanopowders. The different combustion time of these two types of nanopowders indicates that the aluminium with iron burns faster than aluminium without additives. Based on these differences and similarities of time parameter, the determination of each powder type is possible.

Table 1. The time parameters of combustions.

No.	Duration of first wave (s)	Duration of second wave (s)	Induction Time (s)	No.	Duration of first wave (s)	Duration of second wave (s)	Induction Time (s)
	Aluminium nanopowder				Mixture of aluminium and iron nanopowders		
1	1.19	1.92	0.33	1	0.89	1.17	0.15
2	1.31	1.95	0.68	2	0.62	0.90	0.43
3	1.32	0.86	0.65	3	0.88	1.19	0.38
4	1.05	1.72	0.78	4	1.08	0.88	0.18

5. Conclusions

In this work, the combustion processes of aluminium nanopowder and its mixture with iron nanopowder are studied by the method of laser speckle correlation. We obtained the time dependences of the correlation coefficient during the combustion and explained the observation principle using the correlation coefficient for characterizing propagation of the front and heat waves of combustion.

The curves of the time dependences of correlation coefficient and the time parameters, which were used for characterizing the properties of combustion, provided the capability of the method application for determination of nanopowder type. Therefore, the method of laser speckle correlation needs more in-depth research in further to define the time parameters for other kinds of mixtures of aluminium nanopowders.

Acknowledgments
The research is carried out at Tomsk Polytechnic University within the framework of Tomsk Polytechnic University Competitiveness Enhancement Program and supported by The Ministry of Education and Science of the Russian Federation, Project No 11.1928.2017/4.6.

References

[1] Sundaram D S, Yang V, and Zarko E 2015 *Comb. Expl. Shock. Waves* **51** 173–196
[2] Sundaram D S, Puri P, and Yang V 2016 *Combust. Flame* **169** 94–109
[3] Gromov A A and Teipel U 2014 M*etal Nanopowders: Production, Characterization, and Energetic Applications* (Weinheim: Wiley-VCH)
[4] Hunt W H 2000 New directions in aluminum-based P/M materials for automotive applications," Inter *J. Powd. Metal* **36** 50–56
[5] Jeong T, Kim K H, Lee S J, Lee S H, Jeon S R, Lim S H, Baek J H and Lee J K 2009 *IEEE Photon. Technol. Lett.* **21** 890–892
[6] Li L, Gubarev F A, Klenovskii M S and Bloshkina A I 2016 *Proceeding of Int. Siberian Conf. on Control and Communications (SIBCON)* 1–5
[7] Iqbal S M and Gualini M S 2013 *J. Phys.: Conf. Ser.* **439** 012048
[8] Postnov D D, Tuchin V V and Sosnovtseva O 2016 *Biomed. Opt. Express* **7** 1–10
[9] Saceleanu F, Idir M, Chaumeix N and Wen J Z 2018 *Front. Chem.* **6** 465
[10] Figueroa-Labastida M, Badra J, Elbaz A M and Farooq A 2018 *Combust. Flame* **198** 176–185
[11] Bazyn T, Krier H and Glumac N 2006 *Combust. Flame* **145** pp 703–713
[12] Lynch P, Fiore G, Krier H and Glumac N 2010 *Combust. Sci. Technol.* **182** 842–857
[13] Li L, Il'in A P, Gubarev F A, Mostovshchikov A V and Klenovskii M S 2018 *Ceram. Int.* **44** 19800–19808
[14] Gubarev F A, Li L, Klenovskii M and Glotov A 2016 *MATEC Web Conf.* **48** 04003
[15] Il'in A P, Mostovshchikov A V and Timchenko N A 2013 *Comb. Expl. Shock. Waves* **49** 320–324

IOP Conf. Series: Materials Science and Engineering 1019 (2021) 012062 doi:10.1088/1757-899X/1019/1/012062

The choice of feedstock for the biodiesel production with optimal physicochemical and low-temperature properties

N E Belozertseva, I A Bogdanov, A T Balzhanova and M V Kirgina
Division for Chemical Engineering, Tomsk Polytechnic University, Tomsk, Russia
E-mail: belozertsevanatasha@mail.ru

Abstract. The transesterification reaction of sunflower, mustard and linseed oil with ethyl alcohol was studied using potassium hydroxide as a catalyst. The physicochemical properties and low-temperature characteristics of the feedstock and obtained products were studied. The most preferred feedstock for the biodiesel fuel production had been determined.

1. Introduction

In the modern world the motor fuels consumption is increasing every year [1, 2], which leads to an increase in oil production, the development of new fields and, consequently, to reserves depletion and higher prices for hydrocarbons feedstock. With the rise in oil prices interest in alternative energy increases. One of the most popular types of alternative fuel is biodiesel, an environmentally friendly and safe to use energy source, the feedstock for the production of which is renewable. In addition, the biodiesel production allows recycling of food and industrial waste, such as waste oil.

Biodiesel has advantages over conventional diesel in terms of aromatic hydrocarbon content, sulfur, flash point and biodestruction. It can be used on mechanic transport both in pure form and as a mixture with diesel fuel. Biodiesel is the only alternative energy source that is suitable for all mechanic transport with conventional diesel engines [3].

From a chemical point of view, biodiesel fuel is a monoesters mixture of higher fatty acids, obtained by transesterification of vegetable oils and animal fats [4]. Feedstock for the biodiesel production is oilseed lipids. The main crops used are rape (EU countries), sunflower (France, Italy), soybean (USA, Brazil and African countries), canola (Canada), palm oil (Indonesia, Malaysia) and jatropha (India, African countries). The main fatty acids contained in the oilseeds lipids are palm, oleic and linoleic acids.

A large number of works are devoted to the biodiesel fuels production; researchers in this direction are working all over the world. Therefore, the authors of the article [5] describe in detail the various parameters effect on the process of obtaining biodiesel from oilseed plants. To achieve a high biodiesel yield, it is necessary to optimize a variety of catalytic variables, such as catalyst type and composition, carrier type and pretreatment conditions (for example, temperature and calcination time). In addition, many operation conditions require adjustments such as the reaction temperature, alcohol type, molar ratio of alcohol to oil and stirring speed.

A homogeneous alkaline catalyst such as sodium hydroxide (NaOH), sodium methoxide (NaOCH3), potassium hydroxide (KOH), or a homogeneous acid catalyst, such as sulfuric acid (H2SO4) or hydrochloric acid (HCl) can catalyze the transesterification reaction. Homogeneous alkaline catalysts are considered more preferable despite the fact that the transesterification reaction in the presence of this catalysts type causes saponification of the triglyceride or the product of fatty acid methyl ester.

Saponification increases the catalyst consumption, reduces the biodiesel yield and complicates separation processes after the biodiesel production.

Acidic catalysts compared with alkaline do not lead to saponification, but the reaction rate is lower. Therefore, the reaction requires higher temperatures and pressures to achieve the desired conversion. Separation and purification of the product is simple; therefore, the reaction is more economically feasible and the catalyst can be reused after a simple filtration process [6, 7]. In addition to these, heterogeneous catalysts such as alkali metal carbonates (Na2CO3, K2CO3), alkaline earth metal carbonates (CaCO3), oxides of alkaline earth metals (CaO, MgO, SrO, BaO) and other oxides such as ZnO are also used [8].

In addition to alkaline heterogeneous catalysts, green biocatalysts are currently attracting the scientific community attention due to their energy-efficient enzymatic properties. In [9–11], it was found that by using immobilization or in the liquid phase, high yields can be achieved with the help of reliable lipase enzymes. Enzymatic process can be easily combined with the latest technologies (for example, ultrasonic), work in a continuous mode using various reactors designs; also the enzymes lifetime can be improved in order to take further steps towards industrial production. Last but not least, the use of metagenomics in enzyme technologies opens up broad prospects for the development of stable and solvent-resistant biocatalysts for the biodiesel production [9, 12].

For the transesterification reaction, alcohols such as methanol, ethanol, propanol, butanol, pentanol, and amyl alcohol are commonly used.

In the majority of existing methods, methyl alcohol is used as a transesterifying component, which is a toxic liquid of the third hazard class, which has a damaging effect on the human nervous system and vascular system when inhaled through the respiratory system or the skin. Therefore, for safety reasons, it is advisable to use ethyl alcohol, which, in addition, is a more affordable feedstock. However, the biodiesel production using ethanol as a transesterifying agent has not occurrence due to the high solubility of the esters obtained in this alcohol, which makes it difficult to isolate the desired reaction products and leads to a lower yield of the final product.

Both methanol and ethanol have advantages and disadvantages. For example, the reaction with methanol is faster at a lower temperature compared to the reaction using ethanol [12], in addition, the reaction with methanol gives a higher biodiesel yield. However, ethanol obtained by renewable methods, in some cases, for example, by transesterification of soybean oil, increases the reaction rate [13].

To increase the alcohol / oil ratio, there are practically no restrictions, however, the authors of [14] showed that the increase in the ratio is effective only up to a certain limit. However, the authors of article [15] confirm that during the transesterification of soybean oil with methanol with a Li-doped MgO catalyst, the alcohol to oil ratio of more than 15 complicates the processes of heating and separating the product, the optimum ratio is from 12 to 15. The authors of [16] found that the ratio of alcohol to oil, exceeding 30, is harmful due to the increased solubility of the glycerol product in the reaction mixture. However, in the case of the transesterification reaction of rapeseed oil with methanol and Zn / Al catalyst carried out in high-pressure autoclaves at high temperatures, it was found that a higher alcohol to oil ratio is favorable.

One of the important variables that effects on the transesterification reaction rate is temperature. Although the reaction can be carried out at room temperature, which the authors of [17] successfully implement, based on the kinetics, the reaction rate increases with increasing temperature. Moreover, in the case of oils transesterification higher temperatures lead to decrease in oil viscosity and better reagents mixing.

Nevertheless, there is an optimal reaction temperature and its further increase causes a decrease in the biodiesel yield due to increased saponification (in the case of homogeneous alkaline catalysts), as well as the rapid evaporation of the used alcohol [18]. Consequently, the reaction temperature is usually maintained below the boiling point of the alcohol; however, it is possible to carry out the reaction at higher temperatures under reflux conditions or at high pressure.

A literature review showed that the process of biodiesel production from oil plants has a wide range of temperatures, pressures and ratios of raw feedstock. This gives a huge field to study the biodiesel production and the products properties that also change with changing reaction conditions.

The most significant biodiesel characteristics are viscosity, lubricity, flash point, cetane index and low temperature properties. These properties play a decisive role in the equipment design for the biofuels use in diesel engines, in the fuel operation in different climatic conditions.

Thus, the study purpose is to determine the most preferred feedstock for the biodiesel fuel production with optimal physicochemical and low-temperature properties.

To achieve this goal, the following tasks were determined:

- investigate the physicochemical properties and low-temperature characteristics of feedstock for the biodiesel fuel production (vegetable oils);
- choose a method for the biodiesel synthesis;
- investigate the physicochemical properties and low-temperature characteristics of the obtained products (complex mixtures consisting of ethyl esters of fatty acids – biodiesel fuel).

2. Experimental part

Sunflower, mustard and linseed unrefined edible oils from various manufacturers were used as feedstock for biodiesel production.

Ethyl alcohol was used as a transesterifying agent.

The molar ratio "ethanol: oil" is assumed to be 9: 1 based on the molar masses of alcohol and vegetable oil. When the molar ratio is less than 9: 1, the reaction is incomplete, with an increase in the molar ratio, the separation of glycerol becomes more difficult and the yield of ethers decreases.

The reactor was a heat-resistant beaker 4 in the volume of 1 liter, located on the electric stove 5, equipped with a stirrer 1 and a thermometer 3 (Figure 1). The top of the heat-resistant beaker was covered with metal foil to prevent volatilization of alcohol vapors during the reaction.

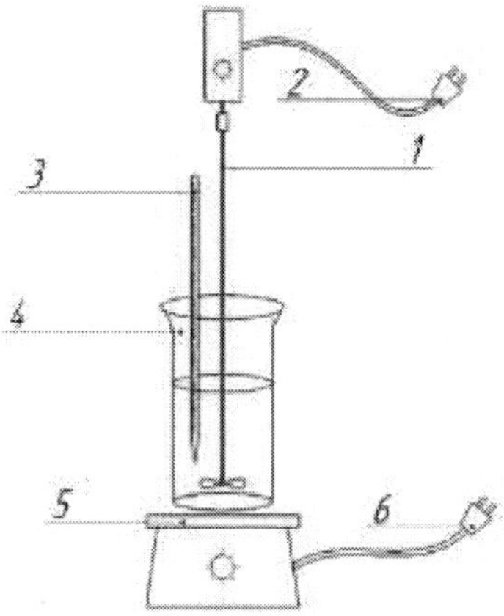

Figure 1. Simplified diagram of the laboratory unit for the transesterification reaction: 1 – automatic blade stirrer; 2 – power-intake connection yoke for blade stirrer to the electricity mains; 3 – thermometer; 4 – heat-resistant beaker; 5 – electric stove with heat control; 6 – power-intake connection yoke for electric stove to the electricity mains.

Vegetable oil was placed in a heat-resistant beaker, slowly heated to 75 °C with constant stirring, then the catalyst was introduced and the reaction time was started. An alcoholic solution of potassium hydroxide was used as a catalyst. The reaction time was 6 hours. At the end of the synthesis, the reaction mixture was cooled to room temperature and glycerol was added, which contributed to the formation of two phases. The beaker's contents were transferred to a separation funnel for further processing. After settling, two phases were formed – the upper one consisting of ethyl esters of vegetable oils and residual ethanol, and the lower one containing glycerin, an excess of ethanol, unreacted potassium hydroxide, formed soaps and products of incomplete transesterification. The obtained ethyl esters were purified from residual ethanol using a rotary evaporator. The catalyst was removed by sequential washing with 1 % citric acid solution (until neutral pH) and then with distilled water.

The use of a rotary evaporator can significantly reduce losses during the distillation of ethyl alcohol and eliminate the stage of drying the obtained product.

3.Results and discussion

The paper was examined various feedstock options (vegetable oil) for the biodiesel fuel production. The physicochemical properties of three different vegetable oils (sunflower, mustard, linseed) were investigated. The results of measurements of the kinematic viscosity [19], dynamic viscosity [19] and density [20] are shown in tables 1–3, respectively.

Table 1. Kinematic viscosity of vegetable oils.

Vegetable oil	Kinematic viscosity at 15 °C (m^2/s)	Kinematic viscosity at 20 °C (m^2/s)
Sunflower oil	$77.76 \cdot 10^{-6}$	$62.46 \cdot 10^{-6}$
Mustard oil	$90.26 \cdot 10^{-6}$	$72.14 \cdot 10^{-6}$
Linseed oil	$62.41 \cdot 10^{-6}$	$51.07 \cdot 10^{-6}$

Table 2. Dynamic viscosity of vegetable oils.

Vegetable oil	Dynamic viscosity at 15 °C (Pa·s)	Dynamic viscosity at 20 °C (Pa·s)
Sunflower oil	$71.77 \cdot 10^{-3}$	$57.43 \cdot 10^{-3}$
Mustard oil	$83.05 \cdot 10^{-3}$	$66.14 \cdot 10^{-3}$
Linseed oil	$58.13 \cdot 10^{-3}$	$47.41 \cdot 10^{-3}$

Table 3. Density of vegetable oils.

Vegetable oil	Density at 15 °C (kg/m^3)	Density at 20 °C (kg/m^3)
Sunflower oil	922.9	919.6
Mustard oil	920.1	916.8
Linseed oil	931.5	928.4

From the data presented in tables 1–3 it can be seen that linseed oil has the lowest both kinematic and dynamic viscosity at temperatures of 15 °C and 20 °C. While the least density at temperatures of 15 °C and 20 °C has mustard oil.

Low-temperature characteristics, such as cloud point [21] and pour point [22], were also determined for feedstock for biodiesel fuel production. The results are shown in table 4.

Table 4. Cloud point and pour point of vegetable oils.

Vegetable oil	Cloud point (°C)	Pour point (°C)
Sunflower oil	-9	-19
Mustard oil	-18	-29
Linseed oil	-13	-16

From the data presented in table 4, we can conclude that mustard oil has the best low-temperature characteristics, sunflower oil – the worst.

The result of triglyceride esters transesterification contained in vegetable oil under the low alcohols action in the catalyst presence is a complex mixture formation consisting of fatty acids ethyl esters – the main synthesis product – biodiesel fuel.

Thus, the overall efficiency of the biodiesel production technology is determined by the conversion degree of the feedstock and the ethyl esters yield of vegetable oils. Product yields for various vegetable oils are presented in table 5.

Table 5. The yields of the mixture of fatty acids ethyl esters obtained from vegetable oils.

Product	Mass of feedstock (g)	Yield (%)
Sunflower oil product	584.40	59.12
Mustard oil product	573.20	50.37
Linseed oil product	567.10	51.28

As can be seen from table 5, which presents the product yields after residual alcohol distillation, the highest yield is observed for the product obtained from sunflower oil, the lowest – for the product obtained from mustard oil.

For the obtained biodiesel fuels were determined physicochemical properties. The results of determining the kinematic viscosity [19], dynamic viscosity [19] and density [20] are shown in tables 6–8, respectively.

Table 6. Kinematic viscosity of obtained biodiesel fuels.

Product	Kinematic viscosity at 15 °C (m^2/s)	Kinematic viscosity at 20 °C (m^2/s)
Sunflower oil product	$20.37 \cdot 10^{-6}$	$17.19 \cdot 10^{-6}$
Mustard oil product	$25.81 \cdot 10^{-6}$	$21.73 \cdot 10^{-6}$
Linseed oil product	$20.96 \cdot 10^{-6}$	$17.88 \cdot 10^{-6}$

Table 7. Dynamic viscosity of obtained biodiesel fuels.

Product	Dynamic viscosity at 15 °C (Pa·s)	Dynamic viscosity at 20 °C (Pa·s)
Sunflower oil product	$18.43 \cdot 10^{-3}$	$15.49 \cdot 10^{-3}$
Mustard oil product	$23.36 \cdot 10^{-3}$	$19.60 \cdot 10^{-3}$
Linseed oil product	$19.18 \cdot 10^{-3}$	$16.29 \cdot 10^{-3}$

When comparing the kinematic viscosity of the feedstock for the biodiesel fuel synthesis (table 1) and the obtained products (table 6), it can be noted that cause of the transesterification reaction, the kinematic viscosity decreased more than three times for each of the products.

Products from sunflower and linseed oil have similar kinematic viscosity values, in contrast to the product obtained from mustard oil, the viscosity of which is the highest among those presented.

Similar trends are observed for dynamic viscosity values (the product obtained from sunflower oil has the lowest dynamic viscosity, the product obtained from mustard oil has the highest viscosity).

Table 8. Density of obtained biodiesel fuels.

Product	Density at 15 °C (kg/m^3)	Density at 20 °C (kg/m^3)
Sunflower oil product	904.7	901.0
Mustard oil product	905.0	901.7
Linseed oil product	915.1	911.4

When comparing the density of feedstock for the biodiesel synthesis (table 3) and the obtained mixtures of fatty acids ethyl esters (table 8), it should be noted that the density of each obtained product decreased by more than 15 kg/m^3. A product obtained from linseed oil characterizes the highest density, both at 15 °C and at 20 °C. While the product obtained from sunflower oil has the lowest density at 15 °C; the product obtained from mustard oil has the lowest density at 20 °C.

A significant decrease in each of the physicochemical characteristics has a positive effect on the possibility of using biodiesel fuel for vehicles, since viscous and heavy fuel makes it difficult for the engine to work.

During the study of the literature, it was found that not many authors pay attention to the low-temperature characteristics of the mixtures of fatty acids ethyl esters. For the cold climate that prevails in most of Russian Federation territory, the issue of compliance of the low-temperature characteristics of the obtained products with the existing standards is most acute.

In the work process, the cloud point [21] and pour point [22] of the transesterification reaction products were determined. The results are presented in table 9.

Table 9. Cloud point and pour point of obtained biodiesel fuels.

Product	Cloud point (°C)	Pour point (°C)
Sunflower oil product	+7	-8
Mustard oil product	-11	-14
Linseed oil product	-10	-12

An overall deterioration in the low-temperature properties of the obtained products is observed in comparison with the corresponding characteristics of the feedstock (table 4). The product obtained from mustard oil has the best cloud point and pour point; the product obtained from sunflower oil – is the worst.

4.Conclusions

Based on the investigation results the following conclusions were made:

- Complex mixture of ethyl esters, obtained from sunflower oil, is characterized by the highest yield, as well as the lowest viscosity values. Thus, based on the physicochemical properties of the obtained products, sunflower oil is the most preferable feedstock for biodiesel production, which is also the most optimal for processing from an economic point of view.
- It was found that products derived from mustard oil have the best low-temperature properties. Thus, mustard oil is the most optimal feedstock in terms of low-temperature characteristics, which plays an important role in cold climate conditions.

References

[1] Belinskaya N 2016 *Pet. and Coal* **58** (1) 126–134
[2] Kirgina M, Bogdanov I, Altynov A and Belinskaya N 2019 *Pet. and Coal* **61** (1) 110–119
[3] Braginskij O 2008 *Ros. him. zhurnal [Russian chemical journal]* **52** (6) 137–146
[4] Dvoreckij S, Zazulya A, Nagornov S, Romancova S and Ryazanceva I 2012 *Vopr. sovr. nauki i*

pr. [Problems of Contemporary Science and Practice] (39) 126–35
[5] Günay M, Türker L and Tapan N 2019 *Fuel* **250** 27–41
[6] Tapan N, Yıldırım R and Günay M 2016 *Biofuels, Bioprod. Biorefin* **10** (4) 422–34
[7] Islam A, Taufiq-Yap Y, Chan E, Moniruzzaman M, Islam S and Nabi M 2014 *Energy Convers. Manage* **88** 1200–1218
[8] Sharma Y, Singh B and Korstad J 2011 *Fuel* **90** (4) 1309–1324
[9] Gusniah A, Veny H and Hamzah F 2019 *Ind. Eng. Chem. Res* **58** (2) 581–589
[10] Hama S, Noda H and Kondo A 2018 *Curr. Opin. Biotechno* **50** 57–64
[11] Tacias-Pascacio V *et al* 2019 *Renewable Energy* **135** 1–9
[12] Meneghetti S *et al* 2006 *Energy Fuels* **20** (5) 2262–265
[13] Ghesti G *et al* 2009 *Appl. Catal.* A **355** (1–2) 139–147
[14] Sai A, Niju S, Meera K and Anantharaman N 2019 *Energy Sources* A **41** (15) 1862–1878
[15] Wen Z, Yu X, Tu S, Yan J and Dahlquist E 2010 *Appl. Energy* **87** (3) 743–748
[16] Xie W and Wang T 2013 *Fuel Process. Technol.* **109** 150–155
[17] Yang X, Wang Y, Yang Y, Feng E, Luo J, Zhang F, Yang W and Bao G 2018 *Energy Convers. Manage* **164** 112–121
[18] Abbah E, Nwandikom G, Egwuonwu C and Nwakuba N 2016 *American Journal of Energy Science* **3** (3) 16–20
[19] ISO 3104: 1994 Petroleum products. Transparent and opaque liquids. Determination of kinematic viscosity and calculation of dynamic viscosity
[20] ISO 12185:1996 Crude petroleum and petroleum products – Determination of density – Oscillating U-tube method
[21] ASTM D2500-05 Standard Test Method for Cloud Point of Petroleum Products
[22] ASTM D97-17b Standard Test Method for Pour Point of Petroleum Products

Analysis of the FACTS devices impact on the transmission line protection operation

N Yu Ruban, A S Gusev, Yu D Bay and I A Razzhivin
School of Energy and Power Engineering, Tomsk Polytechnic University, Tomsk Russia
E-mail: rubanny@tpu.ru

Abstract. The effect of the static synchronous compensator (STATCOM) on distance protection of a high-voltage transmission line is studied in this article. To carry out the research stated in the topic of the paper, an experimental scheme was developed, equipment was selected, the operation settings of the 3 stages of distance protection were calculated and quad characteristics were built on their basis. All listed above, as well as further experiments, were carried out by the real time digital simulator. The STATCOM model was integrated in the scheme and set next. A series of experiments was carried out: two-phase and three-phase short circuits in various nodes of the scheme with and without the compensator and in various connection nodes. After all the experiments were carried out, appropriate conclusions were made about the impact of FACTS devices on the operation of distance protection. Possible ways to solve detected problems were also proposed.

1. Introduction

Currently, flexible alternating current transmission systems (FACTS) – a set of technical and information tools for automatic control of transmission line parameters – are widely spread and developed [1]. Many problems are solved with the help of these devices: increasing the transmission capacity of power lines; ensuring the stable operation of the power system under various disturbances; improving the reliability of energy saving of consumers; reduction of losses in electrical networks. One of such devices is a static synchronous compensator (STATCOM).

As with the putting of any new element into the system, it is necessary to study the effect of STATCOM on network operation in various modes, including faults. The role of automatic devices that detect faults and disconnect equipment or inform personnel and dispatchers about the appearance of an abnormal mode, is performed by relay protection and automation. One of the most widely used basic protection of 110–220 kV lines is distance protection (DP), the principle of this one is to calculate the resistance based on measured voltages and currents. There are guidelines for calculating and configuring relay protection for various equipment and different network modes [2]. Besides each protection manufacturer provides its own recommendations for setting up its own devices. The regulatory documents used in Russia for setting up distance protection do not contain any recommendations for their operation in networks with FACTS devices [3]. Perhaps this is due to the low prevalence of FACTS in Russia, but recently such devices have been shared quite actively.

Thus, there is a question about the qualitative and quantitative influence of FACTS devices on functioning now elements, including the relay protection of transmission lines, as well as the question of taking them into account when calculating the settings and configuring the protection. In the paper a study of the effect of a second-generation FACTS device — a static synchronous compensator — on the distance protection of a 220 kV transmission line was conducted, since it has several advantages over similar compensation devices and is widely used now and also an effective device in terms of its use for maintaining voltage levels or controlling power flows [1, 4, 5]. At the same time, there is a problem of the behavior of FACTS in case of faults, particularly their influence on the operation of relay protection [6–10].

The purpose of the study is to answer the question about the necessity of taking into account the impact of a particular device on a distance protection in Russian networks. Production companies are interested in the results in terms of improving the quality and accuracy of recommendations for protection setting up.

2. Case study

To carry out the study, a circuit in which there is a 220 kV transmission line with a three-stage distance protection installed is required. According to the [2], the first stage is adjusted from the own resistance of the protected line, the second stage is adjusted from the first stage of the previous line, the third stage is adjusted to ensure reliable coverage of the previous line. It is also possible to set from the short-circuit on the low voltage side of the transformer at the end of the protected line. Due to the above, it was decided to use the experimental power grid presented in Figure 1. All the equipment shown in the diagram is made in Russia and its parameters are selected in accordance with [11].

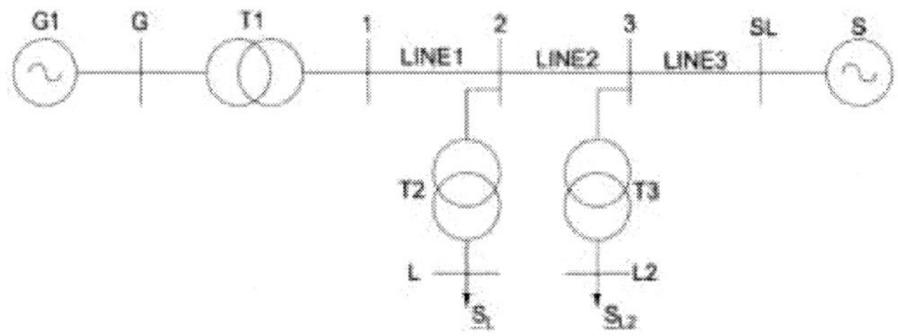

Figure 1. Experimental power grid scheme.

Generator G1 rated power is 100 MW. STATCOM rated power is 100 Mvar. Loads power (S_L) is 40+20i MVA. Lines rated voltage is 220 kV.

Calculation of distance protection parameters is carried out for a set of line 1 installed at substation 1. The model of the DP measuring body in the real-time digital simulator (RTDS) shows the primary resistance. The values of phase voltages and currents are given to the input; it is possible to monitor phase-to-ground resistances, phase-to-phase resistances, and also resistances of direct, inverse, and zero sequences at the output. In the output locuses during modeling data about line resistance are entered, as well as about calculated characteristics in primary values, using this data, the complex draws quad characteristics.

3. Quad characteristics

In general, the quad characteristic of each stage has a form shown in Figure 2.

With the help of the measuring body in the RTDS, the phase resistance is found at the specified fault (node 2), it is also own for the LINE1:

$$\underline{Z}_{LINE1} = 9.2 + j43.19 = 44.159e^{j78^\circ} \text{ Ohm}.$$

The first stage set point calculation for LINE1:

$$\underline{Z}^{(1)}_{set1} = 0.85 \cdot \underline{Z}_{LINE1} = 0.85 \cdot 44.159e^{j78^\circ} = 37.535e^{j78^\circ} \text{ Ohm}.$$

For further calculations LINE2 own resistance is needed. From similar experiment:

$$\underline{Z}_{LINE2} = 11.07 + j51.9 = 53.067e^{j78^\circ} \text{ Ohm}.$$

Matching with the previous line. The current distribution coefficient between LINE1 and LINE2 with a short-circuit at the node 2 equals one, since there are no additional sources from the beginning of the first line to the end of the second one.

The second stage set point calculation for LINE1 with matching:

$$\underline{Z}_{set1}^{(2)} = 0.85 \cdot \underline{Z}_{LINE1} + \frac{0.66}{k_{cII}} \underline{Z}_{LINE2}\ ; \quad \underline{Z}_{set1}^{(2)} = 0.85 \cdot 44.159e^{j78^\circ} + \frac{0.66}{1} 53.067e^{j78^\circ} = 75.56e^{j78^\circ} \text{ Ohm}.$$

The third stage is calculated based on the conditions of reliable coverage of the previous line and the conditions of the required sensitivity. For this, a short-circuit is simulated at the end of the LINE2 (node 3) and the resistance of the is measured by LINE1 protection. The resulting value is:

$$\underline{Z}_{LINE1}^{(3)} = 22.18 + j94.4 \text{ Ohm}.$$

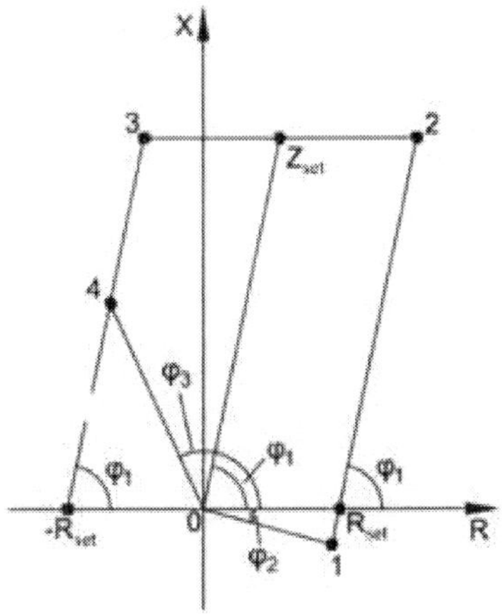

Figure 2. Quad characteristic general form, where Z_{set} – set point of the stage for which the characteristic is built; φ_1 – the angle of maximum sensitivity taken equal to the angle of the set point resistance; φ_2 – the angle of the edge 01 inclination taken to be -15° for 1 and 2 stages, for 3 stage calculated separately; φ_3 – the angle of the edge 34 inclination taken to be 115°; R_{set} – active resistance set point; edge 23 passes through the end of a straight Z_{set} and is parallel to the axis R.

Required sensitivity coefficient for the third stage is $k_s = 1.2$. The third stage set point calculation for LINE1:

$$\underline{Z}_{set1}^{(3)} = k_s \cdot \underline{Z}_{LINE1}^{(3)} = 1.2 \cdot (22.18 + j94.4) = 116.365e^{j77^\circ} \text{ Ohm}.$$

The angle φ_2 is calculated next for the third stage operation characteristic:

$$\cos\varphi_{max} = \frac{\cos\varphi_{nom}}{U_{min} I_{load\,max}} = \frac{0.85}{0.9 \cdot 1} = 0.944\ ;$$

$$\varphi_2 = 5 + \arccos(\cos\varphi_{max}) = 5 + \arccos(0.944) = 24.19^\circ ,$$

where $\cos\varphi_{max}$ – cosine in maximum load mode; $\cos\varphi_{nom}$ – cosine in normal load mode; U_{min} – voltage in maximum load mode, p.u.; $I_{load\,max}$ – current in maximum load mode, p.u.

Active resistance set points:

$$R_{set1} = 13.679 \text{ Ohm};\ R_{set2} = 19.246 \text{ Ohm};\ R_{set3} = 26.208 \text{ Ohm}$$

4. STATCOM model

The STATCOM model in the RTDS is represented by two blocks. Block 1 (STATCOM) is required for elements that are modeled using a small time-step in the program, while a large time-step is used for the main circuit. This block contains DC sources, rectifiers, a reactor, a filter, a special transformer model for connecting parts with different calculation steps, as well as elements generating a sawtooth signal and firing pulses based on the data received from the control system. Experimental power grid scheme with STATCOM connection is shown in Figure 3.

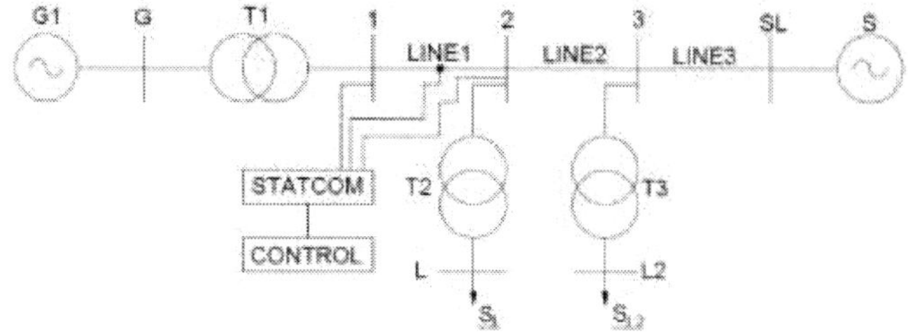

Figure 3. Experimental power grid scheme with STATCOM connection.

In block 2 (CONTROL) the control system is located. The control uses the pulse width modulation method. Based on the measured power flowing between the system and STATCOM, as well as the voltage at the common coupling point (PCC) and the set point, an angle and modulation factor are generated for the reference signal, which is compared with the carrier sawtooth signal in the firing pulse generator. The set points for voltage and active power are put in the control system (when adjusting reactive power and voltage, the set point for active power is equal to zero) in p.u. Base values are set in the measuring bodies. The installed power of STATCOM is regulated by changing the voltage of the element VDC in block 1.

5. Results

The following experiments were carried out during the studies: a two-phase short circuit between phases AB and a three-phase short circuit at the end of the protected and previous lines. Using midpoint STATCOM, additional experiments were carried out with a fault for 25% and 75% of the protected line length. The obtained polygonal operational characteristics of the starting relays of distance protection and the locuses of resistance are presented in Figures 4–10. In the Figures, the red line shows the resistance of the protected line. Blue indicates the response of the first stage, green and purple - the characteristics of the second and third stages, respectively. The graphs in Figures 4, 5 are obtained for the case of connecting STATCOM to node 1.

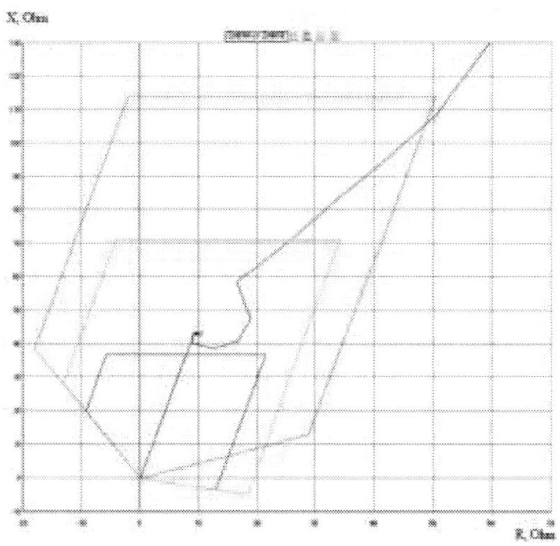

Figure. 4 Phase-to-phase fault, node 2.

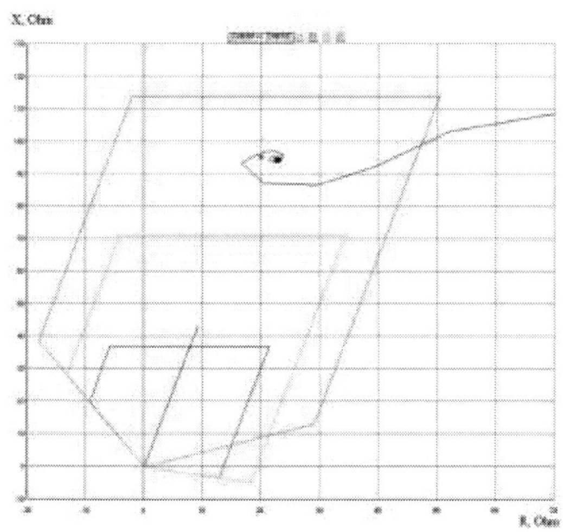

Figure. 5 Phase-to-phase fault, node 3.

The graphs in Figures 6, 7 are obtained for the case of connecting STATCOM to node 2.

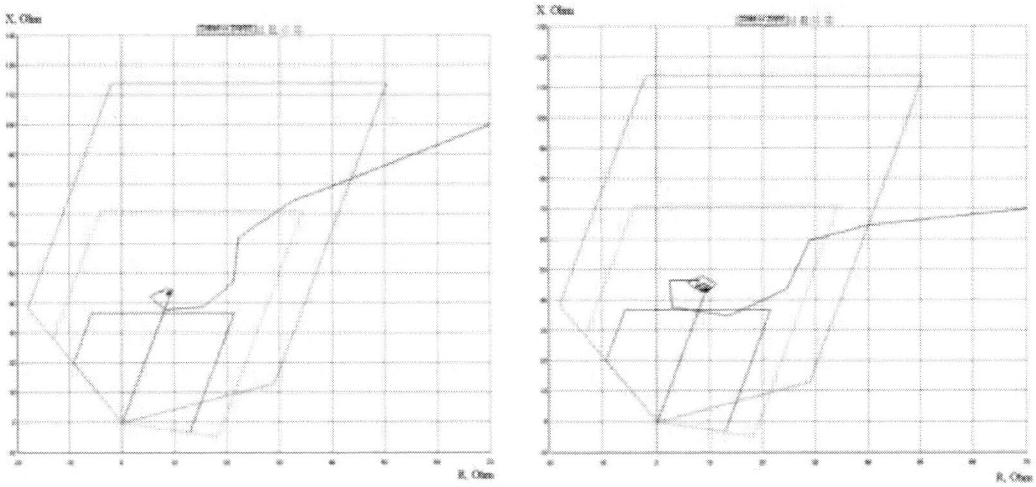

a b

Figure 6. Phase-to-phase (a) and three-phase (b) faults, node 2.

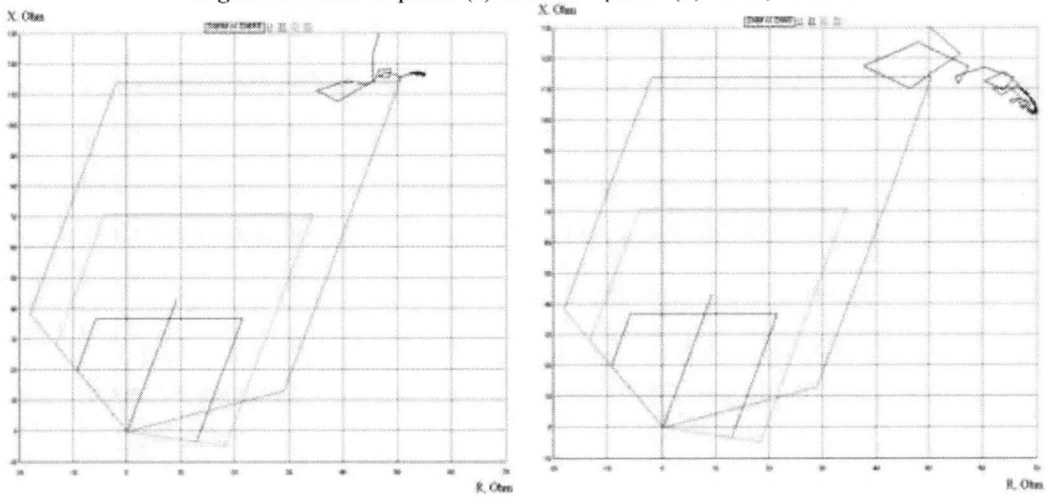

a b

Figure 7. Phase-to-phase (a) and three-phase (b) faults, node 3.

The graphs in Figures 8–10 are obtained for the case of connecting STATCOM to the midpoint of line 1.

IOP Conf. Series: Materials Science and Engineering 1019 (2021) 012063 doi:10.1088/1757-899X/1019/1/012063

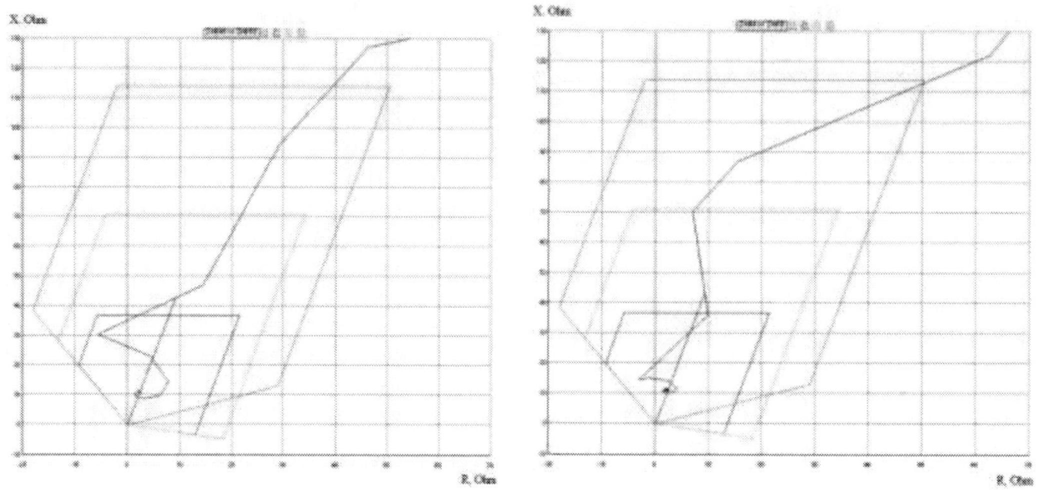

a b

Figure 8. Phase-to-phase (a) and three-phase (b) faults, 25 % length.

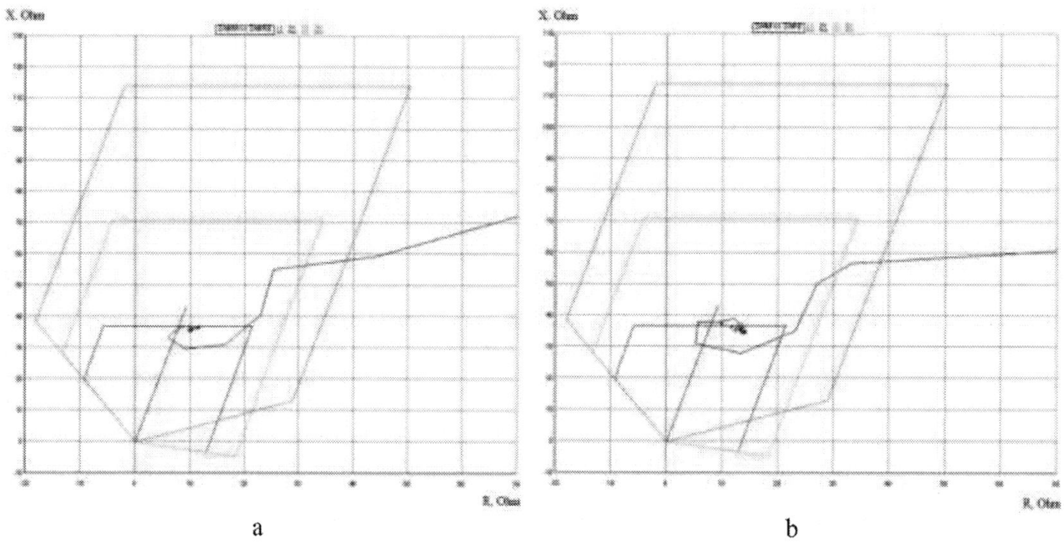

a b

Figure 9. Phase-to-phase (a) and three-phase (b) faults, 75 % length.

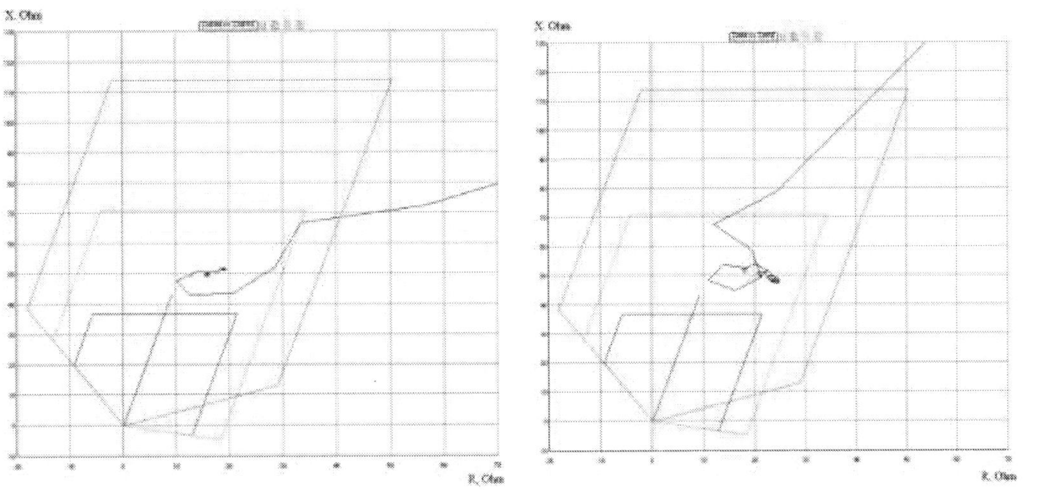

a b

Figure 10. Phase-to-phase (a) and three-phase (b) faults, 100 % length (node 2).

From the results of the performed experiments the fact of the static synchronous compensator influence on the transmission line distance protection operation was established in individual cases. STATCOM does not affect the protection operation if the fault occurs between it and the place where the distance protection is installed, as well as when the compensator is connected at the beginning of the protected line. This is due to the fact that in these cases, the measured current and voltage required to calculate the resistance either do not change or change proportionally. When the STATCOM is located between the distance equipment and the fault place, there is a noticeable influence on the behavior of the protection. For example, the third stage underreaches and does not provide long-distance backup mode with a short circuit at the end of the previous line if the compensator is installed at the end of the protected line, because the locus does not get into the third stage zone of operation. Experiments in which the STATCOM is connected at the midpoint shows that a locus shift relatively to normal operation can lead to the triggering of the second and third stages, when the first and the second ones must operate. This, in turn, leads to a delayed response and may be the result of more serious damage and emergency conditions of the power system.

6. Conclusion

According to the experiments results, it can be concluded that the compensator increases the measured reactive and active components of the resistance. This fact must be taken into account when calculating the settings for the distance protection of power lines and reflected in the relevant regulatory documents. Thus, it is possible to take this into account using the coefficient, which increases the quad characteristic edges. Another problem solution for protection manufacturers may be to develop an adaptive function. The point is to ensure the exchange of information between distance protection and STATCOM and dynamically change the operation characteristics depending on the mode of the compensator and its installed power.

Acknowledgments

The work was supported by Ministry of Science and Higher Education of Russian Federation under the governmental grant "Science" № FSWW-2020-0017.

References

[1] Tiwari R, Niazi K and Gupta V 2012 *Power Energy Soc. Gen. Meet.* 6344702

[2] Hondurov S, Ilyukhin E, Pirogov M and Soloviev A 2013 *Step distance protection of 35–220 kV transmission lines. The method of calculation* (St. Petersburg: PEIPK) p 32

[3] 2015 STO 56947007-29.120.70.200-2015 *Guidelines for the calculation and selection of settings of microprocessor-based devices for relay protection and automation* (Moscow)
[4] Lee C, Leung S and Hui S 2003 *IEEE Trans. Power Electron.* **18**(4) 1084–1092
[5] Parvathy S, Thampatty K C S and Nambiar T N P 2017 *IEEE Asia-Pacific Power Energy Eng. Conf.* 7747905
[6] Samkari H, Allehyani M and Johnson B 2015 *North American Power Symp.* 7335196
[7] Alam M, Leite H, Liang J and da Silva Carvalho A 2017 *Int. J. Electr. Power Energy Sys.* **92** 245–260
[8] Guan R, Xue Y and Zhang X 2018 *The Journal of Eng.* **15** 1038–1042
[9] Bhaskar M and Indhirani A 2014 *Nationl Conf. Emerg. Trends New Renew. Energy Sources Energy Manag. (Chennai, India)* 7088738
[10] Khederzadeh M and Ghorbani A 2014 *IEEJ Trans. Electr. Electron. Eng.* **6**(6) 525–533
[11] Neklepaev B and Kryuchkov I 2014 *Electric part of power plants and substations: Reference materials for course and degree design* (St. Petersburg: BHV-Petersburg) p 542

IOP Conf. Series: Materials Science and Engineering 1019 (2021) 012064 doi:10.1088/1757-899X/1019/1/012064

Engineering a cyclone pre-furnaces' calculating method for combustion fine water-coal suspensions

E E Boyko and Yu V Ovchinnikov
Department thermal Power Plants, Novosibirsk State Technical University, Novosibirsk, Russian Federation
E-mail: e.boyko1991@yandex.ru

Abstract. The article is devoted to the method of engineering calculation of the cyclone pre-furnace for incineration of fine coal-water suspensions. This type of device use for ignition and burning fine water-coal suspension. At present there is still no methods allows to make engineering calculations in order to obtain physical damnations of cyclone pre-furnaces for this type of fuels.

1. Introduction

In the foreseeable future, coal will continue to be one of the main energy resources, both in the world and in Russia. It is the most widely available and cheap resource of fossil fuels. The expansion of the technological base of coal use is relevant. One way of use is to provide a fine coal-water suspension (FCWS) for subsequent burning in power plants of wide application, e.g., in the cyclone pre-furnace (CPF). In previous years, many scientists have developed both technologies for creating CWS and researched the features of burning, including in cyclone-type devices.However, there is problem which hampering the development of this technology. The most important of these are the lack of reliable methodological tools for creating the proper cyclone furnaces with guaranteed power characteristics, process parameters, and the complexity of ignition of fuel with high water content.

2. Engineering calculation method

This article discusses the methods producing cyclone-type devices with reliable ignition and combustion of fine water-coal suspensions using a plasma torch.

The developed method has an engineering focus, which determines the proximity of modelling, which is expressed by the following assumptions: quasistationary process; constancy of heat capacities and heat transfer coefficients (Its values correspond to the average process temperature); fuel particles are isothermal; gas mixture is homogeneous;the ash components are inert; only some chemical reactions of interaction with the oxidizing agent are taken into account; It is considered a one-dimensional flow.

Thermal resistance of the fuel particles can be neglected inasmuch as Fourier criterion value $Fo_o>5$ and at the same time the Bio criterion $Bi<0.5$[1–3].

Such an approach allows analyzing the process and obtaining final expressions suitable for engineering calculations.

The movement of fuel in the form of FCWS inside a cyclone pre-furnace is characterized by physical properties and particle size (δ), cyclone design parameters (L_w, L, D, α), the residence time of the particles in the reaction zone (τ) and wall temperature (T_{wall}), gas temperature (T_{gas}), mixture temperature (T_{mix}), particle temperature (T_{coal}), as well as the initial fuel temperature (T_{fuel})(Figure 1).

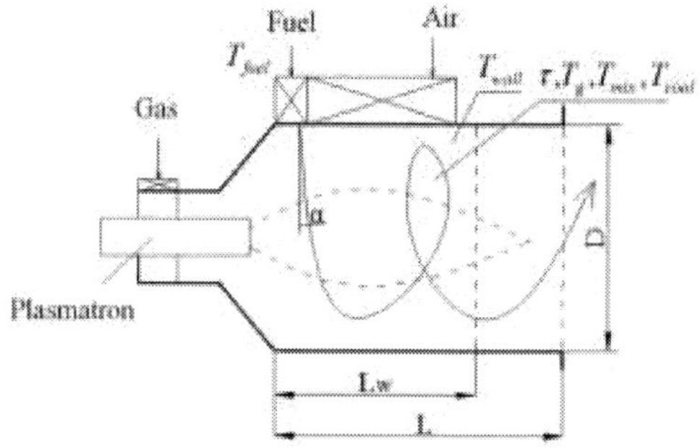

Figure 1. The movement of fuel in a cyclone pre-furnace.

It is assumed that the fuel moves inside some channel. This channel is twisted by forming CPF. In view of this solving a system of equations that takes into account the change in gas temperature, the process of evaporation of moisture from a fuel drop, the temperature change of a fuel particle over time, taking into account the convective-conductive mechanism of heat transfer in the gas mixture, the thermal effect of thermo chemical reactions $\Sigma Q_i f_i$ adopted a methodical approach has the form

$$\begin{cases} T_g = T_{mix}\exp\left(-3BiFo\tau^{-1}\dfrac{c_{coal}}{V_{mix}c_{mix}}\right) + T_{wall}\left[1-\exp\left(-3BiFo\tau^{-1}\dfrac{c_{coal}}{V_{mix}c_{mix}}\right)\right], \\ w = 1.25\dfrac{\psi_{coal}}{\delta\cdot r}\lambda Nu\left(T_{mix}-T_{fuel}\right)\left(1-w_0\right)^{1.67} w_0^{0.33}\tau_w, \\ T_y = T_{mix}e^{-3BiFo} + T_{gas}\left(1+\dfrac{1}{3BiFoc_{coal}T_{gas}\tau^{-1}}\sum Q_i f_i\right)\left(1-e^{-3BiFo}\right). \end{cases} \quad (1)$$

In Eq.(1): T_{wall}, T_{coal}, T_{mix}– CPF wall temperature, fuel particle temperature and mixture temperature,°C; ψ_{coal}- particle surface per 1 kg of fuel; c_{coal}, c_{mix}, – heat capacity of fuel particles and gas mixture, J/K; w_0–fuel initial moisture, %; τ_w– time of evaporation of moisture from the fuel, sec.

The total time of thermo chemical transformations of fuel will be determined by the sum (sec.)

$$\tau_\Sigma = \tau + \tau_w \quad (2)$$

The initial temperature of the particles is taken at the temperature of the mixture (T_{mix}).

In thermo chemical reactions, the equations describing the release of the components (CH_4, CO_2, CO, H_2, H_2O) can be represented as

$$\frac{dC_j}{d\tau} = \left(C_j^0 - C_j\right)a_j = f_j. \quad (3)$$

In Eq.(3) a_j– pre-exponential factor.

$$a_j = K_j\exp\left(-E_j R^{-1} T_y^{-1}\right) \quad (4)$$

K_j– component release rate constant (s^{-1})[4–9];
E_j –corresponding activation energy in the j-th thermochemical reaction (kJ/kμ) [4–9];
R–universal gas constant, J/μK;
C_j^o–initial concentration of components, mg/m^3.
From Eq. (3) find

$$C_j = C_j^0\left(1-e^{-a_j\tau}\right). \quad (5)$$

Another approach to obtain a mathematical model, which will be sufficient for engineering calculations, can be the definition of pyrolysis and combustion processes. These reactions are shown below.

The release of volatile methane is determined by the interaction of the carbon of the coal with the hydrogen released by the heating of the coal:

$$C+2H_2 \rightarrow CH_4 + 6.2\,\text{MJ/kg}.$$

The release of volatile carbon dioxide is determined by the combustion of carbon when interacting with the released oxygen of the coal:

$$C+O_2 \rightarrow CO_2 + 33.8\,\text{MJ/kg},$$

and in following up process using CO_2 for partial carbon dioxide gasification of carbon with the formation of CO:

$$C+CO_2 \rightarrow 2CO - 14.4\,\text{MJ/kg}.$$

In addition, the release of volatile carbon monoxide is determined by the oxidation of carbon in coal by oxygen, taking into account the previous use of oxygen for the combustion reaction:

$$C+O_2 \rightarrow 2CO + 9.4\,\text{MJ/kg}.$$

Also, evaporation of moisture and, in this connection, partial steam gasification of semi-coke carbon should be taken into account:

$$C+2H_2O \rightarrow CO + H_2 - 10.3\,\text{MJ/kg}.$$

The combustion of the formed CO, H_2 and CH_4 is due to the presence of oxygen in the primary and secondary (if necessary) air and passes through the reactions

$$H_2 + O_2 \rightarrow 2H_2O + 17.8\,\text{MJ/kg},$$

$$CH_4 + 2O_2 \rightarrow CO_2 + 2H_2O + 27.8\,\text{MJ/kg},$$

$$CO + 0.5O_2 \rightarrow CO_2 + 8.9\,\text{MJ/kg}.$$

Based on the above approaches and neglect, the calculation algorithm can be represented as follows: first determine the kinetic characteristics of the processes of evaporation of moisture from the FWCS when moving in the pre-furnace in the form of some jet, swirling along the generatrix of the cyclone so that the flame front moves at a linear speed towards the combustion chamber. Then determine the temperature of the fuel particles and the combustion products mixture.At the last stage of the calculations, the design characteristics of the cyclone pre-furnace are determined (Figure 2).

Thus, an approximate physical model of the process of ignition and combustion of FCWS is presented. At first, as the fuel moves inoxidizing environment (air), and then, fuel particles and pyrolysis products ignite and burn.In fact, the cyclone pre-furnace is divided into two parts - the evaporation zone (characterized byL_w) and the zone of ignition and combustion (characterized by L), (Figure 1).

Moisture release time is found to determine the design characteristics of CPF (in sec.):

$$\tau_w = \frac{W^r}{1.25 \cdot w_f \cdot \frac{6}{\rho_p \cdot \delta} \cdot \alpha \cdot \frac{T_{wall} - T_{fuel}}{h_{vap}} \cdot \left(1-W^r\right)^{1.67} \cdot \left(W^r\right)^{0.33}}. \tag{6}$$

Then determine the length of the evaporation section, m

$$L_w = \frac{V_{vap} \cdot G}{\frac{\pi}{4} \cdot d^2} \cdot \tau_w, \tag{7}$$

IOP Conf. Series: Materials Science and Engineering 1019 (2021) 012064 doi:10.1088/1757-899X/1019/1/012064

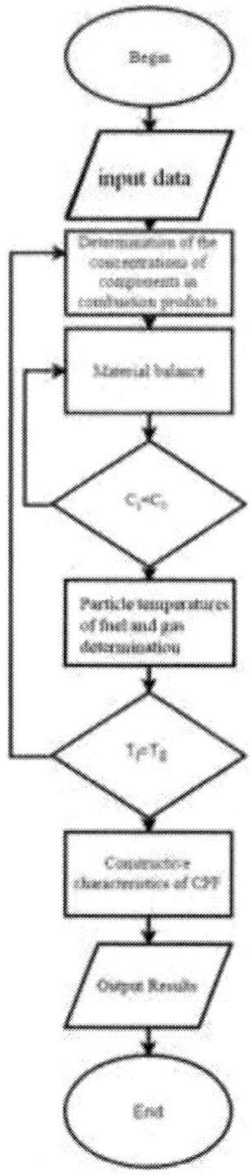

Figure 2. Algorithm engineering calculation of CPF.

also determine the full length of the pre-furnace, given that the flow is twisted

$$L = 2 \cdot d \cdot \frac{\ell_{burn} \cdot \omega_{mix} \cdot \tau + \dfrac{V_{vap} \cdot G}{\dfrac{\pi}{4} \cdot d^2} \cdot \tau_w}{\pi \cdot D \left[\dfrac{\sqrt{(D \cdot tg\alpha)^2 + D^2}}{D} \right]} \quad (8)$$

In these equations (8):

W^r– fuel moisture;

w_f– correction factor for the speed of particles flurries;

ρ_p– particle density, kg/m^3;

δ –characteristic of particle size;

α – flow entry angle;

T_{wall}– CPF wall temperature, °C;

T_{fuel}– fuel temperature, °C;

h_{vap}– vapor enthalpy, kJ/kg;

V_{vap}– volume of steam in the combustion products, m^3/kg;

G– steam consumption, kg/s;
ℓ_{burn} – correction factor to the length of the torch when burning FCWS;
ω_{mix}– velocity of the mixture (combustion) at the flow axis, m/s;
τ – characteristic of length, m;
D– torch twist diameter, m;
tgα – twist angle.

3. The analysis of the combustion characteristics' influence on the cyclone pre-furnace construction

Based on the calculations results the dependences of gaseous products release of pyrolysis and combustion of FCWS are obtained (Figure 3).

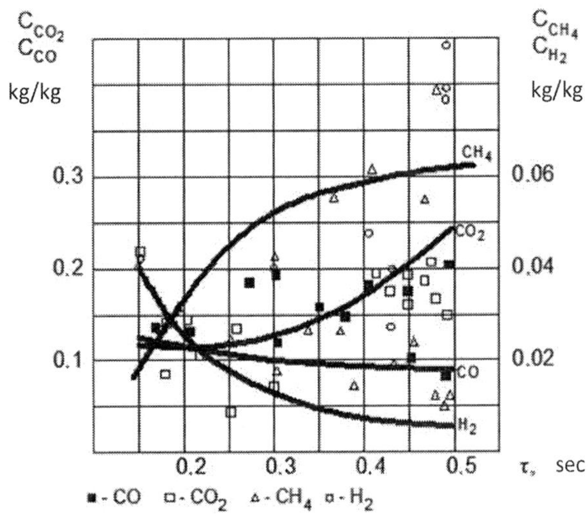

Figure 3. Comparison of the results with experimental and calculated data on the pyrolysis of carbon-containing fuels [10–15].

The adequacy of the model is noted in comparison with the experimental data of other authors [10–15].It was established that due to the complete hydrogen burning into the fuel in the initial part of the process (0…1s), a high temperature of the mixture (T_{mix}) is provided up to 1800–2500 K.The moisture release time for heat-stressed pre-furnaces is less than 2s, and for refrigerated ones it can be up to 5s.

The characteristic area of evaporation, depending on the flow rate, can be 2.5–200 m(Figure 4).

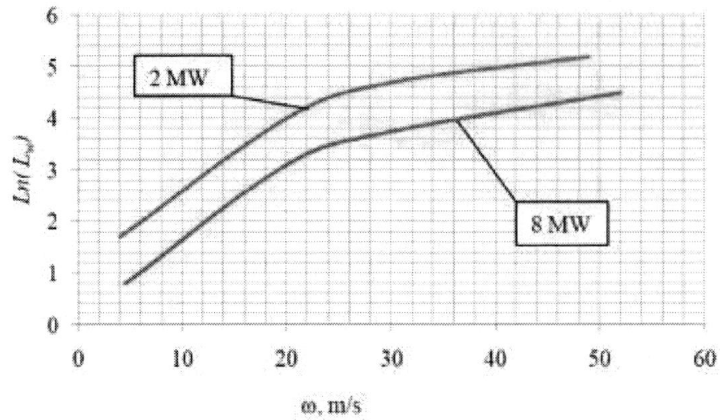

Figure 4. The section length dependence of flow velocity.

Pre-furnace efficiency affects the length of the section.So for pre-furnaces with a capacity of 2 and 8 MW, the length L_w is 2.3–2.4 times less for more powerful CPF (Figure 3). It is shown that the length of the coil inside the cyclone is almost independent of its efficiency (Figure 5).

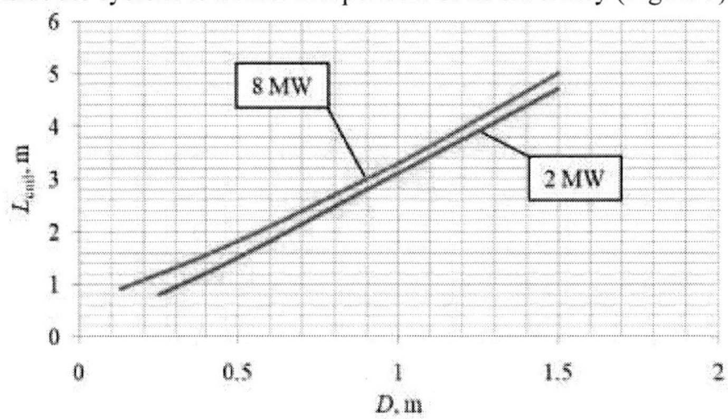

Figure 5. The efficiency effect on CPF size for ω_{mix}=20 m/s.

It can be seen that the influence of temperature is less noticeable than the influence of CPF efficiency on its linear dimensions.If we take the cyclone diameter of 1–1.5 m with burning capacity of 8–16 MW, then the length of the apparatus is 4.5–10m (Figure 7).

In general, the pre-furnace can be designed with different ratios of its length (L) to diameter (D) (Figure 6).

IOP Conf. Series: Materials Science and Engineering 1019 (2021) 012064 doi:10.1088/1757-899X/1019/1/012064

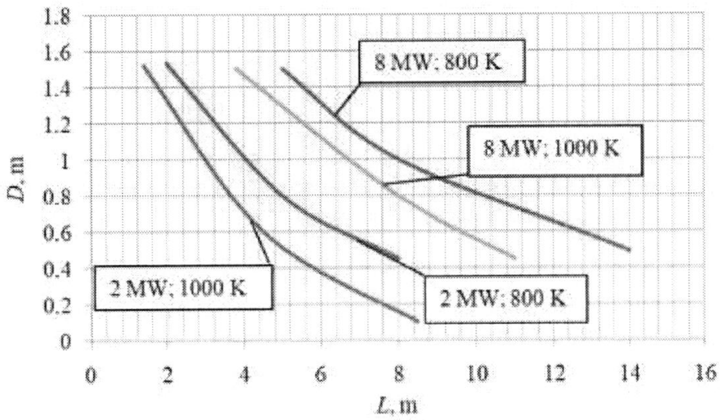

Figure 6. The linear dimensions of CPF depending on its efficiency and reaction for ω_{mix}=20 m/s.

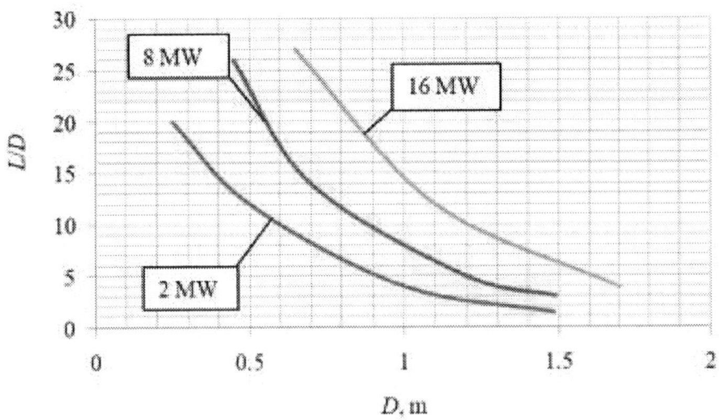

Figure 7. Constructive characteristics of CPF depending on its efficiency (α=6–17°).

4.Summary

Thus, the article presents a technique to calculate the geometric parameters of a cyclone pre-furnace for burning FCWS.The effect of kinetic processes on the physical design of the apparatus is shown.Graphic dependences of various factors are given, such as the speed of the mixture, the angle of fuel injection, the diameter of the flow twist on the diameter and the length of the cyclone pre-furnace. The adequacy of the proposed method is shown by comparison with the calculations and experimental values of other researchers.

References

[1] Kalinenko R A, Levitskii A A, Mirohin Yu A *et al* 1978 *Kinetics and catalysis* **28** (3) 723–729
[2] Pomerancev V V, Shestakov S M, Dudukalov A P *et al* 1985 *Proceedings of Conference Combustion of fossil fuels* 23 122
[3] Pechenegov Yu Ya 1988 *Simulation and calculation of heat transfer of one-and two-phase systems in tubular devices* (Saratov: Saratov State University Press) p 37
[4] Yurenev V N, Lebedev P D 1976 *Thermal reference book* (Moskow: Energy Press) p 896
[5] Karpenko E I 1995*Plasma energy technology integrated use of solid fuels* (Novosibirsk: Novosibirsk State Technical University Press) p 83
[6] Suuburd E M, Peters W A, Howard J B 1987 *Ind.Eng.Chem. Proceses.***17** 37–46
[7] Solomon P R, Hamblen D J, Carangelo R M 1982 *Proceedings of Coal 19thSymp. (Intern.) Combustion (Prol.)*1139–1149

[8] Kosubova M, Krutil Ya, Nevri V 2014 *Physics of combustion and explosion* **50**(4)8–14
[9] Zambon A C, Chelliah H K 2006 *J. Fluid Mech.* **560** 249–278
[10] Volkovinskii V A, Tolmachev IYa 1994 *Themalengyneering* (9) 42–48
[11] Volkovinskii V A, ZholudovYaS, Roskolupa AI *etal*1985 *Burning fossil fuel* 51–55
[12] Ibragimov M H, Marchenko E M, Tuvalbaev B G *etal2011 Heat power engineering. Introduction to the specialty*(Moscow: MGOU)
[13] Kashirskii V G, Vasiliev Yu A, Kovalev E E *et al* 1985 *Proceeding of 5th All-Union Conference 2* 97–102
[14] Kashirskii V G, Mulukova G P 1985 *Proceeding of 5th All-Union Conference* **2** 93–96
[15] Shchinnikov PA 1998 *Choosing an environmentally promising direction for the development of CHP on Kansko-Achinsk coal in modern economic conditions* (Novosibirsk: Novosibirsk State Technical University Press) p158

IOP Conf. Series: Materials Science and Engineering 1019 (2021) 012065 doi:10.1088/1757-899X/1019/1/012065

Critical technologies in the cluster of virtual and augmented reality

V Zhmud[1], A Liapidevskiy [2], V Avrmachuk[3], V Sayapin[4], J N Sedinkin[5] and W Hardt[6]

[1] Department of Automation, Novosibirsk State Technical University, Novosibirsk, Russia

[2]General Director of Novosibirsk Institute of Program Systems, Novosibirsk, Russia

[3] Department of Complex Information Security of Computer Systems, Tomsk State University of Control Systems and Radioelectronics, Tomsk, Russia

[4] Director of R&D Center of Novosibirsk Institute of Program Systems, Novosibirsk, Russia

[5] Vice-Director on Development of Novosibirsk Institute of Program Systems, Novosibirsk, Russia

[6] Department of Computer Science, Technical University of Chemnitz, Chemnitz, Germany

E-mail: oao_nips@bk.ru

Abstract. The program "Digital Economy of the Russian Federation", launched in 2019, contains virtual and augmented reality technologies in its list. Each such technology consists of many assistive technologies called sub-technologies. These assistive technologies make up a large list, while failure to obtain the required results in at least one of these technologies may lead to the fact that the integral technology will not be created. This article discusses a list of only critical assistive technologies, that is, precisely those technologies whose creation is a critical factor for creating a generalized (integral) technology. It is on these technologies that the main emphasis should be placed not only in financing their development, but also in the thorough planning of goals and objectives, in the selection of performers and in the acceptance of research results.

1. Introduction

The implementation of the national research program "Digital Economy of Russia" began in 2019, initially it identified nine areas of technological development [1], of which only six remained later [2]. Initially, the list included the following technologies: virtual and augmented reality technologies, neuro-technologies and artificial intelligence, robotics components and sensor technologies, big data, wireless technologies, distributed ledger systems (blockchain), industrial Internet, new production technologies and quantum technologies [3]. Four roadmaps were adopted in the first reading: 1) artificial intelligence, 2) distributed ledger technologies (blockchain), 3) quantum technologies, 4) virtual and augmented reality technologies. Two more roadways will be improved, these are: 5) new production technologies and 6) wireless communication (5G) [3]. The first year of this program can be considered lost, since the actions to distribute funding were extremely hasty and sometimes unreasonable, perhaps this was the reason for the change of the main responsible government official, therefore it is especially important to correctly develop these technologies in the remaining four years.

2. Statement of the problem

There is no point in explaining the essence of the concept of virtual reality, since nowadays this understanding has been achieved everywhere, and for general education you can refer to the corresponding sites. We will focus on augmented reality and how our understanding differs from the conventional, which is not professional, but has become predominant.

Augmented reality, as you know, is a virtual reality component that is brought into actual reality. The key technologies here are, firstly, the technologies for introducing this additional component into the user's real sensing system, and secondly, the methods of its preliminary transformation so that it is perceived in the most harmonious way.

Effective use of augmented reality does not require the inability to distinguish its elements from reality. This can only matter in the entertainment industry, in some very specific simulators and in some very specific cases. In practice, a person who does not distinguish digital deception from reality can be classified as deceived or mentally ill. If, of course, this deception is so successful that no sense organs can detect it, and a person needs to act in actual reality, and not in virtual and not in augmented reality, it turns out that such augmented reality, instead of expanding his sensory capabilities, takes away him from the real world, that is, hinders his effective activity. Therefore, augmented reality should in most cases clearly differ from physical reality, at least so much that the user does not react to it in an erroneous way. A realistic image of the abyss may add drive to the computer player, but if he, having broken off from it in the game world, dies of a heart attack, since he believes that his life is really in mortal danger, there is nothing good about it. Thus, one of the directions of augmented reality should not replace real images of the external world with virtual fakes, but provide the necessary additional information, explanations, prompts for action.

An example of this would be a navigator hint. Currently, it is a voice, text or graphical prompt, or a combination of these types. Ideally, such a prompter should provide instructions in a form that is most comfortable for the driver, as this is required to improve driving safety. It is useful to visualize the most critical road signs, especially considering the expected actions in connection with its movement along the route. Other clues are also desirable, which are currently not yet available, but which are not prevented from appearing. For example, since GPS and GLONASS systems can see the entire situation on the route, and digital technologies can predict the actions of a driver whose route is known quite well in the next few minutes, such prompts could inform the driver about the danger of his planned maneuver, or vice versa about the safety of the maneuver. In particular, it would be desirable to warn about other vehicles and obstacles that will arise on the way in the very near future. For example, if the driver has planned to overtake, as indicated by his preliminary actions, but he does not see the oncoming vehicle due to poor visibility, or is not aware of the peculiarities of the road surface, the road profile or other obstacles for this maneuver, then a voice message, for example of this type "Attention! Overtaking is dangerous, oncoming traffic is dangerously close!" could in many cases exclude the worst kind of accidents – a head-on collision. It is also known that very often road accidents occur for the reason that the driver believes that he still has time to slip through the green light, and another driver leaves the intersection too early, predicting that the green light has already turned on for him. In other words, both drivers consider the yellow light to be permissive for themselves, while in fact it prohibits all drivers, except for those who stop driving. Occasionally, drivers do not notice vehicles that take advantage of the siren and flashing lights. In such cases, it would be useful to submit a message, for example, "Danger on the right!", or "Car with a siren on the left!" Also, sometimes the cause of an accident is open hatches, ruts, puddles, objects on the road, sharp objects tearing tires, objects that have fallen from previous vehicles, and so on. To detect them, special sensors in cars could serve, GPS and GLONASS systems do not detect them, but such sensors would be very useful to increase traffic safety. Note that there is no emergency notification even in the event of those obstacles that these systems fully recognize. The reason for this is, firstly, the large delay in their detection, which makes such an alert meaningless, since it would still be late, and secondly, apparently, while nobody stated this task.

Similar technical solutions can be offered for training cases, and for helping physicians during an operation, and for specialists who repair and maintain complex equipment. These technologies are already relatively efficiently practiced in gaming applications, in various attractions. In a sense, Disneyland was the prototype for augmented reality platform.

3. The main critical sub-technologies

Augmented reality should have the following features: 1) combine virtual and real; 2) interacts in real time; 3) work in 3D [4].

If you use the image of augmented reality objects on a computer screen or any gadget (the first approach), this does not require any new technologies.

If you use virtual reality glasses (second approach) or turn the walls of the room in which the user is located into screens (third approach), this also does not require substantially new imaging technologies, but this requires technologies for preliminary creation of objects of this virtual reality by means of their appropriate distortion. to obtain such projections that would deceive the user's perception, making the generated reality most similar to volumetric, especially when the user moves and turns the head. Thus, this approach requires only computational methods, which, strictly speaking, is not a new technology, but is a branch of applied mathematics. With a large amount of computation, this is associated with an increase in computing power and the development of more efficient programs, probably also with the development of more efficient algorithms, but these are not production technologies.

The fourth approach is to create a hologram. If a hologram is created on a semitransparent medium, these technologies already exist, but their development is undoubtedly required in the direction of higher resolution, better color reproduction, and higher speed. Without a specially prepared screen, these technologies can exist on random and regular media. An example of a regular medium is the blades of a rotating fan, which creates the effect of transparency in an image that seems to hang in the air. An accidental carrier can be fog, aerosol, or other particles that scatter light at certain points, but generally transmit it. These technologies are very complex, their development seems to be very expedient, it will advance technologies in any case, and not only for augmented reality. In this area, science fiction films demonstrate miracles that, unfortunately, are not yet achievable in practice. This refers to the creation directly in the air of volumetric images that are indistinguishable from images of real objects. We assume that such technologies for creating dense color moving images in an ordinary atmosphere without admixtures of fog, splashes, aerosols, without moving or resting scattering objects within the framework of the discussed program will not be reliably created until 2024. This does not mean that research in this direction is useless, but if the program should fund only those research that in the near future (within a calendar year from the beginning of funding) will yield an output with the production of products using this new technology, then such projects are most likely overly optimistic.

The fifth approach is to project the image onto particles in the air. Such particles are almost invisible in the absence of illumination, and under illumination they can glow so brightly that even short-term flashes, for example, with a duty cycle of $Q = 10$, are perceived by the eye as a constant glow. These are very complex technologies, but they are quite realistic, their development can really be carried out over a period equal to a calendar year. But their application is much narrower than technologies according to the fourth approach.

The sixth approach may be to obtain the same effect as the fifth approach, but with very high laser power. This allows you to create pictures not on special aerosols, but even on natural dust particles present in almost any atmosphere. This method is very dangerous for the eyes of the observer and is characterized by huge energy costs, so it does not seem very promising yet.

The seventh approach is to project images onto objects of reality. For example, an image of a battle panorama is projected onto a number of architectural structures, while such corrections are made to the images that lead to the fact that the structure of real objects is completely masked. This method is based on large calculations and is effective for massive shows in impromptu locations, such as city squares. It is doubtful that it is advisable to develop this method through government funding; it is quite capable of developing through the self-development of enterprises that specialize in mass shows.

4. Possible goals for AR use

Much has been said about using augmented reality to reveal additional information about an object. More than half of these examples are not directly related to augmented reality. The user simply reads some product codes and receives additional information about this object on his gadget. Similar technologies exist for amateur naturalists and professionals. By entering an image into his gadget, the user can get the result of its recognition, the programs recognize animals and plants, report their Latin name, after which the user can find any available information about it by the name of the species. Unfortunately, these programs are not yet reliable enough, it would be rash to use them to distinguish edible mushrooms from poisonous, harmless snakes from poisonous and so on, a program error can cost the user too much.

Technologies combined with smart sensors are the most interesting, but also the most complex. They can offer the user many additional properties, in fact, they expand his sensory capabilities. The user can determine the distance to the object, see the details of its image, can get the capabilities of night vision, thermal imaging, pattern recognition and so on. The prototypes of such devices are binoculars, microscopes, telescopes, range finders and night vision devices. In digital technologies, images are formed not by transforming them by analog means (magnification, amplification), but using digital image recognition and digital formation of new images, and between these two operations any digital image processing can be carried out.

Close to this is the technology of automatic translation of speech from one language to another, but these technologies do not require any new hardware today, they can be developed exclusively by software.

Augmented reality technologies can help people work more efficiently in extreme situations: rescuers, firefighters, cavers, miners, sappers, construction workers, doctors, and so on.

In general, it can be argued that the most critical technologies in this area are, firstly, technologies for collecting a large amount of data on objects of the surrounding reality, secondly, technologies for processing these data, thirdly, imaging technologies and, possibly, others. sensations such as tactile. Sound shaping technologies, although they are adjacent to these technologies, should not be included in this list, since they are already so well developed that their further development in the near future is not critical; special funding for their development is not required.

5. Conclusion

Based on the analysis of the main problems the paper has given the view on the main sub-technologies of VR-AR.

References

[1] "Digital Economy" Program Implementation. Available at: https://ac.gov.ru/en/projects/project/digital-economy-program-implementation-42
[2] TAdviser: Digital_economy_of_Russia. Available at: http://tadviser.com/index.php/Article:Digital_economy_of_Russia
[3] Instructions on implementation of Digital Economy of Russian Federation national programme. Available at: http://en.kremlin.ru/acts/news/60780
[4] What is Augmented Reality (AR) and How does it work. Available at: https://thinkmobiles.com/blog/what-is-augmented-reality/

IOP Conf. Series: Materials Science and Engineering 1019 (2021) 012066 doi:10.1088/1757-899X/1019/1/012066

Visual shape and position sensing algorithm for a continuum robot

R A Manakov, D Y Kolpashchikov, V V Danilov, N V Laptev, I P Skirnevskiy and O M Gerget
Medical Devices Design Laboratory, Tomsk Polytechnic University, Tomsk, Russia
E-mail: ram290495@gmail.com

Abstract. Continuum robots represent an actively developing and fast-growing technology in robotics. To successfully implement control and path planning of continuum robots it is important to develop an accurate three-dimensional shape and position sensing algorithm. In this paper, we propose an algorithm for the three-dimensional reconstruction of the continuum robot shape. The algorithm is performed during several steps. Initially, images from two cameras are processed by applying pre-processing and segmentation techniques. Then, the gradient descent method is applied to compare two-dimensional skeleton points of both masks. Having compared these points, it finds a skeleton of the robot in a three-dimensional form. Additionally, the proposed algorithm is able to define key points using the distance from the robot base along the center line. The latter allows controlling the position of points of interest defined by a user. As a result, the developed algorithm achieved a relatively high level of accuracy and speed.

1. Introduction

Today the continuum robot technologies are actively developing. These robots obtain a higher dexterity compared to the traditional rigid robots. In this regard, they are capable of operating in complex workspaces. Due to this advantage, continuum robots are widely used in medicine for minimally invasive procedures [1]. For instance, robotic catheters are used in endovascular surgery as the main tools for the delivery of instruments and stents. Other fields of application of continuum robots are non-destructive testing and repair [2–5].

In order to successfully implement a closed-loop control and path planning of continuum robots, it is important to develop an accurate three-dimensional shape sensing algorithm. Due to a flexible robot structure, this task is quite challenging and complex. Currently there are several techniques for 3D shape sensing and some of them are fiber-optic sensor-based algorithms (FOS), electromagnetic-tracking (EM), and image-based methods [6].

Image-based methods are less complex, cheaper, and more accessible in comparison with FOS and EM techniques as they do not require any modification of the robot construction and/or sensors. It allows applying them using simple equipment such as webcams. Image-based methods are able to measure the shape of a robot directly and reconstruct it in 3D without additional kinematic modeling.

In connection with the foregoing, we propose an image-based algorithm which is able to estimate a continuum robot shape and its tip position inexpensively. In the algorithm's workflow, the gradient descent method is used for pairwise comparison of the skeleton points obtained by two cameras. Once this comparison is performed, it finds a skeleton of the robot in a three-dimensional form. Additionally, the proposed algorithm is able to define key points using the distance from the robot base along the center line. This algorithm feature allows controlling the position of the points defined by a user.

IOP Conf. Series: Materials Science and Engineering 1019 (2021) 012066 doi:10.1088/1757-899X/1019/1/012066

2. Methods

The workflow of the proposed continuum robot shape sensing algorithm is represented by several steps shown in Figure 1. During the first step, the algorithm simultaneously acquires images using two cameras. At least two cameras are creating a stereoscopic effect and allow us to receive information about the robot's spatial position.

In the second step, acquired images are being segmented and the algorithm recognizes the robot on them. To simplify the image processing operations and improve the detection accuracy, the image with the robot has a monotonous background. The process of image acquisition is carried out with uniform illumination of the scene and adjusted parameters of the cameras (brightness, contrast, saturation, exposure, and clarity). With these adjusted parameters the robot border is not blurred and there are no glare and other undesirable effects. It should also be noticed that the non-linear distortion has been removed from each image. The latter allows reducing the triangulation error. During the pre-processing step the algorithm performs such morphological operations as area opening and morphological closing. Area opening helped us to remove unnecessary small objects and noise, while morphological closing fills the gaps and blurred the edges of the masks. Initial segmentation is performed using RGB thresholding, calculated as follows:

$$\sqrt{(R-R_0)^2+(G-G_0)^2+(B-B_0)^2} > T \quad (1)$$

where R, G and B are the current pixel colors; R_0, G_0 and B_0 are background colors, T is the threshold level.

Once the binary masks are obtained, the algorithm skeletonizes them using distance transform and then computes the bounding boxes. Image skeletonization is required for finding an axial line of the continuum robot on each image. In order not to process the whole image in the following steps, the algorithm defines bounding boxes of the region of interest.

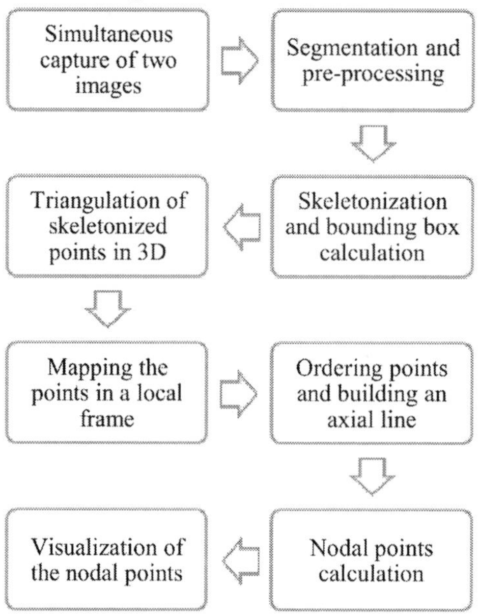

Figure 1. Steps of the proposed algorithm.

Pre-processing and segmentation operations are followed by the triangulation. For the triangulation of the points in three-dimensional space, we implemented the gradient descent method based on the triangulation function. This implementation allows the algorithm to define the pairs of corresponding points on both skeletons. The purpose of an application of triangulation is the error minimization between two rays (Figure 2). The first ray is fixed, and the direction of the second is chosen in such a way that the error of their triangulation is the lowest. The triangulation analyzes the current and the neighboring pixels of the skeleton and iteratively moves along it until the objective function minimum is reached. The initial approximation of the method is the result of the minimization of the previous pair of rays.

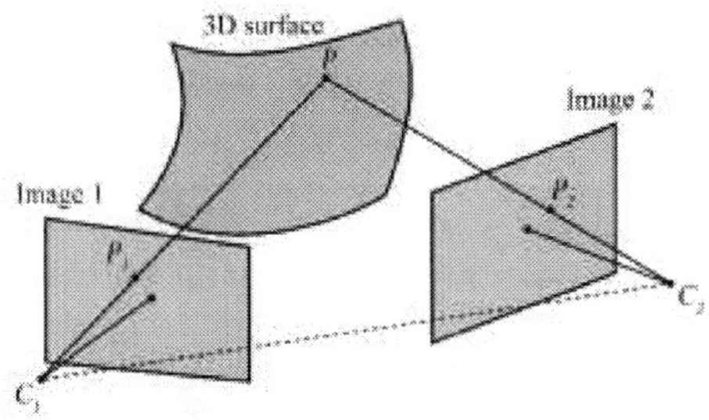

Figure 2. Minimization of the error between two rays.

This implementation allows the algorithm to minimize the number of iterations and speed up the triangulation process.

The described method finds the correspondence between the points of the first and second skeletons. Thus, we can find the pairs of points on two skeletons, triangulate them and receive a set of points corresponding to the robot skeleton in three-dimensional space. In order to obtain the maximum possible number of point pairs, the algorithm is applied to all the pixels of the skeleton on both masks. On the one hand, this may be redundant. On the other hand, such an approach makes it possible to obtain additional points for non-trivial forms of the robot.

For the convenience of further work with the obtained set of points in three-dimensional space, it has to be mapped into the local coordinate system of the robot. In this connection, we use an affine transformation matrix. In the local coordinate system, the robot is located at the frame origin, and its longitudinal axis is directed along the Z-axis. All the further calculations are carried out in this coordinate system.

For the obtained set of points, it is necessary to define the correct order and remove duplicate points. In order to achieve that, the Euclidean metric is used. The process of sorting starts from the robot base. In the beginning, the algorithm takes the point with the smallest Z-coordinate. The next point will be the closest to the selected point. This process is repeated iteratively until the exact order is found for all sets of points. As a result, we obtain a sorted sequence describing the robot center line.

During the last step, the algorithm searches for the robot points defined by a user. The idea behind this step is to set the distance for each point from the robot base along the axial line.

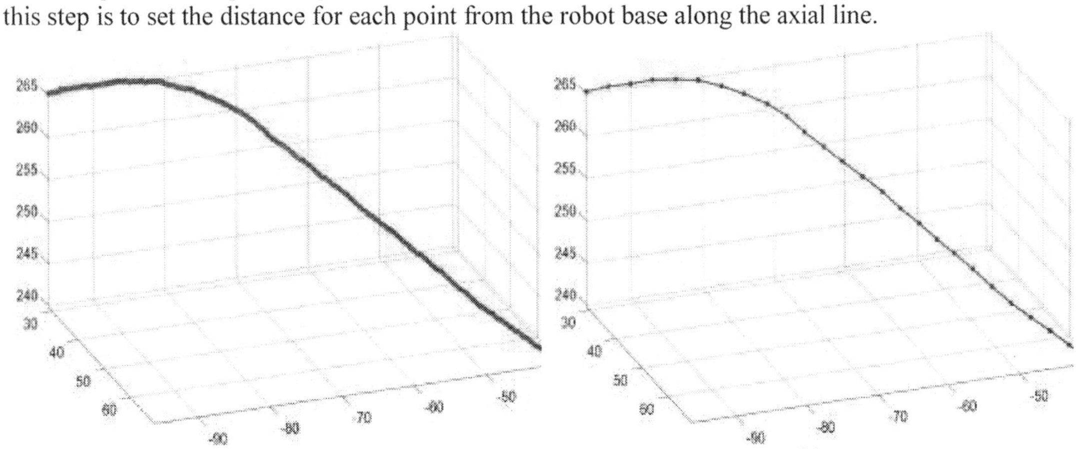

(a) (b)

Figure 3. 3D skeleton and nodal points of continuum robot: a) Triangulated points of the 3D skeleton. b) Calculated nodal points.

Thus, the search for these points is reduced to the integration of the previously obtained sequence along the length of the center line. The beginning of the integration is a point on the center line with some coordinate Z_0, determined by the robot base. It is worth noticing that the linear interpolation is actively used for improving positioning accuracy. As a result of integration, the exact coordinates of the nodal points representing an output of the detection algorithm are defined. If necessary, these points can be converted back to the stereo camera's frame using the affine transformation calibration matrix. Founded axial line and corresponding nodal points are reflected in Figure 3.

3. Testing and Results

The proposed algorithm was tested using a two-section continuum robot. Each section is 40 mm length and has two degrees of freedom. Additionally, the laboratory test bench shown in Figure 4 has five degrees of freedom including the linear movement. As shown above, the laboratory test bench has a monotonous background and two cameras. Image acquisition is carried out with good illumination and well-adjusted parameters.

In order to acquire images in real-time, we used two Logitech C922 Pro Stream cameras at the resolution of 1920x1080. As for the software, the Computer Vision Toolbox of MATLAB R2019a was used for calibration and work with the cameras. Testing was conducted on a computer equipped with an Intel Core i7 CPU and 16 Gb RAM. All the results were obtained at a resolution of 1920x1080. The average error of the triangulation made up 0.215 mm. The final accumulated integration error did not exceed 0.07 mm for every 10 mm of the robot length. Having tested the algorithm's execution time, the average detection time made up approximately 0.7 seconds. Two examples of the source, segmented, skeletonized, and detected images are presented in Figure 5.

Figure 4. Laboratory test bench for continuum robot shape sensing.

The proposed algorithm is compared with two other well-known approaches. One of them is represented by the algorithm of D. Camarillo et al. [7], another is reflected in the paper of A. Vandini et al. [8]. D. Camarillo et al. use the algorithm where images from three cameras are pre-processed by the RGB thresholding obtaining binary masks. Then the voxel carving technique is applied. According to this technique, the volume of interest is divided into a discrete three-dimensional grid of volume elements (voxels). Each voxel is back-projected onto the source images and is classified as part of the 3D shape if it lies within the masks. In related research, A. Vandini et al. conduct experiments with a planar continuum robot attached to the rigid robot. In this study, a single-camera and a naive Bayesian classifier are used to extract the robot center line. The three-dimensional shape is reconstructed from the central line, position, and orientation of the rigid robot. An accuracy comparsion of the proposed algorithm with the above mentioned image-based algorithms is reflected in table 1.

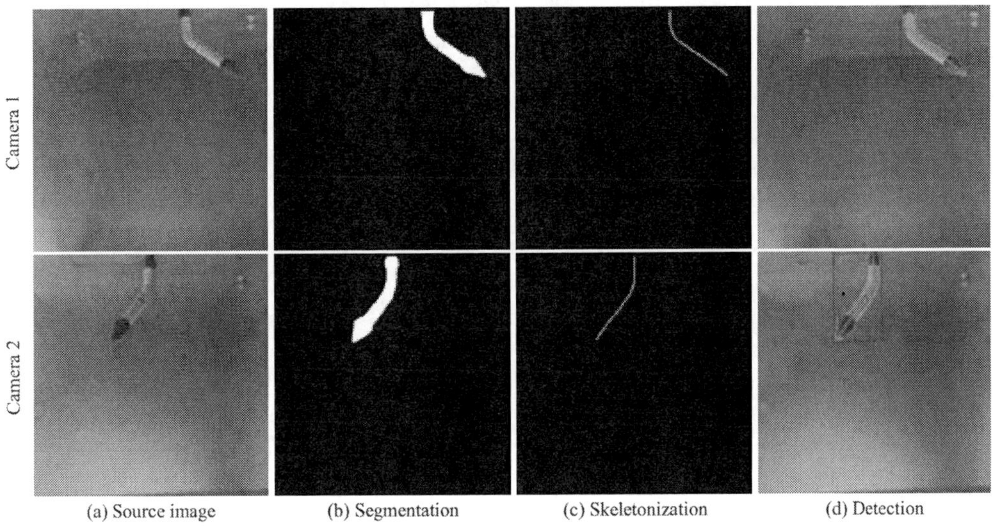

Figure 5. Examples of the algorithm outputs.

Table 1. Comparative Evaluation of the Algorithms.

Algorithm	Error	Reference
Proposed algorithm	0.07 mm per 10 mm of the robot length	–
Vision-based 3D shape sensing of flexible manipulators	8 mm for 160 mm robot	[7]
Vision-based motion control of a flexible robot for surgical applications	1.84 mm	[8]

4. Conclusion

In this paper, we presented the algorithm for the shape and position sensing of continuum robots. The proposed algorithm uses the image-based approach with two stereo cameras. Having tested the algorithm, the accumulated integration error did not exceed 0.07 mm for every 10 mm of the robot length. The average execution time made up about 0.7 seconds. The presented results show that the algorithm is able to estimate the continuum robot shape and define its spatial position. The proposed detection algorithm can be easily implemented and does not require any special expensive hardware. The algorithm is not capable of working in real-time but this issue can be solved by its implementation using C++ and its computer vision libraries, for instance, OpenCV. The final accuracy can be improved by using a higher resolution of the cameras or by increasing the number of cameras.

Acknowledgments
This study was supported by the Russian Federation Governmental Program "Nauka" № 17.8205.2017/БЧ (additional number: 4.1769.ГЗБ.2017).

References

[1] Burgner-Kahrs J, Rucker D C and Choset H 2015 *IEEE Trans. Robot.* **31** 1261–80
[2] Dong X, Axinte D, Palmer D, Cobos S, Raffles M, Rabani A and Kell J 2017 *Robot. Comput. Integr. Manuf.* **44** 218–29
[3] Dong X, Palmer D, Axinte D and Kell J 2019 *J. Manuf. Process.* **38** 313–8
[4] Buckingham R and Graham A 2012 *Ind. Robot An Int. J.* **39** 6–11
[5] Mehling J S, Diftler M A, Chu M and Valvo M 2006 *First IEEE/RAS-EMBS Int. Conf. Biomed.*

Robot. Biomechatronics, 2006, BioRob 2006 **2006** 690–695

[6] Shi C, Luo X, Qi P, Li T, Song S, Najdovski Z, Fukuda T and Ren H 2017 *IEEE Trans. Biomed. Eng.* **64** 1665–1678

[7] Camarillo D B, Loewke K E, Carlson C R and Salisbury J K 2008 *Proceedings - IEEE International Conference on Robotics and Automation* 2940–2947

[8] Vandini A, Salerno A, Payne C J and Yang G Z 2014 *Proceedings - IEEE International Conference on Robotics and Automation* 6205–6211

Information support of stakeholder cooperation: technology for formation of a digital footprint of companies

O V Milekhina [1], I B Adova [2] and V A Fomin [1]
[1] Chair of Computer Science in Economics, Novosibirsk State Technical University Novosibirsk, Russia
[2] Chair of Management, Novosibirsk State Technical University Novosibirsk, Russia
E-mail: olga.milekhina@gmail.com

Abstract. The paper presents a tool for the formation of a digital footprint of companies focused on supporting the processes of strategy with the further search for potential stakeholders, the formation of proposals for cooperation and establishing contacts with them. The world-class head Hunter recruiting company was chosen as the object of the study, its REST API was considered as a data source. The architecture of the system, which provides the ability to collect and store relevant information. It secures the monitoring the digital footprint of personnel needs (vacancies) of companies. Implementation of the proposed tool for the formation of a digital footprint is carried out within the framework of interaction between universities and the business community in the process of formation of professional competencies of students. An interactive map with data on profession vacancies has been developed, which allows supporting the processes of stakeholder interaction.

1. Introduction

Information plays a fundamental role in the future economy of the country. Taking into account the growth of traffic volume by 27.87% (from 1464 to 1872 exabytes per year) in the period 2017–2018 and its increasing projected values to 4752 exabytes per year [1], the principal component of any business is interaction with stakeholders aiming the goal achievement. Indeed, the published research results show a reduction in the life cycle of companies and the positive dynamics of the tempo metrics of successful companies [2]. It is indicated that their age, size, reputation and current sales do not guarantee survival in the future [3]. The authors [4] discuss the fundamental need for systematic work on business development and the business environment [5,6], modeling scenarios of interaction with stakeholders with tracking changes in dynamics [7], mixed types of interaction based on relations with each group [8], the application of the theory of fuzzy sets [9], the need for proactive management [10] based on the use of end-to-end technologies of the digital economy [11].

Despite the fact that various aspects of the interaction of stakeholders and their groups have been discussed for a long time, one of the decisive phases is the phase of strategy with the further search for potential stakeholders, the formation of a proposal for cooperation and establishing contacts with them, is poorly described. The formation of a digital footprint of companies of interest is one of the solutions of the problem. We define the digital footprint as information intentionally or unintentionally left in the digital space with the further possibility of its source identification. It is proposed to use one of the peculiarities of the digital footprint namely if the information was entered it is almost impossible to remove it and it becomes publicly available.

Thus, it is necessary to develop a tool for the formation of a digital footprint of companies, which will allow to collect and accumulate public data on potential stakeholders. This requires the development of appropriate technology and its software realization.

2. Methodology

Development of technological processes requires specification of the application. Implementation of the proposed tool for the formation of a digital footprint is carried out within the framework of interaction between universities and the business community in the process of formation of professional competencies of students. Training of future specialists should be focused on the needs of companies. Taking into account the need to provide personnel for the digital economy, the development of technology for the formation of a digital footprint of personnel needs of companies and its software implementation will allow to organize the educational process at a higher level. Head Hunter, a world-class recruitment company, has been selected as an object under investigation. Its REST API (Representational State Transfer Application Programming Interface) was considered as a data source.

The architecture of the future system, which provides the ability to collect and store relevant information, is formed to monitor the digital footprint of personnel needs (vacancies) of companies. In this study, the problem of modeling the structure of functional subsystems with the subsequent visualization of the digital footprint is solved, and the technology is developed and implemented in the form of several programs.

The structure of the functional subsystems is determined at the level of task groups for the implementation of the digital footprint of companies including creation of a platform to accommodate your digital footprint (creating a website, placing a site on hosting, creation of domain), the platform for the subsystem of data collection, storage and visualization (Figure 1).

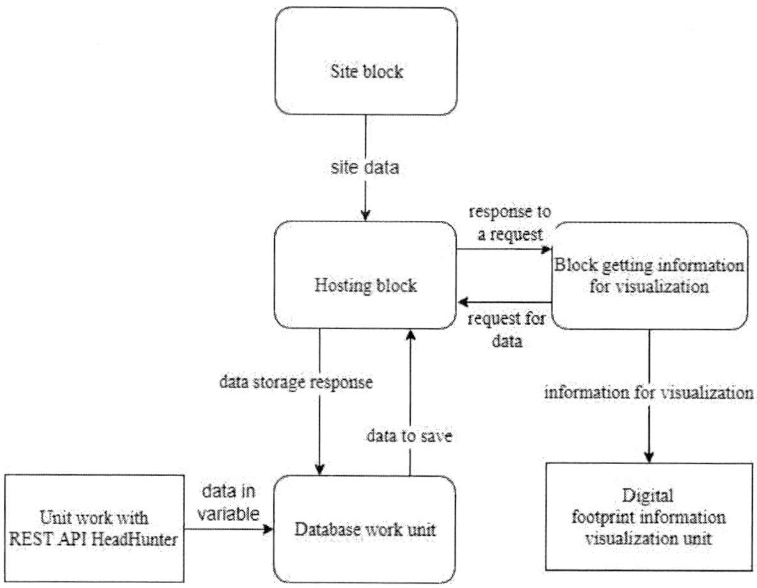

Figure 1. The logical structure of the relationship of functional blocks of the system for formation of the digital footprint of companies.

The structure is divided into six basic blocks:

- site block assuming the style definition of the website, the choice of colors, determining the layout of the pages, the description of the block content, compatibility with various browsers and devices;

- hosting block provides the storage of the site on the server, the rights of the website administration, implementation of backup system, reporting errors in the case of failures of the website, domain name registration, obtaining access to the site by domain name;
- databases (DB) work unit provides a remote connection to DB, creates DB for the site, creates tables in DB, records data on the vacancies in the table, deletes outdated data and records new ones;
- unit work with REST API Head Hunter forms a request to REST API, receives a result on the request, recorders the result of the request to a variable;
- block getting information for visualization creates an interactive map of Russia regions displaying information on vacancies of a certain profession, creates diagrams with the most popular competencies for vacancies of a certain profession.
- digital footprint information visualization unit to forms a request to the database, retrieving data from the database.

The logical structure of the relationship of the functional blocks of the system of formation of the digital footprint of companies includes the following elements:

- the site block is located on the hosting server (Hosting block);
- database work unit creates a database which is stored on the hosting;
- unit of work with REST API Head Hunter provides the data to write to database work unit;
- block of getting information for visualization requests the data for updates from the database stored on the hosting;
- digital footprint information visualization unit receives the data from a block of getting information for visualization;
- hosting block creates domain name for website access of users.

Thus, the system of formation of the company's digital footprint is a set of tools aimed at collecting, processing, storing and displaying information. The choice of information which will be further used, depends on the task assigned to the system. Almost any work with a digital trace can be reduced to the presented logical model, making minor changes according to the chosen subject area.

Information flows of the digital footprint formation system can be classified according to the following features:

- by relation to the digital footprint formation system: incoming, outgoing and internal information flows.
- by degree of openness: public and non-public information.
- by subject area: information for interested persons, including graduates, students, and University professors.

In addition, the system includes functions for storing data with a time stamp on which they were relevant, and the ability to change the structure of the data if necessary.

The software implementation of the technology of forming a digital footprint of companies required the use of several programming languages; query and markup languages in table 1.

Thus, the schematic diagram of the software implementation of the digital footprint involves the sequential realization of two stages: the creation and configuration of the system (deployment, filling data), as well as work with data (updating data) and their visualization. The collected, accumulated and stored data in the system are placed in the hosting storage (in the built-in file system), or in the database created on the hosting by the user.

The final result of the system is the data on vacancies clustered and stored in the database. Built-in visualization tools allow the most current data at the current time to be displayed.

Table 1. Languages used and functions for which they are used.

Block name	Language	Functional
Site block	HTML	Description of the logical structure of the site and its blocks
	CSS	General description and description of individual block ocations
	JavaScript	Connection of modules and scripts, data conversion to the required form
The unit works with the REST API Head Hunter	Python	Sending a request to REST API hh.ru and getting data on vacancies
The database work unit	Python	Connecting to the database and running queries
	SQL	Creation of databases and tables, their filling and clearing
The block getting information for visualization	PHP	Connecting to the database and getting the latest information on the site
The digital footprint information visualization unit	PHP	Calling the data acquisition block
	JavaScript	Displaying the map and data on the mp
	HTML	The logical structure to display on the site
	CSS	Description of of the map and the possibility of increasing or decreasing it

3. Results and summary

Thus, the website development using HTML, CSS, JS, and software implementation for the visualization of digital trace using JS, HTML, CSS, and PHP allowed us to obtain an interactive map of the Russian Federation with the display of relevant data regarding the number of vacancies of a particular specialty in the regions of the Russian Federation, the level of salaries and the demand of professional competences of potential candidates (Figure 2).

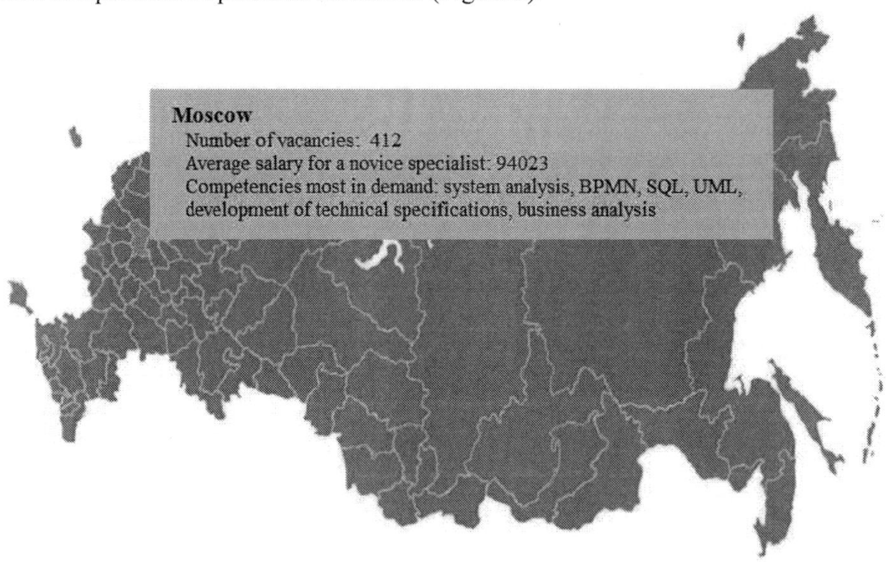

Figure 2. An interactive map with the data on vacancies of the system analyst.

Figure 2 shows the offer for vacancies of the system analyst in Moscow (412 places), the average salary (94023 rubles a month), and also the most demanded competences such as SQL, UML, BPMN, development of specifications.

The presented interactive map is currently created for several professions:

- system analyst;
- business analyst;
- web developer;
- web designer;
- 1C programmer;
- IT project Manager;
- Internet marketer.

These jobs for the software implementation of the digital footprint were chosen according to the following criteria:

- formation of professional competencies for future work should be provided by the current training plan of the program 38.03.05 – Business Informatics faculty of business Novosibirsk state technical University;
- the professional standard for the chosen workplace should be present in the database of the Association of computer and information technology enterprises of Russia.

The proposed version of the interactive map will be beneficial for all stakeholders of the educational process. The state, graduates and their parents (acting as investors in the education of children), as well as junior students have the opportunity of professional orientation and self-determination. Graduates who have already chosen their future profession, as well as senior students can form their own professional educational trajectory through the formation of competencies relevant to the region of the proposed location. The teaching staff of the University will be able to improve the educational process based on requirements for a potential employee, personal characteristics (qualities and abilities), relevant tools, as well as the functional work activities and its context. This will have a positive impact on the reliable provision of the region with really necessary personnel.

Regarding the issues of strategy of universities, the search for potential stakeholders, the formation of proposals for cooperation and subsequent contacts, the proposed interactive map will allow to choose the most promising professions. These professions will allow to purposefully form the required professional competence of teachers (and, consequently, somewhat later, of students) and to carry out technological modernization of the educational process, thereby ensuring the competitiveness of universities in the world market of educational services.

Acknowledgments

The work was carried out according to the plan of research within the program of strategic development of NSTU in the period of 2017-2021 and the thematic plan of research “The formation of the mechanism of interaction of stakeholders in the digital transformation of the business environment”.

References

[1] Barnett T, Jain S, Andra U and Khurana T 2018 *Cisco Visual Networking Index (VNI) Complete Forecast Update, 2017–2022* (Cisco Public) p 78
[2] Kogdenko V, Mel'nik M 2018 *Regional economy: theory and practice* **16** 38–57
[3] Malone M, Ismail S 2017 *Exponential Organizations: Why new organizations are ten times better, faster, and cheaper than yours (and what to do about it)* (Moscow: Al'pina Pablisher Publ) p 391
[4] Timofeeva R 2018 *Beneficium Bulletin of the institute of economics and management NovSU* **3**(28) 80–85
[5] Ivanter V 2016 *Problems of* forecasting **1** 3–7

[6] Gohberg L, Radomirova Ya 2018 *Technological future of the Russian economy* (Moscow: HSE publishing house) p 194
[7] Gorbunova M, Gresko A 2016 *Problems of the modern economy* **2** 215 –218
[8] Gorbunova M, Gresko A 2015 *Fundamental research* **2** 560–565
[9] Gorbunova M, Gresko A 2016 *Vestnik Astrahanskogo GTU Ser: Ekonomika* **4** 46–55
[10] Plotnikov V 2018 *Proceedings of the Saint Petersburg state University of Economics* **4**(112) 16–24
[11] Avdeeva I, Golovina T and Parahina L 2017 *Management issue* **6**(49) 50–56

Investigation of the properties changes observed for plastic samples made by fused deposition modelling under radiation exposure

A A Bulavskaya[1], Yu M Cherepennikov[2], A A Grigorieva[1], I A Miloichikova[2,3], N E Toropkov[4,5] and S G Stuchebrov[1]
[1] Research School of High-Energy Physics, Tomsk Polytechnic University, Tomsk, Russia
[2] School of Nuclear Science & Engineering, Tomsk Polytechnic University, Tomsk, Russia
[3] Cancer Research Institute of Tomsk NRMC RAS, Tomsk, Russia
[4] Institute of Strength Physics and Materials Science of Siberian Branch of Russian Academy of Sciences, Tomsk, Russia
[5] School of Advanced Manufacturing Technologies, Tomsk Polytechnic University, Tomsk, Russia
E-mail: bulavskaya@tpu.ru

Abstract. This work shows investigation of the transmission dynamics of polymer samples made of PLA plastic (polylactide) by fused deposition modelling for electron beam. Besides, results of tests for mechanical destruction (compression) of plastic samples with radiation exposure and without are shown. Article demonstrates that properties of plastic samples are stable up to 1.5 kGy, that proves the suitability of this material for sample production intended for depth dose distributions of electron beams.

1. Introduction

High energy electron beams are used for radiation therapy purposes since 1947 when irradiation by extracted betatron beam was first successfully carried out [1, 2]. Modern medical linear accelerators (linacs) generate electron beams with 4–25 MeV average energy [2, 3]. Advantages of electron using for radiotherapy purposes are possibility to provide higher dose homogeneity in the target volume as well as sufficiently lower dose value in deeper located healthy tissues. The latter leads to achieve the main goal of radiotherapy – maximal damage of malignant tumor while minimizing the impact on normal tissue [1–3].

The common method for radiation field depth dose distribution of radiation beam forming involve using of wedge filter sets, included in the accelerator package. The limited wedge number and shapes cannot provide forming of arbitrary complex dose distributions [1, 2]. These filters are designed for gamma radiation beam and cannot be used for electron one. These limitations significantly affect accuracy of dose delivery to the tumor focus and leads to overexposure of healthy tissue [2]. To form depth dose distributions of electron beam radiation fields and avoid acute post-radiation reactions some medical centers apply either special complex shaped compensators, which are located in the beam path and change its transverse profile, or tissue equivalent boluses located on the patient's body, which provides changes of the depth dose distribution in the patient tissues [3, 4]. Shapes of compensators and boluses is chosen individually for each particular patient, that causes increase the time of preparation for treatment, necessary to produce samples with complex shape [2–4].

In order to increase accuracy and for simplification of patient specific forming samples production, authors of this work propose to use fast prototyping systems. The technical level of these systems allow producing produce custom-made products with submillimeter accuracy in very short time [5, 6]. A number of papers poof the efficiency of this approach [5, 7–10].

The PLA plastic is chosen as investigated material for forming samples production since it is relatively cheap, highly available and strength [11].

Nowadays a number of investigations of physical and mechanical properties of polymers under irradiation by electron beam aimed to improve material parameters are carried out [12–14]. Based on results of a few works [12–14] it can be concluded that physical and mechanical properties of different polymers are stable in radiation dose range equal up to 100 kGy. The total dose in a radiation therapy course is not higher than 100 Gy [15], that 1000 time lower than aforementioned dose range. However, for forming of depth distributions of radiation fields it is necessary to investigate not only mechanical properties, but also ionizing radiation transmission dynamics of this material.

Therefore, the aim of the work is to investigate transmission dynamics of PLA plastic for electron beam and to test this material for mechanical destruction (compression) of plastic samples exposed and non-exposed to radiation.

2. Materials and methods

2.1. *Production of plastic samples*

In a frame of work samples made of PLA plastic is produced using fused deposition modelling by 3D printer Ultimaker2 with modified printhead [16]. The samples are cubic shaped with 2 cm edge length and 100% filling coefficient in accordance with ISO 604:2002 (federal standard 4651–2014) [17]. The number of samples for each type of test equals to 3 (Figure 1). The model of samples for 3D printing is made with Siemens NX software, while Slic3r software is used for layer separation [18, 19].

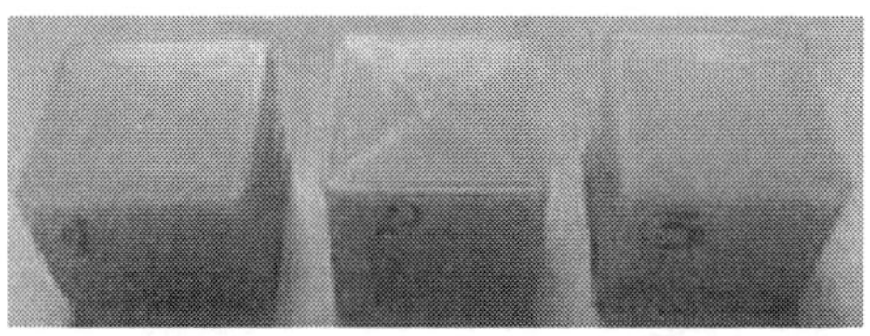

Figure 1. Samples made of PLA plastic using fused deposition modeling technology.

2.2. Experimantal scheme.

The TPU microtron is used as a source of radiation [20]. The PLA samples is placed in the path of the extracted electron beam with 6 MeV energy at 15 cm distance from the exit window (Figure 2).

Dose rate measurement is performed using clinical dosimeter UNIDOS E with plane-parallel ionization chamber Markus type 23343, placed in the special adapter plate of solid phantom RW3 [21–23]. Dose rate of initial beam in absence of sample equals 8.93 Gy/minute (S_0). Current value is measured each minute (S_i). Total dose absorbed by the each irradiated sample equals 1.5 kGy.

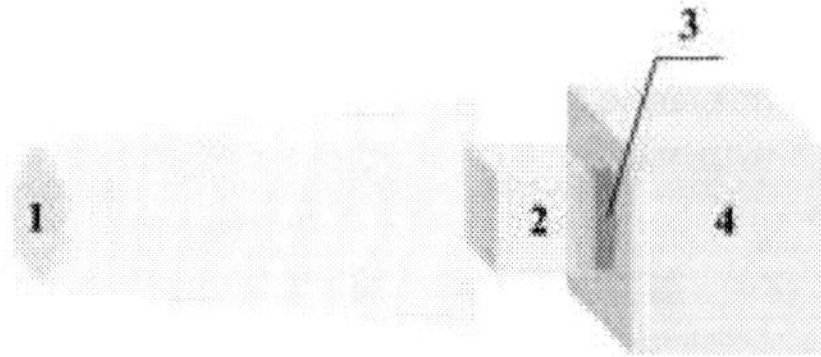

Figure 2. Experimental scheme: 1 – ionizing radiation source; 2 – plastic sample; 3 – ionization chamber; 4 – tissue equivalent phantom.

2.3. Mechanical tests

Investigation of mechanical properties of plastic samples with radiation exposure and without is performed with Instron 5982 setup [24]. Compression tests are performed with 100 kN load sensor and the speed of movement of the loading plate equal to 1mm/minute at room temperature. Load dependence on motion is measured using Bluehill software [25]. Figure 3 shows appearance of the experimental setup.

Figure 3. Mechanical test of plastic samples with Instron 5982 setup.

2.4. Results processing

In accordance with ISO 604:2002 (federal standard 4651–2014) following parameters is determined for both irradiated and non-irradiated samples: compressive yield stress, ultimate compression strength, elastic modulus.

Compressive yield strength was determined as the compression stress value for the sample height reduction on less than twice. On the other hand, PLA plastic is a glass-crystalline polymer, so at yield stress it can be considered as destructed since crystalline part is still stable. Therefore, total fracture was defined as the ultimate compressive strength with a decrease in height of more than twice.

Compressive stress (MPa) is determined by the equation:

$$\sigma = \frac{F}{A} \tag{1}$$

where F – compressive load, (N); A – sample initial cross-sectional area, mm^2.

Elastic modulus (MPa) is determined by the equation:

$$E_C = \frac{\sigma_1 - \sigma_2}{\varepsilon_1 - \varepsilon_2} \tag{2}$$

where σ_1, σ_2 – compressive stress, measured at relative deformation values ε_1, ε_2, (MPa);

Relative compressive deformation value is determined by the equation:

$$\varepsilon = \frac{\Delta L_0}{L_0} \tag{3}$$

where ΔL_0 – reduction of the nominal sample length, mm; L_0 – nominal sample length, mm [17].

3. Results and discussions

The measurement results of transmission dependence of investigated polymer samples for electron beam on absorbed dose are shown in Figure 4. The results of each measurement with investigated sample (S_i) is normalized on value of initial dose rate in absence of sample (S_0), so it shows relative part of radiation passed through the sample $k=S_i/S_0$.

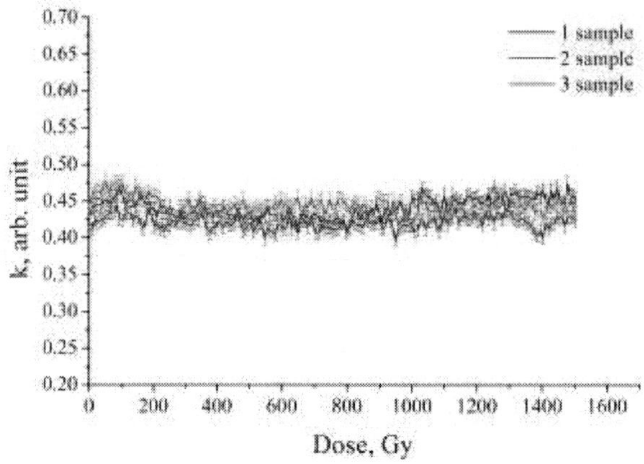

Figure 4. Dependence of 3D printed polymer samples transmission for electron beams on absorbed dose. The samples made of PLA plastic using fused deposition modeling.

Figure 4 shows slight deviation of transmission dynamics for electron beam. This deviation is about ~2% and stable in time, that proves the possibility of investigated material and samples made of it to maintain absorption properties for high energy electrons even with high doses of absorbed radiation (up to 1.5 kGy).

Figure 5 shows appearance of samples made of PLA plastic after compression tests.

Figure 5. Samples made of PLA plastic 3D printed using fused deposition modeling after compression tests.

As one can see in Figure 5, the samples under compression is deformed without gaps or cracks.

The results observed in compression tests of plastic samples are shown graphically: σ_y – compressive stress at material yield strength (Figure 6), σ_M – ultimate stress in compression with the ultimate destruction of the material (Figure 7).

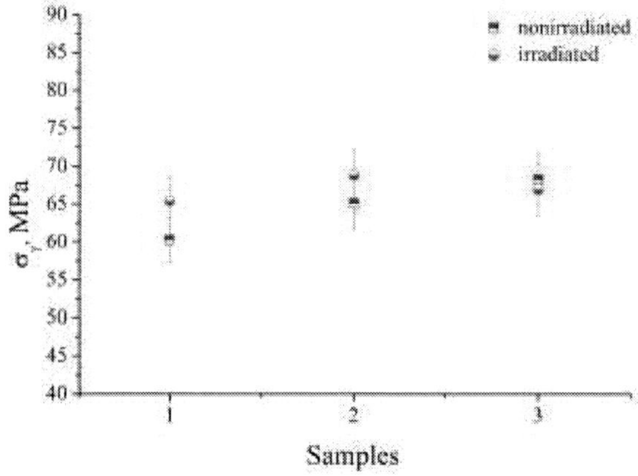

Figure 6. Experimentally observed compressive yield stress for irradiated and non-irradiated samples made of PLA plastic using fused deposition modeling.

IOP Conf. Series: Materials Science and Engineering 1019 (2021) 012068 doi:10.1088/1757-899X/1019/1/012068

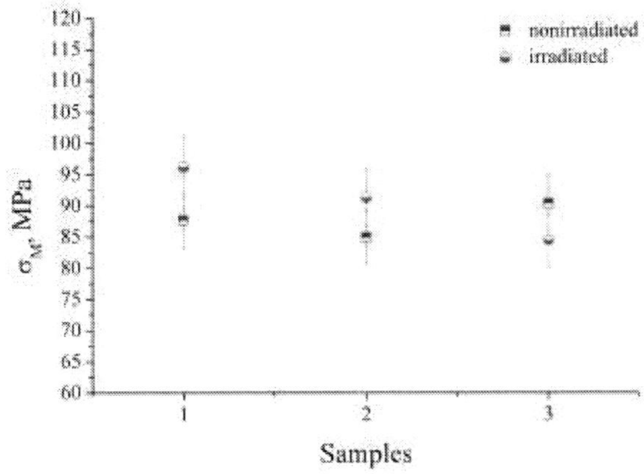

Figure 7. Experimentally observed ultimate compression strength for irradiated and non-irradiated samples made of PLA plastic using fused deposition modeling.

Figure 8 shows results of elastic modulus determining (E_C) for plastic samples.

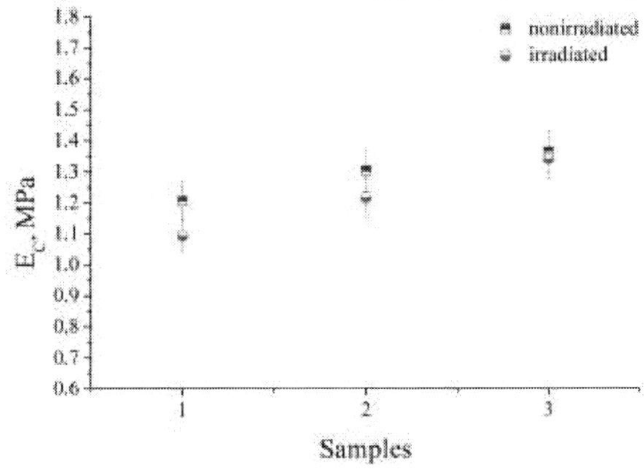

Figure 8. Experimentally observed elastic modulus for irradiated and non-irradiated samples made of PLA plastic using fused deposition modeling.

As one can see from Figures 6–7, the data obtained for irradiated and non-irradiated samples agree within the error. Figure 8 shows that all three irradiated samples have lower strength. As far as elastic modulus is more demonstrative integral characteristic of strength, this effect can be caused by slight structural change in plastic with increasing crystallinity. However, elastic modulus reduction does not exceed 7%. Accounting low statistical sample, this data can relate to the error of the experiment. Summarizing the above, it can be concluded that samples made of PLA plastic using fused deposition modeling maintain mechanical properties in dose range up to 1.5 kGy.

4. Conclusion

The work demonstrates, that samples made of PLA plastic using fused deposition modeling maintain both mechanical and absorption properties, after absorbing of radiation dose up to 1.5 kGy, that proves prospects of this material using for production of samples forming depth dose distribution of electron beams.

Acknowledgment

This work is supported by the Russian Science Foundation, project No. 19-79-10014.

References

[1] Khan F M *et al* 1991 *Med. Phys.* **18**(1) 73–109
[2] Kry S F 2017 *Med. Phys.* **44**(10) 391-429
[3] Thwaites D I and Tuohy J B 2006 *Phys. Med. Biol.* **51** 343–362
[4] Suit H D, Loeffler J S 2011 *Evolution of Radiation Oncology at Massachusetts General Hospital* (Berlin: Springer Science & Business Media*)* p 213
[5] Miloichikova I A, Krasnykh A A, Stuchebrov S G 2016 *AIP Conf. Proc.* **1772**(1) 060016
[6] Ventola C L 2014 *Pharm. and Ther.* **39**(10) 704–711
[7] Su S, Moran K, Robar J L 2014 *J. of Appl. Clin. Med. Phys.* **15**(4) 194–211
[8] Burleson S 2015 *J. of Appl. Clin. Med. Phys.* **16**(3) 166–178
[9] Zhao Y *et al* 2017 *Med. Dosim.* **42**(2) 150–155
[10] Łukowiak M 2017 *J. of Appl. Clin. Med. Phys.* **18**(1) 76-81
[11] France A K 2013 *Make: 3D Printing: The Essential Guide to 3D Printers* (Sebastopol: Maker Media, Inc.) p 230
[12] Murray K A 2013 *Nucl. Instrum. Methods Phys. Res. Sect. B.* **297** 64–74
[13] Linggawati A, Mohammad A W, and Ghazali Z 2009 *Eur. Polym. J.* **45**(10) 2797–2804
[14] Soon Z F 2015 *Investigation of electron beam irradiated polystrene under oven treatment* (Kampar: Universiti Tunku Abdul Rahman) p 99
[15] Gerber D E, and Chan T A 2008 *Am Fam Physician* **78**(11) 1254–1262
[16] Ultimaker 2+ series. Available at: https://ultimaker.com/3d-printers/ultimaker-2-plus
[17] 2002 *ISO B. 604. Plastics–determination of compressive properties* (Geneva: International Organization for Standardization)
[18] Siemens: NX. Available at: https://www.plm.automation.siemens.com/global/ru/products/nx/
[19] Slic3r. Available at: https://slic3r.org/
[20] Miloichikova I A *et al* 2016 *Phys. of Part. and Nucl. Lett.* **13(7)** 890–892 https://doi.org/10.1134/S1547477116070384
[21] UNIDOS E Universal Dosemeter. Available at: https://www.ptwdosimetry.com/en/products/unidos-e/
[22] PTW 23343, Markus Ion Chamber. Available at: https://www.rpdinc.com/ptw-23343-markus-ion-chamber-965.html
[23] Acrylic and RW3 Slab Phantoms. Available at: http://www.equiphos.com/wp-content/uploads/2015/05/3.-Maniqu%C3%AD-de-aguas-s%C3%B3lidas-RW3-Brochure.pdf
[24] Floor Model Adapter Kit for 5984, 5985 and 5988. Available at: https://www.instron.ru/en/products/testing-accessories/grips/other/couplings-adapters/loadcell-mounting-adapters/2501-227
[25] Bluehill Universal: The Next Generation Materials Testing Software. Available at: https://www.instron.ru/ru-ru/products/materials-testing-software/bluehill-universal

IOP Conf. Series: Materials Science and Engineering 1019 (2021) 012069 doi:10.1088/1757-899X/1019/1/012069

Aircraft landing gear with electromagnetic damper

A R Zagidulin[1] and E G Podruzhin[2]
[1] Aircraft Dynamic Strength Research Division, Siberian Aeronautical Research Institute named after S. A. Chaplygin, Novosibirsk, Russia
[2] Aircraft Strength Department, Novosibirsk State Technical University, Novosibirsk, Russia
E-mail: zagidulin@corp.nstu.ru

Abstract. The article describes the design and computation of the aircraft landing gear with an electromagnetic-gas shock absorber. A numerical simulation of the landing impact was carried out at different values of the initial vertical velocity. The dependence of the optimal electric current in the winding of the shock absorber electromagnet on the vertical speed is obtained.

1. Introduction

Flight safety during takeoff and landing largely depends on the aircraft landing gear. Landing gear damping system absorbs the energy of the vertical movement of the aircraft, reduces the overload acting in the airframe at the time of impact during landing, as well as when hitting irregularities during taxiing, takeoff and landing. To it should be attributed pneumatic wheels and shock absorbers. The shock absorber absorbs and dissipates the impact energy (up to 60...80%), so that the overloads occurring in the airframe are within regulated limits. These overloads should be acceptable not only for the structural elements, but also for all that is in the aircraft (passengers, cargo, weapons, avionics, etc.). The shock absorber can be an independent element, or integrated with the landing gear strut. There are many design types of shock absorbers capable of absorbing and dissipating the vertical kinetic energy of an aircraft at the time of landing. In this paper, one of the most promising landing gear schemes with an electromagnetic-gas shock absorber is considered.

2. Description of design and mathematical model

The scheme of the considered landing gear is shown in Figure 1. The landing gear consists of a wheel and a shock absorber. The shock absorber is a cylinder in which the gas is compressed by the piston. The walls of the cylinder are made of copper, and an electromagnet is built into the piston. When a piston moves in a copper cylinder, eddy current arise due to the magnetic field of the electromagnet, creating an induced magnetic field that counteracts the magnetic field of the electromagnet. Thus, during the compression of the shock absorber arises braking electromagnetic force.

The design of the electromagnetic damper is shown in more detail in Figure 2. The damper consists of winding 1 electromagnet, copper cylinder 2 and magnetic core 3. The core of electromagnet together with the steel wall 4 of the shock absorber cylinder forms a toroidal magnetic circuit. The electromagnet is mounted on the rod 5 of the shock absorber. Seals 6 seal the chamber with compressed gas. The electromagnet together with the seals 6 and the axle boxes 7 forms a shock absorber piston.

IOP Conf. Series: Materials Science and Engineering 1019 (2021) 012069 doi:10.1088/1757-899X/1019/1/012069

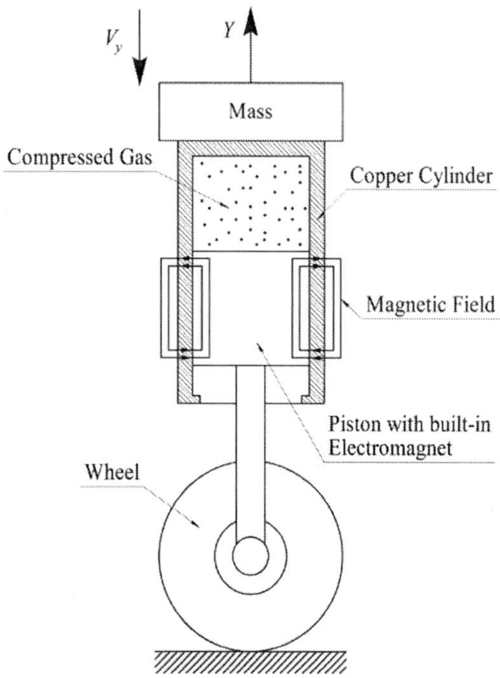

Figure 1. Landing gear scheme.

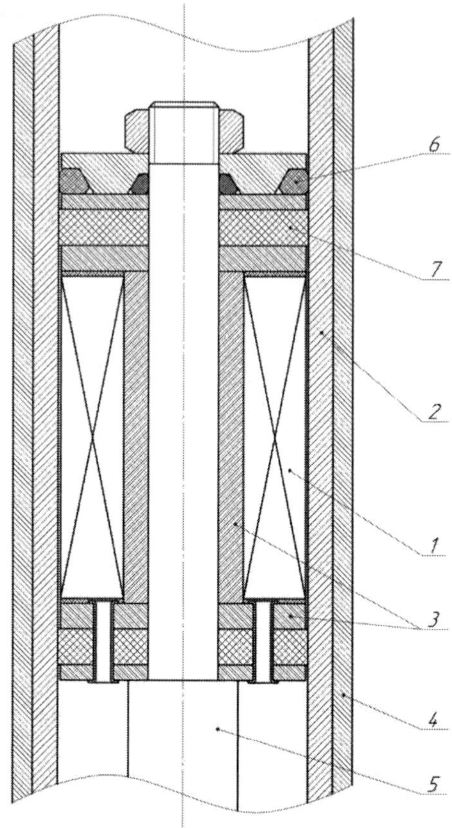

Figure 2. Electromagnetic damper.

At the initial moment of time, the landing gear moves vertically with the mass at a velocity V_y. When the landing gear strut is compressed, a lifting force Y equal in size to 2/3 of the weight of the landing gear with the mass falling on it.

The computation model of the landing gear is presented in Figure 3. Equations of motion:

$$M \cdot \ddot{H} = -P_{sh} - Y + R,$$

$$m \cdot \ddot{h} = P_{sh} - P_w - R,$$

where m is the mass of the moving parts of the landing gear, M is the mass of the load falling on the strut, except for the mass m, H is the vertical displacement, h is the reduction of the wheel, R is the reaction in the piston stop when the shock absorber is not compressed, P_w is the force of compression of the wheel, P_{sh} is the shock absorber compression force:

$$P_{sh} = P_g + P_f + P_e,$$

where P_g is the gas compression force, P_f is the friction force, P_e is the electromagnetic force.

Lifting force:

$$Y = \begin{vmatrix} (M+m) \cdot g \cdot 2/3, \text{ if } H \geq 0 \\ 0, \text{ if } H < 0 \end{vmatrix},$$

where g is the acceleration of gravity.

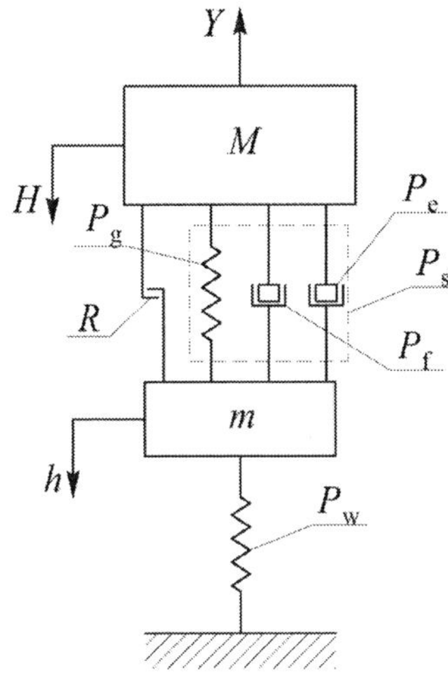

Figure 3. Computation model.

Reaction in the piston stop is defined as follows:

$$R = \begin{cases} R', \text{ if } S = 0 \text{ and } R' > 0 \\ 0, \text{ otherwise} \end{cases},$$

$$S = H - h,$$

$$R' = P_{sh} + (m \cdot Y - M \cdot P_w)/(M + m),$$

where S is the compression of the shock absorber. In addition, if $S = 0$, $\dot{S} < 0$, then $\dot{H}$ and $\dot{h}$ recalculated as $\dot{H} = \dot{h} = (M \cdot \dot{H} + m \cdot \dot{h})/(M + m)$.

The force of gas compression is determined from the expression:

$$P_g = p_0 \cdot F/(1 - S \cdot F/\Omega_0)^{\chi},$$

where p_0 is the initial gas pressure, Ω_0 is the initial gas volume, F is the cross-sectional area of the piston, χ is the polytropic exponent.

Friction force:

$$P_f = \mu \cdot \text{sgn}(\dot{S}) \cdot P_g,$$

$$\text{sgn}(x) = \begin{cases} 1, \text{if } x > 0 \\ 0, \text{if } x = 0 \\ -1, \text{if } x < 0 \end{cases},$$

where μ is the coefficient of friction.

Electromagnetic force [1, 2]:

$$P_e = (1/a^3 - 1/b^3) \cdot 15 \cdot \mu_0^2 \cdot \sigma \cdot q^2 \cdot \dot{S}/1024,$$

where a is the inner radius of the copper cylinder, b is the outer radius of the copper cylinder, $\mu_0 = 1.2566371 \cdot 10^{-6}$ H/m is the magnetic constant, $\sigma = 59500000.0$ S/m is the electrical conductivity of copper, q is the magnetic moment [3]:

$$q = k_m \cdot I \cdot F_m \cdot n,$$

where k_m is the coefficient determined by the parameters of the magnetic core, I is the electric current in the electromagnet winding, F_m is the cross-sectional area of the magnetic core, n is the number of turns of the winding.

Wheel compression force [4]:

$$P_w = k \cdot h/(1 - h/h_{max})^{\alpha},$$

where k is the stiffness, h_{max} is the maximum compression displacement of the wheel, α is the nonlinearity coefficient.

Initial conditions:

$$H = 0,\ h = 0,\ \dot{H} = V_y,\ \dot{h} = V_y.$$

Model parameters: $M = 785.5$ kg, $m = 14.5$ kg, $\Omega_0 = 0.0021$ m^3, $p_0 = 176519.7$ Pa, $F = 0.00785$ m^2, $\chi = 1.35$, $\mu = 0.1$, $a = 0.05$ m, $b = 0.06$ m, $k_m = 16.2$, $F_m = 0.00181$ m^2, $n = 10000$, $k = 201279.584$ N/m, $h_{max} = 0.1158$ m, $\alpha = 0.095$.

3. Simulation results

In Figures 4–7 shows the results of numerical simulation for the parameters: $V_y = 3.05$ м/с, $I = 4.2$ A.

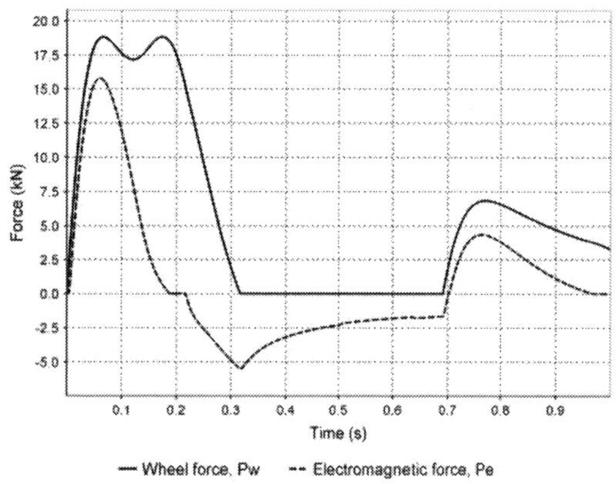

Figure 4. Wheel force and electromagnetic force in time.

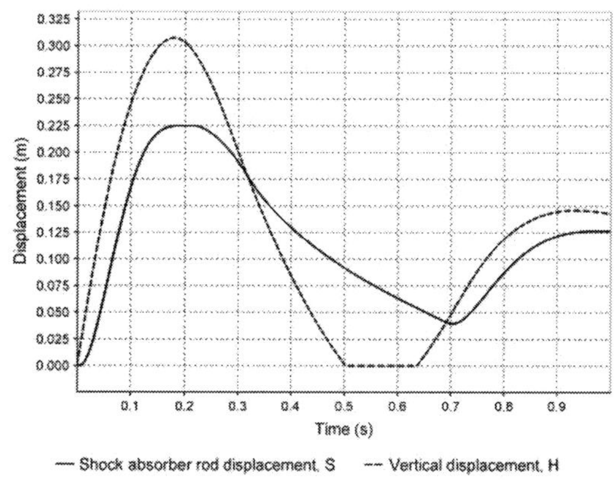

Figure 5. Shock absorber rod displacement and vertical displacement in time.

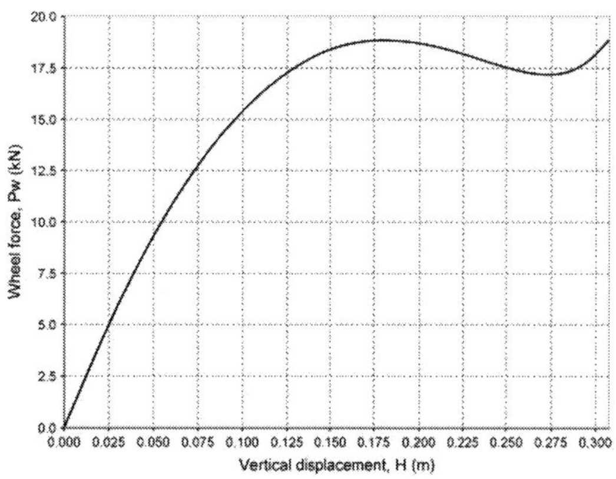

Figure 6. Dependence of wheel force on vertical displacement.

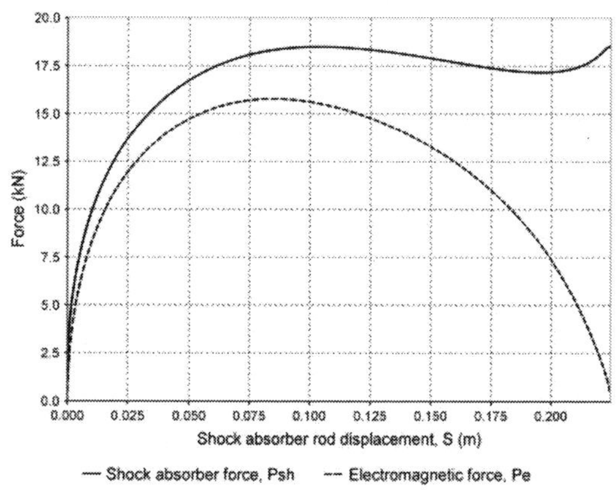

Figure 7. Dependence of shock absorber force and electromagnetic force on shock absorber rod displacement.

According to the results of numerical simulation, the optimum values of the electric current in the electromagnet winding were determined at vertical velocity from 1 m/s to 4 m/s. The optimal electric current value was considered such that the maximum force in the shock absorber during the reduction was the lowest. The diagram of the dependence of the optimal current on the vertical velocity is shown in Figure 8.

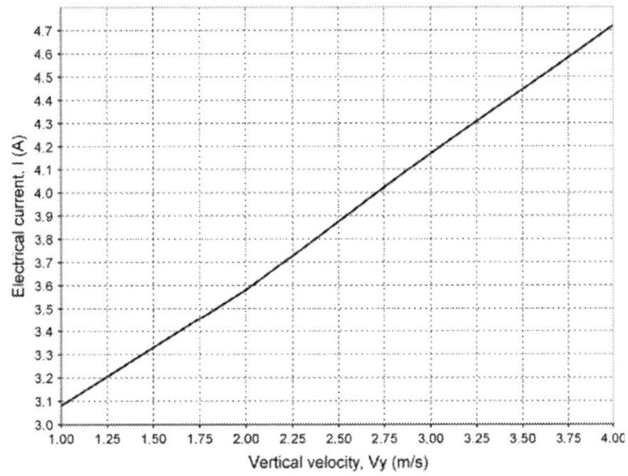

Figure 8. Dependence of the optimal current on the vertical velocity.

The resulting dependence is linearly approximated:

$$I = K_I \cdot V_y + I_0,$$

$$K_I = 0.551 \text{ A·s/m}, I_0 = 2.51 \text{ A}.$$

The results of the calculation of cases of landing using the dependence of the optimal electric current on the vertical velocity are shown in Figures 9–13.

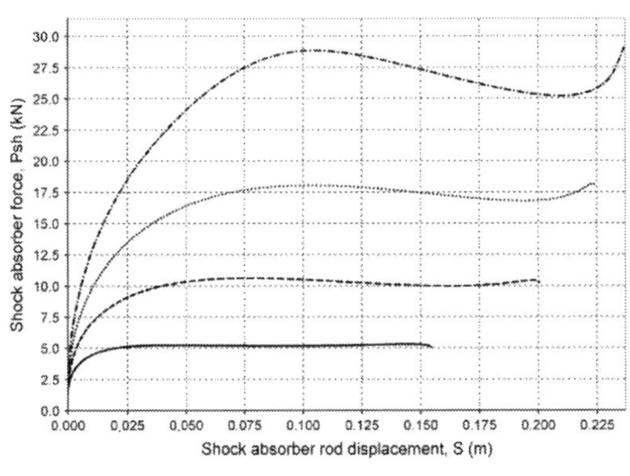

Figure 9. Dependence of shock absorber force on shock absorber rod displacement.

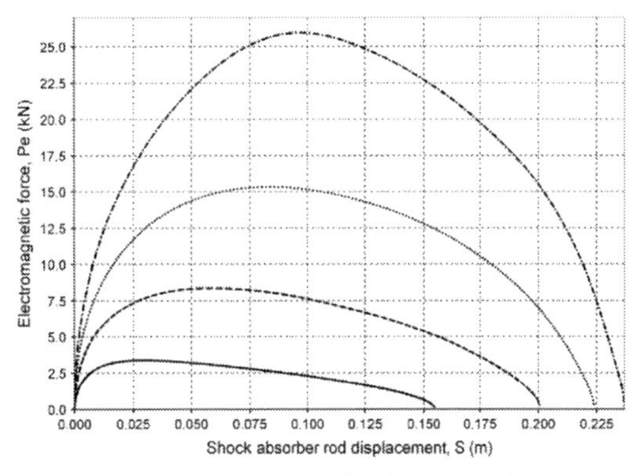

Figure 10. Dependence of electromagnetic force on shock absorber rod displacement.

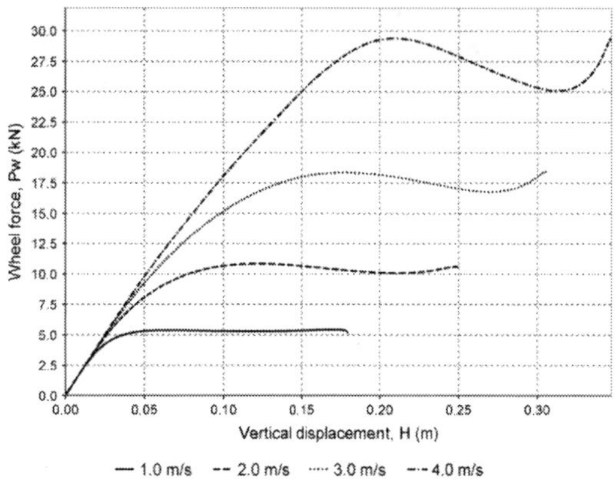

Figure 11. Dependence of wheel force on vertical displacement.

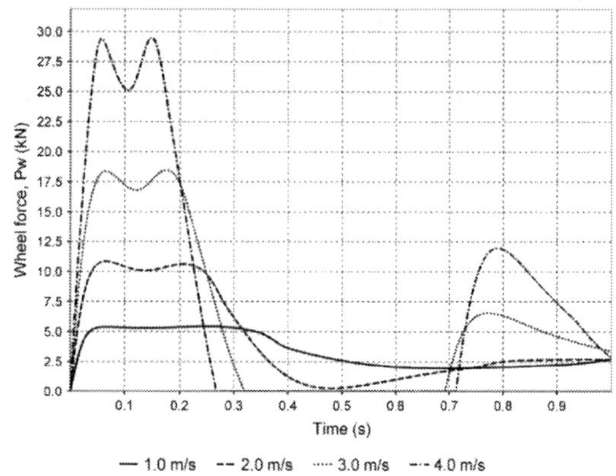

Figure 12. Wheel force in time.

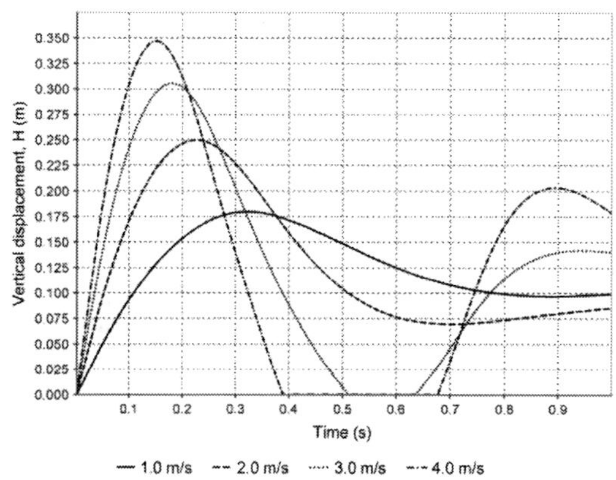

Figure 13. Vertical displacement in time.

4. Conclusion

According to the results obtained, we can conclude about the high efficiency of the electromagnetic-gas shock absorber. Such a shock absorber can be used in adaptive landing gear of aircraft, which allow minimizing the operational load on the airframe of the aircraft during landing. The design of the proposed electromagnetic damper is simple compared to regulated liquid dampers. When changing the control signal, the damper has no inertia, which allows you to fast change the damping force.

References

[1] Knyazev B A, Kotelnikov I A, Tyutin A A and Cherkassky V S 2006 *Uspekhi Fizicheskikh Nauk* **176** 965-974

[2] Saslow W M 1991 *American Journal of Physics* **60** 693–711

[3] Biryukov P V, Borovkov A G, Blaze E S 1976 *Devices and elements of automatic regulation and control systems. Technical cybernetics. Book 3. Executive devices and servomechanisms* (Moscow: Mashinostroenie Publishers)

[4] Podruzhin E G and Rastorguev G I 2002 *Calculation of liquid-gas dampers of the aircraft landing gear* (Novosibirsk: NSTU Publishers)

Technology of partial discharge detection by means analysis of impulse interaction

A Mytnikov and V Strugov
Power Engineering School, National Research Tomsk Polytechnic University, Tomsk, Russia
E-mail:Mytnikov66@mail.ru

Abstract. In this paper the approach based on impulse analysis for partial discharge (PD) control and identification for high voltage insulation are summarized and the experimental research of impulses of different nature for detection some typical PD defects are discussed. Some physical effects at different voltage levels are shown and a method for the representation and characterization of the PD is introduced. Special test cells with different types of defects such as a streamer corona and pore inside oil-paper insulation of condenser type are used for the investigations. The experimental procedure for PD identification by means of impulse form transformation is shown and discussed. Experimental equipment and measurements are described. It is shown that using of impulse of nanosecond duration and commutative impulse allows the characterization and controlling PD in high voltage insulation.

1. Introduction

Partial discharge (PD) can seriously damage the insulation of high voltage power equipment. Power transformers and generators are one of the most expensive and crucial components in any electric power systems, whose reliable operation is most priority for energy generation, transmission sectors and end users. Timely PD control and characterization in inner structure of high voltage insulation is key to stable work providing of power engineering systems [1, 2]. Although many efforts have been concentrated to improve PD control, the failure accidents of high voltage equipment due to insulation damage often take place [3]. Service life of many high voltage apparatus does not reach the designed lifetime. According to a survey [4], it is reported that one third of total transformer failures are attributed to insulation failures. Despite big amount of PD identification techniques, every technology has drawbacks connected with losses problem, complicated measurement process and others. Development PD control technology in high voltage insulation in on-line mode is serious task. Research on the DC voltage-related PD characteristics and PD mechanism has attracted much attention [5]. DC voltage receives special consideration because it has no zero crossing and can cause polarization, there by generating PD characteristics and a PD mechanism that are unique to DC voltage.

It is attractive to use impulse process to control PD in high voltage insulation. These could be external impulses supplying by special generator or switching impulses as result of commutation processes.

The existence in current operation of old and largely worn equipment, simultaneously with new equipment manufactured using present-day technologies and materials, as well as the lack of such modern control technologies for electrical equipment that meet all the criteria and comply with smart grid technologies, causes a decrease in the effectiveness of traditional methods of providing a normal operation of electrical equipment. This factor largely hinders the development of intelligent power management systems.

Due to the aging of a whole generation of power transformers, the present-day electric power industry faces serious problems, since transformer failures, repairs and corresponding costs produce significant economic losses. Transformers have become the most problematic elements of electrical networks. The need for reliable methods of control and diagnostics has led the principal experts in this area to the necessity of developing new technologies that significantly increase the reliability and optimize the performance of each network element. The failure of high voltage insulation of power systems as power transformers or generators often causes accidents with serious consequences. Some of the main reasons for such emergencies are insulation defects caused by PD. The existing diagnostics technologies do not always permit to find out the PD, and hence the number of accidents increases [1–4].

The main purpose of the experiments described in this paper is to study the fundamental possibility of controlling partial discharges (PD) in high-voltage insulation in the ON-LINE mode. The pulse control method for controlling the state of transformer windings, known as the low-voltage impulses, was adopted as the basis for the PD control method.

Attempt to use commutation impulses for PD detection was demonstrated in [6]. It was shown that it is possible to detect PD in insulation of DC motors [6].

Promising results were obtained for winding condition control by using of nanosecond probing impulses [7–9]. This experience is useful to apply impulse processes to PD control. Two ways of controlling the PD were investigated, depending on the form and method of obtaining the pulse. At the first stage, rectangular impulses with duration of 520 and 260 ns were used. These so called external impulses so as they are applied to investigated insulation part from special impulse generator. The second stage of research consisted in the application of inner or switching impulse of a high-voltage network, which was modeled using a circuit based on a pulse capacitor IK-100-0.25. These impulses are generated inside power system due to commutative processes.

2. Experimental research of PD detection in high voltage insulation by means of external rectangular impulses

This approach implies an application of impulse on the part of insulation being under high voltage via the coupling condenser and a pulse signal the containing PD. During the experiments impulses of two different durations 520 and 260 nanoseconds respectively were applied. The cell modeling PD represented the brass sharp inserted into a cell from organic glass. Voltage about 6 kV is applied to a cell from the test transformer. Thus, type PD "incomplete breakdown" on the surface of organic dielectric was organized. Ignition and burning of PD was controlled visually. Block diagram of experiments for PD diagnostics by external nanosecond impulse is shown in Figure 1.

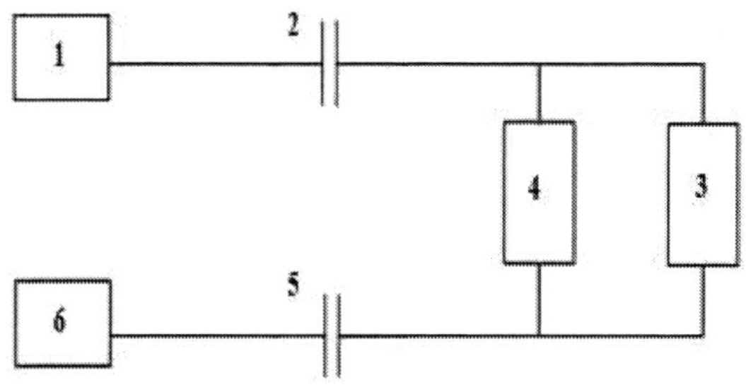

Figure 1. Scheme of experiments. 1 – generator of probing impulses of nanosecond duration, 2, 5 – coupling capacitors, 3 – high voltage transformer, 4 – PD cell, 6 - oscilloscope Tektronix TDS 1012.

Impulse generator forms nanosecond probing impulse which is supplied to detective cell. Oscilloscope measures probing impulse which is applied to cell. At first, probing impulse 520 ns duration is applied to cell without PD. Then cell with PD in streamer corona is placed in the scheme.

IOP Conf. Series: Materials Science and Engineering 1019 (2021) 012070 doi:10.1088/1757-899X/1019/1/012070

Coupling condensers provide safety work of impulse generator and oscilloscope. Waveform of impulse is applied to cell without PD is shown on Figure 2. The same situation with PD as streamer corona is shown on Figure 3.

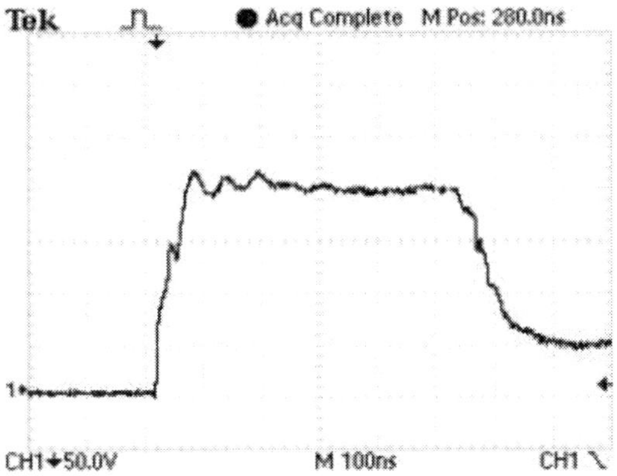

Figure 2. Waveform of impulse 520 ns. No PD.

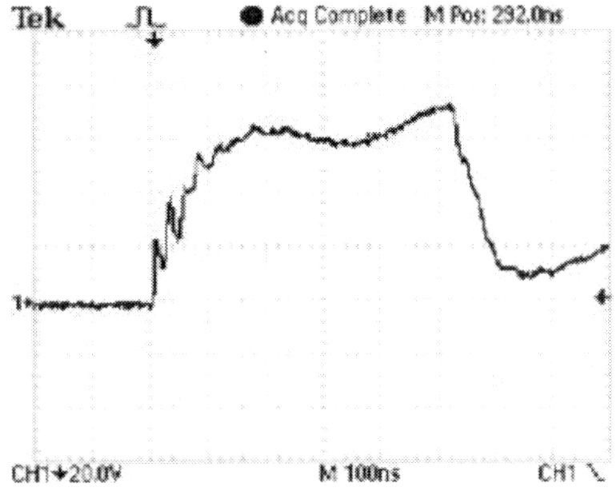

Figure 3. Waveform of impulse 520 ns. PD in streamer corona form is in cell area.

Figures 4 and 5 demonstate waveforms for the same situations, but for impulse duration 260 ns.

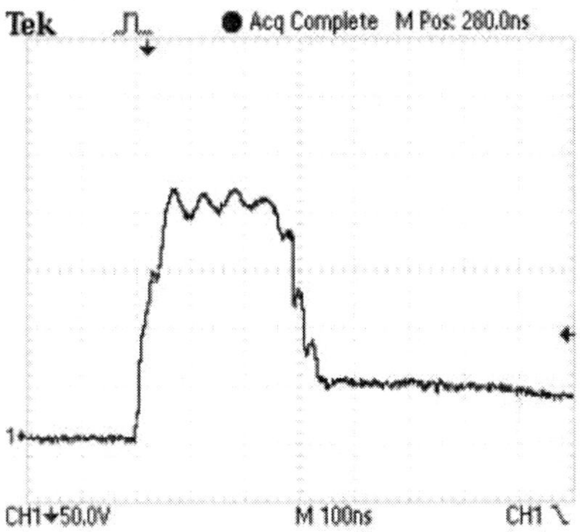

Figure 4. Waveform of impulse 260 ns. No PD.

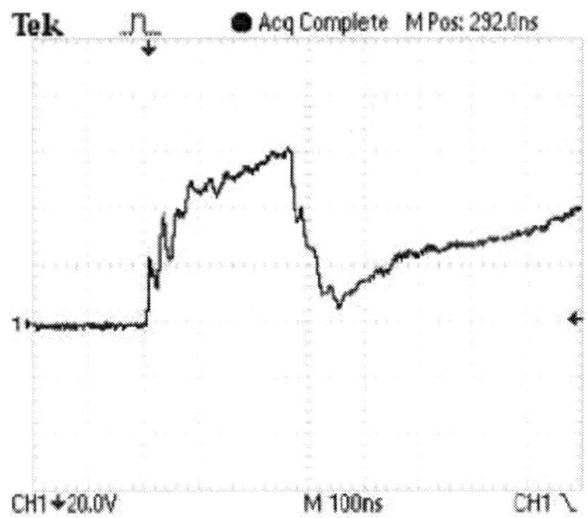

Figure 5. Waveform of impulse 260 ns. PD in streamer corona form is in cell area

Deformation of a form of an impulse comes from an initial form under the influence of PD. The picture observed in experimentscould be explained by following way. A physical basis of similar deformation of a rectangular impulse is interaction of PD pulse waves with the probing impulse ones. Both signals are impulses and contain big amount of high-frequency component. The rectangular impulse at the output of the generator is enriched with high frequencies in the range from hundreds kHz to MHz. The same frequency range has PD arising in internal defects of high-voltage insulation under the influence of operating voltage. Interaction of two frequency ranges when passing an impulse through the part containing PD leads to deformation of an initial form of an impulse and to emergence of the characteristic distortions observed at the Figures 3 and 5. Using a frequency PD-signal analysis is possible to estimate PD level intensity.

IOP Conf. Series: Materials Science and Engineering 1019 (2021) 012070 doi:10.1088/1757-899X/1019/1/012070

3. Experimental research of PD detection in high voltage insulation by means of commutative impulses

The cell of the second type modelled type of PD "PD in Internal Inclusion of Paper-oil Insulation of Condenser Type Bushing". On a metal rod was lapped layers of condenser paper with the leveling plates from thin aluminum foil are reeled up. Two identical cells were made. In one of them the artificial pore with 1mm diameter was made. To model bushing insulation layers of cell were impregnated with transformer oil. The view of a cell is shown on Figure 6.

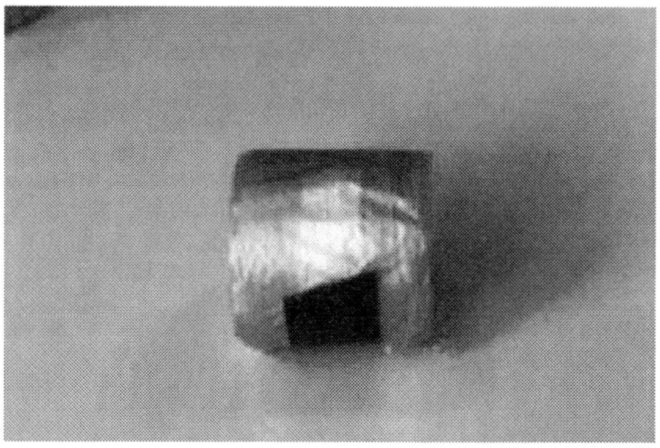

Figure 6. View of PD cell with pore in high voltage insulation of condenser type.

Common scheme of these experiments were following. Pulsed high-voltage capacitor IK-100-0.25 is charged from a direct current source. Then, when the required charge level was reached (in the range of 2–5 kV), a charged capacitor was switched to a cell with PD. The current in the discharge circuit was measured by a specially designed current shunt. The signal from the shunt was measured by an electronic oscilloscope Tektronix TDS 1012. The shape and amplitude of the signals with and without PD are analyzed and compared with each other at different voltages. Three different waveforms were obtained. Without PD current shunt signal has large amplitude in whole spectrum. Voltage on the cell is 2 kV. At the same conditions, but with PD cell (pore inside), amplitude the waves are diminished. Diminish of wave amplitude in more than 3 times is noted for situation with PD cell and 5 kV. A scheme of experiments is shown on Figure 7.

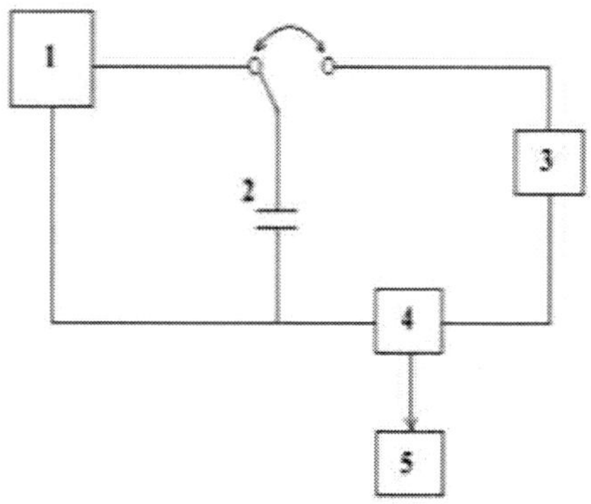

Figure 7. Scheme of PD controlling experiments by means of commutative impulses: 1 – charge source of power supply, 2 – high voltage condensor, 3 – PD cell, 4 – current shunt, 5 – oscilloscope TDS 1014.

Experimental situation without PD corresponds to waveform on the Figure 8. The cell is part of condenser insulation without any fail or pores inside. Figures 9 and 10 show the same experimental situation, but the cell contains a pore and voltage on it 2 kV and 5 kV respectively.

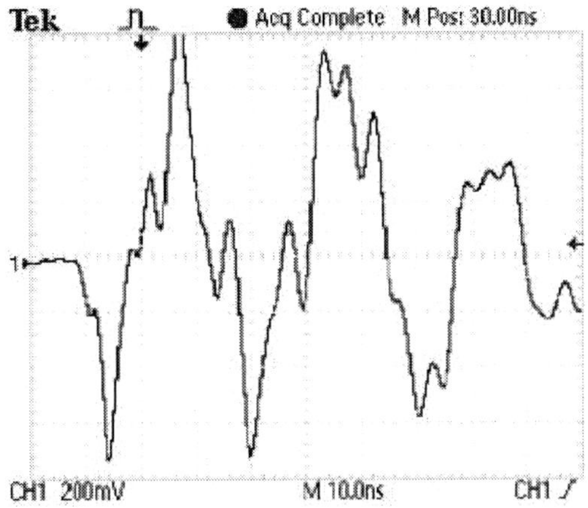

Figure 8. Waveform of current shunt without PD. Voltage on the cell 2 kV.

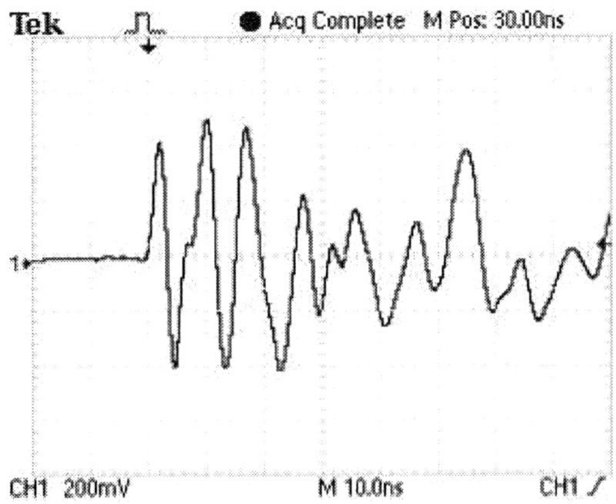

Figure 9. Waveform of current shunt. PD cell contain pore. Voltage on the cell 2 kV.

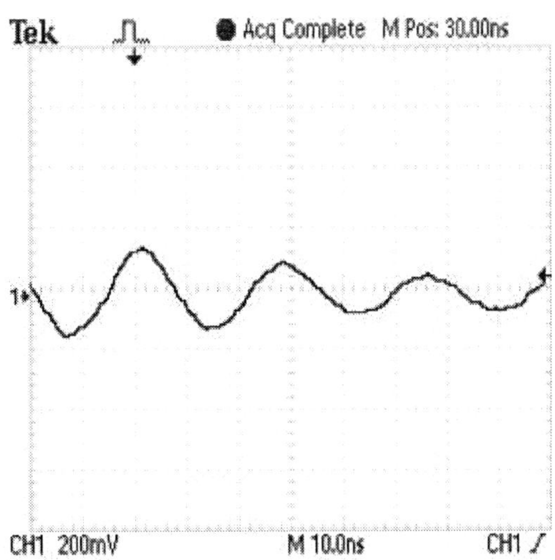

Figure 10. Waveform of current shunt. PD cell contain pore. Voltage on the cell 5 kV.

It is noted that a decrease in the amplitude of oscillations of the current shunt signal is proportional to the intensity of the partial discharge, all other parameters of the measuring system being equal. The decrease is explained by the superposition of the both current PD and initial one.

4.Conclusion

It is established that it is possible in principle to control PD in high-voltage insulation by means of analysis of various types by the impulses. Two different impulses could be appropriate for PD control goal. One of them is impulse of quasi-rectangular form generated by special generator. This PD controlling way is based on technology of consecutive impulse control for transformer winding condition control. Comparison of nanosecond impulses characteristics allows the controlling PD presence on the part on high voltage insulation. Potential application of proposed technology could be realized in ON-LINE regime using measuring sensor of bushing.

Other approach based on using commutative impulses in electric energy system. Consecutive comparison of switching impulses gives information on existence or lack of PD in the investigating high voltage insulation part. In both cases analysis of impulses allow the making of conclusion about presence of absence of PD as well as to make an estimation of PD development degree. Proposed PD-identification could be applied for monitoring for different types of PD - streamer corona, PD in pore in residual atmosphere, PD in inclusion in condenser-type paper-oil insulation.

To fulfill PD control both rectangular pulses created by a special generator and switching pulses of the high-voltage greed can be used. The criterion by which it is possible to state the presence of PD is the reduction of the amplitude of the current signal in the case of switching pulses. When using rectangular pulses, the criterion for assessing the presence and degree of development of the PD is the degree of distortion of the original rectangular shape of the probe pulse proportional to the intensity of the PD.

Potential effectiveness of proposed impulse technology for PD control is high enough, so as it could be used at on-line mode. Analysis of comparison impulse form and amplitude transformation allows the making of conclusion about presence of PD in high voltage insulation. Also, it is possible to make an estimation of PD development degree.

References

[1] Bartnikas R 2002 *IEEE Transactions on Dielectrics and Electrical Insulation* **9** 763–772
[2] Fuhr J Aschwanden T 2017 *IEEE Transactions on Dielectrics and Electrical Insulation* **24** 17–30
[3] Pirker A Schichler U 2018 *IEEE Transactions on Dielectrics and Electrical Insulation* **25** 883–894
[4] Tenbohlen S Coenen S Djamali M Muller A Samim A Siegel M 2016 *Energies* **9** 347–372
[5] Li S Li Q He D 2019 *IEEE Transactions on Dielectrics and Electrical Insulation* **26** 261–269
[6] Strugov V Lavrinovich V 2015 *Bulletin of the Tomsk Polytechnic University. Geo assets Engineering* **326** 72–77
[7] Lavrinovich V Lavrinovich A Mytnikov A 2012 *International Journal on Technical and Physical Problems of Engineering* **13** 149–153
[8] Lavrinovich V Mytnikov A 2015 *IEEE Transaction Dielectrics and Electrical Insulation* **22** 2041–2045
[9] Lavrinovich V Mytnikov A Hongda Li 2016 *Resource-Efficient Technologies* **3** 111–117

InAlN/GaN and AlGaN/GaN HEMT technologies comparison for microwave applications

L E Velikovskiy[1], P E Sim[1], O I Demchenko[1], N E Kurbanova[1], I A Filippov[1], A V Sakharov[2], W V Lundin[2], E E Zavarin[2], D A Zakheim[2], D S Arteev[2], M A Yagovkina[2] and A F Tsatsulnikov[3]
[1]REC Materials and technologies for space applications, Tomsk state university, Tomsk, Russia
[2]Ioffe Institute, St Petersburg, Russia
[3]Submicron Heterostructures for Microelectronics Research & Engineering Center RAS, St Petersburg, Russia
E-mail:velikovsky.1@gmail.com

Abstract. High electron mobility transistors (HEMTs) technologies based on AlGaN/GaN and InAlN/GaN heterostructures have been developed. The research focused on influence of epitaxial growth conditions and buffer doping profiles on electrical properties HEMTs. An output power density of 4W/mm at 17 GHz was demonstrated for InAlN/GaN HEMTs and 7W/mm at 10 GHz for AlGaN/GaN HEMTs.

1. Introduction

GaN-based high electron mobility transistors (HEMTs) have potential for successful application in MMIC technology for various frequencies including mm-wave range. InAlN/GaN and AlN/GaN based heterostructures have a higher, compared to AlGaN/GaN, spontaneous polarization induced electric field and, hence, a higher electron concentration in 2DEG [1–3]. The electron density in 2DEG for InAlN-based heterostructures reaches $2.5\div3\cdot10^{13}$cm^{-3}, more than 2 times of typical AlGaN/GaN 2DEG value. Among them, $In_{0.17}Al_{0.83}N$/GaN heterostructure was of special interest because of its lattice matched, unstrained InAlN barrier, which may lead to improved device reliability [4]. In addition to a higher electron concentration in 2DEG, an electron mobility of 1800cm^2/V·s was reported for InAlN/GaN heterostructure, which is comparable to the values for AlGaN/GaN heterostructures [5]. A high electron concentration in 2DEG in InAlN/GaN heterostructures is expected to extend the frequency range of HEMTs [6, 7]. Reducing the gate length is the main step to increase the operation frequency of the FETs, but this leads to an increase of the electric field in transistor channel and force to decrease operating voltage. To achieve high output power levels for this type of devices one should increase the maximum current range, which is possible with high Al-content InAlN/GaN and AlN/GaN based heterostructures. A thin barrier layer is essential for short gate length high frequency devices, since a high ratio between the gate length and the gate to channel distance is a key factor suppressing the short channel effects in HEMTs [8].

A good confinement of 2DEG in the heterostructure near the threshold voltage even for high drain voltages is important to achieve high power performance of short channel HEMTs because drain leakage at high Vds at |Vgs|≥|VT| affects such an important parameter as Power Added Efficiency (PAE). Possible solutions include AlGaN and InGaN back barriers and adjustment of dopant profiles in the GaN buffer layer [9–15]. Despite the excellent electrophysical properties, some issues regarding the stability of thin InAl(Ga)N barrier layers were reported [16–19]. Most of them are related to traps generated during reliability tests, which are located at the surface or in the InAlN barrier layer causing degradation of the RF performances of the devices. The impact of traps becomes more severe for low thickness barriers, since a decrease in the distance between the 2DEG and traps increases the effect of trap charges on the 2DEG in the channel. Utilizing an optimized SiN in-situ passivation along with other passivation techniques (ALD of Al_2O_3 or SiO_2) is necessary to achieve high DC and RF parameters [20].

The development of GaN-based HEMT and MMIC technologies for frequencies above 30GHz requires several technological and design steps that allows achieving comparable to GaAs-based devices RF parameters by compensating a lower than in GaAs HEMTs electron mobility in the 2DEG. To date, technologies developed by leading companies use gates with lengths of 60-150nm, low resistance non-alloyed ohmic contacts, AlGaN, InAlN and InAlGaN-based heterostructures with in-situ SiN passivation and scaling of device sizes [21–23].

2. Samples preparation and experimental results

AlGaN/GaN and InAlN/GaN heterostructures were grown by MOVPE on 100mm diameter silicon carbide (SiC) substrates. The epi layers included an AlN nucleation layer, an insulating buffer layer, a 1nm thick AlN interlayer, an AlGaN or InAlN layer, and a 5nm in-situ SiN layer. The sheet electron concentration and the electron mobility for the AlGaN/GaN and InAlN/GaN heterostructures were around $1800cm^2/V{\cdot}s$, $1.15cm^{-3}$ and $1100cm^2/V{\cdot}s$, $2.5cm^{-3}$, respectively.

Device fabrication started with alignment marks and ohmic contacts formation. Device isolation was carried out by ion implantation of boron. The alloyed ohmic contacts were formed by evaporation of Ti/Al/Mo/Au-based metallization followed by short (15–20sec) rapid thermal annealing at 780–800°C [24]. 0.1μm-length T-gates were fabricated using electron-beam lithography followed by ICP-RIE of a 0.1μm-thick SiN layer in a fluorine-based plasma. The top of the T-gate was formed by photolithography. Then, a Schottky contact was formed by deposition of Ni/Pt/Au.

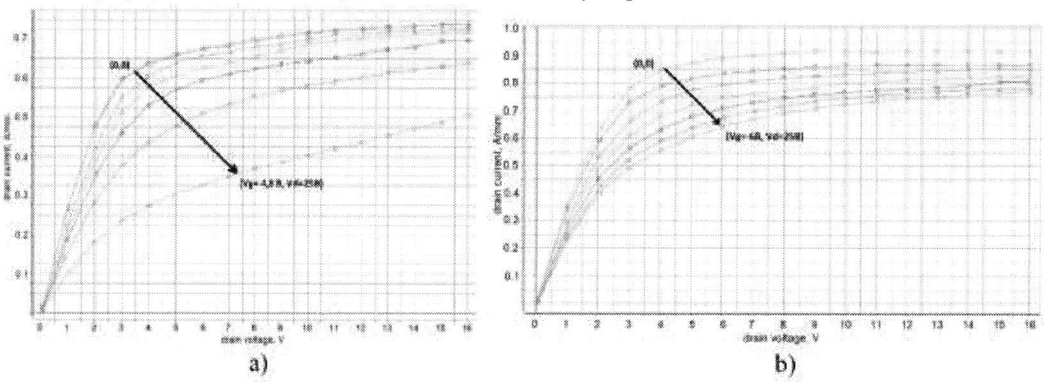

Figure 1. Comparison of the pulsed Id (Vd) characteristics at Vg = 0 V of the AlGaN/GaN HEMT with C- doped (a) and Fe-doped (b) buffer layers at the different quiescent bias points (Vg = 0V, Vd = 0V), (Vg = -6V, Vd = 5 ÷ 25V). The pulse length is 250ns.

Pulse I-V, S-parameters and load pull measurements were used to characterize the obtained AlGaN/GaN and InAlN/GaN HEMTs. AlGaN/GaN and InAlN/GaN HEMTs with a source-drain spacing of 2.5 and 3.5μm and 2x50μm or 2x100μm gate fingers were fabricated and characterized. Short (≤0.15μm) gate lengths required for millimeter wave range devices imposes strict requirements on leakage current values of the GaN buffer layers. To achieve a high breakdown voltage in transistors along with a minimal current collapse effect, an optimization and comparison of buffer layers with different design was carried on, namely, the transistors with undoped, C-doped and Fe-doped buffer layers were fabricated and compared. The highest (≥180V) breakdown voltages were observed in samples with C-doped and Fe-doped buffer layers. To evaluate the current collapse effect, pulsed I-V characteristics were measured on the AlGaN/GaN HEMTs, fabricated on the C- and Fe-doped buffer layers. Curves measured at Vg = 0V, were measured at various quiescent bias points and compared

IOP Conf. Series: Materials Science and Engineering 1019 (2021) 012071 doi:10.1088/1757-899X/1019/1/012071

(Figure 1). The smallest values of current collapse were observed in transistors with Fe-doped buffer layers.

InAlN/GaN is a structure with complex epitaxy requirements (the conditions for the growth of Al-containing and In-containing layers are different - the optimum temperatures for InN are ~ 600°C, and for AlN are ~ 1200°C), which makes an uniform In concentration along the InAlN layer not easily obtained. Also, an unintentional incorporation of Ga atoms into the InAlN layer was reported [25, 26]. Still, a layer-to-layer analysis of the composition of our InAlN/AlN/GaN samples by X-ray photoelectronic spectroscopy (Figure 2) showed that the In content corresponds to the target value of 12% and is almost the same over the entire thickness of the InAlN layer, as well as no Ga incorporation in the InAlN layer is observed.

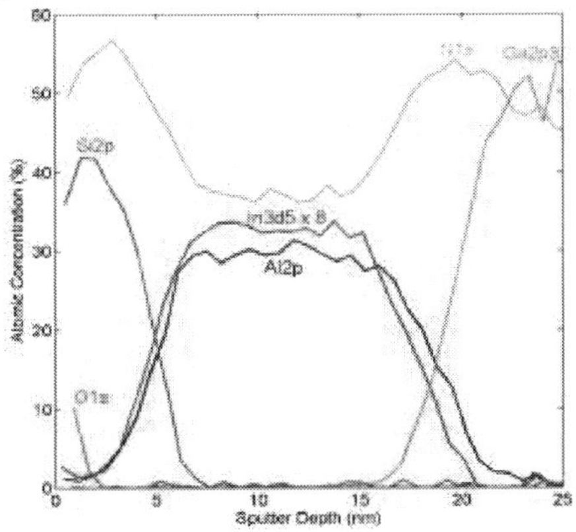

Figure 2. Atomic concentration profile obtained by layer-by-layer etching of SiN/InAlN/AlN/GaN heterostructure samples. The In concentration shown in the graph is 8 times increased.

The investigation of the effect of the InAlN molar fraction on the electrophysical parameters of a two-dimensional electron gas showed the InAlN layer conductivity is highest for indium content of 13÷15% with a relatively good mobility of 1100 $cm^2/V{\cdot}s$, so this content was used in further experiments.

A significant effect of the trimethylindium (TMI) flow rate during InAlN growth on the current collapse effect was also observed (Figure 3), showing a lowest collapse for a low TMI flow rate, which can be attributed to changes in the spectrum and/or concentration of the deep centers in the InAlN layer and on the surface. 0.1μm gate length HEMTs based on this heterostructures demonstrates a maximum drain current $I_{dmax} \geq 1.4$A/mm, a transconductance $g_m \approx 600$mS/mm with $V_{th} \approx -1.5$V.

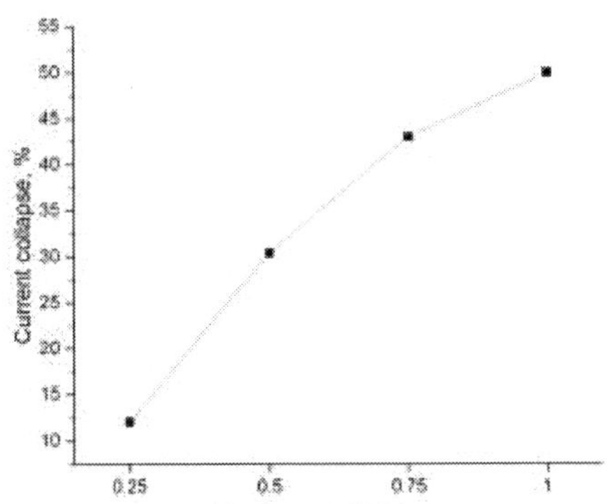

Figure 3. Current collapse effect on InAlN/GaN HEMTs fabricated on InAlN barriers grown with different TMI flow. Current collapse calculated as [Id (0,0) – Id (1.2Vt, 25V)]/ Id (0,0). TMI flow is normalized.

IOP Conf. Series: Materials Science and Engineering 1019 (2021) 012071 doi:10.1088/1757-899X/1019/1/012071

Reducing the gate length without a proportional decrease of the thickness of the wide-gap AlGaN barrier layer of the transistor leads to a significant short-channel effect. Due to the greater than in the AlGaN/GaN heterostructures conduction-band discontinuity and the spontaneous polarization charge density of InAlN/GaN heterostructures, the latter type of heterostructures allows the drain current density to be increased up to 2÷3A/mm, which was confirmed in our experiments. Thanks to this, InAlN-based heterostructures allows to reduce the wide-gap InAlN layer thickness whilst keeping I_{dmax} higher than in AlGaN/GaN heterostructures based HEMTs [23]. As a result, it is possible to compensate for the decrease in the operating voltage of thin barrier layer short-channel transistors by higher maximum current in InAlN/GaN heterostructures and to obtain high output power levels in the millimeter wavelength range [27].

The Ti/Al/Mo/Au-based alloyed ohmic contacts technology, developed for AlGaN/GaN was applied to fabricate contacts for InAlN/GaN heterostructures. This allowed to fabricate low-resistance ohmic contacts ($Rc \leq 0.25\Omega\cdot$mm) on AlGaN/GaN and InAlN/GaN heterostructures along with decreasing the gate-drain distance up to 1.5µm due to the low thickness of contacts metallization (0.12µm) and low surface roughness (RMS≈14nm for AlGaN/GaN and 24nm for InAlN/GaN). The carried out experiments revealed the effect of an electrophysical parameters degradation during rapid thermal annealing at 825°C and above (Figure 4).

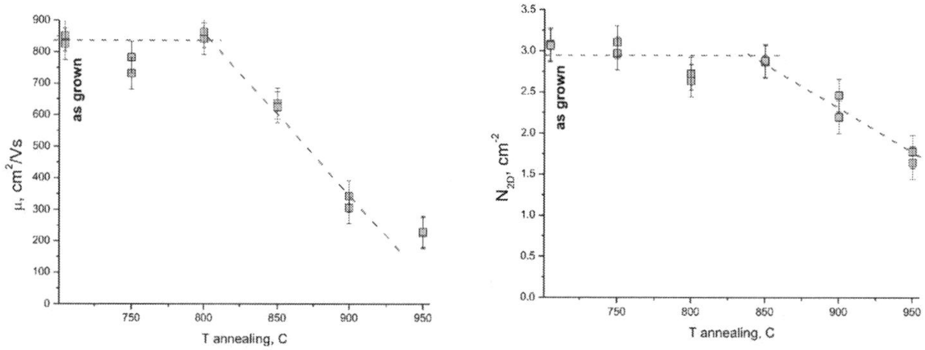

Figure 4. The dependence of carrier mobility (top) and concentration (bottom) in the channel on the annealing temperature (measurements were carried out at room temperature).

InAlN/GaN heterostructures proved out to be sensitive to another process as the gate foot was fabricated. A surface low energy ion treatment during ICP RIE of SiN and an oxygen plasma ashing of photoresist residues may strongly affect the sheet resistance of 2DEG and Ron in fabricated HEMTs. Figure 5 shows the results of InAlN/GaN heterostructure surface treatment. Processing time was 2 minutes. As can be seen from the results, an increase of the plasma power leads to an increase in the semiconductor layer resistance and, consequently, to a decrease of the current in InAlN/GaN-based HEMTs. Therefore, an additional adjustment of the process parameters is required for successful processing of InAlN/GaN-based heterostructures.

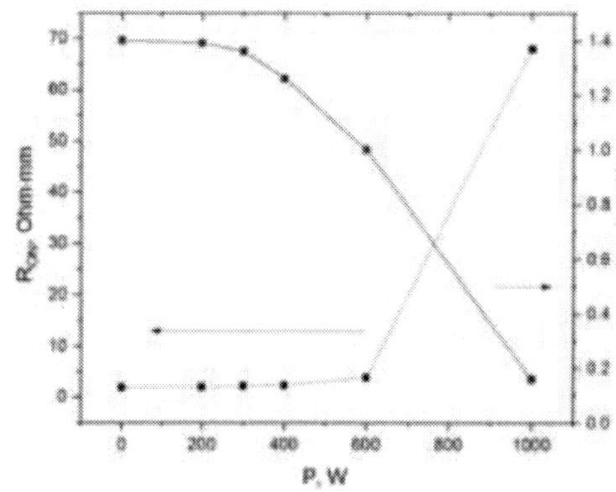

Figure 5. Effect of oxygen plasma surface treatment on InAlN/GaN heterostructure parameters.

Fabricated AlGaN/GaN HEMT and InAlN/GaN HEMT have demonstrated DC characteristics that are in good agreement with their design and 2DEG parameters. AlGaN/GaN HEMT had $I_d \approx 1.0$A/mm, $V_t \approx -3.2$V, $gm \approx 280$mS/mm, while InAlN/GaN HEMT had $I_d \approx 1.3$A/mm, $V_t \approx -1.3$V, $g_m \approx 550$mS/mm. The I-V characteristics of 0.1μm gate length AlGaN/GaN HEMTs shows a low short channel effect which can be attributed to the buffer and T-gate design (Figure 6).

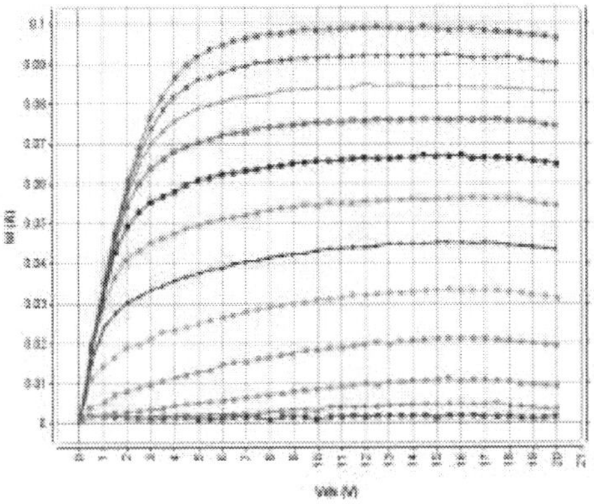

Figure 6. Pulse I-V curves for 0.1x100μm gate length InAlN/GaN HEMT, Vg=+1.5÷-4V.

Comparison of small-signal microwave characteristics of InAlN/GaN and AlGaN/GaN HEMT with a 0.1μm gate length and different source-drain spacing, shown in Figure 7 confirms that, being fabricated in the same process and topology and measured at the same bias and current, AlGaN/GaN HEMT demonstrated a higher gain than InAlN/GaN HEMT. This advantage disappears when AlGaN/GaN HEMT is compared with InAlN/GaN HEMT with a 1μm smaller source-drain spacing, giving similar MaxGain vs frequency curve.

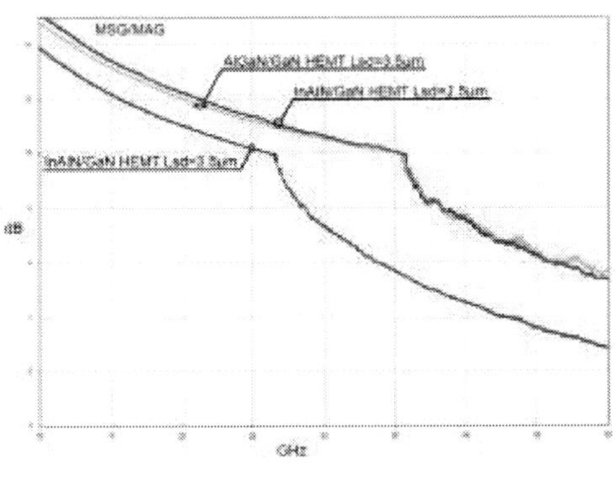

Figure 7. Comparison of MSG/MAG for 0.1μm gate length InAlN/GaN and AlGaN/GaN HEMTs with different source-drain spacing.

A comparison of power characteristics of the devices (Figure 8) using pulsed load pull measurements, carried out at a 17GHz on 100μm gate width for AlGaN/GaN and 200μm gate width for InAlN/GaN devices, showed that both type of devices demonstrated comparable levels of power parameters.

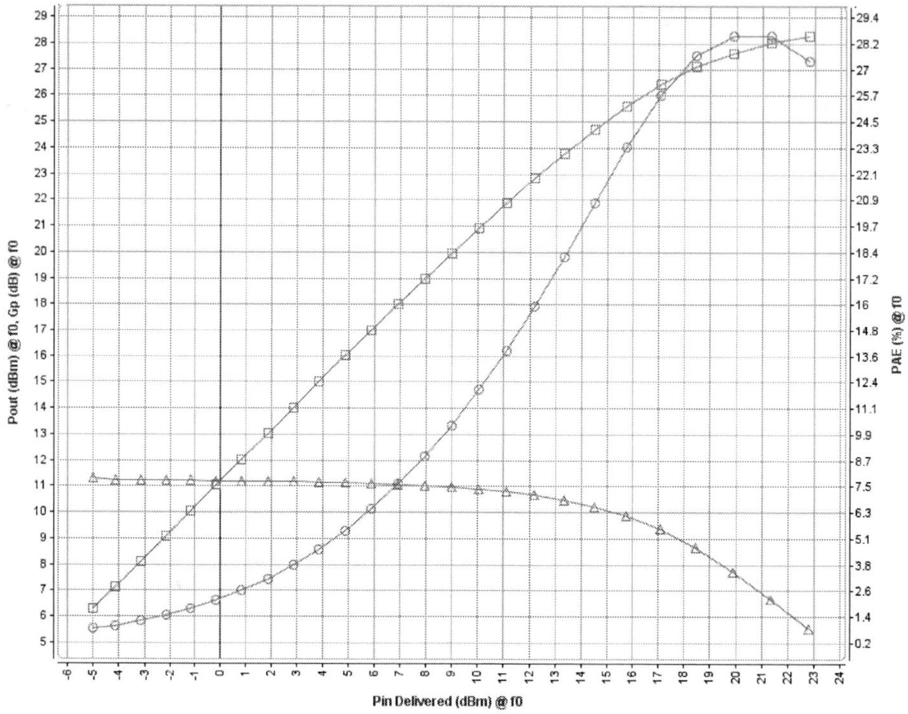

a)

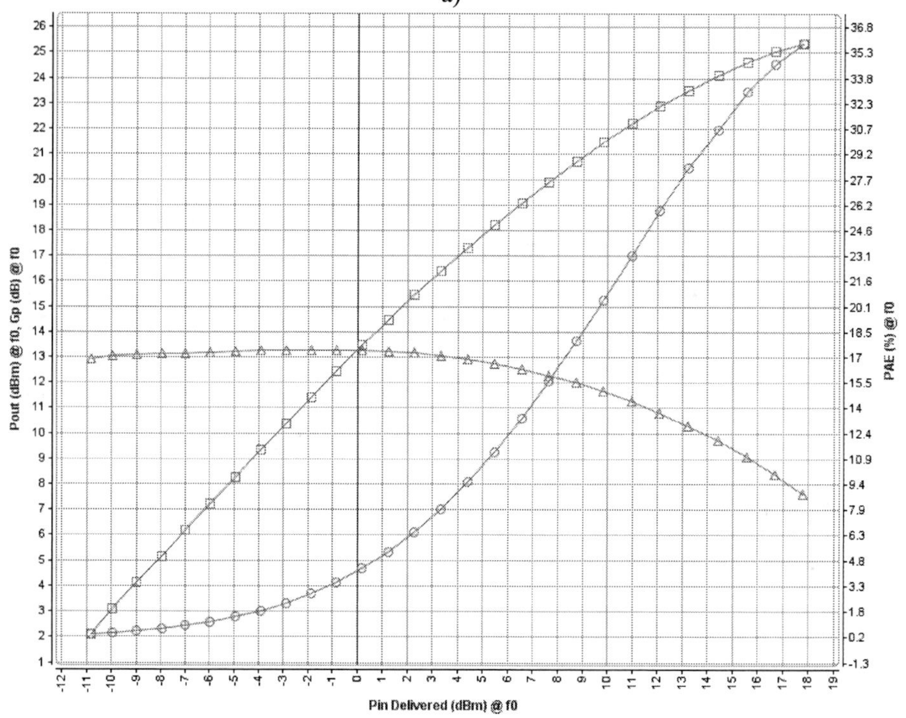

b)

Figure 8. Comparison of loadpull data for 0.1 μm gate length InAlN/GaN (a) and AlGaN/GaN (b) HEMTs.

InAlN/GaN demonstrated a high-power density of 3.5÷4W/mm at a lower V_{ds}=15V supply voltage while the same power level was obtained for AlGaN/GaN devices at V_{ds}=20V. As expected from small signal data, AlGaN/GaN-based transistors have a higher power gain. A lower PAE in InAlN/GaN devices may be attributed to the higher leakage at high drain voltages at pinch-off.

3. Conclusion

The results of the study showed that high quality InAlN/GaN can compete with AlGaN/GaN-based HEMTs for use in millimeter wave range power applications. Using low-resistance alloy contacts in combination with a 0.1μm gate and an optimized design of the buffer layer allows to achieve high microwave parameters required for operation in the selected frequency ranges.

However, further improvement of the RF parameters of the InAlN/GaN transistors is required to reach typical values of small and large signal gain in AlGaN/GaN HEMTs. Besides increasing of 2DEG mobility, an increase in gain can be achieved utilizing short a source-drain and gate length dimensions, non-alloyed regrown ohmic contacts and optimization of the heterostructure. In this case, it will be possible to realize the potential of the InAlN/GaN heterostructure in achieving high values of the output power in the mm-wave frequency ranges.

Acknowledgments

This research is financially supported by the Ministry of Science and Higher Education of the Russian Federation, agreement № 14.578.21.0240 dated 26 September 2017, a unique project identifier УИР RFMEFI 57817X240.

References

[1] Kuzmik J *et al* 2006 *IEEE Trans. Electron Devices* **53** 422–426
[2] Saunier P *et al* 2013 *IEEE Trans. Electron Devices* **60** 3099–3104
[3] Freedsman J J *et al* 2014 *Appl. Phys.Exp.* **7** 104101
[4] Medjdoub F *et al* 2008 *Open Elect. Electron. Eng. J.* **2** 1–7
[5] Gamarra P *et al* 2017 *J. Cryst. Growth* **464** 143–147
[6] Niida Y *et al* 2016 *Proc. of the IEEE Topical Conf. on Power Amplifiers for Wireless and Radio Applications* 24–26 doi: 10.1109/PAWR.2016.7440153
[7] Aubry R *et al* 2016 *IEEE Electron Device Lett.* **37 (5)** 629–632 doi: 10.1109/LED.2016.2540164
[8] Jessen G H *et al* 2007 *IEEE Trans. Electron Devices* **54** 2589–97
[9] Liu B *et al* 2013 *J. Semicond.* **34** 044006
[10] Lu J *et al.* 2014 *Appl. Phys. Lett.* **104** 092107
[11] Han T-C, Zhao H-D, Yang L, and Wang Y 2017, *Chin. Phys. B* **26** 107301
[12] Lee D S, Gao X, Guo S, and Palacios T 2011 *IEEE Electron Device Lett.* **32** 617–619
[13] Lee D S *et al.* 2011 *IEEE Electron Device Lett.* **32** 1525–1527
[14] Killat N *et al.* 2013 *2013 Int. Conf. on Compound Semicond. Manufacturing Technology* New Orleans, LA **4** 113-116
[15] Kabouche R *et al.* 2018 *Proc. of the 13th European Microwave Integrated Circuits Conf.* 5–8
[16] Petitdidier S, Guhel Y, Trolet J L, Mary P, Gaquière C, and Boudart B 2017 *Appl. Phys. Lett.* **110** 163501
[17] Wu Y, Sasangka W A, and del Alamo J A 2017 *IEEE Trans. Electron Devices* **64** 4435–4441
[18] Dutta G, DasGupta N, and DasGupta A 2017 *IEEE Trans. Electron Devices* **64** 3609–3615
[19] Wu Y and del Alamo J A 2016 *IEEE Trans. Electron Devices* **63** 3487–3492
[20] Medjdoub F, Herbecq N, Linge A, and Zegaou M 2015 *IEEE Int. Electron Devices Meeting (IEDM)* 9.2.1-9.2.4
[21] Rocchi M 2017 *IEEE Curran Associates, Inc.* **1** 422
[22] 2015 *Qorvo webinar GaN on SiC* (https://www.everythingrf.com/webinars/details/11-Qorvo-GaN-on-SiC-15-Years-of-Reliability-and-Producibility)

[23] Fujitsu 2017 Achieves World's Highest Output Density with Power Amplifier for W-Band GaN Transmitters, Press release, Fujitsu Laboratories Ltd. Available at: https://www.fujitsu.com/global/about/resources/news/press releases/2017/0724-01.html
[24] Sim P 2018 *Proceedings of TUSUR University*
[25] Dai S et al. 2018 *J. Phys. Appl. Phys.* **51** 035102
[26] Kim J, et al. 2015 *J. Appl. Phys.* **118** 12530
[27] Ohki T et al. 2019 *IEEE Electron Device Lett.* **40** 287–290

IOP Conf. Series: Materials Science and Engineering 1019 (2021) 012072 doi:10.1088/1757-899X/1019/1/012072

The influence of the method of an arc discharge initiation on the product of plasma dynamic synthesis in the W-C system

A A Sivkov, D S Nikitin, I I Shanenkov, A Nassyrbayev and I A Rakhmatullin
School of Energy and Power Engineering, National Research Tomsk Polytechnic University, Tomsk, Russia
E-mail: nikitindmsr@yandex.ru

Abstract. Cubic tungsten carbide WC_{1-x} is distinguished by the difficulty of obtaining in a dispersed and bulk form. The article attempts to plasma dynamic synthesis of cubic tungsten carbide in powder form. The method is based on the use of a coaxial magnetoplasma accelerator for the generating a supersonic plasma jet, which is then rapidly cooled. The possibility of obtaining a powder with a predominant content of WC_{1-x} (up to 95%) is shown. Studying the influence of the method of an arc discharge initiation allowed to reveal the efficiency of using graphitization in comparison with carbon fibers.

1. Introduction

Currently, a particularly pressing problem in the industry is the production of superhard and refractory materials. Carbides of transition d-metals of groups IV-VI have the such properties [1, 2]. Carbides are widely used in the production of materials for work under high loads and temperatures in corrosive environments. One of the notable carbides is tungsten carbide. Tungsten carbide has a high melting point, hardness, wear resistance, chemical resistance [3]. Along with high hardness and melting point, tungsten carbide has a high modulus of elasticity and a low coefficient of thermal expansion than carbides of other metals of the group. The combination of WC properties allows it to be used in the processing of various materials, including as an abrasive material [4].

The dominant phase in the tungsten-carbon system is the hexagonal stoichiometric phase WC. However, according to the state diagram, there are also phases W_2C and WC_{1-x} [5]. The last crystalline modification is characterized by increased electro- and photocatalytic activity [6–9]. The WC_{1-x} cubic phase is known for its narrow range of temperature stability, which makes its production laborious compared to hexagonal phases [10]. The attempts to obtain cubic tungsten carbide in a dispersed and bulk form are known [11–17], however, until now, methods of synthesis of WC_{1-x} are not effective enough. Therefore cubic tungsten carbide WC_{1-x} is the object of active research now. It is known that the cubic phase WC_{1-x} can be synthesized by ultrafast cooling of tungsten carbide melt at a speed of 10^8-10^{11} K/s [18].

Plasma dynamic synthesis of various powder materials and coatings is one of the promising synthesis methods. This method has already been used to obtain carbide compounds in ultrafine form, in particular, silicon carbide [19] and boron carbide [20]. In the present work, plasma dynamic synthesis is used to synthesize cubic tungsten carbide, since it provides the heating to a high temperature required for the formation of this phase and rapid cooling. An important task of the study is to determine the most effective way of initiating an arc discharge, which may influence the precursors sublimation process in the zone of plasma structure formation due to preionization [21].

2. Experimental

The synthesis of ultrafine cubic tungsten carbide was carried out using a high-current pulsed coaxial magnetoplasma accelerator with graphite electrodes (CMPA). The CMPA was powered from a capacitive energy storage device with a capacitor bank capacity of C = 6 mF and a charging voltage of U = 3 kV. Two series of experiments were carried out with different methods for initiating an arc discharge a plasma flow in order to determine the most effective way. Carbon fibers and graphitization were used as an initiating way. Figure 1 shows the assembly of a central electrode using different methods of plasma jet initiation. In the case (*a*) the carbon fibers were stretched from the bottom of the zone of plasma structure formation to the outer cone-shaped surface of the insulator which is the place of contact with the graphite electrode. In case (*b*) a thin layer of graphite aerosol was placed on the inner surface of the zone of plasma structure formation, as well as the surface of the precursor embedded into the zone of plasma structure formation.

A micron tungsten powder with a mass of 0.5 g was used as a precursor, which was placed into the zone of plasma structure formation at the beginning of the accelerator channel. Obtaining carbon was carried out by means of electric erosion when the plasma structure of an arc discharge was moving through a graphite electrode.

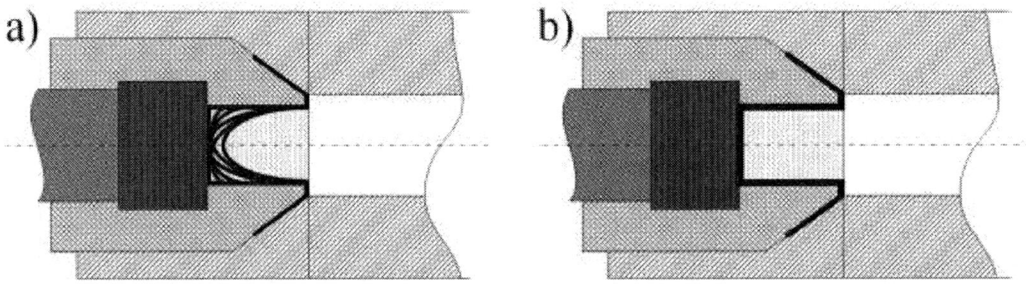

Figure 1. The assembly of a central electrode using different methods of plasma jet initiation by carbon fibers (a) and graphitization (b).

After the power switch was closed, the discharge current of the capacitor battery began to flow through the zone of plasma structure formation, the arc discharge was initiated, and the plasma structure containing the precursors was formed. An electrical erosion of material occurred from walls of a graphite acceleration channel when a plasma flow moved. The plasma flew into the space of the reactor chamber, the synthesized material cooled down and deposited on the walls.

The obtained powder products were analyzed without prior preparation by X-ray diffractometry (XRD) on a Shimadzu XRD7000 diffractometer (Cu-Kα radiation). The quantitative X-ray analysis was carried out by the Rietveld method using the PDF2+ structural data base and the PCW software package. The study of the morphology of the particles and the particle size distribution of the obtained powder was carried out by a transmission electron microscopy (TEM) using a Philips CM 12 microscope with functions of a microdiffraction on a selected area (SAED) and obtaining dark-field images.

3. Results and discussion

During the experiments, it was assumed that the way for initiating an arc discharge affects the phase composition of the synthesis product by changing the nature of the precursor transition to the plasma state. The phase composition of the product is determined by X-ray diffraction. Figure 2 shows the typical XRD patterns of powders obtained by different ways for initiating an arc discharge. The most obvious difference in the XRD patterns is the high intensity peaks of the side phases such as hexagonal carbides and carbon in the case of the product obtained using carbon fibers. The results of the quantitative X-ray diffraction analysis are shown in table 1. The data show that the content of the cubic tungsten carbide is maximum and is about 95% in the product obtained using graphitization. The presence of tungsten carbide W_2C is observed in the product obtained using carbon fiber. In addition to the phase content, the values of coherent scattering areas (CSA) were found, which showed a significant difference in the values of the cubic WC_{1-x} phase from the values of the hexagonal W_2C phase.

Table 1. The results of the quantitative X-ray diffraction analysis.

No.	Phase content, %					CSA	
	WC_{1-x}	WC	W_2C	W	C	WC_{1-x}	W_2C
1	78.4	1.8	18.4	0.8	0.6	35	15
2	94.9	0.4	3.8	0.2	0.7	32	14

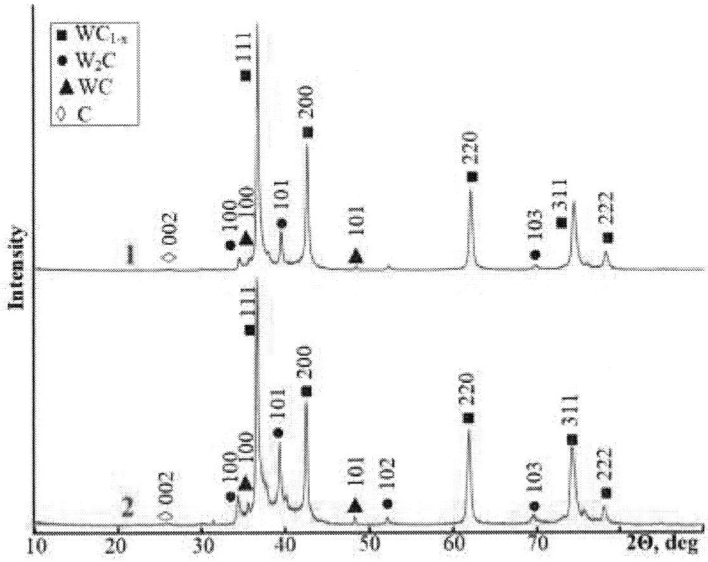

Figure 2. The typical XRD patterns of powders obtained by different ways for initiating an arc discharge: carbon fibers (1) and graphitization (2).

Figure 3*a* shows a bright field TEM image of the product obtained using carbon fibers. The material contains several types of objects. Objects 1 are the largest particles with sizes up to 150 nm. One of such particles in an enlarged form is presented in Figure 3*b*. Analysis of a large number of such objects allowed them to be identified. Figure 4*a* shows a bright-field TEM image of a particle with the corresponding SAED (Figure 4*b*), which is a set of phase reflections of cubic and hexagonal tungsten carbide in the form of Debye rings. The reflexes of the cubic phase have the greatest intensity. When the aperture diaphragm is shifted to the region of tungsten carbide reflexes WC_{1-x}, the reflecting planes of a relative large particle 1 are observed (Figure 4*c*). In addition, there are particles of medium size up to 40 nm in the product (objects 2, Figure 3*b*). In contrast to the sphere-like shape of the particle 1, objects 2 are more pronounced in morphological type and close to hexagons in the plan. It shows the possible belonging of these particles to the hexagonal W_2C phase contained in the product in an amount of ~ 18%. In addition, it is obvious that the large objects 1 and medium particles 2 are in the matrix, which is identified as carbon. The carbon of the synthesized product can also be formed in the form of other forms, including 4 in Figure 3*c*. These are the so-called pineapple-like particles, which are highly deformed particles of graphite [22].

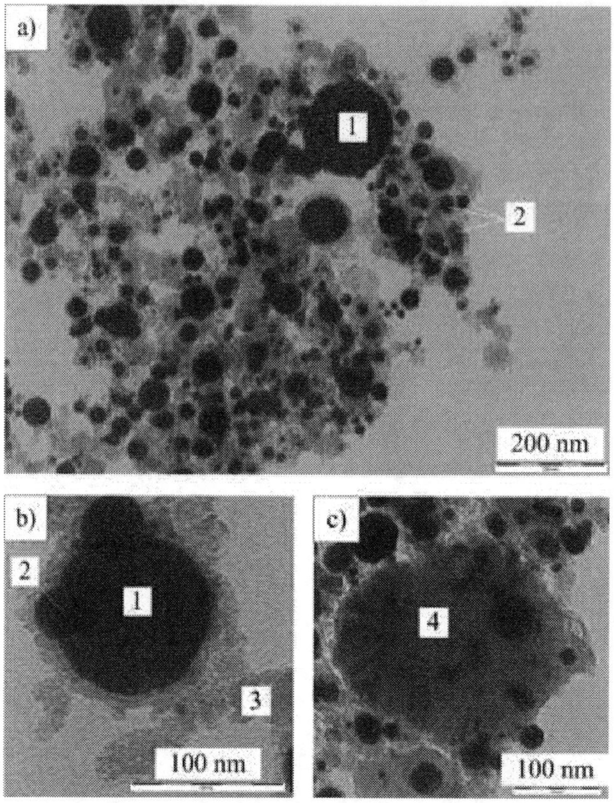

Figure 3. TEM-images of the plasma dynamic synthesis product obtained using carbon fibers: a particle accumulation (a), typical particles (b), carbon structures (c).

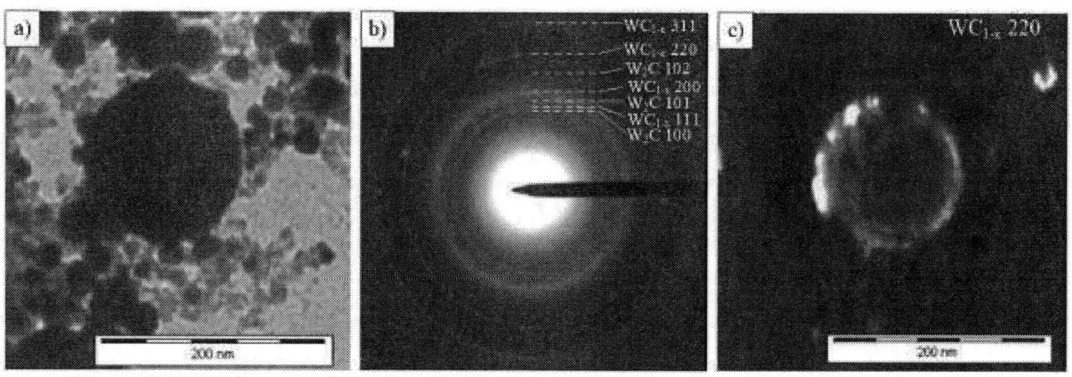

Figure 4. TEM-images of the plasma dynamic synthesis product: bright-field image (a), SAED (b), dark-field image (c).

According to the micrograph of Figure 5, the product obtained using graphitization consists of particles 1 overwhelmingly. A small amount of particles 2 is also contained in the product in the form of smaller particles. An important distinguishing feature is the absence of a carbon matrix, as well as carbon in other morphological forms.

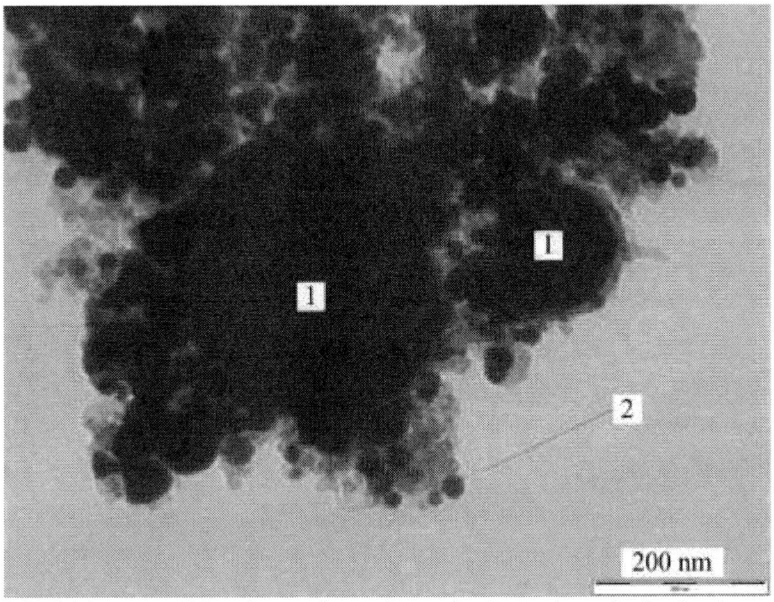

Figure 5. TEM-images of the plasma dynamic synthesis product obtained using graphitization.

According to the combination of X-ray diffractometry and transmission electron microscopy, graphitization is the most effective initiating way of an arc discharge. It can apparently be explained by the physical processes occurring at the initial moment of time. In the case of graphitization of the insulator surface, preionization of the precursor occurs in the zone of plasma structure formation. The activation of the precursor at the preionization stage leads to a smoother entry of precursors into the chemical reaction zone and causes the effectiveness of the chemical reaction and the preferential formation of particles of the cubic phase WC_{1-x}. In the case of using carbon fibers, an electrical explosion of carbon fibers as conductors occurs. It leads to incomplete sublimation of the precursor to the plasma state and the amount of unreacted precursor phases increases in the final product of the synthesis. Also the hexagonal phases of higher and lower carbides (WC and W_2C) are formed. The described process of precursor sublimation in various cases of initiating way of an arc discharge is close to the results of work describing processes in the SiC carbide system [15].

4. Conclusion

The paper discusses the plasma dynamic synthesis in the W-C system. The ways of initiating an arc discharge through the use of carbon fibers and graphitization are considered. The results of analytical studies show that unique phase of cubic tungsten carbide WC_{1-x} is formed as nanoscale particles with sizes up to 150 nm. The most effective way of initiating an arc discharge is the use of graphitization. In this case, the maximum yield of the cubic tungsten carbide WC_{1-x} is up to 95%.

Acknowledgments
The research is carried out at Tomsk Polytechnic University within the framework of Tomsk Polytechnic University Competitiveness Enhancement Program grant and supported by the Russian Science Foundation (project no. 19-13-00120).

References

[1] Pierson H 1996 *Handbook of Refractory Carbides & Nitrides: Properties, Characteristics, Processing and Apps* (New Jersey: Noyes Publications) p 339
[2] Naboychenko S, Murashova I, Neikov O 2009 *Handbook of Non-Ferrous Metal Powders* (Amsterdam: Elsevier) p 995

[3] Kurlov A, Gusev A 2006 *Inorg. Mater* **42** (2) 121–127
[4] Bondarenko V, Andreyev I, Savchuk I. 2013 *Int. J. Refract. Met. Hard Mater* **39** 18–31
[5] Kurlov A, Gusev A 2006 *Russ. Chem. Rev.* **75** (7) 617–636
[6] Kim J, Jang J-H, Lee Y-H 2009 *J. Power Sources* **193** (2) 441–446
[7] Zheng H, Yu A, Ma C 2011 *Chinese Chem. Lett.* **22** (4) 497–500
[8] Bott-Neto J, Beck W, Varanda L 2017 *Int. J. Hydrogen Energy* **42** (32) 20677–20688
[9] Wei X and Li N 2019 *Appl. Surf. Sci.* **463** 1154–1160
[10] Abad M, Muñoz-Márquez M, Mrabet S 2010 *Surf. Coatings Technol.* **204** (21-22) 3490–3500
[11] Gao Y, Song X, Liu X 2013 *Scr. Mater.* **68** (2) 108–110
[12] Carim A, De Jong A, Houdy P 1989 *Thin Solid Films* **176** (2) 177–182
[13] Raekelboom E, Abdelouahdi K, Legrand-Buscema C 2009 *Thin Solid Films* **517** (5) 1555–1558
[14] Reddy K, Rao T, Joardar J 2011 *Mater. Chem. Phys.* **128**(1-2) 121–126
[15] Fenggang Z 2019 *Vacuum* **159** 254–60
[16] Du S, Wen M, Ren P 2017 *Mater. Sci. Forum* **898** 1505–1511
[17] Pawlak W, Kubiak K, Wendler B 2015 *Tribol. Int.* **82** 400–406
[18] Zhang F, Zhu X, Lei M 2017 *Vacuum* **137** 119–124
[19] Sivkov A, Nikitin D, Shanenkov I 2019 *Int. J. Refract. Met. Hard Mater.* **79** 123–130
[20] Sivkov A, Rakhmatullin I, Shanenkov I 2019 *Int. J. Refract. Met. Hard Mater.* **78** 85–91
[21] Kuzenov V, Polozova T, Ryzhkov S 2015 *Probl. At. Sci. Technol.* **98** (4) 49–52
[22] Sivkov A, Pak A, Nikitin D 2013 *Nanotechnologies Russ.* **8** (7-8) 489–494

A computational fluid dynamic approach to validate the wind effect in air pollution of Ulaanbaatar city

B Bulgansaikhan, V Ulziibayar and Ts Munkherdene
School of Applied Science, Mongolian University of Science and Technology
Ulaanbaatar, Mongolia
E-mail: bulgansaikhan@must.edu.mn

Abstract. In the past two decades, inhabitants of Ulaanbaatar city have been increased more than three times. The Air and soil of Ulaanbaatar city contaminated seriously due to the densely populated behavior. We proposed to develop a massive computational model for pollution, which cares about contamination factors. This paper is a part of the purpose. We consider the filtered equation express conservation of mass and momentum in a Newtonian incompressible flow in a computational fluid dynamic model for Ulaanbaatar's wind field. The terrain around the city has been built for numerical computation. The finite element method was used for numerical analysis, from the results, we can characterize the wind field.

1. Introduction

Nowadays, the rapid climate change in the world is a global environmental problem. Thus, every country has been faced with many difficult situations in ecology and economy. With this global reason, most cities have severely deficient air condition problems, especially in developing countries. Mongolia's capital Ulaanbaatar (UB), is one of the world's most polluted towns. Its" harmful dust" is 6–7 times higher than the most lenient World Health Organization standards, [1]. Several factors can be added to the source of UB's air pollution as dust from both insides, outside, partially covered soil surface, less horticulture, smoke from a coal stove, and so on.

Nevertheless, one of the main factors of air pollution is the geographical location of UB. The UB city is built between big mountain [2] , which is the town's unique behavior. Figure 1 shows a real wind field layers and air pollution during wintertime, b denoted surrounding big mountains of the city. It has been a long time to use the wind field's computational fluid dynamics (CFD) for pollution dispersion. Thus, massive studies have been carried on the distribution of air pollution with complex models, including buildings of city and terrain of it [3–8]. Early work was mostly focused on theoretical and numerical aspects in the numerical study: the finite difference method (FDM) was used commonly for less computational time. More precisely, the second-order central difference scheme in space is used to demonstrate the wind flow in the model, which is the obstacles approximated into rectangular shapes [9, 10] and [11]. However, FDM is mostly used for comfortable computing time in computational fluid dynamics (CFD). The method is fragile in modeling specific and complex geometry such as mountains or modern designed buildings.

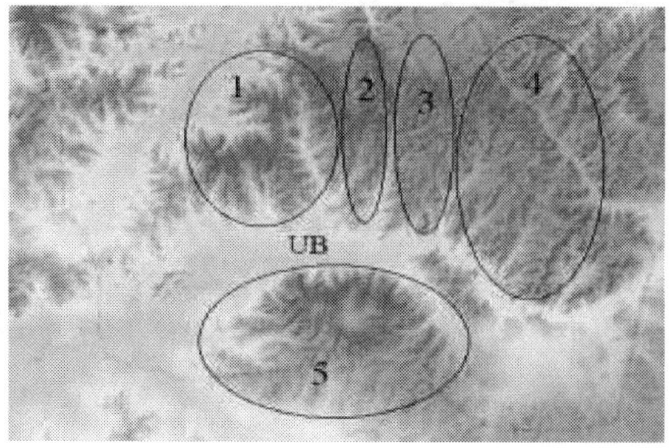

Figure 1. Air pollution and atlas of Ulaanbaatar: a) the air pollution of UB in the winter time; b) elevation map of UB, mountains in the area of 1, 2, 3, 4 and 5 named as Songino Khairhan, Chingeltei, Sukhbaatar, Bayanzurkh, and Bogd Uul. The top of them around 1670m, 1890m, 1981m, 1934m and 2268m, respectively.

Therefore, at some time with FDM, the finite element method (FEM) was developed and mainly used in CFD [12], [13] and [14]. Although FEM requires a particular time for convergence, to avoid the disadvantage, the parallel computation has been developed for large predict flow field and pollutant dispersion over tricky obstacles. This paper is part of massive US air pollution; we will investigate pollution dispersion from the soil and temperature inversion. To do that, we develop a computational fluid dynamic model in the large eddy domain, which can be coupled with the equation of temperature and with particle tracking motion that considered pm type mass. The first approach is considered in this paper as the numerical computation of the wind field in three-dimensional space, based on a finite element model used with an adaptively distributed unstructured mesh for simulation with boundary data as a condition of boundary that simulated to real averaged wind information. It has been noticed intensively in many official reports and papers that the primary source of UB's air pollution is smoke from suburbs, which does not connect to the water pipe system and uses coal for heat. In Figure 3*b* we denote the suburb areas as the red and blue area is the UB city's full area. But real measurement proves that air pollution includes many PM type particles from the soil, and frequent inversion of temperatures must be considered.

2. Model problem

We assume that the observing large domain $\boldsymbol{\Omega}$ (Figure 2) is completely occupied with three-dimensional space R^3 and let the wind velocity potential $\mathbf{u}$ has u, v, w components on Cartesian coordinates. Then following unsteady governing equation holds for large eddies that interpolate the velocity components of u along the direction of the flow in low pass.

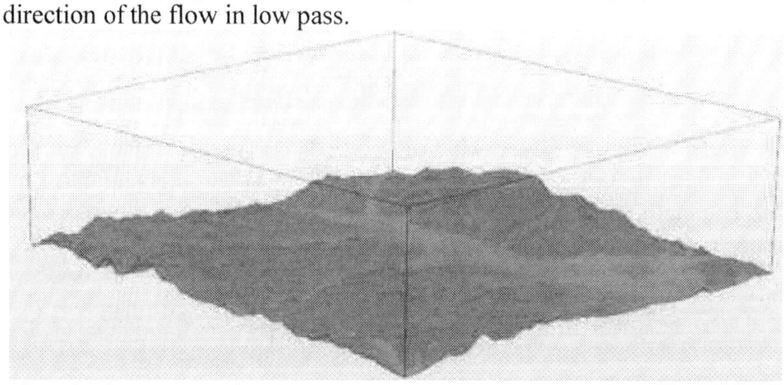

Figure 2. The domain for computation in Comsol Multiphysics, it is a three-dimensional subspace of R^3.

The filtered equation express conservation of mass and momentum in a Newtonian incompressible flow written as in conservative form combined with $k-\omega$ and $k-\varepsilon$ turbulence models as

$$\rho\frac{\partial u}{\partial t}+\rho(\mathbf{u}\cdot\nabla)\mathbf{u}=\boldsymbol{\nabla}\cdot[-p\boldsymbol{I}+\mu(\nabla\mathbf{u}+\nabla\boldsymbol{u}^{\mathrm{T}})]+\mathrm{F} \quad (1)$$

$$\rho\nabla\cdot\mathbf{u}=0 \quad (2)$$

$$\rho\frac{\partial k}{\partial t}+\rho(\mathbf{u}\cdot\nabla)k=\boldsymbol{\nabla}\cdot[\mu+\mu_T\sigma_k\nabla k]+F_k \quad (3)$$

$$\rho\frac{\partial\omega}{\partial t}+\rho(\mathbf{u}\cdot\nabla)\omega=\boldsymbol{\nabla}\cdot[\mu+\mu_T\sigma_\omega\nabla\omega]+F_\omega \quad (4)$$

where ρ is the density of dry air; p is filtered pressure; μ is the viscosity; $(\nabla\mathbf{u}+\nabla\boldsymbol{u}^{\mathrm{T}})$ is scaled strain rate tensor; the variables are defined as $F_k=P-\beta^*\rho\omega k, F_\omega=\frac{\gamma}{v_l}P-\beta\rho\omega^2+2(1-F_1)\frac{\rho\sigma_\omega}{\omega}\nabla k\cdot\nabla\omega$. The turbulence model parameters are can be founded in [15].

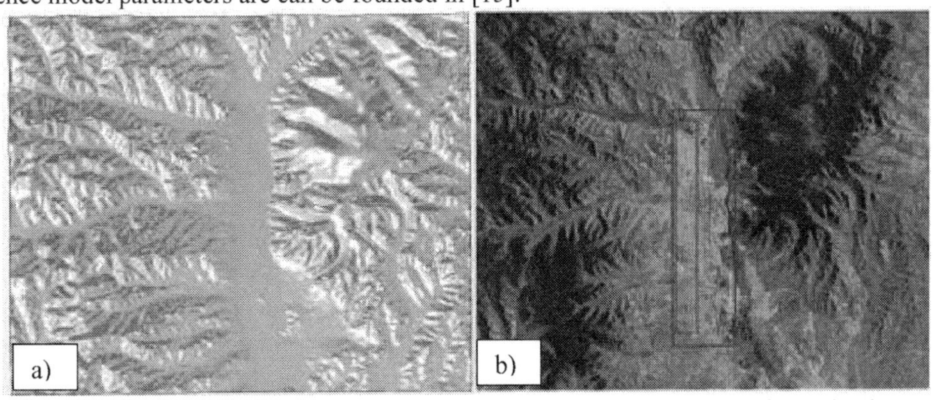

Figure 3. Environment of UB: a) the computational domain of environments of UB city; b) a real 2D view of UB city, the red square is a suburb that does not connect to the main pipe system and uses coal for heat. The area of UB city denoted as a blue square, Comsol Multiphysics.

The computational domain is interpolated with exact behavior real landscape shown Figure 3 and scaled as $10^4m\times 10^4m\times 10^4m$. The geometry is approximated from the information of topography that created from the official source. We consider following boundary conditions to the governing equations. Since the domain is computationally large the top and two besides are set to be non-slipping wall:

$$\mathbf{u}=0, k=0 \text{ and } \omega=\lim_{l_\omega\to 0}\frac{6\mu}{\rho\beta_1 l_\omega^2} \quad (5)$$

In all four seasons, the average UB's wind speed is $6m/s$ and mostly directed from northwest to southeast. So the inlet wind flow boundary condition defined as

$$\mathbf{u}=-U_0\mathbf{n}, U_0=U_{ref}, k=\frac{2}{3}\left(\frac{u_0}{T}\right)^2 \quad (6)$$

$$\omega=\frac{\sqrt{k}}{(\beta^*)^{1/4}L_T} \quad (7)$$

In outlet wind flow the pressure should be free with free divergence for turbulence,

$$[-p\boldsymbol{I}+(\mu+\mu_{\mathrm{T}})(\nabla\mathbf{u}+\nabla\boldsymbol{u}^{\mathrm{T}})]\mathbf{n}=-p_0\mathbf{n} \quad (8)$$

$$\nabla k\cdot\mathbf{n}=0, \nabla\omega\cdot\mathbf{n}=0 \quad (9)$$

where U_{ref} and p_0 are the reference wind flow and pressure, for other turbulence parameters we refer to see [15].

IOP Conf. Series: Materials Science and Engineering 1019 (2021) 012073 doi:10.1088/1757-899X/1019/1/012073

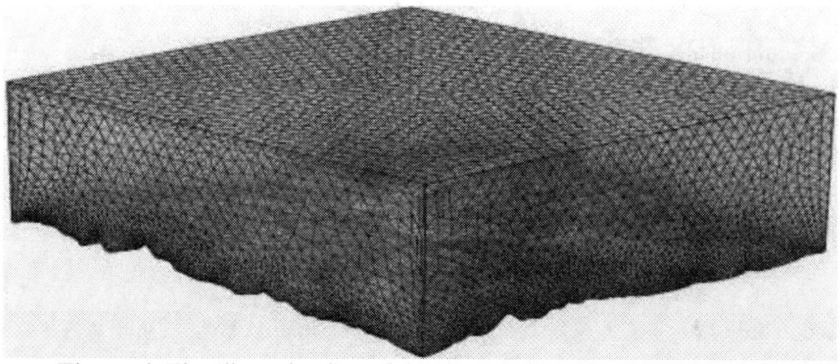

Figure 4. The discretization of the domain into finite tetrahedrons.

3. Experiment and results

In the experimental domain, we set all surrounding five mountains described in Figure 4, the square area (Figure 3*a*). We normally distributed free mesh (tetrahedrons), 205808 number of elements and 19414 node points for computation—the wind inflow and laminar conditions are considered Dirichlet and Neumann boundaries. Figure 5*a*) is the three-dimensional solution of the governing equations defined in (1). The convergence rate is acceptable but not direct as FDM; from the answer, we can visualize many different analysis kinds. For instance, in Figure 5*b*), we show the isosurface images that illustrate the wind field dispersion. Figure 6 shows a cross-sectional view of the solution in streamlines and contour plots that gives information about turbulence and wind field layers.

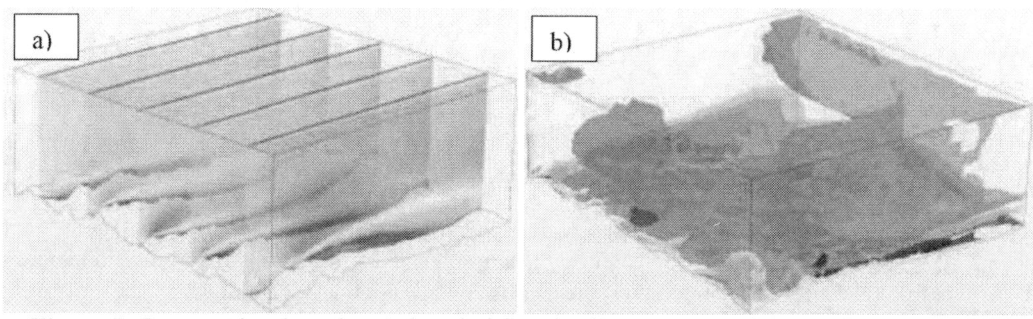

Figure 5. Computational results: a) the wind flow in slice images b) the isosurface graphics in six different levels.

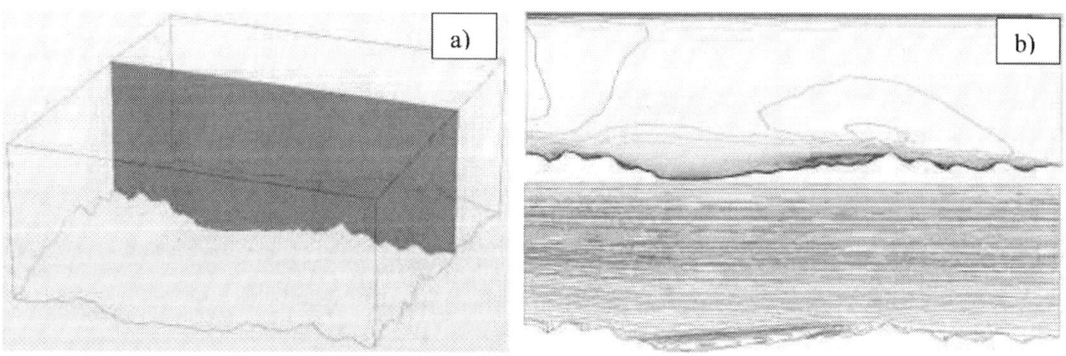

Figure 6. The cross-sectional view of the domain is described in a), contour and streamline representation shown in b).

4. Conclusion

The sizeable computational fluid dynamic model-based numerical setting has been considered. The purpose of the research is completed, and some more management is needing in future work. The computed distribution is used to couple with the thermal flow. From the coupled system PM sized particle motion can be calculated. Some more additional conditions need to consider the smoke distributions of suburban areas. Also, the large eddy simulation would be simple to analyze based on these results.

Reference

[1] WB R 2012 Curbing Air Pollution in Mongolia's Capital *world bank* 1–3
[2] Ulaanbaatar S I Patgaon Map India Google Satellite Maps
[3] Cimorelli A J *et al* 2004 *AERMOD: DESCRIPTION OF MODEL FORMULATION*. Available at: https://www3.epa.gov/scram001/7thconf/aermod/aermod_mfd.pdf
[4] Britter R E and Hanna S R 2003 *Annu. Rev. Fluid Mech.* **35** 469–496
[5] Dobre A *et al* 2005 *Atmos. Environ.* **39** 4647–4657
[6] Kim B-G *et al* 2011 *Boundary-Layer Meteorol.* **139** 457–485
[7] Chavez M, Stathopoulos T and Bahloul A 2012 *Proceeding of seventh international colloquium on bluff body aerodynamics and applications*
[8] Flores F, Garreaud R and Muñoz R C 2013 *Comput. Fluids* **82** 1–13
[9] Tominaga Y and Stathopoulos T 2010 *Build. Environ.* **45** 2231–2239
[10] Tominaga Y and Stathopoulos T 2011 *J. Wind Eng. Ind. Aerodyn.* **99** 340–348
[11] Moon K, Hwang J-M, Kim B-G, Lee C and Choi J 2014 *Environ. Fluid Mech.* **14** 1381–1403
[12] Kawahara M and Hirano H 1983 *Int. J. Numer. Methods Fluids* **3** 137–163
[13] Mccallen R C, Komblum B T and Kollman W 1997 *Proceeding of the 1997 American society of mechanical engineers fluids engineering division summer meeting vancouver* 474666
[14] Ciarlet P G 2002 *The Finite Element Method for Elliptic Problems* (Philadelphia:Society for Industrial and Applied Mathematics)
[15] Wilcox D C 2008 *AIAA J.* **46** 2823–2838

IOP Conf. Series: Materials Science and Engineering 1019 (2021) 012074 doi:10.1088/1757-899X/1019/1/012074

Novel DC arc plasma method for super dispersed metal carbides synthesis

A Ya Pak, G Ya Mamontov and Yu Z Vassilyeva
School of Energy & Power Engineering, Tomsk Polytechnic University, Tomsk, Russia
E-mail: ayapak@tpu.ru

Abstract. This paper presents the experimental results of metal carbides synthesis by DC arc discharge plasma in ambient air. The results indicate that the synthesis takes place in the atmosphere of CO and CO_2, which are generated during the DC arcing process on graphite electrodes and that mixture of CO and CO_2 prevents the reaction zone from air oxygen. The paper introduces experimentally substantiated arguments about the possibility of obtaining metal carbides by the developed arc discharge method. By the moment, the authors have already obtained silicon carbide and titanium carbide in this way.

1. Introduction

Metals and non-metals carbides play a significant role in the scientific and technical development for over 100 years [1]. The scientific community pays much attention to some carbides due to their outstanding properties, such as superhardness, refractoriness, wear resistance, resistance to aggressive media, catalytic activity, etc. In this regard, the number of publications devoted to the carbides synthesis and study of their properties has increased significantly over the past 20 years (Figure 1). The most studied carbides are silicon, boron, tungsten, and titanium carbides, while at the same time more rare carbides are known, such as platinum, palladium, yttrium, uranium, beryllium, cerium, ruthenium carbides, etc. To date, carbides are usually prepared by carbothermic reduction of oxides in furnaces of various types: generation of electron and ion beams, generation of electric discharge and microwave plasma, chemical vapor deposition (CVD), laser ablation, etc. All existing methods are generally implemented in inert gas or liquid, preventing the synthesizing product from air oxidation. At the same time, in the last 5–8 years, there have been some papers in the scientific literature about the possibility of synthesizing carbon nanotubes and graphene by electric arc plasma in the air [2–4], including ambient air [5, 6]. This approach allows significantly simplify the design of electric arc plasma reactors and reduce the cost of carbon nanostructures synthesis [4]. By this moment there is the substantial experience of carbon nanotubes electric arc synthesis by such approach [7]. However, the metal carbides synthesis by electric arc plasma in ambient air practically has not been studied. In particular, the following issues are not resolved: a list of materials, which can be synthesized by such method; an actual gas composition in a reaction zone; operating parameters of a discharge circuit and current-voltage characteristics of arc discharge, which are required to obtain the desired material. Experimental studies with control of electrical parameters of the discharge circuit, the electrodes temperature, and the gas atmosphere can only partially solve the above-mentioned issues.

This paper represents the experimental studies showing the possibility of metal carbides synthesis by DC arc discharge plasma in ambient air. In this case, a DC arc discharge is initiated in the cavity of a graphite crucible, into which a mixture of metal and graphite is placed. The qualitative assessment of the gas composition during the experiments is presented. This result showed that, in fact, synthesis takes place mainly in the atmosphere of CO and CO_2. The data presented in conjunction with our previous papers [8–10] suggest that the synthesis of metal carbides by the non-vacuum method being developed is possible.

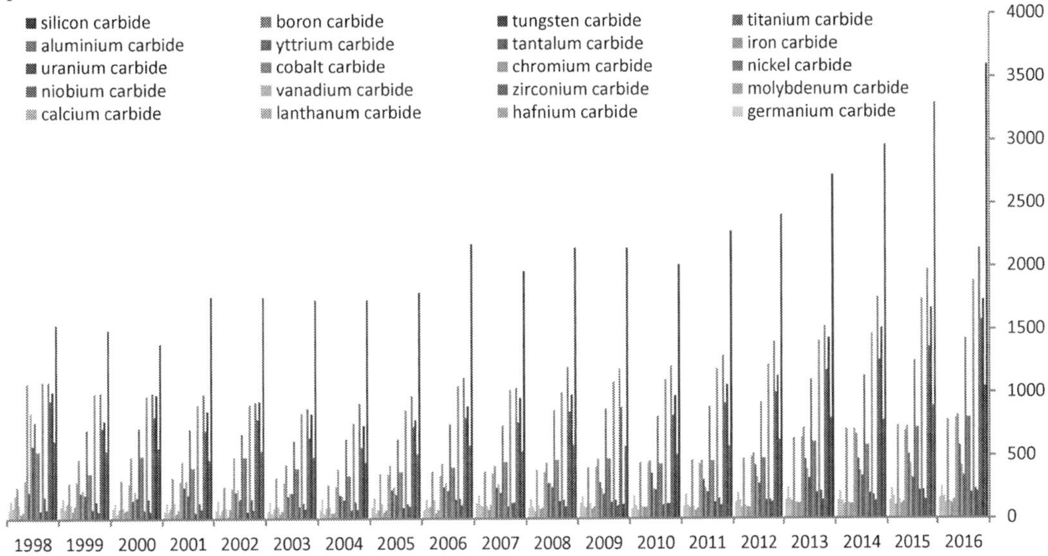

Figure 1. Dynamics of publication activity on metals and non-metals carbides according to sciencedirect.com.

2. Experimental methods

Experiments were carried out using a DC arc discharge reactor (Figure 2) that was created by the authors of this paper [8]. It included graphite electrodes connected to the DC power source. The power source consists of a step-up frequency converter, a step-down transformer, and controlled rectifier. The anode was a graphite cylindrical rod; the cathode was a graphite crucible, filled with a powder mixture of some metal and graphite. In this series of experiments, graphite rods with a diameter of 8 mm were used as anodes; graphite crucibles with a diameter of 30 mm and a height of 30 mm – as the cathodes. The anode and cathode were weighed before and after each experiment. The arc discharge gap was controlled during the experiment by a linear electric drive based on a stepper motor, which was operated by the driver and controller. The arc current and the voltage on the electrodes were recorded using a two-channel oscilloscope through an ohm voltage divider and a current sensor (Hall Effect sensor). The electrodes temperature was controlled by a pyrometer. The gas composition was determined by a gas analyzer, the probe of which was located directly above the arc zone at a distance of about 10 cm. The arcing was generated by a voltage of 30–35 V with DC of 160–170 A.

The general principles of operation of the electric arc reactor being developed and the possibility of synthesizing silicon carbide and titanium carbide by DC arc discharge plasma in ambient air have already been discussed previously [9,10].

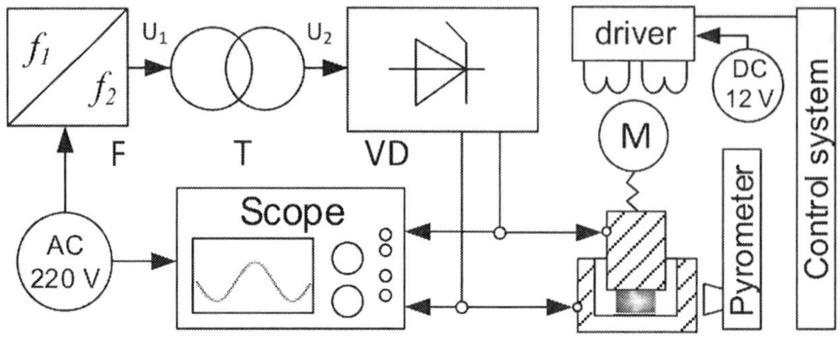

Figure 2. Schematic diagram of the experimental setup.

3. Results and discussion

Figure 3 shows the detected dependencies as a result of a series of experiments in which the duration of arc discharge was varied with constant parameters of the discharge circuit and the current-voltage characteristics. As can be seen from Figure 3*a*, the arc energy grows with increasing synthesis time almost linearly. The energy release lead to the temperature rise in the system, including on a cathode surface (Figure 3*b*). As is evident, the temperature of the cathode surface increases to approximately ~ 1100 °C - 1200 °C over a 20-second interval of the reactor operation. A time of ~15-20 seconds should be considered as the limit for the operation of the reactor without a cooling system since the graphite cathode begins to ignite in the open air. The amount of energy, released during this time in the system, is up to ~ 60-100 kJ.

Figure 3 also presents the dependencies between the anode mass loss (Figure 3*c*) and the mass of cathode deposit (Figure 3*d*), formed during the arcing process, with the synthesis time. The process of mass transferring from anode to cathode with the formation of a cathode deposit is known [7]. In the considered series of experiments, it is clear that as the duration of the synthesis increases, both the anode mass loss increases and the mass of the formed cathode deposit increases, which is logical. In this case, we can see that the anode mass loss is greater than the cathode deposit mass. It indicates that the part of the anode mass is expended on a gaseous product formation.

Gases, formed during the operation of a non-vacuum electric arc reactor, were determined by the gas analyzer. The results of gas analysis should be considered qualitative due to the lack of a hermetic reactor as part of the experimental setup. Nevertheless, the obtained results seem logical.

The formation of CO and CO_2 is uniquely identified as a result of the arc discharge initiation. These gases are products of graphite electrodes combustion when arc discharge initiating and maintaining. As can be seen from Figure 4, the gas analyzer recorded the release of CO and CO_2 gases at different points in time. Unfortunately, we could not synchronize the time of the gas appearance with oscillograms of the reactor operating mode, since it takes some time for the gas to path from the probe to the detector. Nevertheless, the difference in the appearance of CO and CO_2 gases is present in all experiments: first appears CO_2, then CO. At first, oxygen reacts with carbon atoms to form CO_2, and then CO_2 reacts with carbon atoms to form CO [11]. A similar phenomenon is also described in [12] when discussing the four stages of the electric arc process for the silicon carbide synthesis. The synthesis of silicon carbide takes place in a reactor, filled with He, with the CO and CO_2 formation when treated silicon dioxide SiO_2 with carbon plasma. In addition to CO and CO_2, the release of a small amount of hydrogen H_2 was also recorded, which generation often accompanies plasma processes [13,14], and specifically, in this case, it may be the result of the presence of atmospheric humidity and adsorbed moisture in the consumables used. Hydrogen, in turn, is actively used as the atmosphere for electric arc reactors (for example, in [15]), respectively, its presence or absence in the system under consideration is not critical.

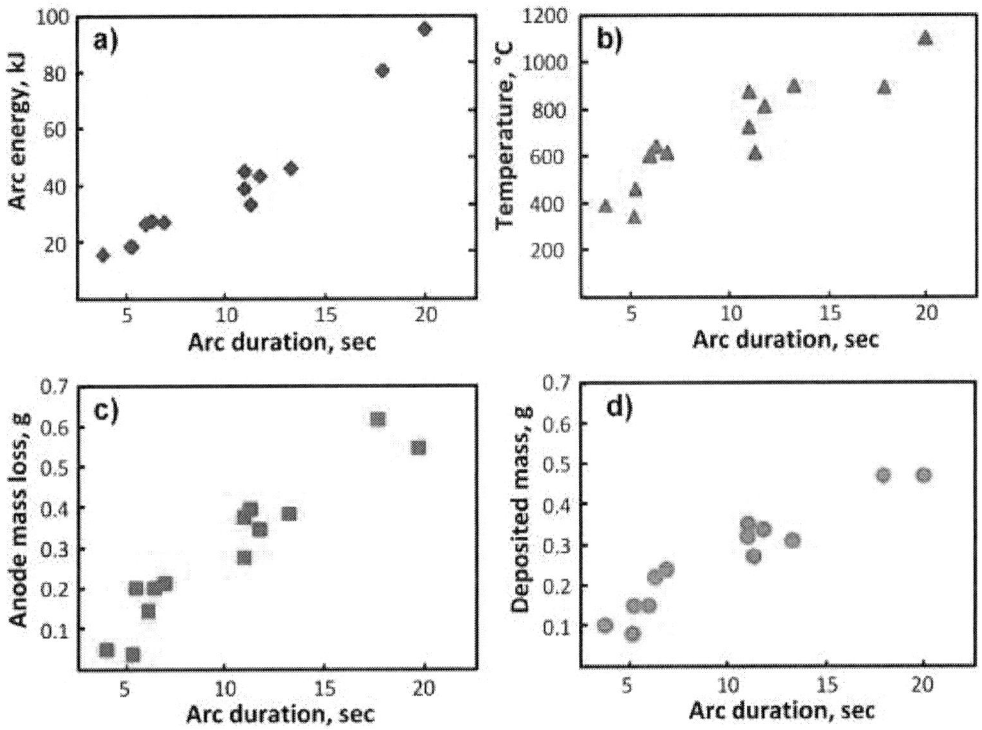

Figure 3. The dependencies between the arc energy (a), the temperature of cathode surface (b), the anode mass loss (c), the cathode deposit mass (d) with the arc duration in a series of experiments.

The obtained result is verified by known data. The possibility of carbon nanotubes synthesis by the arc method in the CO atmosphere is proved [16]. It is also known, that the carbon nanohorns synthesis is possible both in the CO, CO_2 atmosphere, and in ambient air, due to the CO generation [11]. Thus, it is safe to conclude that non-oxide materials can be synthesized by DC arc discharge plasma on graphite electrodes by creating the atmosphere of CO and CO_2. In addition, according to the literature data, oxygen in SiO_2 composition is replaced by carbon atoms in the synthesis process, oxygen combines with carbon and the synthesis of CO and CO_2 occurs during the SiC synthesis from SiO_2 by DC arc discharge plasma in He atmosphere [12]. So, it means the atmosphere in the synthesis process mainly consists of CO and CO_2, both in the case of arc discharge initiation on graphite electrodes in air and inert gases (in the case of the oxygen's presence in the precursor). This means that the presence of a sealed chamber in the reactor is not critical for the synthesis of powder materials by DC arc discharge plasma initiated between graphite electrodes. Since, in any case, oxygen will be in a bound state in the composition of CO and CO_2 during the synthesis process

This principle made it possible to realize the synthesis of silicon carbide and titanium carbide [8–10] by DC arc discharge plasma, initiated on graphite electrodes, in ambient air. This proves the consistency of the above theses. Moreover, according to the well-known state diagrams [17], many metal carbides are formed under relatively similar conditions (in particular, in the temperature range from 1000 °C to 4200 °C). Many materials have already been obtained by the traditional electric arc method in a sealed reactor in the inert atmosphere. Taking into account the carried-out research, it seems possible to synthesize such materials by the non-vacuum method. Accordingly, in addition to the silicon carbide and titanium carbide, that have already been synthesized by the non-vacuum method being developed [8–10], it is also possible to synthesize tungsten, molybdenum, lanthanum, yttrium, zirconium, chromium, and other metals carbides. Thus, the task of future research is to synthesize of the mentioned-above carbides by the non-vacuum method. As evidence of the above assumptions, in the papers [8–10] the results of X-ray diffractometry and electron microscopy are presented. The results in these works indicate that the titanium carbide and the silicon carbide were synthesized and that there was not silicon and titanium oxide in the product.

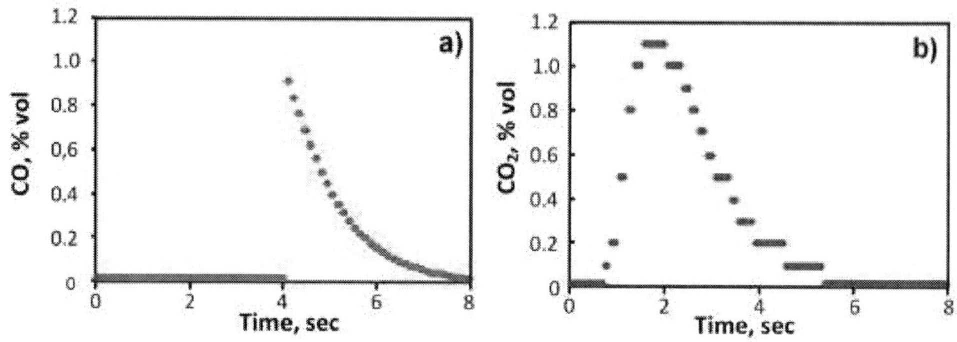

Figure 4. Typical changes in the concentration of carbon monoxide (a) and carbon dioxide (b) during the experiment.

4. Conclusion

The paper presents systematically collected data for the process of electric arc synthesis of powder materials by DC arc discharge plasma in ambient air. As part of the non-vacuum method development, the assessment of the actual gas atmosphere was carried out. For the first time, it was experimentally substantiated that the electric arc synthesis of metal carbides can be implemented in ambient air without a sealed chamber and vacuum equipment. The synthesis process is possible as a result of CO and CO_2 generation during the arcing process on graphite electrodes.

Acknowledgment

This work was supported by the Grant of the President of the Russian Federation for young PhD holders (project No. MK-633.2019) and by Ministry of Science and Higher Education of Russian Federation (projects No. FSWW-0022-2020).

References

[1] Adachi G, Imanaka N, Fuzhong Z 1991 *Handbook on the Physics and Chemistry of Rare Earths* (Amsterdam: Elsevier Science Publishers B V)
[2] Su Y 2013 *Mater Res Bull.* **48** 3232–3235
[3] Kim H H and Kim H J 2006 *Mater Sci Eng B* **133** 241–244
[4] Wang Z 2010 *Nanotechnology* **21** 17
[5] Joshi R 2008 *Diamond Relat Mater* **17** 913–919
[6] Berkmans J A 2015 *Diamond Relat Mater* **55** 12–15
[7] Arora N and Sharma N N 2014 *Diamond Relat Mater* **50** 135–150
[8] Pak A Ya 2019 The setup for titanium carbide synthesis Patent RF
[9] Pak A Ya 2018 *J Superhard Mater* **40** 157–163
[10] Pak A Ya and Mamontov G Ya 2018 *Tech Phys Lett* **44** 615–617
[11] Li N 2010 S *Carbon* **48** 1580–1585
[12] Chiu S, Huang C and Li Y 2007 *J Phys Chem C* **111** 10294–10297
[13] Du C 2015 *Int J Hydrogen Energy* **40** 12634–12649
[14] Wu A 2015 *Int J Hydrogen Energy* **40** 9039–9048
[15] Guo G F 2013 *Surf Coat Technol* **228** S120–S125
[16] Su Y 2011 *Appl Surf Sci* **257** 3123–3127
[17] Landolt H and Bornstein R 2016 *Phase Equilibria, Crystallographic and Thermodynamic Data of Binary Alloys • Ac-Au – Au-Zr* (Berlin: Springer)

IOP Conf. Series: Materials Science and Engineering 1019 (2021) 012075 doi:10.1088/1757-899X/1019/1/012075

The approach to construction and discretization of complex 3D models of geological media in software for airborne geophysical surveys maintenance

Y G Soloveichik, M G Persova, D S Kiselev, O S Trubacheva, N S Kondratyeva and A Grif
Department of Applied Mathematics, Novosibirsk State Technical University, Novosibirsk, Russia
E-mail: mpersova@mail.ru

Abstract. The paper presents a new approach to construction of complex geological models and generation of finite element meshes in order to perform 3D modeling and multidimensional processing of airborne geophysical data. The main principles of geological models construction are considered. They are based on defining geological models in terms of the subject field and take into account their main specifics: complex background media, curvilinear Earth surface relief and borders between the geological layers, complex shape of target objects, etc. We consider the necessary data structures and specifics of finite element mesh generation. The example of generation of the finite element mesh for a geological model typical of airborne geophysical surveys is shown.

1. Introduction

Nowadays, the need for applying 3D approaches to modeling and data processing in airborne geophysical surveys is undeniable [1–3]. In this regard, on the one hand, the requirements for accuracy of representing real geological media grow steadily, and, on the other hand, the requirements for automation of their discretization and modeling accuracy increase.

Together with that, some software applications, which implement 3D modeling and are characterized by sufficiently high degree automation, make it possible to describe a medium only by the objects of simple rectangular shape and use the parallelepiped meshes for discretization. Other software applications allow the description of a medium with complex borders [4–6], but they do not take into account the specifics of the problems, and, correspondingly, do not provide sufficient instruments of automation both for model construction and for optimization of computational cost to obtain the numerical solutions of the required accuracy.

In this paper, we consider the approach, which, on the one hand, allows convenient construction of "continuous" geological media with curvilinear borders and local 3D objects of complex shape, and, on the other hand, makes it possible to automatically generate the optimized non-conforming meshes with hexahedral cells, which ensure both quite an accurate representation of complex borders and the required accuracy of numerical solution approximation and low computational cost.

2. Software structure, principles of model construction and mesh generation

2.1. *The structure of software for airborne geophysical surveys maintenance*

The software includes the following main components:

- computational modules including modules of solution of forward and inverse airborne electromagnetic and magnetic problems;
- graphical user interface (GUI);
- control program for parallel solutions;
- information system for keeping the observed data, geological media recovered, and corresponding numerical data.

In this paper, we consider the principles of construction and discretization of geological models, which are implemented in GUI and corresponding computational modules.

2.2. *Geological model construction*

In the software developed, the geological model is represented by the basis layers the borders between which (including the Earth surface relief) are defined by "heights" $Z_m(x_p,y_p)$ in the sets of points $\{x_p,y_p\}$, $p=1,\ldots,P_m$. Using these sets, the bicubic Hermite splines $Z_m(x,y)$ are constructed. They pass through or as close as possible to these points and ensure the smoothing of corresponding surfaces.

Besides, for each border between the geological model layers, the so-called master coordinate is preset. These coordinates are determined by some middle depth of the corresponding border.

At the next stage of the geological model construction, 3D inhomogeneities are formed within each of the layers.

The 3D objects of parallelepiped shape are represented by six coordinates (minimal and maximal values in each of the coordinate axes). We note that the values of the coordinate in the Z-axis are preset in master coordinates, i.e., in fact, the object location along the vertical axis is determined relative to the top and bottom surfaces of the corresponding layer and its top and bottom borders are curved in accordance with the curvilinear borders of the layer within which this object is contained.

The 3D objects of complex shape in plan are represented with the use of polygons for which we preset the value d defining the accuracy of their borders representation. The positions in the Z-axis of these complex objects are defined in the same manner as for the objects of parallelepiped shape.

The dipped geological structures with possible change of their dimensions at different levels in the Z are defined with the use of the hexahedrons-stencil-plates the top and bottom bases of which are rectangulars with possible different dimensions and locations in plan. These hexahedrons, in fact, define the deformation of 3D objects (structures) located within them.

We note that for 3D objects, the priority can be defined, which enables us to place some objects "over" others (i.e. at the places of their intersections, the properties of the medium are defined according to the object with a higher priority). This is convenient when there is a need for constructing a "continuous" but laterally inhomogeneous medium with curvilinear borders in plan, which, in its turn, includes 3D target objects of complex shape. In this case, the priority of such target 3D objects is set higher than the priority of the objects which represent a laterally inhomogeneous background medium.

For all structural parts considered above, the electrophysical characteristics are set: the values of conductivity (it may be anisotropic) and parameters of induced polarization (chargeability, relaxation time, and the coefficient describing the form of drop).

2.3. *Finite element meshes generation*

The accuracy of the receiver signals obtained as a result of finite element calculation of the 3D electromagnetic field significantly depends on the mesh refinement in small vicinity around the receivers and abnormal 3D inhomogeneties placed quite close to them. Therefore, in order to reduce the computational cost required to obtain the signal with the acceptable accuracy, the non-conforming meshes with local refinements are used [7].

These meshes are generated taking into account the following principles:

- The regular nonuniform mesh with quite a small equal step in the small subdomain defined by transmitter-receiver position (or positions joined into a group) is used as the base mesh. Further, this subdomain is named the *uniform regularity* subdomain.
- The additional local refinements in the vicinities of transmitters and receivers are performed in order to increase the accuracy of the calculated signals.
- The "superfluous" nodes located outside the *uniform regularity* subdomain are deleted, and the corresponding "elongated" cells are joined, which allows decreasing in the number of nodes in the obtained irregular non-conforming mesh approximately by an order.
- 3D objects are embedded with the local refinements of near-border cells of the mesh (in order to take into account the borders of these objects).

In order to implement these principles, before finite element mesh generation, the following specific data structures are formed on the basis of the geological model data considered above.

- The first structure is the structure describing the computational domain with the use of control points, which are put in a regular macro mesh. The control points are put on the plates, which will be transformed into curvilinear borders between the layers, and their coordinates in the Z are defined using master coordinates of the corresponding borders. In plan (at each border between the layers), the control points are put in the following way. Firstly, the control points are put on the remote lateral boundaries of the computational domain, which are defined depending on the time range in which the modeling is carried out. Secondly, the control points are put on the boundaries of the *uniform regularity* subdomain. And finally, the control points are put at the place of vertexes of the hexahedrons-stencil-plates defining the deformation of geological structures.
- The second structure is the structure describing 3D objects locations relative to the control points positions. In order to form this structure, the 3D objects of complex shape are preliminarily approximated in plan (in lateral) using rectangulars taking into account the values *d*, which defines the required accuracy of representation of such objects borders. The approximation is performed in such a way that, on the one hand, the required accuracy would be ensured, and, on the other hand, the number of rectangulars obtained (which are further interpreted as independent 3D objects with the same properties) would be minimized.
- The third structure is the structure containing the information on splines, which represent the curvilinear borders between the geological layers.

On the basis of these structures, the finite element mesh is generated in the following way.

- At the first stage, the base mesh is generated. This mesh takes into account control points positions, the step-size in the *uniform regularity* subdomain, and the factor of increase of the cell size when moving to the remote boundaries of a computational domain. During the mesh generation process, the "superfluous" nodes are deleted from the mesh, and "elongated" cells are joined into more isometrical ones.
- At the second stage, local refinements are performed in the vicinities of transmitter and receiver positions.
- At the third stage, 3D objects are embedded in accordance with their priorities (the objects with higher priorities are embedded over the objects with lower priorities).
- At the fourth stage, the finite element mesh is deformed in accordance with the real coordinates of the control points and splines representing the curvilinear borders between the layers.

3. Examples

We consider a three-layered geoelectrical model the layers of which contain the inhomogeneities of different types. The master coordinates of the borders between the layers are taken equal to -40 m and -240 m. In the top layer of the geoelectrical model, two non-overlapping objects were constructed using the polygons. These objects are shown in Figure 1*a*. Figure 1*b* shows 3D inhomogeneities in the form of polygons in the second layer. The target objects are shown by red color, inhomogeneities simulating geological noise-objects are shown by other colors. The left target object has higher priority, it therefore "slots" noise-objects. The noise-objects also have different priorities (because they overlap one another). So, the objects whose contours are shown by green color have the highest priority, and the objects the contours of which are shown by dark blue color have the lowest priority. Besides, the right target object narrows downward, and this narrowing is defined by the corresponding hexahedron-stencil-plate.

For the Earth surface and surfaces between the layers, the sets of points $\{x_p, y_p\}$, $p=1,\ldots,P_m$ representing the shape of these surfaces were set.

The measurements system is shown by light points, and the points of the current group (for which the calculations are carried out) around which the *uniform regularity* subdomain is constructed are shown by black color.

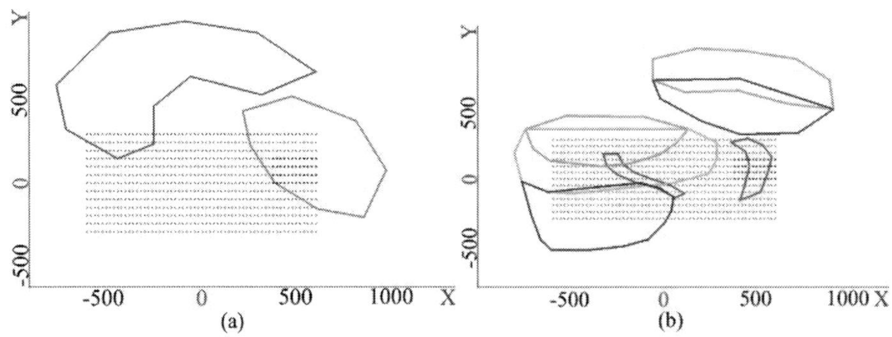

Figure 1. The 3D objects of complex shape in the first (a) and second (b) layers of the geoelectrical model.

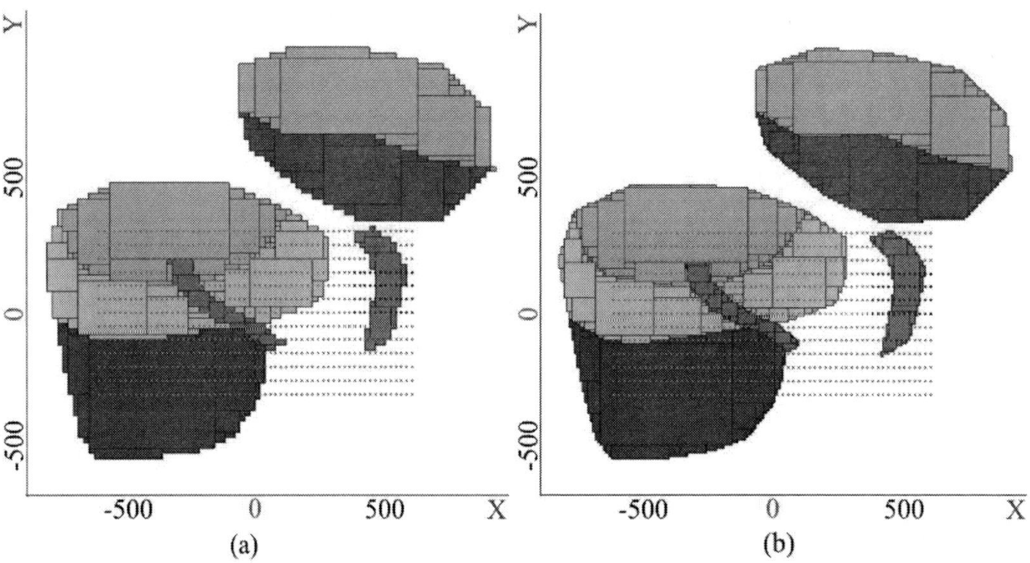

Figure 2. The approximation of objects-polygons using rectangulars for d=20 m (a) and 10 m (b).

As we noted above, before the finite element mesh generation, the objects-polygons are approximated in the plan taking into account the value d, which defines the accuracy of the borders

IOP Conf. Series: Materials Science and Engineering 1019 (2021) 012075 doi:10.1088/1757-899X/1019/1/012075

representations. The result of this approximation is shown in Figure 2 for d=20 m (Figure 2*a*) and for d=10 m (Figure 2*b*).

As previously mentioned, the mesh generation consists of several stages one of which (the third) includes the embedding of 3D objects into the mesh. In order to show the base mesh with local refinements and deformations, we exclude this stage. The finite element mesh without the embedded 3D objects is shown in Figure 3.

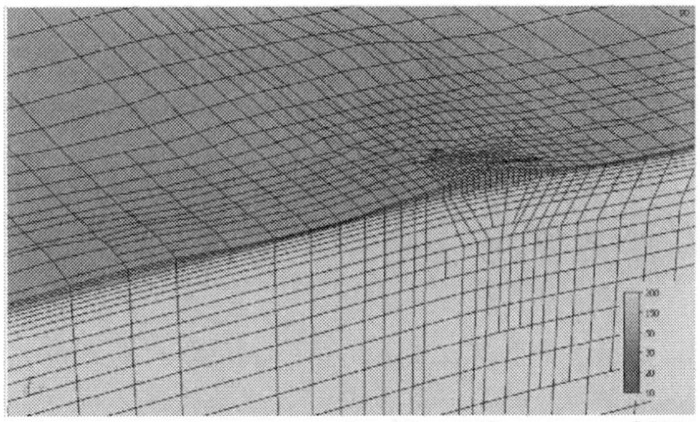

Figure 3. The finite element mesh generated without the stage of 3D objects embedding.

Figure 4 shows the finite element mesh obtained after all four stages (the value d is taken equal to 20 m). The mesh is presented in different slices. The color corresponds to resistivity of structural parts of the geoelectrical model.

In Figures 4*a* and 4*b*, the 3D inhomogeneities in the top and second layers are visible, and in Figure 4*c*, we can see how the sizes of the right target object are reduced in the plan.

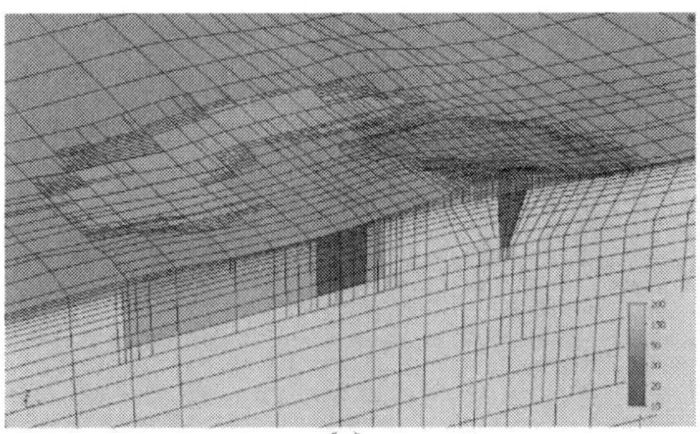

(a)

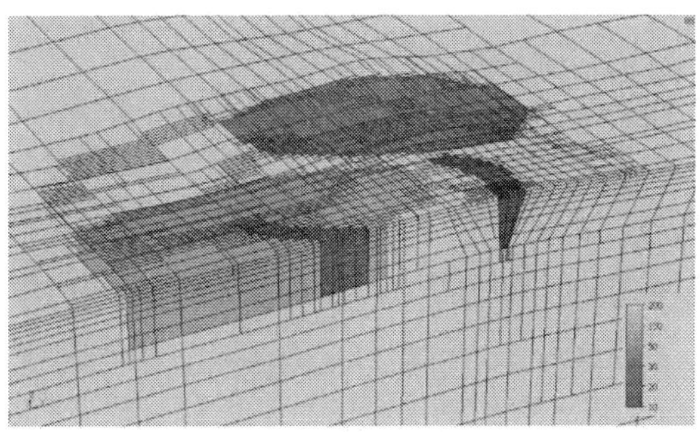

(b)

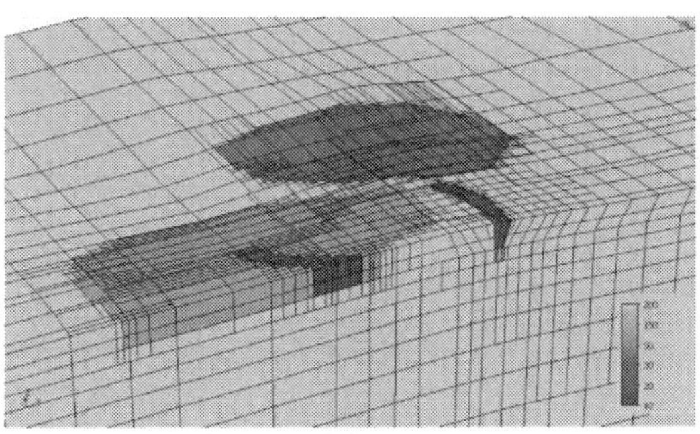

(c)

Figure 4. The view of the finite element mesh after all stages of its generation (the approximation of the objects of complex shape is performed for d=20 m): (a) the slice along the border with air; (b) the slice along the border between the first (top) and second layers; (c) the slice inside the second layer.

Figure 5 shows one of the slices of the finite element mesh, which was generated for the case when the value d was taken equal to 10 m for approximating the polygons.

We can see that the mesh refinements are only in the vicinities of 3D objects borders. This increases the accuracy of borders representation, but the number of nodes in this mesh is bigger by half.

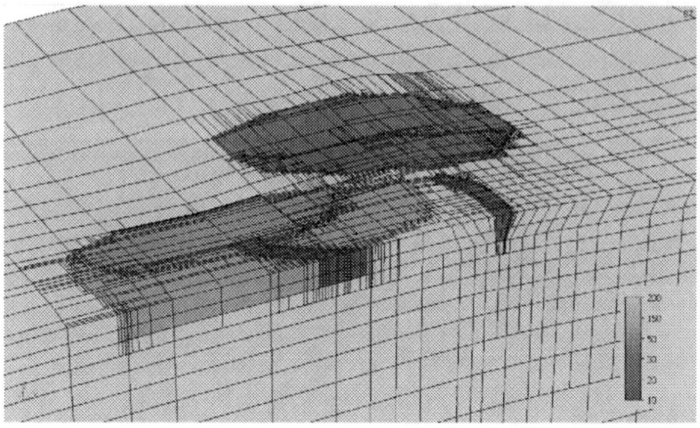

Figure 5. The finite element mesh slice inside the second layer (approximation of the objects of complex shape is performed for d=10 m).

4. Conclusion

The approach developed enables us to construct complex geological models and generate the corresponding finite element meshes. These meshes, on the one hand, include all borders of the structural parts of a geological model and make it possible to obtain the finite element solution with the required accuracy, and, on the other hand, they are optimized and do not contain the "superfluous" nodes, which finally provides the computational effectiveness of 3D modeling and the inversions based on it.

Acknowledgment

This work is supported by the Russian Science Foundation under grant 20-61-47072.

References

[1] Cox L H, Wilson G A and Zhdanov M S 2012 *Geophysics* **77** 59–69
[2] Liu Y and Yin C 2015 *Geophysics* **81** 401–408
[3] Persova M G, Soloveichik Y G, Koshkina Y I, Vagin D V and Trubacheva O S 2016 *Proceeding of 1st Conference on Geophysics for Mineral Exploration and Mining, Near Surface Geoscience* 1–5
[4] Yin C, Qi Y and Liu Y 2016 *J. Appl. Geophys.* **131** 163–178
[5] Yin C, Qi Y, Liu Y and Cai J 2016 *J. Appl. Geophys.* **134** 11–22
[6] Schwarzbach C, Börner R U and Spitzer K 2011 *Geophys. J. Int.* **187** 63–74
[7] Soloveichik Y G, Persova M G, Domnikov P A, Koshkina Y I and Vagin D V 2018 *Geophys. J. Int.* **212** 2159–2193

Whole blood clotting time assessment by method of laser-speckle correlation

I D Liushnevskaya[1], L Li[1], F A Gubarev[1], A V Nosarev[2], V S Gusakova[2]
[1]Research School of Chemical and Biomedical Technologies, Tomsk Polytechnic University, Tomsk, Russia
[2]Department of Biomedicine, Siberian State Medical University, Tomsk, Russia
E-mail: juliasytnik55@yandex.ru

Abstract. The paper presents the results of study of the coagulation process of the whole blood with anticoagulant using the method of laser speckle correlation. The clotting time dependence curves of the correlation coefficient have been presented and discussed. Analyzing the changing trend of these curves is possible to assess the clotting time, which prove the capability of using the method of laser speckle correlation for determining the diseases of blood coagulation.

1.Introduction

As an important adjustable function, the mechanism of blood coagulation plays a key role in protection of biological organs from accidental injuries [1]. The diseases of blood coagulation bring much significant risk to normal life. Severe patients need to take medication regularly to maintain normal mechanism of blood coagulation. More than that, the capability of coagulation mechanism is also a routine indicator of blood tests in clinical practice. Therefore, a faster and more accurate detection of coagulation mechanisms remains an urgent task.

Until now, the most effective method for study the mechanisms of blood coagulation is to measure the clotting time of blood. The coagulation factors that complete the coagulation mechanism are present in the plasma [2]. In this connection, the method of measuring the clotting time requires the plasma extracted from the blood in clinics. Modern coagulometers do not allow to make a quick assessment of blood clotting time. This method is slow and need a large amount of blood for centrifugation to produce plasma. For the analysis carried out using laboratory coagulometers, it is necessary to separate the blood cells from the plasma, place non-magnetic balls. These processes are time consuming. In addition, the device is in contact with the sample, it is necessary to place a non-magnetic ball in the sample that is in sample during the entire study and may make to add any changes in the formation of a fibrin clot. Therefore, researchers started to develop the means of measuring the clotting time using the whole blood for determining the disease of blood coagulation.

The method of laser speckle correlation is a new optical method that can observe or measure many sorts of changes of objects, such as the displacements [3, 4], distorts and the velocity of liquid flow [5], mechanical tension [6], by analyzing the laser speckle patterns. In biomedical flied methods of laser speckle techniques are well applied for measuring the heartbeat pulse [7], the velocity of microvascular blood flow [8], bacteria activity [9].

The correlation coefficient reflects the ratio of two speckle patterns, as a result, reflects velocity of change of two speckle patterns. The higher the rate of change of the image, the less optically dense is the object under study. In the process of blood coagulation, the rate of light scattering changes, corresponding to the rate of formation of a fibrin clot. The laser-speckle methods allow to analyze biological fluids without contacting with the sample. Due to this, non-contact optical analysis of clotting time of whole blood is possible. The blood coagulation process proceeds independently, without prior intervention in the sample under study [10–12].

In [11, 12] we tested the method of laser-speckle correlation using the control plasma. The aim of this work is to test this method with citrate whole blood of the laboratory animals.

2. Experimental details

In this paper, we used whole blood of test systems (rats) for the analysis of prothrombin time and Techplastin reagent. "Techplastin" is thromboplastin-calcium mixture from the rabbit brain. For the preparation of the regents, 5 ml of distilled water was added to Techplastin. Then the mixture was mixed and left for 20 minutes at a temperature of +37 °C. According to the instructions [13] for analysis, whole blood and Techplastin reagent were used in a 1:2 ratio.

Blood was taken from the aorta of animal. To prevent a premature start of the coagulation process, blood was placed in a tube with sodium citrate immediately after collection. Sodium citrate blocks the blood coagulation system allowing.

Figure 1 shows the scheme of the experiment. A He-Ne laser with a wavelength of 650 nm and a power of 5 mW illuminated the container into which the test sample with a volume of 50 μl was placed. The resulting image was recorded with USB digital camera ELP-USBFHD01M-MFV with a Canon Macro Lens EF 100 mm lens. During the process of coagulation a chaotic movement of suspended particles, formed elements, proteins occur in the blood. When the blood is illuminated by coherent radiation, multiple reflections of the beam from moving particles mix and form a speckle pattern. In the proposed method, changes in speckle patterns are recorded, using a camera and then processed in the Matlab software. Images were being recorded during 6 minutes.

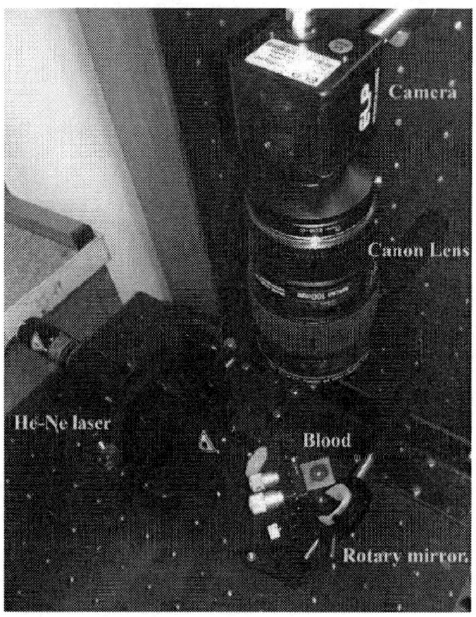

Figure 1. The scheme of the experiments with lens.

Figure 2 shows the experimental setup without lens. In contrast to the experiments presented in [11, 12], in this case we removed the objective keeping the position of the camera approximately the same. Registration of speckle patterns was carried out with 30 fps speed. To determine the blood coagulation time, the calculation of the correlation coefficient during the entire coagulation process were conducted. The correlation coefficient reflects the ratio of two speckle patterns. The closer this value is to 1, the less they differ and the less changes occurs in the blood sample. For the calculation, the formula for the correlation coefficient, given in [14], was used.

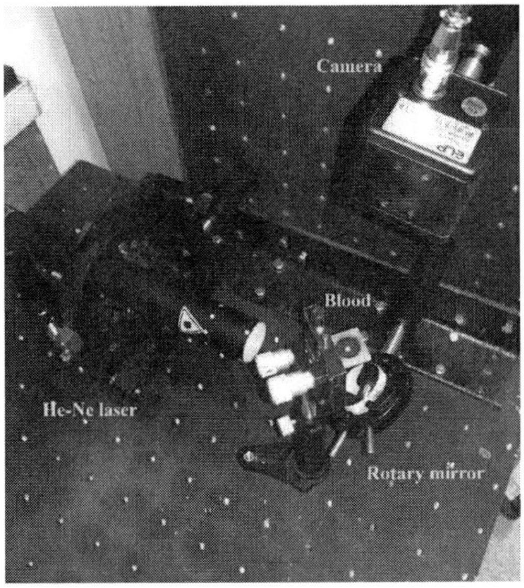

Figure 2. The scheme of the experiments without lens.

3.Results

Figure 3 shows the time dependence of the correlation coefficient for six control experiments conducted on a laboratory setup with a lens shown in Figure 1. Experiments were conducted on whole citrate blood of test systems. To activate the process of clotting in the sample was added reagent Tehplastin. Experiments were carried out to test the applicability of the digital speckle pattern correlation method for analyzing the clotting time of whole blood. In this case, the time of blood clotting is determined by the coagulant Tehplastin. Analyzing the data, we can say that the clotting time in this case is 40–150 s for a series of experiments. The increase in clotting time depends on the concentration of sodium citrate in the sample.

Figure 4 shows a series of experiments conducted on the setup shown in Figure 2. The conditions of the experiments and the ratio of the volumes of the sample-Technical were the same as in the first series of experiments. In this case, we can to see the time of sample activation with Techplastin reagent. This time between 10 and 40 s. from the start of the experiment. Active growth of the fibrin clot is between 40 and 118s. Thus, the generalized growth time of a fibrin clot for a series of experiments was 78s.

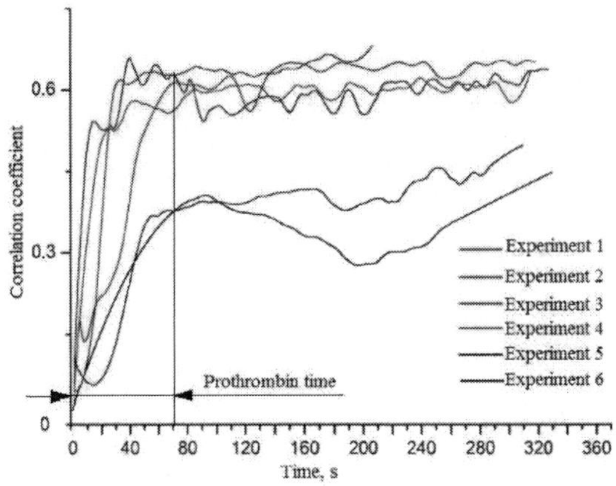

Figure 3. The curves of correlation coefficient dependence on blood clotting time, obtained by recording with a camera with a lens.

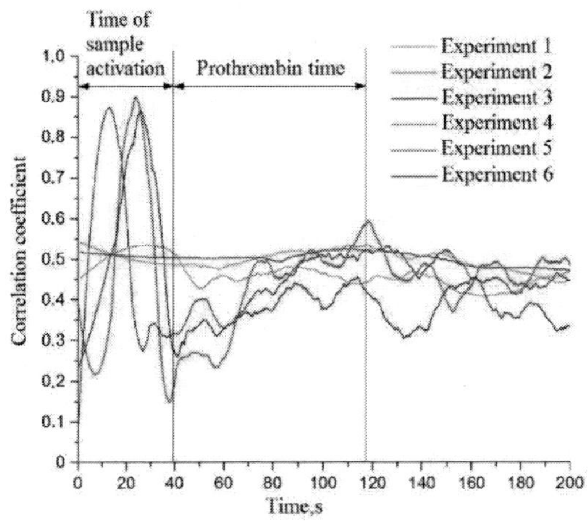

Figure 4. The curves of correlation coefficient dependence on blood clotting time, obtained by recording without a camera with a lens.

4.Conclusion

In this paper, we demonstrated the possibility of using the method of correlation of speckle patterns for analyzing the clotting time of citrated blood of laboratory animals (rats). The results of analyzes carried out on whole blood placed in a tube with sodium citrate are presented. The analysis was carried out on two installations. An objective lens was added to the camera in the first one, and in the second one, it was missing. We can conclude that method of laser-speckle correlation works both with objective and without objective. Due to this fact, we can construct more compact portable device by abandoning the optical elements.

Obtaining the prompt information about the clotting time, the doctor will be able to take the necessary measures immediately to optimize the composition of the patient's blood to avoid thrombosis of the lower veins or large blood losses.

References

[1] Kumar P, Clark M 2005 *Clinical Medicine 6th Edition* (Saunders Ltd.).
[2] Levy J, Dutton R *et al* 2010 *Anesthesia &Analgesia* 354–364

[3] Li L, Gubarev F A *et al* 2016 *Proceeding of International Siberian Conference on Control and Communications (SIBCON)* 16090528
[4] Li L, Bloshkina A I *et al* 2017 *Proceeding of 2017 Progress in Electromagnetics Research Symposium - Spring (PIERS)* 3172–3175
[5] Zhang P, Peterson S D *et al* 2018 *Experimental Thermal and Fluid Science* 207–221
[6] Hild F, BouterfPascal A *et al* 2017 *The Micro-World Observed by Ultra High-Speed Cameras* 185–206
[7] Zalevsky Z, Beiderman Y *et al* 2009 *Optics express* **24** 21566–21580
[8] Khaksari K, Kirkpatrick S J 2016 *of Biomedical Optics* 7 076001
[9] Ramírez-Miquet E E 2017 *Lasers in Medical Science* 1375–1386
[10] Markandey M T, Zeinab H 2014 *Biomed. Opt. Express* **3** 1–15
[11] Li L, Gubarev F A *et al* 2018 *Biomedical Engineering* **3** 177–180.
[12] Li L, Gubarev F A *et al* 2018 *Proceeding of 2018 Progress in Electromagnetics Research Symposium* doi: 10.23919/PIERS.2018.8597803
[13] Technical plate-test. Available at: http://www.tehnologia-standart.ru/catalog/tekhplastin_test/tekhplastin_test_100_opr_.html
[14] Gubarev F A, Li L *et al* 2016 *MATEC Web Conf.* **04003**.

IOP Conf. Series: Materials Science and Engineering 1019 (2021) 012077 doi:10.1088/1757-899X/1019/1/012077

Prospects for the use of wireless reading of prosthetic control signals and methods for generating commands

D Malakhov[1], V Zhmud[1], V Avrmachuk[2], D Gryaznov[1], J Nosek[3], L Dimitrov[4]
[1] Department of Automation, Novosibirsk State Technical University, Novosibirsk, Russia
[2] Department of Complex Information Security of Computer Systems, Tomsk State University of Control Systems and Radioelectronics, Tomsk, Russia
[3] Department of Mechatronics and Computer Engineering (MTI) Faculty of Mechatronics, Informatics and Interdisciplinary Studies Technical University of Liberec Liberec, Czech Republic
[4] Vice rector Technical University of Sofia, Sofia, Bulgaria
E-mail: oao_nips@bk.ru

Abstract. The paper provides a rationale for the use of low-power wireless communication to create non-invasive bionic limb prostheses. The need for such a connection is dictated, on the one hand, by the expediency of using sensors located in various parts of the body, often very remote from each other, on the other hand, by the desire to free the disabled person's body from the wires that would envelop it like a spider web. For each of these positions, arguments are given.

1. Introduction

The Internet of Things implies a big leap forward in the development of Internet technologies, with quite a lot of emphasis being placed on wireless communication systems. Already, wireless systems play a large role in social contacts and information requests of the average user. Wireless communication has organically entered everyday life on the example of Wi-Fi for connecting computers and gadgets with the Internet in homes and public places, wireless manipulators such as "mouse", keyboards, and so on. There are tendencies in the use of sufficiently developed software tools for household appliances (refrigerator, dishwasher, washing machine, coffee maker, and so on) that can be programmed for certain actions remotely. Also, the idea of a smart home is actively developing in the concept of a large number of video cameras and devices to ensure the comfort of the residents of this house. Along with communication by means of radio frequency communication means, optical communication is widely used, in particular, to control a TV set, air conditioner, and some types of other household equipment. The advantage of the optical method is high noise immunity, the advantage of the radio frequency method is that there is no need for optical visibility, since, for example, to successfully control a TV from an optical remote control, if not direct optical visibility, then at least the possibility of light entering due to its scattering from the walls is required or ceiling.

Our concept, which is to use signals that can be read from the human body by non-invasive methods as efficiently as possible to control prostheses, leads to a proposal that, at first glance, seems unexpected and strange, but if it is detailed investigate and test it, it may be quite workable. In this case, we are talking about the use of very unexpected ways of controlling the prostheses, for example, using the toes to control the prosthetic hand.

2. Statement of the problem

In research carried out in collaboration with the Technical University of Liberec (Czech Republic) [1, 2], one of the directions was to create the cheapest hand prostheses and manage them as easily as possible, without implanting electrodes into the human body [3–6]. This study was completed quite successfully, but the obtained control methods turned out to be insufficiently useful, since, for example, control of the left-hand prosthesis with the right hand, on the one hand, gives the disabled person extra opportunities, but on the other hand, it takes away the already existing possibilities of action with the hand that is ... Only a very few actions that require synchronous and completely identical work of two hands can be done in such a case more successfully than without such a prosthesis. In most manual operations, the hands act separately and autonomously, copying the movements of one hand in the movements of the prosthesis of the other hand is in most cases useless.

In addition, the most acute problem of providing hand prostheses is to help such a group of disabled people who do not have both hands. For them, the ability to control the prosthesis must be reliably provided without using the other hand.

The functions of the human legs are significantly less than that of other primates; finger movements are almost unnecessary. At the same time, a person, like any primate, has the ability to move all toes separately. On this basis, we consider it quite expedient to investigate the possibility of controlling prosthetic hands using the simplest commands sent by leg boys. This, in our opinion, will not interfere with movement when walking, since the corresponding buttons can be programmatically blocked when performing certain actions or, for example, by pressing the corresponding key combination. At least most people can use their toes to press three different groups of buttons, which gives six buttons when using two feet. With the appropriate training, apparently ten buttons can be used, moreover, the toes (except for large ones) can also perform a movement similar to scrolling the mouse roller up and down in addition to a simple press, and if the buttons are located not only at the bottom, but at the top, they can be pressed as well. That is, the set of commands that can be sent from the legs is quite large. You can also consider at least such an option when the patient uses this method of controlling the prostheses while at rest, sitting in a comfortable chair. In this case, he can devote all his muscular efforts to precisely controlling these prostheses with the greatest concentration.

3. Implementation of the data transmission system

Of course, wires that would go from shoes to hand prostheses are an unacceptable option, it creates too much inconvenience for the patient. The ideal option is to create special inserts in the socks of the shoes. In this case, it is not necessary to use slightly larger shoes, since these devices can be made quite thin. These tabs can contain individual wireless communications. In this case, contact with the prosthesis is carried out only by wireless methods.

In this case, it is necessary to assess the speed of such a connection, safety for the patient, and sufficiency of power. Through our preliminary research, we have determined that a low-power Bluetooth system is best suited for these tasks. Bluetooth Low Energy – BLE – provides a solution to this problem completely [7].

To test the idea, a program for simulating the work of a hand was developed in the form of displaying a graphic image of a hand on a monitor screen.

This application is developed in the C # language in the Unity3D development environment. A 3D model of the hand has been created and animated using four basic gestures.

Table 1 compares the characteristics of BLE and Bluetooth technologies. In terms of basic parameters, BLE technology is not inferior to Bluetooth, the data transfer rate remains sufficient, the range is not critical. In this case, the low level of radiation is an advantage.

The problem of interpreting signals sent by the patient is the problem of converting the original set of signals into the required set of output signals $\{I_N\} \rightarrow \{O_M\}$.

In the general case, the task is to convert signals from the original set of sensors into signals for controlling the existing set of motion organs (fingers, hand, possibly the elbow joint, and so on).

Table 1. Comparison of BLE and Bluetooth technologies [7].

Data sheet	*Classical Bluetooth*	*Energy efficient Bluetooth*
Radio frequency	2.4 GHz	2.4 GHz
Distance	100 m	>100 m
Airspeed	1 – 3 Mb / s	1 Mb / s
Bandwidth	0.7–2.1 Mb/s	0.27 Mb/s
Minimum total data transfer time (depends on battery status)	100 ms	3 ms
Power consumption	1 W as initial	From 0.01 W to 0.5 W (depending on use cases)
Maximum current consumption	< 30 mA	< 15 mA

For example, when controlling the hand of one hand, in the simplest case, we have three flexor muscles (thumb, forefinger, and a group of middle, ring fingers and little fingers), as well as five extensors. In a more complex version, we have five flexor muscles for each of the fingers separately and five extensor muscles. You can also complement the commands to rotate the brush up and down, right and left, and rotate clockwise and counterclockwise. Thus, in a complex hand, a lot of commands are typed, and this is just a very primitive model of the hand. Indeed, a real hand also allows you to fan out your fingers or squeeze them into a tight palm. There are also other figures that a hand can fold (a boat, a handful, and so on). On this basis, we can say that, in fact, a lot of commands may be required to control the hand, that is, even twenty commands (ten toes with pressing up or down) may not be enough, and besides, such dexterity of the toes is perhaps not no real disabled person can achieve. We do not consider this example with twenty commands as real, realizing that it is unnecessarily difficult for most practical cases. However, it is known that a person who has lost one of the senses becomes more receptive to other senses, a blind person perceives auditory signals better, and so on. Therefore, if a person has set himself the task of achieving the most virtuoso mastery of an artificial hand, then it would be very desirable to provide him with such an opportunity, especially since there are no technical limitations in this case. In general, the equation for transforming body movements (inputs) into commands for a prosthesis (outputs) is as follows:

$$\left(\begin{bmatrix} I_{11} & \cdots & I_{n1} \\ \vdots & \ddots & \vdots \\ I_{1m} & \cdots & I_{nm} \end{bmatrix}\right) \rightarrow \begin{bmatrix} O_{11} & \cdots & O_{p1} \\ \vdots & \ddots & \vdots \\ O_{1q} & \cdots & O_{pq} \end{bmatrix}.$$

In the simplest version, for complete controllability, it is necessary that the number of inputs is not less than the number of outputs

$$m \times m \geq p \times q.$$

Ideally, it is still desirable that the number of possible ways of sending commands be much greater than the required number of elementary movements, since this makes it possible to make redundancy and increase reliability.

For example, it is impossible to exclude the accidental sending of "false" commands, as well as the possibility of not sending a command when it is needed. To protect against false commands, you can use the "AND" logic, that is, only if all variants of this command are generated, it is perceived as actually sent:

$$O_{ij} = I_{ij} \cup I_{2j} \cup \ldots \cup I_{kj}.$$

IOP Conf. Series: Materials Science and Engineering 1019 (2021) 012077 doi:10.1088/1757-899X/1019/1/012077

To ensure the execution of the command in any case, if at least one of the commands is sent, you can use the "OR" logic:

$$O_{ij} = I_{ij} \cap I_{2j} \cap \ldots \cap I_{kj}.$$

Greater redundancy of control commands would be very desirable, since it would allow both methods to be applied simultaneously, that is, the use of duplicate commands, combined through the logic "OR", and the application of the need for confirmation, combined through the logic "AND".

$$O_{ij} = (I_{ij} \cup I_{2j} \cup \ldots \cup I_{kj}) \cap (I_{is} \cup I_{2s} \cup \ldots \cup I_{ks}).$$

The software modules are implemented in Python using the Keras machine learning library.

4. Conclusion

This paper proposes a method for controlling hand prostheses using signals received from the patient's body without implanting invasive sensors. It is proposed to connect sensors with prostheses using a low-power BLE network [7]. Preliminary experiments have been carried out showing the effectiveness of the proposed method. In this case, a virtual image of a hand prosthesis was used.

References

[1] Signal measurement and processing from myo armband device 2019 An assignment for an Erasmus+ student on an intenship (Technical University of Liberec, Faculty of Mechatronics, Informatics ad Interdisciplinary Studies). Available at: https://www.researchgate.net/publication/324889539_Technical_Features_and_Functionalities_of_Myo_Armband_An_Overview_on_Related_Literature_and_Advanced_Applications_of_Myoelectric_Armbands_Mainly_Focused_on_Arm_Prostheses

[2] Zhmud V A, Frantsuzova G A, Dimitrov L, Nosek J 2018 *Journal of Physics: Conference Series* **1015**(3) 032190

[3] Myo Gesture Control Armband – Black 2019 Available at: https://www.robotshop.com/en/myo-gesture-control-armband-black.html

[4] Morais G D, Neves L C, Castro M C 2016 *Proceedings BIOSIGNALS 2016 – 9th International Conference on Bio-Inspired Systems and Signal Processing* 227–231 doi: 10.5220/0005706302270231

[5] Wakita Z, Nagata K, Yamanobe N, Clark M 2019 *The Proceedings of Conference of Kanto Branch* 519–520 doi: 10.1299/jsmekanto.2009.15.519

[6] Park S, Jung Y, Bae J 2018 *Mechatronics* **55** 54–62 doi: 10.1016/j.mechatronics.2018.08.011

[7] Bluetooth low energy overview. Available at: https://developer.android.com/guide/topics/connectivity/bluetooth-le

Generalization of data on heat exchange within a gas mixture with a small value of Prandtl numbers in channels of different forms

S L Elistratov and E Yu Slesareva
Power Engineering, Aircraft Engineering, Novosibirsk State Technical University, Novosibirsk, Russia
E-mail: slesareva@corp.nstu.ru

Abstract. The article presents an analysis of data on heat transfer in the channels of a round and triangular shape with a flow of a gas mixture with a small Prandtl number. It is shown that for the Reynolds number in the interval from 9500 to 49000 the heat-transfer coefficient depends from the length a heat exchange section in tube. Using the dependence of the Nusselt number Nu(Ψ) from a dimensionless parameter $\Psi=1/\mathrm{Pe}\times x/D_h$, we can describe the experimental data for the input region of the round and triangular channels, taking into account the channel length and the distances from the input. The adequacy of the model was checked by comparing the proposed dependence with the experimental data of various authors.

1. Introduction

To improve the recuperative heat exchange equipment in various applications, the development of compact heat exchanges with a complex spatial structure of the heat exchange surface is urgent. In such channels, the heat exchange capacity is much higher, due to the fact that the flow of coolant inside the mini-channels is organized in such a way that heat exchange takes place on the hydrodynamic and thermal entrance length where the heat exchange capacity is much higher than in the fully developed laminar flow. The prismatic shape of the channels, typical of most core fuel assemblies in the nuclear power industry, gas reactors in the chemical industry and other similar devices, does not allow the use of well-known theoretical and experimental dependences obtained for circular channels for evaluation. Numerical studies of such flows are not always adequate, since the number of reliable input parameters is not enough.

Use of new a heat conductor with regulable properties allows improving efficiency of the working behaviours of the heat exchangers. Such heat conductor can be create in form two-component a mixtures of gases or fluids. The special attention has of the mixtures of gases with essentially various molecular weights, in particular such as a mixture of hydrogen or helium with heavy gases. For some concentrations of the heavy gas the Prandtl number Pr for a mixture of the hydrogen or the helium can have small value between 0.1 and 0.3. At such Prandtl numbers, the thermal boundary layers are thicker than the velocity boundary layers, and consequently Thermal boundary layers merge earlier than hydrodynamic. Therefore the thermal stabilization occurs on a shorter channel section than the of flow stabilization length. However, for fluids with small value of the Prandtl number, at calculation of the heat-transfer coefficient there is problem in precision and validity of the computational result [1, 2]. In particular, identification of the influence degree of the duct entrance region for a turbulent flow in the channels of the intricate cross-section form is required [1, 3].

The Nusselt number Nu for a gas laminar flow in the short-length tubes depend on the dimensionless parameter $\Psi = 1/\mathrm{Pe}\times x/D_h$, where Pe – Peclet number, x – length of the duct, D_h – hydraulic diameter of the duct [4]. For $\Psi < 0.01$ dependence Nu (Ψ) asymptotically approaches the solution $\mathrm{Nu} \sim \mathrm{const}\times\Psi^{-1/3}$. For a laminar fluid flow in circular and triangular ducts there are exact solutions for dependence of the Nusselt number Nu (Ψ) from parameter Ψ [4]. In Paper [5] presents the general equations for heat transfer calculations for constant wall temperature in laminar developed flow in ducts of arbitrary cross sections. The results obtained from these equations compare well with the theoretically-calculated values available in the literature for circular, rectangular, triangular, elliptical and parallel plate ducts. But the such dependence Nu (Ψ) for a turbulent fluid flow usually is not used. It is associated with a fact, that as a rule for a turbulent flow a ordinary gases, $\mathrm{Pr} > 0.6$, length the hydrodynamic and thermal entrance regions of the developing flow in pipes x does not exceed $10 \div 20$ hydraulic diameters D_h, and for developed flow state the heat-transfer coefficient α and Nusselt number depend from a Reynolds number Re and not depend from a x/D_h.

In the [6] shows that the experimental heat transfer data for helium–xenon mixture flow in triangular cross-section channel can be described by empirical correlation [7] for the turbulent flow regime:

$$\mathrm{Nu}_{D_h} = 0.022\cdot \mathrm{Re}_{D_h}^{0.8}\cdot \mathrm{Pr}^{\kappa},\quad \kappa = 0.595\cdot \mathrm{Pr}^{-0.126} \tag{1}$$

where Re_{Dh} – Reynolds number. The formula of (1) generalizes adequately the obtained experimental data despite the fact that the recommended range of Prandtl numbers for this formula $0.3 \div 1$, therefore, the range applicability of this formula can be extended to Prandtl numbers equal to 0.2. In this paper, we presented the results of the analysis of experimental heat transfer data for gas mixture with a small Prandtl number at flow in the short channels with circular and triangular form.

2. Data analysis and results discussion

For the analysis heat transfer at gas flow in a short channel, experimental data from the works [6, 8–11] were used. Experiments were carried out with a gas mixture $\mathrm{Pr} \sim 0.2$ and air $\mathrm{Pr} \sim 0.7$ in direct a tubes circular and triangular shape. The cylindrical pipe made from stainless-steel A2 with a wall thickness of 1.3 mm, the length 0.875 m, internal diameter D = 4.6 mm and D = 5.5 mm. The plain triangle channel was made from a stainless-steel thin-wall plain pipe with the outer diameter 5 mm and the wall thickness 0.15 mm. The working length of the triangle heated channel was 670 mm. The hydraulic diameter of the equilateral triangle channel with the side equal to 4.8 mm was D_h = 2.814 mm. The pipe was heated by passing through it an alternating electric current.

The channel inlet temperature was 300 K. The inlet pressure ranged from 392 to 433 kPa, and at the outlet from 382 to 413 kPa. The inlet Reynolds numbers varied over the range from 9500 to 49000, and the x/D_h from 10 to 120. Reynolds number Re_{Dh} and Nusselt number Nu_{Dh} for distance x from the entrance was determined by the hydraulic diameter D_h:

$$D_h = 4\cdot S/\Pi$$

$$\mathrm{Re}_{D_h} = U_{in}\cdot D_h/\nu$$

$$\mathrm{Nu}_{D_h}(x) = 1/x\cdot\int_0^x \alpha(\ell)\cdot D_h/\lambda\cdot d\ell$$

where S – cross section area, Π – length wetted perimeter channel, U_{in} – average rate flow velocity at the entrance to the channel, α – local heat transfer coefficient. The local circumferential-average heat transfer coefficient for the heated walls is defined here as

$$\alpha = (Q/\Pi)/(T_w - T_b)$$

where Q is the rate of convective heat transfer per unit axial length from the heated walls to the fluid. The temperatures T_w and T_b respectively represent the values for the heated wall and the bulk. The values viscosity ν, gas density ρ, thermal conductivity λ of the gas were calculated by data from the works [1, 12, 13].

On Figure 1 Nusselt number Nu_{Dh} varying with dimensionless distance from the entrance x/D_h, the Reynolds number Re_{Dh} and Prandtl number Pr for the gases mix flow in circular tube and triangular duct for various flow regime (9000 < Re_{Dh} < 50000) are shown. As can be seen from Figure 1 the relationship between the value Nu_{Dh} and the parameters Re_{Dh}, Pr and x/D_h has an uncertain correlation. The two vertical lines on Figure 1 are show the range values Nu calculated by formula Tetelbaum (1) [7] for Pr = 0.23, Pr = 0.7 and Reynolds numbers between 9500 and 49000. It is possible see on a graph that the range of a calculated values Nu does not comport with the experimental data.

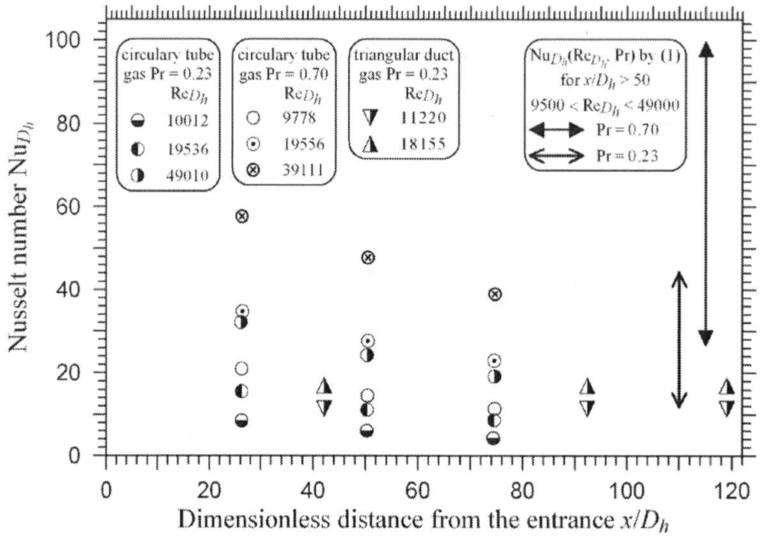

Figure 1. Nusselt number Nu_{Dh} varying with dimensionless distance from the entrance x/D_h for various flow regime (Reynolds number Re_{Dh}) at gasmix flow with different Prandtl number Pr in circular tube [8, 9] and triangular duct [6, 9, 10, 11].

An analysis experimental data showed that dependences spliting for Nusselt number Nu (Re) relative to Prandtl number Pr disappears if use the dependence Nu (Pe) on the Peclet number Pe = Re×Pr. Graph in Figure 2 illustrates examples of the comparisons between a dependence of the Nu from the Re (Figure 2*a*) and dependence of the Nu from the Pe (Figure 2*b*). Examples for the experimental data in a circular tubes for a different distances from an entry *x/D* are shown. Can see that for every one of the arguments x/D_h there exist the unique dependences of a Nu from a Pe for all a Pr.

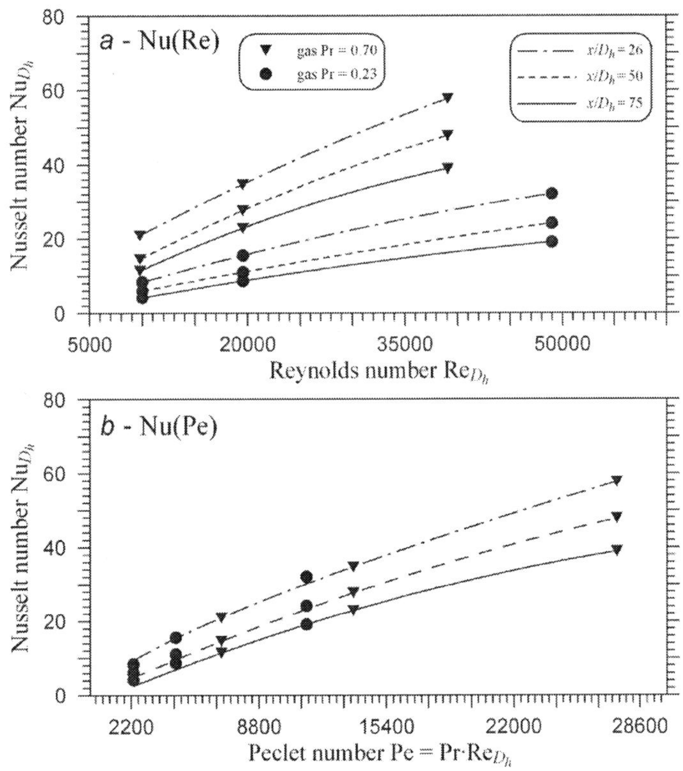

Figure 2. Dependence of the Nusselt number Nu_{Dh} on Reynolds number Re_{Dh} (a) and from Peclet number Pe (b) at gasmix flow with various Prandtl number Pr in circular tube [8, 9] on different dimensionless distances from an entry x/D_h.

Using the hydraulic diameter D_h the effect of channel shape on heat transfer usualy can be taken into account [1, 4, 14–16]. Figure 3 shows the data for Nusselt number Nu_{Dh} at flow of the gas with Pr = 0.23 in circular tube and triangular duct, which have equal wetted perimeter values Π of cross-section. In Figure 3*a*, data for Nu_{Dh} depending on Re_{Dh}, and in Figure 3*b* depending on dimensionless parameter $\Theta = Re_{Dh}\times\pi\times D_h/\Pi$ are presented. It can are seen that for circular tube and triangular duct there is a general dependence for the Nu_{Dh} from the parameter Θ.

Thus, it is quite possible that dependence of the Nu from the dimensionless parameter $\Psi = 1/Pe\times x/D_h$ as well as for the laminar flow of ordinary gases in the short-length tubes, it is valid, and for a turbulent flow of the gas mixes with small a Prandtl number. In [14], an interpolation equation was proposed for heat exchange in a circular pipe with laminar flow and constant temperature of the channel wall

$$Nu = 3.66 + \frac{0.0668/\Psi}{1+0.04\cdot\Psi^{2/3}} \quad (2)$$

By analogy with equation (2), the following equation was used for the generalization dependence:

$$Nu \sim Nu_0 + \frac{1}{(a\cdot\Psi+b)^n} \quad (3)$$

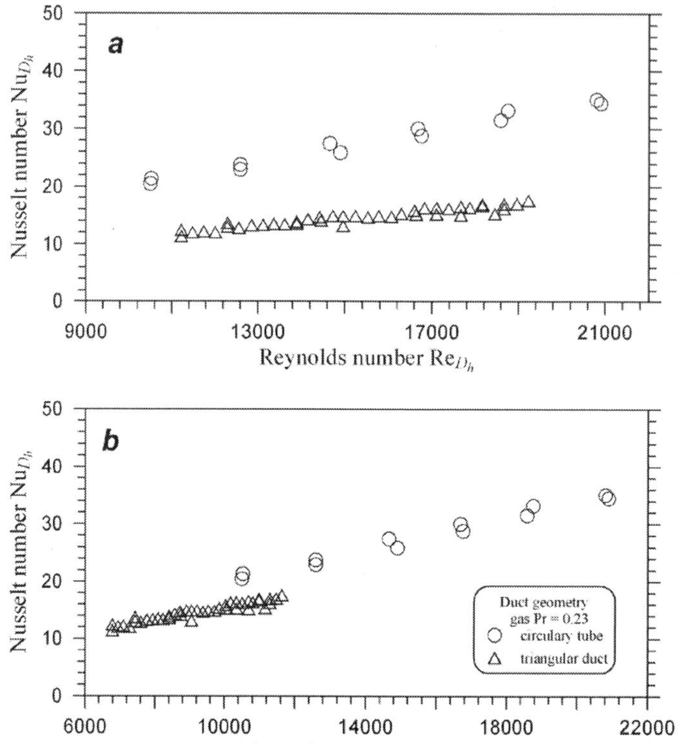

Figure 3. The values of the Nusselt number Nu_{Dh} as a function of the dimensionless parameter Θ for gas mixture flow with the value of the Prandtl number Pr = 0.23 in circular tube [9] and triangular duct [10].

The main difference of equation (3) from equation (2) is that for $\Psi = 0$ Nu $\neq \infty$. To this end, the term b has been added to equation (3). The value of *a* and the exponent *n* must satisfy the condition Nu; const×$\Psi^{-1/3}$ for $\Psi < 0.01$. The value of Nu_0 are chosen such that $\Psi >> 1$ equation (3) would correspond to exact solutions for a laminar flow in circular tube and triangular duct [4]. The coefficients *a* and *b* were founding by reducing the dependence to a linear form, according to the well-known standard mathematical rules. At uniform wall heat flux for the circular tube $Nu_{lam} = 4.364$ and for triangular duct $Nu_{lam} = 2.970$ [1, 4].

Thus, data on heat transfer in a turbulent flow of gas mixtures with a low Prandtl number are described by the dependence (3). In Figure 4 the dependence of the data for Nusselt number on dimensionless parameter Ψ from works [6, 8–11] the are shown.

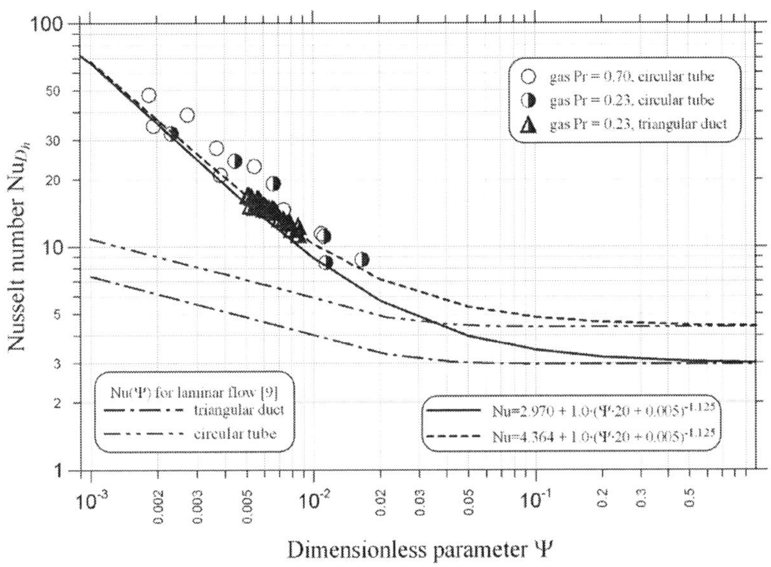

Figure 4. Dependence of the Nusselt number Nu_{Dh} on dimensionless parameter Ψ at gas mixture flow in circular tube [8, 9] and triangular duct [6, 9–11].

The graph also shows the curves of Nu(Ψ) for the developing laminar regime flow [14, 15].Can see that all experimental points is grouping along the common curve. On a graph two fit curves for this data are shown. The resulting expression for the approximation function is given by equation for triangular duct:

$$Nu = 2.970 + 1 / (20 \cdot \Psi + 0.005)^{1.125} \qquad (4),$$

and for the circular tube:

$$Nu = 4.364 + 1 / (20 \cdot \Psi + 0.005)^{1.125} \qquad (5).$$

The dependence Nu (Ψ) (4), (5) differ from the exact analytical solution Nu_{lam} (Ψ) [1, 4, 14, 15] for a laminar fluid flow in smooth tube.

To determine the application range equation (3) in Figure 5 presents the experimental data of different authors [16–19] for laminar flow in ducts of arbitrary cross-sections.

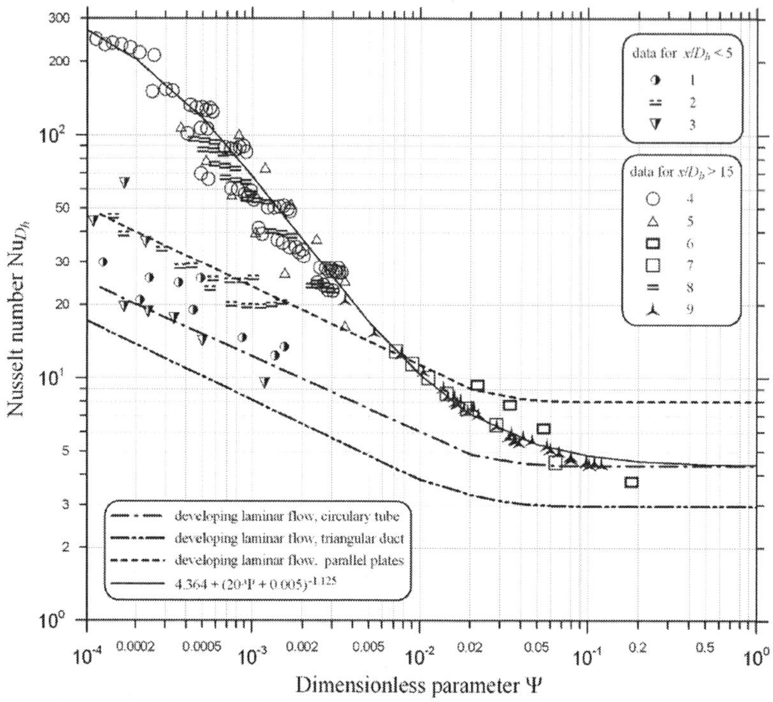

Figure 5. The dependence of the Nusselt number Nu_{Dh} [16–19] on dimensionless parameter Ψ for gas and liquid flow in ducts of arbitrary cross-sections.

Geometric and operational parameters for which experimental data on heat transfer were obtained, used in Figure 5 are presented in table 1.

Table 1. Geometrical and operational parameters at which the obtained experimental data on heat transfer used in Figure 5.

Points	Channel cross-sections	Pr	D_h (mm)	Re_{Dh}	Ref.
1	circular	0.7	37.24	$9.2\times10^3 \div 1.6\times10^5$	[16]
2	parallel plates	0.7	63.75	$1.2\times10^4 \div 6.1\times10^4$	[16]
3	triangular	0.7	22.9	$9.7\times10^3 \div 5.9\times10^4$	[17]
4	circular	0.7	37.24	$9.2\times10^3 \div 1.6\times10^5$	[16]
5	triangular	0.7	22.9	$4.1\times10^3 \div 5.9\times10^4$	[17]
6	rectangular 0.39×1	7.1 ÷ 7.9	0.56	$1.9\times10^2 \div 2.9\times10^3$	[18]
7	rectangular 0.39×1	2.4 ÷ 2.8	0.56	$2.0\times10^2 \div 2.8\times10^3$	[18]
8	parallel plates	0.7	63.75	$1.2\times10^4 \div 6.1\times10^4$	[16]
9	quasi-triangular	0.7	1.31	$7.2\times10^2 \div 5.2\times10^3$	[19]

Experimental data 19 were obtained in the study heat transfer for air flow in a quasi-triangular channel at a constant out wall temperature of 273 K. The walls of the quasi-triangular channel are made from 50-micron copper foil. The channel shape was set so that the inner curvature of the channel walls was 6.4 mm. During the analysis, all data were divided into two groups according to the dimensionless length of zone heat exchange. First group 1 – 3 for $x/D_h < 5$, and second group 4–9 for $x/D_h > 15$. Also, the graph shows curves Nu(Ψ) for the developing laminar flow [1, 4, 14, 15] of channels of round and triangular form , and flat ducts. On the graph you can see that the data for group $x/D_h > 15$ located near the curve (3). The data for group $x/D_h < 5$ is located below curve (3) near the curves for the developing laminar flow. Its can be explained by the fact that for small distances from the entrance transverse and longitudinal pulsations gas (liquid) velocity, typical for developed turbulent flow, were insignificant and did not influence on process heat transfer.

3. Conclusions

Analysing the experimental data on heat transfer for turbulent flow of a gas mixture with small Prandtl numbers may to the conclusion that for the Reynolds number in the interval from 9500 to 49000 the heat-transfer coefficient depends from the length a heating area in tube. For an entrance region of a triangular and circular duct a data on heat transfer may define by the dependence (3) of a value Nusselt number Nu(Ψ) from a dimensionless parameter $\Psi = 1/\mathrm{Pe}\times x/D_h$. The this dependence Nu(Ψ) for a turbulent gas flow differ from the exact analytical solution $\mathrm{Nu}_{\mathrm{lam}}(\Psi)$ for a laminar fluid flow in smooth tube.

References

[1] Kays W 1966 *Convective Heat and Mass Transfer* (New York: McGraw-Hill)
[2] Diaz G and Campo A 2009 *Int. J. Thermal Sci.* **48** 1392–1397
[3] Lebedev V, Lemanov V, Misyura S and Terekhov V 1991 *J. Appl. Mech. Techn. Phys* **32** (3) 360–364
[4] Shah R and London A 1978 *Laminar Flow Forced Convection In Ducts* (New York: Academic Press)
[5] Yilmaz T and Cihan E 1993 *In. J. heat Mass Trunsfer* **36** (13) 3265–3270
[6] Vitovsky O, Elistratov S, Makarov M, Nakoryakov V and Naumkin V 2016 *J. Eng. Thermophys* **25** (1) 15–23
[7] Kirov V, Kozhelupenko Yu and Tetelbaum S 1974 *Inzh. Fiz. Zh.* **26** (2) 226–228
[8] Vitovsky O Nakoryakov V and Slesareva E 2015 *J. Eng. Thermophys* **24** (4) 338–341
[9] Vitovsky O Elistratov S and Slesareva E 2015 *J. Eng. Thermophys* **24** (1) 33–35
[10] Nakoryakov V, Elistratov S, Vitovsky O and Slesareva E 2015 *J. Eng. Thermophys* **24** (2) 139–142
[11] Nakoryakov V, Makarov M, Petukhov Yu, Vitovskiy O and Elistratov S 2015 *Thermal processes in flows of gas mixtures with a small Prandtl number* (Novosibirsk: Akademizdat)
[12] Tournier J and El-Genk M 2008 *Energy Conversion and Management* **49** 469–492
[13] Hewitt G 2002 *Heat Exchanger Design Handbook. Part 5. Physical Properties* (Carlsbad: Hemisphere Publishing Corporation)
[14] Petuhov B 1967 *Heat exchange and resistance under a laminar flow fluid in tubes* (Moscow: Energy)
[15] Kutateladze S 1979 *Fundamentals of the theory of heat transfer* (Moscow: Energy)
[16] Sukomel A, Velichko V and Abrosimov Iu 1979 *Heat transfer and friction for turbulent gas flows in short channels* (Energy Moscow)
[17] Altemani C and Sparrow E 1980 *Transactions of the ASME* **102** 590–597
[18] Sahar A, Özdemir M, Fayyadh E, Wissink J, Mahmoud M and Karayiannis T 2016 *Applied Thermal Engineering* **93** 1324–1336
[19] Slesareva E, Elistratov S and Harisov M 2017 *MATEC Web of Conferences* **92** 01018

IOP Conf. Series: Materials Science and Engineering 1019 (2021) 012079 doi:10.1088/1757-899X/1019/1/012079

Analysis of barriers to the development of Industrial Internet of Things technology and ways to overcome them

V Zhmud[1], A Liapidevskiy[2], V Avrmachuk[3], V Sayapin[2], O Stukach[4] and H Roth[5]
[1]Dept. of Automation, Novosibirsk State Technical University, Novosibirsk, Russia
[2] Novosibirsk Institute of Program Systems, Novosibirsk, Russia
[3]Department of Complex Information Security of Computer Systems, Tomsk State University of Control Systems and Radioelectronics, Tomsk, Russia
[4]Faculty of Computer Sciences of National Research University Higher School of Economics, Moscow, Russia
[5]Faculty of Natural Sciences and Technology, University of Siegen, Siegen, Germany
E-mail: oao_nips@bk.ru

Abstract. As is known, in 2019 the national program "Digital Economy of the Russian Federation" was launched. In the first phase, nine roadmaps were developed, of which six were approved by the Prime Minister. Participation in the development of these roadmaps provides a fresh perspective on some particular problems. The authors reveal some details of the approach to the implementation of the "Industrial Internet of Things" direction. Some specific differences of the industrial Internet from the regular Internet are discussed using examples. Some conclusions are drawn about the directions for further research in this area.

1. Introduction

This paper provides an additional look at the problems of creating new end-to-end technologies using the example of the Industrial Internet of Things (IIoT) technology [1, 2]. It is advisable to pay attention to the barriers to the creation of these technologies. The roadmaps talk about development, but, apparently, in most cases it would be more correct to talk about their creation. This is fully true for the industrial Internet of Things.

2. Statement of the problem

The Internet of Things is the solution to all real, not virtual, problems using Internet technologies. The Industrial Internet of Things means the same when applied not to everyday tasks, but to industrial production. The difference between the industrial Internet of things and the usual one is approximately the same as the difference between an industrial computer and a personal computer, which is intended only for office work. This understanding is not always available even at the highest levels of the organization of these works. This article discusses some specific examples and provides recommendations for the development of these technologies.

For example, let us compare how research and development work should differ from experimental design and serial production of innovative products.

Research works prove the theorem on the possibility of creating at least one sample of a new type of device. In this case, individual elements of the product can be created in any way, including the technology of their creation may be unknown to the creators of a common device, for example, it can be ready-made assemblies from another technical device. For example, to create a laser device for cutting donor tissue using a semiconductor laser at the Institute of Laser Physics SB RAS [3], the mechanical parts of the plotter were used, but due to the additional laser system for accurate measurement of the position of the writing head and due to feedback, the position of the writing unit was controlled much more accurate than a traditional plotter. Also, instead of the writing unit, a semiconductor laser was installed with a power control circuit and an interferometer for measuring distance. Thus, a device was created that can accurately measure the profile of donor tissue, using software, this allows you to create a three-dimensional map of this profile. Next, the software calculates the optimal cutting of this tissue for cutting out the elements of the heart valve. After that, the laser head cuts out the donor tissue, cutting out all the necessary elements of the artificial heart valve. Such a development does not allow launching mass production, but demonstrates that the idea is workable, and that its effective solution is in principle possible.

Research and development requires the availability of technical documentation that ensures the production of all components of the product, or, if the products are purchased, agreements are required on the appropriate supply of these components or an indication of the possibility of purchasing them for the entire period of production of the new product. Research and development is the creation of a technology for the production of these products. Until the completion of the R&D, no product is created for sale, this phase also consists solely of costs. This development can end with the production of several units of the product, which must be investigated to control the quality of the created technology.

Pilot production is a stage of verification of the created technology, at which products are already produced in small quantities. Even if there is no profit from the output, then, in any case, it can be sold - so think most of those who are engaged in research work. But this is wrong. Products resulting from pilot production should be used to investigate their reliability.

Serial production is a stage when all products are produced using a single technology, therefore their quality is determined by the quality of technology. The modern approach is based on such a high degree of production automation that it completely excludes the human factor, therefore, firstly, the products coming off the automatic production line are usually much cheaper and much more reliable than the products of the pilot batch, and even more so since product obtained as a result of research work.

Taking into account this difference between different stages, we will consider the stages of creating technologies for the industrial Internet of things.

3. The difference between IoT and IIoT

Most of the authors writing about the Internet of Things on the Internet are specialists in economics, business organization, or even writers specializing in science fiction. The idea is being introduced that it is enough just to entrust the solution of all tasks to computers, and they will be solved by orders of magnitude more accurately, better, faster, more reliably. These writers are inspired by the results of using traditional Internet technologies for a variety of tasks. They see new opportunities in increasing the range of tasks solved with the help of Internet technologies, but they often lose sight of the need to create fundamentally new Internet technologies, industrial rather than household ones.

Let's define what distinguishes the regular Internet from the industrial one. For this purpose, let's draw an analogy between a personal computer and an industrial one.

A personal computer is used to facilitate the work of an office worker, it allows you to speed up the creation of texts, the calculation of mathematical relationships, the modeling of simple tasks, the calculation of statistical models, and so on. Since personal computers, due to their extreme efficiency, became widespread, many researchers began to use personal computers for professional control purposes, that is, as control machines. However, personal computers were not originally designed for these purposes, so they must be seriously reprogrammed, and in some cases hardware modifications are necessary. One of the main problems is the condition for the inheritance of all the capabilities of the previous series by new computers, compatibility with them, both software and hardware. This forces us to use far from optimal algorithms for computer actions. If an industrial computer in multitasking mode completely obeys the priorities set by the user, then a personal computer operates according to its own rules, which are far from optimal when solving problems of managing production processes or scientific research. A professional computer is focused primarily on working with devices that it controls, a personal computer is focused on working with a user, for it the perception of a task from a person and the most convenient presentation of information for a person is a priority. When receiving tasks for irreversible actions, the personal computer asks again whether this action is exactly required, meaning the possibility of an erroneous command. In an industrial computer, the possibility of an erroneous command must be excluded in principle. An industrial computer must operate according to a different logic, it must have fundamentally different interface capabilities, a different speed, a different memory size, it needs many ADC and DAC channels with high speed, bit depth, rigidly connected with the machine time, and the machine time must be strictly synchronized with a unified system of precise time. A personal computer is considered secure enough if it has anti-virus programs and is powered by an uninterruptible power supply. For an industrial computer, these measures are not enough: it should, in principle, be invulnerable to viruses, and uninterruptible power supply should work not only for a time sufficient to save the created files and close running programs, but to continue all functions without disrupting the rhythm of work.

The industrial Internet should differ in approximately the same way from the usual one, that is, it should be different in everything. Thus, the creation of the industrial Internet of things technology is the creation of a new software and hardware environment and new communication networks.

Traffic volumes of such networks are estimated at up to 1 PB (petabyte) per day.

Very often, authors of articles and even developers of development programs confuse the use of the Internet for industrial purposes and the creation of special Internet technologies for industry. The Internet has long been used for industrial and especially for commercial purposes, with its help you can order and pay for many services and goods, but this is the usual Internet of things, not the industrial Internet of things. Unfortunately, officials from the management of these programs often forget about this. We met the opinion that the funds of the Program should be spent as soon as possible on the creation of a smart housing and communal service, that is, a system of smart meters for the consumption of water, heat, electricity and gas, as well as a system for automatic calculation of rent. Perhaps such an investment can pay off most quickly, but this has nothing to do with the industrial Internet.

4. The main problems

Standardization is one of the biggest challenges. The creation of a whole variety of mutually incompatible technical solutions should not be allowed. The compatibility of technical solutions would greatly simplify the implementation of this technology. But history shows that compatibility is achieved in two cases: firstly, if a standard has been created, the observance of which is necessary for all participants, and secondly, if there is a production leader, whose technical solutions are the standard for everyone who intends to join the market for these products ... As a rule, standards are developed with a great delay after many developers have already created a variety of hardware. It is certainly reasonable to follow the leader's wake, but such an approach cannot be state policy. At the same time, the Russian Federation is facing extensive sanctions actions that have already begun to affect the cyber space. Some technical opposition explains the need to create a GLONASS system as opposed to the already existing GPS system. Similarly, the United States has launched several parallel projects to create rocket engines as opposed to Russian engines. Therefore, the creation of a Russian standard for the industrial Internet of things, apparently, may be quite reasonable. If possible, you should use the available world developments to achieve maximum compatibility with foreign technical solutions. However, it may turn out that it is not so necessary to obey ready-made foreign solutions in this case. The answer to this question should be given by a consultation of specialists.

In our opinion, the main problem in solving all problems in this direction is the lack of a systematic approach to their solution. The nominated and selected experts are most concerned with how to get the largest possible share of funding, rather than how to avoid misallocation of funding. This is also manifested in the fact that instead of a single plan for creating technologies, a matrix of projects is created, into which not elementary technologies but projects fit into.

Also, one of the mistakes is the requirements for individual projects, according to which the results should be obtained as early as possible, that is, within one calendar year. If the creation of fundamentally new technologies is required, then the results should not be required for some holiday dates. The timing of obtaining results should depend on the complexity of technologies, on the necessary components for their creation, on the availability of relevant technical knowledge at the disposal of developers (including world-class technology).

5. Smart sensors

Currently, smart sensors are understood to be those sensors that are most convenient to use, since they give a ready-made measurement result, that is, they are optimized for user-researchers. The requirements from a person to the sensors that he is going to use are radically different from the requirements of the control system. The technical device requires the highest speed, highest reliability, a digital output is also desirable, but in its absence, this problem is solved with the help of high-speed ADCs. Consequently, the development of the industrial Internet requires the development of sets of different sensors.

In some cases, sensors should not provide the highest accuracy, but the greatest amount of complex information, such as video cameras with image recognition. One of the key solutions can be distributed computing, in which sensors equipped with intelligence carry out primary information processing. Distributed computing can dramatically improve performance.

In general, it seems that for the industrial Internet one should move away from centralized data processing, preferring decentralized and distributed processing. The principle of operation of such systems is based on the complete interchangeability of individual processors by their functions, as is done when implementing swarm algorithms.

6. Conclusion

This paper cannot address all the problems of the industrial Internet development, but it does discuss some aspects that, in our opinion, were outside the sphere of roadmap developers. In particular, it is a mistake to equate the use of existing Internet technologies for industrial purposes with the creation of industrial Internet technologies. Industrial Internet technologies imply the creation of technologies for remote highly reliable and high-speed industrial production management by methods and technologies that are, in fact, close to Internet technologies, but at a fundamentally new technical level. This is not a simple quantitative increase; it is a qualitatively different technology.

References

[1] Available at: https://ac.gov.ru/en/projects/project/digital-economy-program-implementation-42
[2] TAdviser: National_program_Digital_economy_of_the_Russian_Federation. Available at: http://tadviser.com/index.php/Article:National_program_Digital_economy_of_the_Russian_Federation.
[3] Institute of Laser Phycics. Directions and results. (In Russian) Available at: http://www.laser.nsc.ru/napravleniya-i-rezultatyi/

Model of on-line control for the smart object by communication technologies

Y A Maksimova[1], D Barchukov[2] and S Slobodyan[2]
[1]Department of Earth Sciences & Engineering, National Research Tomsk Polytechnic University, Tomsk, Russia
[2]Department technology metals & materials, Tver State Technical University, Tver, Russia
E-mail: sms_46@ngs.ru

Abstract. Based on analysis of opportunities of usage existing models, methods and techniques for on-line monitoring and control of unmanned dynamic objects, a mathematical model of remote estimation of distance to wire-controlled unmanned mobile object is proposed. The presence of model and techniques of remote estimation of distance to wire-controlled unmanned object increases the reliability of control for exclusion of critical situations in the strict radio counteraction conditions and powerful electromagnetic jamming setting during control.

1. Introduction

The development of capabilities of the element base and video equipment allows to implement new principles of construction of autonomous mobile transport [1–6], information and unmanned systems (US) [7–12] with more efficient algorithms for space review, detection, recognition, information processing, coordinate estimation for navigation and control in almost all environments of the Earth: air, water, ground and underground [13–19]. Mobile objects are generally considered to be objects moving at a relatively low speed (about tens of kilometers per hour or less) both on a two-dimensional plane, cross-country, and in three-dimensional space. Developers of unmanned vehicles are actively conducted research aimed at ensuring the autonomy of using US in the conditions of exclusion of the satellite navigation signal based on alternative data sources. These data include, first of all: synthesized image formed by phased antenna arrays of radars; video information taken by onboard photo and video cameras of optical and infrared ranges; digital lay of land data; space photos and aerial photographs, and "signals of natural influence": gravity vector, the Earth's magnetic field, stars position of the Universe [20–24].

Generally, applications of US are wide and varied.US as a universal platform for a wide range of application, among the tasks to be solved, except direct purpose, may include the solution of side tasks, such as automatically patrolling protected areas (for example, their perimeter) [25–30]. An applications of US allows he following: reduce the reaction time to an emergency situation and the number of false alarms; patrol the perimeter (including in the dark); eliminate the influence of the human factor; reduce the time of detection of violations; perform aerial photography and mapping of hard-to-reach areas, while the possibilities of obtaining information is much wider than using airplanes; search people at any time of the day; delivery of individual means of rescue or other small size of cargo in cases where the use of manned vehicles is impossible or impractical (for example, at a great distance from existing airports); it is also possible to use as a communication relay with ensuring restoration, reservation or rapid expansion of deployed networks (lines); high-speed communication with groups; link in hard-to-reach areas (including infected, mountainous, marshy, in wooded areas, etc.); communication at high rates of movement and much more [1–13,14–32].

The task of determining the coordinates of the location, range of distance and angular position of US occurs with each case of practical application system. Determination of the coordinates of US is advisable to calibrate at the start before using them, when the elements and units of US came to a steady state, characterized in that all transients from switch-on time are completed. Reliability of control of an Autonomous mobile object is based on a systematic approach, more fully considering all the factors acting in solving the problem – from the method of navigation and features of means ensuring security to the psychological perception of the actions of the system by the operator. The dynamics, mobility and speed of movement of unmanned vehicles, often controlled by man, increase requirements for accuracy of navigation and visualization of vehicle control for operator. Thus, studies aimed at improving the quality of control for navigation of a mobile object in difficult navigation and weather conditions are relevant. Determining coordinates of Autonomous US is an extremely important task, which is solved by using different navigation systems. The most popular navigation systems for determining coordinates of the locations of Autonomous US are [13–18] inertial navigation systems (INS), which are peculiar by characteristic features – increasing coordinate measurement errors as time passes, as well as a sufficiently significant mass and cost. These factors [1–19, 25, 29] determine the limits of their use by aviation and sea transport. For reduce these very significant errors, are used additional means ensuring correction and initial installation of INS. Satellite navigation systems (SNA) [1], which have a number of advantages, such as an unlimited area of application, small size and cost, high enough accuracy, are also widely used. However [2, 13], well-known fact, when hijacking American US RQ-170, carried out by Iranian militaries, through action of external signal formed false coordinates and ensuring landing, shows that development of new means of navigation is relevant. This and other similar facts suggest that increasing resistance to interference is one of the priorities in control of US. The greatest noise immunity to external influences hold principle of control of US by wired or cable line [1, 20], ensuring a solution to a wide range of problems in a localized area of action.

The purpose of article is to create model and method of operational on-line determination range of distance Autonomous mobile object controlled by wire line for improve accuracy of determining coordinates of its position in the conditions of possible interference in the local control area.

2. Method of research

The method used by authors can be attributed to the method of precise reference. This method provides high accuracy of determination absolute coordinates on range even at absence of inertial navigation systems, at zero range of visibility in terrestrial environments and at bad meteorological conditions of application of US. It also allows you to find the absolute position of US with camera even when its approximate coordinates are unknown. This is achieved by scanning entire database with images of the area and comparing them with current image. In addition, having "referenced" picture to range can be determine with high accuracy coordinates of ground objects had detected by optical and electronic video equipment [1,12–27]. The essence of considered method of estimation coordinates on range of distance of US as a method of precise reference of coordinates on range is based on application of model and algorithm of wave method of estimation of properties and parameters of control line of US as a line with dispensed parameters [28–34]. To do this [33–37], we accept in the model that geometric center of US coincides with remote end of control line, with possibility of providing a controlled short circuit (SC) of end of the line. In principle, the method of standing waves is suitable for determining location of control lines damage of US.

3. Equations

The essence the method of standing waves is [32–37] that a high frequency harmonic signal source is connected to disconnected control line of US. In area of high frequencies, the electric control line of US can be considered as electric long line with dispensed parameters. If frequency of the test signal generated by the source and f_0 being the own resonance frequency of electric long line will be coincided,

IOP Conf. Series: Materials Science and Engineering 1019 (2021) 012080 doi:10.1088/1757-899X/1019/1/012080

the amplitude value of voltage (current) at the beginning of the control line of US will be maximum. With a sinusoidal test signal of frequency ω, the voltage at any point in the long control line of US can be represented as sum of two terms [33, 38]:

$$U = U_{+m}\exp(i\omega t) + U_{-m}\exp(i\omega t + \gamma x), \quad (1)$$

then U_{+m} is a complex amplitude of direct (+) wave of the test signal, V; U_{+m} is a complex amplitude of back (–) wave of the test signal, V; i is an imaginary number; γ is a propagation constant; x is a distance from beginning of control line to being measured location point of US, m.

The wave propagation in long electrical line is characterized by propagation constant, which generally has a complex form:

$$\gamma = \alpha + i\beta, \quad (2)$$

then α is an attenuation coefficient of wave in the control line of US, characterizes the change amplitude of wave per unit length of the line; β is a coefficient of phase change of an electrical signal in long control line of US [1, 33–35].

The point, in which value of phase of the test electric signal remains unchanged, moves in electric line with v is a phase velocity equal to:

$$v = 2\pi f/\beta, \quad (3)$$

then f=ω/2π is a linear frequency of test signal change in the control line. If we consider the initial time of test signal input into control line to be fixed (for example, t=0) and to consider the change of instantaneous value of test signal along control line depending on x being the coordinate of distance of US, then each of terms in (1) corresponds to harmonic wave. The wave described by the first term is a straight or falling wave. The wave represented by the second term is inverse or reflected [33].

If incident sinusoidal wave attenuates (decreases in amplitude) when moving from beginning of the control line of US, combined with the origin of coordinate system of the reference range to US, to end of the control line, combined with center of US, as well as reflected wave attenuates when moving from end of the control line, that is from center of US, to beginning of the control line, that is to origin of coordinate system of the reference range to US.

4. Results of research and discussion

For an example of the algorithm and method of application of the standing wave method [33–38] consider calculation algorithm for control line of US, parameters of which, generally, are given below. The length of the control line of US $l=l_k$, km; specific resistance of the line r_k, Ohms/km; specific inductance of line L_0, H/km; phase speed (wave propagation speed) v=297000 km/s. Suppose that at end of the control line of US (at point of location of US) occurred on single-phase short circuit to ground (the casing of US) [38]. Let determine f_l– frequency of the test signal, when standing waves effect occurs and estimate dispensing of voltage and current in the control line of US at a given frequency of the test signal. Considering this long control line of US as electric line without loss, let us determine f_0 as an own frequency of resonance such control line [33, 38]:

$$f_0 = v/4l_k, \quad (4)$$

Knowing the own resonance frequency of the electric control line of US, you can determine frequency of the test signal emitted into this line:

$$f_1 = 2f_0, \quad (5)$$

then the phase coefficient in the line will determine [33,38]:

$$\beta = \omega\sqrt{(L_0 C_0)} = \omega/v. \tag{6}$$

Figures 1 demonstrate the behavior of the amplitude damping coefficient α and the phase damping β, where α is an attenuation coefficient and β – coefficient of phase change; eq. 2 and 6); Figures 2 – dependence the phase velocity (v) on the distance (l_k) from measured location and f – linear frequency of test signal (eq. 3).

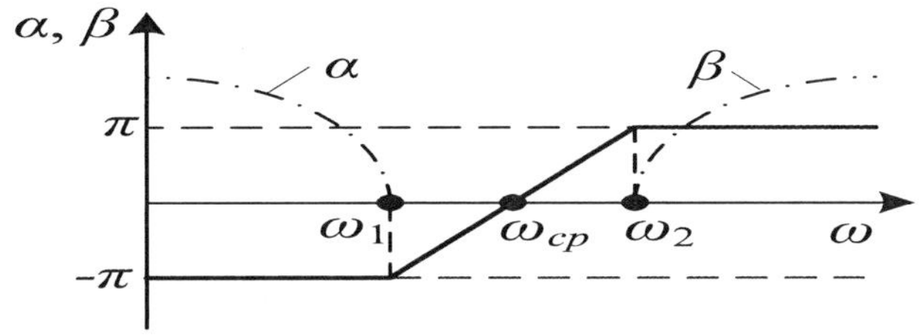

Figure 1. The structure damping (α –attenuation coefficient and β – coefficient of phase change) of the output signal in long electrical line is characterized by propagation constant: α – attenuation coefficient of wave in the control line of US, characterizes the change amplitude of wave per unit length of the line; β – coefficient of phase change of an electrical signal in long control line of U (eq. 2 and 6).

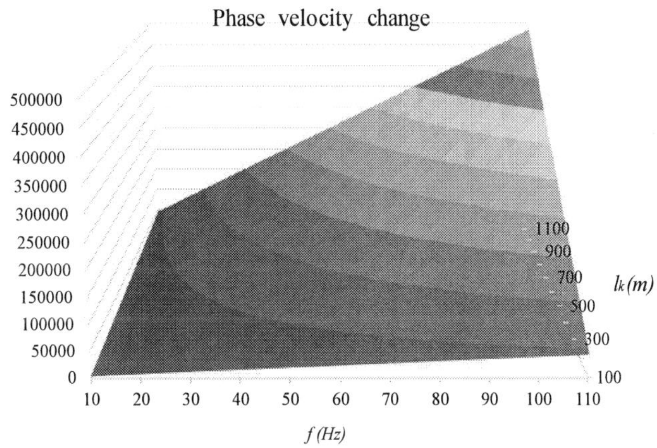

Figure 2. Dependence the phase velocity (v) on the distance (l_k) from measured location and linear frequency of test signal (f) (eq. 3).

The wave resistance of the control line of US will determine [33, 34]:

$$Z_0 = \sqrt{(L_0 C_0)}\,. \tag{7}$$

The length of the test signal wave in the control line of US will be equal to [33, 38]:

$$\lambda = 2\pi/\beta. \tag{8}$$

In accordance with (8), the length of the test signal propagation wave in the control line of US is proportional distance to damage site, in our case, to the geometric center of US. Note that the method of standing waves has been known for a long time, but it hasn't been widely used, because of the hardware complexity.

The modern level of development of electronics, radio, nanophotonics and microprocessor technology allows virtually and hardware with high speed to implement the on-line algorithm and the technique of high-precision remote determination of coordinates on range of distance of US by simply using the existing control line of US and elementary algorithm of the method for determining the location of line damage based on the effect of propagation of standing waves when short circuit in the line occurs [15–18, 21, 39].

The standing waves method supposes measuring total input resistance of damaged control line in a wide frequency range. The input impedance is most easily determined by measuring the current in line at a constant input voltage. Therefore, for determination distance as coordinate to the line closure point, that means to the center of the location of US by this method would requires a high-frequency sinusoidal signal generator, a meter of the input resistance of control line and microcontroller (MC), which in real time measures parameters of the electrical control line. The estimation technique is based on described above control model and distance to the site of damage determination algorithm or point of short circuit of the control line of US is combined with geometric center of US. When connecting to an undamaged line in MC by its definition, the length of this control line defined at the moment is automatically entered. Microcontroller is calculating initial frequency of measurement and forms remote monitoring algorithm and determining coordinate on range of distance of US. At the same time, the input impedance of the control line of US is determined by measuring the current when constant input voltage.

Changing the frequency of the test signal introduced into the control line, the MC finds, at maximum current, resonance frequency of undamaged line [33,34]. Subsequent connection to other lines associated with investigated US, their resonance frequencies are calculated [38]. By the ratio of frequencies of resonance at the length of the control line, MC is calculating distance to US, which corresponds to the point of given short circuit or damage of the control line of US [1, 33, 34]. Their coincidence with geometric center of US corresponds to coordinate on range of distance of US; non coincidence indicates problem with this control line.

The simulation results, numerical results and analytical description [35–38] statistically with a satisfactory accuracy of ~10% of the linear size of US coincide with data obtained on above mathematical model [35–41].

5. Summary

The method of determining coordinates of the location of US, controlled at the communication line, by remote on-line estimation on range of distance of US with knowledge of the angular position are similar to the methods used by man for orientation in three-dimensional space. The application of the above method in the management of deep-sea US during study of the sea depths, as well as US designed to work in hard-to-reach, man-made dangerous or closed earth and water spaces: dungeons, catacombs, caves and labyrinths of any type of environments is especially efficiently. Computer simulation of algorithm for remote on-line estimation of US's removal in closed earth space proved principal possibility of achieving position coordinates estimation of US by the wave method with an error not exceeding 0.01% of resonance frequency of US's used control line. When using the remote method for estimating on range of distance of US together with video navigation, which ensure provides an image of the terrain by camera, with subsequent analysis of this image by a special processor or computer, an exhaustive complete set of coordinates can be obtained for ensure Autonomous orientation of US. Based on the summary, the following conclusion can be drawn. The considered model, algorithm of estimation and technique of realization of the remote method for determining on range of distance of US by the wave method based on standing waves effect used for monitoring the state of electric lines and determining places of damage or short circuit in the lines is highly effective for remote monitoring of US state in critical and life-threatening situations, in hard-to-reach and rough terrain.

Acknowledgments

This work was supported by the Ministry of Education and Science of the Russian Federation in the framework of the implementation of the Program «Research and development on priority directions of scientific-technological complex of Russia for 2014–2020», and is funded from Tomsk Polytechnic

University Competitiveness Enhancement Program grant. Assistance in the publication of the manuscript was provided by Tomsk Polytechnic University within the framework of TPU DRIaP-79/2018.

References

[1] Tsupin A and Slobodyan S 2013 *Laser tools of navigation equipment for orienteering mobile objects* (Moscow: Meiler)
[2] Chatterji G *et al* 1997 *IEEE Trans on Aerospace and Electronic Systems* **33** 1012
[3] Liu T *et al* 2006 *AIAA Paper* **2006** 1437
[4] Koch A *et al* 2006 *AIAA Guidance Navigation and Control Conference* 2006
[5] Volkov V *et al* 1981 *Instruments and Experimental Techniques* **24** 1522
[6] Barber D B *et al* 2007 *J of Aerospace Computing Information and Communication* **4** 770
[7] Slobodyan M *et al* 2008 *Measurement Techniques* **51** 798
[8] Jones C *et al* 2006 *Proc of the AIAA Guidance Navigation and Control Conf* 2006
[9] Saripalli S *et al* 2003 *IEEE Transactions on Robotics and Automation* **19** 371
[10] Yakimenko O *et al* 2002 *IEEE Trans on Aerospace and Electronic Systems* **38** 1181
[11] Bol'shanin A A *et al* 1987 *Measurement Techniques* **30** 954
[12] Bondarev V G 2015 *Nautchnyi vestnik MGTU GA* **213** 54
[13] Kaiser M K *et al* 2010 *IEEE Trans on Aero & Electr Syst* **46** 1064
[14] Deeva V *et al* 2018 *International Journal of Engineering & Technology* (UAE) **7** 167
[15] Hu G *et al* 2007 *Proc of the IEEE Multi-conference on Systems and Control* 1474
[16] Deeva V *et al* 2019 *Journal of Tribology-Transactions of the ASME* **141** 031602
[17] Slobodyan S 2008 *J of Optical Technology* (A Translation of Opticheskii Zhurnal) **7** 301
[18] Sharp C S *et al* 2001 *Proc of the IEEE Int Conf on Robotics and Automation* 1720
[19] Slobodyan S 2003 *Measurement Techniques* **46** 28
[20] Eren F *et al* 2017 *Sensors* **17** 1741
[21] Slobodyan S 2006 *Measurement Techniques* **49** 1
[22] Bryson M *et al* 2008 *IEEE Trans on Aerospace and Electronic Systems* **44** 261
[23] Sorensen A J *et al* 1996 *Control Eng Practice* **4** 359
[24] Roberts J *et al* 2002 *Proc of the Australasian Conference on Robotics & Automation*
[25] Arutyunov V *et al* 1985 *Instruments and Experimental Techniques* **28** 176
[26] Prazenica R *et al* 2005 *AIAA Guidance Navigation and Control Conference*
[27] Trisiripisal P *et al* 2006 *AIAA Aerospace Sciences Meeting and Exhibit*
[28] Lefeber E *et al* 2003 *IEEE Trans Contr Sys Tech* **11** 52
[29] Arutyunov V *et al* 1983 *Measurement Techniques* **26** 641
[30] Alsaraj A *et al* 2015 *IEEE Adv Inf Technol Electron & Autom Contr Conf* 704
[31] Skjetne R *et al* 2004 *Modeling, Identification & Control* **25** 3
[32] Khamis A *et al* 2014 *International Journal of Aerospace Engineering* 1
[33] Beletsky A 1967 *Fundamentals of the theory of linear electrical circuits* (Moscow: Svyaz')
[34] Vashchuk S *et al* 2018 *Measurement Techniques* **61** 1148
[35] Vashchuk S *et al* 2019 *Atomic Energy* **126** 56
[36] Sorensen R *et al* 2004 *Automatica* **40** 373
[37] Cheng Y *et al* 2018 *Neurocomputing* **272** 63
[38] Whitehead A N 2016 *An Introduction to Mathematics* (New-York: Create Space Ind)
[39] Deeva V et al 2017 *Proc of the 27th European Safety and Reliability Conf* ESREL 2577
[40] Yan Z *et al* 2017 *Sensors* **17** 1599
[41] Deeva V *et al* 2020 *Applied Surface Science* **500** 143999

Study on interfacial tension of surfactant and its oil displacement performance

Risu Na and V I Erofeev
School of Earth Science & Engineering, Tomsk Polythetic University, Tomsk, Russia
E-mail:narisu2011@yandex.ru

Abstract. Tuha oil field is located at the Turpan basin in China. Currently, the water content of the main reservoir of the section Y-6 of the Tuha oil field is more than 93% and the recovery rate is less than 20%. At present, an efficient oil displacing agent must be selected for this oil field, which can be used in reservoirs with a high degree of mineralization. Based on the results of experiments measuring the interfacial tension of the surfactant solution and oil, a non-ionic surfactant was selected. Non-ionic surfactant can reduce interfacial tension to the ultra-low level (10^{-3} mN/m), and when the amount of adsorption by oil sand is 3, interfacial tension also can maintain at the level of 10^{-2} mN/m, so the solution of non-ionic surfactant has good anti-adsorption ability. The results of the oil displacement experiments show that the solution of a non-ionic surfactant can increase the efficiency of oil displacement. But under condition of high water cut, the dominant flow channel in the core is formed, and the surfactant solution flows into the main channel preferentially and reduces the residual oil saturation in the main channel, which leads to a decrease in the resistance coefficient and injection pressure. Therefore, after water flooding, take measures to control the fluid profiles firstly so that the following surfactant solutions change direction in areas not exposed to water in order to achieve the best effect of the increase in oil recovery.

1. Introduction

Oil displacement agent refers to the chemical substances used to improve oil recovery, mainly refers to the functional compounds that can reduce residual oil saturation through chemical reaction, namely surfactants [1, 2]. The surfactant molecule has two functional groups, the hydrophilic (water soluble) and hydrophobic (oil soluble) group [3, 4]. Two functional groups are easily absorbed at the liquid-liquid or gas-liquid interface and form a single-molecular film. Therefore, surfactant can reduce surface tension and change the surface condition, as a result, change the wetting angle, emulsification increase and others [5–7]. Depending on the characteristics of the hydrophilic group, the surfactants are divided into four (anionic, cationic, zwitterionic and non-ionic) groups. The commonly used surfactants for oil displacement are anionic, non-ionic and zwitterionic surfactants [8–11].

Tuha oil field is located at the Turpan Basin in China. Section Y-6 is the main water injection development area in the oilfield. The degree of mineralization of the formation water ranges from 115.5 to 191.8 g/l, and the concentration of Ca^{2+} and Mg^{2+} exceed 7 g/l. Currently, the water content of the main reservoir of the section Y-6 of the Tuha oil field is more than 93% and the recovery rate is less than 20%, so it is essential to use the enhance oil recovery technology to increase the oil recovery rate. At present, an efficient oil displacing agent must be selected for this oil field, which can be used in reservoirs with a high degree of mineralization. In this paper, based on the geological characteristics and fluid properties of the oil reservoir of the Y-6 section of the Tuha oil field, the suitable surfactant solution was selected, the interfacial tension of the solution with oil was measured, the adsorption characteristics and the oil displacement ability of the surfactant solution were evaluated.

2. Experiment description

2.1. Experimental materials
Surfactants include petroleum sulfonate produced by Daqing Refinery and Chemical Company (effective content 50%), heavy alkylbenzene sulfonate produced by Donghao Company of Daqing Oilfield (effective content 50%) and non-ionic surfactant (DWS), produced by Dalian Davis Technology Co., Ltd. (effective content 35%). Alkali include weak alkali sodium carbonate (Na_2CO_3, effective content 98%) and strong alkali sodium hydroxide (NaOH, effective content 96%). The experimental water is simulated injection water of the target oilfield, mass concentrations of ions (K^++ Na^+), Ca^{2+}, Mg^{2+}, Cl^-, SO_4^{2-}, CO_3^{2-} and HCO^{3-} are 51.433, 6.915, 0.393, 92.524, 0.129 and 0.059 g/l respectively. The overall degree of salinity of simulated injection water is 151.453 g/l. The oil used in the experiment is the crude oil of the Tuha oil field. Oil sands are obtained by crushing and sieving the natural cores of the target reservoir, and their particle size range from 40 meshes to 80 meshes. Cores are natural cores from the Tuha oil field.

2.2. Instruments and equipment for experiments
Interfacial tension between the displacing agent and oil was tested using the Spinning Drop Interfacial Tensiometer Model 500. Experiments were performed at a reservoir temperature of 55 °C. The main equipment used for the experiment for the oil displacement in heterogeneous cores includes a core holder, an advection pump, a hand pump for generating rock pressure, a pressure sensor and two intermediate containers. With the exception of the advection pump and the hand pump, the remaining parts were placed in a thermostat oven at 55 °C. Figure 1 shows the experiment-process diagram.

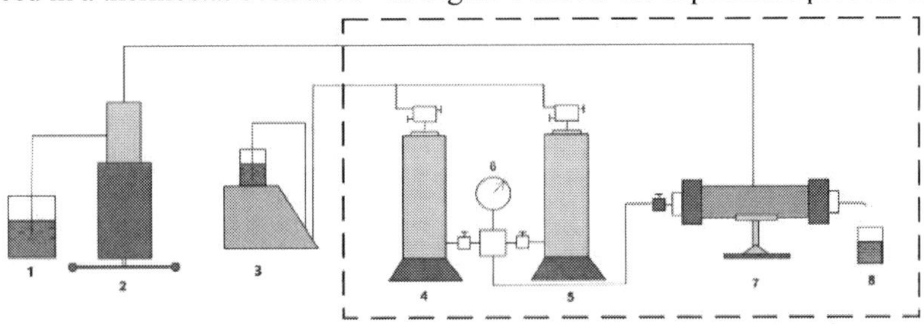

Figure 1. Experiment-process diagram: 1 – beaker, 2 – hand pump, 3 – advection pump, 4 – the container of water, 5 – the container with surfactant, 6 – sensor, 7 – core holder, 8 – measuring cup.

3. Results and discussion

3.1. Interfacial tension and adsorption characteristics of the solution of non-ionic surfactant and oil
The solution of strong alkali and heavy alkylbenzene petroleum sulfonate (1.2% NaOH + 0.3% heavy alkylbenzene petroleum sulfonate), weak alkali and petroleum sulfonate (1.2% Na_2CO_3 + 0.3% petroleum sulfonate) and non-ionic surfactant solution (0.3%) were prepared with 100 ml simulated injection water. The interfacial tension between the oil displacement agent and crude oil in different solution was measured by interfacial tensiometer. The results of measurement of interfacial tension are presented in table 1.

It can be seen from table 1 that the solution of strong alkali + heavy alkylbenzene sulfone and the solution of weak alkali + petroleum sulfonate can slightly reduce the interfacial tension of the solution and oil, comparing simulated injection water and oil. The non-ionic surfactant can reduce interfacial tension to the very low level of 0.0021 mN/m. In conjunction with this, the non-ionic surfactant (DWS) was selected for the next study.

Table 1. The results of measuring the interfacial tension (mN/m).

Composition	Interfacial tension
simulated injection water	2.95
1.2% NaOH + 0.3% heavy alkylbenzene petroleum sulfonate	2.91
1.2% Na_2CO_3 + 0.3% petroleum sulfonate	1.50
non-ionic surfactant (0.3%)	0.0021

Different concentrations of non-ionic surfactant (DWS) solutions were prepared with simulated injection water. The experimental data of the relationship between interfacial tension and concentration and time of DWS are shown in table 2.

Table 2. Relation between DWS concentration and interfacial tension (mN/m).

Concentration (%)	Time（min） 0	4	8	12	16
0.025	2.470	0.55000	0.1400	0.0590	0.04600
0.05	1.900	0.30000	0.0260	0.0044	0.00023
0.1	0.950	0.20000	0.0026	0.0012	0.00260
0.2	0.730	0.01400	0.0016	0.0050	0.00690
0.3	0.037	0.00036	0.0090	0.0036	0.00110
0.4	0.050	0.00160	0.0029	0.0021	0.00290

From table 2, it can be seen that the interfacial tension shows a downward trend as the test time prolongs, and some samples show the "downward, upward and downward" trend. The interfacial tension decreases with the increase of surfactant concentration. When the surfactant concentration reaches or exceeds 0.05%, the interfacial tension reaches ultra-low interfacial tension (10^{-3} mN/m).

Under the condition of 0.3% DWS concentration, the surfactant solution was contacted with fresh oil sand for many times, and the interfacial tension was measured 24 hours after each contact. The experimental data of interfacial tension are shown in table 3.

Table 3. The results of measurement of interfacial tension after adsorption (mN/m).

Parameters t	The amount of adsorption initial	1	2	3
Non-ionic surfactant (0.3%)	0.0011	0.0021	0.0130	0.0240

It can be seen from table 3 that with the increase of the amount of contact between the solution of non-ionic surfactant and fresh oil sand, the interfacial tension increases. When the amount of adsorption is 3, the interfacial tension can be maintained at level 10-2 mN/m, which indicates that the solution of non-ionic surfactant has good anti-adsorption ability.

3.2. Oil displacement efficiency of nonionic surfactant solution.

The experimental results of the effect of surfactant solution concentration and injection timing on oil displacement efficiency (oil recovery) are shown in table 4. The dynamic characteristics of injection pressure, water cut and oil recovery during core displacement experiments are shown in from Figure 2 to Figure 4.

Table 4 shows that surfactant solution can effectively reduce the interfacial tension between oil and water and improve the oil washing efficiency. Compared with experiment No.1 and No.2, experiment No.3 and No.4 have greater increase in recovery and higher final recovery. It can be seen that the earlier the injection time of surfactant solution, the higher the oil saturation of core, the higher the seepage resistance and the injection pressure, so the area affected by surfactant larger, and the increase of the recovery greater.

Table 4. Experimental data of oil recovery.

Experiment number	The content of the experiment	Core length (cm)	Permeability Kg (10^{-3} μm^2)	Recovery (%)	
				Water flooding	Final
1	Water flooding to water cut 98% + 0.2PV (pore valume of the core) DWS + subsequent water flooding to water cut 98%	6.9	110	39.47	43.86
2	Water flooding to water cut 98% + 0.4PV DWS + subsequent water flooding to water cut 98%	7.1	117	38.88	49.48
3	Water flooding to water cut 60% + 0.4PV DWS + subsequent water flooding to water cut 98%	6.2	131	27.23	56.68
4	Water flooding to water cut 0% + 0.4PV DWS + subsequent water flooding to water cut 98%	6.3	143	-	68.74

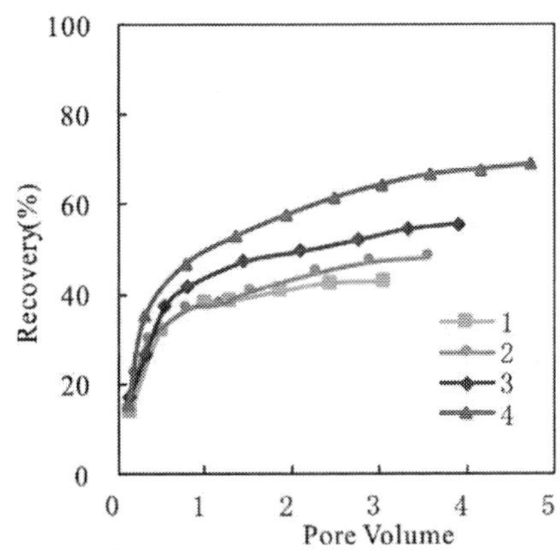

Figure 2. The relationship between injection pressure and pore volume.

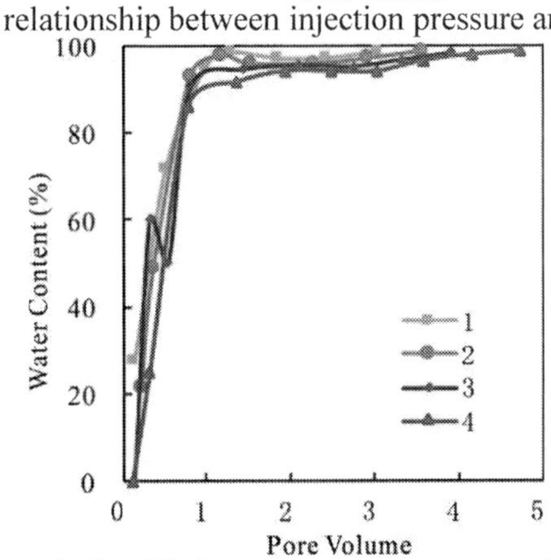

Figure 3. The relationship between water content and pore volume.

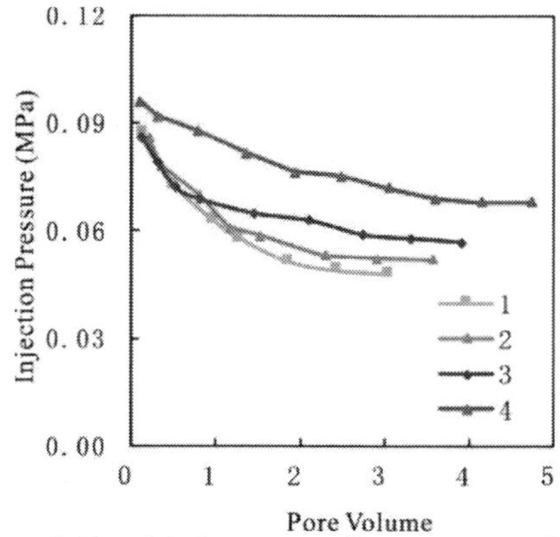

Figure 4. The relationship between oil recovery and pore volume.

From Figure 2 to Figure 4, it can be seen that the injection pressure of surfactant solution in displacement process decreases due to the poor retention ability of surfactant itself in core pore. Compared with experiment No. 1 and No. 2, injection pressure of experiment No. 3 and No. 4 is obviously higher, which indicates that displacement pressure gradient is larger, capillary diameter of the pore that surfactant solution can enter is smaller, so the effect of sweep volume enlargement and recovery increase is better. Thus, under the condition of high water cut, the dominant channel of flow has been formed in the core, and the surfactant solution will enter the dominant channel preferentially, further reduce the remaining oil saturation in the dominant channel, resulting in the decrease of seepage resistance and injection pressure. Therefore, after water flooding, profile control measures should be taken first to make the following surfactant solutions divert into areas not affected by water, so as to achieve better oil recovery increase effect.

4. Conclusion

The solution of strong alkali +heavy alkylbenzene sulfonate and the solution of weak alkali +petroleum sulfonate can slightly reduce the interfacial tension of the solution and oil compared with formation water and oil. The non-ionic surfactant DWS can reduce interfacial tension to the ultra-low level (10^{-3} mN/m), and when the amount of adsorption is 3, interfacial tension can be maintained at 10^{-2} mN/m level, so the solution of non-ionic surfactant has good anti-adsorption ability. The solution of non-ionic surfactants can effectively reduce the interfacial tension between the oil and the solution, and increase the efficiency of oil displacement. But under conditions of high water cut, the dominant flow channel in the core is formed, so the surfactant solution flows into the dominant channel preferentially and reduces the residual oil saturation in the dominant channel, which leads to a decrease in the seepage resistance and injection pressure. Therefore, profile control measures should be taken first after water flooding to make the following surfactant solutions divert into areas not affected by water, so as to achieve better oil recovery increase effect.

References

[1] Olajire A A 2014 *Energy* **77** 963–982

[2] Zhu Y Y, Zhang Y, Niu J L, Liu W D, Hou Q F 2016 *Petrol. Explor. Dev.* **39** 346–352

[3] Zheng L, Yang T, Liu X, Zhang R, Wang X 2014 *J. Shanxi Univ. Sci. Technol.* **32** 97–100

[4] Yang J, Li G, Liu Z, Feng Y, Zhou J, Wang Q 2015 *Unconventonal oil gas* **2** 46–49

[5] Xie K, Lu X, Pan H, Han D, Hu G, Zhang J, Zhang B, Cao B 2018 *SPE Prod. Oper.* **33** 1–11

[6] Narisu, Erofeev V I, Lu X, Lv J, Wang X, Zhang L 2019 *Bull. Tomsk Polytechnic University. Geo Assets Eng.* **330** 136-145

[7] Sheng J J, Leonhardt B, Nasser A 2015 *J. Can. Pet. Technol.* **54** 116–125

[8] Li R *et al* 2014 *Petrochem. Ind. Appl.* **33** 55–62
[9] Narisu, Erofeev V I, Lv J, Wang W 2019 *Bull. Tomsk Polytechnic University. Geo Assets Eng.* **330** 147–157
[10] Ma S *et al* 2009 *Fuel* **88** 1049–1056
[11] Narisu, Erofeev V I, Lu X, Tian Z, Zhang L 2019 *Bull. Tomsk Polytechnic University. Geo Assets Eng.* **330** 59–68

IOP Conf. Series: Materials Science and Engineering 1019 (2021) 012082 doi:10.1088/1757-899X/1019/1/012082

Varying spatial orientation of photoelectric panels as method of managing graph of electric energy generation

Y Daus[1], I Yudaev[1] and D Desyatnichenko[2]
[1]FSBEI HE «Saint-Petersburg State Agrarian University», Saint-Petersburg, Pushkin, Russia
[2]FSBEI HE «Don State Agrarian University», Zernograd, Russia
E-mail: zirochka2505@gmail.com

Abstract. One of the main issues in the determining layout of photovoltaic installations is choosing optimal parameters for photovoltaic panels and auxiliary equipment and duplicating energy sources, which makes it possible to justify economically the electrical energy generation. Regularity of solar radiation variation for the studied geographical point throughout the year, and the influence of atmospheric phenomena, as well as the dynamics of daily and seasonal electrical energy consumption must be considered, which will help to predict the operation mode of the plant, and match generation and consumption schedules. Variation of tilt angle of the photovoltaic module receiving surface relatively to horizon and cardinal points makes it possible to manage the schedule of power generation at solar plant and to match it with the load curve, due to the maximum possible convergence of them, as change in the spatial orientation of the photovoltaic panel surface makes it possible to shift the maximum of solar radiation flux in time. The results of the experimental research also revealed the dependence of the current produced by the photoelectric panel on its spatial orientation due to the change of the incoming solar radiation flux on it.

1. Introduction

The widespread usage of photovoltaic plants is due to first of all infinity of the Sun as renewable energy source, ecological purity of the production of electric energy, the simplicity in the implementation and operation of equipment. During last 6 years prompt and high evolution of solar power plant engineering became evident even under conditions of availability of fossil raw materials and fuel.

Such dynamics is primarily associated with decrease in the capital expenditures of projects and equipment of photovoltaic installations, decreasing their payback period. This is result of mostly of reduction in the specific primary expenditures and increase of total efficiency ratio of solar modules [1].

Most of the information presented in the open press is presented in general outlooks describing general indexes of the energy sector functioning. First of all, the utility photovoltaic plants that are connected to the National Grid of the country, and produce electric energy in large volumes, are described there. Most authors name high price of generated electricity to be the basic problem in the ubiquitous installation of solar systems, and argue the application of network and autonomous low-power plants to supply consumers of housing and communal sector, small-power production facilities, social and health sector, resort zones, outlying and rural territories to be one of the methods to solve it [2]. The listed categories also include objects of distributed generation, the projected growth of which for the next nine years in the world is 200% [3]. In [4] it is noted that number studies how to increase efficiency and decrease expenditures reduces and most of them are dedicated to the design of new layouts of low-power photovoltaic installations that would meet the need for electric power of small-capacity consumers. Introducing such low-power photovoltaic installations decrease both in the load of the power supply system and in power losses in the components of power grid at electric energy transportation to the consumer can be achieved.

One of the main issues in the determining layout of photovoltaic installations is choosing optimal parameters for photovoltaic panels and auxiliary equipment and duplicating energy sources, which makes it possible to justify economically the electrical energy generation [5. Regularity of solar radiation variation for the studied geographical point throughout the year, and the influence of atmospheric phenomena, as well as the dynamics of daily and seasonal electrical energy consumption must be considered, which will help to predict the operation mode of the plant, and match generation and consumption schedules. Such multicriteria selection of the required power of photovoltaic modules allows to reduce the costs of equipment, installation and maintenance of the system by reducing the number and power of photovoltaic modules and duplicating energy sources, and to justify their operation modes. Without changing design and configuration of projected solar power installation, increase of indicators of its efficient functioning and coordination of the schedules of electric energy generation and consumption by determining the optimal spatial orientation of the photovoltaic modules with respect to horizon and the sides of the world can be achieved.

2. Background

The variation of the tilt angle of coming the sun's rays on the receiving module surface has direct influence on the intensity of its irradiation, which allows the most complete utilization of the incoming solar radiation flux by means of trekking systems at power stations. The installation structure that observes the Sun, depending on the day time and season is 28–35% more efficient than the fixed installation [6]. However, this approach requires the presence of a mobile base for photovoltaic panels oriented for the Sun position in the sky, which is associated with the manufacture of a more complex system of their constant spatial displacement, and also the use of high reliability attachments requiring special maintenance. For the adjustment and operation of the tracking sensor for un movement in the sky, special high-precision means of software and electronic control of this process are necessary, which would have individual settings considering both level of actinometrical conditions, various influencing climatic factors [7, 8]. As a result, tracking systems have a high cost [9], and the character of the generation schedule corresponds only to the schedule of the solar radiation arrival during the day.

This leads to difficulties in matching the modes of electrical energy generation and consumption. The solution of such issues is carried out, most often, on the basis of monthly or annual amounts of electricity production, without taking into account the fact that the selection of spatial orientation, first of all, affects the justification and forecasting of operating modes of power equipment [10].

Therefore, it is advisable to study the effect of the tilt angle relatively to horizon and cardinal points of the unregulated fixed photovoltaic module, characterized by the simplicity of the system design and its relatively low cost, on the position and value of the maximum of the solar radiation that comes on it.

Analysis of scientific literature has revealed that large number of such studies is most often aimed at increasing the level of incoming insolation upon module by varying its tilt angle by means of analytical and experimental approaches.

The existing computational methods for determining the optimal inclination angle on horizon and orientation along the sides of the world are designed to solve the problem of maximizing the arrival of direct component of solar radiation at any time and are intended for systems for tracking the Sun [11]. But this approach cannot be applied to fixed photovoltaic panels, since all the components of solar radiation are not taken into account. For the south of Russia, a high proportion of diffused radiation in the radiation balance is typical, which is within 25% in summer and 60% in the winter months [10]. Neglecting the diffused and reflected components of solar radiation will lead to an overestimation of the generating equipment power and low accuracy in predicting its operation mode. To maximize the arrival of total solar radiation on receiving surface, there are no strict analytical dependences.

In [12–14], the effect of tilt angle relatively to horizon and cardinal points of the stationary module on the form schedule and value of solar radiation maximum is widely studied.

In [12], the values of the cosine of the incidence angle of the sun's rays on the differently orientated in space areas were analyzed. The regularities of the variation in the maximum value of insolation at diverse tilt angles relatively to horizon for 4 directions along the sides of the world were noted. It can be concluded from the analysis results that at the oriental orientation the maximum is reached at inclination angles of more than 40° for 900 and 60° for 830. It is not possible to make inferences about the evening time, since there are no corresponding data, because the studies were carried out only for a period of time from 900 to 1500.

In [13], the average monthly values of hourly values of the total and diffused solar radiation, arriving at the horizontal and vertical surfaces with different orientations around the world for June and January, were studied. It was revealed that in June, due to the high altitude of the Sun in comparison with January, the horizontal surface gets much more insolation than the vertical one. It was also noted that in the morning and in the evening a large irradiation was detected at the vertical platform, oriented to the east and west respectively. And although the direction of the receiving surface oriented to the south gives maximum at noon, for January, greater decrease in its value compared with other orientations was found. In work [13], the influence of only the extreme values of the inclination and orientation angles of the photovoltaic panel on the amount and regularities of the arrival of solar radiation was studied.

In the last three above-mentioned works, the issues of the effect of tilt angle relatively to horizon and cardinal points of the stationary module on the position and maximum value of the insolation arriving at it under different angles are considered. However, the dependence between spatial orientation and maximum of the incoming insolation, are not fully considered, since in most cases only strict orientation along cardinal points and/or the vertical and horizontal position of the receiving surface are taken into account. This is caused by the fact that attention has been paid to the research of irradiance of buildings, rather than photovoltaic panels in a solar power plant.

Thus, in order to justify the parameters of the photovoltaic installation components and also to harmonize electric energy generation and consumption schedules by choosing the operation mode of installation, it is necessary to evaluation of spatial orientation effect of the stationary module on the position and maximum value of insolation coming on them must be studied.

3.Methods of Research

3.1. Method of experimental research

Photo of designed and manufactured experimental set-up for experimental studies is shown at Figure 1.

The installation consists of photovoltaic panel 1 ($Pmax$ = 5W, Ump = 9V) and control photovoltaic element 2 (8.84 W/m²/mA for a horizontal surface) fixed on a movable base allowing to adjust their inclination angle relatively to horizon and cardinal points with the help of rotary-tilt mechanism 7 providing two degrees of rotation freedom, which makes it possible to model change in spatial position of module by means of mechanism levers. To perform precise adjustment of the tilt angle relatively to horizon electronic protractor 3 is used, and to the sides of the world - a scale with a calibration step of 1°, located at the fastening base. The adjustment is from 0° to 85° in the vertical plane and from 0° to 360° in the horizontal plane.

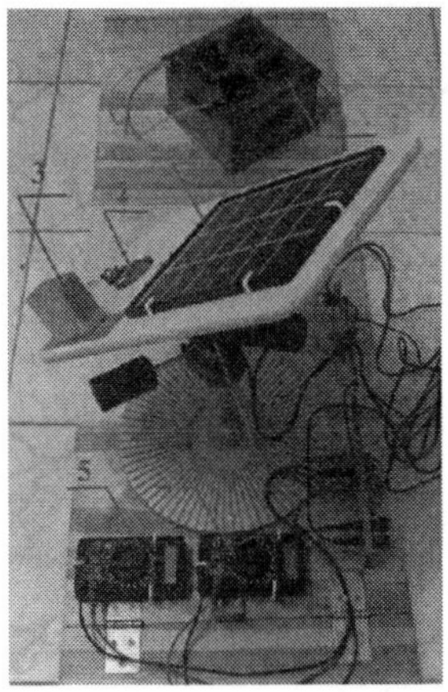

Figure 1. External view of the experimental installation.

To simulate the connected load, laboratory resistance 6 from 1 ohm to 9999 ohms is used. Fixation of current and voltage at the terminals of the photovoltaic panel and the control element is carried out with the help of multimeters 4 and 5. All the components of this experimental installation are fixed rigidly to the wooden base of the folding structure, which makes it possible to move, deploy, adjust and operate it at any geographical point.

The method of experimental research was as follows. The volt-ampere characteristics of the photovoltaic panel and the determination of its output power were taken at natural sunlight on a site unshadowed by neighboring building structures and green plantations according to [15, 16] at a geographic coordinate of 46.8°N. 40.3°E. (Zernograd, Rostov Region).

3.2. Method of computational research

In order to solve the problem of estimating tilt angle effect on the schedule graph and maximum value of output of electric energy during the day, there must be detected orientation (p) of module, in which, at a constant power and a specific number of photoelectric panels, the amount of insolation coming on it at considered time period will be the maximum as the generation directly depends on the flux of the sun's rays. The theoretical thesis to solve this task is the mathematical formulation of the optimization issue using the criterion – the maximum of the hourly values of total insolation at time t, with the objective function being:

$$\begin{Bmatrix} R^p_{sum} = R^p_{dr} + R^p_{df} + R^p_{rf} \rightarrow \max \\ tn \\ for \quad t = t_1..t_2 \end{Bmatrix} \tag{1}$$

where $R^p_{sum}, R^p_{dr}, R^p_{df}, R^p_{rf}$ are hour sums of the total, direct, diffused and reflected insolation components, respectively, falling on the receiving surface at position p corresponding to the angle β to horizon and γ to the sides of the world, kWh/m2. Calculation of the hourly sums of solar radiation is carried out using express methodology to calculate the solar energy potential at set geographic location [17, 18].

Following limitations were adopted solving above optimization issue:

- values of inclination angle of module relatively to horizon equal 0° – 90°;
- values of tilt angle of module by cardinal points equal -90° – 90°;
- considering technical implementation possibility of the obtained solutions calculations accuracy is limited by β and γ integer values.

The solution to the objective function is in fact a solution to the search for the optimal orientation of the PV panels for each hour of the day in question, the purpose of which is to increase the utilization of the solar potential of the area, as well as the efficiency of the photovoltaic panels used.

IOP Conf. Series: Materials Science and Engineering 1019 (2021) 012082 doi:10.1088/1757-899X/1019/1/012082

The solution to the objective function is actually solution of the search for the optimal orientation of the PV panels for each hour of the considered day, the purpose of which is to increase the solar potential utilization of the area, as well as the efficiency of the used photovoltaic panels.

4.Results and their discussion

Figure 2 presents graph of the solar radiation flux on the receiving surface at the selected optimal positions for each hour by the example of June 22.

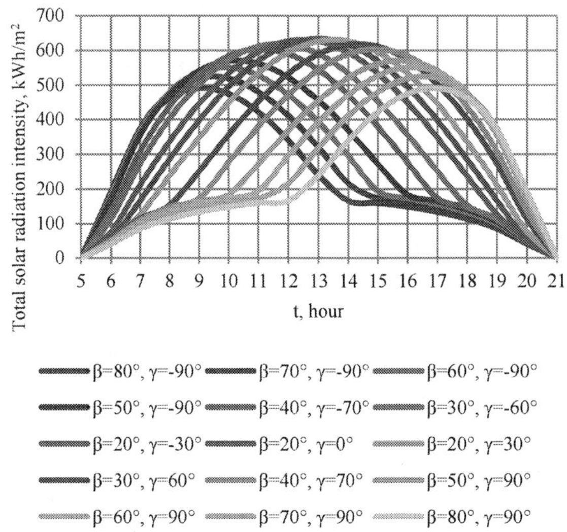

Figure 2. Hourly values of total insolation on the module at hourly optimal tilt angles.

By varying the tilt angle of the photovoltaic panel of experimental setup, shift in the maximum of the solar radiation flux in time was also revealed. The results of the analysis of the nature of the coming solar radiation on the differently oriented receiving surfaces of photovoltaic panels have revealed that due to different combination of the spatial placement of module surface respectively to cardinal points and horizon, it is possible to shift the maximum power take-off relatively to noon for the morning and evening time by about four hours.

Figures 3, 4 present the daily energy production curves obtained by calculation and experimentation for 4 variants of tilt angle of module – in this case of the photovoltaic panel of the experimental setup.

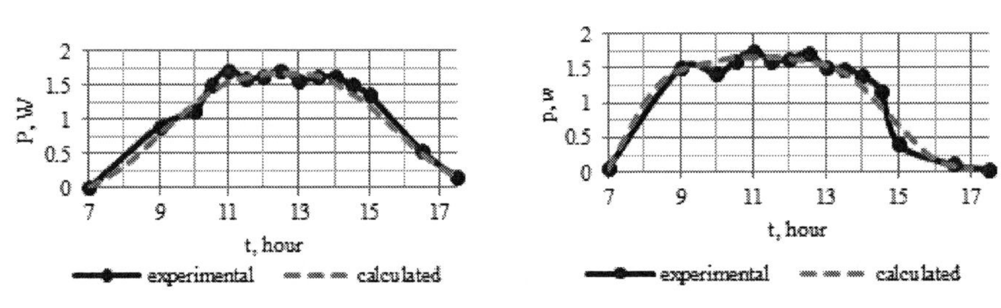

a) b)

Figure 3. Daily graphs of electric power generation by a photovoltaic panel, obtained by calculation and experimental way for the following spatial orientations: $\beta = 20°$, $\gamma = 0°$ (a); $\beta = 40°$, $\gamma = -45°$ (b).

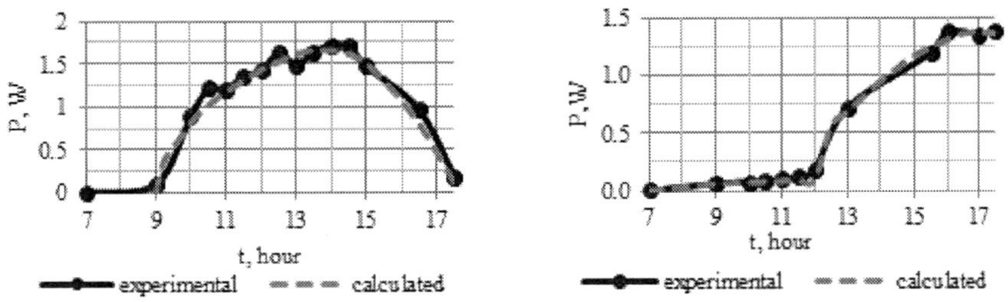

a) b)

Figure 4. Daily graphs of electric power generation by a photovoltaic panel, obtained by calculation and experimental way for the following spatial orientations: $\beta = 40°$, $\gamma = 45°$ (a); $\beta = 60°$, $\gamma = 90°$ (b).

The results of experimental and computational research have shown that the obtained error value between them is acceptable and does not exceed 10 %. The difference between the actual actinometric and the calculated data is caused by the fact that the calculation model is based on data averaged over 22 years, and the experiment was carried out for a day of one year. The instantaneous total insolation value is affected by cloudiness, which can change dynamically for short time periods (even few minutes), and be different for the one and the same day of various years.

5. Summary

Thus, it can be stated that the variation of photovoltaic module's tilt relatively to horizon and cardinal points makes it possible to manage schedule of power generation at the solar power plant and to match it with the load graph, due to their maximum possible convergence, as variation in the spatial orientation of the photovoltaic panel surface makes it possible to shift the maximum of the coming solar radiation in time. The results of the experimental research also revealed the dependence of the current produced by the photoelectric panel on its spatial orientation due to the change in the incoming solar radiation flux at it.

Acknoledgements

This research hasn't received grants from funding agencies in the public, commercial, or not-for-profit sectors. Nonetheless, authors of this work would like to acknowledge associate professor of the Operation of Power Equipment and Electrical Machines department G. V. Stepanchuk for help in organizing experiment. Without his support, conducting full study would not have been possible.

References

[1] Kharchenko V and Daus Yu 2015 *Industrial technology and engineering* **3** 1–6
[2] Belenov A T , Daus Yu V, Rakitov S A, Yudaev I V and Kharchenko V V 2016 *Applied Solar Energy* **52** 105–108
[3] *Distributed Generation. An Overview of Recent Policy and Market Developments.* Available at: https://www.publicpower.org/files/ PDFs/Distributed%20Generation–Nov2013.pdf.
[4] *Next Generation Wind and SolarPower. From cost to value.* Available at: https://www.iea.org/publications/freepublications/publication/NextGenerationWindandSolarPower.pdf.
[5] Markvart T and Castaner L 2003 *Practical handbook of photovoltaics: fundamentals and applications* (New–York: ELSEVIER) p 998
[6] Dahecha K, Allouchea M, Damaka T and Tadeob F 2017 *Electric Power Systems Research* **143** 182–188
[7] Patkó I, Szeder A, and Patkó C 2013 *Energy Procedia* **32** 222–229

IOP Conf. Series: Materials Science and Engineering 1019 (2021) 012082 doi:10.1088/1757-899X/1019/1/012082

[8] Yurchenko A V and Kozlov A V 2007 *Proc. of XXII European PV Solar Energy Conference and Exhibition* 3019–3022
[9] Corrada P, Bell J, Guan L, and Motta N 2014 *Energy Procedia* **48** 806–812
[10] Mahjoub Essefi R, Souissi M, and Hadj Abdallah H. 2014 *International Journal of Modern Nonlinear Theory and Application* **3** 53–65
[11] Vissarionov V I *et al* 2008 *Solar energy: manual for universities* (Moscow: Publishing house MPEI) p 276
[12] *Matching Utility Loads with Solar and Wind Power in North Carolina: Dealing with Intermittent Electricity Sources*. Available at: www.ieer.org/reports/NC–Wind–Solar.pdf
[13] Jamil Ahmad M, and Tiwari G N 2010 *International Journal of Energy and Environment* **1** 257–276
[14] Chandel Shyam S , Rajeev K Aggarwal 2011 *Smart Grid and Renewable Energy* **2** 45–52
[15] *Standart GOST R 51597–2000 Alternative energy. Modules solar photo-electric. Types and basic parameters* 2000 (Moscow: IPK Publishing house of Standards) p 6.
[16] *Standart GOST 28977 – 91 (904–1–87). Photovoltaic devices. Part 1. Measurements of photoelectric volt-ampere characteristics* 2004 (Moscow: IPK Publishing house of Standards) p 7.
[17] Daus Yu V, Kharchenko V V and Yudaev I V 2016 *Applied Solar Energy* **52** 124–129
[18] Daus Yu V and Yudaev I V 2016 *Proc. of EMASS2016* (Beijing: DEStech Publications Inc) 306–309

Method for numerical optimization of the parameters of regulator from unstable condition

A A Voevoda, D O Romannikov
Automation Department, Novosibirsk State Technical University, Novosibirsk, Russia
E-mail: dmitry.romannikov@gmail.com

Abstract. The task of selecting the values of coefficients for a given controller structure from the initial unstable position is considered using the methods of numerical optimization. A method is proposed which consists in correcting the error function which is used for optimization. The main idea is multiplication by the exponent in the degree of the simulation step which the convergence of the transition process is achieved. After reaching convergence, you can perform numerical optimization. It is recommended to reduce the value of the exponent so that the figure of the transition process remains convergent. Further, when a stable transition process is reached, the value of the exponent should be zeroed out and then the optimization procedure for the error of the transition process should be performed without adjusting it until the change in the value of the objective function changes less than the specified threshold value. An example of such optimization is considered on the example of the transfer function of a second order oscillator.

1. Introduction

At present, the task of developing regulators for closed systems is relevant and is used in a variety of automatic and automated industrial systems. During the development and research of the theory of automation, a number of methods have been created that allow the structure and values of the controller coefficients to be obtained in [1, 2] proposed. All the above methods allow to obtain a regulator for the linearized model.

Another approach to the creation of control systems is the direction of learning with the confirmation of [3, 4], which assumes the absence of detailed information about the object and the formation of the control law on the basis of the obtained statistics of the object's responses to specified control actions. But reconciliation of the teaching direction methods with confirmation faces the problem that for some objects the formation of the control law takes quite a long time (there is a theoretical justification for the convergence of training with confirmation for time tending to infinity, which is hardly applicable in practice).

On the other hand, numerical optimization [3, 5] methods are often used for a stable closed system, which allow adjusting the controller coefficients depending on the optimization criterion used. At the same time, the authors did not encounter papers that propose the use of numerical methods for obtaining controller coefficients for an initially unstable system.

A closed system is considered with the object W(s) and the regulator $W_r(s)$. The structure of the regulator is obtained by one of the existing methods for the synthesis of regulators, for example [1, 2]. In the problem under consideration, according to the known structure of the regulator, the values of the initial coefficients are arbitrarily chosen in such a way that the closed system is in an unstable state. In this case, the task is to apply numerical optimization methods in order to obtain controller values to ensure the minimum error of the optimization criterion function.

The above problem was considered for the transfer function of the object $W(s) = \frac{1}{s^2}$.

In conclusion, we propose the idea of how to switch to neural networks from the obtained regulator structure. The resulting structure was also applied to the proposed method of stabilizing a closed system.

2. An optimization method with known regulator structure

In the article, for convenience of simulation, discrete models of transfer functions are used with a sampling time of 0.1 s, which have the following form:

$$W(z) = \frac{0.005z + 0.005}{z^2 - 2z + 1}$$

$$W_r(z) = \frac{az + b}{z + c}$$

The optimization criterion is Q = | v – y |, where the value of the setpoint v takes the value of one. The initial values of parameters a, b, c are chosen in such a way that a closed system is unstable, for example, a = -5, b = 10, c = -10. With the selected initial conditions, the simulation time of 3 seconds, optimization did not produce any results, and the optimization criterion did not change its value. Thus, we can conclude that the optimization by numerical methods of an unstable system is a rather complicated task. Figure 1 shows a graph of the divergent transition process, which is obtained as a result of performing an optimization procedure with an error module as an optimization criterion (similar results were obtained for the sum of squares of error as an optimization criterion).

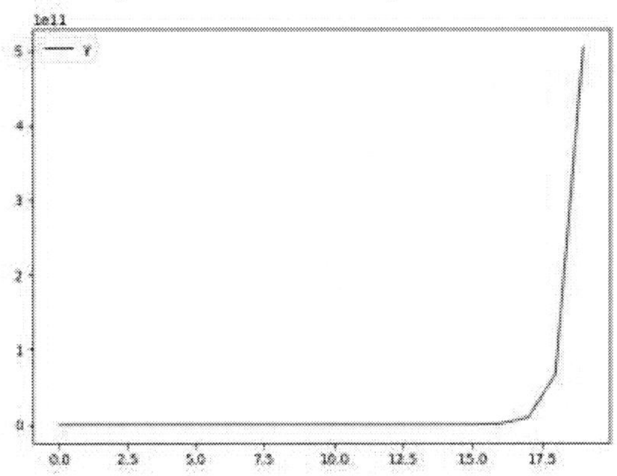

Figure 1. Divergent transition process of the studied closed system, obtained as a result of the optimization process.

To optimize the unstable closed dynamic system, the authors propose to additionally multiply the error by the factor e^{-k*i}, where i is the iteration number in the simulation. The above multiplier allows you to make a graph of errors converging and allows you to perform optimization. The purpose of this multiplication is to obtain a converging error graph and, as a consequence, the possibility of optimizing it. At the same time, in the course of research, it was found that for the convergence of the optimization process it is necessary to adjust the value of the coefficient k so that the error graph remains convergent, but not too close to zero.

Thus, for it is possible to formulate an algorithm for the method of finding parameters of a regulator according to a known structure.

1. For the object W, specify the structure of the regulator W with unknown values of the coefficients.
2. To obtain the value of the transient error during the simulation time T.
3. Select the value of the coefficient k so that the error graph, corrected by the factor e^{-k*i}, is convergent.

IOP Conf. Series: Materials Science and Engineering 1019 (2021) 012083 doi:10.1088/1757-899X/1019/1/012083

4. Perform the optimization procedure until the value of the optimization coefficient changes by less than e.

5. If the graph of the transition process has become stable, then set the value of the coefficient k to zero. Otherwise, go to step 3.

6. Perform the optimization procedure without considering the correction of the error value (k = 0) until the change in optimization criterion will be less than e.

The authors used the simulation to optimize the stochastic optimization Adam [6, 7] with a learning parameter value of 0.05, without regularization, with exponential decay rates of the first and second moments of 0.9 and 0.999, respectively.

3. An example of applying an optimization method with a known regulator structure

Consider the application of the optimization method on the example of the transfer function of the object $W(s) = \frac{1}{s^2}$. In the discrete representation, it has the form: $W(z) = \frac{0.005z + 0.005}{z^2 - 2z + 1}$ with a sampling time of 0.1 seconds. According to the above method, it is necessary to specify the structure of the regulator, which can be selected as (as + b) / (s + c). Under the initial conditions of -5, 10, -10 for the coefficients a, b, c, respectively, the graphs of the error are presented in Figure 2.

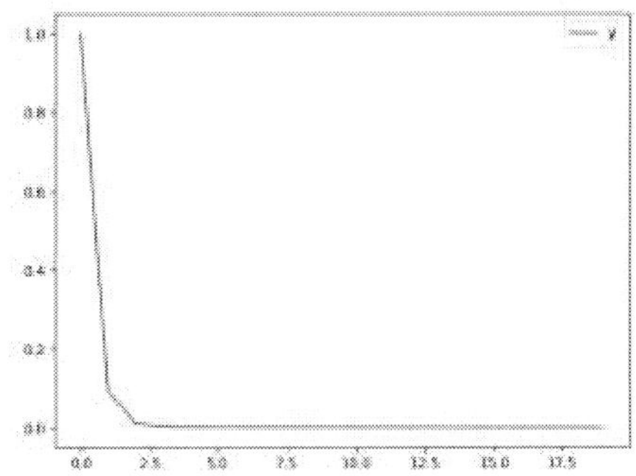

Figure 2. The initial figure of the transition process error of the studied closed system, corrected by the factor e^{-k*i} with k = 2.4.

Further, according to the method proposed above, the value of the coefficient k was corrected several times and consistently took the values 0.4, 0. In this case, the graph of changes in the values of the coefficients has the following form (see Figure 3).

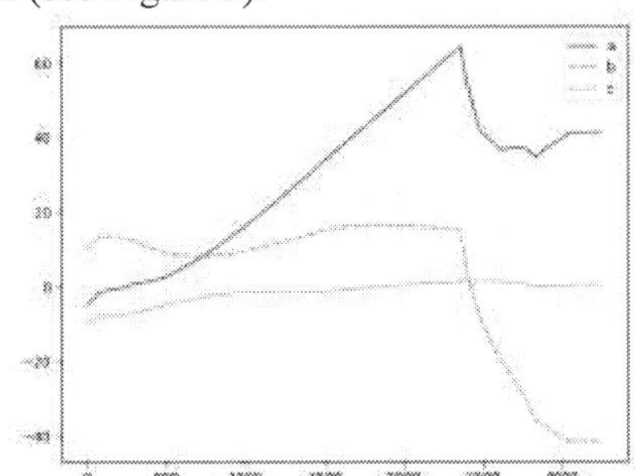

Figure 3. Changes in the values of the coefficients a, b, c.

In Figure 3, the number of simulations is shown on the abscissa axis, and the values of the coefficients a, b, c are plotted on the ordinate axis. The change of the coefficients occurred at 2300 and 2900 iterations.

The graph of changes in the values of coefficients in simulations from other initial conditions (the values of the coefficients a, b, c are equal to -40) is shown in Figure 4.

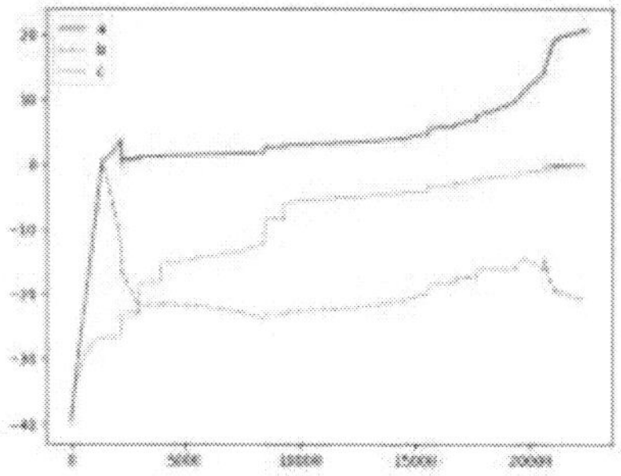

Figure 4. Figures of changes in the values of the coefficients during optimization starting from -40.

From Figure 4 it is clear that the simulation required a greater number of iterations. In this case, abrupt changes in the values of the coefficients on the graph also correspond to differences and coincide with the moments of the change in the value of the coefficient k.

A further area of research for this task is to replace the controller coefficients with neural networks while maintaining the controller structure.

4. Using Neural Networks with a Known Regulator Structure

Using the structure of neurons, which allows to divide the input signal into ranges (it was originally used for static tasks [8–10], for which the values of the coefficients of neurons could be calculated) did not give a positive result due to too sharp transition between the shared ranges. The values of the coefficients of the above structure for the replaced coefficient c from the structure of the regulator are shown in Figure 5.

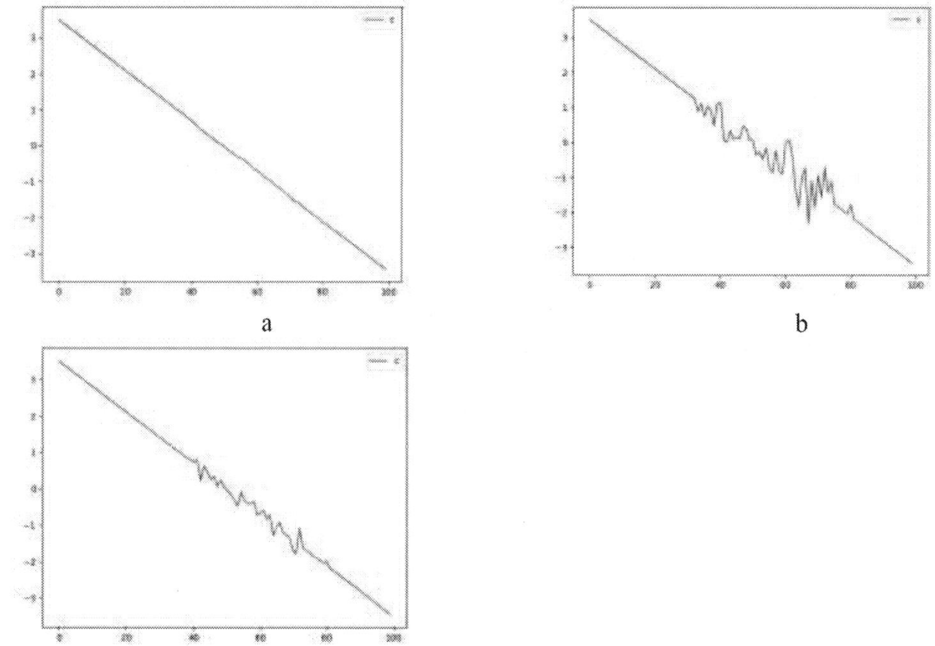

Figure 5. Figures of the neural structure used instead of the coefficient with: a) after 100 iterations; b) after 1000 iterations; c) after 10,000 iterations.

To prevent significant differences in values, it is proposed to use an approximation of the value between two neurons instead of "hard" partitioning. For example, when dividing a range from minus one to three using four neurons, as shown in Figure 6.

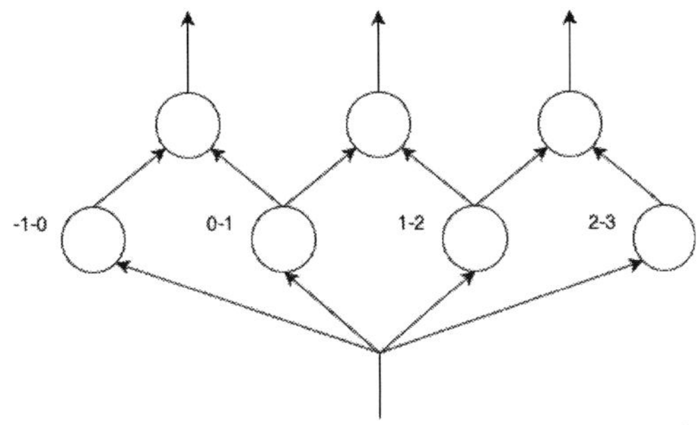

Figure 6. The structure of the neural network for the range from -1 to 3.

In Figure 6, when the value of the input value falls into one of the ranges, the interpolation of a straight line between neighboring neurons is taken to form the output value of the considered neural structure, which allows optimization of the resulting structure.

5. Conclusion

The article proposes a method for optimizing the controller values by its given structure from the initial unstable position, which is based on stabilizing the error graph by multiplying it by e^{-k*i}, where i is the iteration number of the closed system simulation, k is a coefficient to ensure the convergence of the graph mistakes. With a stable graph of errors, it is proposed to perform the optimization procedure gradually reducing the value of the coefficient k. An example of the application of the proposed method on the object $W(s) = \frac{1}{s^2}$ and the structure of the regulator W(s) = (as + b) / (s + c) is considered.

References

[1] Besekersky V A, Popov E P 1975 *Theory of automatic control systems* (Moscow:Nauka) p 768
[2] Kim D P 2003 *Automatic control theory* (Moscow: Fizmatlit) p 203
[3] Bishop C 2007 *Pattern Recognition and Machine Learning Information Science and Statistics* (Berlin: Springer) p 738
[4] Mnih V, Kavukcuoglu K, Silver D, Graves A, Antonoglou I, Wierstra D, Riedmiller M 2013 *Playing Atari with Deep Reinforcement Learning*. Available at: https://arxiv.org/pdf/1312.5602.pdf
[5] Voevoda A A, Romannikov D O 2017 *IEEE Transl. Proceedings of SPIIRAS* **54** 205-223
[6] Hastie T, Tibshirani R, Friedman J 2001 *The Elements of Statistical Learning: Data Mining, Inference, and Prediction* (New York: Springer) p 745
[7] Hagan M, Demuth H, Jesús O 2003 *Engineering Applications of Artificial Intelligence* **16**(2) 121–129
[8] Voevoda A A, Romannikov D O 2018 *Transaction of scientific papers of the Novosibirsk state technical university* **1**(91) 104–111
[9] Voevoda A A, Romannikov D O 2018 *Transaction of scientific papers of the Novosibirsk state technical university* **1**(91) 86–94
[10] Voevoda A A, Romannikov D O 2017 *Proceedings of SPIIRAS* **54** 205–223

Implementation of vibration signals receiving unit on Raspberry single-board computers

V A Faerman[1], A V Tsavnin[2] and S A Andreev[2]
[1]Institute of System Integration and Security, Tomsk State University of Control Systems and Radioelectronics, Tomsk, Russia
[2]School of Computer Science & Robotics, Tomsk Polytechnic University, Tomsk, Russia
E-mail: fva@fb.tusur.ru

Abstract. The paper dedicated to vibration measurement system development on the Raspberry Pi basis. Main features of the solution are low-cost, easy access to components and functionality and flexibility provided by Raspberry Pi usage. Circuit solution and configuring procedures are presented and described. In particular, block diagram of the vibration measurement system with piezo-electric sensors. Two variants of the vibration measurement system, on the Raspberry Pi B+ and Raspberry Pi 3B correspondingly, were implemented and tested as well. In addition, two variants of impedance-matching device were implemented. Conducted and described experiments confirm performance both separate components and the whole solution. Presented results are applicable for correlation leak detectors new algorithmic solutions and in educational process.

1. Introduction

The main trend in modern instrument manufacturing is microprocessor devices usage as for different signal processing algorithms as for diagnostic purposes. Implemented algorithms determine operability and performance of microprocessor units generally [1].

Nowadays, new algorithms development and testing are main trend in the field of advanced facilities of technical diagnostics. New algorithms testing requires hardware and software solutions application for signal processing [2–4]. These solutions are applicable in educational process as a part of software-oriented education of engineers [5, 6].

In this paper vibration measurement system implementation on the low-cost Raspberry Pi computer basis. Developed solutions are used for time-delay estimation algorithms testing. The algorithms are used in leak detection correlation devices software, described in [6, 7].

2. Components and computers

2.1. Concept of the solution

The solution comprises four interconnected blocks: vibratacoustic sensors, interface module, ADC and single-board computer. Structural diagram of the solution depicted in Figure 1.

Interface module task is analog sensors output and ADC input impedance and level matching. Single-board computer that calculates and processes the data controls ADC.

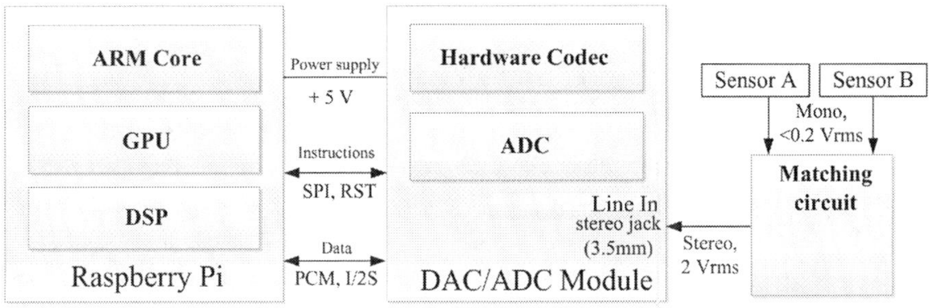

Figure 1. Structural diagram of Raspberry based receiver for vibration signals.

2.2. Vibroacoustic Sensors
Piezo-electric accelerometers are used as a vibroacoustic sensors.

Presented devices has the following features: low level of output voltage, that proportional to vibro acceleration of the oscillating object; sensor output circuit impedance influence on precision and frequency measurement range [8].

In the research accelerometers DN-3 and DN-4 [9] are used.

2.3. Hardware
According to Figure 1, Raspberry Pi B+ [10] and compatible Wolfson audio card is used for receiving, discretization and collecting signal records in stereo format [11]. Despite the presented devices are assumed to be obsolete and are not manufactured nowadays, they are still applicable for signal processing problem. In particular, in [12] said that Raspberry Pi B+ is powerful enough for time-delay estimation problem in real-time.

2.4. Raspberry Pi 3B
Alternative version of signal receiver implemented on the most-recent Raspberry Pi 3B+ basis [13]. The most significant feature is almost three times more powerful CPU in comparison to Raspberry Pi B+ [14]. HiFiberry DAC+ ADC module [15] is used for signal discretization.

3. Impedance and level matching circuit
Necessary condition for proper signal receiving is output circuit level and impedance matching with ADC. According to sensors specification [9], in order to fix the lower bound of frequency range at level of tens of Herz, it is necessary for input impedance to be not less than 100 kOhm with input capacitance about 1μF.

Since sensors have different sensitivity and thus provide different level of output voltage, matching device gain has to be adjustable in wide range. As a result two devices were developed. Circuits of the devices are shown in Figure 2.

3.1. Matching circuit without filtering
Matching devices are constructed on the basis of the cascade of two serial-connected instrumentation and differential amplifiers. Instrumentational amplifiers application is native to measurement systems since it combines differential amplifiers advantages (high input impedance, coherent noise rejection) but it has stable gain value with low level of self noise [16].

Output amplifier provides necessary level of differential input impedance 200 kOhm with 1 μF capacitance. That yields that $f_{\min} = 7Hz$ according to [9].

The GPU_FFT library allows unrelated data arrays to be grouped into batches immediately before calling the calculation function. It reduces the amount of time spent on context switches when processing large data sets [12]. A practical implication in our research is that when calculating the correlation, you can produce two forward FFTs with a single function call.

$$f_{\min} = \frac{0.79\left(R_S + R_3\right)}{R_S \cdot R_3\left(C_S + C_2\right)}, \tag{1}$$

where R_S – sensor insulation resistance (10 GOhm);

C_S – sensor self capacitance with cable (2000 pF).

The second amplifier comprises dimmer for gain adjusting in the range from 1 to 3, in order to work with both DN-3 and DN-4 sensors. The output of the circuit supplied with Zener diode based limiter for avoiding ADC failure [17].

3.2. Matching circuit with low-pass filtering

The second circuit differ from the first by low-pass filter with cutoff frequency $f_c = 50\ Hz$

$$f_c = \frac{1}{2\pi\left(R_3 + R_4\right)C_2} \tag{2}$$

The filter reduces electric pickups influence [18]. Although, low-pass filter, according to (1), increase $f_{\min}$ value to $f_{\min} = 577$ Hz, experiment shows operability of the device with frequencies from 100 Hz.

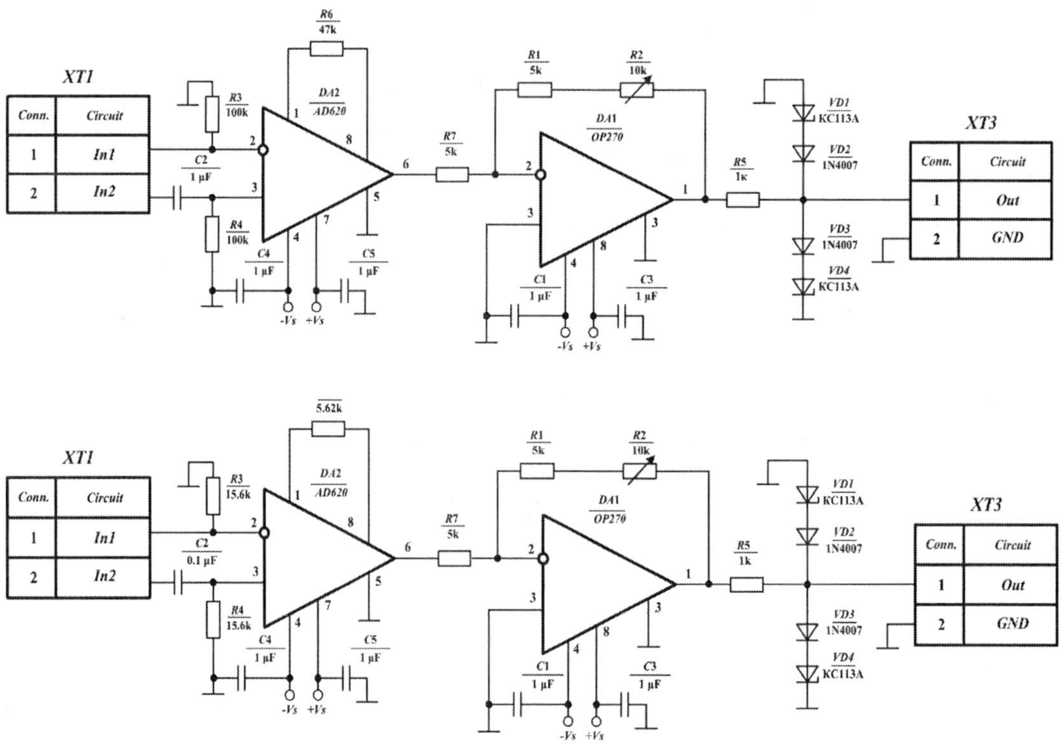

Figure 2. Electrical circuits of the matching devices variants (in the top – the one without filtering, in the bottom – the one with low-pass filtering).

4. Implementation of the receiving unit

As well as aforementioned hardware software essential. Since the research conducted in Raspberry Pi basis oriented way, software development comes down to single-board computer configuration.

4.1. Raspberry Pi B+ configuring

Single-board computer Raspberry Pi B+ with mounted Wolfson audio card configuring process includes the following steps: driver installation, record and playback interface choice and record format setting. In order to provide proper interfacing between Raspberry and audio card, the driver installation should be performed using "apt-get" utility for UNIX-based operating systems. The next step is record device choice. Since Wolfson audio card has several onboard interfaces such as Sony/Philips Digital Interface (SPDIF), digital microphones, etc. In the research 3.5 Line in connector is used and is activated via running Bash script "Record_from_lineIn.sh". The final step is proper recording. In order to get record with required quality and properties, one has to form terminal command of the following format:

"arecord –c 2 –f S16_LE –r 44100 <filename.wav>"

where –c defines number of channels, in our case 2,
–f defines recording format (16-bit, Little Endian) and
–r defines sample rate.

4.2. Raspberry Pi 3B+ configuring

Since both Raspberry Pi B+ and Raspberry Pi 3B+ use the same UNIX-based operating system Raspbian Jessie, configuring process comprises almost the same steps as a Raspberry Pi B+. The main difference is that Hifiberry allows users to adjust inner gain value for input signal with hardware onboard jumpers. Possible values are 0dB, 12dB and 32db. For the experiment, 0 dB value was chosen in order not to inject additional disturbances.

4.3. Interfaces and wires

Stereo jack 3.5 mm was chosen as a connectors in order to simplify wiring and experiment conduction. Mono signals channels were soldered with each other.

System shown in Figure 3. The second variant of matching device was implemented in metal case for electromagnetic pickups reduction.

5. Operability test

Experiment was conducted in laboratory conditions. Experiment includes receiving two time-shifted copies of vibroacoustic signals. Through stereo amplifier, PC-plugged vibrodynamics utilized as a playback device. More detailed experiment description presented in [19]. Experimental setup shown in Figure 3.

In Figure 4 actual and reference signals spectrum characteristics are depicted. Spectrum difference is due to non-homogeneous surface where sensors and dynamics were positioned and time-delay between channels as well.

During the experiment it was discovered that time-delay estimation can be perform using received vibroacoustic signal copies. Time-delay estimation precision is 0.5 ms (1.5% of relative length of correlation function time scale, shown in Figure 4). This fact allows system to be utilized for leak detectors software testing in laboratory conditions [3].

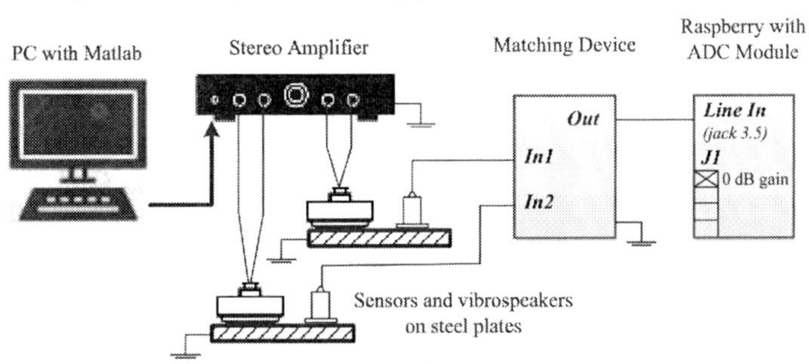

Figure 3. Experimental setup.

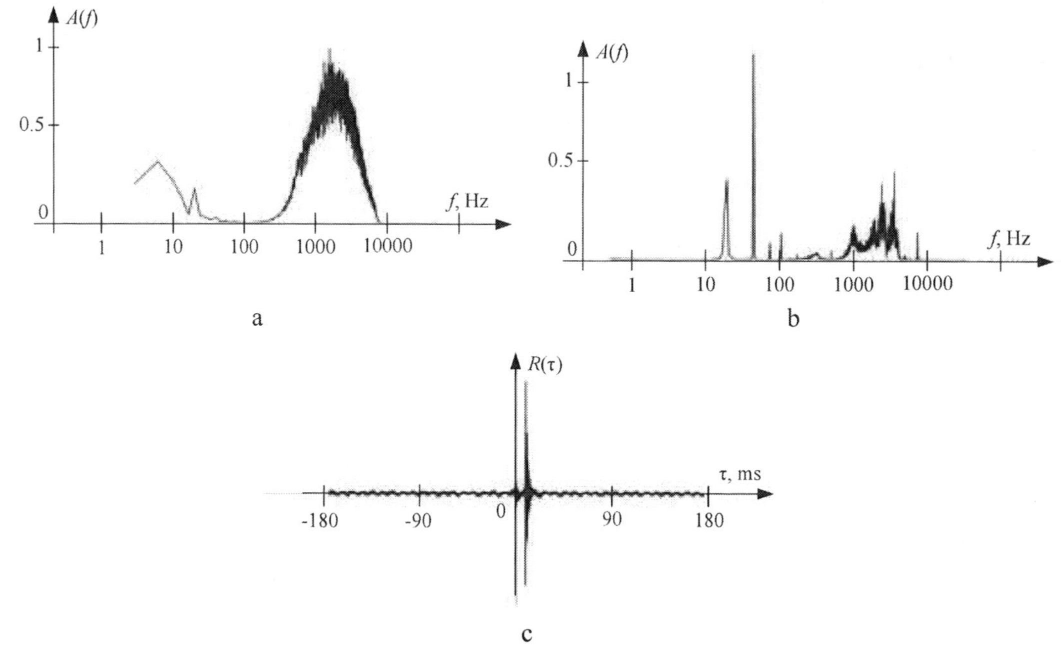

Figure 4. Specters of the: a) power cross spectrum of the reference signal; b) power cross spectrum of the received signal; c) cross-correlation between channels of the received signals.

6. Conclusion

Developed and tested vibroacoustic signals processing on the Raspberry Pi single-board computer basis device is properly functioning and applicable as for research leak detector model as for educational purposes. One of the main advantages of Raspberry Pi-based solutions is high-performance (Raspberry Pi 3B+ in particular). It allows not only performing software signal processing including electrical pickup reducing, but also time-delay estimation in real-time scale.

The next step of the development is deploying software algorithms directly on single-board computers and their performance research. It gives the opportunity to get correlation leak finder functioning prototype that can be used with pipe virtual model and in field conditions as well. Another important direction of the future study is applying the receiving unit with the software [20] to developed lab bench [19] to use it in education as cyberphysical system. For this purpose, we have plans to develop a Matlab library to simulate a pipe rig in a bench using the methods applied in [21].

References

[1] Chen Y-K, Chakrabarti C, Bhattacharyya S, Bougard B 2009 *IEEE Signal Proc. Mag.* **26**(6) 24–25
[2] Chen J, Benesty J, Bhattacharyya S, Huang G 2009 *J. Acoust. Soc. Am.* **138** 1737
[3] Brennan M J, Kroll de Lima F, de Almeida F C L, Joseph P F, Paschoalini A T 2016 *Appl. Acoust.* **102** 137–145
[4] Holovatyy A, Tesluyk V, Iwaniec M, Mashevska M 2017 *Eastern-European Journal of Enterprise Technologies* **6** 52–62
[5] Fuchs H V, Riehle R 1991 *Appl. Acoust.* **33** 1–19
[6] Ionel R 2015 *Proc. of International conference of the IEEE Industrial Electronics Society* 2465–2471
[7] Pasolini G 2017 *IEEE Signal Proc. Mag.* **34**(4) 151–158
[8] Steps to selecting the right accelerometer 2017 *Endevco Newsletters* 7. Available at: https://endevco.com/news/newsletters/2012_07/tp327.pdf

IOP Conf. Series: Materials Science and Engineering 1019 (2021) 012084 doi:10.1088/1757-899X/1019/1/012084

[9] Charactristics of the vibration sensors DN-3, DN-4 2014 Datasheet in Russian. Available at: http://izmeri.ru/08.html
[10] Introducing the Raspberry Pi Model B+ 2015 Blog record. Available at: https://cdn-learn.adafruit.com/downloads/pdf/introducing-the-raspberry-pi-model-b-plus-plus-differences-vs-model-b.pdf
[11] Wolfson audio card user documentation 2014 Official datasheet. Available at: https://www.farnell.com/datasheets/1805130.pdf.
[12] Dalskov D 2014 *Locating acoustic sources with multilateration:* master thesis (Aalborg university)
[13] Raspberry Pi 3 model B+ 2018 Official datasheet. Available at: https://static.raspberrypi.org/files/product-briefs/Raspberry-Pi-Model-Bplus-Product-Brief.pdf
[14] Raspberry Pi 3 B+ specs and benchmarks 2018 Official release. Available at: https://www.raspberrypi.org/magpi/raspberry-pi-3bplus-specs-benchmarks/
[15] HiFiberry DAC + ADC Datasheet 2018 Official datasheet. Available at: https://www.hifiberry.com/build/documentation/datasheet-dac-adc/
[16] Tietze U, Schenk C, Schmidt E 2008 *Electronic circuits* (Berlin: Springer-Verlag)
[17] Andreev S A, Kim V L 2017 *Proc. of the 2016 Int. Conf. on Information technologies in science, management, social sphere and medcine* **1** 803 doi: 10.2991/itsmssm-16.2016.70
[18] Kay A 2012 *Operational amplifier noie: Techniques and tips for analyzing and reducing noise* (Newnes: Elsevier)
[19] Faerman V, Tsavnin A 2019 *AIP Proceedings* **2195** 020006
[20] Faerman V, Tsavnin A, Shvetsov M 2020 *Journal of Physics Conference Series* **1615** 012004
[21] Tsavnin A, Efimov S, Zamyatin S 2020 *Symmetry* **12**(7) 1092

Wettability inversion of aluminum-magnesium alloy surfaces

A Islamova and E Orlova
School of Energy and Power Engineering, Tomsk Polytechnic University, Tomsk, Russia
E-mail: agi2@tpu.ru

Abstract. The paper presents the experimental results on the use of low-temperature heating to reduce time of wetting inversion (from superhydrophilicity to hydrophobicity) of aluminum-magnesium alloy surfaces textured by laser radiation. Stable growth of the contact angle to 137.3–144.2° after heating surfaces (wettability properties deteriorate) was recorded. Wetting inversion from superhydrophilicity to hydrophobicity occurs in 2–3 hours of low-temperature heating of textured samples. The wettability inversion time depends on the type of texture. A significant increase in carbon content of elemental composition of the near-surface layer of samples after their low-temperature heating was registered.

1. Introduction

Laser processing is the most promising way to obtain surface structures with unique functional properties, as well as controllable wetting, spreading and evaporation of droplets [1,2]. Such surfaces are applied in aviation, energy, shipbuilding, medical equipment, biotechnology, microelectronics. It is used for non- and metallic surfaces, has low capital and operating costs, a fast and simple process easily scaled for details of a complex profile.

Many studies showed that immediately after laser texturing, the surface of the freshly treated metal was hydrophilic or superhydrophilic with the presence of micro / nanostructures [3–5]. When the laser-textured surface was exposed to ambient air for a relatively long time, it was possible to observe a transition of wettability from superhydrophilicity to superhydrophobicity [5–10]. Consequently, superhydrophobicity can be achieved on laser-textured metal surfaces when stored their in the ambient condition. The inversion time is different for different metals. For example, copper or brass textured by a nanosecond laser takes about 11–14 days to become superhydrophobic [11,12]. Jagdheesh et al. [13] reported that wettability inversion of laser ablated aluminum takes about 40 days. Whereas the wettability change of femtosecond laser ablated stainless steel requires more time than other metals (52–60 days) [14, 15].

To reduce the time required for the transition metal wettability from hydrophilic to superhydrophobic without using any chemical coating, it is proposed to use low-temperature annealing. D.-M. Chun et al. [16] reduced wettability transition time of pure copper plate treated by nanosecond pulsed laser from 2 weeks to several hours using low-temperature annealing (100 ° C).

In this work, we strive to find out the influence of low-temperature heating on the wetting inversion from super-hydrophilicity to hydrophobicity of laser-textured aluminum-magnesium alloy surfaces

2. Methods and materials

At present work, the aluminum-magnesium alloy is selected because of its widespread use in industry. Samples made of aluminum-magnesium alloy (Al 91.2, Mg 6.8, Mn 0.8, Fe 0.4, Si 0.4, Zn 0.2, Ti 0.1, Cu 0.1 in wt%) with a thickness of 5 mm and a diameter of 50 mm were used. Before texturing, the surfaces were polished using the "Grinding Polishing Machine MP1B". We used ultrasonic cleaner with distilled water and chemically pure isopropyl alcohol (C_3H_8O) for removing contaminants. Three-dimensional non-contact optical profilometer "MicroMesure 3D station" was used to investigate the texture of samples.

IPG-Photonics ytterbium nanosecond pulsed fiber laser (MiniMarker-2 Laser Center, 1064 nm wavelength) was used for laser texturing procedure. The procedure was carried out in air at a temperature of 20–22 °C, relative humidity of 35–40% and atmospheric pressure. The samples were affected by single laser pulses: duration is 200 ns, high average power – 20 W and frequency – 20 kHz. By varying the beam linear speed v (mm/s) and the number of lines n (mm^{-1}), three samples with different texture were created (table 1). The texture of sample No 3 formed at n = 7.1 mm^{-1} and v = 2800 mm/s is characterized by crater-like elements located at some distance from each other.

With such values of linear speed and number of lines, the light spots do not overlap on the surface. When v was reduced and n was increased, the nearest craters edges on the surface contact with each other (sample No 2). A partial overlap of the light spots occur with a further change in these parameters, and a texture of sample No 1 is formed with randomly located drops and jets of molten and solidified metal. Thus, two different types of textures were created: periodic (samples No 2 and No 3) and anisotropic (sample No 1).

Table 1. Laser parameters.

Parameters of Laser Beam Spatial Displacement	**Sample No**		
	1	*2*	*3*
n (mm^{-1})	20	15	7.1
v (mm/s)	1000	1320	2800

Images of the microstructure of aluminum-magnesium alloy surfaces textured with laser radiation were obtained using a scanning electron microscope (SEM) and are presented in Figure 1.

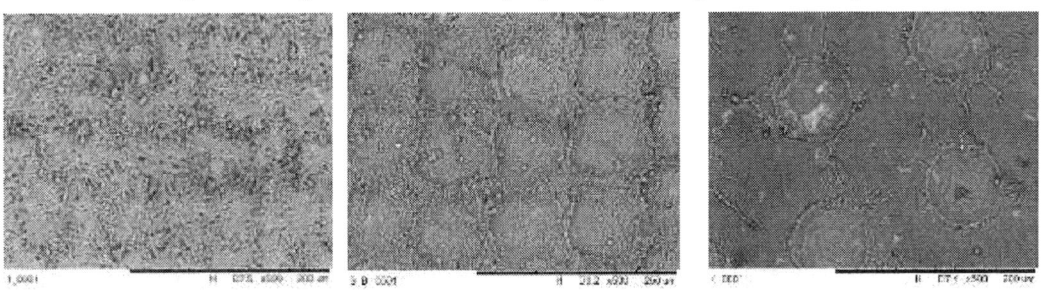

(a) (b) (c)

Figure 1. Surfaces microstructure images obtained with a scanning electron microscope. Sample Number: (a) No 1; (b) No 2; (c) No 3.

Experimental studies of wetting inversion were conducted on the setup with the use of shadow optical system (Figure 2).

We used low-temperature heating (heating of samples from their lower side by thermal conduction) for accelerating the wetting inversion process. This heating differs from low-temperature-annealing [17] implemented in the furnace where heat is transferrred by conduction, convection and radiation.

A polymer glass laboratory box (3 mm thick) was used to isolate experimental setup from external influences (convection). The substrate was mounted on the working cell, which consisted of a goniometer and a silicone heater connected to a laboratory autotransformer. The surface was heated to 100 ° C for six hours. The temperature under the substrate and on its surface was recorded by chromel-alumel thermocouples (with a measurement error of ± 0.1 °C). The temperature difference did not exceed 0.1 °C in the longitudinal coordinate direction. In addition, humidity inside the box was recorded.

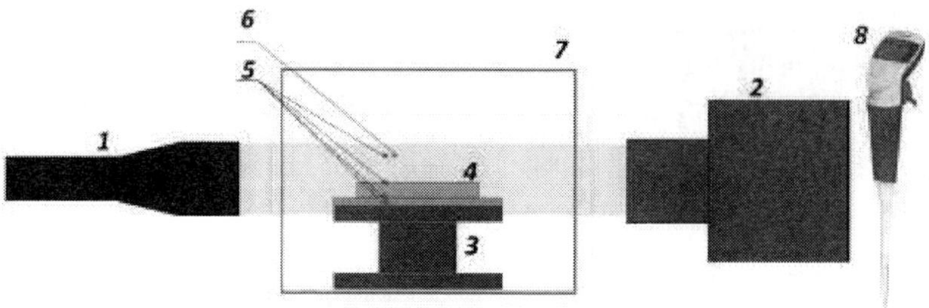

Figure 2. Experimental setup: 1 – light source; 2 – high-speed video camera; 3 – goniometer; 4 – substrate; 5 – thermocouples; 6 – humidity sensor; 7 – transparent box; 8 – dispensing device.

To measure the static contact angles, the shadow method was used. A beam of light from source passed through a fiber-optic illuminator, falling into a telecentric tube, where it was transformed into plane-parallel light. This light illuminated a droplet formed on the surface of the substrate with a dispensing device. Video camera with a macro lens was used to obtain photographs of droplets. The latter were processed by goniometry methods for determining contact angles. The random error of measuring contact angle did not exceed 5 %.

The elemental composition of the samples was analyzed on the Hitachi S-3400N scanning electron microscope using energy dispersive spectroscopy equipped with a Bruker XFlash 40 EDS chemical analysis unit. Studies were conducted before and after heating the samples.

3. Result and discussion

The contact angle was measured immediately after laser texturing. Then samples were heating for 6 hours at 100 °C, and contact angles were measured every hour. For each measurement, a droplet was placed on a previously unwetted surface area. The obtained results are presented in Figure 3. The static contact angle of polished surface of aluminum-magnesium alloy (before laser texturing) was 88.1°.

Figure 3 shows hydrophilicity of all samples that immediately after texturing ($\tau = 0$). At $\tau = 0$, the static contact angle increased in the sequence from sample No 1 to No 3. After two hours of heating, the contact angles on all samples increased: samples No 2, 3 showed hydrophobicity with $\theta = 142.9°$ and $\theta = 133.3°$, respectively, and sample No 1 remained hydrophilic ($\theta = 22.5°$).

At $\tau = 3$ hours, we registered the wetting inversion of sample No 1 from hydrophilicity to hydrophobicity. The wettability properties on sample No 1 stabilized more slowly than that on other one. Ngo and Chun obtained similar results that the decrease in the distance between the effects of light spots while texturing leads to an increase in the stabilization time of the static contact angle [17].

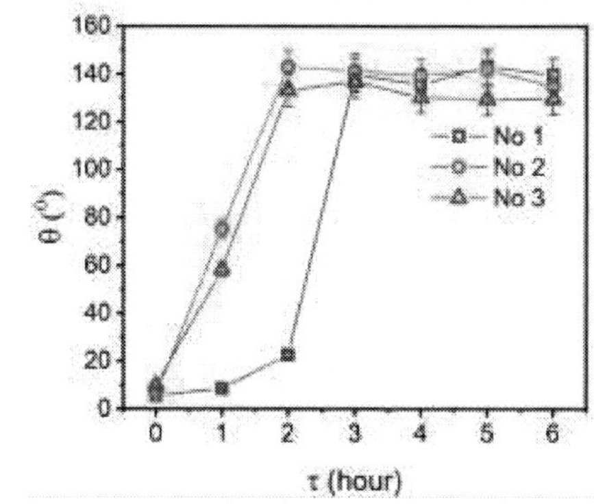

Figure 3. The static contact angle of textured samples at the heating time.

When all surfaces showed hydrophobic properties, in the range from 3 to 6 hours of heating, contact angle increased in the sequence from sample No 3 to sample No 1. In addition, 3–6 hours range is characterized by constant contact angles over heating. In Figure 4 shows typical photographs of water droplets before and after heating the substrate for six hours.

(a) (b)

Figure 4. Typical photographs of distilled water droplets on the surface immediately after texturing (a) and after six hours of heating (b).

In addition, for 1 month after heating the static contact angles on samples No 1–3 were measured. Table 2 shows the results of measurements of the static contact angle immediately after texturing with laser radiation, after 6 hours of heating and at 1 and 30 days after heating.

Table 2. Static contact angles measured on textured surfaces before and after heating.

Sample No	After Laser Texturing	After 6-hour Heating	One Day after Heating	Thirty Days after Heating
1	4.0±0.3	137.3±6.7	135.7±5.4	137.2±5.0
2	7.3±0.3	144.2±6.0	144.6±3.5	142.4±3.2
3	11.8±0.5	130.8±5.5	132.6±3.3	121.3±4.3

All samples retained hydrophobicity after heating, confirming the possibility of using low-temperature heating of an aluminum-magnesium alloy textured by laser radiation in order to reduce the time of wetting inversion from superhydrophilicity to hydrophobicity.

In [17], the change in the wettability properties of stainless steel treated with laser radiation after 6–20 hours is explained by an increase in carbon in the elemental composition of the samples. The latter is adsorbed from CO_2, which is part of the air. When samples are stored in the atmosphere, the reaction of CO_2 decomposition with carbon adsorption occurs slowly. Heating intensifies the decomposition process; the wetting inversion is accelerated.

Thus, the carbon adsorption can lead to a change in the wettability properties. To confirm this assumption, the elemental composition of the near-surface layer of anisotropic (sample No 1) and periodic (sample No 3) textures was analyzed using the EDS method. Table 3 presents the mass elemental composition in percent for samples No 1 and 3 before and after heating.

Table 3. Mass element composition.

	Al	Mg	O	C	C/Al
			wt %		
No 1 (before heating)	52.92	4.74	38.86	3.48	0.066
No 1 (after heating)	44.19	4.29	28.50	23.03	0.521
No 3 (before heating)	88.44	5.78	4.14	1.64	0.0185
No 3 (after heating)	55.18	4.60	13.21	27.01	0.489

The elemental composition of the near-surface layer of sample No 1 was obtained by averaging the percentage of Al, Mg, O, C over the area. The data presented in table 3 for sample No 3 correspond to the measurements of the elemental composition in the center of the crater (Figure 5). The elemental composition of sample No 2 was not analyzed since it was similar with that of sample No 3 since the conditions of the texture element formation was similar, the difference was in the parameters of laser beam spatial displacement.

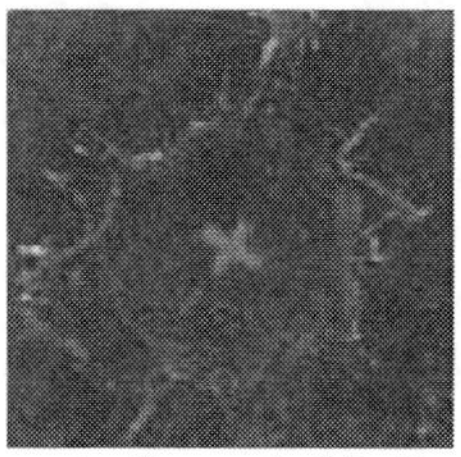

Figure 5. SEM image of the texture element of sample No 3 with the designation of the point at which the elemental composition was determined by EDS method.

Table 3 presents that after heating the samples, the carbon content in the elemental composition increases. The C/Al ratio for samples No 1 and 3 after heating increased significantly. It can be assumed that the adsorption of carbon leads to a significant change in the wettability properties in the process of heating the textured surfaces of the aluminum-magnesium alloy.

It should be noted that after heating the samples leading to the wetting inversion, their surface showed ultrahigh adhesion. It was not possible to register the roll-off angle (the droplet did not roll when turning the samples upside down).

4. Summary

It was established that after laser texturing all samples showed hydrophilic properties. With a decrease in the density of the arrangement of texture elements, which is controlled by the parameters of the spatial displacement of the laser beam, the static contact angle after texturing increases.

Wetting inversion from superhydrophilicity to hydrophobicity occurs in 2–3 hours of low-temperature heating of textured samples. The change time of wetting properties depends on the type of texture.

The samples with the static contact angles up to 137.3–144.2° were obtained after laser texturing and low-temperature heating.

A steady increase in the contact angle after low-temperature heating of samples was recorded (wettability properties deteriorate).

A significant increase in the carbon content in the elemental composition of the surface layer of the samples after their low-temperature heating was recorded.

Acknowledgments

The reported study was funded by RFBR, research project No. 19-38-90136.

References

[1] Toosi S F, Moradi S, Ebrahimi M, Hatzikiriakos S 2016 *Appl. Surf. Sci.* **378** 426–434
[2] Wang M 2019 *Surf. Coatings Technol.* **364** 57–62
[3] Meijer J 2004 *J. Mater. Process. Technol.* **149** 2–17
[4] Meijer J *et al* 2002 *CIRP Ann-Manuf. Technology* **51**(2) 531–550
[5] Jagdheesh R 2014 *Langmuir* **30**(40) 12067–12073
[6] Wu R *et al* 2012 *Appl. Surf. Sci.* **258** 5933–5937
[7] Ta V D *et al* 2016 *Appl Surf Sci.* **365** 153–159
[8] Vorobyev A Y, Guo C 2015 *J. Appl. Phys.* **117** 033103
[9] Gamaly E G 2011 *Femtosecond Laser-Matter Interaction: Theory, Experiments and Applications* (Singapore: Pan Stanford Publishing)
[10] Ahuir-Torres J I *et al* de 2017 *Surf. Coat. Tech.* **321** 279–291
[11] Jagdheesh R, García-Ballesteros J J, Ocaña J L 2016 *Appl. Surf. Sci.* **374** 2–11
[12] Ta D V *et al* 2015 *Appl. Surf. Sci.* **357** pp 248–254
[13] Tang M, Shim V, Pan Z Y, Choo Y S, Hong M H 2011 *J. Laser Micro/Nanoeng.* **6** 6–10
[14] Kenar H *et al* 2013 *Colloid Surf. B* **108** 305–312

[15] Kietzig A-M, Hatzikiriakos S G, Englezos P 2009 *Langmuir* **25** 4821–4827
[16] Chun D-M, Ngo C-V, Lee K-M 2016 *CIRP Ann. Manuf. Technol.* **65** 519–522
[17] Ngo C-V, Chun D-M 2017 *Appl. Surf. Sci.* **409** 232–240

IOP Conf. Series: Materials Science and Engineering 1019 (2021) 012086 doi:10.1088/1757-899X/1019/1/012086

Automation of the monitoring system for surface contamination with α-active radionuclides

V A Khan[1,2], V F Myshkin[2], D M Khorokhorin[3], D A Izhoykin[4], M S Kuznetsov[2], A N Panamaryova[5] and L A Rakhimzhanova[6]

[1]Laboratory of Optical Location, V.E. Zuev Institute of Atmospheric Optics SB RAS, Tomsk, Russia
[2]Division for Nuclear-Fuel Cycle Tomsk Polytechnic University, Tomsk, Russia
[3]FSUE «Mining and Chemical Combine», Zheleznogorsk, Russia
[4]Department of Applied Mechanics and Materials Science, Tomsk State University of Architecture and Building, Tomsk, Russia
[5]Division for Foreign Languages, Tomsk Polytechnic University, Tomsk, Russia
[6]Department of Physics and Chemistry, Kazakh Agro Technical University, Astana, Kazakhstan
E-mail: gos100@tpu.ru

Abstract. The paper presents research on radiation control at nuclear enterprises working with α-active radionuclides. The focus of the research is on the registration of ^{239}Pu radionuclides on hands surface. The experimental results show the possibility to create a turnstile for automatic monitoring the presence of α-active radionuclides on hands. Such a device can be used with a minimum helium consumption of 66 liters per hour in a measuring chamber with an open bottom of 5.4 liters.

1. Introduction

The radiation safety maintenance and monitoring personnel in terms of radionuclides contamination are integral parts of the nuclear fuel cycle enterprise. The radiometric control systems of the RZB, RZA, RZBA series are currently manufactured in Russia. They are made as a frame with built-in detection units and a measurement processing unit. Some of these devices are equipped with a turnstile. Existing devices, as a rule, have β-particles detection units, fixed at various points of the turnstile and external α-particles detection units. Monitoring β- and γ-emitting radionuclides is carried out on existing devices with a turnstile in automatic mode. Such contamination is also possible with the disposal of radioactive waste [1, 2].

It is known that the α-particle path in the air does not exceed 2–3 cm [3]. Therefore, it is difficult to create a device for automatic control of the radiating nuclide presence, for example, 239Pu, at stationary positioning of a detector. This is due to the fact that stationary α-particles detection units may not register the radionuclides out-of-the-way places for the detector flat surface, for example, in the recesses between fingers or under nails. Therefore, the remote detection units are used in a manual mode for more accurate measurement. In this case, a human factor appears in the process of monitoring radioactive contamination, the detector damage risk, for example a scintillator film, increases, and the detection unit may fall.

The work of stationary radiometric control devices becomes ineffective with the nuclear industry development and the creation of new industries using radionuclides emitted mainly α-particles [4, 5].

Therefore, the development of stationary automated devices for monitoring α-active contamination of personnel in nuclear fuel cycle enterprises are topical.

2. Gas Mixture Search For α-Particles Recording

The device in which the source and detector of α-particles is in a single hermetic chamber is assembled to substantiate the possibility of creating an automated monitoring system of surface contamination with α-emitting radionuclides (Figure 1). This allows controlling a gas mixture composition in which α-particles are distributed. The α-particles source is installed thereby an external magnetic manipulator can change the distance from the detecting unit to an α-particles source. The source installation on the rail runner allows experimental determination of the α-particles path length.

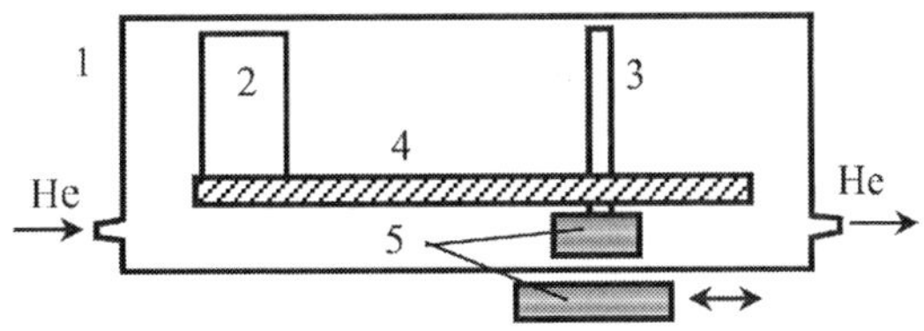

Figure 1. Experimental facility: 1 – chamber, 2 – sensor α-particles, 3 – source α-particles, 4 – rail, 5 – permanent magnet.

The counting unit is located outside the hermetic chamber. The chamber is equipped with two fittings for changing the gas mixture. An MKS-AT1117M radiometer with a remote BDPA-01 detection unit was used. An α-particles source is performed in the tablet form with a diameter of 30 mm and a thickness of 4 mm. ^{239}Pu (activity 3.4 kBq) was chosen as a α-particles source, because this isotope makes the main contribution to the surface contamination of personnel [6, 7].

The ^{239}Pu activity is associated with α decay with the probability close to 100% [8]. In this case, α-particles are emitted with energy: 5105.5 keV – 11.94%; 5144.3 keV – 17.11%; 5156.59 keV – 70.77%. Spontaneous fission with the probability not exceeding 3.2 10^{-10} % is also possible. The particle energy is about 5.1 MeV at the α-decay of ^{239}Pu. The path of the α-particles in the air at 760 torr and at 15°C depends on the energy of α-particles, respectively, on the initial velocity, and is determined as:

$$L = K \times E_a^{1,5}.$$

The K-value depends on the units of measurement and the propagation medium [9]. If the energy of α-particles (E_a) is in MeV, at a temperature of 15°C in air, the α-particle path can be determined by the equation:

$$L_0 = 0.316\ E_a^{1,5}.$$

For example, the mean path of α-particles with an energy of 5.1 MeV in air is L_v = 3.63 cm. The α-particles path in a given medium can be determined from the known path in the air with an accuracy of ±15% using the equation:

$$L = L_0\left(\rho_0 : \rho\right)\sqrt{A : A_0},$$

where L is a path of α-particle in substance; L_0 is path of α-particle in air; ρ_0 and A_0 are air density and atomic mass (ρ_0 = 1.29 g/l, atomic mass is 14.4); ρ, A are density and atomic mass of substance.

It can be seen that the α-particle path will be maximal in gases which have a minimum density and maximum atomic mass. Therefore, to increase the α-particle path and the number of pulses, it is necessary to create a helium atmosphere in the space between the detector and the controlled surface. In the helium atmosphere, the path of α-particles has a maximum value.

The α-particles flux at different distances from the detector and the maximum path of α-particle in the air and a mixture of helium with air are experimentally determined. Experimental results are

IOP Conf. Series: Materials Science and Engineering 1019 (2021) 012086 doi:10.1088/1757-899X/1019/1/012086

shown in Figure 2. It can be seen that the minimum distance at which it is possible to register of α-particles in air is 2.5 cm. In the helium, the path of α-particles does not exceed 12 cm. The obtained path of the α-particles allows the measuring unit formation for an automated complex for the radioactive contamination control.

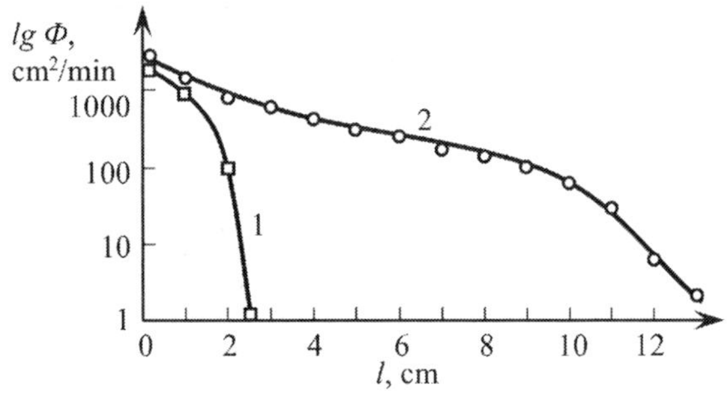

Figure 2. α-intensity at various distances from the source: 1 – Air, 2 – Helium.

The source was installed at a distance of 6 cm from the detector, and helium was introduced into the chamber for additional verification of the obtained results. The flux density was 250 α-particles. After an air blow of the chamber, the detector stopped detecting α-particles.

3. Development of Measuring Chamber for a Turnstile

It is necessary to place two detectors of α-particles, designed to monitor contamination of the inner and outer sides of a palm into the chamber. The requirements for the measuring chamber of the automated device are the following: the chamber should provide free and unobstructed movement of a hand during the measurement, maintain constant measurement conditions - keep the gas mixture with helium content above a certain specified value. Helium is lighter than air, therefore the hermetic chamber window for hand insertion should be located in its lower part (Figure 3).

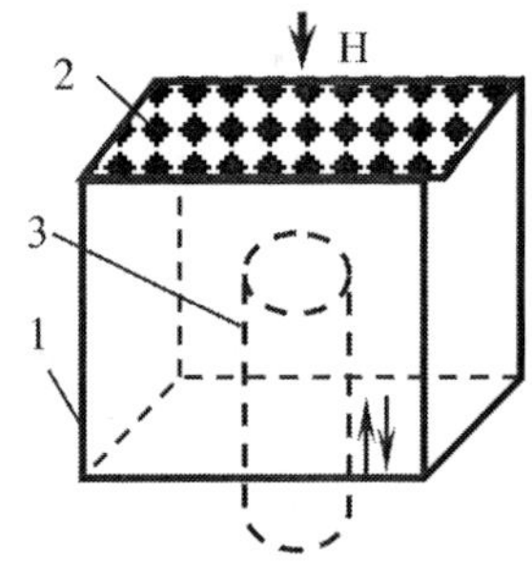

Figure 3. Measuring chamber scheme: 1 – chamber, 2 – perforation, 3 – mobile cylinder.

When a hand is put into the chamber, the working gas mixture is forced out, and when the hand is removed, air is drawn into the counting volume. As a result of a hand movement in the chamber, turbulence is created, which also draws air into the counting volume, reducing He concentration. Also, to maintain He concentration above a predetermined value, a continuous supply of helium is necessary in the upper part of the box. This is due to the fact that the attenuation of the α-radiation flux in helium is much less than in air.

The gas volume in the chamber and the He feed rate are the main parameters providing, under a given people flow through the turnstile, required helium content in the gas mixture of the measuring chamber. To estimate these parameters, we calculated the gas flows in a helium-air mixture with periodical hand moving inside the chamber.

In the simulation, the chamber has a parallelepiped shape (18 × 25 × 50 cm) oriented vertically with the open part down. An 8 cm diameter cylinder with a rounded upper end is inserted every 10 seconds for a time of 2 seconds into the chamber along the longitudinal axis of symmetry. The cylinder stops at a distance of 5 cm from the chamber top. Helium is fed into the rectangular chamber from the upper end over the entire area uniformly. The simulation results of the gas flows inside the chamber obtained using the Comsol are shown in Figure 4.

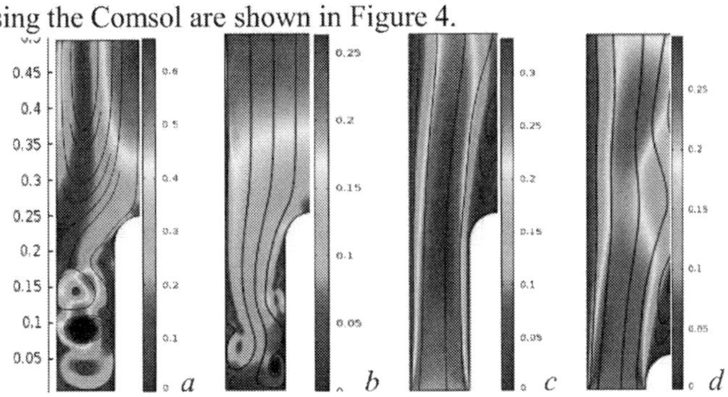

Figure 4. Gas flow diagrams in the chamber at the cylinder moving: *a*, *b* – up; *c*, *d* – down.

The calculated values given in Figure 4 allow us to state the following: the cylinder movement up through the chamber leads to the intensive formation of a turbulent flow in the space between the cylinder and the body. A laminar flow is maintained in the volume of the chamber above the cylinder. Therefore, the mixing of helium with air is insignificant. The cylinder moving downward, gas vortices are formed mainly when it leaves the chamber cavity.

Therefore, each time the cylinder exit from the chamber forms a turbulent flow, the vortex of which replaces the gas mixture with clean air at the chamber bottom. Thus, two processes contribute to the helium dilution in the chamber: displacing a helium with a cylinder followed by replacing this volume with air and replacing helium with air as a result of cylinder removing from the chamber, which triggers the vortex gas movement. The scale near to each Figure 4 shows the linear velocity in m/s.

4. Experimental Tests of the Turnstile Chamber Model

For experimental studies, a 5.4-liter parallelepiped chamber (18 × 25 × 50 cm), a rotameter RM-06 with a maximum air flow of 83 *l*/h, and a MS-200 mass-spectrometer were used. Helium was supplied from a balloon through a rotameter. The mass spectrometer MS-200 is designed for analysis with constant pumping at a sample flow rate of 110 m*l* / min through a 5 m long capillary. Therefore, sampling on mass-spectrometer was carried out for less than 2 minutes at regular intervals. For periodic introduction of the capillary into the chamber on its side wall, a tube was installed vertically, ending at a distance of 5 cm from the upper end.

First, the helium natural leakage rate from the chamber was evaluated. For this purpose, the mass-spectrometer capillary was installed at the 5 cm height from the lower section of the chamber. Then the helium balloon was opened. After the helium steady state at the chamber cut, the mass-spectrometer sampling capillary was transferred to the upper part of the chamber through a tube, and the supply of helium was stopped. Sampling S was carried out periodically. The measurement results are shown in Figure 5.

It can be seen from Figure 6 that the helium concentration is about 2 times greater than in the section at the initial moment in the center of the chamber. It is caused by the constant dilution of the working mixture due to the mutual diffusion of air and helium. The comparison of the first two peaks shows that the helium concentration decreased 1.55 times in 6 minutes. The estimated helium leakage rate is 0.5 *l* / min. Therefore, to maintain a constant composition gas mixture in the chamber, it is necessary to continuously feed He into the chamber.

Then, the He concentration change was experimentally determined with periodic cylinder (5 cm diameter) bringing into and out the chamber at different helium consumption values. First, He was fed into the chamber until the stationary concentration establishing (the first 1.5 minutes) and then the feed was blocked. The cylinder was stopped at a distance of 5 cm from its upper end. Manipulation with the cylinder in the chamber was repeated 4 times with an interval of 20 seconds. The He content change graph at a distance of 5 cm from the upper end of the chamber is shown in Figure 5.

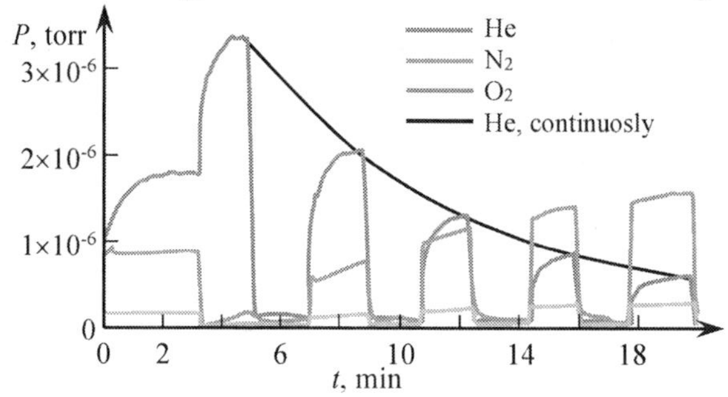

Figure 5. Time dependence of the gas mixture components.

The graphs in Figure 6 show four small peaks associated with the cylinder insertion into the chamber. At the beginning of the cylinder movement, the partial pressure of helium increases, and when the cylinder is removed, air is sucked into the released volume, so the He concentration decreases. The downward shift of the curve is seen by about 5% after each peak. The estimate of this helium loss is 0.27 liters. This time period on Figure 6 is highlighted with a circle. However, a more significant decrease in He concentration is observed due to leakage.

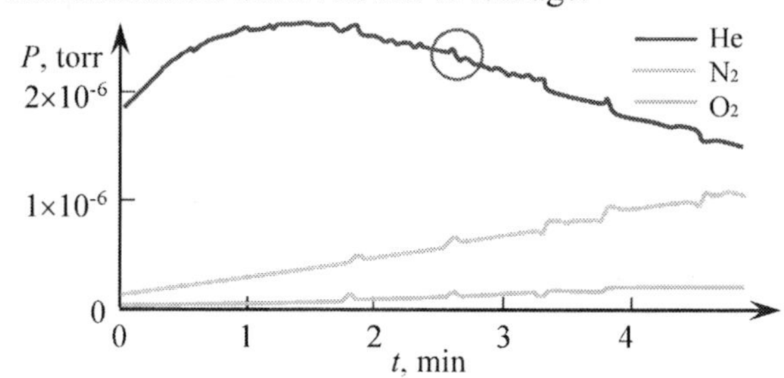

Figure 6. Gas mixture composition changes in the chamber lower part.

To design a turnstile, it is necessary to estimate the minimum helium flow rate at which its concentration fluctuations will be negligible. Figure 7 shows helium concentration changes at a consumption of 33 *l*/h, 50 *l*/h, 66 *l*/h and 83 *l*/h supplied from the upper end of the chamber.

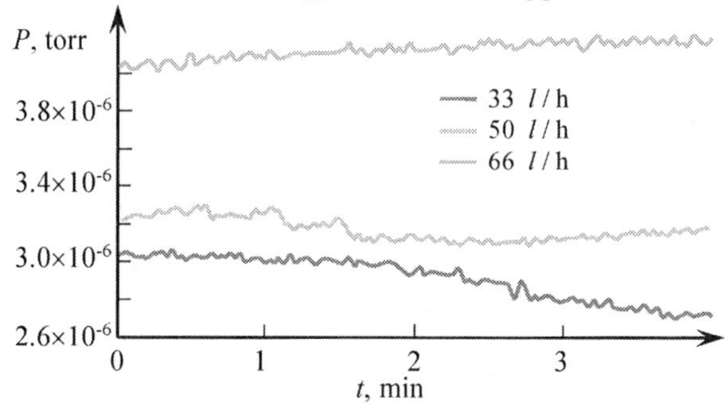

Figure 7. Helium concentration changes.

Figure 7 shows that the steady state of helium with the concentration of 84% of the maximum possible is achieved at a flow rate of 33 liters per hour. The helium concentration decreases with several manipulations with a moving cylinder inside the chamber. The steady state of helium is achieved at 90% of the maximum concentration at a flow rate of 50 liters per hour. The helium concentration decreases slightly with cylinder multipassing re-insertion into the chamber. The helium concentration in the chamber is maintained at the maximum level at a flow rate of 66 and 83 liters per hour. The concentration of helium does not change, and concentration fluctuations are not noticeable during several manipulations with the cylinder in the chamber.

5. Summary

The experiments showed that it is possible to implement an automated system for monitoring hand surface contamination with α-active radionuclides. For this purpose a rectangular chamber blown with helium is used in which the α-particles sensor is installed. It is necessary to maintain a helium consumption of at least 66 liters per hour when using a 5.4 liter-chamber with an open bottom.

References

[1] Antonenko M V, Myshkin V F, Grigoriev A S *et al* 2018 *AIP Conference Proceedings* **1938** 020002 doi:10.1063/1.5027209

[2] Dorofeeva L I, Orekhov D 2018 *AIP Conference Proceedings* **1938** 020011 doi:10.1063/1.5027218

[3] Mashkovich V P, Kudryavtseva A V 1995 *Protection against ionizing radiation. Directory* (Moscow: Energoatomizdat) p 496 in Russian

[4] Kotelnikova A A, Karengin A G, Mendoza O 2018 *AIP Conference Proceedings* **1938** 020015 doi:10.1063/1.5027222

[5] Myshkin V F, Khan V A, Plekhanov V G *et al* 2016 *J of Industrial Pollution Control* **32** (2) 603–607

[6] 2009 *SanPiN 2.6.1.2523-09. Radiation Safety Standards. NRB-99/2009. Sanitary rules and regulations* (Moscow: Rospotrebnadzor) p 101 in Russian

[7] 2014 *Radiation protection and safety of radiation sources: International Basic Safety Standards. General safety requirements. IAEA Safety Standards, N GSR Part 3* (Vienna: IAEA) p 471

[8] Mukhin K N 1993 *Experimental Nuclear Physics. In 2 books, Book 1. Nuclear physics* (Moscow: Energoatomizdat) p 137 in Russian

[9] Mayles P, Nahum A, Rosenwald J C 2007 *Handbook of radiotherapy physics. Theory and practice* (New York, London: Taylor & Francis) p 1452

High-level architecture for a digital oilfield: features of the transition to data-driven decision management

A O Savelev
School of Computer Science & Robotics, Tomsk Polytechnic University, Tomsk, Russia
E-mail: sava@tpu.ru

Abstract. This paper is devoted to the design of a distributed heterogeneous data warehouse of a digital oilfield. With increasing amounts of data collected from intelligent controllers and sensors, the lack of mechanisms for combining data from different sources and providing them to consumers affect the overall management efficiency. In addition, without it is impossible making the next logical step – the effective application of intelligent analysis methods. The paper describes the high-level architecture, as well as subsystems of operational management and decision support. The presented data are intermediate results of the project "Digital oilfield heterogeneous distributed data warehouse for informational support of decision-making processes".

1. Introduction

Today, due to the development of digital oilfield (DOF) technologies, there are virtually no production and management tasks in the field of oil production, the solution of which would not be automated to one degree or another. At the same time, there is a growing interest in the study and implementation of data mining technologies to support decision-making and improve the overall management processes efficiency. However, the effectiveness of implementing these depends on the quality of the previous automation and control stages solutions. However, the distribution, implementation, and use of individual information solutions often occur outside the framework of a rational automation strategy. Due to the complexity of the digital oilfield, due to the nature and composition of the appropriate management tasks, the lack of integration between the multitudes of existing software solutions prevents the formation of single information space of the enterprise. Thus, the effective use of data analysis tools and the consequent transition to data-based management are complicated. Separately, we should note that the oil production companies recognize the data integration problem. Several software solutions automate the main tasks of oilfield development management. The main applied field of research in the digital oilfield area is solving problems of forecasting and modeling the state of development, including the current monitoring of oil well equipment [1–5].

Authors of papers [6,7] considered the existing groups of Energetics consortium international standards as a way to ensure the uniformity of data presentation. The advantage of these standards is their data schemes, covering a wide range of oil industry elements. With an increase in the volume of data generated by the industry-recognized by most authors, various ways are offered to increase the efficiency of working with them. The authors of [8,9] proposed the use of cloud technologies subject to the availability of reliable communication channels with high bandwidth. The use of Big Data technologies, as well as machine learning and artificial intelligence concerning specific problems and tasks of the oil industry, are discussed in [2, 4, 10–13].

Despite the progress in this research area of, there are several unsolved problems, including data integration, knowledge extraction from the existing information volume, complexity, and efficiency of processing the available information [14]. In other words, there is no unified approach to storing and managing large amounts of data from heterogeneous, distributed sources and then extracting knowledge from them to support decision making and creating appropriate analytical problem-oriented algorithms and systems.

Summarizing the above, we can identify the following most pressing problems in the application of data mining and artificial intelligence methods in oil field development management:

- The lack of uniform software solutions for the digital oilfield distributed heterogeneous data sources integration.
- The volume of data collected in the field and the mining process has moved into the Big Data category, the effective use of which in the management process requires the development of specialized software tools.

As a solution to these problems, we propose heterogeneous distributed data warehouse creation to unify the processes of interaction between the components of the digital oilfield infrastructure, ensuring the unity of the data collection, processing and storing processes, as well as intellectualizing the oilfield development management.

2. Digital oilfield infrastructure modeling

Within the framework of the research, we created a software-hardware stand for simulating the infrastructure of a digital oilfield. A hardware-software stand is a system of three main subsystems: data generation, data aggregation, data processing, and storage. The stand includes the following data sources:

- Sources of low-level data collected from cluster controllers and sensors. Data source include four main data sets (5560, 3368, 15040 and 22144 records) both in the two most commonly used formats in the industry – wellsite information transfer standard markup language (WITSML) and log ASCII standard (LAS).
- Instrumentation and automation equipment database. Information is presented as a Microsoft SQL database and contains records of 4000 models of equipment, as well as of the regulated time for maintenance and repair work.
- Reference and reporting information stored as .doc, .pdf files. This set of files is the “passport” of the oil field and contains an extensive list of textual, tabular, and graphical information describing the oil fields starting from the exploration stages.

Low-level data sources support an improved drilling data transfer protocol – Energestics transfer protocol (ETP), which implements a more reliable data transfer channel at the program level. The stand generally allows studying the peculiarities of the technical interaction between DOF information infrastructure components. The stand is applicable in the tasks of specialists training, as it allows to visually demonstrating the interaction of different software and technical solutions within the infrastructure of one oil company, including the associated features of synchronization, integration and data import/export. Besides, the stand provides flexible options for setting up a test environment for testing the interaction with elements of the external infrastructure, load modeling, system testing.

3. Attribute dependency model

The problem of efficient use of heterogeneous and distributed sources of information is the difficulty of ensuring timely access to the full list of data, the result of processing and analysis of which is necessary for management processes.

As already noted, one of the problem solutions is the creation of systems and integration tools that provide a unified presentation of data. The amount of information generated by the elements of the digital oilfield infrastructure makes it virtually impossible to integrate at the physical level. Integration at the logical level requires a single, global scheme that allows combining different ways of defining dependencies between data, as well as implement different strategies for finding solutions.

To form a unified scheme for presenting digital oilfield data, we propose to use the attribute dependency model, which is part of the object-oriented system analysis methodology. The network of functional dependencies describes the links between data located in various sources. The network is a non-strict hierarchy, which may have several roots. The attribute dependency model is used to determine the values of some attributes based on the values of other attributes. There are two main classes of tasks to be solved using the model:

- The interpretation task. The initial data are the values of the primary attributes. It is necessary to find the values of some target attributes that correspond to the given source data. This type includes the tasks of medical diagnostics, diagnostics of technical equipment, forecasting tasks. The usage of direct output methods can solve such problem.
- The task of finding a feasible solution (inverse problem). The source data is the specific (specified) value of a non- primary attribute. It is necessary to find the corresponding values of the primary attributes. This type includes the tasks of planning, target management. To solve the inverse problem, use methods of reverse output.

The task of finding the optimal solution can be solved in two ways. In one way, as a series of solutions to the problems of interpretation for various alternative variants defined by multiple combinations of primary attributes. Alternatively, as a series of solutions to the problem of finding feasible solutions for different values of the efficiency criterion and target constraints. Methods for solving the optimization problem use the techniques of direct and inverse derivation, complementing them with procedures for generating variants and checking restrictions. Figure 1 describes an example of the attribute dependency model structure.

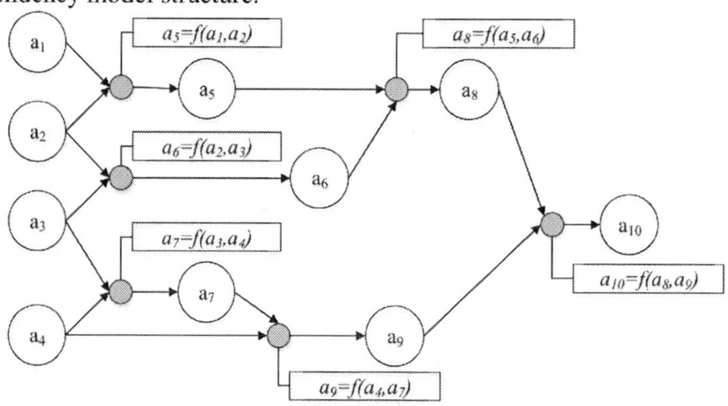

Figure 1. An example of the attribute dependency model structure.

Thus, the attribute dependency model makes it possible, first, to create a diagram of data interconnections stored in various sources. Secondly, the meta description of model attributes – data types – makes it possible to automate the tasks of searching and providing data access, as well as evaluating the feasibility of storing attribute values that are calculated (dependent on primary ones).

4. Heterogeneous data warehouse high-level architecture

The study of the structure and types of data located in the different information sources described in section 2, as well as scientific, technical and regulatory documentation analysis, made it possible to formulate the following main theses:

1. Technical interaction unification of heterogeneous data sources, subsystems, and software solutions today are the focus of the DOF-research. There are several issues of data integration at the physical level. The WITSML and PRODML standards are considered as the basis for integration.
2. In other areas of activity that are not related to oil production, in particular, in the framework of the ATLAS experiment at CERN, working with heterogeneous sources is based on the development of solutions for the collection, storage, and analysis of metadata.

3. Solutions and approaches that integrate technical cooperation standardization tools for production systems and decision support tools based on the analysis of information from distributed heterogeneous sources have not been found within the framework of the DOF-concept.

Partly, we can explain the current situation by the following: the use of data mining tools for decision support purposes requires well-functioning solutions in the areas of data integration, including low-level ones, within a unified information environment.

Thus, we formulate the following basic principles for the development of the heterogeneous distributed data warehouse architecture for digital oilfield:

- The principle of technical interaction unification – the architecture should provide for the availability of physical integration solutions for low-level data obtained using industry-standard solutions and equipment. The integration of low-level data is necessary to provide information support for the processes of operational accounting and control over the state of oilfield development.
- Principle of single information space – the primary tool for the tasks of strategic management should be an appropriate decision support system, based on the knowledge base, describing the relationship of all data from heterogeneous and distributed sources. The unity of the information space should be ensured by an appropriate logical model in combination with the principles of its change and maintenance.

Figures 2–4 show the first version schemas of the heterogeneous distributed data warehouse architecture for digital oilfield consisting of operational management subsystem architecture (Figure 2), decision support subsystem architectures (Figure 3) and high-level representation of the heterogeneous distributed data warehouse architecture (Figure 4).

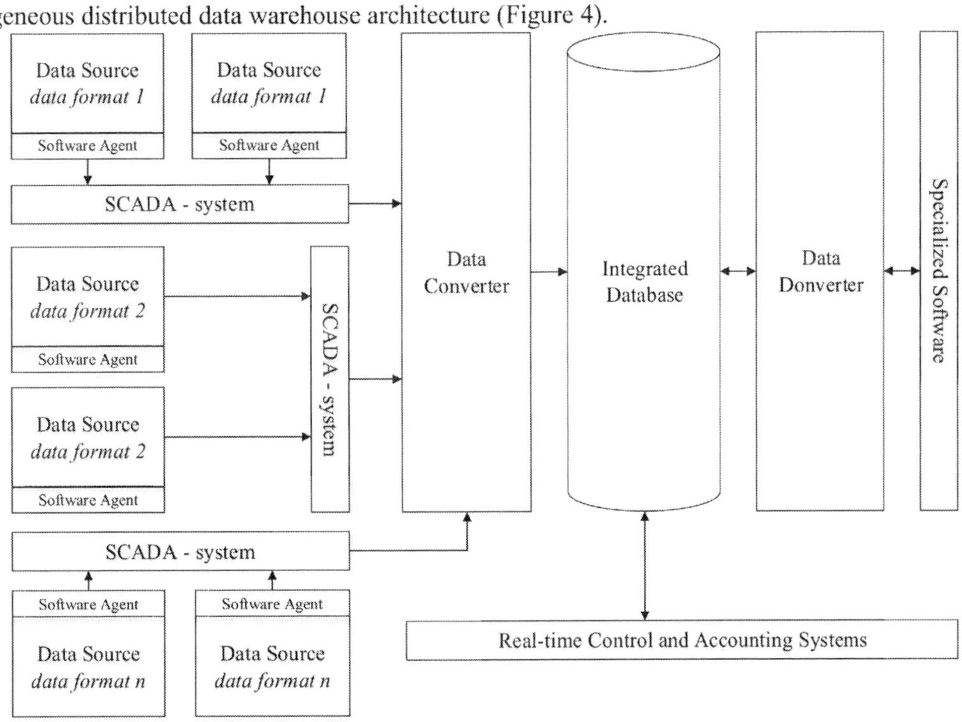

Figure 2. High-level architecture of the operational management subsystem.

Thus, a mixed approach to the organization of a heterogeneous data warehouse, which consists in sharing the operational management subsystem and the decision subsystem, while maintaining the existing infrastructure of the digital field, improve the quality of decisions and implement the transition to a data-driven decision management model.

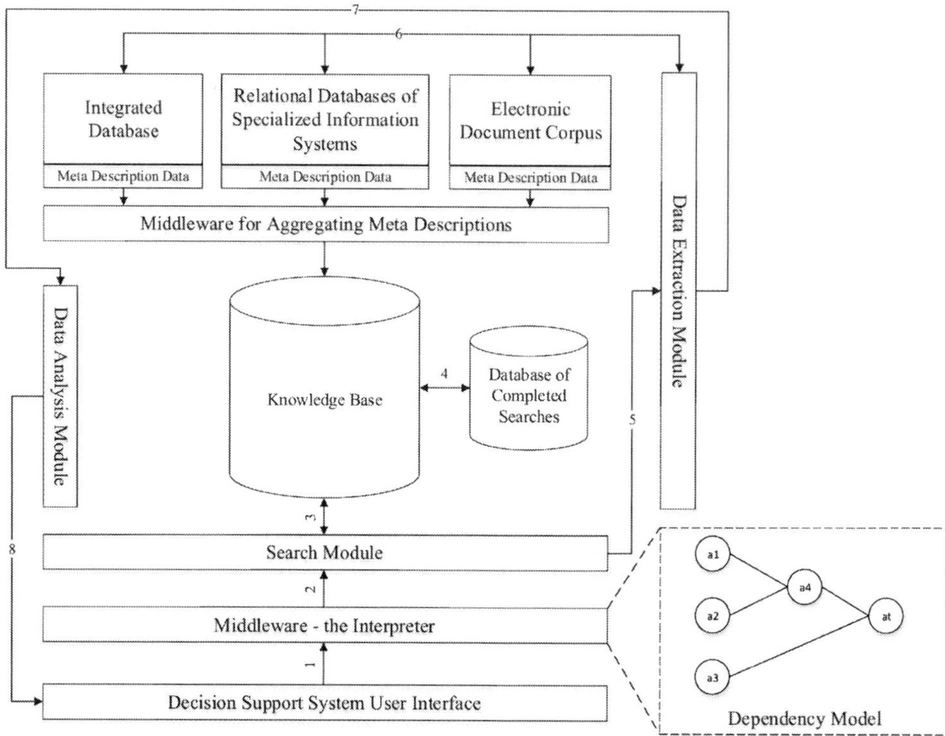

1 – The list of target attributes that determine the desired state of the system
2 – List of primary attributes defining the value of target attributes
3 – Request for a list of data sources that store basic attribute values
4 – Check request whether the search query is new
5 – List of data sources and base attribute value extraction rules
6 – Extract the values of the primary attributes that determine the value of the target
7 – List of base attribute values
8 – Calculated values of target attributes and estimation of their errors

Figure 3. High-level architecture of the decision support subsystem.

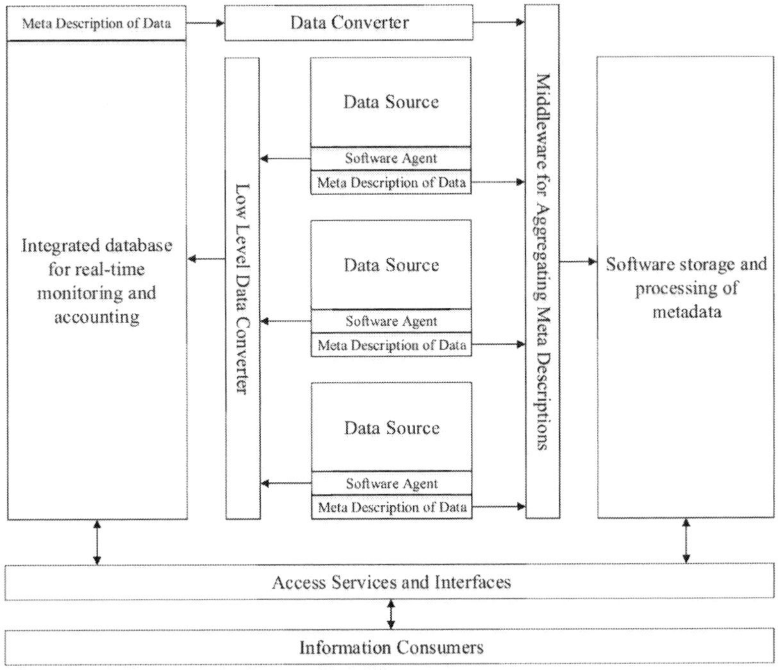

Figure 4. High-level architecture of a heterogeneous distributed data warehouse for digital oilfields.

5. Summary

The transition to a data-driven digital oilfield management model implies an analytical module using machine learning tools. The development of such a module is a complex task that requires a comprehensive integration of data sources to build the most accurate model of the analyzed processes. Because DOF data sources usually exist autonomously from the analytical module, the priority is to create a system for integrating data from heterogeneous sources into a single repository for quick and efficient search and retrieval of data needed by an expert. Within the framework of the current stage of the work, we propose a high-level architecture of a heterogeneous data warehouse for a digital oilfield, combining subsystems of operational management and decision support. The implementation of the software prototype according to the developed architecture will allow going over to the tasks of designing an analytical module for the data coming into the knowledge base using machine learning tools.

Acknowledgements

The reported study was funded by Russian foundation for basic research (RFBR) according to the research project № 18-37-00311.

References

[1] Qin H and Han Z 2016 *IEEE Systems Journal* **12** 1295–1306
[2] Hu H, Fan L and Guan X 2017 *Proceedings of the 2nd IEEE International Conference on Cloud Computing and Big Data Analysis* 533–537
[3] Alenezi F and Mohaghegh S 2016 *Proceeding of 4th Saudi International Conference on Information Technology (Big Data Analysis) (KACSTIT)* 1–6
[4] Preveral A, Trihoreau F and Petit N 2014 *Proceedings of the SPE Intelligent Energy Conference & Exhibition* 295–304

[5] Patri O, Panangadan A, Chelmis C *et al* 2014 *Proceedings of the SPE Annual Technical Conference and Exhibition* 1313-1326
[6] Zhong Q and Du H 2015 *Proceedings of the 2015 International Industrial Informatics and Computer Engineering Conference* 139-142
[7] Mckenzie W, Morandini F, Verney P *et al* 2012 *SPE Intelligent Energy International* 350–360
[8] Zhao G, Di W and Wang R 2017 *Proceedings of the 2nd International Conference on Cloud Computing and Big Data Analysis (ICCCBDA)* 201–205
[9] Alguliyev R, Ramiz M and Hajirahimova M 2016 *Proceedings of the IEEE 10th International Conference on Application of Information and Communication Technologies (AICT)* 1–5
[10] Brule V 2015 *Proceedings of the SPE Digital Energy Conference and Exhibition* 547-553
[11] Cameron D 2014 *Proceedings of the SPE Intelligent Energy Conference & Exhibition* 219-227
[12] Perrons R and Jensen J 2015 *Energy Policy* **81** 117–121
[13] Chanana P, Son T and Bhakne U 2016 *Proceedings of the Offshore Technology Conference* 1487–1499
[14] Crompton J 2015 *Proceedings of the SPE Digital Energy Conference and Exhibition* 500–518

Study of enzymatic hydrolysis parameters for reducing anti-nutritional properties of flour

N Ashirova, E Bychkova and D Gurova
Department of Technology and Organization of Food Production, Novosibirsk State Technical University, Novosibirsk, Russia.
E-mail: bychkova.nstu@gmail.com

Abstract. The paper addresses the problem of production of foodstuffs for people suffering from diagnosed celiacia. There are no ways of pharmacological treatment of this group of people. The only possibility to maintain their optimal living standards is a life-long gluten-free diet. Grain-based proteins (wheat gliadins, rye secalines, barley hordeins, and oat avenines) and products of their processing are the disease triggers. Most often one generally accepted term "wheat gluten" is used instead of all these species. Celiacia is one of the urgent social problems in many countries which can be explained by its wide morbidity rate, difficulties in its diagnostics and a lack of experience of its prophylaxes and treatment. Strict requirements to the chemical composition of gluten-free flour-based foodstuffs dictate their high retail prices. Market analysis has shown that gluten-free flour-based products on average are 240% more expensive than usual flour-based products. So crackers, butter biscuits, pizzas and pasta made from gluten-free flour are the most expensive products. However the market of gluten-free flour-based products is one of the most fast growing markets. In this connection there is a need to provide people suffering from diagnosed celiacia with inexpensive domestic gluten-free products of high quality including products suited for public catering facilities such as pre-school, school and higher education institutions, military army facilities, canteens and sanatorium-and-spa resorts. Developing specialized foodstuffs based on nontoxic raw materials having a positive effect on human organisms and able to decrease risks of disease recurrence is a challenge to the scientific community.

1. Introduction

Celiacia has long been considered a rare disease from which little kids suffer. In 2014, National Recommendations on diagnostics and treatment of celiacia of adults were adopted in Russia. In 2015, Federal Clinical Recommendations on the delivery of health-care services to children suffering from celiacia were approved. Application of advanced diagnostic methods in recent decades resulted in a considerable growth of celiacia patients due to the identification of latent or dormant forms of the disease. Lately the perception of celiacia has drastically changed. At present it is a disease with a high incidence rate (1-3% of the European population) and a chance of occurrence at any age [1]. Gastroenterologists develop algorithms of laboratory examination of patients with suspected celiacia and risk groups. These algorithms make it possible to timely diagnose the disease. Dietitians and food technologists try to monitor the results of diet therapy of patients sensitive to gluten enteropathy. The population groups mentioned above join in societies of celiacia patients in such cites as St.Petersburg, Moscow, Novosibirsk and other cities. The authorities of the Novosibirsk Oblast render financial assistance to such families with children. This money is used to purchase expensive foodstuffs.

It is known that for celiacia to actively show itself three factors are necessary, namely, heredity, the availability of gluten in food and a trigger mechanism, e.g. emotional stress, pregnancy, surgeries, virus

IOP Conf. Series: Materials Science and Engineering 1019 (2021) 012088 doi:10.1088/1757-899X/1019/1/012088

infections or early ablactation and passing to infant formulas, etc. A damage of the small bowel caused by gluten is a key factor of the disease. When gluten enters the digestive tract of a genetically predisposed person, antigliadin antibodies began to develop in the digestive tract. These antibodies join with a part of a gluten molecule, which causes damages and destruction of villi of the intestinal mucosa. As a result of an enzymatic deficiency the surface of the small bowel becomes smooth and practically unable to absorb nutritious substances necessary for normal functioning of the organism [2].

The elaboration and application of biotechnological approaches in food production is a dynamically developing area. Scientists are making attempts to find and study safe alternatives to the existing celiacia treatment, i.e. a gluten-free diet. One of them is selection of some high-yielding variety of wheat not containing toxic gluten fractions. Another direction is the development of processes of complete proteins degradation by enzymes which lead to hydrolysis of immunogenic peptides to biologically available amino acids and short peptides [3]. An effectively selected and stabilized enzyme will stimulate gluten digestion or can be used as a food additive [4].

To implement the research objective an analytical review of Russian and foreign literature was made. The RF patent No. 2057458 describes the method of obtaining a carbohydrate protein module from grain variety flour for children nutrition. In 1996, scientists from the Research Institute for Baby Food carried out amilosubtiline enzymatic hydrolysis G10x. The use of enzymes as technological additives in the biotechnological industry to detoxify gluten is investigated at the Chemical Faculty of the University of Salerno (Italy) and the European laboratory of food-borne disease study in the University of Naples Federico II (Italy) [5]. María Cristina Sotomayor Grijalva studies enzymatic hydrolysis of wheat gluten at different protein concentrations [6]. Doctor Atze Jan Van Der Goot describes research on determining an intensified process of wheat gluten hydrolysis. He argues that enzymatic hydrolysis is a process in which "soft" conditions are used i.e. pH from 6 to 8 and the temperature from 40 to 60° C. Their studies prove that understanding the enzymatic hydrolysis reactions can help to realize various functional properties of hydrolyzates of wheat gluten, for example solubility, emulsion and foam formation.

In 2017 and 2018, research results obtained by R.G Kondratenko and E, N Urbanchik and other scientists from the Mogilev State University for Food (Belarus), were published. Their papers address the problem of the chemical composition of flour made from enzymatic wheat grains. Their colleagues E.V. Nelyubina, E.V. Zakharova and others are involved in developing gluten-free products made from new bioavailable vegetable raw materials. A range of pea hydrolizate crisps was developed at the department of technology and organization of food production in the Novosibirsk State Technical University. The complex enzymatic preparation Protosubtiline G3x was used.

It is known that wheat proteins are hydrolyzed with the use of acids, bases or enzymes. Compared to this, enzymatic hydrolysis of wheat flour is a softer process with a more selective hydrolysis. The development of specialized enzymatic gluten-free foodstuffs made from cheap raw materials available in the Novosibirsk Oblast for celiacia patients is quite relevant.

The objective of the research is to study the parameters of the raw material enzymatic hydrolysis to reduce anti-nutritional properties of wheat flour based on gluten degradation.

2. Materials and metods

Local flour of various varieties from NPO Avangard and the enzyme preparation Feedbest VGPro from OOO PO Sibbiofarm were use to prepare the samples. All the ingredients are produced in the Novosibirsk Oblast, Russia.

Research objects are:

- patent wheat flour (top grade), GOST 26574-2017, the gluten content is no less than 28%;
- light wheat flour (first grade), GOST 26574-2017, the gluten content is no less than 30%;
- second grade wheat flour, GOST 26574-2017, the gluten contentis no less than 25%.

Feedbest-VGPro is an enzyme complex for feed production. It is used in rations with a high content of oil-bearing seed by-products (oilseed residues, oil cake) and grain legumes (soya peas, chicl-peas, lupine, etc.).

Feedbest-VGPro contains such enzymes as xylanase, β-glucanase and pectinase (table 1). They provide hydrolysis of non-starch polysccharides and proteolytic enzyme complexes which destroy high molecular weight proteins to available peptides and amino acids. Enzymes contribute to the neutralization of anti-nutritional substances in grain fodders (non-starch polysaccharides), which ensures an increase in fodder nutrient digestibility. This preparation is the most available on the market and economically more profitable as compared to foreign and Russian counterparts.

Table 1. Feedbest-vgpro composition.

Characteristic	Enzyme activity
Xylanase activity (XA),no less than U/g	10000
β-glucanase activity (β-glA), no less than U/g	3500
Pectinase activity, no less than U/g	5000
Proteolytic activity, no less than U/g	120

Advantages:
- Feedbest-VGPro increases the level of the ration metabolizable energy of by 4–7%;
- animal and poultry products can be used for food after using Feedbest-VGPro without limitation;
- Feedbest VGPro is compatible with all ingredients of mixed fodder, all drugs and feed additives,
- the preparation is not toxic for people and does not pollute the environment.

The introduction of the enzyme complex into raw materials makes it possible to:
✓ accelerate hydrolysis of high molecular weight proteins and polysccharides;
✓ increase nutritional substance utilization in the digestive tract;
✓ offset the deficit of digestive enzymes.

General sequence of research stages is shown in Figure 1.

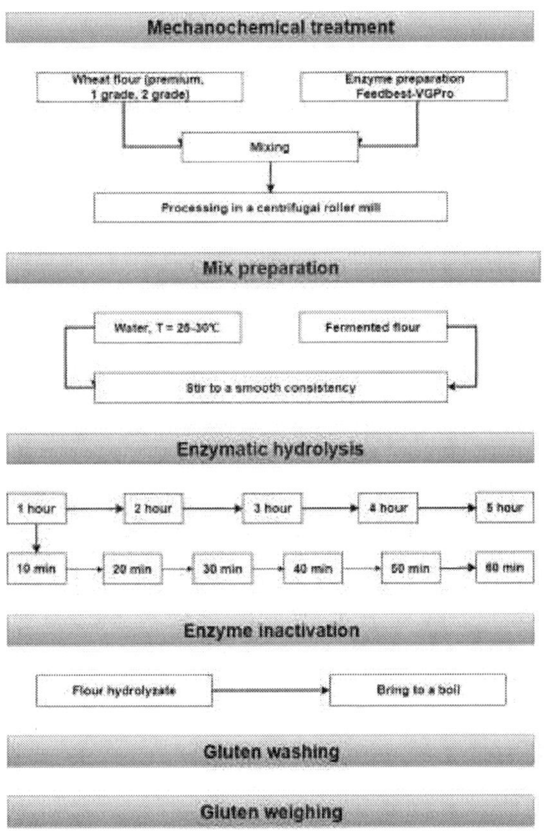

Figure 1. Diagram of research stages.

The first stage of sample preparation was carried out in the Institute of Solid State Chemistry and Mechanochemistry, SB RAS.

- enrichment by Feedbest-VGPro at the rate of 20.0 units per 1kg of flour;
- further mechanochemical treatment of the mix was carried out by the procedure [7] in the TM-3 centrifugal-roll mill.

The following samples were obtained:

Sample No. 1 –enzymes with wheat flour, top grade;

Sample No. 2 –enzymes with wheat flour, first grade;

Sample No. 3 –enzymes with wheat flour, second grade.

At the next stage the samples were sifted through a sieve with meshes of 1mm in diameter. After that they were diluted with warm water in the proportion of 1:1.692 until receiving a homogeneous consistency.

Enzymatic hydrolysis was carried out at 50-55oC in the proofing oven during 5 hours with pH=6 and subsequent enzyme inactivation. The hydrolizate was tested for the residue of crude gluten in the samples by washing them with water at 16-18°C. Washing was not interrupted; washing water was changed several times and was filtered though a fine silk sieve. Residues of gluten on the sieve were weighed and photographed.

3. Resultes and discussion

Table 2 presents the content of crude gluten in the samples after 5 hours of enzymatic hydrolysis.

Table 2. Content of crude gluten in samples after 5 hours of enzymatic hydrolysis, %.

Content of crude gluten	Samples		
	No.1	No. 2	No. 3
	0.116±0.015	0.061±0.001	0.049±0.005

The results in the samples after 5 hours of enzymatic hydrolysis and further washing of gluten are shown in Figures 2 and 3.

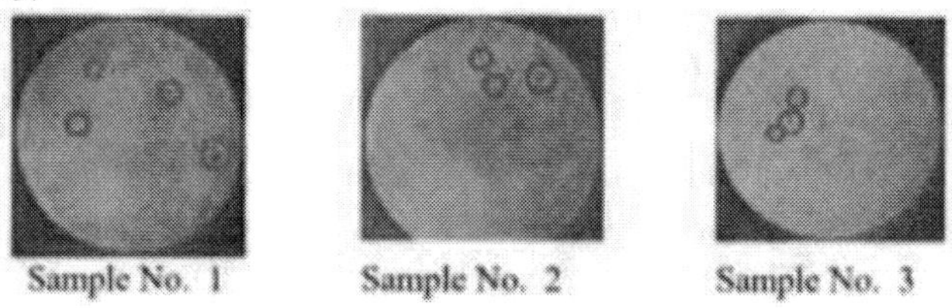

Figure 2. Traces of residual gluten.

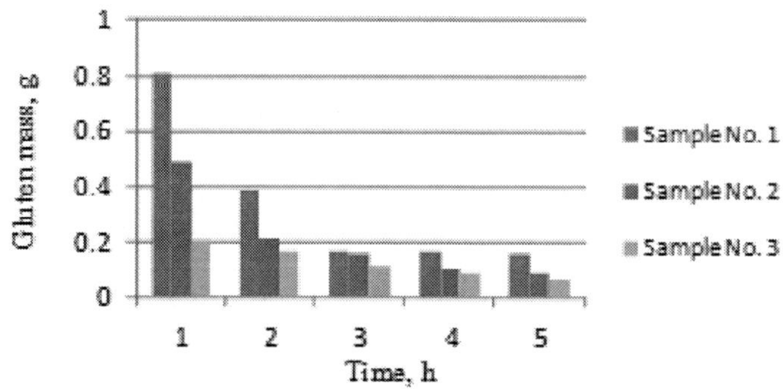

Figure 3. Results in the samples.

The results given in table 3, Figures 2 and 3 allow us to assume that hydrolysis occurred at earlier stages and it was decided to control its occurrence every hour (table 3).

Table 3. Grude gluten content in the samples by hours after enzymatic hydrolysis, %.

Samples	**Time of enzymatic hydrolysis, hours**				
	1	*2*	*3*	*4*	*5*
No.1	0.605± 0.015	0.281± 0.003	0.121± 0.002	0.118± 0,004	0.116± 0.008
No. 2	0.356± 0.011	0.155± 0.005	0.113± 0.001	0.073± 0.001	0.061± 0.009
No. 3	0.150± 0.002	0.119± 0.007	0.078± 0.001	0.064± 0.001	0.049± 0.003

Pictures with traces of residue gluten after washing every hour of enzymatic hydrolysis are shown in Figure 4.

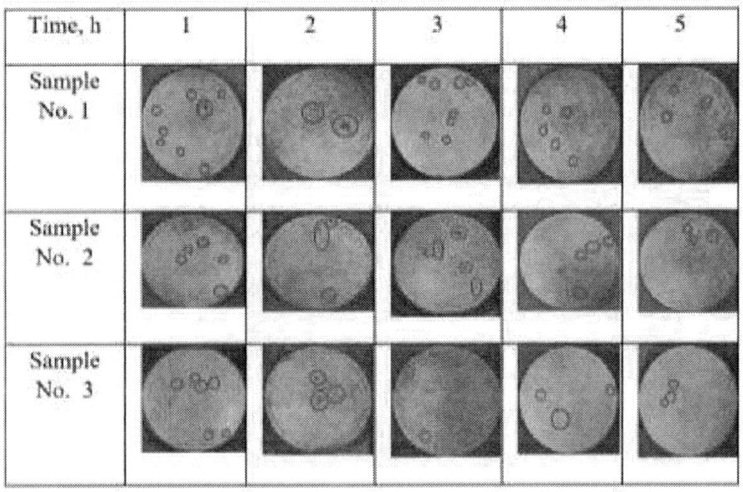

Figure 4. Traces of residual gluten after enzymatic hydrolysis during 5 hours.

It can be seen that after the first hour of hydrolysis the quantitative content of crude gluten in the rest samples did not change. Further studies were carried out every 10 minutes during one hour. The results are presented in Figure 5.

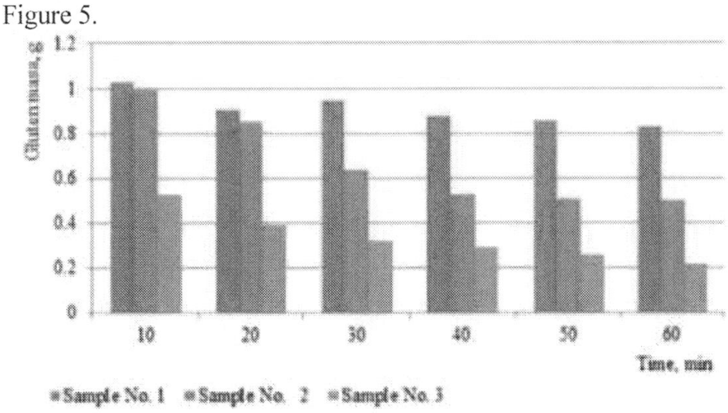

Figure 5. Results of enzymatic hydrolysis of samples every 10 minutes.

4. Conclusions

It can be concluded that the protein content in the reaction, the enzyme-substrate ratio, pH, the temperature and the tine of the reaction are the most important parameters for carrying out enzymatic hydrolysis. Based on the results of the research done the enzyme-substrate ratio of 1:1.962 was determined and the mechanochemical treatment of wheat flour of different grades was carried out. Enzymatic hydrolysis was completed by using the Feedbest-VGPro. complex enzyme preparation As a result wheat protein which is the key factor in gluten intolerance was degraded during 10 minutes into low molecular weight compounds.

References

[1] Presutti R J, Cangemi J R, Cassidy H D and Hill D A 2007 *Am Fam Physician* **76** 1795–1802
[2] Bychkov A, Podgorbunskikh E, Bychkova E, Lomovsky O 2019 *Biotechnology and Bioengineering* **116** 1231–1244
[3] Bychkov A L, Lomovskii O I 2010 *Russian Journal of Applied Chemistry* **83** 1106–1108
[4] Caputo I, Esposito C, Lepretti M and Esposito C 2010 *Enzyme Research* 174354
[5] Cristina M *Evaluation of Enzymatic Hydrolysis of Wheat Gluten at Different Protein Concentrations.* Available at: http://repositorio.educacionsuperior.gob.ec/bitstream/28000/1199/1/T-SENESCYT-000330.pdf
[6] Atze Jan van der Goot *Enzymatic Hydrolysis of Wheat Gluten* Available at: https://www.wur.nl/en/show/Enzymatic-Hydrolysis-of-Wheat-Gluten.htm
[7] Bychkov A L, Buchtoyarov V A and Lomovsky O I 2014 *Cell Chem Technol* **48** 545–551

IOP Conf. Series: Materials Science and Engineering 1019 (2021) 012089 doi:10.1088/1757-899X/1019/1/012089

Characteristic barriers to the implementation of the roadmap for augmented reality technology development program

V Zhmud[1], **A Liapidevskiy** [2] **and V Avrmachuk**[3]
[1] Department of Automation, Novosibirsk State Technical University, Novosibirsk, Russia
[2]General Director of Novosibirsk Institute of Program Systems, Novosibirsk, Russia
[3] Department of Complex Information Security of Computer Systems, Tomsk State University of Control Systems and Radioelectronics, Tomsk, Russia
E-mail: oao_nips@bk.ru

Abstract. The paper gives the author's vision of what the main barriers to the creation and implementation of new technologies in the field of virtual and augmented reality may exist during the implementation of the Digital Economy of the Russian Federation program. Since some of the authors took part in the development of roadmaps for this program, there is a two-sided vision of the problem. Authors can put themselves in the shoes of both the funding organization for this project and the campaign that is developing concretely in this area. Almost two years have passed since this program was launched. The conclusions of the article can be useful both for organizations funding this program and for grant recipients

1. Introduction

The program "Digital Economy of the Russian Federation" was launched in 2019 [1]. At the same time, nine roadmaps were being developed in accordance with the nine main areas of technology [2]. One of the most relevant areas is the direction of virtual and augmented reality. These two terms traditionally go together, although it seems that it is high time to separate them despite the fact that they have much in common [3].

The fact is that virtual reality has long entered a stage of development that does not require centralized government support. This is because the industry is already making a steady profit. Another explanation is that even small results in this area have independent commercial significance, therefore, there is no need for relatively large investments in order to overcome such a barrier that cannot be overcome gradually.

In this article, we will consider the main problems of the development of augmented reality in our opinion, and we will talk about virtual reality only because of the close connection of some technological solutions with this main and most difficult task.

2. Statement of the problem

Virtual reality is now known to almost every child, even not yet a schoolboy. The abundance of games played on the phone, tablet and computer allows you to keep your child busy and thus free up time for parents for personal affairs. This, of course, is a great advantage of our time, but there is also a disadvantage of this situation. Although now it is very often said that children have become addicted to these computer games, as well as from virtual communication with real or virtual friends, but few people notice that it was the parents who were the first to get into this addiction. Indeed, it is much easier for them to give the child a game and get the desired free time than to entertain the child personally. Further, the parents believe that the process of preschool education can also take place without their substantial help; it is enough to address the child to the corresponding Internet pages. The problem is getting out of control due to the fact that school teachers themselves often use sources from the Internet when preparing for classes, and schoolchildren do the same. This process is carried over to universities: teachers have tremendous opportunities to download textbooks, including those subjects in which they need to teach, and students have the opportunity to download ready-made abstracts, term papers, and diploma works. The humorous forecast of the Polish satirist Stanislaw Lec that "humanity will soon be able to do well without itself" is being implemented in practice.

Along with this unpleasant feature of the present time, which consists in the fact that people can receive all the necessary forms of communication without the help of other people, and with the help of Internet technologies and virtual reality technologies, one cannot deny the benefits of this. Indeed, in our time the world is faced with the fact that in critical situations, professional help may be delayed, or even be inaccessible to most of those people who urgently and urgently need it. So, for example, the arrival of ambulance teams, which according to the regulations should take place within 20 minutes, in a pandemic is delayed by a day or more. But the arriving doctors often carry out a standard survey and prescribe standard sets of drugs, so it turns out that in some cases there was no need to wait for real doctors, virtual doctors could cope with this task, who would question the patient about the symptoms and prescribe at least harmless treatment. Also, some examination methods can be done remotely, since there are fully automated devices not only for measuring temperature, but also for measuring pressure, pulse, apparently, it is not difficult to make the same devices for measuring other most important indicators of health. For example, it is not necessary to be present in person to see a white coating on the tongue or an icteric skin color, and so on. Thus, we can say that the use of virtual reality technologies could go much further if it were combined with technologies for remote data collection and processing, with technologies for measuring parameters of physical objects. And such a connection is element of augmented reality technologies.

Augmented reality technologies are that they not only allow the user to immerse the user in a fictional world, which is formed by the methods of computer creation of sounds, images, tactile sensations, possibly also heat, aroma, air movement, changes in humidity (including splashing water), a change in the sense of gravity by accelerating the cockpit with the user, (which makes you feel like you are in almost any fictional world). Augmented reality technologies should allow this to be done not under the condition of complete disconnection of the user from reality, but by combining his sensations of reality along with the sensation of the generated virtual reality. Such technologies do not completely deceive the user's feelings, but only partially, adding additional sensations to the correct ones. Science fiction has shown a lot of examples of augmented reality, but practice is still far from achieving such effects. Some effects, perhaps, will remain fiction, others are likely to be widely implemented in the near future.

For the correct choice of financed areas, it is advisable to understand, firstly, what exactly we want to finance, and secondly, how exactly promising projects should be selected, to distinguish them from unpromising ones.

3. Some non the main directions

We believe that government funding for the development of science is not always necessary. Apparently, if a direction can develop from itself, independently, then financing from outside, of course, can accelerate this development, but it is not at all a necessary condition for such development. For example, when the United States channeled money to pay unemployment benefits in an attempt to quell the first wave of the crisis in the early twentieth century, it not only did not help, but only exacerbated the crisis, because free money never promotes the development of industry and technology. Because people who were not employed in production began to receive money for nothing, people employed in production had to create material values both in order to feed themselves, their families and their employers, but also in order to feed these unemployed. Napoleon Bonaparte approached this problem in a completely different way. In order to develop the textile industry in France, he supplied money not to manufacturers, but to customers, that is, he allowed the upper classes of society to enrich themselves, at the same time obliging them to spend huge sums on luxurious clothes, carriages and other items that were products of the domestic industry. Apparently, this approach has a right to exist, despite the obvious violation of the ideals of social equality and justice. Transferring this experience to the modern situation, it can be argued that for the development of certain technologies, instead of giving gifts to those who can develop these technologies, it is necessary to create conditions for those who can buy services obtained with the help of these technologies to buy such services would be more and more.

As for augmented reality technologies, the difficulty of their development is that they use too many other related technologies, so they cannot be sufficiently developed by individual small groups of enthusiasts, representatives of small and medium-sized businesses. Big capital will flow into this area only when it is obvious that profits in this sector have become greater than profits in other sectors. If virtual reality technologies can develop gradually, then augmented reality technologies in such a state, when they will bring obvious profit, are at a distance from the current situation in Russia by such an amount that can be figuratively described as an "abyss". And as you know, it is impossible to jump over the abyss in two jumps. One "leap" is needed, one breakthrough stage, after which these technologies will start to pay off steadily. It is for such disruptive technological developments that government funding is the only effective way to solve the problem. Only the state can afford to finance the development of such technologies that cannot pay off in the next two or three years, but will pay off only after large capital investments. In this area, we see some not entirely correct approaches that are found in the statements of some experts who participated in the creation of roadmaps on virtual and augmented reality.

3.1. Display manufacturing

An erroneous emphasis on the quality of the visual picture, that is, attempts to drag funding to the production of higher-quality displays for displaying generated images. Undoubtedly, the quality of video image formation is a rather important property of available technologies, but this direction has been successfully developing for many years; for its further development, there is no need to class it as a specific digital technology, which should be financed from funds for the development of virtual and augmented reality. Billions of people use televisions and computers, phones and tablets, all of these devices use displays created using approximately the same technologies. Thus, there are already those who will pay for the best quality of these products. There is a market, there is a consumer, there is a sufficient supply of funds to purchase better quality goods. These technologies can be developed without government support.

3.2. Production of new audio systems

There are also quite a lot of consumers of sound equipment in the world, and there are enough connoisseurs of high-quality sound. Therefore, if the manufacturer creates loudspeaker systems of higher quality, there will be no problems with the market. Therefore, the inclusion of sound generation technologies in the list of virtual reality technologies is a mistake.

3.3. Wireless data transmission technologies

These technologies are a separate area of the Program development. Of course, they will be in demand in the direction of the industrial Internet of things, and in the direction of robotics and sensorics, and in the field of computer intelligence and neural computing. Virtual and augmented reality technologies should be developed autonomously, considering that the technologies listed above will develop on their own, and by a certain meeting point in the future they will already be at a new technical level that will be used in these technologies too.

4. Some main goals

Augmented reality technologies should be distinguished by the feature that they allow the formation of some virtual reality objects on real objects, which are not, for the most part, objects specially prepared for the formation of this virtual reality. Along with the fact that you can look at reality through some special glasses, or a screen, or, for example, aiming the smartphone camera at this object and observing it directly on the smartphone screen, virtual reality technologies in the daring dreams of science fiction writers present completely different possibilities.

On the Internet, you can already find video images that demonstrate technologies that do not yet exist in such a form as if they already exist. Such videos are fabricated by computer methods of combining reality with digital animation. Almost all fantastic films, and not only fantastic ones, have been shot using these methods. These films show images that are formed simply in the air, without the use of any additional media. These can be images of computer diagrams, drawings, interactive screens, as well as supposedly holographic images of living people, which outwardly do not differ from real people. Unfortunately, these technologies are still unrealistic. Some kind of light reflectors are needed to form an image in air. These can be splashes of water, blades of a fast-rotating fan, smoke particles, dust particles, aerosols, and so on. Since the human eye has a short-term memory, it perceives a short but bright flash as a constant glow, reacting to the average value of the brightness of the light. Therefore, an effect can be obtained in which the eye sees a real situation, against the background of which it also sees a picture, consisting of many bright points, from which an integral image is formed. But in order to make a certain point in space glow, it is necessary that it contains some object reflecting or scattering light, even a point one. In clean and transparent air, a luminous point can be created only due to very powerful radiation, which is very sharply focused only at this point, but this method of forming a hologram already poses a serious danger to the eyes. Another way is to use virtual reality glasses. But the method with glasses, screens, fountains, propellers and aerosols has long been implemented in many mass-produced devices. Therefore, before financing developments in this direction, it is necessary to understand exactly what it is proposed to finance, and what result is expected to be obtained, how breakthrough it is.

Sound shaping technologies can also contain fundamentally different methods. For example, sound can be shaped using light, remotely, and not only shaped, but also transmitted over long distances at the speed of light, and not at the speed of sound. In this case, no speakers will be used, the sound can be formed directly in air, or on the surface of a solid or liquid body due to periodic thermal expansions and contractions of individual volumes of the medium or surface. If we talk about the formation of sufficiently loud sounds directly in the air, then these technologies can also pose a danger to human health. In addition, the technologies discussed above, if they are associated with the use of laser technology, require ideal visibility from the transmitter to the place of image or sound formation, that is, any shielding will disrupt the formation of augmented reality objects.

Measurement and perception of the surrounding reality is very important. In this case, it is necessary to additionally process the actual picture of reality in order to obtain a digital twin of the subjective picture of reality perceived by the subject, since the virtual additional image must be superimposed on this real picture so harmoniously that the perception of these two different types of reality is carried out as a single reality. With the inept creation of augmented content, it is possible to overload the user's psyche so that, instead of helping, such a combined reality will destroy his psyche. There will be many technical and psychological problems on the way to implementing augmented reality. The developers of the roadmap paid great attention to legislative problems, but it can be argued that to eliminate legal "slingshots" does not require a lot of funding. These problems should be solved with the help of management resources, and not with funding. After all, the state will not buy off its own legal services to eliminate obstacles to economic development using digital technologies. And if the solution to the problem does not require state funding, then this problem is not a task that should be solved with the help of a program whose only toolkit is targeted financing.

5. Preferences for small businesses

The program was initially aimed at executors concentrated in small and medium-sized enterprises. The initial conditions of the competition did not allow large enterprises to participate in the competition. This, of course, is not correct. Later, the competitive conditions were adjusted. Small businesses are usually created with the aim of improving the welfare of their creators. The development of science-intensive technologies, bringing them to real products, making a profit is too long a process in which small enterprises, as a rule, do not participate.

Most often, small businesses are created on the basis of an existing scientific groundwork with the aim of commercializing it. Often, the core of these enterprises is a group of developers who have spun off from a large research and development enterprise on the grounds that they have all the necessary knowledge to use the idea that was generated during their work in the enterprise. This enterprise paid for both the first experiments and the creation of the first product samples, but it is not aimed at commercializing such ideas. In addition, the commercialization of ideas in a large enterprise does not bode well for the team of authors, and therefore it is separated into a separate small enterprise. At least in the economy of the Russian Federation, such a scheme was carried out quite often. A small business cannot provide co-financing for its development. A small business cannot provide all the staffing requirements for the development of a new technology. The only thing a small business can most often do is to quickly gather a sufficient team of programmers around it, provided that they are provided with higher salaries than in a large campaign.

The condition of obligatory co-financing in the amount of at least 50% of the cost of the work seems unreasonable. If the research results are a new technology that is in demand by the state economy, then it is not clear why to limit the selection of performers only to such teams that can cover half of the costs from some other sources. Apparently, this requirement is a desire to hedge, which is natural in the event of a shortage of independent and highly qualified experts.

It also seems strange to require that the team of performers have publications in high-ranking journals, and that the main performers have scientific titles and scientific degrees. In addition, one of the requirements that is almost always set for competitive funding is a small average age of the performer team. It is clear that if the average age of performers is 65 years old, one can hardly expect breakthrough digital technologies, but what does not suit the team in which, mainly, for example, programmers aged 45-55, work, it is not clear.

It is also not serious to demand that the team, which received funding, began to issue products by the end of the current calendar year. It turns out that in a few months the recipient of financing must not only create a fundamentally new technology, but also introduce it into production, and get some kind of fundamentally new product. It seems to us that the average statistical team aimed at creating a truly new technology, most likely, will not be able to provide co-financing, will not meet the requirements for publication activity, for age criteria, for scientific titles and degrees of performers. This means that in order to win, this team will make a deal with an enterprise that can provide financing, will include people with appropriate indicators in the team.

That is, we see that these requirements, at first glance, are aimed at saving public funds, because if the winner of the competition adds their own funds in the amount of 50% of the cost of the work, then the state will receive research results for half the price. But in fact, the process will go in the opposite direction, because in order to create a tandem with an organization that will provide co-financing, the real performer will have to attract her with something. And it can be attracted only by the fact that this organization will receive more money from the project than it will provide. That is, the condition for providing financing does not actually increase the amount of money for development, but decreases it. For example, if the cost of development is indicated as 200 million rubles, and at the same time the contractor indicates that he will receive half from the co-financing organization, then this means that in fact the contractor intends to spend on the work itself an amount that is less than half of the indicated cost, since half , received as co-financing, he will definitely have to return, as well as add some kind of "bonus", for example, he will attract 50% of the amount, and return 60% to the benefactor, that is, only 40% will remain for work, which is 80 million rubles. So, wouldn't it be easier to allocate 100 million rubles to this enterprise without any requirement for co-financing?

If an organization that has entered into a competition for funding includes some doctors of sciences with important publications in the list of executors, and some students to reduce the average age of the team, then these extra people will have to receive their share of funding at least for agreement to be listed in this team. This again means that some of the money will not go to the purposes for which the program was created, but wasted. Consequently, the wrong list of achievements and indicators of the contestants is simply unnecessary costs, unnecessary obstacles for a truly effective team and unnecessary opportunities for various combinator-manipulators to get involved in the division of money.

6. Creation of leading centers

The creation of leading centers assumes that these centers will carry out competitive selection locally in the regions. If we talk about science-intensive technologies, this suggests that there is an understanding that there is, for example, Siberian science, or Far Eastern science, or Ural, or Moscow, and so on. Apparently, the attraction of regional experts to finance the development of technologies for regional projects was meant to maintain some semblance of fairness. Apparently, not all the money should be left in the two capitals, this fact has already been understood. But instead of changing the conditions of the competitive selection, excluding corruption components, it seems that it was easier to split funding into territorial quotas. This gives hope that some funds will be shared impartially.

7. Conclusion

This paper provides some alternative views on funding research in virtual and augmented reality. Probably, one of the main problems in the implementation of the Program is not so much a lack of competent experts in this area, but a lack of trust in truly competent experts from among domestic specialists; instead, high-rated journals enjoy trust. We must first answer the question of what we want to receive as a result of financing the Program - an additional number of articles with a high quartile, or actually working technologies for the production of new products, even if not a single new paper is written about them.

References

[1] "Digital Economy" Program Implementation. Available at: https://ac.gov.ru/en/projects/project/digital-economy-program-implementation-42

[2] Report on the digital economy development in Russia. Competition in the digital age: strategic challenges for Russia. Available at: https://roscongress.org/en/materials/doklad-o-razvitii-tsifrovoy-ekonomiki-v-rossii-konkurentsiya-v-tsifrovuyu-epokhu-strategicheskie-vyz/

[3] Russia Moves to Fund Digital Economy in Five Sectors. Available at: https://www.russia-briefing.com/news/russia-moves-fund-digital-economy-five-sectors.html/

IOP Conf. Series: Materials Science and Engineering 1019 (2021) 012090 doi:10.1088/1757-899X/1019/1/012090

Application of the probabilistic technologies to power plant design

L V Krivova and A V Shmoilov
Department of Electrical Engineering, National Research Tomsk Polytechnic University, Tomsk, Russia
E-mail: krivova_lv@tpu.ru

Abstract. In the report the developed approach for choice and substantiation of power equipment, current-carrying parts and switching devices using existing and proposed probabilistic method of borders selection of input and output data (SBID) is presented. SBID method allows to receive complete probabilistic characteristics or probability distribution laws (PDL) of output data as functional dependence from the arguments (input data) by probabilistic characteristics of arguments. Any task, including electrical values in operating modes and transient conditions in power system can be exposed as dependence on input or output data. Due to SBID method, the PDL of input data processing results of these dependencies may be received. PDL allows to calculate risks of overload of the power components in the operating modes and destruction, also power equipment parameters, current-carrying parts and switching devices are going to be selected on the basis of the minimum specified risks.

1. Introduction

Questions of reliability, efficiency, safety, maintenance, other functional specifications for the design and operation of any objects are always relevant. These issues even more updated in the case of electrical plants - large and complex facilities in the power sector, such as substations, power plants, or the whole power grid. The importance of these issues constantly accompanies all stages of development, design and operation of electric power systems.

According to this, the aim to assess the functional reliability of the main parts of power system as power plants and substations is stated. Thus, the probabilistic approaches applied to equipment choice for power plant with renewable resources and power system stability assessment are presented in [1–7].

Reliability is a property of an object (element, a component of the system) to perform specified functions under specified conditions.

Structural reliability – reliability designs and schemes without taking into account their operational modes (downloads).

Functional reliability is filling sensitive component properties and their aggregates included in designs and schemes of objects that are not included in the structural indices of reliability.

The structural and functional reliability of systems are the tasks of determining the reliability of interesting elements, objects, and whole systems on indices of reliability of all system components.

2. The proposed methodology

Performance of objects and components of the systems can be different and similar. Depending on different methods for determination of reliability parameters of objects, systems and their components can be formed. Experience shows that rational methods of reliability calculation are the methods that allow to estimate the reliability of any units (individual items, different sets of them) and other objects with similar indices.

At the Department of Power Stations, Tomsk Polytechnic University the method of evaluating of functional reliability indices in the form of probabilities or risks of overload and destruction of the switching devices, current-carrying parts and equipment that are included in the above mentioned objects was developed [8, 9]. Risks are defined, respectively, in the operational and emergency modes according to the probability distribution laws (PDL) of electrical quantities (currents and capacity in the branches and voltage at nodes) in the choice and substantiation of the power components. Methods and indices of functional reliability are important and necessary, in addition to choice and substantiation of power objects in power supply systems and for many other engineering tasks: definition of imbalances and deficits of active power at switchgear, power plants, substations, and, more generally, power systems, determine the settings and the effectiveness of relay protection and automatics, etc.

Feature of the proposed methodology application for assessment of functional reliability in contrast to expert guidance methods for recommending solutions, is that it allows not only to implement an objective choice of interesting parameters based on certain criteria, but also to assess the effectiveness of this choice, and therefore, to adjust the choice, to achieve the relevant maximum efficiency. These methods can be realized only when known complete probabilistic characteristics of interesting parameters, i.e. the way to obtain these characteristics have to be known.

3. SBID method

Discussed methodology of functional reliability is based on the method of boundaries selection of input and output data (SBID), also developed at the Department of Power Stations, Tomsk Polytechnic University. SBID method allows for complete probabilistic characteristics (probability distribution function (PDF) or probability distribution density) of arguments of any deterministic (non-random) or random function to determine almost exactly or inexactly the PDF of function results. Knowing the same complete probabilistic characteristics of interesting parameters or values, simple and objective solutions to many practical problems can be easily found.

The essence of the method is based on the following SBID:

1. Any task can be represented as an ordinary nonrandom or random function. With nonrandom functions the arguments or the original (input) data are arbitrarily (randomly) change, functions remain the same and the results or output are formed by transformation of input data by using nonrandom functions (NF). Therefore, the output data are also random, but dependent on the input data through a rigid (deterministic) conversion of the NF. In case of random functions (RF) arguments are arbitrarily (randomly) change, and transformation of random functions and results or output when will be formed by a random transformation of random input data.

2. Integral probability characteristic – the function of probability distribution (PDF) or the probability that each random variable will not exceed the given values has such great properties as taking their own measure in the range from zero to unity throughout the range of the specified arguments from the minimum possible to the maximum possible, and at any interval is not declined. In other words, PDF is the no decreasing function.

These properties of PDF may it possible to decline that if all arguments of NF (the input data) accept as the values corresponding to the same value of their PDF, in case of increasing NF, the result of NF transformation or output data should match the same value PDF of nonrandom function.

PDF value corresponding to the value of its argument, called quantiles of given order to be determined by the value of the NT. In other words, the transformation of input data, quantiles of the same order of increasing SF gives the result as a quantile of the same order (as the order of the quantiles of the input data). In case if input data are transformed by decreasing NF, quantiles of the same order of decreasing NF gives the result as quantile of order equal to the difference between the unity and the order of the quantiles of the input data. sign of increasing or decreasing NF is that if the difference between the previous and subsequent orders of input data, have the same polarity, the NF is increasing, otherwise the NF is decreasing. Thus, starting with the second version of the input data as quantiles of

IOP Conf. Series: Materials Science and Engineering 1019 (2021) 012090 doi:10.1088/1757-899X/1019/1/012090

the same order, based on the results of each NF it is possible to specify increasing or decreasing character of the NF and, consequently, to identify specific orders of output data (results) as the quantiles of the same order as the order of the input data quantiles, or about equal to the difference between the unity and the order of the input data quantiles.

In this paper materials on the development and application of algorithms, sequences, calculations and procedures for the choice and substantiation of the different components of the power stations and substations are presented, some of them tested for specific components by using named methods and procedures.

4. Example of SBID method application

For illustration of SBID method the CAG - 24 - 30 / 30000UZ circuit breaker in the output circuit of the generator 200 MW at Surgut GRES – 1 was chosen. In case of retrofitting the power plant equipment the methodology may be also applied for choice and substantiation of SF6 and vacuum circuit breakers.

The electrical quantities in the calculated operating and emergency modes are determined for specified areas of the Tyumen energy system, the databases of which are formed in the computing complexes (CC) of the TKZ-3000 (forced sinusoidal electrical quantities with short-circuits and other disturbances), Dakar (operating modes).

Since the circuit breaker switches the generator of the G-T unit, the maximum observed value of the current in operating modes should be taken equal to the maximum operating current of the generator.

The minimum observable current of the operating mode is determined by the switching state of the area of the network where the G-T unit is located and the minimum mode of the generators. The minimum observed operating current is determined by the minimum mode of the generator itself.

By means of computer program TKZ - 3000 it is possible to find the maximum and minimum observed values of the fault current flowing through the circuit breaker. The maximum observed value is the periodic component of the super-transient short-circuit current from the generator and the power system when short-circuit takes place on the periphery of the circuit breaker and the mode of the sources providing the maximum values.

The minimum observable value in accordance with the prescribed recommendations will be found in two stages. First, the minimum observable value due to the switching state of the network and the type of short circuit damage is determined at the same periphery of the circuit breaker.

Then the minimum observable value is determined taking into account 30% of power source load and type of short circuit, which is considered as a quantile of order p_2=0.0013.

Next, the ME and MSD of the normal PDL of the short circuit current through the circuit breaker are determined:

$$m(I_f) = \frac{i''_{p1} + i''_{p2}}{2} = \frac{65.492 + 35.254}{2} = 83.119 \quad \text{kA}$$

$$\sigma(I_f) = \frac{i''_{p1} - i''_{p2}}{6} = \frac{65.492 - 35.254}{6} = 5.033 \quad \text{кA}$$

Similarly, in emergency conditions, quantiles of orders and thermal effects are calculated:

$$b_{kp_1 main} = i''^2_{p_1 main}(t_{ter} + \tau) = 65.492^2(3 + 0.15) = 13511 \quad \text{kA}^2 \cdot \text{c}$$

$$b_{kp_2 main} = i''^2_{p_2 main}(t_{ter} + \tau) = 35.254^2(3 + 0.15) = 3915 \quad \text{kA}^2 \cdot \text{c}$$

$$m(B_{k\ main}) = \frac{b_{kp_1\ main} + b_{kp_2\ main}}{2} = \frac{13511 + 3915}{2} = 8713 \quad \text{kA}^2 \cdot \text{c}$$

$$\sigma(B_{k\ main}) = \frac{b_{kp_1\ main} - b_{kp_2\ main}}{6} = \frac{13511 - 3915}{6} = 1935.3 \quad \text{kA}^2 \cdot \text{c}$$

IOP Conf. Series: Materials Science and Engineering 1019 (2021) 012090 doi:10.1088/1757-899X/1019/1/012090

In a static state, the load circuit breaker must withstand the emergency short circuit currents arising from disturbances in the power plant and in the power grid. In this regard, for this device, the amplitude and effective values of the limiting test short circuit current for the main circuit and grounding conductor are set. Assuming a possible random change of these values is similar to other short-term allowable test values, we define the parameters of the normal PDL specified random changes of the short circuit current $I_{s\,main}$ given in catalogue for main circuit. For grounding circuit, the named values are found according to similar equations.

Similarly, a previously possible predetermined random change in the test current of thermal withstanding allows forming the parameters for a normal PDL in main circuit.

Based on the results of the calculations, the calculated operating (i_{co}) and emergency electrical quantities are accepted. It is advisable to take nominal values as initial calculated operating quantities, and as similar emergency values can be recommended as a half-sum of mathematical expectations of normal PDL actual emergency (I_e) and short-term acceptable (I^{sa}) values. It is possible to determine the risks accordingly:

1) r_{ol} overload in operating conditions:

$$r_{ol} = 0.5 - \Phi\left[\frac{i_{co} - m(I_e)}{\sigma(I_e)}\right] + 0.5 + \Phi\left[\frac{i_{co} - m(I^{la})}{\sigma(I^{la})}\right],$$

where $i_{co} = i_r$

i_r - rated current,

I^{la}- long-term allowable current.

2) damage under emergency conditions:

$$r_d = 0.5 - \Phi\left[\frac{i_{dem} - m(I_{max})}{\sigma(I_{max})}\right] + 0.5 + \Phi\left[\frac{i_{dem} - m(I^{sa})}{\sigma(I^{sa})}\right] \text{ where } i_{dem} = \frac{m(I_{max}) + m(I^{sa}_{s\,main})}{2}$$

where Φ - Laplace's function.

Finally, may be found:

1) the risk of overload in operating conditions:

$$r_{ol} = r_o + r_{la} = 0{,}5 - \Phi\left[\frac{8.625 - 6.911}{0.715}\right] + 0.5 + \Phi\left[\frac{8.625 - 30}{0.6}\right] = 0.5 - 0.490613 + \frac{x_{-\infty} + x_{п} - x_{\Phi_2}}{-100} 0.000000000000032 =$$

$$= 0.009387 + \frac{-100 - 7.5 - (-35.625)}{-100} 3.2 \cdot 10^{-14} = 0.009387000000023$$

i_{pp}=8.625 kA,

x_{tL}– table values of Laplace's function,

x_{pL} – positive values of Laplace's function,

x_{nL} – negative values of Laplace's function.

Calculating of extremely small values of Laplace's function is presented in [9].

2) the risk of destruction in emergency conditions:

$$r_d = 0.5 - \Phi\left[\frac{218.36 - 50.318}{5.033}\right] + 0.5 + \Phi\left[\frac{218.36 - 353.6}{7.07}\right] = \frac{x_{\infty} + x_{п} - x_{\Phi_1}}{100} 3.2 \cdot 10^{-14} +$$

$$+ \frac{x_{-\infty} + x_{п} - x_{\Phi_2}}{-100} 3.2 \cdot 10^{-14} = \frac{100 + 7.5 - 33.388}{100} 3.2 \cdot 10^{-14} + \frac{-100 + (-7.5) - (-19.129)}{-100} 3.2 \cdot 10^{-14} = 5.2 \cdot 10^{-14},$$

$$i_{dem} = \frac{83.119 + 353.6}{2} = 218.36 \text{ kA}.$$

Further, it is possible to optimize the calculated values by changing their magnitudes and based on the minimum risks of overload and destruction respectively.

5. Conclusions

The specified positive and unique opportunity to substantiate the power equipment, switching devices and current-carrying parts by means of the probabilistic technology allows not only selecting interesting parameters and electrical quantities characterizing these objects, but immediately and objectively assessing the total technical losses - risks of overload and accidental destruction. This substantiation can be implemented if risks of overload and destruction due to exceeding the selected values of the parameter, respectively, in operating and abnormal conditions, other risks of overload and destruction due to the reduction of the permissible limits long respectively and short time limits on selected values for the specified parameter or parameters can be adjusted to the same observation conditions. The criterion of synergies can be considered and used as a tool to optimize the choice of values of the specified parameter or parameters of the power components. Finding the optimal values of selected parameters is the subject of a task or group of tasks that are not finally solved, but it is possible.

References

[1] De Caro *et al* 2012 *Proceeding of 21st International Symposium on Power Electronics, Electrical Drives, Automation and Motion.* 474–476 doi:10.1109/SPEEDAM.2012.6264605

[2] Dorman L I 1975 *Variations of Galactic Cosmic Rays* (Moscow: Moscow State University Press) 103 p.

[3] Malley M *et al* 2011 *IEEE Transactions on Power Systems* **26** 564–572

[4] Rueda J. L *et al* 2011 *IEEE Transactions on Power Systems* ***26*** 1733–1743.

[5] Byun J 2017 *Springer Series in Reliability Engineering* (Cham: Springer) 27–46 doi:10.1007/978-3-319-52425-2_2

[6] Straub D, Papaioannou I, Betz W 2017 *Reliability Updating in the Presence of Spatial Variability. In: Gardoni P. (eds) Risk and Reliability Analysis: Theory and Applications.* (Cham: Springer) 365–383 doi: 10.1007/978-3-319-52425-2_16

[7] Varde P V, Pecht M G 2018 *Probabilistic Approach to Reliability Engineering. In: Risk-Based Engineering* (Singapore: Springer) 31–69 doi: 10.1007/978-981-13-0090-5_3

[8] Krivova L V, Shmoilov A V 2007 *Proceeding of International Forum on Strategic Technology* 408–410

[9] Krivova L V, Schmoilov A V 2014 *The choice and substantiation of the power equipment, switching devices and current-carrying parts with application of probabilistic technologies: monography* (Tomsk: TPU Publishing) 106 p.

IOP Conf. Series: Materials Science and Engineering 1019 (2021) 012091 doi:10.1088/1757-899X/1019/1/012091

Glass composite modified with silicon carbide and gallium arsenide, that absorbs electromagnetic radiation

O V Kazmina[1], V I Semenova[1] and K V Dorozkin[2]
[1]Scientific and educational center of N. M. Kizhner, National Research Tomsk Polytechnic University, Tomsk, Russia
[2]Laboratory for Terahertz Research, Tomsk State University, Tomsk, Russia
E-mail: kazmina@tpu.ru

Abstract. The possibility of using waste of semiconductor manufacturing containing silicon carbide and gallium arsenide as an effective additive in radio absorbing material preparation was established as a result of research. The resulting material meets the requirements of effective porous electromagnetic radiation absorbers of extremely high frequency range: maximum absorption and minimum reflection of electromagnetic radiation, fire safety, the minimum allowable level of toxic and hazardous substances.

1. Introduction

The relevance of research on the porous glass composites creation due to the demand for radio absorbing materials with improved characteristics that differ from their counterparts in environmental and fire safety. The glass composite considered in this paper is a porous inorganic material consisting of a vitreous matrix and micro-sized particles of crystalline phase in the interporous septum. The particles of crystalline phase with high dielectric and magnetic permeability, give to material ability to absorb electromagnetic radiation. Such materials been used as radio absorbers of electromagnetic radiation of extremely high frequency range in various areas: shielding of premises, protection of mobile objects from detection, when detecting dangerous objects and substances. The requirements of functional properties for this material such as high absorption coefficient and low reflection coefficient of electromagnetic radiation, as well as reduced density and increased mechanical strength. Achieving these properties increases the material efficiency and has great prospects, but it is rather difficult to implement in practice, therefore it is of scientific interest for researchers.

Foam glass materials and composites obtained on the basis of glass microspheres are analogous to porous glass composite. Many foreign scientific publications contain the results of development and use of foamed glass materials in a wide variety of field. The use of foam glass as an absorber of electromagnetic radiation is a current new direction [1]. Broadband, efficient, lightweight, and relatively inexpensive radio absorbing materials are required in this area. Broadband is achieving due to the multi-layer structure, for example, sandwich panels, laminates and epoxy composites [2–5]. This approach greatly complicates the technology of radio absorbers and requires high-tech production. The material effectiveness is proving by the additional introduction of such absorbing additives as carbon nanotubes, carbon fibers, graphene, ferrites, magneto dielectrics [6, 7]. Carbon nonmaterial interacts with electromagnetic radiation and are effective modifiers, including the extremely high frequency range 53–73 GHz [8]. At the same time, the price of carbon nonmaterial is currently still quite high. Therefore, effective and relatively inexpensive additives are of practical interest.

Previously obtained experimental data showed that electronic manufacturing waste, consisting of silicon carbide and gallium arsenide, could be using as an effective modifier of porous silicate system [9]. Material safety is confirming by biotesting results and atomic emission spectroscopy of aqueous extract. According to the measurements, the amount of arsenic, transferred from the material to water, does not exceed the maximum permissible concentrations and amounts to 9.35 mg/l. This indicates the possibility of using this type of waste to obtain a safe material [10]. The waste transfer to safe state and its use as a component for production radio-absorbing material is an extremely important task. Gallium arsenide is supposed to improve the electrophysical properties of radio absorbent, since it has a number of structural features: a wide band gap (1.43 eV), high electron mobility (8500 $cm^2/(V \cdot s)$), direct interband transitions. Thus, without the use of expensive components, solved problems of environmental improvement and improvement functional properties of material.

The aim of the work is to develop a porous glass composite modified with silicon carbide and gallium arsenide at temperatures not exceeding 400 °C, which has a high radio absorbing ability in the extremely high frequency range (from 30 to 300 GHz).

Due to high permeability, electromagnetic waves of this range have great prospects and are using in such fields as medicine, safety systems, the study of fundamental characteristics of materials [11, 12].

2. Materiaks and experimental method

Liquid glass, perlite microspheres, hardener, semiconductor manufacturing waste containing silicon carbide and gallium arsenide using to obtain a radio absorbing material. Industrial sodium liquid glass with a silicate module 3 and a density of 1.36 g/cm^3 using in the work. Na_2SiF_6 with a content of sodium fluorosilicate of 98% been chosen as a hardener of the liquid glass composition. Perlite hollow microspheres with a bulk density of 150 kg/m^3 provide a lower material density. Perlite microspheres have the following chemical composition, wt. %: SiO_2 – 74.7; Na_2O – 5.3; CaO – 0.6; Fe_2O_3 – 1.2; K_2O – 5.1; Al_2O_3 – 12.8; MgO – 0.3. The phase composition of perlite microspheres is represented by orthoclase and glass phase. The average size of microspheres is 125 μm with a wall thickness of about 2 μm according to electron microscopy data (Figure 1), the shape is uneven, elongated, asymmetric. Figure 1 shows that microspheres have a polycameral cellular structure.

Figure 1. Electronic micrograph of perlite microspheres.

Waste of semiconductor manufacturing is forming in the process of cutting plates of gallium arsenide and polishing. The work investigated the sludge formed after polishing plates using silicon carbide. By particle size, waste refers to finely dispersed materials. The studied waste is 90% represented by particles with a size of less than 5 microns, and 30% of the grains have a size of less than 3 microns according to the results of laser diffraction (Figure 2). The bulk density of the waste powder, dried at room temperature, is 730–760 kg/m^3. The phase composition of waste is representing by silicon carbide and gallium arsenide. Reflexes corresponding to gallium arsenide are more pronounced and are consistent with the main established reflection maxima (d = 0.3263; 0.1998;

0.1708 nm). Less intense reflections correspond to silicon carbide (d = 2.5162; 2.5208; 2.6276 nm). The content of gallium arsenide is about 55%, and silicon carbide in pure form is 40% according to the results of a quantitative X-ray structural analysis carried out using the Match! Program. Less intense reflexes correspond to minor impurities in the form of gallium and boron carbide.

The radio absorbing properties were evaluating by measuring the reflection, transmission, and absorption coefficients using the "free space" method on a STD-21 terahertz spectrometer in the 120–260 GHz range. To study the electromagnetic response, flat samples with a size of 30×50 mm^2, of various thicknesses were using.

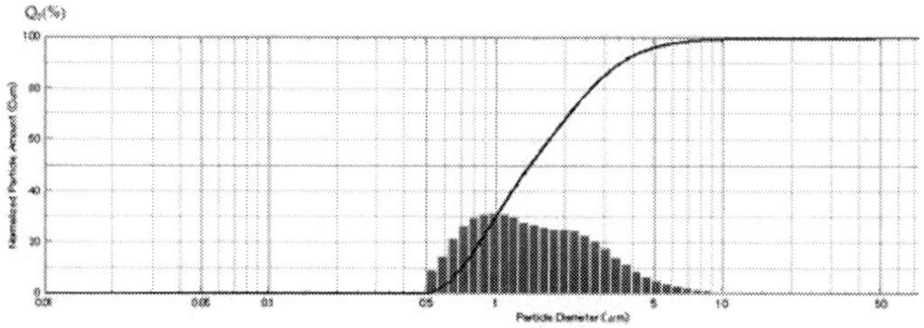

Figure 2. Electronic micrograph of perlite microspheres.

The absorption coefficient was calculated (1):

$$A = 1 - R - T \tag{1}$$

where R – reflection coefficient, T – transmission coefficient, rel. unit.

3. Results

Production of porous glass composite was caring out according to the developed technological scheme presented in Figure 3. The technology includes the stage of waste preparation, mixing of components and heat treatment of liquid glass composition to a solid state. Waste preparation involves drying, since the sludge formed after grinding plates has a moisture content of about 15%. Granulation was caring out with the addition of a solution of liquid glass, which leads to the coating of the granules with a strong film of sodium silicate and transfers the waste to a safe state. On the other hand, the contact of the granules with the liquid glass composition is improved. The use of liquid glass due to its ability to foam at a temperature of from 100 °C to 550 °C, which gives the material additional porosity and lightness. A hardener converts sodium silicate to an insoluble state. Perlite microspheres contribute to reducing mass of finished absorber, reacting with liquid glass and promoting its silicatization.

The temperature of the synthesis of the material does not exceed 400 °C. Gallium arsenide decomposes with arsenic starting with 600 °C.

It is knew that electro physical properties of porous material depend on its composition, structure and thickness. The presence in the material composition additives in the form of dispersed particles with high dielectric constant leads to increase absorption coefficient. In the case of waste, the role of such additives is performing by silicon carbide and gallium arsenide. The influence of waste amount on the efficiency of electromagnetic radiation absorption of the considered range was carried out on samples containing 10, 20 and 30 wt. % waste. A higher amount of waste leads to increase material density. The results of all experiments are giving for samples with a thickness of 1 cm.

Figure 4 shows the dependences the values of absorption, reflection and transmission coefficients of electromagnetic radiation for a sample with waste (10%) and without. The results show that the value of reflection coefficient is close to zero, including the material obtained without waste. This indicates that the mechanism for reducing intensity of electromagnetic radiation for these porous samples lies in the wave absorption and not in its reflection. When waste content at composite in the amount of 30%, the value of the absorption coefficient increases at a frequency of 22 GHz by 28 times, and at a frequency of 40 GHz by 4 times. The transmission coefficient in this case is regularly reducing by 2 times at 22 GHz and 5 times at a frequency of 40 GHz.

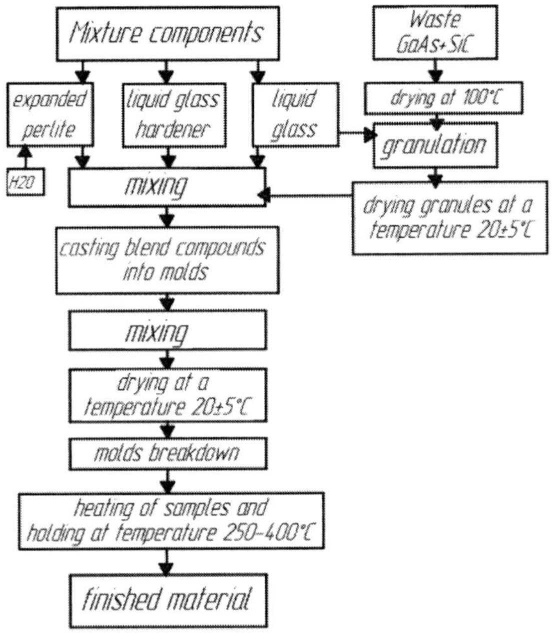

Figure 3. Technological scheme for obtaining electromagnetic radiation absorber with waste of semiconductor manufacturing.

The similarly situation in the high-frequency range of 120 – 260 GHz. The reflection coefficient is zero for all samples. With an increase the waste amount in the composite, the transmission coefficient decreases to zero when the waste content is 30%. The maximum value of absorption coefficient equal to 100% has a sample with a maximum content of waste at a frequency of 240 GHz (Figure 5). The value of the sample wave transmission coefficient with 30% waste compared to the sample without waste decreases at a frequency of 120 GHz by 15 times, and at a frequency of 240 GHz by 33 times.

For the obtained samples, the main physic mechanical properties were determined, the results of which are giving in table 1. The apparent density of the material increases, which also leads to an increase in the mechanical strength of the material with an increase waste content. Material corresponds to the class of heat-insulating structural materials according to the values of these indicators.

Table 1. Physical and mechanical properties of the composite.

The amount of waste in the composite, wt. %	Apparent density, kg /m^3	Compressive strength, MPa	The porosity of the composite,%
0	330	0.3	87
10	350	1.5	86
20	510	4.7	80
30	560	6.0	78

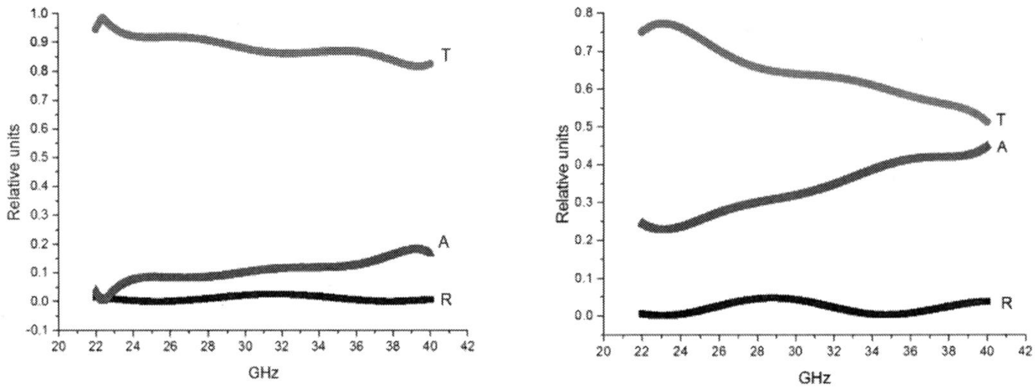

Figure 4. The dependence of absorption (A), reflection (R) and transmission (T) coefficients at a frequency of 20-40 GHz for a sample without waste (from the left) and with a waste of 10% (from the right).

Because of the research conducted, it was establishing that semiconductor production wastes could be using as an effective additive in obtaining a composite that absorbs electromagnetic radiation, including the extremely high frequency range.

4. Discussion

The developed mixture composition, comprising up to 30 wt.% of semiconductor manufacturing waste containing silicon carbide and gallium arsenide provides a composite that absorbs electromagnetic radiation in the extremely high frequency range of 30–300 GHz.

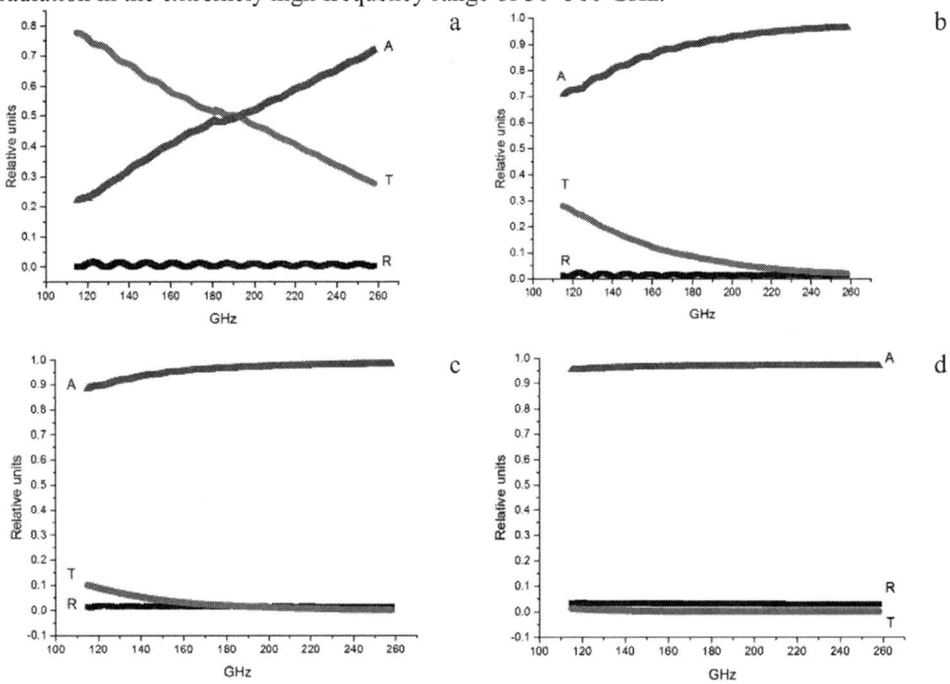

Figure 5. The dependence of the absorption (A), reflection (R) and transmission (T) coefficients of the frequency 120–260 GHz for the sample without waste (a) and with a waste of 10% (b), 20% (c), 30% (d).

A low-temperature (up to 400 °C) technology for producing a porous electromagnetic radiation absorber is proposing. The reflection coefficient equal zero. The absorption coefficient is 100% at a frequency of 260 GHz and 84% at a frequency of 40 GHz. The developed technology is a resource-saving and energy-efficient solution.

It is established that with an increase the waste amount in material from 10 to 30%, the material apparent density increases from 350 kg/m^3 to 560 kg/m^3 and compressive strength from 1.5 MPa to 6.0 MPa. This increases absorption efficiency of electromagnetic radiation, which is due to the influence of silicon carbide and gallium arsenide on the processes of polarization and conductivity in the composite during the absorption of electromagnetic radiation.

The resulting material meets the basic requirements for effective porous electromagnetic radiation absorbers of extremely high frequency range: maximum absorption and minimum reflection of electromagnetic radiation, fire safety, the minimum acceptable level of emission of toxic and harmful substances. The composite is recommending for use in anechoic chambers and high frequency devices.

Acknowledgments

The reported study was funded by RFBR, project number 19-33-90099.

References

[1] Ostergaard B M, Cai B and Petersen R R 2019 *Materials Letters* **250** 72–74 doi:10.1016/j.matlet.2019.04.106
[2] Ki-Yeon P, Sang-Eui L and Jae-Hung H 2006 *Composites Science and Technology* **66** 576–584 doi: 10.1016 / j.compscitech.2005.05.034
[3] Bollena P, Quiévy N, Huynena I *et al* 2013 *Scripta Materialia* **68** 50–54
[4] Volski V, Vandenbosch A E 2009 *Compos Sci Technol* **69** 161–168 doi:10.1016 / J.COMPSCITECH.2008.09.040
[5] Jung-Hoon O, Chun-Gon K *et al* 2004 *Composites Part B: Engineering* **35** 49–56 doi:10.1016/J.COMPOSITESB.2003.08.011
[6] Kolanowska A, Herman A P *et al* 2018 *Carbon* **126** 31–52 doi:10.1016/j.carbon.2017.09.078
[7] Dong J, Wancheng Z P *et al* 2018 *Applied Physics A: Materials Science and Processing* **124** 693 doi:10.1007 / S00339-018-2117-0
[8] Zakharychev E A, Ryabov S A *et al* 2013 *Inorganic Materials: Applied Research* **4** 405–409 doi: 10.1134 / S2075113313050201
[9] Stebeneva V I, Dorozkin K V and Cibo Z 2018 *High technology in modern science and technology* **1** 92-93
[10] Stebeneva V I, Kazmina O V 2018 *Conf. materials "National Development"* **1** 176–178
[11] Dréan Y L, Yonis S M *et al* 2013 *Comptes Rendus Physique* **14** 402–411
[12] Homenko A, Kapilevich B *et al* 2009 *Bioelectromagnetics* **30** 167–175 doi: 10.1002/bem.20466

IOP Conf. Series: Materials Science and Engineering 1019 (2021) 012092 doi:10.1088/1757-899X/1019/1/012092

Research of the operation voltage limiter in pulsed mode

I Krasniy and S Akimov

Department semiconductor and microelectronics, Novosibirsk State Technical University, Novosibirsk, Russia

E-mail: kr.ivan.work@gmail.com

Abstract. This paper presents the results of investigations on the matter of withstanding power and switch-on time of the voltage limiter operating in a pulsed mode. The voltage limiter is based on a silicon $p^{+}\!-\!n\!-\!n^{+}$ structure obtained using the «batch» method of simultaneous two-way diffusion of boron and phosphorus out of solid phase.

1. Introduction

The transistor, diode or thyristor semiconductor structures utilized for data transmission and processing systems produce a capacitive current while operating. If no proper elimination measures are taken, this current leads to a deterioration in noise immunity and reliability of the entire electronic circuit. As reported in the work [1], a capacitive current increases the duration of the output voltage transition to a high level state and increases the resorption time of carriers. To suppress the influence of capacitive current, the authors of work [2] propose to create a low-impedance capacitive current tap circuit using a voltage limiter instead of traditional resistor. The advantage of the voltage limiter lies in its very low impedance in switch-on mode, as well as its high frequency range. However, these parameters differ from each other depending on the technology and design of a particular voltage limiter. The shorter the carrier resorption time, the less time it takes for an output voltage to rise and the greater the frequency range, thus expanding the range of voltage limiter application [3].

This paper presents the results of investigations on the matter of withstanding power and switch-on time of the voltage limiter operating in a pulsed mode. The voltage limiter is based on a silicon p+–n–n+ structure obtained using the new «batch» method [4] of simultaneous two-way diffusion of boron and phosphorus out of solid phase.

2. Materials and Experimental Procedures

The voltage limiters based on silicon p^{+}–n–n^{+} structures utilized in the study are produced using diffusion technique. n-type silicon of «КЭФ» grade (i.e. phosphorous doped silicon with electronic conductivity) with an initial thickness of 200 μm and a resistivity of 0.02 Ohm×cm was chosen as the base region. The diffusion process was carried out simultaneously on several samples placed in a quartz cartridge («batch» method). A microrelief was created by means of grinding on the surface of the initial crystals with a given resistivity to improve adhesion with the diffusant. The first diffusant, «КБК-45» (containing 45% of boric acid), which forms a p–n junction with the initial crystal, is applied to the surface of the crystal twice, while the second diffusant, «КФК-54» (containing 54% of phosphoric acid) is applied as a single layer. In order to provide infinite source conditions, the n^{+}- and p^{+}-type gaskets matted with corundum, 800 μm in diameter and 1 mm in thickness, are placed between the initial crystals. The initial crystals are placed inside the cartridges in such a way as to enclose gaskets between the homogeneous sides, as it is shown in Figure 1.

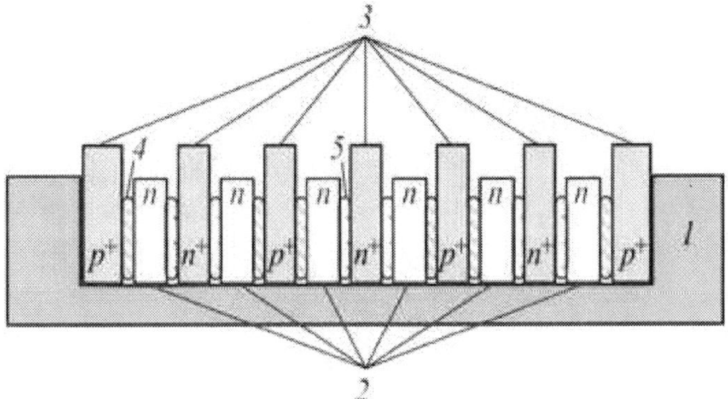

Figure. 1. Layout of components for diffusion process: 1 – quartz cartridge, 2 – base crystals, 3 – gaskets, 4 – p^+-type diffusant, 5 – n^+-type diffusant.

The diffusion process itself is carried out in a furnace chamber. The temperature profile of the diffusion process comprises the heating of the system from 650 °C up to 1250 °C in a nitrogen atmosphere, holding for 48 hours and then cooling to 650 °C in a dry oxygen environment. The use of the «batch» diffusion method in incorporation of boron and phosphorus impurities into silicon made it possible to reduce the surface resistance of the p^+-layer to 1.1–1.2 Ohm/square and that of the n^+- layer to 0.3–0.4 Ohm/square, in contrast to the values achieved for diffusion out of the gas phase, which are 2.2 - 1.8 Ohm/square. Specific to the proposed method, in the area with a depth of up to 25 μm the resistivity close in value to the surface resistivity is obtained, which is then rather sharply increases by 3–4 orders of magnitude, creating an abrupt p–n junction.

The distribution of the impurity over the thickness of the base region, as well as the occurrence depth of p–n junction were determined by analyzing the structure of the transverse section of the crystal using scanning electron microscope (SEM).

The calculated value of the diffusion coefficient was equal to 7.47×10^{-12} cm^2/s. The depth of boron diffusion into silicon increases in a sublinear manner depending on the diffusion time and at 48-hour point reaches the values of 50–70 μm. A similar dependence also takes place in the diffusion process of phosphorus. Thus, we can assume that the thickness of the diffusion regions on one side, p^+ -type, is about 60 μm in depth, while the thickness of those on another side, n^+ -type, is 70 μm.

Taking into account the low rate of thermal diffusion and electrodiffusion, as well as low penetrating ability into silicon, the following systems of metals were chosen as the contact material for the p^+–n–n^+ structures obtained: V – Ag, Ti – Ni – Au, in which case Ag and Au provide reliability of soldering process. Contact materials were deposited sequentially with thickness in the range of 0.1–0.65 μm. Then, Ag layers with a thickness of 5 μm were galvanically deposited on their surfaces, the p^+–n–n^+ structures have a hexagonal crystal shape (3.4 mm 3.0 mm) with an area of 0.0742 cm^2. Afterwards, the thermal expansion joints were formed on both surfaces of the crystals by means of fusing with a silver solder alloy, 2–3 μm in thickness. Thermal expansion joints are made of a 380 μm thick copper disc with a diameter of 0.33 cm, on both surfaces of which layers Ni, 6 μm in thickness, are applied.

The structures prepared this way were installed in plastic shells with metal leads (Figure 2), soldered using the soft solder of «ПСр-2.5» grade with a melting point of 295–300 °C. On the basis of technology described above, samples of voltage limiters with different resistivity of the base area and breakdown voltages ranging from 12 to 340 V were manufactured.

Figure 2. The structures assembled in a plastic shell.

3. Experimental

Investigations of the capacity-voltage and current dependences, as well as the study of the dependence of the maximum pulse power on the pulse duration, were carried out on the structures obtained.

As shown in Figure 3, the obtained dependence of the barrier capacity of the p+–n–n+ structure on the blocking voltage is in accordance to the power law 1/C2 = V.

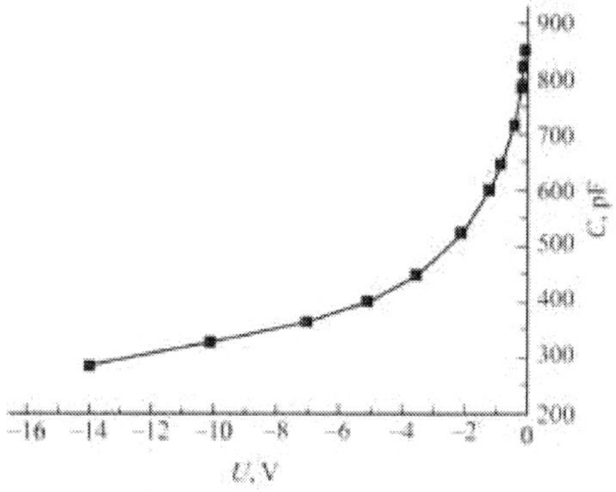

Figure 3. Dependence of barrier capacity on blocking voltage.

The thickness of the space charge layer of the p^{+}–n junction is derived based on data of the barrier capacity values in the specified voltage range:

$$W_{p\text{-}n} = \frac{\varepsilon\varepsilon 0 S}{Cv} \tag{1}$$

which provides us with a value of 0.09914 μm at zero offset and 0.27 μm at a voltage of 14 V. The concentration of charge carriers at the boundaries of the depleted layer, valid for any distribution of charge density at small increments of the offset voltage, can be determined from the equation [5]:

$$N_n(W_{p-n}) = -\left(\frac{2}{q\varepsilon\varepsilon_0 S^2}\right)\left(\frac{1}{\frac{d[C^{-2}]}{dV}}\right) \tag{2}$$

which provides us with a value of 1.3 10^{15} – 1.2 10^{16} cm^{-3} (Figure 4).

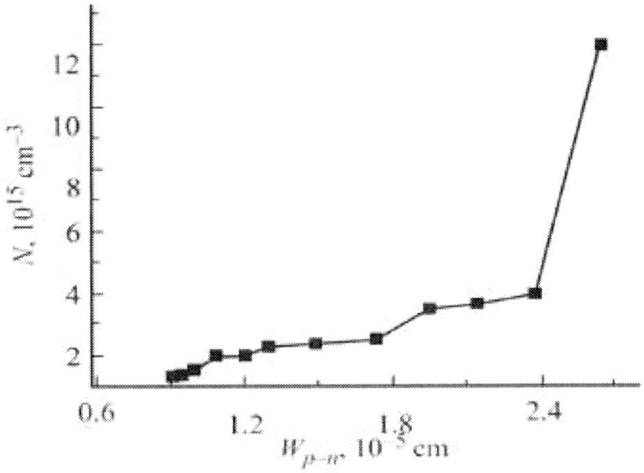

Figure 4. Distribution of the carrier concentration at the various distances from the p^+–n junction boundary.

Thus, the structure is divided into four layers with different charge concentrations (Figure 5):
1. diffusion layer p^+ - 10^{19} cm^{-3};
2. partially compensated layer n_0 - 1.3 10^{15}...4 10^{15} $cм^{-3}$;
3. base material n with resistivity - 0.02 Ohm×cm;
4. diffusion layer n^+ - 10^{18} cm^{-3}.

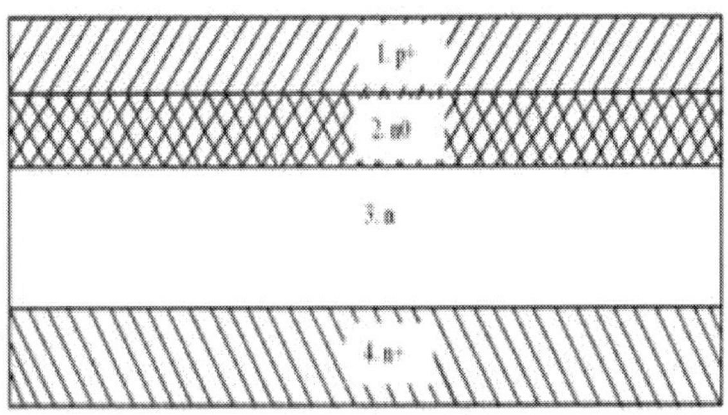

Figure 5. Structure of the crystals obtained

4. Discussion of the results

The current–voltage characteristics of the p^+–n–n^+ structures, obtained using the measuring bench by setting the reverse current and voltage fixation, allowed us to prevent the transition to the mode with a steep rise in the reverse current. Investigations of the current-voltage characteristics showed that their breakdown voltages increase within the temperature range of –60 ± 2 °C to 125 ± 2 °C (Figure 6), which is typical of avalanche breakdown [5]. Concurrently, the electric field intensity in case of the observed breakdown voltage of 23 V equals 8.5×10^5 V/cm.

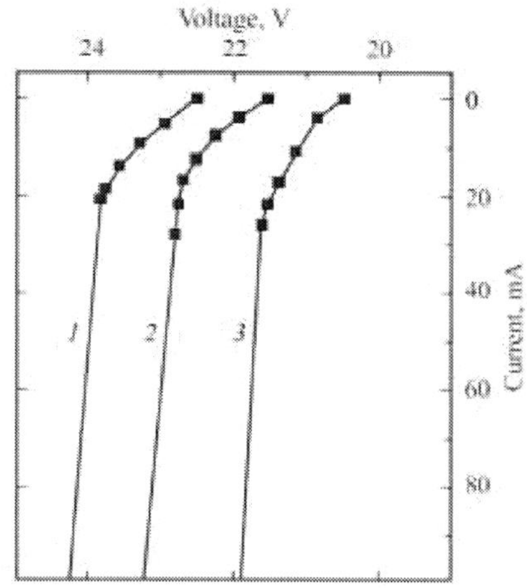

Figure 6. Dependence of the reverse current on the voltage of the silicon $p^+–n–n^+$- structure at different temperatures, C: 1 — 125 ± 2, 2 — 25 ± 2, 3 — -60 ± 2.

In the avalanche breakdown mode, the newly generated avalanche shock front passing through the diode leaves a populated plasma with a weak electric field. In this case, the current flowing through the diode is increased significantly due to the additional multiplication of carriers in the base, while the produced plasma reduces the voltage on the diode to near-zero values. Furthermore, the holes situated in the base region drift to the p^+ region, whereas the electrons drift to the n^+ region at a speed significantly lower than the drift saturation velocity. The plasma gradually dissolves keeping the current unchanged and the process repeats. The process of carrier multiplication is characterized by a high value of the diode field intensity and precedes the new formation of the avalanche shock front, which possesses the longest duration [6]. In this mode, the longer the pulse duration is, the higher the temperature will be and the less withstanding power will be.

The behavior of the withstanding power dependence on the pulse duration is characterized by equal decrease as the pulse duration increases in case of voltage limiters of 0.5 kW and 1.5 kW values (Figure 7). In particular, as the pulse duration increases, the switch-on time will increase, as well, leading to a rise in the temperature of the limiter.

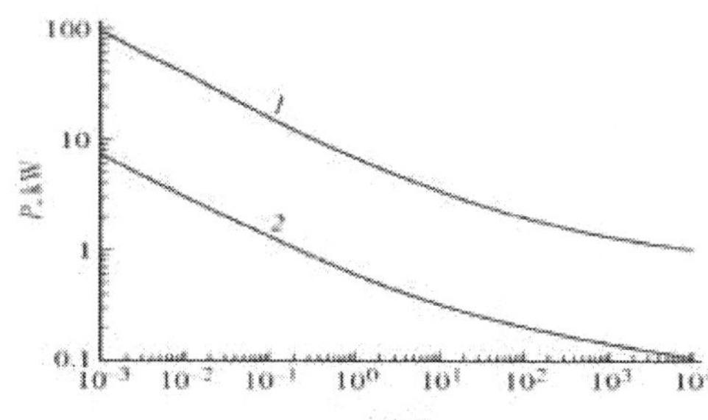

Figure 7. Dependence of the maximum pulse power on the pulse duration for the voltage limiters of the following power: 1 – 1.5 kW, 2 – 0.5 kW.

Since the steepness of reverse current growth in the breakdown mode is rather sharp and the voltage increment is very small, the switch-on time value will be very small, as well (the frequency range value will be high, correspondingly).

When a sharply increasing reverse current pulse is applied to the limiting diode, a capacitive current will flow through it before reaching the breakdown voltage. In p–n junction structures, the main condition for the existence of an inductance is the presence of a voltage drop in the high-resistivity region, as well as the injection of minority carriers into this region, which leads to an increase in its conductivity. The proportionality factor connecting the voltage and the rate of capacitive current change dI/dt is the inductance. The magnitude of the current is proportional to the barrier capacitance and the rate of voltage change, which is also associated with the base series resistance. The electric field will increase across all cross-sections at rate

$$\frac{dE}{dt} = \frac{j(t)}{\varepsilon\varepsilon 0} \tag{3}$$

The field will proceed to build up as such until the electric field at some point reaches the maximum value of E_m. When the field exceeds this value, an avalanche shock front is generated, which passing through the diode leaves a populated plasma with a weak electric field. The current build-up occurs so fast that the process of carrier drift can be neglected. The cross section in which the field reaches the maximum value ($E > E_m$) moves rapidly in the direction of the n^+ region where the shock wave propagates (Figure 8).

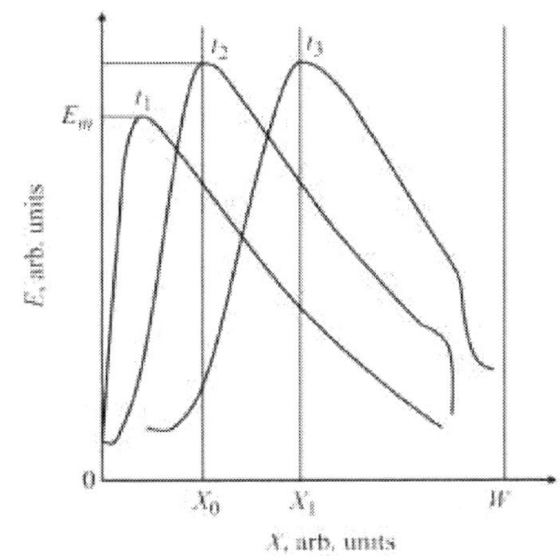

Figure 8. Shock wave propagation in the region of the p–n junction.

The slope of the field to the right of the X_0 section is determined by the equation

$$\frac{dE}{dt} = \frac{q}{\varepsilon\varepsilon 0} N_b \tag{4}$$

where N_b is the impurity concentration in the depletion region. The velocity of shock wave propagation can be calculated if we divide equation (3) by (4), which results in the following

$$V_{sw} = \frac{dx}{dt} = \frac{j(t)}{aN_b} N_b \tag{5}$$

The maximum carrier velocity is $v_s \ll v_{sw}$, but even assuming that $v_s < v_{sw}$ the generated plasma will not have enough time to dissipate and the shock wave will be developed as a result.

The ratios obtained allow one to calculate the total switch-on time of the limiting diode t_{on}. It comprises the sum of the time required to charge the barrier capacitance of the p–n junction t_{RC} within the voltage range of V_m to V_b with the initial current $I = jS$, where S is the area of the p–n junction,

$$t_{RC} = \frac{C_b(V_b - V_m)}{I} - y\frac{\varepsilon\varepsilon 0 S}{W}\frac{E_m W}{2}\frac{1}{jS} = y\frac{\varepsilon\varepsilon E_m}{j} \tag{6}$$

C_b – the p–n junction capacitance, C_b – 0.01, and the shock wave propagation time

$$t_{sw} = \frac{W}{2V_{sw}} = \frac{qWN_b}{2j} \tag{7}$$

It is necessary to add the energy relaxation time $t_e \approx 10-12$ s to these components, i.e. the time required for the energy to transfer from carriers to the lattice.

Accordingly, the total switch-on time of the limiting diode will be determined by the following equation

$$t_{on} = Y \frac{\varepsilon\varepsilon 0 E_m}{j} + \frac{qWN_b}{2j} + t_e \tag{8}$$

As can be deduced from this equation, after the shock wave occurs the diode dynamics is largely defined by the switching circuit, which determines the current density j. The diode is turned off shortly after the impact of the current pulse. The time period is determined by the resorption time of the plasma arising in the depletion region.

The distinctive feature of the diodes under study is that the current increases with a low limiting coefficient (1.13 as opposed to 1.18), in comparison with the conventional alternatives [7], while the current density is very high, predisposing the voltage limiter to low switch-on values in accordance with equation (8). In case of the experimental data $\gamma \approx 0.01$, $E = 8.5 \times 10^5$ V/cm, Wp–n = 0.27 µm and current density (100 mA/0.0742 cm^2) = 1.35 10^{-3} A/cm^2 the switch-on time t_{on}, equals 13 µs, according to (8).

5. Conclusions

1.It has been experimentally demonstrated that the presented technological and design solutions allows one to obtain semiconductor structures that, acting as voltage limiters, possess an increased resistance to impulse loads and a shorter operating switch-on time. The data on the withstanding impulse power obtained over the course of the experiment showed that the nominal power decreases according to the same law with an increase in the pulse duration, which indicates that the parameters of the designed structures remain the same in the limiter obtained.

2.The relationship between transition times (switch-on and switch-off) and the characteristic parameters of the voltage limiter structure is presented. The possibility that some part of power might dissipate at the resonating frequency acts as an additional stimulus for increasing the withstanding power values.

Acknowledgment

The authors of this paper would like to thank the research laboratory at «Institute of semiconductor physics after A.V. Rzhanova SB RAS».

References

[1] Belous A I, Silin A V, Ponomar V N 1998 *Circuitry of bipolar microcircuits for high-performance information processing systems* (Minsk: Polyfact) (in Russian)
[2] Belous A I, Soloduha V S, Sokol V A, Syakersky V T 2009 *Electronic components* 9 50–53 (in Russian)
[3] Korshunov F P *et al 2005 FTT* **47** 419 (in Russian)
[4] Muratov A F, Rakhmatov A Z, Merkulov A A, Ismoilov I R Patent RUS 05328
[5] Farenbruch A, Byube R 1987 *Solar cells* (Moscow: Energoatomizdat) pp 99–100
[6] Kozlov V A, Cardo-Sysoev A F, Brylevsky V I 2001 *FTP* **35** 629–631 (in Russian)
[7] Skornyakov S P, Pavlov V A, Rakhmatov A Z 2008 *Components and technologies* 3 70–71 (in Russian)

Dielectric particle-based strategy to design a new self-bending subwavelength structured light beams

I V Minin and O V Minin
School of Nondestructive Testing, Tomsk Polytechnical University, Tomsk, Russia
E-mail: prof.minin@gmail.com

Abstract. During last 2 years it was shown that an electromagnetic field can be made to curve after propagation through a simple dielectric material of special shape, which adds a new-found degree of simplicity. This effect was termed 'photonic hooks' – it is an unique electromagnetic beam configuration behind a mesoscale dielectric particle with a broken symmetry and differ from Airy-family beams. PH features the radius of curvature, which is about 2 times smaller than the electromagnetic wavelength – this is the smallest curvature radius of electromagnetic waves ever reported. The nature of a photonic hook is in dispersion of the phase velocity of the waves inside of particle, resulting in interference. Here, we report a new dielectric particle-based strategy to design self-bending subwavelength structured light beams.

1. Introduction

The word "optics" is derived from the ancient Greek philosophers term meaning "appearance, look" and was based on the idea that light propagates along straight lines expanding as they travel. For example, Euclid wrote his "Optica" in about 300 BC [1] in which he postulated that light travelled in straight lines. The development of Maxwell's electrodynamics further reinforced these notions by ensuring the conservation of electromagnetic momentum. The possibility that a wave packet can freely accelerate even in the absence of an external force was first discussed four decades ago [2]. As indicated in [2], this is only possible as long as the quantum wave function follows an Airy-function profile. The most unusual aspect of Airy beams is that their paths seem to bend: whereas the axis of a Gaussian beam always follows a perfectly straight line (in homogeneous propagation medium).

Due to the center of such beams seems to follow a nonlinear path, accelerating perpendicular to the initial beam direction, despite the absence of external forces acting on the moving photons, such beams are referred to as self-accelerating. In 2007, this Airy self-acceleration process was suggested and experimentally observed for the first time in optics [3, 4]. It is important to note that for finite power Airy beams, while the local intensity features do self-bend in a self-similar fashion, the Ehrenfest theorem still holds thus preserving the balance of the transverse electromagnetic momentum. Ever since, this class of accelerating or self-bending beams has attracted considerable attention and found applications in many and diverse fields, especially in the paraxial domain. In the past few years, other types of Airy-like accelerating curved beams have been intensely explored; among them: "half Bessel" [5], Weber (travel along parabolic) and Mathieu (travel along elliptic curves) beams [6, 7]. These beams are by nature non-paraxial and hence can bend at larger angles. In all cases, these Airy-like wavefronts propagate on a curved trajectory over many Rayleigh lengths while defying diffraction effects. Until recently, they provided the only example of "curved light transport" in nature with curvature more than several wavelength.

IOP Conf. Series: Materials Science and Engineering 1019 (2021) 012093 doi:10.1088/1757-899X/1019/1/012093

2. Photonic nanojet

In 2004, Chen et al. coined a new term "photonic nanojet" (PJ) for the sub-wavelength-scale near-field focusing at the shadow side of a mesoscale (with dimensions of several units of wavelength) dielectric particle [8]. By increasing the dimensions of a spherical particle the electromagnetic field structure evolves and tends to be more localized and directed forward. Despite the beneficial performance of PJs in several applications, up to date, all photonic nanojet beams are fundamentally similar – single-colored and straight [9].

3. Photonic nanojet based airy-like beams

From the geometrical optics point of view the rays of light in an Airy beam cross each other, staying on one side of an enveloping curve. Each of the rays is briefly tangent to the envelope at some point. Such an envelope of a rays system is called caustics, which are common features in everyday life. In some papers, curvilinear caustics were studied in the diffraction of an off-axis limited size wave (the size of the beamwaist is less than the particle diameter) on a dielectric cylinder. However, no one argued that the caustic curve in this case is an Airy-like beam (Figure 1*a*) formed near the shadow surface of dielectric cylinder or hemicylinder (Figure 1*b*, Figure 2*a*). As Bernhard Baruch told in New York Post (24 June 1965): "Millions saw the apple fall but Newton was the one who asked why".

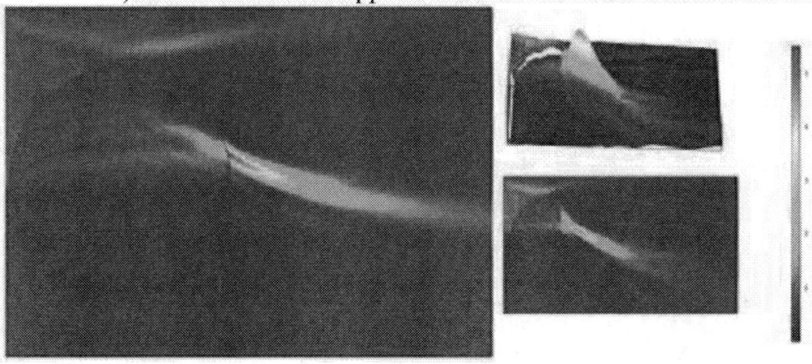

a) b)

Figure 1. Formation of near field Airy-like beam by dielectric particles: cylinder (*a*, left) and by transversely oriented dielectric hemicylinder (*b*, right).

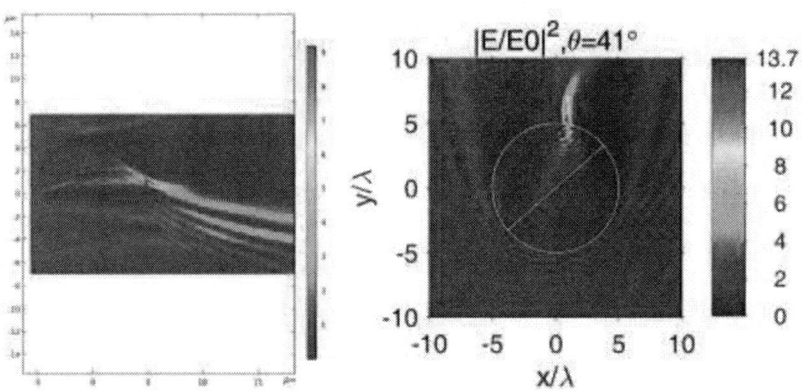

a) b)

Figure 2. Formation of near field Airy-like beam by longitudinally oriented dielectric particles: hemicylinder (*a*, left) and photonic hook (see Paragraph 5 below) by two hemispherical particles with different refractive index (*b*, right).

Nevertheless, the acceleration of photonic jets on the basis of the approaches considered above is much less than on the basis of a cuboid with broken symmetry. Moreover, the subdiffraction value of the main beam lobe was not observed yet.

4. Double curved jet based on edge diffraction

It is well known that under diffraction on dielectric edge in the Fresnel approxiation the edge wave has the eikonal S ~ $Z+X^2/2Z$) [10] and a solution of the diffraction at the edge of the semi-infinite opaque screen [11] is a paraxial 2D light for which the argument of the complex amplitude function is depends on variables, like X^2/Z, where X is the transverse coordinate, and Z is the longitudinal coordinate.

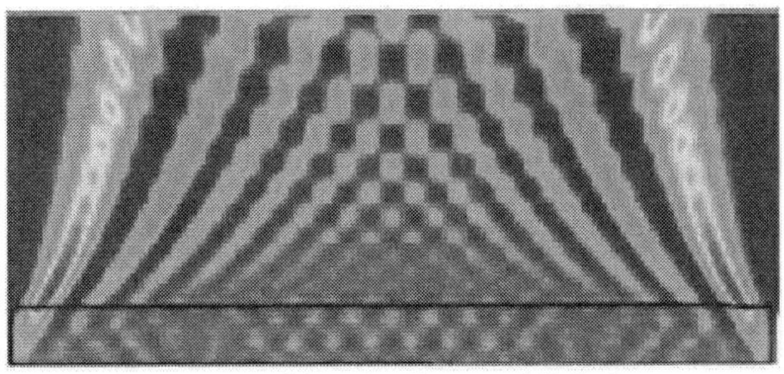

Figure 3. Double-edged curved light beams.

It means that instead of the Airy beams which have a parabolic ray path $y=x^2$, whereas the considered edge beams in question propagate along a root-parabolic path $y=sqrt(x)$ (Figure 3).

5. Photonic hook

Recently, a new family of near-field localized curved light beams, which differs from the family of Airy beams and is formed in the near-field zone: the so-called "photonic hook" (PH) has been discovered by I.V. Minin and O.V. Minin [9].

In a photonic hook, individual light waves propagate along a straight line, but their relative phases and amplitudes are chosen so that the maximum energy in space forms a constant curved shape in time. Since the electromagnetic waves themselves propagate along linear paths, the energy density in the curved beam decreases with distance from the source.

Our recent research shows [12] that it is possible to produce a localized optical beam whose intensity peaks move along curved trajectories in the near-field. Unlike ordinary optical localized beams (like PJ), the photonic hook self-bends (i.e., transversely accelerates) throughout propagation in the near-field. PHs appear even in free space and do not require any waveguiding structures or external potentials. They are unique in the sense that their radius of curvature is substantially smaller than the wavelength, meaning that such structured light beams have the maximum acceleration among the known curvilinear beams. The curvature of electromagnetic waves with such a small radius was described for the first time in [12].

The realization of such curved near-field beams is possible based on the interaction of a plane wave with a mesoscale dielectric particle with broken symmetry. In [9, 12] we theoretically studied a PJ formed behind the symmetric particle with an equal rib and discussed some examples of PHs behind the mesoscale particles with broken symmetry. It is important to note that such a particle may be considered as a Janus particle since a PH is formed only when the particle is irradiated in one direction (and not in the opposite) [9]. The curve of the brightest region results from a complex near-field interference pattern inside the particle as the phase velocity disperses. Because of this shape, the time of the full oscillation phase of an optical wave varies in the particle unevenly. As a result, a curved light beam is produced at the exit from the particle (near its shadow surface). Besides, the curvature of

IOP Conf. Series: Materials Science and Engineering 1019 (2021) 012093 doi:10.1088/1757-899X/1019/1/012093

the beam can be adjusted by changing the wavelength, polarization of the incident light, as well as the geometric parameters of the dielectric particle.

One of an idiom supported by people since ancient times is ''Seeing is believing''. So the experimental verification of this accelerating near-field beams would be an additional proof of the PH concept, which can be used to control light at the new level. Recently, the existence of Minins' PH phenomenon has been verified experimentally for the first time using continuous-wave scanning-probe microscopy, operating at 0.25 THz due to the scalability of Maxwell's equations [13].

In Figure 4, we show the results of experimental PH visualization [13]. It can be seen that curvature of PH is about $\alpha = 148^0$, the length of PH is less than two wavelengths with inflection point position near Z=1.2λ from the shadow surface of the particle. In contrast to traditional Airy beams (generated using a complicated optical element with cubic phase or with a spatial light modulator behind the focus of a spherical lens [4–6]), a PH can be created using a compact mesoscale dielectric particle-lens. Moreover, the Airy beam consists of the main lobe and a family of side beamlets whose intensity decay exponentially [4–6]. Interestingly, in the case of PH, only the main lobe has a curved shape, and the family of curved sidelobes is absent. It was also shown that PH phenomenon is observed on a scale much smaller than Airy beams. In addition, it was shown for the first time that when switching two orthogonal states of linear polarization of the incident radiation, the photonic hook changes the direction of its curvature to the opposite. Note that the mesoscale dimensions and simplicity of the PH concept open a way for the practical integration of PH elements into lab-on-a-chip platforms and indicating their large scale potential applications.

Figure 4. Experimental visualization of a photonic hook in terahertz

These particles are unique among micro objects because they provide asymmetry and can thus impart drastically different physical properties and directionality even with in a single particle. The broken symmetry offers efficient and distinctive means to realize the emergence of properties in unconceivable for homogeneous particles or symmetric particles. Moreover, since the formation of a photonic hook occurs on the basis of a particle with broken symmetry, such a particle may be considered as a Janus particle. The curved electromagnetic beam caustic is formed only when the asymmetric particle is irradiated in one direction - if the particle is pumped from the side of the prism (and not in the opposite) due to broken symmetry [9]. Also the shape of the dielectric particle considered above is not the only possible one.

At the same time, the asymmetry of particles can be not external, but internal (Figure 5).

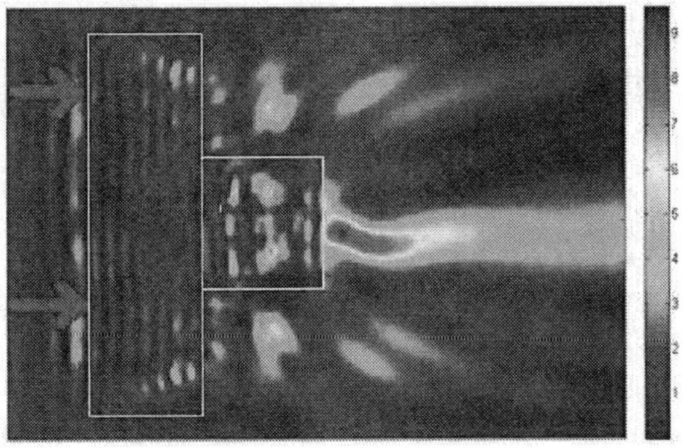

Figure 5. Formation of curved focusing beams by cubic particle with internal broken symmetry placed on dielectric substrate.

PH beam waist also has a cross-dimension smaller than the wavelength that makes high resolution possible. Thereby, the observed PH phenomenon can provide advantages over common photonic jets in imaging applications, similar to the beneficial character of the Airy beams over the Gaussian and Bessel beams reported in [14]. Moreover, it turned out that the photonic hook concept can be used to implement a manipulator to move particles along a curved path around transparent obstacles [15].

6. Photonic loop

Due to the phenomenon of focus bending is caused by the interference of waves inside the dielectric particleas the phase velocity disperses, the shape of the photon hook and the characteristics of curved near-field beams, depending on the specific application, can be quite exotic. For example, by choosing the shape of mesoscale particles we may control of phase delays across the wavefront and, thus, control a shape of localized near-field (Figure 6).

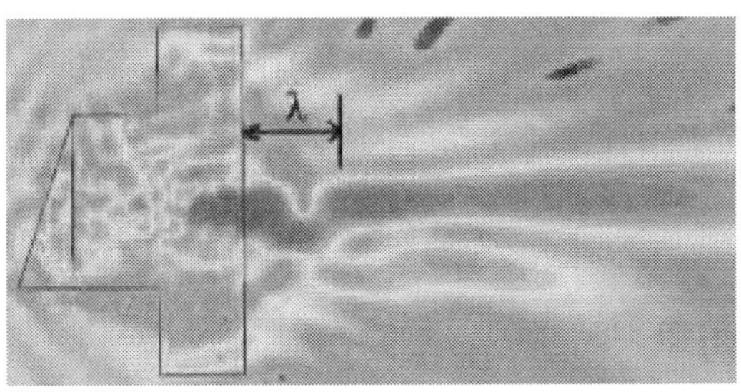

Figure 6. Example of a subwavelength photonic loop.

7. Plasmonic hook

The concept of photonic hook has gone beyond optics and penetrated other fields, including plasmonics. In 2010, the idea of Airy surface plasmon polaritons (SPPs) was introduced [16], and experimental observations soon followed [17]. However, in low-dimensional systems (in which at least in one of the three dimensions of the electronic state wave function is confined) until recently the families of Airy plasmon beams were the only beams that have a curved trajectory. But for Airy-type SPPs high beam acceleration is required to achieve significant curvature of the beam over the propagation length.

Our previous studies have shown that dielectric particle with broken symmetry can modulate the wave fronts of plane wave at will, achieving interesting physical effects such as photonic hook formation. The basic idea is to use a carefully designed particle to locally control the phases of the

IOP Conf. Series: Materials Science and Engineering 1019 (2021) 012093 doi:10.1088/1757-899X/1019/1/012093

impinging waves, thus, reshaping the wave fronts of refracted beams based on interferences. Inspired by this observation, we have propose to extend the idea of the dielectric particle with broken symmetry to control the wave fronts of surface wave. We can put a carefully designed particle on a plasmonic surface, which can also locally control the phase of an impinging surface wave. When a surface beam is launched on the plasmonic surface and strikes the particle, the incident surface wave gain different local phases from the refractions in the particle, thus, generating a refracted surface wave beam with the desired wave front.

Recently, the phenomenon of the photonic hook plasmon (PHP) was also introduced for SPPs propagating along metal-dielectric interfaces [18]. This concept is fundamentally simpler than the generation of the SPP Airy-family beams. It also demonstrated for the first time a simple and reliable real-time technique to dynamically control the curved trajectories of light beams over metallic surfaces by simply changing the illumination wavelength (without the need of any permanent guiding structures). The PHP propagates along wavelength scaled curved trajectory with a radius less than the SPP wavelength, which represents the smallest radius of curvature ever recorded for SPP beams, and can exist despite the strong energy dissipation at the metal surface (Figure 7). It is worth to note that the plasmonic PH is formed in the spatial region where the effects of evanescent fields play a significant role. The use of overlapping evanescent fields from several active radiators causes the specific interference interaction between the radiators which allow breaking the fundamental Rayleigh criterion.

Figure 7. Example of plasmonic photonic hook.

From an applications perspective, photonic hook plasmons are important for the following main reasons. First, it demonstrates that the Airy plasmons now are not the only self-bending beams in two dimensions. Second, the photonic hook plasmons are self-bending localized field, giving us the means to control the flow of light on the surface in the near-field. Moreover, electromagnetic waves at the surface of a metal can be channeled into circuit components smaller than the diffraction limit. Importantly, these new possibilities prove that the concept of wave acceleration appears in general electromagnetism and can be observed in non-paraxial approximation in the near-field. Another important outcome is that the localized near-field beams no longer have to propagate only in a straight line - they can be self-bending.

8. Acoustical hook

The observation of a new type of near-field curved acoustic beam (acoustic hook) different from the Airy-family beams both through simulations and experiment recently was reported [19]. This new self-bending acoustical beam also was formed from a rectangular trapezoid of a dielectric material immersed in water. The origin of this curved beam is in the vortices of intensity flow that appear inside the solid due to the conversion of the incident longitudinal wave mode to a shear wave in a solid. These vortices redirect the intensity flow resulting in a bending of the beam.

9. Passive amplifier of airy beams

Detailed studies of the localization of the electromagnetic field by dielectric particles of arbitrary shape [9], and in particular, by cuboids, allow us to propose a method of amplifying of Airy-family beams. In fact, such dielectric structures play the role of a passive repeater (Figure 8).

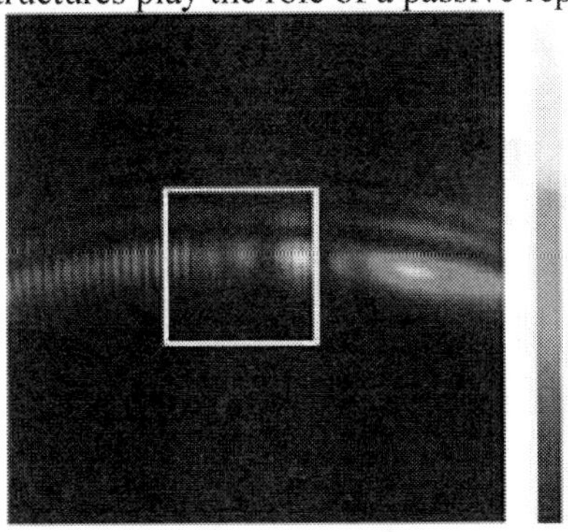

Figure 8. Example of Airy beam passing through dielectric cube.

10. Dynamic formation of Janus particles

The dynamic formation of Janus particles allows both control of the shape of the radiation localization region and opens the new way for obtaining new functional properties of structured fields. This same property can be used for optical switches, commutators, and various sensors. Similar effects can be observed in atmospheric physics, dew in nature, etc. For example, when a water drop freezes, its phase state changes from liquid to solid (ice). These materials have different properties and, in particular, different refractive index. As an example, Figure 9 shows the formation of a structured field behind a drop, consisting of water and ice. Similar effects can be observed upon evaporation of liquid droplets.

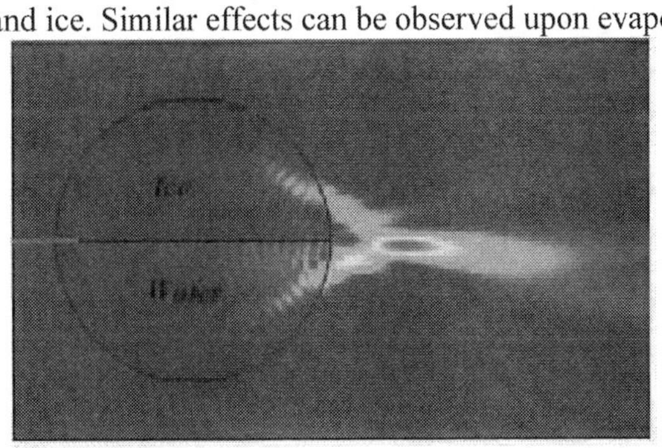

Figure 9. Structured light formation by water-ice droplet.

11. Conclusions

Finally, there are a few reasons that the concept of photonic hook [9,13,21,22] has penetrate other fields of modern physics (i.g., acoustics, plasmonics and THz) both in CW and pulsed [23] radiation as in transmitted as in reflection modes and has gone above and beyond optics. First, the PH phenomenon occurs under non-paraxial conditions and so the concept is common, thus opening interesting opportunities in other physical systems, i.g., acoustics, and ultrasonics. Second, the early experiments have revealed that PH beams are exceptionally easy to realize and manipulate.

We believe that the curved trajectory of PH is beneficial to photonics and the related disciplines. It could be used for advanced manipulation of nanoparticles, biological systems, material processing, and surgical systems, where curved sub-wavelength-scale optical beams are of great interest. The method could accommodate other applications, including particle acceleration and separation, short-range micromanipulation, terahertz generation, near-field spectroscopy, laser machining of various guiding structures to include wavelength scaled division multiplexers, and interferometer beam-splitting and beam-coupling, to redirect an optical signal in meso- and nano-scale or even as a scalpel tips for potential use in ultraprecise laser surgery.

On a more fundamental level, the introduction of the photonic hook concept into the field of optics will help the scientific community to better understand the physical phenomena of accelerating near-field waves and to manipulate light in unusual ways. As it was mentioned in [22] "such elegant examples of structured light beams deepen our understanding of light propagation, and will fuel anticipation and excitement for the next generation of imaging and manipulation". Moreover, it has opened up new paths for new applications.

Acknowledgments
This work was partially supported by the Russian Foundation for Basic Research (Grant No. 20-57-S52001).

References

[1] Burton H E 1945 *Journal of the OSA* **35**(5) 357
[2] Berry M and Balázs N 1979 *American Journal of Physics* **47** 264
[3] Siviloglou G, Broky J, Dogariu A and Christodoulides D 2007 *Physical Review Letters* **99** 213901
[4] Siviloglou G and Christodoulides D 2007 *Optics Letters* **32** 979
[5] Kaminer I, Bekenstein R, Nemirovsky J and Segev M 2012 *Physical Review Letters* **108** 163901
[6] Aleahmad P, Miri M, Mills M, Kaminer I, Segev M and Christodoulides D 2012 *Physical Review Letters* **109** 203902
[7] Zhang P, Hu Y, Li T, Cannan D, Yin X, Morandotti R, Chen Z and Zhang X 2012 *Physical Review Letters* **109** 193901
[8] Chen Z, Taflove A and Backman V 2004 *Optics Express* **12** 1214
[9] Minin I V and Minin O V 2016 *Diffractive Optics and Nanophotonics: Resolution below the Diffraction Limit* (Cham: Springer)
[10] Kopylov Y and Popov A 1996 *Radio Science* **31** 1815
[11] Born M and Wolf E 1986 *Principles of Optics* (Pergamon Press Ltd.: Maxwell House, N.Y.)
[12] Yue L, Minin O V, Wang Z, Monks J, Snalin A and Minin I V 2018 *Optics Letters* **43** 771
[13] Minin I V, Minin O V, Katyba G, Chernomyrdin N, Kurlov V, Zaytsev K, Yue L, Wang Z and N. Christodoulides D 2019 *Appl. Phys. Lett.* **114** 031105
[14] Vettenburg T, Dalgarno H, Nylk J, Coll-Llado L, Ferrier D, Cizmar T, Gunn-Moore F and Dholakia K 2014 *Nature Methods* **11** 541
[15] Ang A, Karabchevsky A, Minin I V, Minin O V, Sukhov S and Shalin S 2018 *Sci. Rep.* **8** 2029
[16] Salandrino A and Christodoulides D 2010 *Opt. Letters* **35** 2082
[17] Li L, Li T, Wang S, Zhang C and Zhu S 2011 *Phys. Rev. Lett.* **107** 126804
[18] Minin I V, Minin O V, Ponomarev D and Glinskiy I 2018 *Ann. Phys.* **530** 1800359
[19] Rubio C, Tarrazó-Serrano D, Minin O V, Uris A and Minin I V 2020 *Results in Physics* **16** 102921
[20] Minin I V, Minin O V, Yue L, Wang Z, Volcov V and Christodoulides D 2019 *ArXiv:* 1910.09543
[21] Minin I V and Minin O V 2019 *Proceeding of Fourth Russian-Belarusian Workshop "Carbon nanostructures and their electromagnetic properties"* 52–57
[22] Dholakia K and Bruce G 2019 *Nature Photonics* **13** 229
[23] Spector M, Ang S, Minin O V, Minin I V and Karabchevsky A. 2020 *Nanoscale Advances* **2(6)** 2595

IOP Conf. Series: Materials Science and Engineering 1019 (2021) 012094 doi:10.1088/1757-899X/1019/1/012094

The possibility of obtaining materials by plasma dynamic synthesis in the Ti-O system

A A Sivkov[1,2], Yu N Vympina[2], I A Rakhmatullin[2], A S Ivashutenko[2] and Yu L Shanenkova[1,2]
[1] College of Communication Engineering, Jilin University, Changchun, PR China
[2] School of Energy & Power Engineering, National Research Tomsk Polytechnic University, Tomsk, Russian Federation
E-mail: xyulyashax@mail.ru

Abstract. The possibility of obtaining a dispersed material by the method of plasma dynamic synthesis in the Ti-O system is shown. The plasma dynamic synthesis is based on the operation of a pulsed high-current coaxial magnetoplasma accelerator with titanium electrodes. The obtained product was investigated by X-ray diffraction method and transmission electron microscopy. The results showed the presence of two main titanium oxide (IV) crystalline modifications: anatase with tetragonal syngony and rutile also with tetragonal syngony. Moreover, the dominant modification is anatase in synthesized material.

1. Introduction

The problem of obtaining bulk ultrafine-grained materials and fine-dispersed powders of metals, alloys and compounds designed for various areas of technology has been discussed for a long time in the literature. Bulk ultrafine-grained materials can be obtained directly from bulk coarse-grained and amorphous materials or by powder metallurgy methods from fine-dispersed powders. In these days, interest in methods for obtaining superfine-grained bulk and dispersed materials has increased significantly, since it was found that a size decrease of structural elements (particles, crystallites, grains) below a certain threshold value can lead to a noticeable change in properties [1–5]. Such effects appear when the average size of crystalline grains does not exceed 100 nm.

The field of metal oxides is an important area in nanotechnology. Metal oxide nanostructures are applied in different fields of material science and engineering. For example, the main advantage of sensing and optical devices based on these materials arises from the high surface to volume ratio, favorable chemical stability, altered physical properties, confinement effects resulting from the nanoscale dimensions, and remarkable resistivity variation in gaseous environments. Additionally, the electrical conduction behavior of semiconducting metal oxide nanostructures and the proton conductivity of insulating metal oxide nanostructures commonly changes at different gaseous environments and by their dissociation of protons, making them promising materials for a wide range of eco-friendly and “green” applications [6].

From the discovery of superconductive oxides [7] and oxides with large magnetoresistance [8] the interest in these materials has increased even more, especially those of transition metal oxides [9]. Transition metal oxides are also commonly used as catalysts and in electronics, with applications varying from semiconductors, dielectrics, and conductive electrodes, which clearly demonstrates the importance and versatility of these materials [10].

The most common and widely used as sorbents and photocatalysts is titanium (IV) oxide due to a combination of properties such as low cost, chemical stability, catalytic activity. Moreover, in the nanoscale state, the particles of this material have an increased specific surface area, which positively affects its quantum efficiency [11–13]. There are many ways to synthesize nanoscale titanium dioxide, but none of the methods is universal in terms of the cost of precursors, time and simplicity of the process. For example, the well-known sol-gel method allows to obtain materials with both simple and complex chemical composition, in addition to regulate the particle size at any stage of the process [14–16]. However, the raw material has a high cost, the process of synthesizing the material is multistage. The solvothermal method also allows to control the particle size, the crystalline phases of the synthesized product [17, 18], but the rare material requires careful preparation. It is worth nothing that the whole process is strongly dependent on the reaction temperature, time, organic solvents.

The purpose of this work is plasma dynamic synthesis in the Ti-O system. It is assumed that it is possible to obtain titanium oxide (IV) with structural modification of anatase, which is more preferable for photocatalytic applications, in such a system. Unlike existing methods, plasma dynamic synthesis also allows to obtain nanoscale materials, while it is environmentally safe, one-step, fast (less than 10^{-3} seconds) and does not require additional preparation of precursors.

To achieve this goal, the following tasks have been set: 1) to obtain material by plasma dynamic synthesis in the Ti-O system; 2) to analyze the material by X-ray diffraction; 3) to analyze the product using transmission electron microscopy.

2. Experimental part

Plasma dynamic method is generally based on a system with a high-current (~ 100 kA) pulsed (~ 1 ms) coaxial magnetoplasma accelerator (CMPA) with metal titanium electrodes. A sketch-map of the coaxial magnetoplasma accelerator is shown in Figure 1.

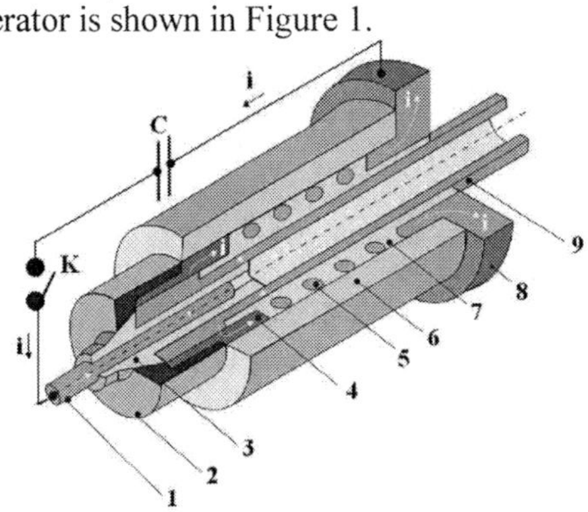

Figure 1. Sketch map, elements of coaxial magnetoplasma accelerator: 1) central electrode; 2) plug; 3) central electrode insulation; 4) contact cylinder; 5) solenoid; 6) case; 7) solenoid insulation; 8) contact flange; 9) electrode-barrel.

Titanium BT-1-O was used. Its phase composition was determined by X-ray fluorescence analysis. The results showed the presence, in addition to titanium, oxygen in the amount of 9%, chromium – about 1%. The presence of other elements (iron, silicon, copper, etc.) was less than 0.2%. The power supply of the CMPA is provided from a pulse capacitive energy storage device with a capacity of up to C_{ch} = 14.4 mF and a charging voltage up to U_{ch} = 2.5 kV. The energy stored by the capacitive energy storage device W_c, the amplitude of the discharge current of the power supply to the accelerator I_m, the developed power in the accelerating channel P_m, the energy W supplied to the accelerator can vary over wide ranges depending on U_{ch}. The initiation of the arc discharge between the central electrode and the electrode-barrel at the beginning of the accelerating channel was carried out by using of an electro-explosive carbon bridge deposited on the surface of the fiberglass insulator and electrically connecting the electrodes in the initial state.

After closing the power keys (K), a rising current begins to flow along the circuit. After some time, it reaches its critical value, which leads to the formation of a plasma structure inside the accelerating channel. As the jet moves, the electroerosion wear of the barrel occurs, the eroded material enters the plasma, thereby producing the base material, titanium. The expiration of a hypersonic plasma jet occurs in a sealed volume of a chamber-reactor filled with oxygen/argon gas (O_2/Ar). The synthesized product settles down to the wall of the chamber. After one hour, the chamber is opened and the product is collected. The experiment was carried out under the following conditions: C_{ch} = 14.4 mF, U_{ch} = 2.5 kV, accumulated energy W_c = 45 kJ, barrel length ℓ = 230 mm, barrel diameter d = 12 mm, O_2/Ar ratio – 1: 1, pressure in the chamber-reactor P = 1 atm., room temperature $t_0 \approx 20$ ° C.

To measure the rapidly changing voltage on the electrodes of the CMPA, ohmic voltage dividers are used. The discharge current characteristic is removed using a Rogowski transformer. Current and voltage signals are recorded on Tektronix TDS2012 oscilloscopes. Figure 2 shows the current and voltage waveforms recorded by digital oscilloscopes. It is seen that the peak values of voltage and current reach high values, up to 1.05 kV and 112 kA, respectively. These values are not constant, they depend on the capacitance and voltage of the capacitive energy storage device: the higher the charging voltage, the greater the discharge current and the voltage on the electrodes of the accelerator. By changing the capacity value, it is possible to regulate the time of the impulse process.

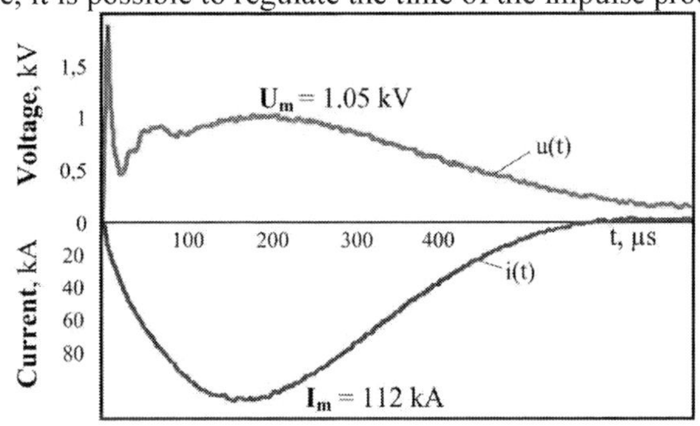

Figure 2. Wareforms of current i(t) and voltage u(t) on electrodes.

The power curves of the plasma discharge and the applied energy are shown in Figure 3. The maximum power value is 112 MW, and the sum of the energy supplied is 28.8 kJ.

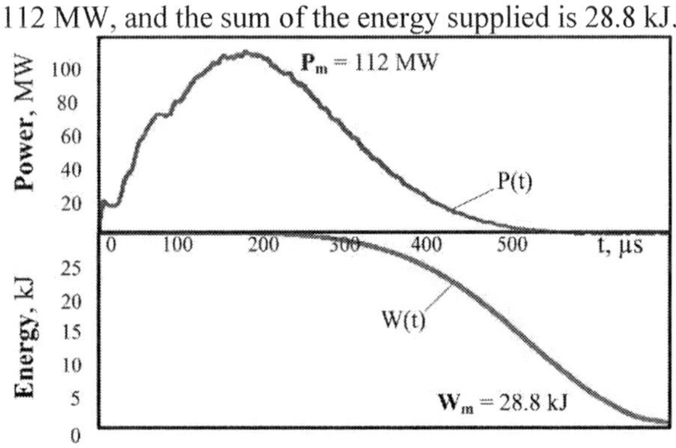

Figure 3. Power P(t) and energy W(t) curves.

3. Results and discussion

The product of plasma dynamic synthesis without additional preparation was investigated by X-ray diffractometry using Shimadzu XRD 7000S X-ray diffractometer (Cu-K_α). Figure 4 shows the X-ray diffraction pattern of the obtained product. The captured diffraction pattern is a set of coherent reflections of two titanium dioxide crystalline phase modifications: anatase aTiO_2 and rutile rTiO_2.

The structural models corresponding to the cards 21-1272 and 21-1276 are the closest to the found phases. Card numbers are taken from the PDF 4+ database.

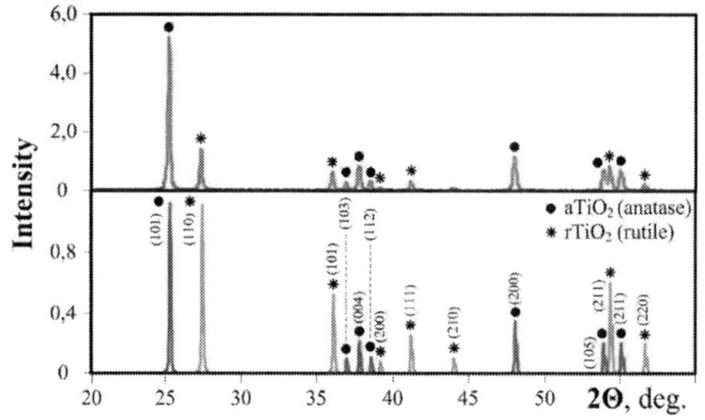

Figure 4. XRD patterns of product, synthesized by plasma dynamic method, and reference phases.

Processing and analysis of the XRD pattern was performed in PowderCell 2.4 software using the PDF 4+ database. The program made it possible to determine the quantitative ratios of the found phases, the size of the coherent scattering regions (CSR), and the levels of internal micro-distortions Δd/d. Table 1 shows the results of X-ray phase analysis.

Table 1. Results of X-ray diffraction analyze.

Phase, Space group	Content, (mass%)	CSR, nm	$\Delta d/d \cdot 10^{-3}$	Lattice parameters, Å Experimental /PDF, Å
$aTiO_2$ (anatase) 141/amd	61.7	97	1.3	a: 3.7844/3.7852 c: 9.5024/9.5139
$rTiO_2$ (rutile) P42/mnm	38.3	111	0.3	a: 4.5920/4.5933 c: 2.9578/2.9592

It is seen that the dominant modification is anatase. The difference in the lattice parameters of the identified phases from the standard values can be explained by the non-equilibrium and highly dynamic crystallization of the material. The consequence of this is quite high levels of internal micro-distortions. It can be concluded that materials are synthesized with defects. This fact can play a positive role in the processes of photocatalysis. According to numerous literatures, scientists are aware of the desire of scientists to obtain highly defective TiO_2 materials. Defects in the crystal lead to a decrease in the band gap and, as a consequence, a shift in the spectral absorption region of titanium dioxide from the ultraviolet region towards visible light [19].

To confirm some of the conclusions obtained from the results of X-ray phase analysis, the product of plasma dynamic synthesis was investigated by transmission electron microscopy. Figure 5 shows bright field TEM images of a cluster, a picture of selected area electron diffraction (SAED), and dark field TEM images of the same cluster. The specifics of sample preparation of highly dispersive materials for TEM exclude micron-sized particles from entering the sample.

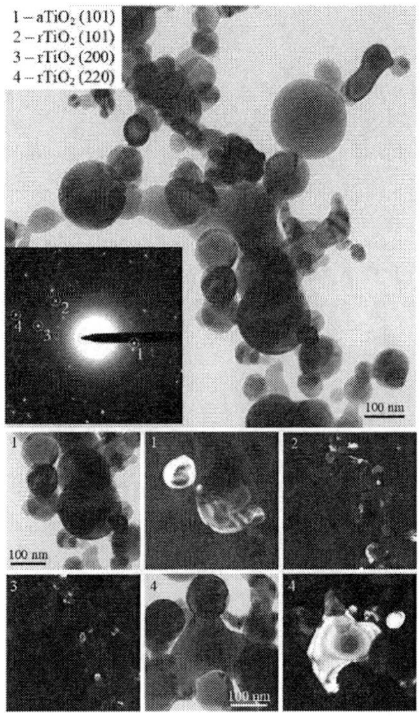

Figure 5. TEM images of synthesized product.

The material contains spherical particles with their size in the range of 27 ÷ 200 nm. At the same time there are shapeless objects whose length can exceed 100 nm. The SAED picture has a point-like appearance and contains reflections of two polymorphic phases of anatase and rutile.

It can be seen that some particles consist of several grains that belong to the structures of both anatase and rutile.

4. Conclusion

The paper presents the results showing the possibility of obtaining a nanostructured material by the plasma dynamic method in the Ti-O system. According to the results of X-ray diffraction analysis, 2 crystalline modifications of titanium dioxide were identified: anatase with tetragonal syngony and rutile with tetragonal syngony. The difference in the lattice parameters of the detected phases from the standard values can be explained by the non-equilibrium and highly dynamic crystallization of the material. The results of transmission electron microscopy showed that the minimum particle size is 27 nm. Moreover, some particles consist of several grains that belong to the structures of both anatase and rutile.

References

[1] Gleiter H 1992 *Nanostruct. Mater.* **1** 1–19 doi: 10.1016/0965-9773(92)90045-Y

[2] Siegel R W 1991 *Annu. Rev. Mater. Sci.* **21** 559–578 doi: 10.1146/annurev.ms.21.080191.003015

[3] Siegel R W 1993 *Nanostruct. Mater.* **3** 1–18 doi: 10.1016/0965-9773(93)90058-J

[4] Mayo M J 1997 *Nanostruct. Mater.* **9**(1–8) 717–726 doi: 10.1016/S0965-9773(97)00158-X

[5] Edelstein A S and Cammarata R C 1998 *Nanomaterials: Synthesis, Properties and Applications, Second Edition* (Florida: CRC Press) p 618

[6] Gangwar J, Gupta B K and Srivastava A K 2016 *Def. Sci. J.* **66** 323–340 doi: 10.14429/dsj.66.10206

[7] Cava R J 2000 *J. Am. Ceram. Soc.* **83** 5–28 doi: 10.1111/j.1151-2916.2000.tb01142.x
[8] Sun J Z and Gupta A 2003 *Annu. Rev. Mater. Sci.* **28** 45–78 doi: 10.1146/annurev.matsci.28.1.45
[9] Rao C N R, Muller A and Cheetham A K 2004 *The Chemistry of Nanomaterials: Synthesis, Properties and Applications* (Weinheim: WILEY-VCH Verlag GmbH & Co. KGaA, Weinheim) p 761
[10] Barquinha P, Martins R, Pereira L and Fortunato E 2012 *Transparent Oxide Electronics: from materials to devices* (Chichester: John Wiley & Sons) p 312
[11] Chung I 2012 *Nat.* **485**(7399) 486–489 doi: 10.1038/nature11067
[12] Huang H H 2004 *Scr. Mater.* **51**(11) 1017–1021 doi: 10.1016/j.scriptamat.2004.08.017
[13] Lilja M 2012 *Biotechnol. Let.* **34**(12) 2299–2305 doi: 10.1007/s10529-012-1040-2
[14] Trino L D 2018 *Coll. and Surf. A* **546** 168–178 doi: 10.1016/j.colsurfa.2018.03.019
[15] Attia S 2002 *J. Mater. Sci. Technol.* **18** 211–218
[16] Vorkapic D and Matsoukas T 2005 *J. Am. Ceram. Soc.* **81**(11) 2815–2820 doi: 10.1111/j.1151-2916.1998.tb02701.x
[17] Li G *et al* 2012 *J. Alloys Compd.* **532** 98–101 doi: 10.1016/j.jallcom.2012.03.050
[18] Liu J *et al* 2011 *ACS Appl. Mater. Interfaces* **3**(4) 1261–1268 doi: 10.1021/am2000642
[19] Amor C O *et al* 2019 *Phys. B: Condens. Matter* **560** 67–74 doi: 10.1016/j.physb.2019.02.017

IOP Conf. Series: Materials Science and Engineering 1019 (2021) 012095 doi:10.1088/1757-899X/1019/1/012095

Automated anomalies detection in the work of industrial robots

A Goncharov, A Savelev, N Krinitsyn and S Mikhalevich
School of Computer Science & Robotics, Tomsk Polytechnic University, Tomsk, Russia
E-mail: asg19@tpu.ru

Abstract. This article describes the results of the anomalies automated detection algorithm development in the operation of industrial robots. The development of robotic systems, in particular, industrial robots, and software for them is ahead of the tracking and managing technologies development. The operation of the digital production system involves the generation of a large amount of various data characterizing the state of both the specific equipment and the industrial system as a whole. Such a system produces a sufficient amount of data to develop machine learning models to analyse this data to solve problems such as forecasting and modelling. As part of the study, an experiment was conducted based on the equipment of the laboratory of industrial robots of Tomsk Polytechnic University. In the course of the research, the industrial manipulator moved loads belonging to different classes by weight. An algorithm was developed for the automated analysis of the values of the parameters of the consumed current and the position of the manipulator.

1. Introduction

In connection with the intensive development of information technologies and the onset of the 4th industrial revolution [1,2], the number of robotic industries is steadily increasing. The production and use of robots are also growing. At the same time, the direction of maintenance and management of digital production is developing [3]. Most robotic systems need human presence, as they need to be fine-tuned, timely serviced, and monitored [4,5]. One of the areas of work to eliminate unexpected failures is preventive maintenance. These works are carried out according to the equipment maintenance regulations, which are being developed by the enterprise. The regulation is based on a variety of factors. One of the factors is the cost of the product life cycle, assessment of the condition of the equipment, and recommendations for maintenance from equipment manufacturers.

Modern digital production environments generate massive datasets from the many areas and sources involved. Data sources may include sensors, optical sensors, drive controllers, tracking cameras, and other similar equipment. These ever-increasing amounts of data contain useful knowledge and information that can improve the performance of the entire production process, due to their subsequent analysis. For example, this knowledge is a valuable asset to support decision-making in various areas, especially in the field of equipment monitoring and maintenance.

There are several approaches to the maintenance of industrial equipment – reactive, proactive, preventive, and prognostic. The main difference of the reactive approach is the production of repairs after an equipment failure. For the formation of the maintenance rules with the prediction of the operation mode is the more suitable prognostic approach. The implementation of proactive maintenance allows to predict and manage equipment failures before they occur. This approach, based on data analysis, is implemented to form a maintenance plan for robotic systems [6]. Data analysis is

IOP Conf. Series: Materials Science and Engineering 1019 (2021) 012095 doi:10.1088/1757-899X/1019/1/012095

carried out on a large amount of historical data of equipment operation using statistical and machine learning methods [7,8]. The software architecture for data collection and analysis includes five main modules: collection, transmission, storage, analysis, and visualization [9,10]. The relevance of this work is to develop an approach to managing the means of production using the methods of analyzing data on the state of industrial equipment.

Timely acquisition and processing of data allow to most effectively develop strategies for the operation, support, maintenance, and repair of fixed assets. When analyzing the software for monitoring the state of industrial equipment from leading manufacturers of robotic systems, we identify the following general functionality:

- Monitoring of current equipment parameters.
- Archiving data.
- Assessment of equipment condition.
- Remote access to equipment functions/modes.

The disadvantage of these software products is the lack of data mining tools, which made it possible to estimate the boundaries of the standard equipment operation parameters values to identify abnormal values. At the same time in these counterparts, there is an alarm function, in the event of a malfunction of the equipment or deviations from typical values. Also, disadvantages include the fact that each software product is compatible only with the equipment type for which the product was developed. Different systems of monitoring in their functions, possibly, can complement each other's work and work more efficiently. However, in practice, this approach can be implemented only with the addition of third-party software that would provide data exchange.

The primary purpose of this work is to develop software that allows to automatically detect anomalous values in the process of operation of robotic manipulators. The research component of the work is to choose the most optimal statistical method for finding abnormal values in the data array. Different statistical methods have various limitations and accuracy of determining anomalies, so it is necessary to evaluate their effectiveness in the framework of this work.

2. Automated anomalies identification in the work of industrial robot

Abnormal values in the data set can be represented as emissions, shifts, changes like the distribution, or deviations from the "seasonal" value. Equipment operation values that go beyond the norm are often the result of possible malfunctions.

We experimented based on the KUKA robotic manipulator (Figure 1) with a loading capacity of 40 kilograms. The purpose of the manipulator is the cyclical movement of cargo from point A to point B with the operation of lifting the load with the help of vacuum suction cups. Table 1 describes the load types.

Table 1. Classes of the lifted load.

Type of the load	Weight (kg)
0	0
1	2.5
2	5
3	7.5
4	10

For the experiment, a software instruction was written consisting of 6 basic commands. The list of basic robot actions:

- Rotate the robot to point A.
- Capture cargo at point A.
- Moving the load to an intermediate point.
- Moving the weight to point B.
- Unloading the robot at point B.
- Return to the intermediate position

Figure 1. Industrial KUKA robot used in the experiment.

We select the following data values to analyze the anomalous values in the robot's work:

- The current value along the axes.
- The number of performed operation.
- Mass of additional load of cargo.

To understand how the current value changes over time as a whole, when operating with a certain mass of load, we calculate the average value of the currents:

$$\bar{I}_{O^k,M_x}{}^{n} = \frac{\sum_{i=O_{t_1}{}^k}^{O_{t_2}{}^k}(I_{i,Mx}{}^{n})}{O_{t_2}{}^{k} - O_{t_1}{}^{k}} \quad (1)$$

where:

n – the axis number,

O^k – operation number k,

$I_{i,Mx}{}^{n}$ is the current value for the n axis, at time i, when the load moves Mx.

To obtain the upper and lower limit of current values when the robot is working with loads of different mass is necessary to get a list of maximum and minimum average values of current for operations:

$$Max\left(\bar{I}_{O_j{}^k,M_x}{}^{n}\right) \quad (2)$$

$$Min\left(\bar{I}_{O_j{}^k,M_x}{}^{n}\right) \quad (3)$$

where:

j – the iteration number on proceeding operation O^k with the load M_x.

Next, we need to obtain the maximum and minimum values of currents for operations with all values of the mass of the additional load:

$$Max\left(Max\left(\bar{I}_{O_j{}^k,M_x}{}^{n}\right)\right) \quad (4)$$

$$Min\left(Min\left(\bar{I}_{O_j{}^k,M_x}{}^{n}\right)\right) \quad (5)$$

This operation will allow you to build an upper and lower limit of values for all cycles of the manipulator. Figure 2 shows an example of construction for the axis №3.

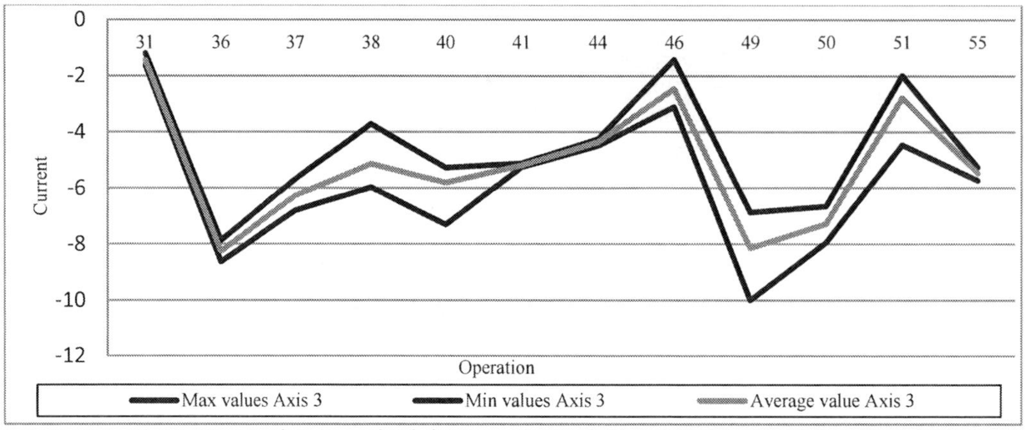

Figure 2. Maximum and minimum values of currents when operating on axis 3.

A robotic arm during an operating cycle may suddenly be exposed to external influences. For example, an unforeseen/abnormal increase in load will occur. A robot set up by an operator to work with cargo weighing 5 kilograms will suddenly accept a shipment weighing 10 kilograms. Figure 3 demonstrates how this reflected in the parameters of the equipment.

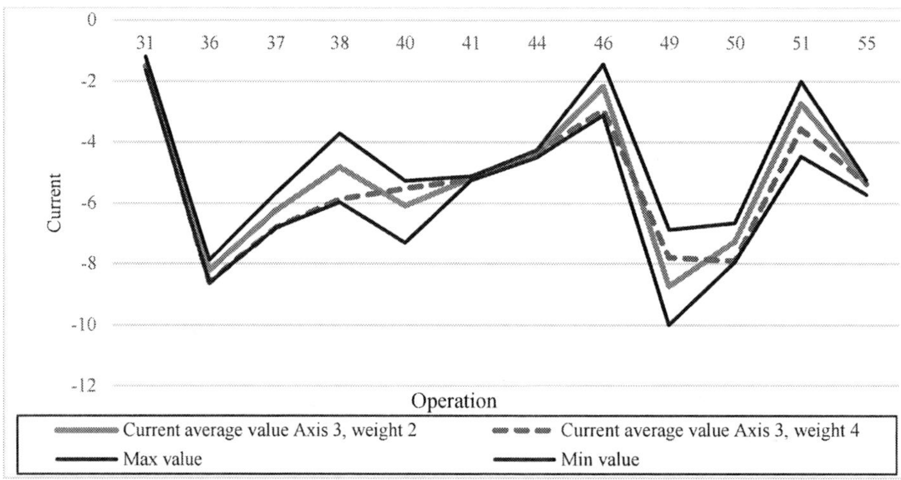

Figure 3. The values of currents when working with weights of types two and four.

The graph clearly shows the deviations in the average current parameters for operations performed with different loads. We should note that even with a massive load, the parameters do not go beyond the limits of the maximum and minimum values since these values are determined for working with all loads. Figure 4 shows a more detailed demonstration of deviations from the border of the maximum and minimum values.

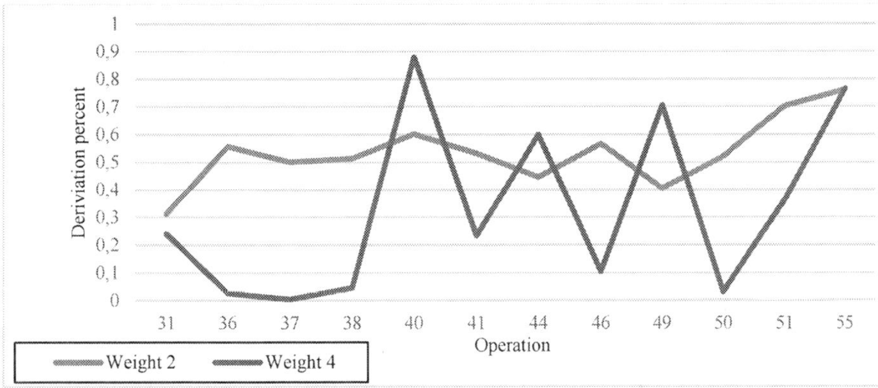

Figure 4. The percentage deviation of the average current parameters relative to the minimum values of the equipment.

An algorithm has been developed for automating operations with experimental data (Figure 5). The algorithm consists of three main stages: obtaining the boundaries of typical values during the initial setup of the equipment, calculating the permissible deviation of the equipment parameters during the operation, monitoring data during the equipment operation.

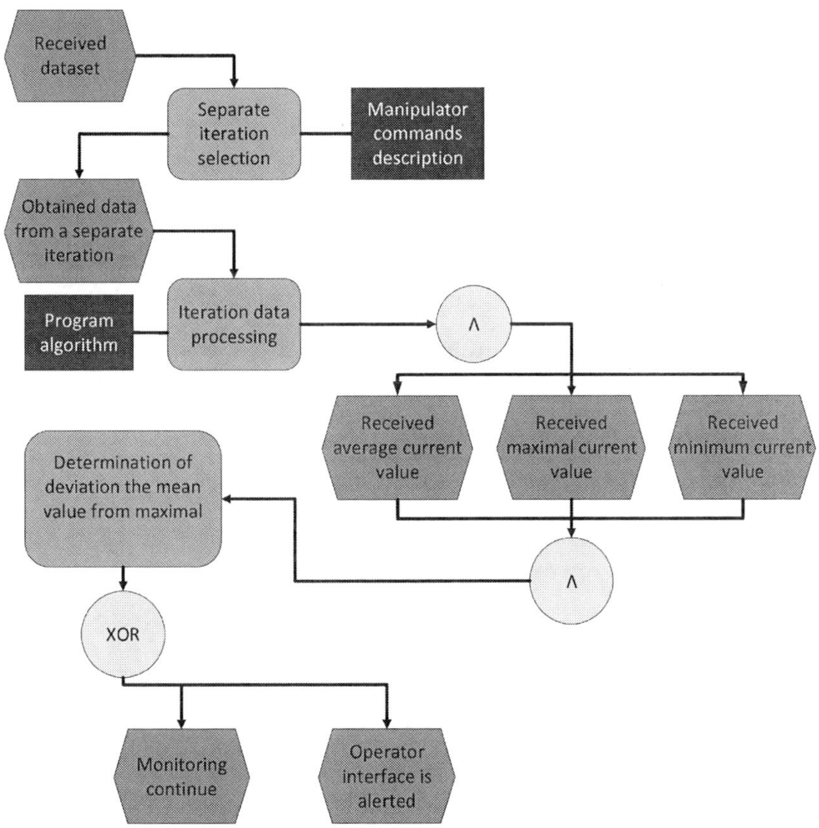

Figure 5. The eEPC diagram of the automated anomalies detection process for the industrial manipulator.

3. Summary

This paper presents an analysis of the current state and approaches to the processing of technical data used in the problem of predicting equipment maintenance, in particular, industrial robotic manipulators. Experiments were carried out on the KUKA robotic manipulator. An algorithm was developed and applied to monitor the deviations of the equipment operation parameters in different modes (under normal and excessive load).

The possibility of automated anomalies detection in the industrial robot operation based on real-time analysis of the operating parameters is demonstrated. The results obtained are intermediate in nature and will be used as the basis for the development of algorithms for diagnosing problems and predicting the state of industrial robots.

Acknowledgments

This work was funded as part of Federal government-sponsored program “Science” number 2.5760.2017/БЧ by Tomsk Polytechnic University.

References

[1] Hermann M, Pentek T, Otto B 2016 *Proceeding of the 49th Hawaii International Conference on System Sciences (HICSS)* 3928-3937
[2] Lasi H 2014 Industry 4.0 *Business & Information Systems Engineering* **6** (4) 239–242
[3] Marc Ingo Wolter 2015 Industrie 4.0 und die Folgen für Arbeitsmarkt und Wirtschaft *IAB Forschungsbericht* **8**
[4] Medha S, Rahul K 2017 *Proceeding of the International Conference on Industrial Engineering, Management Science and Application (ICIMSA)* 1–4
[5] Rastegari A, Mobin M 2016 *Proceeding of the Annual Reliability and Maintainability Symposium (RAMS)* 1–8
[6] Borgi T, Hidri A, Neef B, Naceur M S 2017 *Proceeding of the International Conference on Advanced Systems and Electric Technologies (IC_ASET)* 412–417
[7] Sathish V, Ramaswamy S, Butail S 2016 *Proceeding of the 4th IFAC Conference on Advances in Control and Optimization of Dynamical Systems ACODS* **49** (1) 385–390
[8] Eski I, Erkaya S, Savas S, Yildirim S 2011 *Robotics and Computer-Integrated Manufacturing* **27** (1) 115–123
[9] Karim R, Westerberg J, Galar D, Kumar U 2016 *Proceeding of the 3rd IFAC Workshop on Advanced Maintenance Engineering, Services and Technology AMEST* **49** (28) 214–219
[10] Ferreira L L 2017 *Proceeding of the 13th International Workshop on Factory Communication Systems (WFCS)* 1–9

IOP Conf. Series: Materials Science and Engineering 1019 (2021) 012096 doi:10.1088/1757-899X/1019/1/012096

Studying the influence of supply pulse duration on the phase composition of iron oxides obtained by the plasma-dynamic method

I I Shanenkov, A A Sivkov, A S Ivashutenko, M I Gukov and Y L Shanenkova
School of Energy and Power Engineering, National Research Tomsk Polytechnic University, Tomsk, Russia
E-mail: nikitindmsr@yandex.ru

Abstract. Magnetic materials and in particular iron oxides are of a great practical interest. The magnetite phase and the unique epsilon phase of iron oxide can be especially pointed out. The main difficulty in the synthesis of the epsilon phase is connected with the fact that it can exist only in a nanoscale state and is extremely difficult to obtain. We used the method of direct plasma dynamic synthesis, which allows obtaining multiphase powders of iron oxides containing both the epsilon phase and magnetite. It was found that by varying the initial parameters of the power system, namely the pulse duration by increasing the capacitance of the capacitive energy storage, it is possible to influence the phase composition of the obtained products and to achieve the preferential output of the epsilon phase. In addition, in the mode with the maximum pulse duration, when the best product is obtained from the point of the epsilon phase output, the system efficiency of converting the stored energy into released energy significantly increases. In general, it has been established that such a regime is most favorable for the system operation for the purpose of the iron oxides synthesis.

1. Introduction

Magnetic materials are the most popular and used in various fields of science and technology. Among them iron oxides can be mentioned as the simplest compounds with high magnetic characteristics [1]. There are 7 non-hydrated phases of iron oxides with different structure and physical properties: α-Fe_2O_3 (hematite), β-Fe_2O_3, γ-Fe_2O_3 (maghemite), ε-Fe_2O_3, z-Fe_2O_3, FeO and Fe_3O_4 [2]. All of them have their own features, but the ε-Fe_2O_3 and Fe_3O_4 phases are of the greatest practical interest, due to their unique magnetic properties. For example, particles of epsilon phase have the highest value of coercive force among all known metal oxides (~23 kOe) and ferromagnetic resonance in the terrahertz range ~ 190 GHz [3–5]. Particles of the magnetite phase have a maximum saturation magnetization (92 G·cm^3/g) among ferrites at room temperature [1]. These properties can be useful for creating on their basis modern permanent magnets used for storing information and other electronics. Also, particles of the ε-Fe_2O_3 and Fe_3O_4 phases are non-toxic and highly resistant to corrosion. Despite a number of known method for magnetite synthesis, obtaining the epsilon phase is a rather difficult scientific task, due to the fact that it can exist only in a nanoscale state, as well as it is thermodynamically unstable [3, 6].

To date, many different methods have been developed that allow obtaining the magnetite phase (for example, solid-state reactions [7]; high-energy mechanosynthesis [8]; sol-gel method [9]; chemical coprecipitation [10]; microwave sintering; automatic combustion, conventional ceramic method, two-stage synthesis , etc.), however, the range of methods for obtaining the thermodynamically unstable ε-Fe_2O_3 phase is rather limited due to the fact that most synthesis methods also lead to the appearance of the hematite or maghemite phases. The most commonly used method is a sol-gel method, which is based on the following mechanism of phase transformations γ-Fe_2O_3 → ε-Fe_2O_3 → α-Fe_2O_3 when heated in the presence of a protective matrix. Despite its advantages and widespread occurrence, the sol-gel method also has a number of significant drawbacks, which mainly boil down to the need to use expensive precursors and complex chemical reactions, which leads to an increase in cycle time up to several days or even weeks [11].

It is known that the method of plasma dynamic synthesis has the following advantages: high reaction rate, low energy consumption, high achievable energy parameters in the synthesis process and high cooling rate, in a system based on a pulsed high-current coaxial magnetoplasma accelerator (CMPA). The possibility of using this method to obtain a multiphase product consisting of several modifications of iron oxide, including the rare epsilon phase and magnetite phase [12], has already been shown. In addition, in [13, 14], the possibility of controlling the phase composition by changing the process energy and oxygen concentration in the working chamber was shown.

In this regard, the aim of this work was to estimate the influence of the supply pulse duration on the phase composition of the iron oxides synthesized by plasma dynamic method. It was found that an increase in the pulse duration leads to obtaining the product with the predominant content of epsilon phase. More over, at longer pulse duration the overall efficiency of energy conversion in the system drastically rises up.

2. Experimental part

The studies were carried out in the known system based on a coaxial magnetoplasma accelerator (CMPA) used for generating and accelerating the plasma structure [13, 15]. The system of pulsed power supply from a capacitive energy storage device can be considered a sequential RLC-circuit. The capacitance (C) can be varied from 1.2 mF to 28.8 mF. The total inductance (L) of all in-series connected elements in the discharge circuit (inductance of capacitors, feeding coaxial cables, busbars), including the CMPA solenoid, does not exceed 2.0 μH. Active resistance (R) after beginning the arc stage does not exceed the order of 10-3 Ohms. With such values of the parameters, the following condition is satisfied:

$$R < 2\sqrt{L / C} \quad (1)$$

This condition indicates the oscillatory-damped nature of the change in the discharge current in the short circuit mode. This is confirmed by experimentally obtained current oscillograms, one of which is shown in Figure 1.

However, the current attenuation coefficient, determined from real oscillograms, is significantly higher than the theoretical one, estimated by the expression:

$$b = R / 2L \quad (2)$$

This difference is due to a significant increase in the resistance of the arc discharge in the ignitron dischargers when the voltage polarity changes in an oscillatory mode. Under the conditions of the plasma-dynamic synthesis process in the full-wave mode, the increased resistance of the re-forming arc discharge in the accelerating channel of the CMPA will be included in the series circuit. This, naturally, will lead to a limitation of the current amplitude and a strong decrease in the efficiency of the process. Therefore, taking into account the actual operating conditions of the energy storage device with ignitron dischargers, the half-wave power supply mode of the CMPA is used. The duration of the current rise stage in these conditions up to a maximum, during which the CMPA consumes more than half of the supplied energy, and the maximum discharge power develops, can be estimated from the Thomson equation:

$$\frac{T}{4} = \frac{\pi}{2}\sqrt{LC} \quad (3)$$

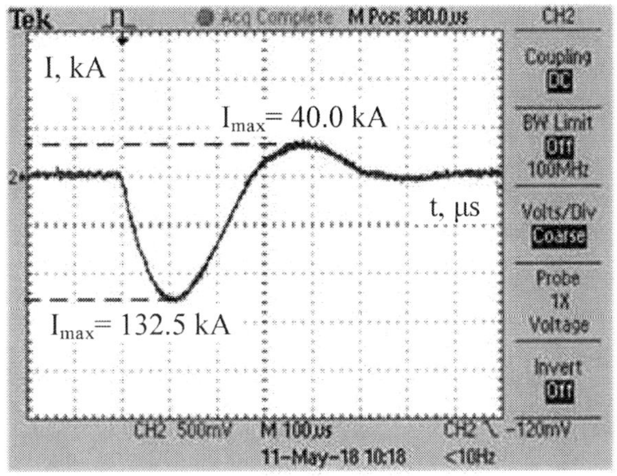

Figure 1. Short-circuit current oscillogram in the system at charging voltage of 500 V and capacitance of 28.8 mF.

From the formula 3 it becomes obvious that the most acceptable way to change the duration of a power supply pulse while maintaining high efficiency in the use of stored energy is to change the capacitance value of the capacitive energy storage that can be provided by its partitioned circuit. On this basis, experimental studies with the different supply pule duration were carried out at the following values of the capacitance: C_1 = 7.2 mF, C_2 = 14.4 mF, C_3 = 21.6 mF, C_4 = 28.8 mF. To save the value of stored energy W_c at the same level of 64-65 kJ in every experiment the charging voltage U_c was changed from 4.25 kV to 2.10 kV, respectively. The working chamber was filled with the technically pure oxygen at a pressure of 10^5 Pa.

3. Results and discussion

In order to study the influence of the supply pulse duration t_{pul} on the phase composition of the plasma dynamic synthesis products, a series of experiments were carried out with the varied capacitance value from 7.2 mF to 28.8 mF. The main data for a series of experiments are given in table 1. It should be noted that with an increase in the pulse duration t_{pul} both the efficiency of energy conversion and the mass of obtained powder m enhance almost twice. It indicates that system work with the enhanced capacitance positively affects the process productivity at the same initial parameters. Moreover, the decrease in the charging voltage U_c positively influence the lifetime of capacitor banks.

Table 1. Data of experimental series.

No.		1	2	3	4
Charging voltage, U_c	kV	4.25	3.00	2.45	2.10
Capacitance, C	mF	7.2	14.4	21.6	28.8
Charging energy, W_c	kJ	65.0	64.8	64.0	64.8
Released energy, W	kJ	30.0	43.0	49.5	58.0
Energy efficiency, W/W_c	%	46.1	66.3	77.3	89.5
Pulse duration, t_{pul}	µs	280	400	500	550
Powder mass, m	g	4.5	7.6	8.0	8.9

The obtained samples of powdered materials were investigated by the methods of X-ray structural and phase analysis. Typical XRD-pictures of the plasma dynamic synthesis products in comparison with the reference cards from the PDF4+ database are shown in Figure 2.

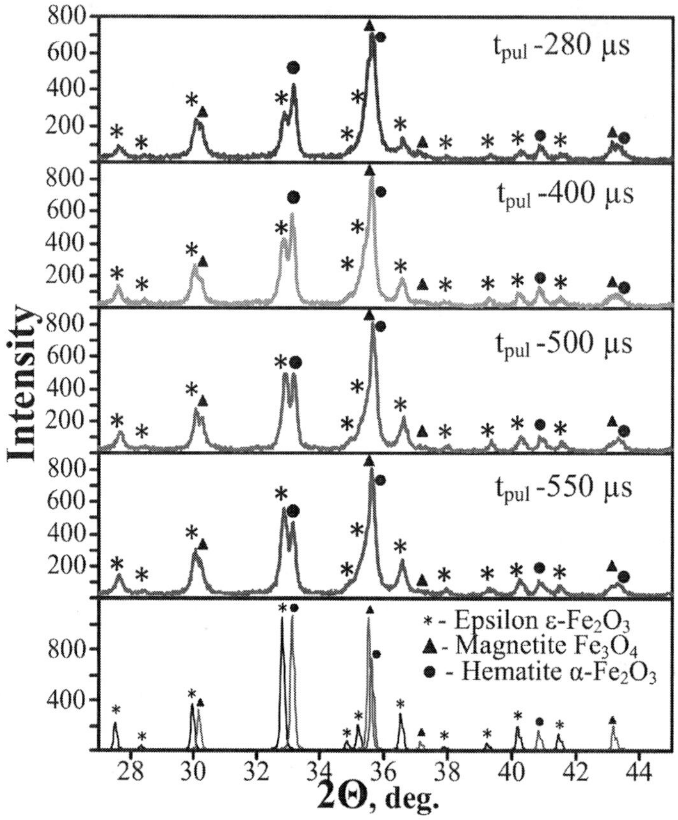

Figure 2. XRD-pictures of plasma dynamic synthesis products at various supply pulse durations t_{pul}.

An analysis of XRD pictures allowed us to establish that the powder synthesized with a minimum supply pulse duration t_{pul} = 280 μs is characterized by the presence of strong reflections corresponding to the phases Fe_3O_4 and ε-Fe_2O_3. An increase in t_{pul} leads to an increase in the relative intensities of the ε-Fe_2O_3 reflections at practically unchanged intensities of the reflections corresponded to the α-Fe_2O_3 phase. Thus, it may be concluded that a possible increase in the ε-Fe_2O_3 content occurs due to a decrease in the content of the Fe_3O_4 phase. This is unambiguously confirmed by a change in the numerical estimates of the percentage and the values of the main parameters of the crystalline phases, depending on their type (table 2).

The obtained results, given in table 2, make it possible to establish a pattern for increasing the yield of the ε-Fe_2O_3 phase with an increase in the pulse current flow duration in the power supply circuit. To determine the main factor affecting the increase in the output of epsilon phase with a change in the pulse duration, the oscillograms of the arc discharge currents, shown in Figure 3, were analyzed.

Table 2. Results of full-profile structural and phase analysis of synthesis products.

Phase	t_{pul}, μs	Mass content, wt. %	Lattice parameters, Å		CSR, nm	Δd/d ·10^{-3}
			Exp.	PDF4		
ε-Fe_2O_3	280	31.3	a: 5.10 b: 8.79 c: 9.46	a: 5.0950 b: 8.7900 c: 9.4400	36.0	1.0

	400	40.4	a: 5.09 b: 8.78 c: 9.47		43.5	0.6
	500	46.1	a: 5.08 b: 8.77 c: 9.46		48.0	0.2
	550	50.1	a: 5.08 b: 8.78 c: 9.46		56.0	1.2
Fe_3O_4	280	45.7	a: 8.37	a: 8.3700	29.5	1.9
	400	36.6	a: 8.36		31.0	1.7
	500	31.4	a: 8.33		36.0	2.4
	550	27.9	a: 8.35		40.0	2.3
α-Fe_2O_3	280	23.0	a: 5.03 c: 13.75	a: 5.0200 c:13.7350	124.5	1.3
	400	23.0	a: 5.03 c: 13.74		104.0	0.3
	500	22.5	a: 5.03 c: 13.71		122.5	1.6
	550	22.0	a: 5.03 c: 13.74		110.0	1.1

As noted in earlier works [13,14], the formation of the epsilon phase, apparently, should occur when a liquid-phase material is sputtered from the boundary of a shock wave, in the front of which high pressures and temperatures occur. The formation of a zone with high P,t-parameters occurs during the formation of the classical form of a shock-wave structure with a region of the Mach disk. When the iron-containing material passes through the boundaries of the Mach disk, it is subjected to high pressures and temperatures and starts reacting with oxygen. It was previously found that in the system under consideration the formation of a Mach disk occurs when the arc discharge current approaches its maximum. Thus, the time Δt_{max}, when the current is near its maximum value, can characterize the quasi-stationary state of the shock-wave structure of a supersonic flow [12]. With the existence of a quasi-stationary state, the most favorable conditions are created for the occurrence of a plasma dynamic reaction at the boundary of a shock wave with the formation of epsilon nanocrystallites.

Taking into account the system features described above, it was proposed to estimate the time Δt_{max} in each of the experiments performed. To do this, taking into account the known relationship between the formation time of a shock-wave structure stable form and the nature of the arc discharge current oscillogram [12], the conditional ten-percent zone is selected, marked with a dotted line in Figure 3, which is used to estimate the lifetime of the quasi-stationary expiration mode. As can be seen from Figure 3, with an increase in the pulse duration, the time Δt_{max} also increases. This indicates that the quasi-stationary state of the shock-wave structure is maintained for a longer time period, and the probability of the epsilon phase formation rises up. The objectivity of these judgments is confirmed by the results of full-profile X-ray phase analysis, indicating that the ε-Fe_2O_3 yield directly depends on the duration of the quasi-stationary supersonic plasma jet flow, which is determined by the duration of the pulsed power supply.

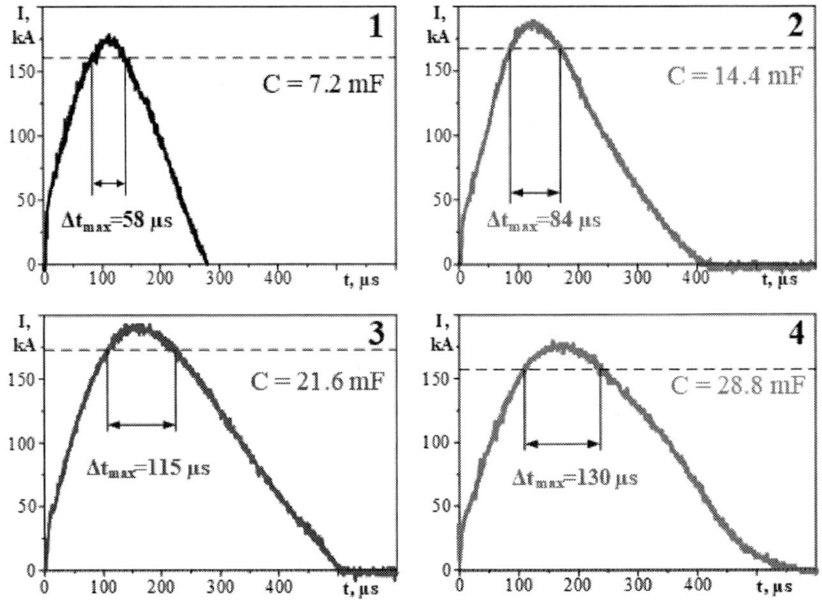

Figure 3. Oscillograms of arc discharge currents at different pulse durations.

4. Conclusion

Thus, evaluating the complex of the conducted studies to estimate the influence of supply pulse duration on the work of the system based on coaxial magnetoplasma accelerator the following conclusions can be made:

- The pulse duration in the considered system depends on the initial parameters of the RLC-circuit and can be changed by increasing the capacitance of the capacitive energy storage;
- The output of ε-Fe_2O_3 phase in the plasma dynamic synthesis products directly depends on the duration of the power supply pulse and the time when the system is at the maximum of its energy parameters;
- The efficiency of the energy conversion drastically rises up from 45% to ~90% in the case of system work with the higher pulse duration.

Acknowledgments
This work was supported by the Russian Foundation for Basic Research and Tomsk Region (Project No. 19-42-703007).

References

[1] Cornell R, Schwertmann U 2003 *The iron oxides: structure, properties, reactions, occurrences and uses* (Weinheim: Wiley-VCH) p 617
[2] Tuček J, Machala L, Ono S 2015 *Scientific reports* **5** 15091
[3] Tucek J, Zboril R, Namai A *et al* 2010 *Chemistry of Materials* **22** 6483–6505
[4] Ohkoshi S, Kuroki S, Sakurai S 2007 *Angewandte Chemie International Edition* **46** (44) 8392–8395
[5] Namai A, Sakurai S, Nakajima M 2008 *Journal of the American Chemical Society* **131**(3) 1170-1173
[6] Gich M, Frontera C, Roig A 2006 *Nanotechnology* **17**(3) 687
[7] Schneeweiss O, Zboril R, Pizurova N 2006 *Nanotechnology* **17**(2) 607
[8] Marinca T, Chicinaş H, Neamţu B *et al.* 2016 *Materials Chemistry and Physics* **171** 336-345
[9] López-Sánchez J, Serrano A, Del Campo A 2016 *Chemistry of Materials* **28**(2) 511–518

[10] Han D, Yang S, Yang J 2016 *Nanoscience and Nanotechnology Letters* **8**(4) 335–339
[11] Popovici M, Gich M, Nižňanský D 2004 *Chemistry of materials* **16**(25) 5542–5548
[12] Sivkov A, Naiden E, Ivashutenko A 2016 *Journal of Magnetism and Magnetic Materials* **405** 158–168
[13] Shanenkov I, Sivkov A, Ivashutenko A 2019 *Journal of Alloys and Compounds* **774** 637–645
[14] Shanenkov I, Sivkov A, Ivashutenko A 2017 *Solid State Phenomena, Trans Tech Publications* **265** 652–656
[15] Kuzenov V, Ryzhkov S, Frolko P 2017 *Journal of Physics: Conference Series* **830** 012049

IOP Conf. Series: Materials Science and Engineering 1019 (2021) 012097 doi:10.1088/1757-899X/1019/1/012097

The high-level overview of social media content search engine

A O Savelev[1], A Yu Karpova[2], D V Chaykovskiy[2], A D Vilnin[1], A Yu Kaida[1], S A Kuznetsov[1], L O Igumnov[1] and N G Maksimova[3]
[1]School of Computer Science & Robotics, Tomsk Polytechnic University, Tomsk, Russia
[2]School of Core Engineering Education, Tomsk Polytechnic University, Tomsk, Russia
[3]School of Engineering Entrepreneurship, Tomsk Polytechnic University, Tomsk, Russia
E-mail: ayk13@tpu.ru

Abstract. An increasing amount of social networks users-generated data is the most remarkable research challenge nowadays. Despite the progress in the field of semistructured data processing algorithms creation, even initial data collection could not be treated as issues that have been optimally solved. The paper covers a high-level overview of the automated social media content search system. The proposed structure enables to implement instruments for multisource content extraction tasks as well as supporting of identification processes of new patterns, which describe a certain type of content. Issues of Search engine organization, logically unified extracted data repository and possible content classification techniques with the appropriate knowledge base's application are considered. Under the work, existing approaches and automated web-data extraction methods have been analyzed; social media API's functions and limits, as well as ways of semistructured data storage system organization, have been studied. The planned result's application area is automation and informational support of sociological research based on the social media content analysis techniques namely a content propagation simulation in interconnected groups; social and personal anomy study; clarification of the weak linkage's strength concept.

1. Introduction

The modern sociology which has entered in the age of digitalization focuses on the understanding of how to operate with various digital devices and modern data aggregation and analysis approaches for the human behavior diagnosis in a context of social networking. Modern digital technologies enable sociologists to evaluate the level of digital content influence on the human being: views, perspectives, and beliefs [1]. The "Digital Sociology" is inextricably linked to exploring the Internet space whose scale and speed of growth as well as the level of diffusion into the daily life may cause a serious potential risk to spread the radical views and beliefs across the society [2]. It is important to note that the most vulnerable group for producing social tensions is the youth. The existing sociological approaches that are used by leading social science services are deprecated for the "digital age"[3]. The scientific community had started to pay attention to social networks analysis about ten years ago. This kind of analysis is considered as a multidisciplinary research area that gathers to aggregate, extend and adapt methods of social media content data analysis [4–6]. It was extended through the years, and now, in a context of the research to define the impact of the disruptive content driven to promote radical ideas, has evolved to a milestone where the social media has a great impact. Internet users leave loads of data about its activity surfing on web pages – a digital footprint [7]. The digital footprints analysis based on the aggregated preprocessed data from social networks may let to define a particular behavior of potentially dangerous groups and users that develop view related to extremism. To study the users' behavior in social networks, the set of particular tools created on the basis of artificial intelligence implementation models and methods is required. The core idea is to define the networks as dynamic social groups driven to make social tensions and to be able making a prognosis of the consequences of the actions within the Internet [8]. The solution can be implemented only as a result of the collaboration between sociologists and IT-specialists.

2. The general issue

Today, Internet users that produce disruptive content use modern digital technologies, as well as common users do. This fact makes a new challenge in front of social security services. Tracking the digital footprints of Internet users is still a pressing issue that requires a complex solution that can be used for social media data aggregation, analysis, and visualization. The proposed solution is a particular search engine that able to find a user or a group of users by a context of their posts and comment as well as analyze the activity on a particular page. The aim of the developing engine is to provide a set of tools for defining objects (users or groups) that are producing disruptive content in social media. The general schema of the engine is presented in Figure 1.

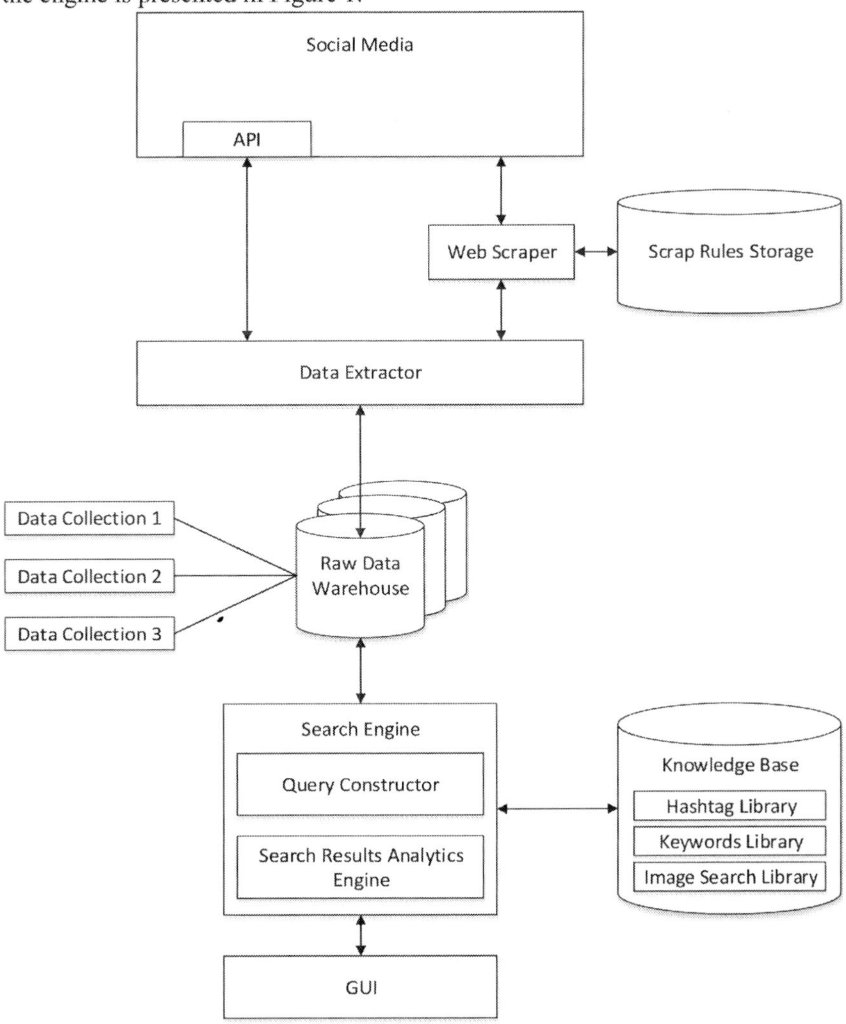

Figure 1. The high-level overview of the search engine.

3. Web scraping engine

Before among solutions for data acquiring from social networks, it is possible to distinguish such services as that could communicate on API (Application Programming Interface) and the rest that could not. The only way to get data from the last ones is a direct way also known as Web scraping.

API is an interface for interaction between a site and external applications including a complete set of classes, procedures, data structures, etc. which is needed for connection. With API an external application could get and use for its purposes a user's public data if user permission is provided, for instance, list of friends, photos, audio and video files.

VKontakte API allows acquiring data from vk.com database over HTTP requests. An applying of such interface avoids the necessity of detailed understanding about internal database structure namely, what tables and field types it consists of. Request's syntax and a type of the response data are strictly specified into the relevant technical documentation.

Despite the benefits of API, one limit still exists: a social network could not provide all details visible while site browsing. It depends on that at first social networks try keeping the privacy of its users and secondly some functions make a too high load for a database [9]. To surmount the limit the web scraping should be used.

Web scraper - an algorithm for collecting data from the network. A web crawler is the very first part of the data extraction process. The crawler is a web robot also known as an Internet bot or bot that systematically scans the World Wide Web with the aim of web spidering. Web crawlers are used by some sites and search engines to update their contents. The procedure of crawling starts with the list of URLs to crawl, these URLs are called seeds URLs. After visiting these seed URLs, crawler identifies all the links that exist on the page and store them into the list of URLs to visit called Crawl Frontier. Now URLs from the crawl frontier are recursively visited. This process goes on until all URLs get visited.

The data needed for analysis process are present on the Internet, this data contains a large amount of information and it is not possible for the analyst to manually take that data from any World Wide Web to perform data extraction. The website data is unstructured, which means it can contain noise or unwanted data. To extract large and unstructured data from the web there are some data extraction techniques and tools that can easily extract large unstructured data and convert them into a meaningful and structured format.

Web data extraction is basically a technique of extracting user-required information from the websites. The extraction process starts with indexing or crawling. It crawls data from the web using a web crawler. In the crawling process, it also transforms unstructured data on the web page into structured data. After crawling, the list of crawled links will be available to us. In these web pages, a lot of junk and useful data is mixed, So we have to extract the useful data and convert that data into needed format. The Data extractor extracts the needed information from the crawled web pages. To separate out the used one from the mixture we target only that part of the page in which information is present. To target that specific part, we will use the CSS Selectors [10].

4. Data warehouse

The aggregated data from social media – extracted using API or Web Scraping method – are related to unstructured or semi-structured data that causes a set of requirements for the data warehouse. The unstructured data is represented as a text that may include the other data fragments for future analysis. At the same time, semi-structured data is represented in some formats that do not require a particular schema (e.g. XML and JSON). Due to the uncertain complexity of analyzed text data, JSON format of files seems as the most suitable ones across serializable formats. Moreover, it is suitable for complex data structures processing.

The raw data warehouse is a key element of the system that keeps raw data aggregated from social media for future analysis. The search engine considers a flexible data model because users can fulfill their own account in social networks partially according to their own decision. Moreover, during the analysis, the number of attributes may be changed because of the results of the previous steps. It is worth noting, however, that the major problem of using database management systems is normalization large volumes of data. The normal form, in a relational data model, considers data representations with the minimal level of superfluous data; however, because of semi-structured form, this way of data representation is a time-consuming task with a high risk of data schema scalability failure or that a large number of empty cells occur. In this case, the normalization process also leads to the enormous growth of the total number of tables, so it is too difficult to define the intersections between analyzed objects (groups or users). NoSQL storages do not use a relational data model (so they may be schemaless), so it rids the requirement to store semi-structured data using relational way.

Today, a huge variety of NoSQL storages exist, so it is essential to define some additional requirements for the system. The developing system is defined as a distributed ones. According to the CAP theorem [11], all data storages can satisfy only two of three requirements at the same time: consistency, availability and partition tolerance. In general, the system consists of a number of standalone independent components where any system element is able to obtain the same set of relevant data – in other words, it is possible to get access to any node and obtain the data without any contradictions. MongoDB has been chosen as a raw data warehouse cause it ensures consistency and partition tolerance. The MongoDB architecture is able to provide distributed data storage using a sharding option. It allows decreasing the load level of a single server operating with the large volumes of data as well as it prevents exhausting CPU capacity. MongoDB storage data on a hard drive source in JSON-like files with a flexible schema. Moreover, MongoDB has a Python distribution containing tools, PyMongo, that makes a process of extraction and analysis faster and easier. PyMongo is able to extract data as JSON files, so then all data can be transfer easily to any other tool for future operations.

5. Knowledge base

The search engine checks the raw data using a knowledge base that contains predefined sets of keywords, hashtags, and images. The main task of the knowledge base is content classification, for the implementation of which statistical correlations are used to build rules for placing objects in specific content classes. The classification issue is the task of recognition, where, for a control sample, the system assigns a new object to a particular category.

Methods based on machine learning to create a training set and build a classifier model, as a rule, require the formation of a dictionary with marked keywords. The tagged (marked) keywords are considered a positive example. The remaining words are considered a negative sample. Next, the relevance of each word of the training text is calculated by comparing the vector of values of various parameters, for example, TF-IDF measures, word lengths, parts of speech, word positions in the headline, word positions in the first paragraph, last paragraph, in literature lists. Further, the difference between the values of the vectors of these parameters for keywords and not important ones (not keywords) is recorded. The next step is to calculate the probability of assigning each word to a group of keywords and setting its threshold, thus learning the model. The extraction of keywords from the new document occurs by calculating the relevance of the words and their probability of attributing to the key following the constructed model [12].

Among the various classification methods based on machine learning, researchers distinguish the naive Bayes classifier and the support vectors as the least complex in their implementation and at the same time the most effective in solving many practical problems.

Naive Bayes classifier

A Bayes classifier is a method that assigns class labels to observations represented by feature vectors. Each attribute independently affects the probability that the observation belongs to the class. For example, an object can be considered an apple if it has a round shape, red color, and a diameter of about 10 cm. The naive Bayes classifier "believes" that each of these attributes independently affects on the probability that this object is an apple, regardless of any possible correlations between the characteristics of color, shape, and size [13].

A Naive Bayes classifier based on a supervised learning method. An additional advantage is the small number of examples required for training.

Support-vector machine

The support vector method is a set of supervised learning algorithms based on the translation of the original vectors of parameter values into a higher dimension space and the search for a separating hyperplane with a maximum gap in this space. This approach involves obtaining some rules that allow distinguishing objects of two classes.

The method of support vectors belongs to the family of linear classifiers used for the problems of classification and regression analysis, one of the main properties of which is the continuous decrease in the empirical classification error. When the method works, the summation is performed not over the entire sample, but only over the reference vectors. This property of sparsity that distinguishes the method of support vectors from other linear delimiters — the Fisher discriminant, logistic regression, and the single-layer perceptron [14].

6. Summary

«Digital sociology» specializing in people's behavior diagnosis questions, informational content influence to opinions, views, and beliefs forming based on the information presented on the Internet and particularly in social media has completely developed as a scientific direction. Research tasks' complexity also depends on the availability of automation tools for big amount processing of initial information. The main task considered in the paper is an adaptation of the smart content analysis methods to meet the diagnosis of Informational influence in social media challenges. The automated social media content search system high-level overview has been developed. The proposed overview has several major advantages as follows:

- On the one hand a hybrid approach to the initial data extraction implementing by the combined use of social networks' API and web-scraping allows to organize a unified data collection from a number of heterogeneous information sources, on the other hand the system sustainability against changes due to API functionality reconsideration by the social media owners or HTTP structure updating is increasing.
- Knowledgebase availability formalizing attribute space for content classification in addition to supportive mechanisms development for the identification of the new attributes on the search module's results enables to identify semantic communication among groups and communities while absence from cross-references as well as new communities, categories, and users' interaction language timely tracking.

The presented results are interim ones and have been achieved with a project devoted to the creation of the software tools and new content promotion communicative technologies' examination method.

Acknowledgments

The reported study was funded by RFBR according to the research project № 19-011-31535.

References

[1] Holt T, Freilich J, Chermak S 2018 *American Journal of Criminal Justice* **44** 83–105
[2] Porta D 2018 *Annual Review of Political Science* **21** 461–474
[3] Ali K, Dong H, Bouguettaya A 2017 *Proceeding of the IEEE International Conference on Web Services (ICWS)* 660–667
[4] Zeng D, Chen H, Lusch R 2010 *IEEE Intelligent Systems* **25** 13–16
[5] Chen X, Madhavan K and Vorvoreanu M 2013 *Proceeding of the International Conference on Cloud and Green Computing* 383–388
[6] Jensen A, Seate A and James P 2018 *Terrorism and Political Violence* 1–24
[7] Gao C and Liu J 2017 *IEEE Transactions on Systems, Man, and Cybernetics: Systems* **47** 171–183
[8] Chung W, Mustaine E and Zeng D 2017 *Proceeding of the IEEE International Conference on Intelligence and Security Informatics (ISI)* 191–193

[9] Sukhanov A and Maratkanov A 2017 *Proceeding of the International Scientific Review of the Problems and Prospects of Modern Science and Education: International Scientific and Practical Conference* 22–25
[10] Parvez M, Tasneem K, Rajendra S *et al* 2018 *Proceeding of the International Conference on Smart City and Emerging Technology (ICSCET)* 1–7
[11] Gilbert S and Lynch N 2002 *ACM SIGACT News* **33** 51–59
[12] Sheremetyeva S, Osminin P and Kokhanova L 2015 *Bulletin of the South Ural State University* **12** 76–81
[13] Kotelnikov E and Klekovkina M 2012 S *Computational Linguistics and Intellectual Technologies* 27–37
[14] Volik A and Murlin A 2014 *Scientific works of KUBSTU* 63–68

IOP Conf. Series: Materials Science and Engineering 1019 (2021) 012098 doi:10.1088/1757-899X/1019/1/012098

On the self-testing (m,n)-code checker design

N Butorina[1], Yu Burkatovskaya[2] and E Pakhomova[1]
[1] Institute of Applied Mathematics and Computer Science, National Research Tomsk State University, Tomsk, Russia
[2] School of Computer Science and Robotics, National Research Tomsk Polytechnic University, Tomsk, Russia
E-mail: tracey@tpu.ru

Abstract. We propose an approach to a self-testing (m, n)-code checker design, based on subdividing the set of all code words into special subsets called segments. The checker circuit is constructed by using one- and two-output configurable logic blocks (CLB). Previously, in each output of a CLB, a function representing exactly one segment was implemented. In the proposed approach, at each CLBs output, it is possible to implement functions that represent several segments and to provide the self-testing property. It allows reducing the number of CLBs and simplifying the circuit of the checker.

1. Introduction

In self-checking circuits, code checkers are used to provide the self-testing property. As a rule, the checker is designed on the same element base as the circuit. In the checker, the weight of an input codeword is usually calculated. For this purpose, threshold elements or parallel counters are used [1–6]. This paper proposes a design method of the checker based on configurable logic blocks (CLBs). In papers [7, 8], the authors propose a method for design a self-checking finite-state machine. In the case of correct operation of the circuit, the code words of some unordered code are implemented at the output of the circuit. Usually, the (m,n)-code is used, where n is the length of a code word, m is the number of unity components (the weight). The papers also develop a method for self-testing (m,n)-code checker design based on the repeated use of a special decomposition formula for a set of the (m,n)-code words and implementing the obtained formula by CLBs. In the method of self-testing checker design proposed in [8], the presence of two-output CLBs is not necessary. This means that the method allows building self-testing checkers using any modern CLBs (Field Programmable Gate Arrays (FPGA), manufacturers Xilinx, Altera, Achronix, Actel, Atmel, Lattice semiconductor, etc.).

There is a special requirement for the implementation of the formula: the checker of the (m,n)-codes must be self-testing for a given set of faults, say V. The set V includes all multiple stuck-at faults occurred at CLBs inputs and outputs. In this case, only one CLB in the checker can be defective. It is assumed that in a system consisting of a self-testing circuit and a (m,n)-code checker, either the circuit or the checker can be faulty, but not both.

The self-testing checker should satisfy the following requirements:

1) when a non-code word appears at the circuit output (or at the checker input), the checker should issue a corresponding signal;

2) in the checker itself, a fault from the considered set of faults V may occur, which must be detectable in the working area of the detector, i.e. on the set of all its code words. This means that there must be an (m,n)-code word for which this fault appears at the outputs of the checker.

The self-testing checker has two outputs with the following combinations of signal values:
a) (01) or (10) mean that the input word is a word of the (m,n)-code and the checker is faultless;
b) (00) or (11) mean that either the input word is non-code or the checker is faulty.
The checker is designed according to a formula representing the set of all (m,n)-code words, being the implementation of this formula.

2. The decomposition method for the set of (m,n)-code words

The number of all (m,n)-code words is equal to C_n^m, i.e., the number of combinations of n elements taken m at a time. The code words can be represented by a disjunction of conjunctions of the rank n. We denote this disjunctive normal form (DNF) as $D_n^m(X)$, where X = {x1, x2, ..., xn} is the set of variables. If, for example, n = 10, m = 5, then $D_{10}^5(X)$ includes $C_{10}^5 = 10!/(5!\cdot 5!) = 252$ conjunctions of the rank 10 and 2520 symbols. The DNF can not be reduced: $D_n^m(X)$ is both the perfect DNF and the sum of prime implicants, since any two conjunctions of $D_n^m(X)$ are orthogonal by at least two variables.

To represent all (m, n)-code words, a special decomposition formula was proposed in [9]. We divide the set of variables X into two subsets X^1, X^2, where $X^1 = \{x_1, ..., x_g\}$, $X^2 = \{x_{g+1}, ..., x_n\}$. The whole set of the (m,n)-code words of can be represented by the formula

$$D_n^m(X) = \sum_{i=0}^{m} D_g^i(X^1) D_{n-g}^{m-i}(X^2), \tag{1}$$

where the $\wedge$ between $D_g^i(X^1)$ and $D_{n-g}^{m-i}(X^2)$ is omitted.

Here $D_g^i(X^1)$, $D_{n-g}^{m-i}(X^2)$ are the decomposition functions; we call the cardinality of variable subsets after the last use of formula (1) as the decomposition base and denote it as k.

In paper [8], we propose to choose g as the least integer number which is greater than or equal to $n/2$, i.e., $g = \lceil n/2 \rceil$. If $g > k$ and/or $n - g > k$, then formula (1) is used again for any decomposition function $D_g^i(X^1)$, $D_{n-g}^{m-i}(X^2)$, $i = \overline{0,m}$ etc. Upon doing that, we obtain the formula for all (m,n)-code words, where $p \le k$ for all $D_p^q(X^r)$. Multiple decomposition is applied both for the first and to the second multipliers in formula (1).

In this paper, we use only one decomposition step. For example, for D_{14}^7 and $k = 7$, where k is the number of CLBs inputs, we have two subsets of variables

$$X^1 = \{x_1, x_2, x_3, x_4, x_5, x_6, x_7\}, \quad X^2 = \{x_8, x_9, x_{10}, x_{11}, x_{12}, x_{13}, x_{14}\}.$$

Formula (1) has the form (2):

$$\begin{aligned} D_{14}^7 = {} & D_7^0(X^1)D_7^7(X^2) \vee D_7^1(X^1)D_7^6(X^2) \vee D_7^2(X^1)D_7^5(X^2) \vee D_7^3(X^1)D_7^4(X^2) \vee \\ & \vee D_7^4(X^1)D_7^3(X^2) \vee D_7^5(X^1)D_7^2(X^2) \vee D_7^6(X^1)D_7^1(X^2) \vee D_7^7(X^1)D_7^0(X^2). \end{aligned} \tag{2}$$

To ensure the self-testing property of the checker, special requirements will be used for the functions implemented by the CLBs. Without loss of generality, we will consider single-output CLBs with not more than 7 inputs and two-output CLBs with not more than 6 inputs.

To provide the self-testing property in the considered class of faults, we use special functions.

Let us denote the set of Boolean vectors where the function $f(x_1, \ldots, x_n)$ is equal to 1 as $S_1(f)$. Let us represent the Boolean function by the table where the columns correspond to Boolean variables, and the rows present Boolean vectors from the set $S_1(f)$.

Definition 1. We call a function f as *function of type 2,* if in the table representation, every column contains exactly one unity component, and the number of unity components in all rows is the same.

Table 1 demonstrates an example of a type 2 function.

Theorem 1. For a CLBs output, implementing a type 2 function, and for any input multiple-stuck-at fault, either a test from $S_1(f)$ exists, or the multiple fault manifests itself as an output stuck-at-1 fault at the CLB.

Theorem 2. If a type 2 function is implemented on one of the outputs of a two-output CLB, then whatever the second function associated with the CLB, an input multiple-stuck-at fault of the CLB is either detected on the corresponding output type 2 function when the CLB receives a vector from $S_1(f)$ in the input, or the fault manifests itself as a stuck-at-1 fault.

Theorems 1 and 2 are proven in [9]

Table 1. Example of a type 2 function.

x_1	x_2	x_3	x_4	x_5	x_6
1	1	0	0	0	0
0	0	1	1	0	0
0	0	0	0	1	1

3. Identifying the properties of configurable logic blocks implementing the (m,n)-code words

Definition 2. We call as a *segment* $F_k^q(x_1,...,x_k)$ the set of all (q,k)-code words, i.e., all Boolean vectors where the function $D_k^q(x_1,...,x_k)=1$.

We denote as $FrSet(x_1, ..., x_k)$ the set of all segments $F_k^q(x_1,...,x_k)$, i.e., the set $FrSet(x_1, ..., x_k)$ includes all Boolean vectors of the length k and the weight q, where $0 \leq q \leq k$.

We denote as $FrSubset(x_1, ..., x_k)$ a subset of the set $FrSet(x_1, ..., x_k)$ satisfying the following conditions:

1) It contains at least two segments;

2) if we arrange the weights of the segments in ascending order then in the sequence of weights there will be two neighboring elements with the difference at least two.

Examples. A union of segments with the weights 2, 4, 5 is a $FrSubset(x_1, ..., x_k)$, a union of segments with the weights 2, 3, 4 is not a $FrSubset(x_1, ..., x_k)$.

Definition 3. Let us call a function whose domain of unity values coincides with a subset $FrSubset(x_1, ..., x_k)$, *as* $Fsubset(x_1, ..., x_k)$.

Table 2 demonstrates an example of such a function.

Table 2. Example of a function Fsubset(x_1,x_2,x_3).

x_1	x_2	x_3
0	0	0
0	1	1
1	0	1
1	1	0

The subset $FrSubset(x_1,x_2,x_3)$ for the function from table°2 consists of two segments: $F_3^0(x_1,x_2,x_3)$, $F_3^2(x_1,x_2,x_3)$.

Subsets $FrSubset(x_1, ..., x_p)$ can be of different cardinality.

Let a CLB implement a function $Fsubset(x_1, ..., x_k)$. Here k is the number of CLBs inputs.

Further we represent an input multiple stuck-at fault by a *ternary vector* with the components from the set {0, 1, – }. Components equal to 0 or 1 are called as *outer components*, they correspond to inputs with stuck-at-0 or stuck-at-1 faults. Components equal to – are called *inner components,* they correspond to faultless inputs.

Example. Let a CLB have 6 inputs. If it has stuck-at-0 fault at the inputs $\{x_1,x_4\}$ and stuck-at-1 fault at the input $\{x_3\}$ then this multiple fault is described by the vector $0-10--$.

Theorem 3. If a CLB implements a function $Fsubset(x_1, ..., x_k)$, where k is the number of the CLBs inputs, then for a multiple input stuck-at fault, either there is a test from the set $FrSubset(x_1, ..., x_k)$, or a multiple fault manifests itself as a stuck-at-1 fault at the CLBs output.

Proof. Let the CLB implement the function $Fsubset(x_1, ..., x_k)$, k be the number of the CLBs inputs. Let the fault be described by a ternary vector β and the vector has r outer components, $r \leq k$. Consider all possible cases.

1. Let $r = k$, i.e., all the components of β are outer. So,

a) if β is a code word from $FrSubset(x_1, \ldots, x_k)$ then the fault manifests itself as the output stuck-at-1 fault;

б) otherwise, if β is not a code word from $FrSubset(x_1, \ldots, x_k)$, then the fault is detected by any vector from $FrSubset(x_1, \ldots, x_k)$; the output value is equal to 0 instead of 1.

2. Let $r < k$, s be the number of unity components of the vector β. There can be several cases:

a) if $0 < s \leq r < k$, then any vector α from $FrSubset(x_1, \ldots, x_k)$ having at least one zero component with the same number as a unity component of β, is a test for the fault. Such vector α exists, because for any component, there is a vector from $FrSubset(x_1, \ldots, x_k)$ in which the component is equal to zero;

б) if $s = 0$, then any vector α from $FrSubset(x_1, \ldots, x_k)$, having at least one unity component with the same number as a zero component of β, is a test for the fault. Such vector α exists, because for any component, there is a vector from $FrSubset(x_1, \ldots, x_k)$ in which the component is equal to unity.

The theorem is proved.

Theorem 4. If at one of the outputs of two-output CLB a function of the type $Fsubset(x_1, ..., x_k)$ is implemented, whatever the second function associated with the CLB, a multiple input stuck-at fault of this CLB is either detected on the corresponding vector α from $FrSubset(x_1, ..., x_k)$, or it appears on this output as a stuck-at-1 fault.

Proof. At one of the CLBs outputs, a function of the type $Fsubset(x_1, ..., x_k)$ is implemented. According to Theorem 1, for a multiple input stuck-at fault , either there is a test from the set $FrSubset(x_1, ..., x_k)$, or the multiple fault manifests itself as an output stuck-at-1 fault. Both functions implemented by the CLB depend on the same sets of variables , and the functions themselves are implemented on separate memory blocks (LUT). At the second output of the CLB, the multiple stuck-at fault may not appear at all. It depends on the type of function implemented by the second output. The theorem is proved.

Consider the subcircuit 1 (Figure 1). The lower level of this subcircuit consists of a one- or two-output CLB (CLB1), which implements the function (one of the outputs) $Fsubset(x_1, ..., x_k)$, and the output of this CLB corresponds to the input of several CLBs implementing the function of type 2 (CLB2 and CLB3). The set of CLBs variables that implement the function of type 2, are different. The outputs of the subcircuit are the outputs of the CLB, implementing a function of type 2.

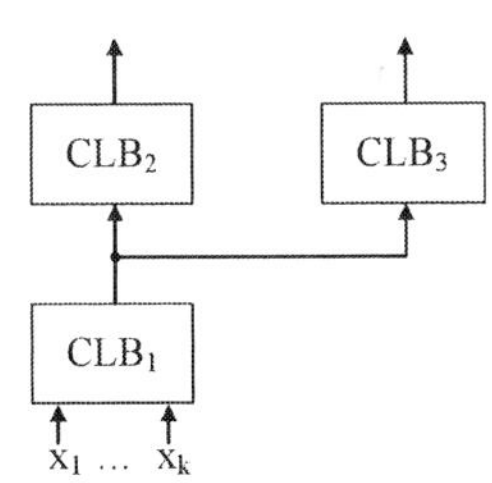

Figure 1. Subcircuit 1.

Theorem 5. For a multiple input stuck-at fault of one of the CLB of subcircuit 1, either there is a test from the set $FrSubset(x_1, ..., x_k)$, or the fault manifests itself as an output stuck-at-1 fault of subcircuit 1.

The statement follows from Theorems 1 – 4.

The detector circuit has a tree structure, on the lower level of which only functions $D_p^q(x_1, \ldots, x_p)$ are used, where $0 < q \le p$, $p \le k$, k – is the number of the CLBs inputs. Earlier, in [7, 8], at every CLBs output exactly one function $D_p^q(x_1, \ldots, x_p)$ was implemented. Theorem 5 allows simplifying the self-testing-checker circuit and to reduce the number of CLBs in it. For example, Figure 2 demonstrates "old" checker D_{14}^7, as Figure 3 shows a "new" circuit of the same checker D_{14}^7.

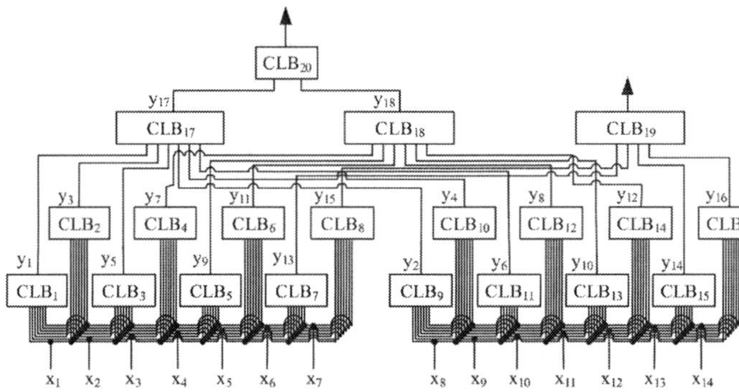

Figure 2. The checker of the (7, 14)-code.

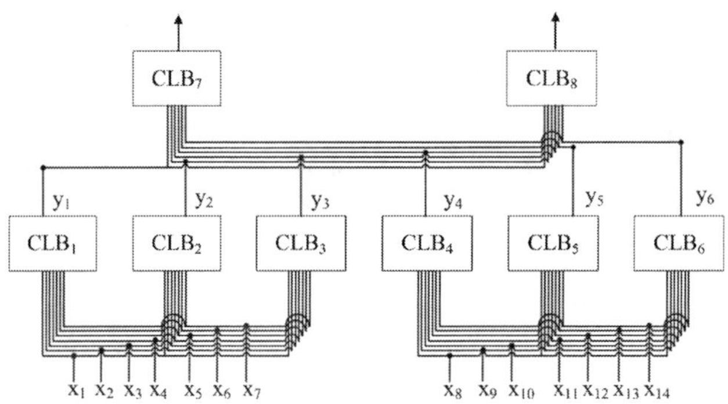

Figure 3. The new checker of the (7, 14)-code.

Consider the "old" circuit. Let us construct a checker for D_{14}^7, using subcurcuit 2 if $k = 7$. At the first step of decomposition one can obtain.

$$D_{14}^7 = D_7^0(X^1)D_7^7(X^2) \vee D_7^1(X^1)D_7^6(X^2) \vee D_7^2(X^1)D_7^5(X^2) \vee D_7^3(X^1)D_7^4(X^2) \vee \\ \vee D_7^4(X^1)D_7^3(X^2) \vee D_7^5(X^1)D_7^2(X^2) \vee D_7^6(X^1)D_7^1(X^2) \vee D_7^7(X^1)D_7^0(X^2). \quad (2)$$

There are 8 Boolean functions depending on the variables x_1, x_2, x_3, x_4, x_5, x_6, x_7. For their implementation one needs 8 CLBs. Their characteristics are listed in table 3.

Table 3. Boolean functions depending on the variables x_1, x_2, x_3, x_4, x_5, x_6, x_7.

№	Implemented function	Output	№	Implemented function	Output
1	$D_7^0(x_1,x_2,x_3,x_4,x_5,x_6,x_7)$	y_1	5	$D_7^4(x_1,x_2,x_3,x_4,x_5,x_6,x_7)$	y_9
2	$D_7^1(x_1,x_2,x_3,x_4,x_5,x_6,x_7)$	y_3	6	$D_7^5(x_1,x_2,x_3,x_4,x_5,x_6,x_7)$	y_{11}
3	$D_7^2(x_1,x_2,x_3,x_4,x_5,x_6,x_7)$	y_5	7	$D_7^6(x_1,x_2,x_3,x_4,x_5,x_6,x_7)$	y_{13}
4	$D_7^3(x_1,x_2,x_3,x_4,x_5,x_6,x_7)$	y_7	8	$D_7^7(x_1,x_2,x_3,x_4,x_5,x_6,x_7)$	y_{15}

There are 8 Boolean functions depending on the variables x_8, x_9, x_{10}, x_{11}, x_{12}, x_{13}, x_{14}. For their implementation one needs 8 CLBs. Their characteristics are listed in table 4.

Table 4. Boolean functions depending on the variables x_8, x_9, x_{10}, x_{11}, x_{12}, x_{13}, x_{14}.

№	Implemented function	Output	№	Implemented function	Output
9	$D_7^7(x_8,x_9,x_{10},x_{11},x_{12},x_{13},x_{14})$	y_2	13	$D_7^3(x_8,x_9,x_{10},x_{11},x_{12},x_{13},x_{14})$	y_{10}
10	$D_7^6(x_8,x_9,x_{10},x_{11},x_{12},x_{13},x_{14})$	y_4	14	$D_7^2(x_8,x_9,x_{10},x_{11},x_{12},x_{13},x_{14})$	y_{12}
11	$D_7^5(x_8,x_9,x_{10},x_{11},x_{12},x_{13},x_{14})$	y_6	15	$D_7^1(x_8,x_9,x_{10},x_{11},x_{12},x_{13},x_{14})$	y_{14}
12	$D_7^4(x_8,x_9,x_{10},x_{11},x_{12},x_{13},x_{14})$	y_8	16	$D_7^0(x_8,x_9,x_{10},x_{11},x_{12},x_{13},x_{14})$	y_{16}

To accomplish formula (2) one needs 4 CLBs. Their characteristics are listed in table 5.

Table 5. Boolean functions for accomplishing formula (2).

№	Implemented function	Output
17	$y_1y_2 \vee y_3y_4 \vee y_5y_6$	y_{17}
18	$y_7y_8 \vee y_9y_{10} \vee y_{11}y_{12}$	y_{18}
19	$y_{13}y_{14} \vee y_{15}y_{16}$	Outputs of the checker
20	$D_2^1(y_{17},y_{18})$	Outputs of the checker

The outputs of CLB_{19} and CLB_{20} are the outputs of the checker. Hence, 20 CLBs are used to implement the checker. The circuit is represented in Figure 2.

At the "new" circuit, we use the following functions (tables 6 and 7).

Table 6. Boolean functions depending on the variables x_1, x_2, x_3, x_4, x_5, x_6, x_7.

№	Implemented function	Output
1	$D_7^0(x_1,x_2,...,x_7)$, $D_7^3(x_1,x_2,...,x_7)$, $D_7^6(x_1,x_2,...,x_7)$	y_1
2	$D_7^1(x_1,x_2,...,x_7)$, $D_7^4(x_1,x_2,...,x_7)$, $D_7^7(x_1,x_2,...,x_7)$	y_3
3	$D_7^2(x_1,x_2,...,x_7)$, $D_7^5(x_1,x_2,...,x_7)$	y_5

Table 7. Boolean functions depending on the variables x_8, x_9, x_{10}, x_{11}, x_{12}, x_{13}, x_{14}.

№	Performing function	Output
4	$D_7^7(x_8,x_9,x_{10},x_{11},x_{12},x_{13},x_{14})$, $D_7^2(x_8,x_9,x_{10},x_{11},x_{12},x_{13},x_{14})$, $D_7^1(x_8,x_9,x_{10},x_{11},x_{12},x_{13},x_{14})$	y_2
5	$D_7^6(x_8,x_9,x_{10},x_{11},x_{12},x_{13},x_{14})$, $D_7^4(x_8,x_9,x_{10},x_{11},x_{12},x_{13},x_{14})$ $D_7^0(x_8,x_9,x_{10},x_{11},x_{12},x_{13},x_{14})$	y_4
6	$D_7^3(x_8,x_9,x_{10},x_{11},x_{12},x_{13},x_{14})$, $D_7^5(x_8,x_9,x_{10},x_{11},x_{12},x_{13},x_{14})$	y_6

To accomplish formula (2) one needs 2 CLBs. Their characteristics are listed in table 8.

Table 8. Boolean functions for accomplishing formula (2).

№	Performing function	Output
7	$y_1y_2 \vee y_3y_4 \vee y_5y_6$	outputs of the checker
8	$y_1y_4 \vee y_3y_6 \vee y_5y_2$	outputs of the checker

The outputs of CLB_7 and CLB_8 are the outputs of the checker. Hence, to perform checker 8 CLBs are used. The circuit of the detector is represented in Figure 3.

4. Conclusion

Thus, we propose an approach to design of a self-testing (*m*,*n*)-code checker, which allows reducing the number of CLBs and simplifying the checker circuit. Previously, at each output of the CLB, functions representing exactly one segment. In the proposed approach, at each CLBs output, it is possible to implement functions that represent several segments.

References

[1] Parag K. Lala 2000 *Self-Checking and Fault-Tolerant Digital Design* (San Francisco: Morgan Kaufmann) p 400
[2] Anderson D A and Metze G 1973 *IEEE Trans. Computers* C **22** 263–269
[3] Marouf M A and Friedman A D 1978 *IEEE Trans. Computers* C **27** 482–490
[4] Ubar R, Raik J and Vierhaus H-T 2011 *Design and Test Technology for Dependable Systems-on-Chip* (New York: IGI Global) p 578
[5] Göessel M, Ocheretny V, Sogomonyan E and Marienfeld D 2008 *New Methods of Concurrent Checking, Edition 1* (Dordrecht: Springer Science+Business Media B. V.) p 182
[6] Sapozhnikov V V, Sapozhnikov V V, Efanov D V and Pivovarov D V 2017 *Radioelectronics and Computer Science (Radioelektronika i informatika)* **3** 15–22
[7] Matrosova A Yu and Nikitin K V 2003 *Tomsk State University Journal, Supplement* **6** 124–136
[8] Butorina N B and Tsidendorzhieva S R 2008 *Proceeding of 7°Russian conference with international participation "New information technologies in the study of complex structures* (Tomsk: TSU) p 44
[9] Burkatovskaya Yu B, Butorina N B and Matrosova A Yu 2006 *Tomsk State University Journal, Supplement* **17** 190–197

IOP Conf. Series: Materials Science and Engineering 1019 (2021) 012099 doi:10.1088/1757-899X/1019/1/012099

An investigation of novel active phased array components

S A Alekseytsev, A P Gorbachev and Y N Parshin
Department of Radio Resievers and Radio Transmitters, Novosibirsk State Technical University, Russia
E-mail: alekseytsev.94@mail.ru

Abstract. This paper deals with a novel approach for designing dual-band dipole radiators. The brief explanation of the basic antenna characteristics obtained is given in the paper. The approach proposed is based on the exact analytical solution for the overall input impedance of a system which consists of two ends-fed dipoles. Then, based on this solution the comparison of the analytical and the simulated results is derived and shown in the paper. In the printed model of the system proposed we compensate huge capacitance occurring in the input impedance of a stand-alone ends-fed dipole by decreasing the gap between the dipoles and the ground plane. Besides, the mathematical model of the phase shifter for phased antenna arrays designing, which is designed for improving the linearity of the phase shifting performance is presented below. Thirdly, the crossed dipole-like antenna fed by rectangular waveguide is also proposed in the paper.

1. Introduction

Wireless communication systems as well as large industrial antenna arrays, which are of the great importance in the last few decades, cannot be further developed without deep study of its radiating and feeding elements. Modern tendency in the miniaturization of all kinds of such devices leads investigators to search for new ideas how to achieve desirable radiation characteristics as well as lightweight and low profile of newly designed antennas. These demands are often controversial.

Another subject to point on is that the research into the properties of nontraditional antenna structures is of a great importance today, because it may contribute to discovering of fundamentally new approaches when designing and implementing radiating structures for modern communication systems.

2. Problem statement

2.1. Dual-band ends-fed dipole-like antenna.

In this section, we present a novel dual-band radiator, to design which we use ends-fed technic for excitation of a pair dipole-like radiators (Figure. 1). The design is based on a split power divider [1], which is placed on the bottom surface of the dielectric substrate with permittivity 1.5 and permeability 1. The power divider in tandem with a microstrip lines on the top side of the substrate mentioned form out-of-phase balancing unit (balun). In [2] and [3] authors use kindred approach to excite dual-band structure. However, in contrast to the design mentioned in [2] and [3], where one-side printed balun is used, in this project the balun is formed by both the bottom and the top side components. Also, in this case we use ends-fed excitation technique.

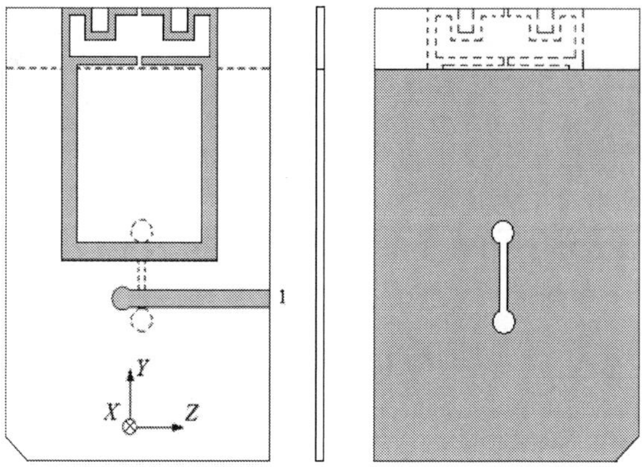

Figure 1. The topology of the dual-band antenna proposed.

As can be seen, the low-frequency dipole-like radiator is meandered along z-axis for reducing the overall structure size.

To explain briefly the approach one should start with theoretical results when deriving a double-dipole structure input impedance. Based on the exact analytical solution for surface current distribution along a linear dipole (along z-axis on Figure 1) when it excided at its separated end points [4],

$$I_z(z) = I_m \sin(\pm kz), \tag{1}$$

the longitudinal component of the electric field induced by one radiator to another can be derived [5]:

$$E_z = \frac{1}{j4\pi\omega\varepsilon_r\varepsilon_0}\left\{ I_m \sin(kl)\left[\frac{e^{-jkR_{1l}}}{R_{1l}^{3}}(z-l)(1+jkR_{1l}) - -\frac{e^{-jkR_{2l}}}{R_{2l}^{3}}(z+l)(1+jkR_{2l})\right] - -kI_m\left[\cos(kl)\left[\frac{e^{-jkR_{1l}}}{R_{1l}} + \frac{e^{-jkR_{2l}}}{R_{2l}}\right] - 2\frac{e^{-jkR_0}}{R_0}\right]\right\}, \tag{2}$$

where $R_{1l} = \sqrt{r^2 + (z-l)^2}$, $R_{2l} = \sqrt{r^2 + (z+l)^2}$, $R_0 = \sqrt{r^2 + z^2}$ – arguments of the free space Green function. The formula in closed form for the input impedance of a stand-alone ends-fed dipole radiator is to be found in [6]. From results in [6] and expression for the mutual impedance the overall input impedance can be derived.

Now, to keep the report brief, we are about to derive equation in closed form for mutual impedance of a system of two ends-fed dipole-like radiators being excited simultaneously with skipping intermediate steps of integral calculations to further formulate the expression for overall input impedance of the system under investigation:

$$Z_{12} = \frac{CI_2}{2i}\begin{bmatrix} -e^{ikl_2}\left(\Upsilon(s^{\bullet}{}_2) - 2\Upsilon(s^{\bullet}{}_1) + \Upsilon(s_2')\right) + \\ +e^{-ikl_2}\left(\Upsilon(s_2) - 2\Upsilon(s_1) + \Upsilon(s^{*}{}_2)\right) \end{bmatrix} -$$

$$-Ae^{ikl_2}\Gamma(0,ikl_2+ik(\sqrt{d^2+(l_1-l_2)^2}-l_1))+2Ae^{ikl_2}\Gamma(0,ikl_2+ik\sqrt{d^2+l_2^2})-$$
$$-Ae^{ikl_2}\Gamma(0,ikl_2+ik(\sqrt{d^2+(l_1+l_2)^2}+l_1))-Ae^{-ikl_2}\Gamma(0,-ikl_2+ik(\sqrt{d^2+(l_1-l_2)^2}+l_1))+$$
$$+2Ae^{-ikl_2}\Gamma(0,-ikl_2+ik\sqrt{d^2+l_2^2})-Ae^{-ikl_2}\Gamma(0,-ikl_2+ik(\sqrt{d^2+(l_1+l_2)^2}-l_1))+$$
$$+B\Gamma(0,ik(\sqrt{d^2+l_1^2}-l_1))-2\Gamma(0,ikd)+B\Gamma(0,ik(\sqrt{d^2+l_1^2}+l_1)),$$
$$A=-0.5i\,\mathrm{ctg}(kl_2)kCI, B=-\frac{iCl_2}{\sin(kl_2)}k, \quad (3)$$

where $\Gamma(0,x)$ – incomplete appear gamma function, and $\Upsilon(\gamma)$ - an auxiliary function, defined as:

$$\Upsilon(\gamma)=\frac{2\sqrt{\gamma}e^{-ikd\sqrt{\gamma}}}{d(\gamma+1)}+ikCi(kd\sqrt{\gamma})+kSi(kd\sqrt{\gamma}). \quad (4)$$

In (3) l_1, l_2, d – lengths and spacing of the dipoles, respectively. The integration limits in (3) are following:

$$s^{\bullet}_1=\frac{1-\sin(arctg(-\frac{l_2}{d}))}{1+\sin(arctg(-\frac{l_2}{d}))},\ s^{\bullet}_2=\frac{1-\sin(arctg(\frac{l_1-l_2}{d}))}{1+\sin(arctg(\frac{l_1-l_2}{d}))}; \quad (5)$$

$$s_2'=\frac{1+\sin(arctg(\frac{l_2+l_1}{d}))}{1-\sin(arctg(\frac{l_2+l_1}{d}))},\ s^{*}_2=\frac{1+\sin(arctg(\frac{l_1-l_2}{d}))}{1-\sin(arctg(\frac{l_1-l_2}{d}))}; \quad (6)$$

$$s_1=\frac{1-\sin(arctg(\frac{l_2}{d}))}{1+\sin(arctg(\frac{l_2}{d}))},\ s_2=\frac{1-\sin(arctg(\frac{l_2+l_1}{d}))}{1+\sin(arctg(\frac{l_2+l_1}{d}))}; \quad (7)$$

When both the intrinsic and the mutual impedances are formulated, it is possible to derive input reflection coefficient of the system of two ends-fed dipole-like radiators.

2.2. Improved phase shifter

It is known that quarter-wave directional couplers are an elementary base when building wide-band differential phase shifters. The closest such analogue is the Schiffman phase shifter [7]. The phase shifter under investigation was patented in 1987 [8], but so far the results of his research have not been published anywhere.

The strip phase shifter [8] consists of two segments of connected lines, the two adjacent ends of which are connected, and the other two ends are respectively the input and output of the strip phase shifter, characterized in that, in order to reduce distortion of time intervals in a pulsed mode, a circular conductor is additionally connected the entire perimeter with segments of connected lines, which in turn is shown in Figure 2.

Strip phase shifter works as follows. The microwave signal passing from the input to the output undergoes a phase shift determined by the electrical length of the segments of the connected lines. Additionally introduced annular conductor due to electromagnetic coupling with segments of connected lines along the entire perimeter provides additional electromagnetic coupling between segments of connected lines, as a result of which the linearity of change in phase shift with frequency changes and the distortion of time intervals decreases when operating in pulsed mode.

To draw the topology, it is required to choose coupling coefficients k in the phase shifter so that at the frequency f_0 it is provided strictly -45°, and the non-uniformity of the phase shift ψ in the frequency band would be ±3°.

For reference: the phase shift ψ takes the following values at specific points for any coupling coefficients k:

- at zero frequency $f/f_0=0$: $\psi=0°$;
- at the center frequency $f/f_0=1$: $\psi=-45°$;
- on the second harmonic $f/f_0=2$: $\psi=-90°$.

Equation transfer function:

$$S_{21}^{GORB}=S_{13}+\frac{S_{12}^2}{1-S_{13}}, \tag{8}$$

where S_{12} and S_{13} are the voltage transmission coefficients.

$$S_{12}=j\frac{k\sin\theta}{\sqrt{1-k^2}\cos\theta+j\sin\theta}, \tag{9}$$

$$S_{13}=\frac{\sqrt{1-k^2}}{\sqrt{1-k^2}\cos\theta+j\sin\theta}, \tag{10}$$

where θ is the electrical length of the transmission line

$$\theta=\frac{\pi f}{2f_0}, \tag{11}$$

where f is the current frequency, f_0 is the center frequency.

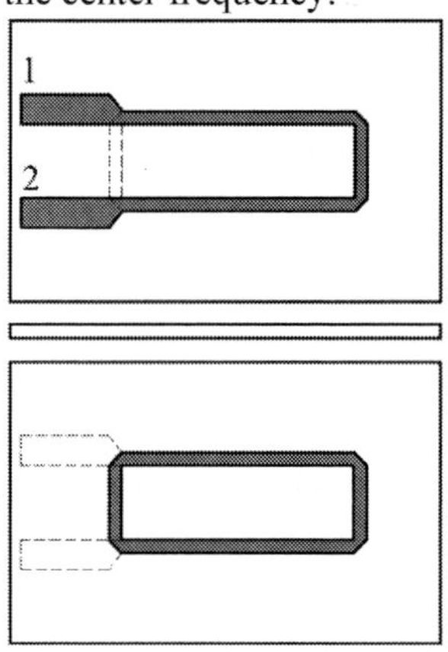

Figure 2. The topology of the dual-band antenna proposed.

Required phase shift:

$$\Psi=-1.25\theta-\text{argument}\left(S_{21}^{GORB}\right). \tag{12}$$

The use of the mathematical model is presented in part B of Section "Results".

2.3. Crossed dipole-like antennas fed by waveguide

It is known that investigation of waveguide-fed radiators is necessary when constructing wireless communication systems. Another important area of using these radiators is radiolocation. The dipole and dipole-like antennas are often used for creating many complex radiators with the appropriate radiation pattern directivity. Waveguides are the traditional feeding structures for stand-alone and phased array antennas due to their low losses at high frequencies that results in high radiation efficiency. Then, if a classical center–fed dipole antenna is excited by a rectangular guide operating at the dominant TE_{10} mode, the split coaxial balun is positioned perpendicularly to a broad waveguide wall onto its centerline. The input port of the balun is excited by a coaxial probe placed inside the waveguide. The balun itself is composed of a coaxial line whose outer conductor, at one end, has two opposite open quarter-wave slots. The two sides of the split coaxial line form the electrically balanced nodes with the center conductor connected to one side. However, the use of split coaxial balun with corresponding slots as the dipole support above the guide often leads to unacceptable manufacturing complexity.

In the present study, to simplify radiator fabrication, the ends-fed dipole antenna (EFDA) and the center-end-fed dipole antenna (CEFDA) are used. The whole radiator structure is excited without any baluns through the rectangular waveguide using four coaxial probes only. This is the novel approach related to derivation of such the arrangement and excitation to date.

There are many radiators constructed on a base of turnstile antenna which proposed in [9]. The name *'turnstile'* is applied to an antenna made up of two dipoles normal to each other and arranged so that their axes intersect at their midpoints. A turnstile antenna can operate in any of several modes, depending on the relative magnitude of the currents at the dipole inputs. When the dipoles of a turnstile antenna are energized by currents of equal magnitude and in phase-quadrature, the spatial field-strength radiation pattern has the omnidirectional (also known as all-directional or non-directional) shape. There are no nulls (or zeros) of radiation, and the maxima occur in directions normal to both dipoles, that is, along the positive and negative z -directions. In the plane xoy containing the dipoles, the radiation is linearly polarized, whereas in all other directions it is elliptically polarized. In the z-direction, the polarization of the radiation is purely circular. In the two half-spaces separated by the plane xoy, the polarization vector rotates in opposite directions. Thus, the radiation field contains all likely states of polarization.

Omnidirectional radiation is produced by horizontally polarized transmitting TV antennas set up on tall masts. Along with the usual cylindrical dipoles, it is convenient in such antennas to employ what are known as batwing dipoles first proposed by B.V. Braude [5]. The batwing element is a number of parallel metal tubes taking up the space within the enclosing rectangular frame. The dipole arms are shorted to the mast for lightning protections at upper points, and the excitation voltage is fed at the lower points. The input impedance is then nearly constant and equal to 140 Ohms over a bandwidth of 7.5%.

Obviously, by adjusting the amplitudes and phases of the currents fed to the crossed dipoles, it is possible to obtain any desired state of polarization. This offers an additional degrees of freedom to effect polarization selectivity in a radio-communication system.

Another way to realize controlled polarization operation at UHF frequencies is the use of waveguide-fed CFDA and CEFDA first described in Russian Patents [4, 10].

To excite the arms of CFDA and CEFDA four probes are used inside the rectangular waveguide (Figure 3; the corresponding Cartesian coordinate system differs from that mentioned above in this subsection). These probes are characterized by corresponding depths inside the waveguide and are positioned perpendicular to a broad wall of that. Both the ends-fed and center-end-fed dipole antennas are placed parallel to a broad wall of the guide at the distance from that (Figure 3). This distance is equal to one quarter wavelength in free space. To connect the arms of the crossed antennas with corresponding probes four coaxial stubs are used. The filling of the stubs is the material with relative permittivity equal to nearly unity but greater than that.

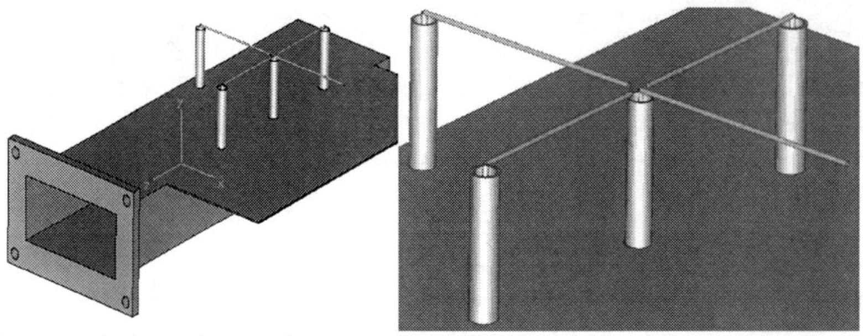

Figure 3. General view of the radiator proposed (large scale) and proposed arrangement of the crossed antennas (fine scale).

3. Results

3.1. Dual-band ends-fed dipole-like antenna

For the structure (Figure 1) we simulated return losses and radiation patterns at both central working frequencies. To compensate extremely great capacitance which occurs when exciting ends-fed dipole [6], we decrease the gap between high-frequency dipole and the ground plane. Here we also bring analytical results for input reflection coefficient when the reactive (image) part of the overall input impedance of the system proposed is artificially canceled (Figure 4).

Figure 5 reveals that when a printed model of the structure presented is designed the way the extremely great reactive part of the overall input impedance is compensated by proper setting the crucial antenna sizes it is able to radiate over two separated frequency bands. It seems that the shift of the low central frequency of the simulated model appears due to presence of complicated structure balun, which affects the all system resonance frequencies.

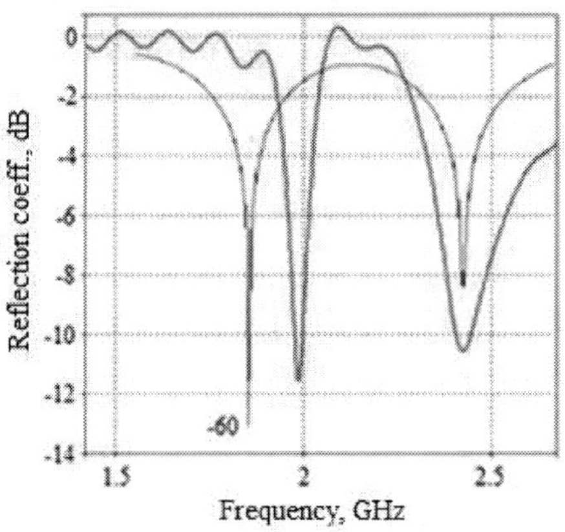

Figure 4. Analytical (blue) and simulated (red) input reflection coefficients (dB) of the antenna proposed.

Figure 5 shows the radiation patterns obtained.

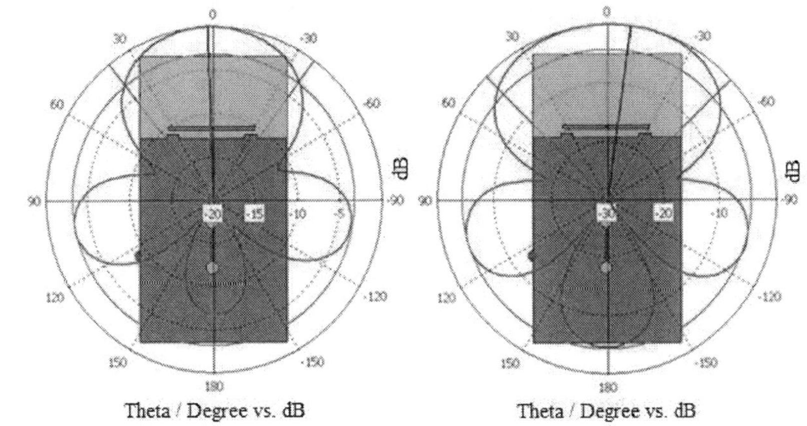

Figure 5. Radiation pattern at low central frequency in E-plane (left) and Radiation pattern at appear central frequency in E-plane (right).

3.2. Improved phase shifter

According to the formulas from the previous section, the mathematical model was calculated. When $k = 0$, we obtain the dependence graph presented in Figure 6. The task in relation to the required values was achieved with k = 0.68, the graph with the result is shown in Figure 6.

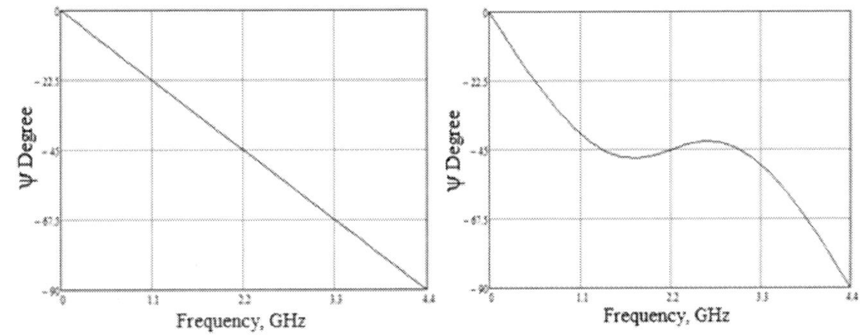

Figure 6. Phase shift graph when k = 0 and phase shift graph when k = 0.68.

3.3. Crossed dipole-like antenna

The radiation pattern of the whole aerial has been investigated through full-wave simulation in free space. The corresponding rectangular waveguide [11] has been used at the central frequency 1.75 GHz. The steepest-descend method [12] was then used to determine the coaxial probe location. Figure 7 displays the dependence of the return loss versus frequency. The plots of the principal cuts of 3D radiation pattern at the frequency of 1.75 GHz are plotted in Figure 8. Thus, it can be seen that the proposed whole aerial has the fan–beam radiation pattern. Good agreement between the measured plots (green curves) and simulation-derived results (red markers) can be observed. Some discrepancy is attributed to impedance discontinuities in the test arrangement, the finite accuracy of the optimization procedure, and attainable resolution of the manufacturing.

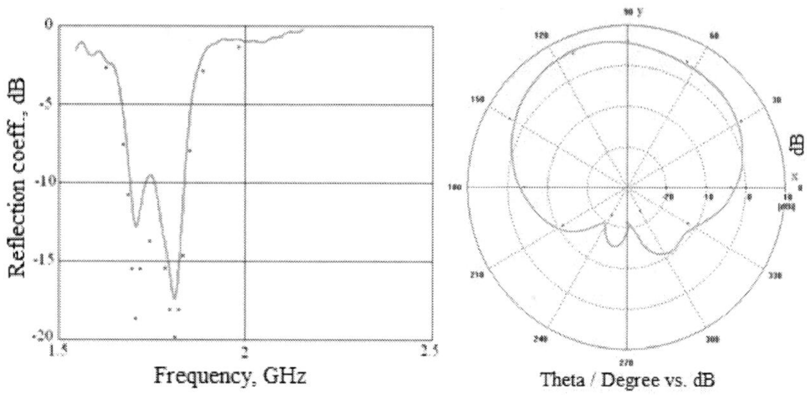

Figure 7. Measured (green curve) and simulated (red markers) return losses of the whole aerial against frequency (left) and measured (green curve) and simulated (red markers) polar x-y radiation patterns (right).

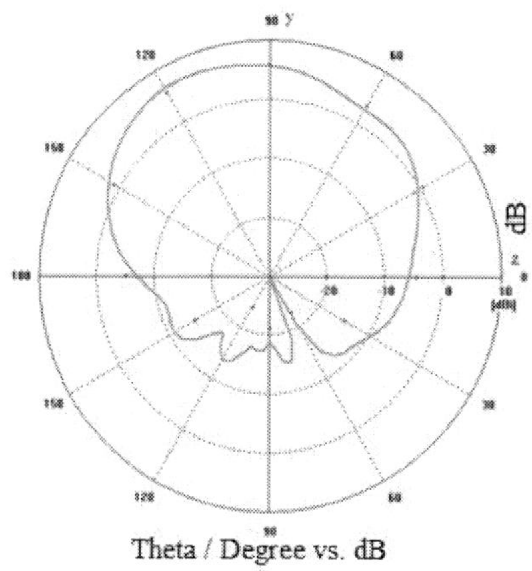

Figure 8. Measured (green curve) and simulated (red markers) polar z-y radiation patterns.

4. Conclusion

The one of the possible approaches for designing dual-band antennas is presented in the communication. This approach is based on use of a printed balun, which is formed by a split power divider in the ground plane, and the microstrip lines on the top side of the substrate. The analytical and simulated results of the structure designed are compared and examined. From this it follows that when using ends-fed radiators the dual-frequency operating might be achieved.

As a result of the investigation of the phase shifter proposed, it appears to be possible to achieve quite acceptable results, namely: at specific points, the phase shift corresponds to the required values and the non-uniformity of the phase shift in the 1.1÷3.3 GHz frequency band did not exceed ±3°.

A new implementation of the crossed dipole-like antennas has been also proposed. To realize the necessary pattern the four coaxial probes placed inside the rectangular waveguide have been manufactured. The presented investigation paves the way for next generation of the circularly polarized aerials

References

[1] Abbosh A, Bialkowski M, Mazierska J 2006 *Proceedings of Asia-Pacific Microwave Conference* **1** 905–908.

[2] Tarasenco N, Gorbachev A 2014 *Proceedings of 15th International conference on micro/nanotechnologies and electron devices (EDM)* 161–163
[3] Alekseytsev S, Gorbachev A 2018 *Proceedings of 19th International conference on micro/nanotechnologies and electron devices (EDM)* 241–243
[4] Buhtiyarov D, Gorbachev A, Filoimonova J 2013 Dipole emitter *Patent RU № 2472261 C1*, Bull. 1
[5] Markov G, Sazonov D 1975 *Antennas* (Moscow: Energiya Publ.)
[6] Buhtiyarov D, Gorbachev A, Atuchin V 2015 *Journal of Electromagnetic Waves and Applications* **29** 1720–1727
[7] Schiffman B 1958 *IRE Transactions on Microwave Theory and Techniques* **6** (2) 232–237
[8] Gorbachev A 1987 Strip phase shifter *Patent SU № 1334221 A1*, Bull. 32.
[9] Brown G 1936 *Electronics* **9** 14–17
[10] Borejchuk A, Gorbachev A, Kirillova N, Shvedova A 2015 Dipole antenna *RU Patent no. 2571156,* Bull. No. 35.
[11] Volakis J 2007 *Antenna engineering handbook* (New York: McGraw Hill)
[12] Gill P, Murray W, Wright M 1985 *Practical Optimization* (New York: Academic Press)

IOP Conf. Series: Materials Science and Engineering 1019 (2021) 012100 doi:10.1088/1757-899X/1019/1/012100

System for automated environmental monitoring using remote sensing data of the Earth from open data sources

K A Maslov and O S Tokareva
Division for Information Technology, Tomsk Polytechnic University, Tomsk, Russia
E-mail: kam20@tpu.ru

Abstract. Environmental monitoring using remote sensing data requires an analyst to perform a large amount of routine work related to downloading, processing and analyzing data, especially in cases when the study area is covered with a large number of satellite imagery. The paper presents the results of the design and software implementation of the system that automates downloading and processing of remotely sensed data according to developed scenarios and, thus, greatly simplifies the processing of satellite imagery. It provides the description of tools for accessing data from the archive of the United States Geological Survey (USGS) and describes the data flow in the system. The paper gives an analysis of results obtained using the developed system on the example of monitoring the state of Siberian pine forests of the Tomsk region.

1. Introduction

Monitoring of land cover change using remotely sensed data helps to make an objective assessment of the state of environmental objects and to identify changes occurring because of natural processes or anthropogenic impact.

Methods of the remote sensing of the Earth are widely used in solving problems of environmental monitoring [1–4]. However, processing of the Earth remote sensing data by an analyst is slow and can lead to errors due to inattention or loss of the researcher concentration, and it consists of steps that can be automated, therefore, the issue of automating this process is relevant [5].

To solve the problems of monitoring of land-cover change, index images based on combination of pixel values from different electromagnetic spectrum regions are widely used [6–8]. Examples of such index images are vegetation index maps, e.g. normalized difference vegetation index (NDVI) or normalized difference water index (NDWI), standing for the amount of photosynthetically active phytomass and water content in leaves and needles of vegetation, respectively [9, 10].

To provide continuous monitoring of the land cover using remotely sensed data of the Earth, it is necessary to have a system that simultaneously automates both the processes of downloading data from external sources and processing of the fetched data according to specific scenarios.

2. System design

2.1. General architecture of the system

Figure 1 demonstrates the diagram showing the general architecture of the designed system for automated environmental monitoring using remote sensing data of the Earth from open data sources. The user controls the system by interacting with the console interface of the monitoring system. Through network requests, the system interacts with data from the external sources, downloads them onto the local storage and processes them.

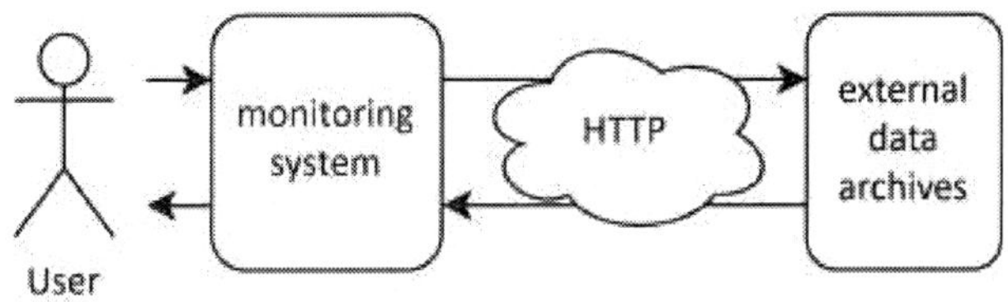

Figure 1. Architecture of the system.

The system provides a framework focused on development and execution of the scenarios, which are sequences of instructions that automate data acquisition and processing. The scenarios are run at regular intervals according to a predefined schedule.

Note that the architecture of the system does not depend on terms of the remote sensing domain, so the system can be used with any kind of data obtained from external sources, e.g. financial, meteorological, etc.

2.2. Data access tools

Nowadays network resources allow end users to get access to remotely sensed imagery via public application program interfaces (APIs). These tools automate data searching and downloading processes.

In this paper, the United States Geological Survey (USGS) data archive was used as a source of external data. To automate the process of obtaining the correct identifiers of satellite images EarthExplorer API was used [11]. The EarthExplorer API provides a flexible and powerful mechanism for searching remotely sensed data and their products in the USGS data archive, allowing to search across hundreds of data collections from various spacecraft and sensors [12], e.g. Landsat, MODIS, ASTER, VIIRS, etc. It also gives an opportunity to search satellite imagery using complex queries. Earth resources observation and science (EROS) center science processing architecture (ESPA) on demand interface API [13] was used to order scenes and get the URLs to download the scenes.

2.3. Data flow in the system

The designed architecture and selected data access tools allow to automate the processes of acquisition and processing of the satellite imagery scenes. Figure 2 shows the corresponding data flow diagram.

The user specifies search parameters for the particular area of interest. Then the system obtains the appropriate scenes identifiers via the EarthExplorer API, after that it orders scenes and receives URLs to download them via the ESPA ordering system API. Finally, the system downloads the scenes and processes them, it puts the scenes in the DATA directory and places the results of the processing in the PRODUCTS directory.

2.4. Scenario algorithm

Below is an example of a scenario for monitoring of vegetation, which allows to trace changes in its condition and locate areas of its degradation and restoration.

At each iteration, the scenario waits for its next run, and then it acquires new satellite imagery and processes each image. For each image it performs the calculation of open area mask, cloud mask and maps of vegetation indices and calculates the zonal statistics (mean values of vegetation indices and percentage of cloud pixels) for the study area, which is represented in a vector format as a set of polygons.

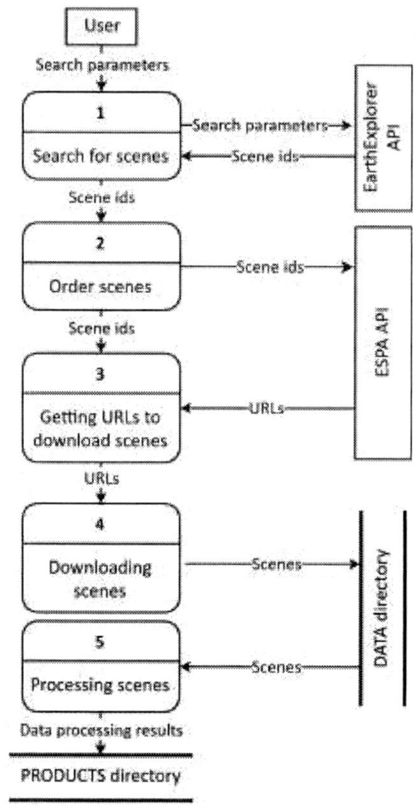

Figure 2. Data flow diagram for a scenario of downloading and processing satellite imagery from the USGS data archive.

The following is the algorithm in pseudo code that demonstrates the steps performed by the designed scenario:

```
BEGIN
 WHILE running
  Wait for the next run
  Search for scenes
  Order new scenes
  Get URLs to download available scenes
  FOR each URL in URLs
   Download scene archive by URL
   Unzip the archive
   Calculate open area mask, cloud mask, NDVI map and NDWI map
   Calculate zonal statistics for the studied polygons
   Write zonal statistics to a .csv file
   IF no clouds within the polygons
   THEN
    Write zonal statistics to a separate .csv file
    Calculate new dNDVI and dNDWI maps
   END IF
  END FOR
 END WHILE
END
```

Vegetation indices NDVI and NDWI are calculated by the formulas [9, 10]:

$$\mathrm{NDVI} = \frac{\mathrm{NIR} - \mathrm{RED}}{\mathrm{NIR} + \mathrm{RED}}, \tag{1}$$

$$\mathrm{NDWI} = \frac{\mathrm{NIR} - \mathrm{SWIR}}{\mathrm{NIR} + \mathrm{SWIR}}, \tag{2}$$

where NIR, RED and SWIR are spectral reflectance measurements acquired in the near infrared, red and short wave infrared regions of the electromagnetic spectrum, respectively.

To assess changes in the state of vegetation, the differenced indices dNDVI and dNDWI are calculated by the formulas:

$$\mathrm{dNDVI} = \mathrm{NDVI}_{\mathrm{post}} - \mathrm{NDVI}_{\mathrm{pre}}, \tag{3}$$

$$\mathrm{dNDWI} = \mathrm{NDWI}_{\mathrm{post}} - \mathrm{NDWI}_{\mathrm{pre}}, \tag{4}$$

where $\mathrm{NDVI}_{\mathrm{post}}$ and $\mathrm{NDWI}_{\mathrm{post}}$ are vegetation index values for imagery with a later acquisition date, $\mathrm{NDVI}_{\mathrm{pre}}$ and $\mathrm{NDWI}_{\mathrm{pre}}$ are vegetation index values for imagery with an earlier acquisition date.

The csv files obtained during the execution of the scenario can be opened in Microsoft Excel or other software for further analysis, including, for example, plotting the dynamics of the values of vegetation indices. Calculated open area masks and cloud masks, as well as maps of NDVI, NDWI, dNDVI, dNDWI indices are raster images with spatial reference and, accordingly, can be opened and visualized in geographic information system (GIS), used for more complicated geoanalysis, etc.

3. System implementation

To implement the system, the object-oriented approach was chosen. Python was used as a programming language; rasterio and rasterstats libraries were used to deal with raster files and to calculate zonal statistics.

The system and the scenario have been implemented. Figure 3 shows the corresponding class diagram.

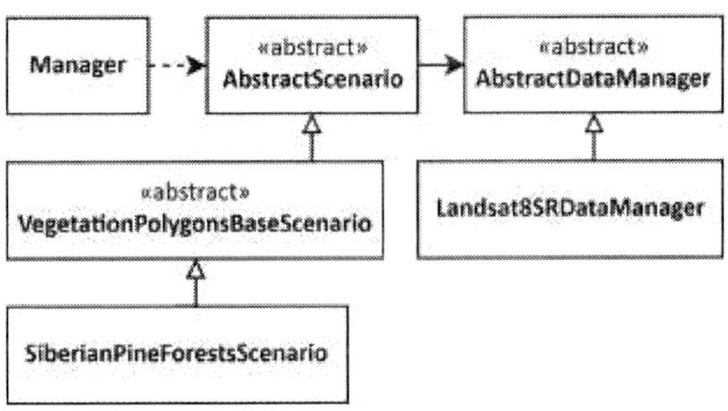

Figure 3. Classes of the developed system and scenario (only essential classes and their relationships are shown).

The *Manager* class implements the console user interface and is responsible for the interpretation of user commands. The *AbstractScenario* and *AbstractDataManager* classes are abstract and represent, respectively, a generalized scenarios classes and classes for receiving data from external sources.

The *Landsat8SRDataManager* class implements a mechanism for searching, ordering and downloading satellite images acquired by the Landsat 8 satellite of the L2 processing level (surface reflectance) from the USGS archive [14]. This class can be reused by other scenarios to access data from the same collection of the USGS data archive, but for other spatial and temporal parameters.

The *VegetationPolygonsBaseScenario* class implements the algorithm described previously, and the *SiberianPineForestsScenario* class implements the scenario for monitoring the state of the Siberian pine forests, which are specially protected natural areas.

4. Results

This section gives the description of the results obtained with the *SiberianPineForestsScenario* scenario for monitoring of the state of the Siberian pine forests of the Tomsk Region, which are specially protected natural areas (Aksenovskiy, Belousovskiy, Bogashevskiy, Loskutovskiy, Luchanovo-Ipatovskiy, Magadaevskiy, Nizhne-Sechenovskiy, Petrovskiy, Petukhovskiy, Plotnikovskiy, Protopopovskiy, Trubachevskiy, Voronovskiy Siberian pine forests and forest park near the village Yar). Reference [1] gives a more detailed description of the Siberian pine forests.

The mean values of the NDVI and NDWI indices were calculated within the boundaries of the Siberian pine forests of the Tomsk region. We compared the values of vegetation indices calculated automatically with values obtained with GIS, in this way the correctness of the scenario implementation was shown.

The scenario automatically determined cloudless scenes within the boundaries of all Siberian pine forests. Table 1 contains the description of these scenes. Using data from the file with zonal statistics for these scenes, plots were built showing the changes of mean values of the NDVI and NDWI indices (Figure 4) by year for mid-July.

Table 1. List of the used cloudless Landsat 8 satellite imagery.

WRS2 path/row	Acquisition date	Identifier
148/21	14.07.2013	LC08_L1TP_148021_20130714_20170503_01_T1
148/21	22.07.2016	LC08_L1TP_148021_20160722_20180525_01_T1
148/21	26.08.2017	LC08_L1TP_148021_20170826_20170913_01_T1
148/21	12.07.2018	LC08_L1TP_148021_20180712_20180717_01_T1

As depicted in Figure 4, the Luchanovo-Ipatovskiy Siberian pine forest has shown significant negative state dynamics, not correlating with other Siberian pine forests. In July 2018, the mean NDVI and NDWI indices of this forest decreased.

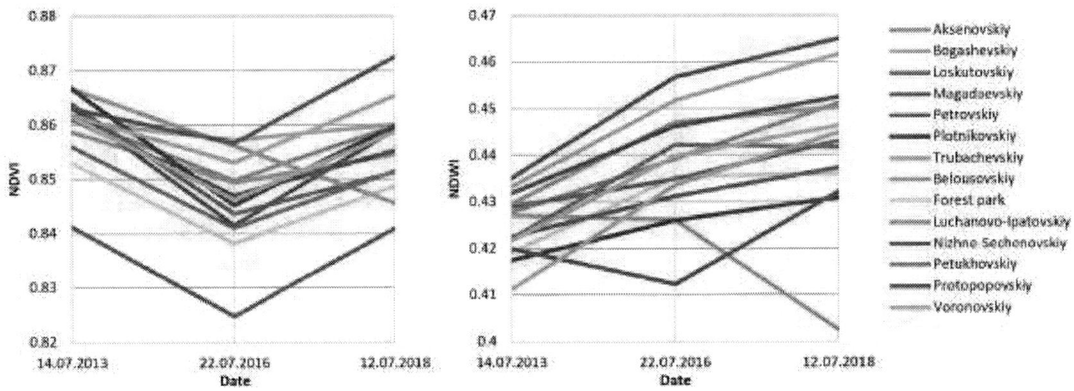

Figure 4. Mean values of the vegetation indices NDVI (left) and NDWI (right) by years for the Siberian pine forests.

According to the scenario dNDVI and dNDWI maps were calculated for the scenes listed in table 1. Figure 5 demonstrates fragments of these maps for the Luchanovo Ipatovskiy Siberian pine forest, boundaries of the Siberian pine forest are shown with black line. According to these fragments, areas of the most negative state dynamics were established: those are forest areas near the village Luchanovo and the village Ipatovo, these areas are shown in red shades; and the period of negative impact is the period from 2016 to 2017.

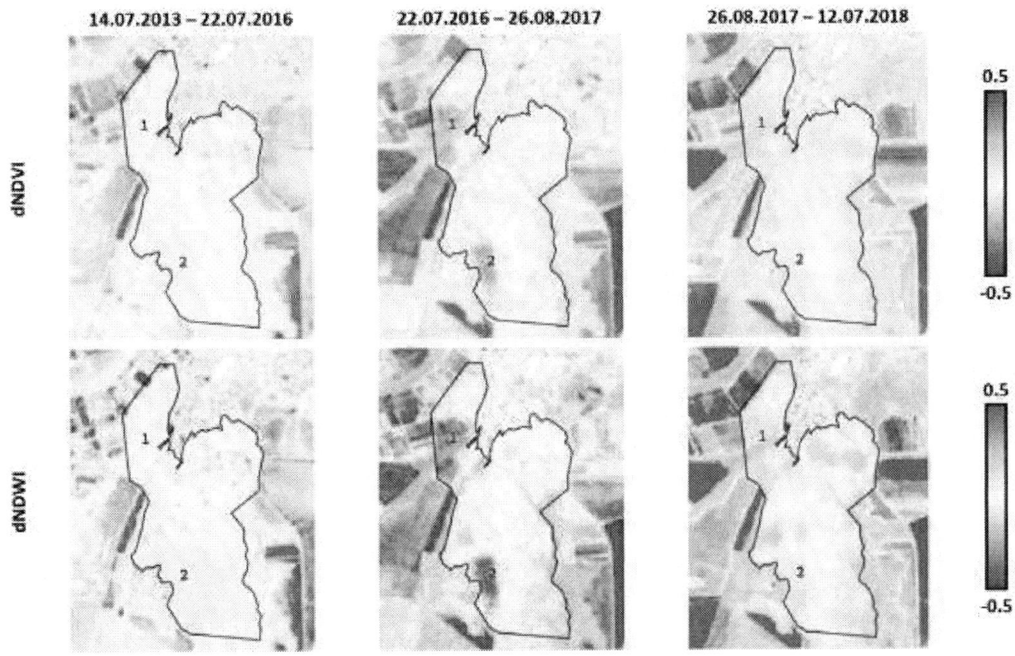

Figure 5. Fragments of the obtained dNDVI and dNDWI maps for Luchanovo-Ipatovskiy Siberian pine forest. Location of the damaged areas: 1 – the area near the village Luchanovo, 2 – the area near the village Ipatovo.

The analysis of high spatial resolution imagery and ground surveys of the territory showed that these areas were affected by Siberian silkworm larvae. It is worth noting that the NDWI index is more sensitive to this kind of forest damage, as Figure 5 shows.

5. Conclusion

This paper presented the results of the design and implementation of the automated system for monitoring of land-cover change using Earth remote sensing data obtained via public APIs. A scenario was designed and implemented to monitor the state of the Siberian pine forests of the Tomsk region.

The functionality of the developed system was successfully tested. The results of the testing showed the negative dynamics of the state of the Luchanovo-Ipatovskiy Siberian pine forest and allowed to locate its damaged areas.

In the future, the developed system may be modified and extended for environmental monitoring of other objects or phenomena using a wider range of Earth remote sensing data.

Acknowledgements

The research has been carried out with the financial support of the RFBR (Project 18-47-700010).

References

[1] Pasko O A, Tokareva O S, Alshaibi A, Chernikova T Y and Cabral P 2019 *Bull. TPU Geo Assets Eng.* **330** 98–109

[2] Maslov K A 2019 *Proc. XVI Int. Scientific and Technical Conf. of students, graduates and young scientists «Youth and modern information technology»* 98–99

[3] Mjachina K V, Baynard C W, Chibilyev A A and Richardson R D 2018 *Int. J. Sust. Dev. World Ec.* **25** 6 541–553

[4] Lamqadem A A, Afrasinei G M and Saber H 2019 *J. App. Rem. Sens.* **13** 1 180590 doi: 10.1117/1.JRS.13.014517,

[5] Balashov I V, Khalikova O A, Burtsev M A, Loupian E A and Matveev A M 2013 *Curr. Prob. Rem. Sens. Earth Space* **10** 3 9–20
[6] Liu L, Niu Q, Heng J, Li H and Xu Z 2019 *Rem. Sens.* **11** 10 1254 doi: 10.3390/rs11101254
[7] Albarakat R and Lakshmi V 2019 *Rem. Sens.* **11** 10 1245 doi: 10.3390/rs11101245
[8] Quintero N, Viedma O, Urbieta I R and Moreno J M 2019 *Forests* **10** 518 doi: 10.3390/f10060518
[9] Cherepanov A S and Gruzhinina E G 2009 *Geomatics* 3 28–32
[10] Bannari A, Morin D, Bonn F and Huete A R 2009 *Rem. Sens. Rev.* **13** 1 95–120
[11] United States Geological Survey, EarthExplorer – Service Documentation. Available at: https://earthexplorer.usgs.gov/inventory/documentation
[12] United States Geological Survey, EarthExplorer – Home. Available at: https://earthexplorer.usgs.gov/
[13] United States Geological Survey, API for the ESPA ordering system, Github. Available at: https://github.com/USGS-EROS/espa-api
[14] United States Geological Survey, Landsat Surface Reflectance. Available at: https://www.usgs.gov/land-resources/nli/landsat/landsat-surface-reflectance

IOP Conf. Series: Materials Science and Engineering 1019 (2021) 012101 doi:10.1088/1757-899X/1019/1/012101

Compatibility of the dimensions of polymer molecular aggregates to the pore throat of a reservoir

Risu Na and V I Erofeev
School of Earth Science & Engineering, National Research Tomsk Polythetic University, Tomsk, Russia
E-mail: narisu2011@yandex.ru

Abstract. The compatibility of the dimensions of the polymer molecular aggregates and the pore throat of the reservoir were studied. The W section of Tuha oilfield was the study area and polymers produced by Daqing Refining and Chemical Company were used. The permeability limit of the polymer molecules with different molecular masses and concentrations, matching relationship between the dimension of polymer molecular aggregates and pore throat were obtained by experiments. The results of the research are important for the development and implementation of a polymer flooding technical scheme in the middle and late stages of the operation of the Tuha oilfield.

1. Introduction

In recent years, polymer flooding has become important in the development of oilfields and in optimizing and improving productivity [1]. Polymer flooding can improve oil recovery primarily by increasing the viscosity of the injected water resulting a reduction in the permeability of the water phase. Polymers with a high relative molecular mass or concentration tend to be viscose and have a low water phase permeability [2,3]. Hence, if a polymer has a higher relative molecular mass and concentration it will be more viscous and the polymer hydration molecule should go through the natural selection of pore throat structure when passing through porous media of reservoir [4]. With the increase of relative molecular weight or concentration of polymer, the size of molecular aggregates becomes larger, the injectability becomes worse, and the mechanical trapping phenomenon of polymer in the reservoir becomes more obvious [5,6]. When the dimension of molecular aggregates is larger than the rock pore throat, it is difficult for the polymer to pass through the pore throat under normal injection pressure. When an external force is applied to push the polymer through the pore throat its molecular structure is destroyed and it loses its primary function of displacement [7,8]. The selection of polymers with regards to their molecular mass and concentration is a challenge at oilfields. Studies have been conducted on the relationship between the relative molecular mass of polymers, their permeability and their oil displacement effect, however, the results did not define the matching relationship between relative molecular mass and permeability on oil displacement effect [9,10].

This paper focuses on the evaluation of a polymer produced by Daqing Refinery and Chemical Company so as to be used at the W section of Tuha Oilfield in China. The dynamic light scattering technique was used to study the dimensions of the polymer molecular aggregates in an aqueous solution. A number of core seepage experiments were conducted to determine the core permeability limit of the polymer molecules with different molecular masses and concentrations as well as taking into consideration the relationship between the median pore radius and the dimensions of the polymer molecular aggregates. The results of the research are important for the development and implementation of a polymer flooding technical scheme in the middle and late stages of the operation of the Tuha oilfield.

2. Experiment description

2.1. Experimental materials

Polymers (Partially hydrolyzed polyacrylamide (HPAM)) produced by Daqing Refinery and Chemical Company were used. The relative molecular mass of the HPAM polymers used were $400x10^4$, $800x10^4$, $1400x10^4$ and $2100x10^4$ respectively, and the solid content of the polymers was 88%. The experimental water which was used in place of the injection water from W section of the Tuha oil field had a mass concentration of (K^++Na^+), Ca^{2+}, Mg^{2+}, Cl^-, SO_4^{2-}, HCO_3^- of 4342, 7935, 437, 20561, 1152, 1026 mg/l respectively. The total salinity of the experimental water was 35453 mg/l. Artificial columnar cores made of quartz sand and clay cemented with epoxy resin we created and used [11]. The permeability of the artificial cores was measured using a gas and cores with different permeabilities were used. The cores were made by mixing quartz sand of different grain sizes with epoxy resin and they were all the same size i.e. they had a diameter and a length of 2.5cm and 10cm, respectively. The permeability of the cores is presented in the table 1.

2.2. Instruments and equipment for experiments

The viscosity of the polymers was measured using a DV-II Brookfield Viscometer at a shear rate of 7.34 s^{-1} at 75°C and the dimensions of the polymer molecular aggregates were measured using a Malvine NanoZS90 laser particle size analyzer system. The equipment used for core flooding tests was used to determine the seepage behavior of the polymers in the reservoir. The experimental setup consisted of an advection pump, pressure sensor, core holder, hand pump and two intermediate containers and other parts. All the equipment, except the advection pump and hand pump were placed in a thermostat with the reservoir temperature of 75 °C. The schematic diagram of the experimental process is given in Figure 1.

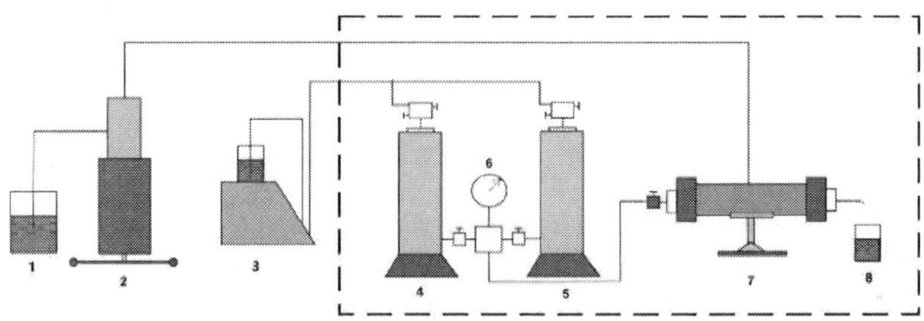

Figure 1. Experiment-process diagram: 1 – beaker, 2 – hand pump, 3 – advection pump, 4 – the container of water, 5 – the container with surfactant, 6 – sensor, 7 – core holder, 8 – measuring cup.

3. Results and discussion

3.1. Compatibility between the HPAM molecular aggregates and the pore throat

When a polymer solution passes through a core pore throat, the relationship between the injection pressure and the pore volume multiple (PV), the resistance coefficient and the residual resistance coefficient reflect its level of retention level in the porous media. It also represents its compatibility or the adaptability between polymer solution and pores of the rock. We divided the polymers into two categories based on their trends in response to the injection pressure. An increase in the injected polymer solution resulted in a stable injection pressure and was represented by a horizontal segment on the pressure curve of the seepage experiment. It showed that the polymer solution and the core pore throat were highly compatible. The lowest permeability at which the polymer solution does not plug and move through the core is called the permeability limit of the polymer solution. For the other polymer solutions an increase in the injected polymer solution resulted in the injection pressure increasing. The increase in the injection pressure suggested that the polymer had blocked the core pores and the compatibility between the polymer and the core pore throat was low.

The seepage characteristics of the polymer solution are generally evaluated using the resistance coefficient and the residual resistance coefficient which are the technical indicators for describing the retention of polymers in porous media. The resistance coefficient (F_R) and the residual resistance coefficient (F_{RR}) were determined using the equations below [12, 13]:

$$F_R = \frac{\partial P_2}{\partial P_1} \tag{1}$$

$$F_{RR} = \frac{\partial P_3}{\partial P_1} \tag{2}$$

Table 1. Resistance coefficient (F_R) and residual resistance coefficient (F_{RR}).

M ($\times10^4$)	C_p (mg/L)	Viscosity (mPa·s)	Permeability Kg ($\times10^{-3}\mu m^2$)	F_R	F_{RR}
400	400	2.6	15	9.71	5.22
			20	8.03	5.13
	600	3.4	25	14.15	9.89
			30	10.59	8.25
	800	4.1	30	14.47	11.03
			35	12.33	9.74
	1000	4.8	35	15.90	11.29
			40	13.33	9.80
800	400	3.7	20	14.82	11.80
			25	10.76	8.17
	600	4.3	30	17.50	14.03
			35	13.52	9.38
	800	5.5	35	19.12	14.83
			40	13.69	11.12
	1000	6.7	40	20.53	16.12
			45	15.76	12.78
1400	400	4.2	30	19.71	15.99
			35	15.33	12.29
	600	5.8	40	23.73	17.96
			45	17.29	12.42
	800	7.9	45	24.33	20.02
			50	17.51	14.51
	1000	10.3	50	24.88	20.41
			55	17.78	13.92
2100	400	5.1	40	12.94	18.51
			45	15.96	13.33
	600	7.2	50	25.61	20.59
			55	18.76	14.54
	800	9.3	50	26.34	20.73
			60	19.44	14.65
	1000	12.4	70	29.00	22.66
			80	21.72	16.48

IOP Conf. Series: Materials Science and Engineering 1019 (2021) 012101 doi:10.1088/1757-899X/1019/1/012101

In the above equation, P_1 is the drop in pressure during the water flooding process, P_2 is thedrop in pressure during the chemical flooding process and P_3 is the drop in pressure during subsequent water flooding.

Experiments were conducted on the seepage characteristics of the polymer solutions using the experimental setup given in Figure 1, using polymers with different relative molecular masses (M) and polymer concentrations (Cp) to determine the permeability limit of the HPAM solution. The injection speed of the experiment was 0.3 mL/min and the pressure were measured every 30 minutes. The resistance coefficient and the residual resistance coefficient of the HPAM solution are presented in table 1. The relationship between the injection pressure of the polymer solution and the PV is given in Figure 2.

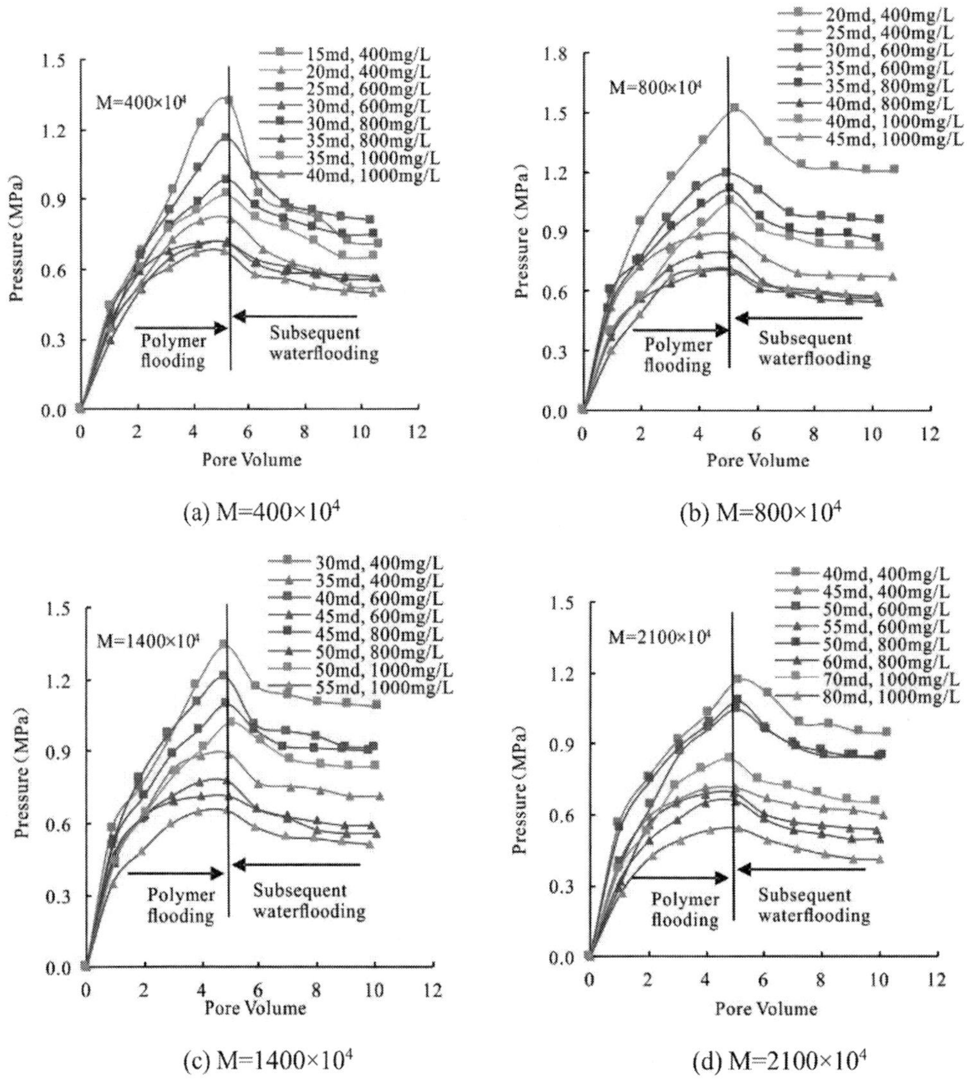

(a) $M=400\times10^4$ (b) $M=800\times10^4$

(c) $M=1400\times10^4$ (d) $M=2100\times10^4$

Figure 2. Relationship between the polymer solution injection pressure and the pore volume.

The results showed that the polymer concentration and the core permeability had a strong influence on the resistance coefficient and the residual resistance coefficient as shown in table 1. When the relative molecular mass and the concentration of polymers were not changed, the resistance coefficient and the residual resistance coefficient of the polymer increased with a decrease in the permeability of the rock. An increase in the relative molecular mass and the concentration of the polymer solution resulted in gradual increase in the permeability limit of core. Under the same permeability conditions, the polymer solutions viscosity increased as the polymer concentration increased resulting in an increase in its resistance coefficient and residual resistance coefficient. The high residual resistance coefficient suggests that the compatibility between polymer solution and pores in the rock was high.

The dimensions of the polymer molecular aggregates were not compatible to the size of the pores of the rock when the injection pressure increased as the injection volume increased as shown in Figure 2. This is in contrast to the previous results which showed that a high compatibility resulted in a stable injection pressure as the injection volume increased as shown in Figure 2. Polymer concentration of 400 mg/L, 600 mg/L, 800 mg/L and 1000 mg/L had permeability limits of 20md, 30md, 35md and 40md, respectively, for the polymer solutions with relative molecular mass of M=400×104. The permeability limits for the polymer solutions with relative molecular mass of M=800×104 were 25md, 35md, 40md and 45md, respectively. The permeability limits of the polymer solutions with a relative molecular mass of M=1400×104 were 35md, 45md, 50md and 55md, respectively. The permeability limits of the polymer solution with a relative molecular mass of M=2100×104 were 45md, 55md, 70md and 80md, respectively.

3.2. Compatibility of the dimensions of the molecular aggregates of the polymers to the pore throat of reservoir

The dimensions of the polymer molecular aggregates (D_h) were measured using the Malvine NanoZS90 laser particle size analyzer system and the results are presented in table 2.

Table 2. D_h test results (nm)

C_p (mg/L)	Relative molecular mass (M×10^4)			
	400	800	1400	2100
400	105.5	162.5	197.3	324.7
600	130.8	187.4	242.8	378.8
800	168.3	221.3	278.3	463.5
1000	186.5	259.1	336.4	541.9

An increase in the concentration of the polymer solution resulted in an exponential increase in the dimensions of its polymer molecular aggregates. This resulted in a gradual increase in the range of D_h as the relative molecular mass of the polymer solution remained unchanged. Under a certain polymer concentration, the dimensions of the polymer molecular aggregates increased as its relative molecular mass increased, and the overall range of D_h was between 105.5 nm and 541.9 nm.

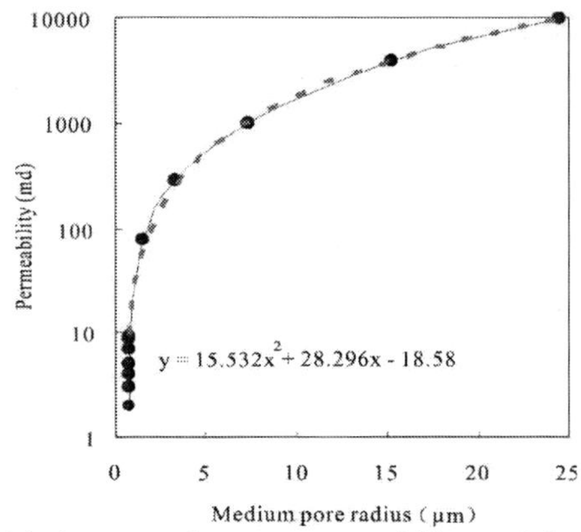

Figure 3. Relationship between the core permeability and the median pore radius.

The relationship between the reservoir rock permeability and the median pore radius (R_m) is presented in Figure 3 including the equation. The results show that an increase in the median pore radius resulted in an increase the logarithm of the permeability of the rock.

The median pore radius (R_m) of the reservoir rock corresponding to each permeability limit is presented in Figure 3. The relationship between the dimensions of the polymer molecular aggregates of the polymer solution and the median pore radius are presented in Figure 4. The relationship between the ratio of the dimensions of the polymer molecular aggregates of the polymer solution to the median pore radius (R_m/D_h) and polymer solution concentration are presented in table 3 and Figure 5.

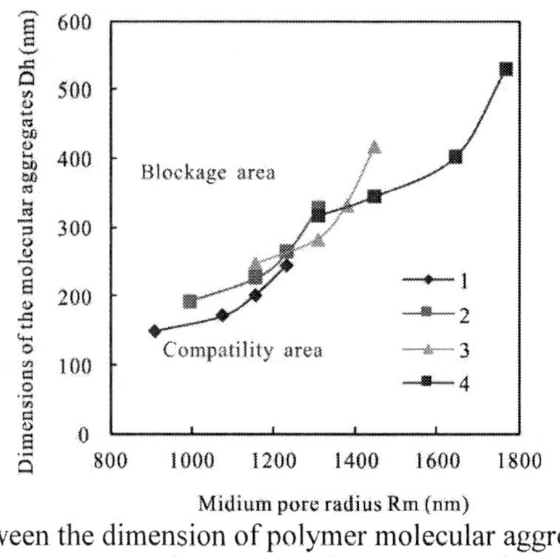

Figure 4. Relationship between the dimension of polymer molecular aggregates and the medium pore throat radius, 1 –400×10^4, 2 - 800×10^4, 3 - 1400×10^4, 4 - 2100×10^4.

The corresponding equation of line 1–3 and 4 of Figure 4 are shown in (3), (4), (5) and (6) respectively:

$$y=1\times10^{-7}x^3+0.0007x^2-1.6171x+937.11 \quad (3)$$

$$y=4\times10^{-6}x^3-0.0125x^2+13.049x-4381.2 \quad (4)$$

$$y=5\times10^{-6}x^3+0.0174x^2+19.987x-7423.4 \quad (5)$$

$$y=4\times10^{-6}x^3+0.0189x^2+27.506x-13027 \quad (6)$$

Table 3. Relationship between R_m/D_h, relative molecular mass and concentration of the polymer

C_p (mg/L)	Relative molecular mass (M×10⁴) 400	800	1400	2100
400	6.14	6.31	5.73	5.05
600	5.17	5.14	4.70	4.01
800	4.72	4.66	4.16	3.48
1000	4.13	4.20	4.09	3.35

The area with coordinates is divided into two parts i.e. the compatibility area and blockage area as shown in Figure 4 and Figure 5. In summary, the compatibility between the polymer and reservoir rock is determined by the permeability of the relative molecular weight and concentration of the polymer solution. An increase in the concentration of the polymer solution resulted in a gradual decrease in the ratio R_m/D_h for polymers with the same molecular weight and it ranged between 3.35 and 6.14. The cumulative percentage of the injected reservoir thickness and its corresponding reservoir permeability can be determined based on the status of the oilfield field. The injectable D_h and R_m/D_h can be determined using the corresponding equation obtained. These data have important reference value for field production.

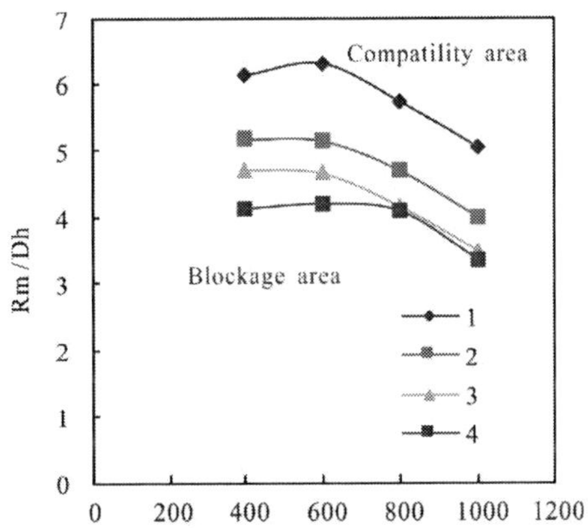

Figure 5. Relationship between R_m/D_h and the concentration of the polymers: 1 – 400×10⁴, 2 – 800×10⁴, 3 – 1400×10⁴, 4 –- 2100×10⁴.

The corresponding equation of line 1,2,3 and 4 of Figure 5 are shown in (7), (8), (9) and (10) respectively:

$$y=1\times10^{-8}x^3-3\times10^{-5}x^2+0.0244x+0.9177 \quad (7)$$

$$y=-4\times10^{-6}x^2+0.0038x+4.3114 \quad (8)$$

$$y=5\times10^{-9}x^3+1\times10^{-5}x^2+0.0106x+2.5212 \quad (9)$$

$$y=-9\times10^{-9}x^3+1\times10^{-5}x^2-0.007x+5.216 \quad (10)$$

4. Conclusion

The polymers used had concentrations of 400 mg/L, 600 mg/L, 800 mg/L and 1000 mg/L although they had different relative molecular masses. The permeability limits of the polymer solutions with a relative molecular mass of $M=400\times10^4$ were 20md, 30md, 35md and 40md, respectively. The permeability limits of the polymer solutions with a relative molecular mass of $M=800\times10^4$ were 25md,35md,40md and 45md, respectively. The permeability limits of the polymer solutions with a relative molecular mass of $M=1400\times10^4$ were 35md, 45md, 50md and 55md, respectively. The permeability limits of the polymer solutions with a relative molecular mass of $M=2100\times10^4$ were 45md, 55md, 70md and 80md, respectively. An increase in the polymer concentration resulted in a gradual decrease in the ratio R_m/D_h for the polymer solutions with the same molecular weight and it range between 3.35 and 6.14.

Acknowledgment
The work was financially supported by China National Found for studying abroad.

References

[1] Standnes D C, Skievrak I 2014 *J. Pet. Sci. Eng.* **122** 761–775
[2] Algharaib M, Alajmi A, Gharbi R 2014 *J. Pet. Sci. Eng.* **115** 17–23
[3] Erofeev V I, Lv J, Wang W 2019 *Bull. Tomsk Polytechnic University. Geo Assets Eng.* **330** 147–157
[4] Jiang W, Zhang J, Tang X 2012 *Oilfield Chem.* **29** 446–451
[5] Lu X, Zhou Y, Sun Z, Xie K, Yang H, Zhang J 2015 *Pet. Geol. Oilfield Dev Daqing* **34** 88–94
[6] Lu X, Wang X, Wang R, Wang H, Zhang S 2011 *Pet. Explor Dev* **38** 576–582
[7] Jiang X, Wang R, Lu X, Deng Q, Xiao L 2014 *Oilfield Chem* **31** 269–273
[8] Xie K, Lu X, Pan H, Han D, Hu G, Zhang J, Zhang B, Cao B 2018 *SPE Prod. Oper* **33** 1–11
[9] Narisu, Erofeev V I, Lu X, Lv J, Wang X, Zhang L 2019 *Bull. Tomsk Polytechnic University. Geo Assets Eng* **330** 136–145
[10] Xie K, Lu X, Li Q, Jiang W, Yu Q 2016 *SPE J* **21** 1–9
[11] Lu X, Gao Z, Yan W 1994 *Pet. Geol. Oilfield Dev. Daqing* **13** 53–55
[12] Liu J, Lu X, Zhou Y, Hu S, Xue B 2014 *J. China Univ.Pet* **38** 171–179
[13] Narisu, Erofeev V I, Lu X, Tian Z, Zhang L 2019 *Bull. Tomsk Polytechnic University. Geo Assets Eng.* **330** 59–68

IOP Conf. Series: Materials Science and Engineering 1019 (2021) 012102 doi:10.1088/1757-899X/1019/1/012102

Adaptive frequency islanding system for distributed generation energy districts

R B Abeuov, E A Aryutina and Yu Z Vassilyeva
School of Energy & Power Engineering, Tomsk Polytechnic University, Tomsk, Russia
E-mail: Abeuov_RB@list.ru

Abstract. This paper presents the developed Adaptive frequency islanding system for distributed generation energy districts. The research is initiated due to the changes of power system structure, increasing the share of distributed generation. At the same time the existing islanding systems have a number of disadvantages, which leads to generation-shedding of distributed generation objects while allocating them for isolated operation. This is primarily due to the low speed of existing islanding systems. The solution to this problem can be an Adaptive frequency islanding system, which is high-speed and evens out the power balance in the distributed generation energy districts at the time of its allocation to isolated operation. The operating principle of this system is to determine the amount of control actions and their issuance at the substation of distributed generation energy districts based on the assessment of the electric power system mode and energy district real-time parameters.

1. Introduction

According to world practice, recently there has been a tendency of rapidly growing world energy consumption [1]. To solve this problem, in addition to large hydraulic and thermal power plants, distributed generation (DG) objects are actively used. In developed countries, among all types of distributed generation resources (DER) much attention is paid to renewable energy sources (such as solar, wind, geothermal, mini-hydro, etc.) due to the positive environmental impact [2]. However, in Russia the DG development is carried out mainly due to thermal power plants with efficient installations based on gas or diesel fuel. These include power plants with gas turbines and gas-piston engines.

There exist a number of features of parallel operation of the DER and the electric power system (EPS) that are not typical for such operation of large power plants and EPS [3]. These features have an impact on solving problems of ensuring the power supply systems' reliability, maintaining operating parameters within acceptable limits, as well as preventing and eliminating violations of the normal mode, both in energy districts with DER and in the adjacent EPS network.

The key indicator of power quality and the most important parameter of the power system mode is frequency. The electric current frequency regulation in the Russian power system is carried out in accordance with the requirements [4]. A significant reduction in frequency leads to emergency generation-shedding and mass disconnection of electrical energy consumers.

The under-frequency load-shedding (UFLS) is used in the power industry to rescue systems facing extreme disturbances to avoid system collapse. This system disconnects consumers to restore the balance of active power in the energy system [5]. This method of frequency control is quite effective, however, it can lead to a massive disconnection of consumers in distributed generation energy districts.

Another way to frequency control in distributed generation energy districts during system faults is to allocate such energy districts for isolated operation. This method allows to avoid mass disconnection of consumers by the action of UFLS, however, it quite often leads to shutdown of gas turbines and gas-piston power plants at isolating of DG energy districts with power unbalance. The reason for this is the low speed of frequency islanding system and slow load-shedding by UFLS, aimed at equalizing the power balance.

Studies have shown that the allocation of the distributed generation energy districts to the isolated operation can be an effective way to ensure uninterrupted energy supply in case of system fault, leading to decrease frequency, but it requires the development of a new high-speed frequency islanding system.

In this paper, we report a development Adaptive frequency islanding system (AFIS), which is high-speed and aligns the power balance in the distributed generation energy districts at the time of its allocation to isolated work.

2. Requirements to developing system

We have carried out the analysis of the effectiveness of existing islanding systems [6–8]. According to it, currently operated systems implement the allocation of DG energy districts only after the frequency changes in the EPS, without current monitoring the mode parameters in the distributed generation energy districts and not adapting to their change.

The solution of this problem can be an Adaptive frequency islanding system, which capable of carrying out the following:

- continuous measurement of real-time mode parameters of the EPS and DG energy districts;
- continuous calculation of the amount of control actions (CA);
- issuing commands for load-shedding (LS) and transmitting information about the volumes of CA to the power substation when it is allocated from the EPS with the subsequent control of their implementation through the Telemetry and SCADA system.

The AFIS development will allow solving the following tasks:

- allocation of DG energy districts for isolated operation in case of emergency frequency reductions in the power system while maintaining stable operation of generators;
- maintenance of the power supply of the motor load of the distributed generation energy districts while allocated to isolated work.

To solve the tasks posed, the AFIS must meet current requirements for speed and adaptability. The speed is achieved by the fact that the amount of CA is determined before the emergency mode and, accordingly, before the command of the network division (ND). In this case, the formation of control actions and the command to load disconnection is carried out at the same time as ND. Adaptability is provided by the function of automatic CA recalculation based on real-time measurements of parameters from the EPS and distributed generation energy districts and its processing in the pre-emergency mode. This eliminates the possibility of over-load-shedding.

3. Adaptive frequency islanding system

3.1. Main units

The developed Adaptive frequency islanding system is shown in Figure 1. The system consists of following elements: measuring unit (MU), transmitter/receiver of SCADA, starting unit (SU), computing unit (CU), control actions distribution unit (CADU), and operating unit (OU) implementing of CA.

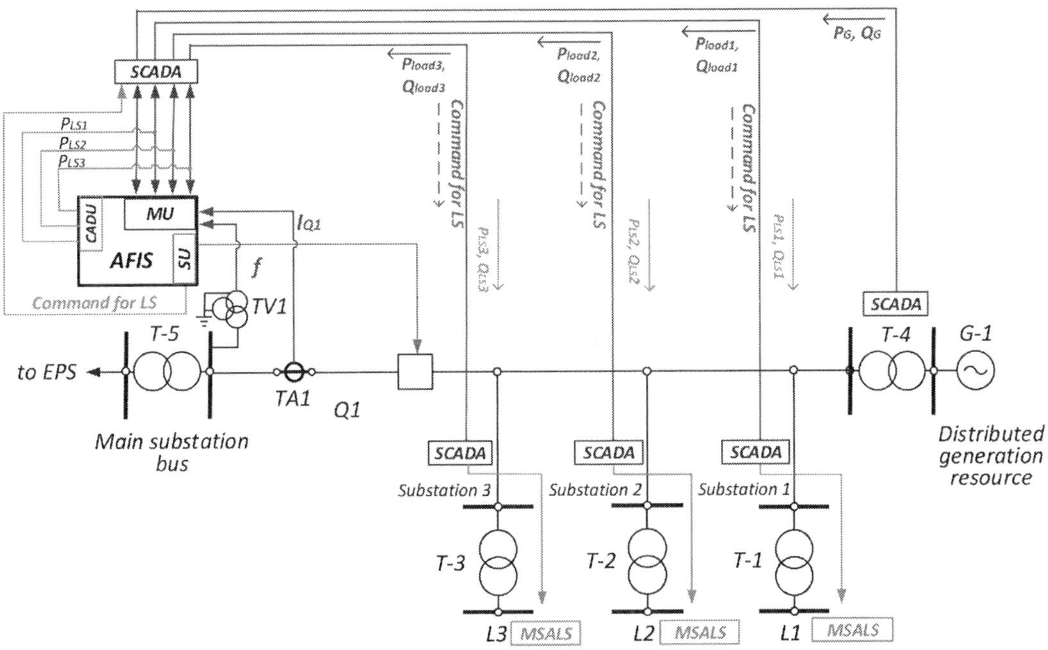

Figure 1. The schematic structure of the Adaptive frequency islanding system.

In case of abnormal frequency decreasing AFIS gives a signal to turn off the switch by the starting unit. The real-time values of frequency on the main substation bus, active and reactive power transferring from the EPS to the DG energy district, as well as the values of consumed and generated power at the stations and substations are measured in MU. For this, it is necessary to ensure the interaction of the ADIS with the SCADA.

A feature of the developed system is the presence of the computing unit. The CU task is to continuously calculate the amount of control actions according to real-time measurements of the mode parameters of the EPS and the DG energy district. The calculated amount of CA is sent to the CADU, and then redirected them to the SCADA, which issues a command to load-shedding of the corresponding substation by multifunctional system for automatic load-shedding (MSALS). The MSALS receives two signals from the SCADA: the first is the system actuation signal and the second is signal characterizing the amount of load shedding.

The use of Adaptive frequency islanding system with Multifunctional system for automatic load-shedding will allow to more effective load-shedding in cases of emergency reduction of the frequency in the EPS, and avoiding the excessive effect of load-shedding by emergency control system.

The MSALS can carry out disconnection of consumers on the fact of frequency and voltage reducing on low voltage buses of substation, as well as on the fact of receiving the actuation signal from the external emergency control system. The multifunctional system for automatic load-shedding can work both independently and as part of an emergency control system as an actuator. The MSALS is a separate complete actuator. The development of this system is a different issue and is not considered in this paper.

For more detailed description, Figure 2 shows the structural-functional scheme of AFIS.

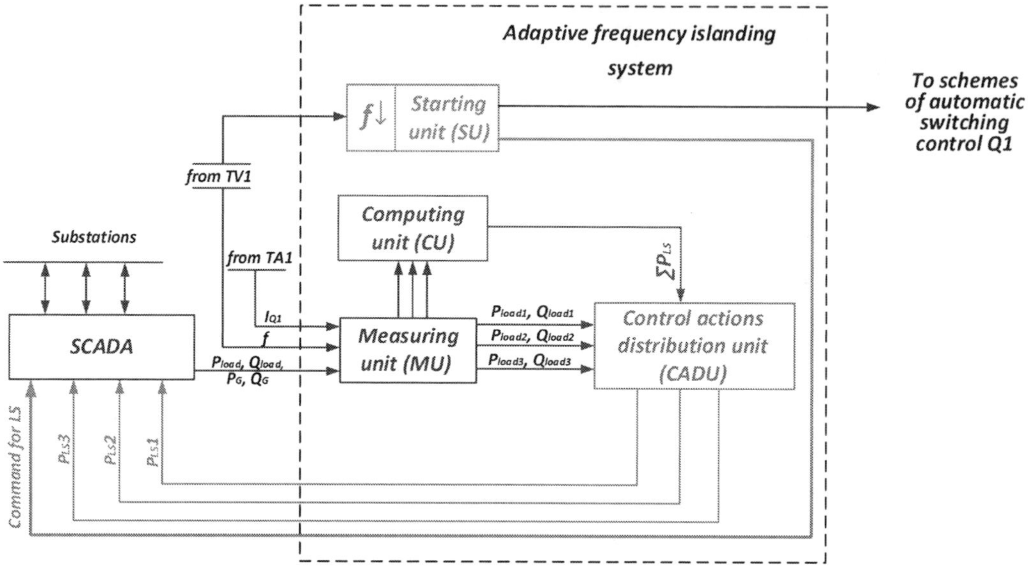

Figure 2. The structural-functional scheme of the Adaptive frequency islanding system.

3.2. Operating Principle

The developed Adaptive frequency islanding system works as follows:

1. A system fault occurs in the EPS.
2. The frequency decreases below rated level.
3. The SU on the main substation bus issues a command to allocate (split) the DG energy district for isolated work.
4. The command of power system splitting is issued to schemes of automatic switching control.
5. Simultaneously with the previous action, the amount of CA is calculated on the basis of the last measured parameters by the MU for load-shedding in the isolated DG energy district.
6. The distribution of total CA to the substations of the DG energy district is carried out by the CADU.

In general, the total active power, which is to be shut down, is calculated by the following equation:

$$\Delta P_{LS} = \frac{\Sigma P_G}{1,05} - \Sigma P_{load} \cdot (1 + \frac{(f_{rated} - f_{previous})}{f_{rated}} \cdot K_f) \ , \qquad (1)$$

where $\sum P_G$ – is the total active power generated by sources of the DG energy district at the time preceding the isolation, MW;
$\sum P_{load}$ – is the total active load power at the time preceding the isolation, MW;
$f_{\text{previrated}}$ – is the rated voltage frequency, Hz;
f_{previous} – is the voltage frequency on the main substation bus at the time preceding the isolation, Hz;
K_f – is the self-regulation of the load in the interconnection, relative unit.

As mentioned before, the distribution of the total CA at the substations of the DG energy district is carried out by CADU. The distribution of the CA should occur in proportion to the measured values of consumers' power loads. For optimal distribution of the CA between substations, the coefficient of participation of the substation load in the total energy district consumption should be taken into account:

$$k_i = P_{load\ i} \ / \ P_{load\ \Sigma}, \qquad (2)$$

where $P_{\text{load i}}$ – is the active power of *i*-substation, MW; $P_{\text{load}\ \Sigma}$ – is the total active power of energy district, MW.

IOP Conf. Series: Materials Science and Engineering 1019 (2021) 012102 doi:10.1088/1757-899X/1019/1/012102

The obtained coefficients are multiplied by the value of the total amount of the CA ($\Delta P_{LS} \cdot k_i$) and, in the form of digital information, are distributed through the SCADA channels at the substations of the DG energy district.

As a result, load-shedding and restoring of the power balance in the isolated DG energy district is occurred at the time of the power system islanding. The DG energy district reliably stands out for isolated work.

4. Conclusion

Thus, this paper presents the developed Adaptive frequency islanding system, which meets the modern requirements of speed and adaptability. The implementation of the AFIS will significantly improve the efficiency of the allocation of the DG energy districts for isolated work, as well as increase the reliability of power supply to consumers of the DG energy districts in cases of unacceptable reduction in the frequency in the EPS. Due to the actions of the Adaptive frequency islanding system, it is possible to ensure stable generators operation and to preserve the power supply of the responsible motor loads while isolating the DG energy district.

References

[1] Wang W and Chen J 2019 *Energy Procedia* **158** 3559–3564
[2] Peças Lopes J A 2007 *Electric Power Systems Research* **77**(9) 1189–1203
[3] Wei G, Wu W, Hu D and Li Z 2007 *High Voltage Engineering* **33** 36–40
[4] Standard no. 59012820.27.100.003-2012. 2012 *Regulation of frequency and active power flows in the EPS of Russia. Standards and requirements* (Moscow: System Operator of the United Power System Joint-stock Company)
[5] Karimi M, Mohamad H, Mokhlis H and Bakar A H A 2012 *Electrical Power and Energy Systems* **42** 127–138
[6] Tolnar J, Caroline H, Hymer L and Bauer M 2012 System and method for adaptive islanding for stored/distributed energy devices. Patent USA no. US8335595B2
[7] Rabinovich R S 1984 The dividing automation system of under-frequency of the power station. Patent RF
[8] Shabad M A 2006 *Dividing protection - automatic division in case of accidents* (Moscow: Energoprocess)

Concept of the comprehension level of situation awareness using an expert system

J Nine, S Manoharan and W Hardt
Department of Computer Engineering, Technische Universität Chemnitz, Chemnitz, Germany
E-mail:julkar.nine@informatik.tu-chemnitz.de

Abstract. One of the major hurdles in high level autonomous driving is the integration of software that has the ability to understand its surrounding situations. Situation awareness is to perceive information, take actions based on that perception and to predict the future events based on the taken actions. Mica R. Endsley introduced the term "Situation Awareness" where the concept was applied only for humans. Expert Systems, on the other hand is a rule based system that is used to make critical and fast decisions. In this paper a concept on the comprehension (understanding) level of situation awareness using an expert system is presented for high level autonomous vehicles. This concept will be realized on the "CE-Box" component of the hardware demonstrator "BlackPearl" from the department of Computer Engineering in TU Chemnitz. Perception is done with the help of sensing data from the sensors from the ECU's of "BlackPearl". This data is provided to the comprehension level to make an assessment and take a decision. The goal of this paper is to provide a conceptual method to extract these decisions based on an expert system.

1. Introduction

Millions of people around the world have lost their lives or became physically disabled as a consequence of road accidents. One of the purposes of making driving autonomous is to eradicate the number of accidents and improve road safety. As per the report was given by the European Transport Safety Council (ETSC) every one million inhabitants around 51 lost their lives due to road accidents in the year 2016 [1]. According to the statistical report in the European Union between 2015 and 2016 (figure 1), the total number of road deaths was decreased by 2% [1]. The human driver being unaware of the situation as result in majority of the road deaths. With the advancement in the automated driving, rate of death can be reduced every year if situation awareness is integrated properly. In the future, death rates can be reduced even further with the help of autonomous driving. Finally, at Level 5 of the autonomous driving death rates due to situation unawareness of drivers will be nullified [1].

As we move up the ladder of autonomous driving that is, from Level 0 to Level 5 the complexity of the software increases. Mica. R. Endsley highlighted in her paper [3], automotive companies have a software failure rate of 82 percent. Most of this failure lies due to the lack of understanding of surrounding situations of the vehicle. Collecting the information to identify the driving situation will further intensify the software's complexity. This has been the motivation to develop the concept of the classification of the driving situations. The classification of the situation might help to decrease software complexity. The Situation Awareness [2] is defined as how the environmental elements or events are perceived, comprehended the meaning and future actions are projected with respect to space and time.

IOP Conf. Series: Materials Science and Engineering 1019 (2021) 012103 doi:10.1088/1757-899X/1019/1/012103

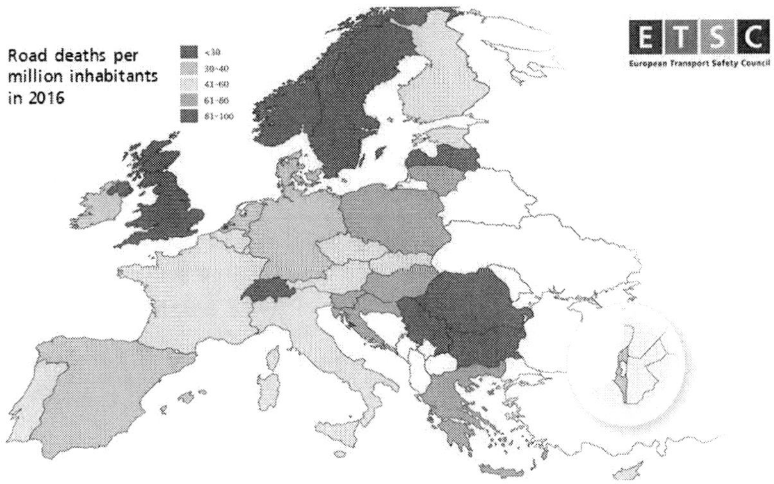

Figure 1. Road Deaths in 2016 (per million).

There are three levels in situation awareness [4]:

(i) Perception: Observing and attention to the information such as dynamics, features, and status which are important and relevant to the situation [4].

(ii) Comprehension: All the information are processed. This also involves the integration of the situation's complete picture from different elements which results in the understanding the importance situation [4].

(iii) Projection: Reacting according to the result of comprehension in a timely manner. This also predicts future actions on the source of the situation comprehension [4].

Disengagements from the autonomous driving include (1) detect a problem in the software of automation and itself gets disengaged, passes the control back to the driver, (2) driver disengages the automation and takes over manually [3]. The reasons for disengagements of the autonomous driving includes failures of software (81.6%), issues in the hardware (14%), conditions of the road surface (0.4%) and weather (4 %) [3]. From the table 1 [3], Waymo [3] is the best performing technology in autonomous driving. In Waymo vehicles, intervene of the human driver needed on an average of 5,600 miles approximately, or before some detected on vehicle and gets disengages by itself.

Table 1. Disengagement of automonous driving for average miles driven [3].

Manufacturer	Miles per Disengagement		
	2015	2016	2017
Google/Waymo	1244	5128	5596
Nissan	14	146	208
Delphi	41	17.5	22.4
Mercedez-Benz	1.5	2	4.5

The system should automatically detect the objects and surrounding features, also analyzing the features and predicting the near future development from the current situation data. These methods can be briefed as situation awareness, which was proposed for the aviation domain. According to the traffic situation, the system have to identify and interpret the objects correctly and also have to react to these elements by predicting and giving respective command to the ECU. The system have to understand other elements such as traffic signs, signal or weather conditions act to their relevance for safe driving.

An expert system [4] is a computer program which pretends as the decision and action are made by a human. The system has expert understanding and knowledge in a specific field. The inputs are provided to the end user by entering the data or selecting from a list which has one or more answers. The system which makes the decision and solves the problem based on the (1) logical rules, (2) task's knowledge and, (3) procedural knowledge.

2. Literature Survey

2.1. *Situational Awareness*

Endsley, M.R. [5], situation awareness in the system operation is one of the major concern in which decisions are made based on an eloquent view of the different state is proposed. Here initially, several individual and environmental factors are explored and their relation between situation awareness. The situation awareness deals with the complexity of system, features and operator automation. The model of the situation awareness introduces the classification of errors and design implications are generated to enhance the operator situation awareness.

2.2. *Levels of Situation Awareness*

Rashaad E.T. *et al.* [6], presented a paper on the situation awareness levels. The first level perceives the important factors in the environment, the second level tries to recognize the meaning of the perceived factors, and finally, the third level will predict the near future with the perceived situation.

2.3. Expert System

Niehaus A. *et al.* [7] assumed that vehicle consists of sensors for the detection of traffic, signs, geometry of the road, and also actuators and control logic for controlling the brakes, steering wheel's angle and throttle. The expert system is given with the situation of the traffic, signs and driver's strategy such that commands are issued to the controllers (Figure 2). The system comprises of a rule base that provides the required knowledge for driving. The reasoning was performed by backward chaining inference, and the reasoning process is optimized by the compiler. The vehicle is controlled either by the expert system or preset strategy.

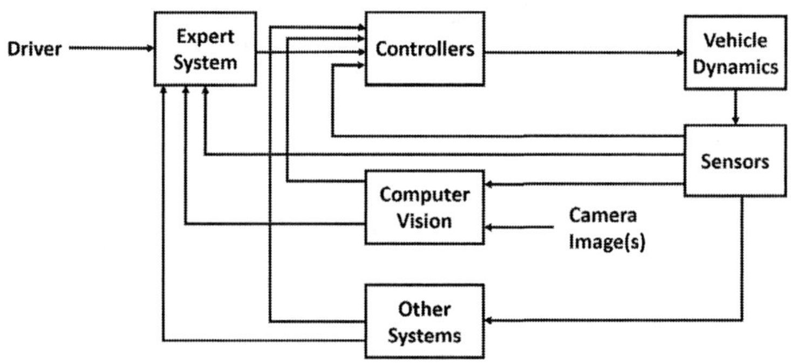

Figure 2. Road Deaths in 2016 (per million).

2.4. *Backward Chaining Algorithm*

Yuliadi E. [8] used a backward chaining algorithm to start with a goal and end with fact leading to a new goal. This algorithm starts with the last element in the chain and continues until reaches the first element. Depth- and breath-first search algorithms are used by backward chaining method. The depth-first algorithm is used to search the data in the deeper level of the tree before backtracking. The breath-first algorithm is used to search the data in adjoining nodes and then moves to a deeper level of the tree. This method provides answers dynamically for the inference engine, computation effort, etc.

2.5. *Rete Algorithm*

Charles L. Forgy [9] presented about Rete Match Algorithm in this article. This algorithm is very efficient in comparing the massive collection of patterns and objects as not iterating over the sets. The iteration over the patterns does not occur in this program as it contains a network which works on tree-structure sorting or pattern index and efficiently matches all the objects with the pattern. The algorithm is used as the interpreters in the production system and used for the systems which consist of hundreds or thousands of objects and patterns.

In this study, a conceptual method for the comprehension level is highlighted, focusing on an expert system which uses Rete algorithm as its inference engine.

3. System Design

3.1. *System Architecture*

Videos are capture from Raspberry Pi camera which is provided as the software architecture layer of the Raspberry Pi. The application layer will provide image processed data as input whereas sensory data through CAN message are also taken as the input. Different image processing algorithms can be used to detect traffic light, traffic signs, pedestrian, curvature, and lane. Once the process of the assembler and the expert system is done, the respective CAN message command is given to the ECU of the vehicle. With the help of PiCAN 2 mounted on the Raspberry Pi, this can be achieved. The architecture of the system and data flow control is shown in Figure 3 and Figure 4.

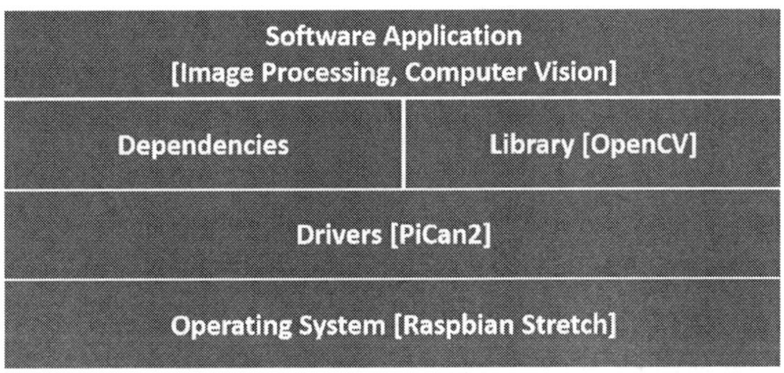

Figure 3. System Architecture Layer.

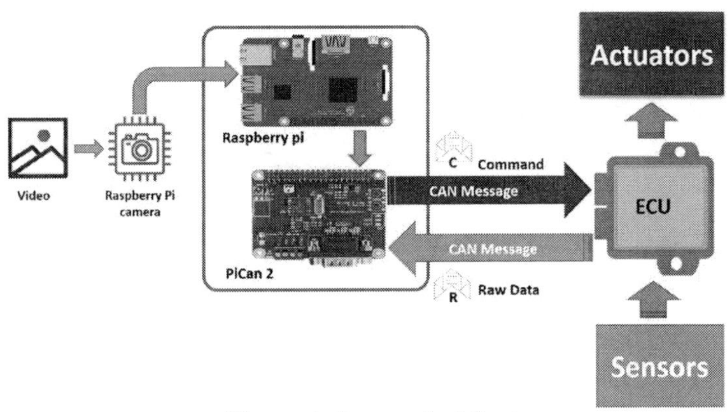

Figure 4. System Architecture.

3.2. *Software*

Raspbian Stretch operating system is used in each Raspberry Pi of the CE-Box. Open CV library is installed on Raspberry Pi with Python and C++ dependencies for performing the computer vision and image processing applications. The C Language Integrated Production System which in short known as CLIPS. The CLIPS is used for the implementation of the expert system.

IOP Conf. Series: Materials Science and Engineering 1019 (2021) 012103 doi:10.1088/1757-899X/1019/1/012103

3.3. *Hardware*

The "BlackPearl" is one of the hardware demonstrator of Department of Computer Science in TU Chemnitz is used for the applications. The hardware unit of CE-Box consists of multiple racks of Raspberry Pi mounted with the PiCAN 2. The CAN messages are sent and received using the PiCAN 2. The Raspberry Pi of the CE-Box is programmed with the assembler and expert system. Through the CAN message, the raw data are taken as input from the ECU, generate the fact list and corresponding action commands are given back to the ECU. The hardware setup is shown in (Figure 5).

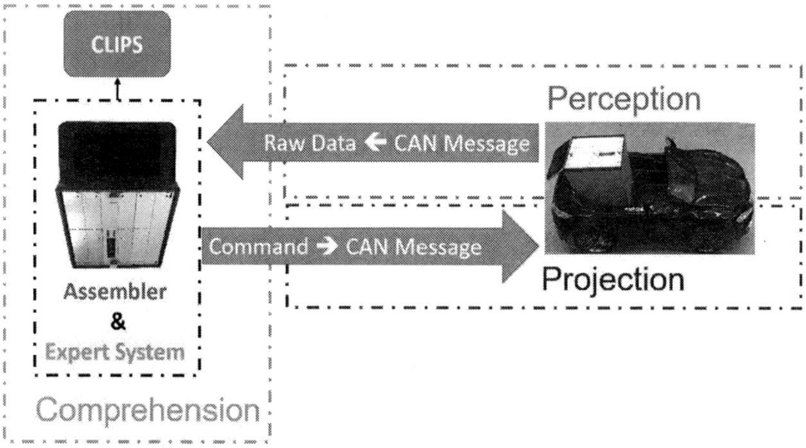

Figure 5. Hardware Setup using CE-Box (left) and BlackPearl (right).

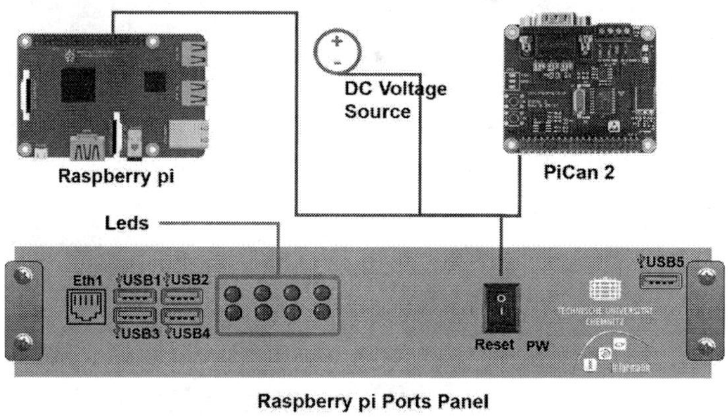

Figure 6. CE-Box Connection (Front panel).

There are 8 LEDs, 1 Ethernet port, 5 USB ports, and 1 switch for power and reset option (Figure 6). The power and reset functionality are designed using the Finite State Machine (FSM) (Figure 7). For the safety restart of the Raspberry Pi, the switch need to be pressed and hold for more than 5 seconds. The single press of the switch will disconnect the power from the source and shut down the Raspberry Pi.

The Assembler code and expert system are coded onto the Raspberry Pi. The CAN message from ECU received and transmitted using PiCAN 2 which mounted on the Raspberry Pi. The CAN messages are perceived and converted into raw data. The raw data is given as an input to the assembler. In assembler, data is changed into fact and given to the expert system which in turn give back command (action) to control the operation of the ECU.

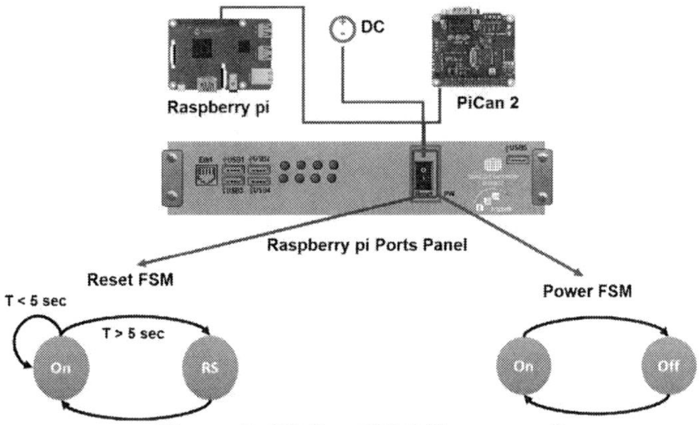

Figure 7. CE-Box FSM (Front panel).

The capability of the CAN Bus is given to the Raspberry Pi via PiCAN 2 board. It uses Microchip MCP2515 and MCP2551 for the CAN controller and transceiver. The PiCAN Board can be powered via DB9 connector or screw terminal. This can be programmed using C or Python.

4. Conceptualization

The situation awareness deals with the complexity of system, features and operator automation. Several individual and environmental factors are collected in the perception level. From the perception level, collected elements and converted into facts which are used as the input for the comprehension level. The comprehension level will consist of a rule-based system which will match the element with the set of rules. The comprehension level will also predict the near future action. Finally, action commands are used to control the operation of the vehicle in the projection level.

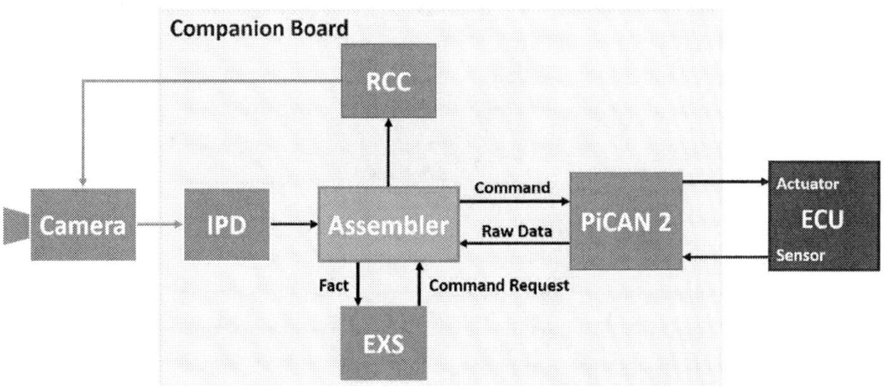

Figure 8. System Setup.

The main steps of the conceptualization are: (1) getting the sensory data from the ECU through PiCAN 2, (2) getting the Image Processed Data (IPD) from the Raspberry Pi camera which is controlled by the Raspberry Pi Camera Controller (RCC), (3) the assembler will generate fact list from the raw data, (4) the facts are matched with the predefined rules in the Expert System Shell (EXS), (5) the expert system will update the fact and give the command request to the assembler, (6) the assembler will send the command as CAN message to the ECU, (7) ECU executes the command. All these steps will run continuously in a cyclic manner.

The ECU / Raspberry Pi will perceive the required data or element from the vehicle sensor and environment factors. These elements are used to classify the situation. The collected data is provided to the assembler where the understanding of the data will take place. Once, all the elements are recognized by the assembler, the assembler will generate the initial fact list. The generated facts are given to the expert system. The expert system will match the facts with the predefined set of rules. Then, action commands are given back to assembler from the expert system. Upon receiving the command from the expert system, the assembler will convert this command into CAN message and send it to the ECU and action will take place.

The main role of the assembler is to act as the interface between the ECU and expert system. The raw data from the ECU will be received by the assembler via PiCAN 2. The received raw data as CAN message will be processed and changed into facts. These facts are sent to the expert system. Then, the commands are received back from the expert system. The received command is converted back into CAN message by the assembler. Finally, the command as CAN message is given back to the ECU and the ECU will act on it.

The main functionality of the assembler is: (1) Fusing all the sensory data received via CAN bus from the ECU, (2) With the data, initial fact lists are generated and sent to the expert system, (3) Receive an action request from the rule-based expert system, (4) Send the received command from expert system to the ECU via CAN message, (5) Then acknowledge the received command, (6) Finally, update the fact list.

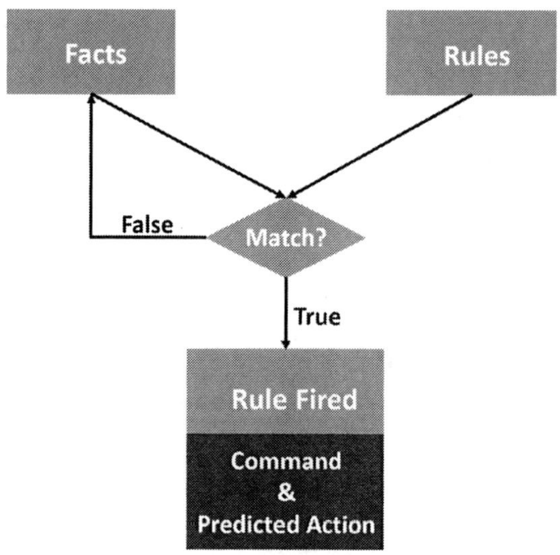

Figure 9. Flow Chart of Expert System.

The rule-based expert system (Figure 9) does the process of situation comprehension. This system will receive the fact list from the assembler. Upon the received fact, the expert system will match the fact with the predefined set of rules. When there is a match between the fact and rule, then the rule is fired. As per the fired rule, the projection of command and predicted action will be generated. The generated command is given to the assembler. The predicted action for the near future will be based on the data collected from the current situation.

The complete set of rules are sub-categorized. The generated facts are given as input to every rule of one sub-category. The facts are matched with the rules and commands are given back to the assembler. On acting upon the rules, we may transit between Rule 1 to Rule 2 for the criteria 1 ➔ 2 and Rule 1 to Rule 3 for the criteria 1 ➔ 3. This transition between rules takes place based on the different situation criteria which are depicted as FSM (Figure 10).

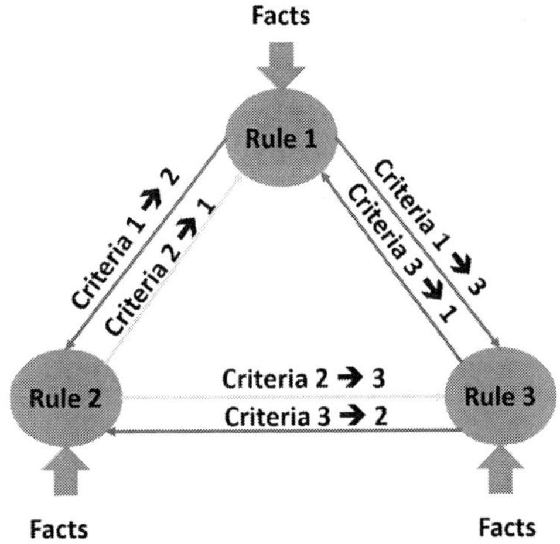

Figure 10. Rule's FSM.

5. Future Work

Implementation of the presented concept using Raspberry Pi, PiCAN 2 and checking the output. In order to increase the performance, use a Jetson Nano board which comes with an in-built GPU. Using Assembler to compare the sensory data and generate the fact list may not be realistic. Hence, assembler and expert system design can be modified such a way that assembler will convert the CAN message into data and send the data to an expert system. The expert system will compare and give the command to the assembler. Finally, assembler sends the command to the ECU. Therefore, most of the computation and comparison part is moved to the expert system which makes the overall design feasible for the real-time applications.

6. Conclusions

In this paper, a new concept for the situation awareness in the autonomous driving is proposed. This concept can make the system aware of the situations by collecting the required data, recognizing the data and giving the command to control the ECU according to the current situation. This setup can be carried on Raspberry Pi 3B model with a PiCAN 2 mounted. The hardware demonstrator called BlackPearl with the CE-Box of the department of computer engineering in TU Chemnitz can be used. Once the implementation is realized, an evaluation can be done and presented as well.

As part of the future aspects of this study, a concept for the projection level can also be carried out, completing the three levels of situation awareness.

References

[1] ETSC (European Transport Society Council). Available at: https://etsc.eu/category/publications/
[2] Krems J F and Baumann M R K 2009 *Proceeding of International Conference on Human Centered Design* 986–994
[3] Endsley M R 2018 *Proceedings of the 20th Congress of the International Ergonomics Association)* pp 303-309
[4] Kumar Y and Jain Y 2012 *International Journal of Computing & Business Research* 2229-6166
[5] Endsley M R 1995 *Toward a Theory of Situation Awareness in Dynamic Systems (Human Factors The Journal of the Human Factors and Ergonomics Society)* vol 37(1) pp 32-64
[6] Jones R E T and Connors E S and Endsley M R 2009 *Incorporating the Human Analyst into the Data Fusion Process by Modeling Situation Awareness Using Fuzzy Cognitive Maps (12th*

International Conference on Information Fusion Seattle (USA)) pp 1265-1271
[7] Niehaus A and Stengel R F 1991 *An expert system for automated highway driving (IEEE Control Systems Magazine)* vol 11(3) pp 53-61
[8] Erdani Y 2012 *Developing Backward Chaining Algorithm of Inference Engine in Ternary Grid Expert System (International Journal of Advanced Computer Science and Applications)* vol 3(9) pp 241-245
[9] Forgy C L 1982 *Rete: A Fast Algorithm for the Many Pattern / Many Object Pattern Match Problem (Department of Computer Science, Carnegie-Mellon University, Pittsburgh, PA 15213, U.S.A)* pp 17-37

AUTHOR INDEX